T0377769

COSMIC MASERS: PROPER MOTION TOWARD THE NEXT-GENERATION LARGE PROJECTS

IAU SYMPOSIUM 380

COVER PICTURE: The Sakura-jima volcano viewed from Kagoshima-city.

The Sakura-jima mountain is one of the most active volcanos in Japan, whose altitude is 1117 m. It is located ~10 km east of the downtown Kagoshima-city on the opposite side across the Kinko-wan Bay. Although it was quiescent during the IAU Symposium 380 held in Kagoshima-city from March 20 to 24, 2023, one can usually see a variety of interesting phenomena analogous to astrophysical maser activities: Intermittent eruptive events are observed frequently at the southern peak of Sakura-jima. In some cases, a large amount of volcanic dust and smoke are ejected from the crater up to a few 1000 m above the top of the mountain, and obscure the surrounding regions. Thanks to its activity, there are a number of hot springs in/around Sakura-jima including the central part of Kagoshima-city. As such, Sakura-jima is recognized as a symbolic landmark of Kagoshima-city.

The photograph was taken by Hiroshi Imai.

IAU SYMPOSIUM PROCEEDINGS SERIES

INTERNATIONAL ASTRONOMICAL UNION

UNION ASTRONOMIQUE INTERNATIONALE

COSMIC MASERS: PROPER MOTION TOWARD THE NEXT-GENERATION LARGE PROJECTS

PROCEEDINGS OF THE 380th SYMPOSIUM OF THE INTERNATIONAL ASTRONOMICAL UNION KAGOSHIMA, JAPAN 20–24 March, 2023

Edited by

TOMOYA HIROTA

National Astronomical Observatory of Japan, Japan

HIROSHI IMAI

Kagoshima University, Japan

KARL MENTEN

Max-Planck-Institut für Radioastronomie, Germany

and

YLVA PIHLSTRÖM

University of New Mexico, USA

CAMBRIDGE UNIVERSITY PRESS
University Printing House, Cambridge CB2 8BS, United Kingdom
1 Liberty Plaza, Floor 20, New York, NY 10006, USA
10 Stamford Road, Oakleigh, Melbourne 3166, Australia

First published 2024

Printed in Great Britain by Henry Ling Limited, The Dorset Press, Dorchester, DT1 1HD

Typeset in System LaTeX 2ε

A catalogue record for this book is available from the British Library of Congress Cataloguing in Publication data

This journal issue has been printed on FSC™-certified paper and cover board. FSC is an independent, non-governmental, not-for-profit organization established to promote the responsible management of the world's forests. Please see www.fsc.org for information.

ISBN 9781009398923 hardback
ISSN 1743-9213

Table of Contents

Chapter 4: Dynamics of Formation of Massive Stars

Chapter 6: Theory of Masers and Maser Sources

Chapter 7: New Projects and Future Telescopes

Chapter 8: Concluding Remarks

Preface

The International Astronomical Union (IAU) Symposium 380 entitled *Cosmic Masers: Proper Motion toward the Next-Generation Large Projects* (IAUS 380) was held from March 20 (Mon) 2023 to March 24 (Fri) 2023 in Kagoshima, Japan. Kagoshima-city is located in the south-western region of Japan, and provides grand views of an active volcano, Sakura-jima, which was the location of the IAUS 380 excursion. At the local Kagoshima University, there is an active and large astronomy community which is involved in many aspects of maser research. Staff members and students from this community contributed to the IAUS 380 by serving on the LOC and as symposium volunteers.

Considering the unexpected situation due to the world-wide COVID-19 pandemic, the IAUS 380 was planned as a hybrid conference. At the time of the registration in late 2022, both in-person and online participants were accepted. Finally, because the COVID-19 situation eased in Japan in early 2023, in total 102 people participated in-person in Kagoshima and 70 participated online. In total 28 countries were represented. Among the 172 registered participants, 43 (25%) were at an early career stage before having received their PhD degrees.

Since the discovery of the strong molecular lines of OH and H_2O in 1960s, cosmic masers have been employed as unique probes of various astronomical objects, ranging from newly born stars and evolved stars, the interstellar medium to active galactic nuclei. The maser scientific community is diverse and multidisciplinary but has long been tied together through the common background physics and observational techniques. To connect and build new collaborations, international meetings focusing on masers have been organized regularly since 1992 in US, including the past IAU symposia 206 in Brazil (2001), 242 in Australia (2007), 287 in South Africa (2012), and 336 in Italy (2017). The IAUS 380 is the 6th big international maser conference and the first one in Asia. It took place about 5.5 years after the last meeting and thus filled in the final gap in global coverage and time.

In the science sessions we discussed seven major themes of maser research, from maser theory, cosmology, galaxies, the Milky Way Galaxy, star-formation, evolved stars, to future projects. In order to allow online speakers from outside of Japan to join the meeting during their convenient time zones, we divided each science topic into 2 or more sessions at time ranges. There were 8 review talks including a summary talk of the IAUS 380, 19 invited talks, 37 contributed talks, and 55 poster presentations including 1-minute flash talk for every poster.

As explicitly defined in the sub-title, we organized intensive discussion sessions for currently on-going and future projects related to most of the maser science topics. Taking the opportunity of the IAUS framework, several informal satellite meetings were held during lunch breaks to discuss international collaborations for future studies. An emphasis was on time-domain studies from daily to decadal monitoring of maser sources that were reported using a variety of telescopes from many different research teams in all regions of the world. Furthermore, multi-wavelength studies of maser sources thrived over the last decade exploiting synergies with large facilities such as ALMA, JVLA, Gaia and various VLBI networks.

It is a great pleasure to acknowledge the financial support of all the sponsors listed on the next page of these Proceedings. We would like to thank the active support of the LOC and SOC members to prepare and realize the IAUS 380 under the difficult conditions of the COVID-19 pandemic. In particular, we are grateful to volunteer members of the IAUS 380, and local staff members related the conference venue with organizing the logistics, excursion, coffee/tea breaks, and banquet during the meeting.

Tomoya Hirota, Hiroshi Imai, Karl Menten, & Ylva Pihlström

The Organizing Committee

Scientific

A. Bartkiewicz (Poland)
C. L. Brogan (USA)
J. O. Chibueze (South Africa)
C. J. Cyganowski (UK)
G. Garay (Chile)
Y. Hagiwara (Japan)
T. Hirota (co-chair, Japan)
K. Immer (The Netherlands)
J. -h. Kang (Republic of Korea)
S. Leurini (Italy)
K. M. Menten (co-chair, Germany)
Y. M. Pihlström (co-chair, USA)
M. J. Reid (USA)
M. Rioja (Australia)
B. Zhang (China)

Local

R. A. Burns
T. Handa
M. Honma
H. Imai (co-chair)
A. Imakado
A. Nakagawa (co-chair)
H. Nakanishi
H. Shinnaga
S. Takakuwa
Y. Tsukamoto

Acknowledgements

The symposium is sponsored and supported by the IAU Divisions H (Interstellar Matter and Local Universe), B (Facilities, Technologies and Data Science), G (Stars and Stellar Physics), and J (Galaxies and Cosmology).

Funding and support by the
International Astronomical Union,
Amanogawa Galaxy Astronomy Research Center, Kagoshima University,
National Astronomical Observatory of Japan,
KAKENHI, Grants-in-Aid for Scientific Research,
Foundation for Promotion of Astronomy,
Inoue Foundation for Science,
Kajima Foundation,
and
Kagoshima Convention & Visitors Bureau,
are gratefully acknowledged.

Participants

Susanne **Aalto**, Chalmers University of Technology, Sweden
Artis **Aberfelds**, Ventspils University of Applied Sciences, Latvia
Zulema **Abraham**, University of Sao Paulo, Brazil
Takuya **Akahori**, National Astronomical Observatory of Japan, Japan
Kei **Amada**, Kagoshima University, Japan
Kitiyanee **Asanok**, National Astronomical Research Institute of Thailand, Thailand
Anna **Bartkiewicz**, Torun Institute of Astronomy, Poland
Alain **Baudry**, University Bordeaux, L.A.B., France
Olga **Bayandina**, INAF - Osservatorio Astrofisico di Arcetri, Italy
Rajorshi **Bhattacharya**, University of New Mexico, USA
Shuaibo **Bian**, Purple Mountain Observatory, China
James **Braatz**, National Radio Astronomy Observatory, USA
Jan **Brand**, INAF - Istituto di Radioastronomia, Italy
Shari L. **Breen**, SKA Observatory, UK
Crystal L. **Brogan**, National Radio Astronomy Observatory, USA
Andreas **Brunthaler**, Max-Planck-Institut fur Radioastronomie, Germany
Ross Alexander **Burns**, National Astronomical Observatory of Japan, Japan
Roldán A. **Cala**, Instituto de Astrofisica de Andalucia (IAA-CSIC), Spain
Paola **Castangia**, INAF - Osservatorio Astronomico di Cagliari, Italy
Priscila **Chacón**, Universidad de Guanajuato, Mexico
James Okwe **Chibueze**, North-West University, South Africa
Se-Hyung **Cho**, Seoul National University, Republic of Korea
Mark J. **Claussen**, National Radio Astronomy Observatory, USA
Anca **Constantin**, James Madison University, USA
Claudia J. **Cyganowski**, University of St Andrews, UK
Joanne R. **Dawson**, Macquarie University / CSIRO Space & Astronomy, Australia
Deepshikha **Deepshikha**, Deepshikha, Japan
Jean-Francois **DESMURS**, Observatorio Astronomico Nacional, Spain
Philip **Diamond**, SKA Observatory, UK
Richard **Dodson**, ICRAR/UWA, Australia
Tânia Pereira **Dominici**, National Institute for Space Research (INPE/MCTI), Brazil
Le Thong **Duc**, Institute for Computational Science, Vietnam
Michał Tomasz **Durjasz**, Nicolaus Copernicus University, Poland
Andrews Mawuli, Kodzo **Dzodzomenyo**, North-West University, South Africa
Simon **Ellingsen**, University of Tasmania, Australia
Dieter **Engels**, Hamburger Sternwarte, Universitat Hamburg, Germany
Sandra **Etoka**, Jodrell Bank Centre for Astrophysics, University of Manchester, UK
Kenta **Fujisawa**, Yamaguchi Univeristy, Japan
Marcin **Glowacki**, CIRA (Curtin Institute of Radio Astronomy), Australia
Sharmila **Goedhart**, South African Radio Astronomy Observatory, South Africa
José-Francisco **Gómez**, Instituto de Astrofisica de Andalucia (IAA-CSIC), Spain
Mark Daniel **Gorski**, Chalmers University of Technology, Sweden
Malcolm David **Gray**, National Astronomical Research Institute of Thailand, Thailand
Jimi A. **Green**, SKA Observatory, UK
Yoshiaki **Hagiwara**, Toyo University, Japan
Chaojie **Hao**, Purple Mountain Observatory, China
Tomoya **Hirota**, National Astronomical Observatory of Japan, Japan
Peter **Hofner**, New Mexico Tech & NRAO, USA
Mareki **Honma**, National Astronomical Observatory of Japan, Japan
Martin **Houde**, University of Western Ontario, Canada
Elizabeth **Humphreys**, ESO / ALMA, Chile

Todd **Hunter**, National Radio Astronomy Observatory, USA
Lucas Jordan **Hyland**, University of Tasmania, Australia
Nao **Ikeda**, Kagoshima University, Japan
Hiroshi **Imai**, Kagoshima University, Japan
Masatoshi **Imanishi**, National Astronomical Observatory of Japan, Japan
Katharina **Immer**, Leiden University / ESO, The Netherlands
Violette **Impellizzeri**, Leiden Observatory, The Netherlands
Carme **Jordi**, University of Barcelona (ICCUB-IEEC), Spain
Jihyun **Kang**, Korea Astronomy and Space Science Institute, Republic of Korea
Rina **Kasai**, Kagoshima University, Japan
Kaito **Kawakami**, Kagoshima University, Japan
Dongjin **Kim**, MIT Haystack Observatory, USA
Jungha **Kim**, Korea Astronomy and Space Science Institute, Republic of Korea
Kee-Tae **Kim**, Korea Astronomy and Space Science Institute, Republic of Korea
Mikyoung **Kim**, Otsuma Women's University, Japan
Soon-Wook **Kim**, Korea Astronomy and Space Science Institute, Republic of Korea
Keita **Kitaguchi**, Yamaguchi University, Japan
Agnieszka **Kobak**, Nicolaus Copernicus University in Torun, Poland
Hideyuki **Kobayashi**, National Astronomical Observatory of Japan, Japan
Toktarkhan **Komesh**, Nazarbayev University, Kazakhstan
Busaba **Kramer**, Max Planck Institute for Radio Astronomy, Germany
Martin Gustav Heinrich **Krause**, University of Hertfordshire, UK
Jayender **Kumar**, University of Tasmania, Australia
Cheng-Yu **Kuo**, National Sun Yat-Sen University, Taiwan
Tomoharu **Kurayama**, Teikyo University of Science, Japan
Dmitry A. **Ladeyschikov**, Ural Federal University, Russia
Elisabetta **Ladu**, Universita degli studi di Cagliari - Osservatorio Astronomico di Cagliari - INAF, Italy
Boy **Lankhaar**, Chalmers University of Technology, Sweden
Megan Olivia **Lewis**, Nicolaus Copernicus Astronomical Center, Polish Academy of Sciences, Poland
Jingjing **Li**, Purple Mountain Observatory, China
Yingjie **Li**, Purple Mountain Observatory, China
Zehao **Lin**, Purple Mountain Observatory, China
Dejian **Liu**, Purple Mountain Observatory, China
Michael **Logue**, University of St Andrews, UK
Xiaofeng **Mai**, Shanghai Astronomical Observatory, China
Lynn D. **Matthews**, MIT Haystack Observatory, USA
Karl M. **Menten**, Max Planck Institute for Radio Astronomy, Germany
Ryo **Miyamoto**, Gifu Univerisity, Japan
Hafieduddin **Mohammad**, ITB, Indonesia
James Michael **Moran**, Center for Astrophysics — Harvard & Smithsonian, USA
Ren **Moriizumi**, Ibaraki University, Japan
Luca **Moscadelli**, INAF - Osservatorio Astrofisico di Arcetri, Italy
Kazuhito **Motogi**, Yamaguchi University, Japan
Akiharu **Nakagawa**, Kagoshima University, Japan
Naomasa **Nakai**, Kwansei Gakuin University, Japan
Hiroyuki **Nakanishi**, Kagoshima University, Japan
Jun-ichi **Nakashima**, Sun Yat-sen University, China
Keisuke **Nakashima**, Kagoshima University, Japan
Katarzyna **Nowak**, University of Hertfordshire, UK
Keiichi **Ohnaka**, Universidad Andres Bello, Chile
Mateusz **Olech**, University of Warmia and Mazury, Poland
Gabor **Orosz**, JIVE, The Netherlands
Gisela **Ortiz-Leon**, Universidad Nacional Autonoma de Mexico, Mexico

Tomoaki **Oyama**, National Astronomical Observatory of Japan, Japan
Jennie E. **Paine**, University of Colorado Boulder, USA
Dominic **Pesce**, Center for Astrophysics — Harvard & Smithsonian, USA
Anita **Petzler**, CSIRO, Australia
Montree **Phetra**, Chiang Mai University, Thailand
Ylva **Pihlström**, University of New Mexico, USA
Bannawit **Pimpanuwat**, University of Manchester, UK
Luis Henry **Quiroga Nunez**, Florida Institute of Technology, USA
Mark J. **Reid**, Center for Astrophysics — Harvard & Smithsonian, USA
Anita M. S. **Richards**, Jodrell Bank Centre for Astrophysics, University of Manchester, UK
Maria J. **Rioja**, ICRAR/UWA,CSIRO,OAN, Australia
Hayley **Roberts**, University of Colorado Boulder, USA
Tatiana Magali **Rodríguez**, New Mexico Institute of Mining and Technology, USA
Zulfazli **Rosli**, International University of Malaya Wales, Malaysia
Kazi L. J. **Rygl**, INAF - Istituto di Radioastronomia, Italy
Raghvendra **Sahai**, Jet Propulson Laboratory, USA
Daisuke **Sakai**, National Astronomical Observatory of Japan, Japan
Nobuyuki **Sakai**, National Astronomical Research Institute of Thailand, Thailand
Yuichi **Sakamoto**, Kagoshima university, Japan
Alberto **Sanna**, INAF - Osservatorio Astronomico di Cagliari, Italy
Hidetoshi **Sano**, Gifu University, Japan
Satoko **Sawada-Satoh**, Osaka Metropolitan University, Japan
Mavis **Seidu**, North West University, South Africa
Nadezhda N. **Shakhvorostova**, Astro Space Center of the P.N. Lebedev Physical Institute of RAS, Russia
Yosuke **Shibata**, Kagoshima University, Japan
Hiroko **Shinnaga**, Kagoshima University, Japan
Ka Yiu **Shum**, Kagoshima University, Japan
Loránt **Sjouwerman**, National Radio Astronomy Observatory, USA
Ivars **Smelds**, Ventspils University of Applied Sciences, Latvia
Andrey M. **Sobolev**, Ural Federal University, Russia
Bringfried Gerhard Werner **Stecklum**, Thueringer Landessternwarte, Germany
Georgina **Stroud**, University of Manchester, UK
Hiroshi **Sudou**, Gifu Univerisity, Japan
Koichiro **Sugiyama**, National Astronomical Research Institute of Thailand, Thailand
Kazuyoshi **Sunada**, National Astronomical Observatory of Japan, Japan
Gabriele **Surcis**, INAF - Osservatorio Astrofisico di Arcetri, Italy
Zsófia Marianna **Szabó**, Max Planck Institute for Radio Astronomy, Germany
Hiroshi **Takaba**, Gifu University, Japan
Daisuke **Takaishi**, Kagoshima University, Japan
Shigehisa **Takakuwa**, Kagoshima University, Japan
Yoshihiro **Tanabe**, Ibaraki University, Japan
Koki **Tanaka**, Kagoshima university, Japan
Andrea **Tarchi**, INAF - Osservatorio Astronomico di Cagliari, Italy
José-María **Torrelles**, Institut de Ciencies de l'Espai (CSIC/IEEC), Spain
Sascha **Trippe**, Seoul National University, Republic of Korea
James Stuart **Urquhart**, University of Kent, UK
Lucero **Uscanga**, University of Guanajuato, Mexico
Stefanus Petrus **van den Heever**, South African Radio Astronomy Observatory, South Africa
Huib **van Langevelde**, JIVE, The Netherlands
Sophie Maria **Venselaar**, Leiden Observatory, The Netherlands
Maxim **Voronkov**, CSIRO Space & Astronomy, Australia
Jakobus Marthinus **Vorster**, University of Helsinki, Finland
Dylan **Ward**, New Mexico Institute of Mining and Technology, USA
Ryosuke **Watanabe**, Kagoshima University, Japan

Gwenllian **Williams**, University of Leeds, UK
Pawel Izydor **Wolak**, Nicolaus Copernicus University in Torun, Poland
Ka Tat **Wong**, Uppsala University, Sweden
Gang **Wu**, Max Planck Institute for Radio Astronomy, Germany
Yuanwei **Wu**, National Time Service Center of Chinese Academy of Sciences, China
Shuangjing **Xu**, Korea Astronomy and Space Science Institute, Republic of Korea
Ye **Xu**, Purple Mountain Observatory, China
Ci **Xue**, Massachusetts Institute of Technology, USA
Wenjin **Yang**, Max Planck Institute for Radio Astronomy, Germany
Yoshinori **Yonekura**, Ibaraki University, Japan
Youngjoo **Yun**, Korea Astronomy and Space Science Institute, Republic of Korea
Ingyin **Zaw**, New York University Abu Dhabi, UAE
Bo **Zhang**, Shanghai Astronomical Observatory, China
Jianjun **Zhou**, Xinjiang Astronomical Observatory, China

smic
sers
Motion
rd the
neration
Projects
Astronomical
posium 380
March 20-24
Kagoshima, Japan
AGARC
James Urquhart

Chapter 1
Cosmic Distance Scale and the Hubble Constant

Cosmic Masers: Proper Motion toward the Next-Generation Large Projects
Proceedings IAU Symposium No. 380, 2024
T. Hirota, H. Imai, K. Menten, & Y. Pihlström, eds.
doi:10.1017/S1743921323002521

Megamaser Cosmology Project II : The prospects for measuring a 1% H_0 and distances to high-z galaxies

Cheng-Yu Kuo[1,2], Dominic Pesce[3,4], Violetta Impellizzeri[5], James Braatz[6] and Mark Reid[3]

[1]Physics Department, National Sun Yat-Sen University, No. 70, Lien-Hai Road, Kaosiung City 80424, Taiwan, R.O.C. email: cykuo@mail.nsysu.edu.tw

[2]Academia Sinica Institute of Astronomy and Astrophysics, PO Box 23-141, Taipei 10617, Taiwan, R.O.C.

[3]Center for Astrophysics — Harvard & Smithsonian, 60 Garden Street, Cambridge, MA 02138, USA

[4]Black Hole Initiative at Harvard University, 20 Garden Street, Cambridge, MA 02138, USA

[5]Leiden Observatory, Leiden, the Netherlands

[6]National Radio Astronomy Observatory, 520 Edgemont Road, Charlottesville, VA 22903, USA

Abstract. H_2O megamaser emission from sub-parsec circumnuclear disks at the center of active galaxies allows a single-step, direct distance measurements to galaxies in the Hubble flow without any external calibration. Based on accurate distance determinations of six maser galaxies within 150 Mpc, the Megamaser Cosmology Project (MCP) team recently obtained $H_0 = 73.9 \pm 3.0$ km/s/Mpc ($\sim$4% accuracy), independent of distance ladders and the cosmic microwave background. To further applying the megamaser technique to attain a 1% H_0 measurement, detecting more high-quality disk maser systems is crucial. In this conference proceeding, we update the status of the MCP and discuss strategies of detecting additional high-quality disk maser galaxies within z $\sim$ 0.1. In addition, we show the prospects of reaching a 1% H_0 measurement with the supreme sensitivity of the ngVLA. Finally, we demonstrate that applying the maser technique to distance measurements of high-z galaxies with future submm VLBI systems is promising and this will allow for investigation of the new tension between the ΛCDM model and the high-z Hubble diagram.

Keywords. H_2O maser, megamaser, cosmology, quasar

1. Introduction

Astrophysical water masers are natural microwave amplifiers by stimulated emission of radiation. Observations in the past three decades have shown that maser emissions from the $6_{16} - 5_{23}$ transition of ortho-H_2O at 22.23508 GHz are abundant in star-forming regions and in the nuclear regions of some galaxies hosting active galactic nuclei (AGNs). When detected at galaxy centers, the 22 GHz H_2O masers are typically millions of times more luminous than those associated with typical star-forming regions in the Milky Way galaxy. They are therefore termed "megamasers."

H_2O megamasers provide a direct determination of the Hubble Constant, H_0, independent of standard candles (Reid et al. 2013), and enable precise measurements of supermassive black hole (BH) masses, M_{BH} (Gao et al. 2017), via sub-milliarcsecond

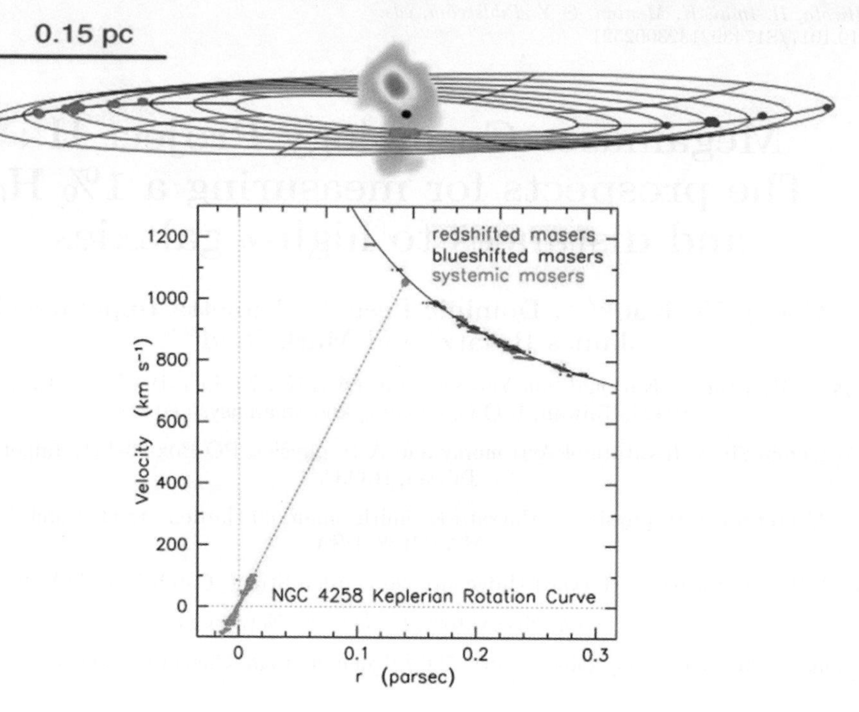

Figure 1. The distribution and Keplerian rotation curve (Argon et al. 2007) of the 22 GHz H_2O masers in NGC 4258.

imaging of H_2O maser emission from disk maser systems, such as NGC 4258 (see Figure 1; Herrnstein et al. 1999; Argon et al. 2007). In such a system, the masing gas resides in a subparsec-scale thin disk viewed almost edge-on and following near Keplerian rotation. These disk properties support M_{BH} measurements to percent-level accuracy (Kormendy & Ho 2013), and the disk maser geometrical and kinematic information can be modeled to provide a standard ruler for measuring an accurate angular-diameter distance to a galaxy well into the Hubble flow (Kuo et al. 2013) without relying on distance ladders and the cosmic microwave background (CMB).

2. The Hubble Tension Problem

From distance measurements of six maser galaxies within 150 Mpc, the Megamaser Cosmology Project (Reid et al. 2009; Braatz et al. 2010) team obtained $H_0 = 73.9 \pm 3.0$ km s^{-1} Mpc^{-1} (Pesce et al. 2020), well consistent with the majority of direct, *late universe* measurements of Hubble constant, including the H_0 obtained from Type 1a supernovae ($H_0 = 74.03 \pm 1.42$ km s^{-1} Mpc^{-1}; Riess et al. 2019) and strong gravitational lensing systems ($H_0 = 73.3 \pm 1.8$ km s^{-1} Mpc^{-1}; Wong et al. 2020), but is in tension with indirect, *early universe* H_0 predictions based on CMB data assuming a flat ΛCDM universe (e.g. $H_0 = 67.4 \pm 0.5$ km s^{-1} Mpc^{-1}; Planck Collaboration 2020). The discrepancy between the early- and late-universe determinations of H_0 is known as the "*Hubble tension problem*", which suggests new physics beyond the standard ΛCDM model (Adballa et al. 2022).

To resolve the Hubble tension, *numerous* models (i.e. >30) have been proposed to modify or replace the ΛCDM model, such as dynamical dark energy, early dark energy, phantom crossing, and modified theory of gravity (see a thorough review in Di Valentino et al. 2021). Many of these models have been shown to resolve the Hubble

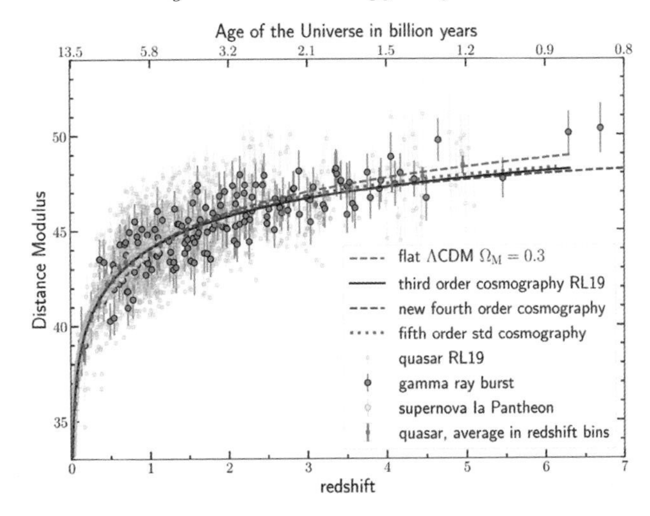

Figure 2. Hubble diagram showing a distance modulus, μ, normalized to that expected for a ΛCDM universe. The yellow dots indicate distance measurements using the quasar technique (Lusso et al. 2019).

tension if one only uses CMB data for modeling. However, when Baryon Acoustic Oscillation (BAO) data is included in the analysis, tension is restored for some models (e.g. Yang et al. 2021), making these models less favorable. This suggests that more cosmological datasets from various different probes are needed to break degeneracies between models (King et al. 2014) and to identify the best solutions that could resolve the Hubble tension.

3. The Need for New High-z Distance Indicator

In recent years, accurate measurements of distances to high-z galaxies have been proposed as an important probe that could break degeneracies between cosmological models. To demonstrate an example, King et al. (2014) show that the differences between distance moduli predicted from the standard ΛCDM and various models of dynamical dark energy are large enough to be distinguishable only at $z \gtrsim 1$, which is probed only by a small fraction of Type Ia supernovae. One needs new distance indicators beyond z∼1 to differentiate between the dynamical dark energy and ΛCDM models.

To reach the high-z universe, Lusso et al. (2019) use luminous quasars as new standard candles for measuring galaxy distances up to z∼5 based on the quasar X-ray/ultraviolet luminosity correlation. Their work shows that the best fit to the high-z Hubble diagram indicates a $\sim 4\sigma$ deviation from the ΛCDM prediction (see Figure 2), suggesting a tension with the standard model at high redshifts. However, recent studies argue that the quasar luminosity correlation is redshift-dependent, and the $\sim 4\sigma$ deviation seen in Lusso et al. (2019) could be a result of redshift evolution of quasar properties (e.g. Li et al. 2022). To avoid the impact of galaxy evolution on distance measurements, we need an alternative tool such as the H_2O maser technique, which is based on well-known gas dynamics and does not suffer from the redshift-dependent effects. When applied to high-z galaxies, this technique is promising to provide substantially more accurate distance measurements than the quasar method. In addition, it will enable direct, dynamical measurements of BH masses at $z \gtrsim 1$ for the first time, enhancing studies of the black hole/host-galaxy coevolution (Kormendy & Ho 2013) in the early universe.

4. The Importance of Improving the Hubble Constant Measurements

While resolving the Hubble tension may require new cosmological probes of the high-z universe, continuing to improve the accuracy of the Hubble constant is equally important. The long-term goal of the observational cosmology community is to attain a 1% H_0 measurement with agreement across multiple, independent observational approaches, including the H_2O megamaser technique. Having an H_0 with such an accuracy would be essential to check whether the current discrepancy in different H_0 measurements could partly or entirely result from unrecognized systematic uncertainties in underlying observations. In addition, it can provide the theoretical community a more precise determination of the difference between H_0 obtained from direct and indirect methods, giving a tighter constraints on various models of cosmology. It is likely that future joint analysis of cosmological datasets that include a 1% H_0 and accurate distances to high-z galaxies could substantially deepen our understanding of the Hubble tension problem.

5. The 2nd Phase of the Megamaser Cosmology Project : MCPII

To fully harness the potential of using the H_2O megamaser technique for studying cosmology, we attempt to apply the maser technique to both low-z and high-z galaxies in the next stage of the Megamaser Cosmology Project (MCPII). In the "low-z MCPII" project, we continue to use the H_2O megamaser technique to measure accurate distances to additional maser galaxies at low redshifts (i.e. $z \lesssim 0.1$) to achieve a $\sim$1% H_0. For the "high-z MCPII", we aim to apply the H_2O megamaser technique to high-z galaxies and use their maser-based distances as a new cosmological probe to identify the origin of the Hubble tension. The details of the two aspects of the MCPII are described as follows:

5.1. *Accurate H_0 measurements with the ngVLA at Low Redshifts*

From the latest H_0 determination from (Pesce et al. 2020), the current accuracy of H_0 achieved with the H_2O maser technique is $\sim$4%. To improve the accuracy from $\sim$4% to $\sim$1%, we need accurate distances (i.e. accuracy $\sim$ 7%) to additional 50 maser galaxies (Braatz et al. 2019). Before making new distance measurements for additional galaxies, it is essential to first looking for new, high-quality disk maser galaxies within $z \sim 0.1$ with improved survey strategy. The number of disk maser systems that have been found by previous single-dish H_2O maser surveys is $\gtrsim$30, insufficient for a 1% H_0 measurement. To find out additional >20 new disk megamasers, a blind survey on all Seyfert 2 galaxies and Liners as done in the MCP is no longer practical because the detection rate of disk megamasers is found to be only $\sim$1% based on this survey strategy. To improve the detection rate of disk maser systems, Kuo et al. (2020) found that it is easiest to find disk masers if a survey targets *Compton-thick* (CT) AGNs (i.e. column density of X-ray obscuring medium $N_H \gtrsim 10^{24}$ cm^{-2}). As shown in Figure 4, the detection rate of disk masers reach $\sim$20% if an AGN has $N_H \geq 10^{24}$ cm^{-2}. In contrast, the disk maser detection rate drops rapidly when $N_H < 10^{24}$ cm^{-2}. This suggests that targeting Compton-thick AGNs can boost the disk maser detection rate by a factor of $\sim$20, allowing one to identify the required number of disk megamasers by surveying a much smaller sample of galaxies.

In addition to searching for additional disk maser systems with new survey strategy, improving the sensitivity of current Very Long Baseline Interferometry (VLBI) facilities is also critical. Among the $\sim$30 disk megamasers that have been discovered, only $\sim$10 have sufficiently strong maser lines and the required spectral characteristics† for accurate

† An accurate distance measurement with the H_2O megamaser method requires detections of multiple blueshifted and redshifted maser features to accurately characterize the Keplerian rotation curve of a disk maser system. In addition, the systemic masers also need to be strong enough for precise acceleration measurements (e.g. Kuo et al. 2015).

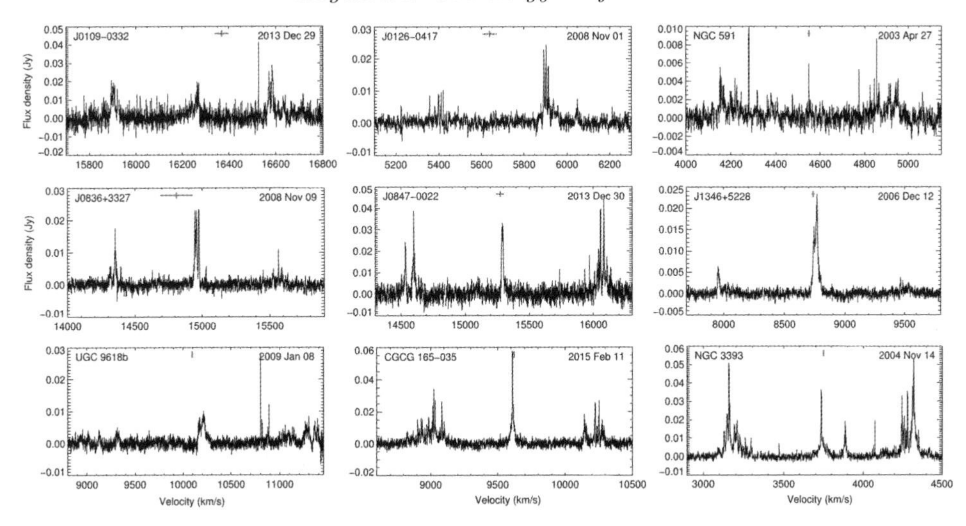

Figure 3. The 22 GHz H_2O maser spectra of nine disk maser systems discovered by the MCP (Pesce et al. 2015). These systems are a subset of the candidate disk megamasers in which the maser features are either too faint or the velocity ranges covered by the systemic masers are too narrow to allow for accurate distance measurements with the sensitivity of current VLBI facilities such as the High Sensitivity Array.

distance measurements. For the rest of ~20 disk maser systems (see Figure 3; Pesce et al. 2015), the maser emissions are too faint to enable accurate distance measurements. To measure accurate distances to these maser galaxies, the sensitivity of current VLBI facilities needs to be improved significantly. This will be realized soon with the advent of the next-generation Very Large Array (ngVLA), which would bring about an order of magnitude improvement in sensitivity. The enhanced sensitivity of the ngVLA will not only allow for efficient detections of maser systems in a ~30 times larger volume, it will also permit accurate distance measurements for the fainter disk maser systems, making a 1% H_0 determination possible with the maser technique in the future.

5.2. *Reaching the High-z Universe with Submillimeter H_2O Gigamaser Emissions*

Based on a relatively recent modeling of H_2O maser emissions (Gray et al. 2016), it has been shown that some of submillimeter H_2O maser lines can be significantly stronger than the well-known 22 GHz water maser, suggesting that submillimeter water masers may be a better tool to access the high-z universe than the 22 GHz maser given the availability of Atacama Large Millimeter/submillimeter Array (ALMA). Among the most luminous submillimeter water masers that could be present in the circumnuclear gas disks in high-z galaxies, the H_2O maser emissions $4_{14} - 3_{21}$ at 380.19736 GHz is one of the most promising lines that would allow for distance measurements of high-z galaxies. The 380 GHz transition corresponds to the back-bones levels, i.e. levels with the lowest energy for a given rotational quantum number (e.g. de Jong 1973). Members of the back-bone group are expected to display large line strengths resulting in large optical depths, slowing down radiative decay and thus carrying the bulk of the population. Based on modelling done by (Gray et al. 2016), the 380 GHz maser transition can be intrinsically brighter than the 22 GHz water maser line by a factor of $\gtrsim$10 under certain physical conditions, making high signal-to-noise detections of high-z maser emissions possible.

Despite the line strength of the 380 GHz maser could be significantly higher than that of the 22 GHz H_2O maser, its frequency falls into a deep absorption trough in the Earth's atmosphere. As a result, it can only be observed from space if the maser emission

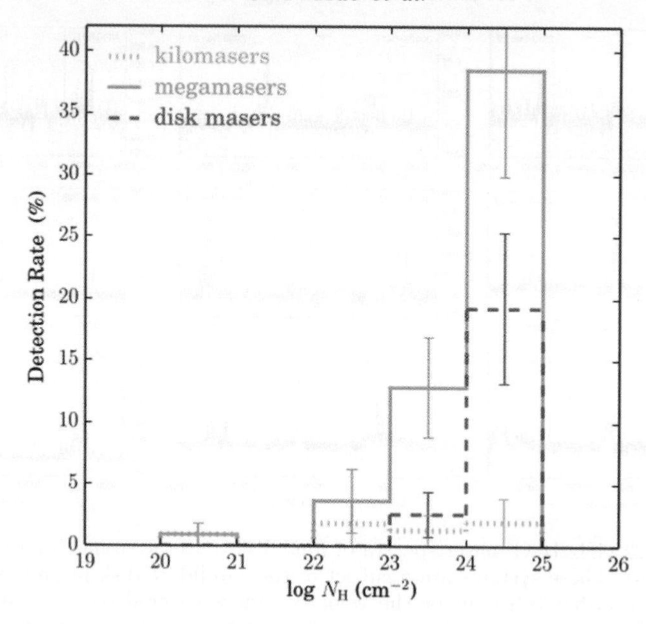

Figure 4. Detection rates of megamasers and disk masers as a function of column density N_H (Kuo et al. 2020).

is from a source in the local universe. However, since we are targeting high-z galaxies, the 380 GHz lines are redshifted to frequency bands that can be easily accessible for follow-up by existing VLBI facilities, which can provide extremely high resolution maser mapping (e.g. ~20−40 μas). Based on our recent model prediction of the angular sizes of H_2O maser disks at z > 1 (i.e. ~3.4 mas or ~30 pc in physical size; Kuo et al. in prep.), we infer that if $\gtrsim 5\sigma$ detections of the 380 GHz maser emissions can be achieved in future follow-up VLBI observations that include ALMA, the maser maps would have fractional uncertainties (i.e. maser position uncertainties $\lesssim 10$ μas) nearly as small as those in the maser disk in NGC 4258 (Figure 1), making highly accurate distance and BH mass measurements feasible at high redshifts.

The 380 GHz H_2O maser line has been detected toward ~10 starburst galaxies and 1 type-1 quasar at z $\gtrsim$ 2 in recent ALMA cycles (Yang et al. in prep.), but robust detections of disk maser systems have not yet been made. To optimize the chance of detecting disk maser systems, it is beneficial to first target *Compton-thick* quasars at z ~ 2 (i.e. the peak of quasar space density) in future ALMA observations.

One expects higher maser detection rates in these systems because the fraction of nearly edge-on disk masers is known to be highest in CT AGNs (see Figure 4; Kuo et al. 2020). In addition, a recent modeling of the physical conditions in disk maser systems (Kuo et al. in prep.) suggests that the maser amplification path length $L_{\rm amp}$ for an edge-on maser disk is strongly correlated with BH mass. Given that the BH mass function of luminous quasars at z ~ 2 peaks at $M_{\rm BH} \sim (1-3)\times10^9 M_\odot$ (i.e. $\gtrsim$100 times more massive than BHs in local H_2O megamasers; see Figure 5; Vestergaard & Osmer 2009), this model predicts that $L_{\rm amp}$ for a quasar maser disk at z ~ 2 would be significantly greater than that for a low-z disk maser, leading to a boost in maser intensity by a factor of $\gtrsim$100 for saturated masers.

Finally, since the 380 GHz maser transition could be intrinsically brighter than the 22 GHz water maser line by a factor of $\gtrsim$10 (Gray et al. 2016), the total maser luminosity of the 380 GHz line of a quasar maser disk at z ~ 2 is likely to be $\gtrsim$1000 times greater than that of a low-z 22 GHz water maser disk, making detections of high-z 380 GHz

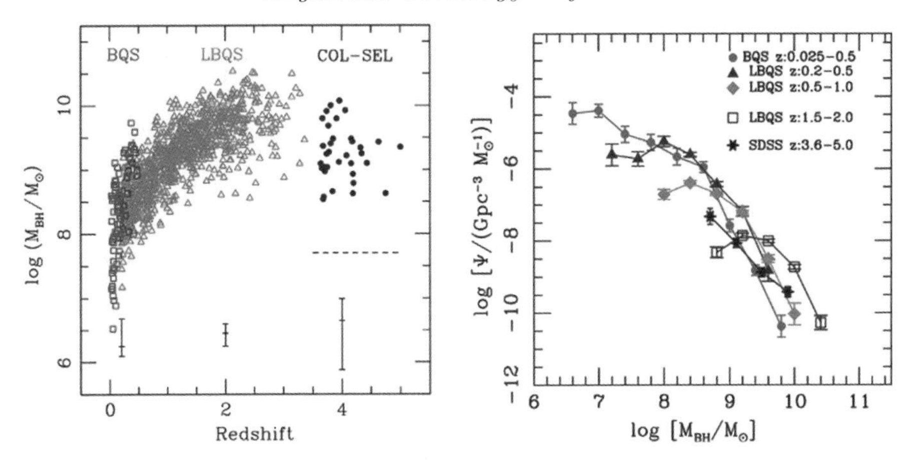

Figure 5. The BH masses distributions of luminous quasars obtained from Vestergaard & Kumar (2009). The left panel shows the BH mass as a function of redshift and the right panel shows the BH mass functions for various redshift bins. These figures show that at the redshift around z $\sim$ 2, the BH mass function peaks at $M_{BH} \sim 10^9\ M_\odot$.

maser emissions possible without gravitational lensing. We can thus call these high-z 380 GHz water maser systems "*Gigamasers*". Depending redshifts (i.e. $1.2 < z < 3.5$; set by the spectral coverage of ALMA bands 3 through 5), the peak flux densities of the 380 GHz maser emissions from CT quasars are expected to be $\sim$1.5$-$4 mJy, estimated based on the typical peak luminosity densities of the 22 GHz water maser lines L_ν in the low-z disk megamaser systems ($L_\nu \sim (2-4) \times 10^{29}$ erg/s/Hz; see Figure 6) plus the expectation that the 380 GHz maser emissions from high-z quasar maser disks could be $\gtrsim$1000 stronger than the local 22 GHz maser disks. Given the sensitivities of ALMA, the 380 GHz lines of high-z CT quasars can be detected within integration times of $\sim$1$-$5 hours.

In the past two decades, deep X-ray observations of high-z galaxies have led to detections of more than 100 CT quasars at $z > 1$ (e.g. Lanzuisi et al. 2009; Gilli et al. 2022). Table 1 shows some CT quasars discovered by (Brightman et al. 2014). If our modeling of disk maser systems is correct, it is promising that ALMA will detect H_2O gigamasers from the high redshift universe in the near future.

6. Conclusion

The studies of H_2O disk megamaser systems in the past two decades have demonstrated that the H_2O megamaser technique enables an accurate, robust determination of the Hubble constant and provide an important consistency check for H_0 values obtained across multiple, independent observational approaches. Comparisons of H_0 from different methods have clearly indicated a significant tension between direct measurements of H_0 and predictions based on the CMB assuming the ΛCDM model. To understand the origin of the Hubble tension problem, we have shown that achieving a 1% H_0 measurement is important. In addition, accurate distance measurements of high-z galaxies would be a new probe that could break degeneracies between models that could resolve the Hubble tension. For the purpose of improving the accuracy of H_0 and measuring precise distances to high-z galaxies, the H_2O megamaser technique will play an important role in the future. At low redshifts, the supreme sensitivity of the ngVLA will allow for efficient detections of disk megamaser systems within $z \lesssim 0.1$ and enable accurate distance measurements for $\gtrsim$50 galaxies, making a 1% Hubble constant measurement possible. For the high redshift universe, we predict that some of the submillimeter H_2O maser lines such as the 380 GHz

Table 1. Compton-thick Quasars in the Chandra Deep Field South and the COSMOS field.

Name	z	F_ν	band	$\log N_{\rm H}$	$\log L_{\rm bol}$	Field
GOODS-CDFS-MUSIC 07810	2.578	106.26	3	25.7	45.7	CDFS
VCDFS 69258	3.153	91.54	3	24.53	45.7	CDFS
[BMP 2010]2575	2.563	106.71	3	24.05	45.25	CDFS
ACS-GC 20094946	2.704	102.65	3	25.55	47.6	COSMOS
COSMOS2015 0663735	3.175	91.07	3	24.14	46.2	COSMOS
COSMOS2015 0441487	3.471	85.03	3	24.95	46.5	COSMOS
ACS-GC 90044755	2.026	125.63	4	24.12	45.7	CDFS
ACS-GC 90018695	1.374	160.42	4	26.0	45.7	CDFS
ACS-GC 20040645	1.796	135.98	4	24.05	46.2	COSMOS
NMBS C30828	1.692	141.63	4	24.8	46.2	COSMOS
ACS-GC 20125049	1.933	128.97	4	24.8	46.5	COSMOS
COSMOS2015 575250	1.407	157.95	4	24.1	45.4	COSMOS
ACS-GC 20088692	1.272	167.34	5	24.5	45.9	COSMOS
ACS-GC 20116538	1.265	167.86	5	24.6	46.4	COSMOS
COSMOS2015 797617	1.244	169.43	5	24.2	46.4	COSMOS

Note. CT quasars found by Brightman et al. (2014). Column (1): source name; Column (2): redshift; Column (3): observing frequency; Column (4): observing ALMA band; Column(5): absorption column density $N_{\rm H}$ (cm^{-2}); Column(6): the bolometric luminosity of the quasar estimated from absorption-corrected intrinsic X-ray luminosity.

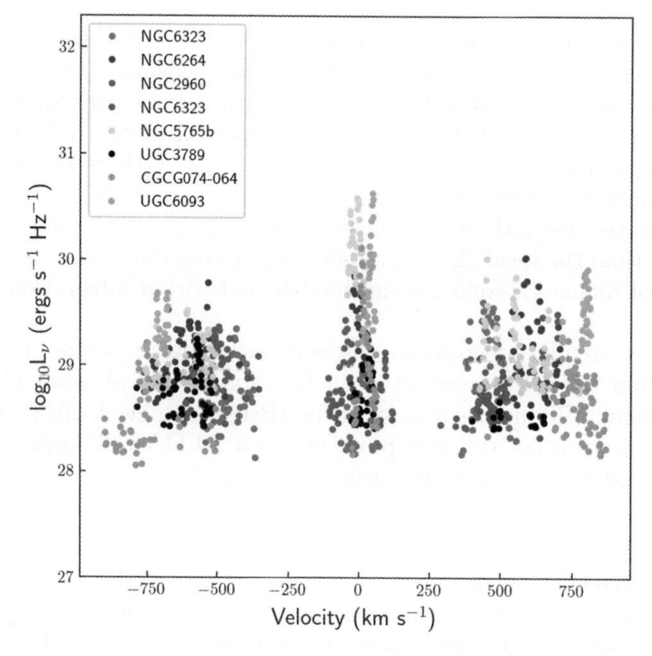

Figure 6. The luminosity densities L_ν of the 22 GHz H_2O maser emissions of eight edge-on disk maser systems discovered by the MCP (Kuo et al. in prep). The median luminosity densities of these systems, especially their redshifted and blueshifted maser components, are all around $L_\nu \sim 10^{29}$ erg s^{-1}Hz^{-1}, suggesting that edge-on maser disks behave like standard candles. This property enables predictions of the flux densities of maser emissions from edge-on H_2O maser disks at different redshifts.

maser transitions from highly obscured quasars could be $\gtrsim$1000 times more luminous that local 22 GHz disk megamasers. This makes the detections of these systems possible with ALMA without strong gravitational lensing. The follow-up VLBI observations of these water "gigamasers" are promising to give maser maps that are as accurate as the prototypical maser system NGC 4258, making precise distance measurements of high-z galaxies feasible. By combining the low-z and high-z distance measurements, the H_2O megamaser technique would provide a unique probe of the cosmic expansion history over a wide range of redshifts, permitting independent constraints on new models of cosmology.

References

Abdalla, E., Abellán, G., Aboubrahim, A. *et al.* 2022, *JHEAp*, 34, 49

Argon, A. L., Greenhill, L. J., Reid, M. J., Moran, J. M., Humphreys, E. M. L. 2007, *ApJ*, 659, 1040

Braatz, J. A., Reid, M. J., Humphreys, E. M. L., Henkel, C., Condon, J. J., Lo, K. Y. 2010, *ApJ*, 718, 657

Braatz, J., Pesce, D., Condon, J., Reid, M. 2019, *BAAS*, 51, 446

Brightman, M., Nandra, K., Salvato, M., Hsu, L.-T., Aird, J.; Rangel, C. 2014, *MNRAS*, 443, 1999

de Jong, T. 1973, *A&A*, 26, 297

Gao, F., Braatz, J. A., Reid, M. J. *et al.* 2017, *ApJ*, 834, 52

Gilli, R., Norman, C., Calura, F. *et al.* 2022, *A&A*, 666, 17

Gray, M. D., Baudry, A., Richards, A. M. S. *et al.* 2016, *MNRAS*, 456, 374

Herrnstein, J. R., Moran, J. M., Greenhill, L. J. *et al.* 1999, *Nature*, 400, 539

Karwal & Kamionkowski 2016, *PhRvD*, 94, 103523

King, A. L., Davis, T. M., Denney, K. D., Vestergaard, M., Watson, D. 2014, *MNRAS*, 441, 3454

Kormendy & Ho 2013, *ARA&A*, 51, 511

Kuo, C. Y., Braatz, J. A., Reid, M. J. *et al.* 2013, *ApJ*, 767, 155

Kuo, C. Y., Braatz, J. A., Lo, K. Y *et al.* 2015, *ApJ*, 800, 26

Kuo, C. Y., Braatz, J. A., Impellizzeri, C. M. V. *et al.* 2020, *ApJ*, 892, 18

Li, Z., Huang, L., Wang, J. 2022, *MNRAS*, 517, 1901

Lanzuisi, G., Piconcelli, E., Fiore, F. *et al.* 2009, *A&A*, 498, 67

Lo 2005, *ARA&A*, 43, 625

Lusso, E., Piedipalumbo, E., Risaliti, G. *et al.* 2019, *A&A*, 628, 4

Pesce, D. W., Braatz, J. A., Condon, J. J. *et al.* 2015, *ApJ*, 810, 65

Pesce, D. W., Braatz, J. A., Reid, M. J. *et al.* 2020, *ApJ*, 891, 1

Planck Collaboration *et al.* 2020, *A&A*, 641, 6

Raveri 2020, *RhRvD*, 101, 083524

Reid, M. J., Braatz, J. A., Condon, J. J., Greenhill, L. J., Henkel, C., Lo, K. Y. 2009, *ApJ*, 695, 287

Reid, M. J., Braatz, J. A., Condon, J. J. *et al.* 2013, *ApJ*, 767, 154

Riess, A. G., Casertano, S., Yuan, W. *et al.* 2019, *ApJ*, 876, 85

Di Valentino, E., Mena, O., Pan, S. *et al.* 2021, *Class. Quantum Grav.* 38 153001

Vestergaard & Osmer 2009, *ApJ*, 699, 800

Wong, K. C., Suyu, S. H., Chen, G. C.-F. *et al.* 2020, *MNRAS*, 498, 1420

Yang, W., Di Valentino, E., Pan, S., Wu, Y., Lu, J. 2021, *MNRAS*, 501, 5845

Cosmic Masers: Proper Motion toward the Next-Generation Large Projects
Proceedings IAU Symposium No. 380, 2024
T. Hirota, H. Imai, K. Menten, & Y. Pihlström, eds.
doi:10.1017/S1743921323002971

Distance of the Seyfert 2 galaxy IC2560 and the Hubble constant

Naomasa Nakai[1], Aya Yamauchi[2], Madoka Yamazaki[3] and Reo Harada[1]

[1]Kwansei Gakuin University, Sanda-shi, Hyogo 669-1330, Japan. email: ilo77771@kwansei.ac.jp

[2]National Astronomical Observatory of Japan, Oshu-shi, Iwate 023-0861, Japan

[3]University of Tsukuba, Tsukuba-shi, Ibaraki 305-8571, Japan

Abstract. Water vapor maser emission from a Seyfert 2 galaxy IC2560 was observed with sensitive VLBI. The accurate black hole mass was determined by separating the mass of the disk surrounding the black hole. In addition, the distance of the galaxy was directly measured from the maser disk. The Hubble constant determined using the distances and recession velocities of IC2560 and othere megamasers is inconsistent with that determined with CMB by 3–6 km s^{-1} Mpc^{-1}.

Keywords. Active galactic nuclei, Angular diameter distance, Hubble constant, Water vapor maser

---

1. Introduction

IC 2560 is a SBb galaxy located at the distance of 44 Mpc. Its nucleus is classified as a Seyfert 2. Braatz *et al.* (1996) detected water vapor maser emission at the systemic velocity of the galaxy, Ishihara *et al.* (2001) detected blue- and red-shifted high-velocity features (Figure 1) and made VLBI observations of the systemic velocity features, and Yamauchi *et al.* (2012) made VLBI observations of the systemic features and a part of red-shifted features. In additon, Yamauchi *et al.* (2012) detected the velocity drift of the systemic features.

We made new sensitive VLBI observations of both the systemic velocity features and all blue- and red-shifted high-velocity features in 2010, using VLBA and GBT. The synthesized beam was about 1.5×0.35 mas. Here we show results of the observations with VLBI, determine the distance of the galaxy and estimate the Hubble constant. Details of the results will be presented by Yamauchi *et al.* (in preparation).

2. Maser distribution and rotation curve

Figure 2a shows the distribution of the maser spots on the sky which exhibits an edge-on maser disk. Figure 2b is the velocity of the maser spots along the disk. By fitting the velocity distribution with the power low of the rotation curve profile, α, the systemic velocity $V_{\rm sys}$ and the center position of the galaxy were determined. Figure 2c is the resultant rotation curve of the maser spots. If we fit the rotation curve with $V_{\rm rot} \propto r^{\alpha}$, the power is $\alpha = -0.415 \pm 0.022$. If $|\alpha| = 0.5$, the rotation is Keplerian, i.e., only point mass exists at the center. The lower power of $|\alpha| < 0.5$ however indicates non-negligible mass distribution around the central mass. Thus we adopt two mass components model of the central mass and disk mass rotationg around the center showed in Figure 3. In

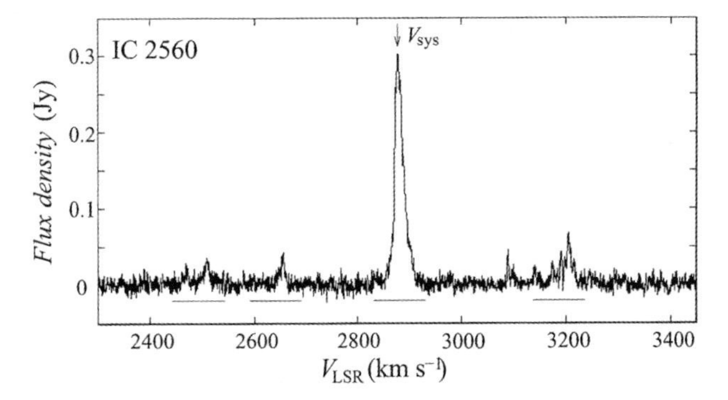

Figure 1. Maser spectrum of IC2560.

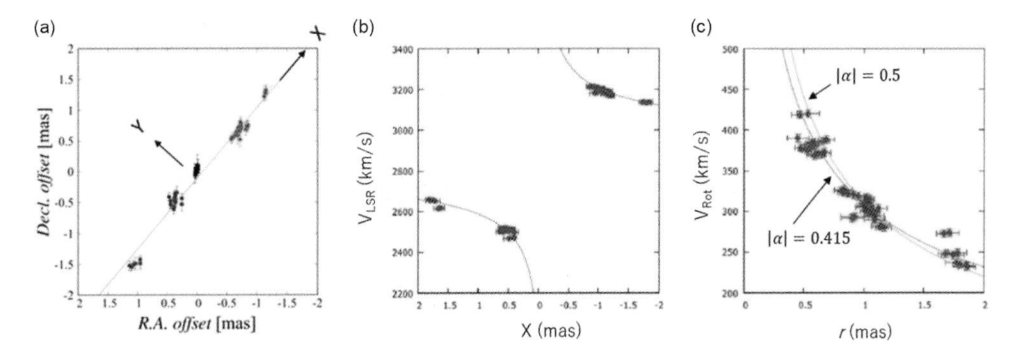

Figure 2. (a) Distribution of maser spots on the sky. (b) Velocities along the maser disk. (c) Rotation curve of the maser disk.

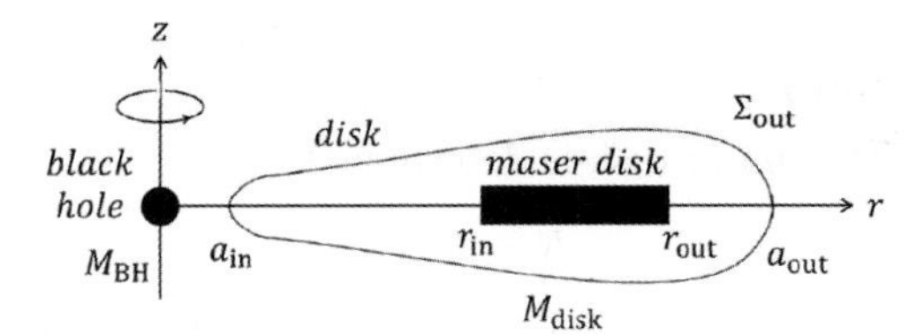

Figure 3. Schematic view of the central black hole and the gas disk rotating aroud the black hole.

this case, the centripetal force gives the eqation,

$$\frac{V_{\rm rot}^2}{r} = G\frac{M_{\rm BH}}{r^2} + \frac{\partial \Psi_{\rm disk}}{\partial r} \tag{1}$$

where $M_{\rm BH}$ is the central mass (i.e., black hole mass) and $\Psi_{\rm disk}$ the potential of the disk at r. We follow the mass model of the disk by Hure (2002), Hure & Hersant (2007) and Hure *et al.* (2008), adopting the distribution of the surface density in the disk,

$$\Sigma_{\rm disk}(r) = \Sigma_{\rm out}\left(\frac{r}{a_{\rm out}}\right)^{\beta}. \tag{2}$$

Adopting the Mestel disk ($\beta = -1$) and assuming $a_{\rm in} \ll a_{\rm out}$ and $r_{\rm out} \ll a_{\rm out}$,

$$\frac{\partial \Psi_{\rm disk}}{\partial r} = 2\pi G \Sigma_{\rm out}\frac{a_{\rm out}}{r}. \tag{3}$$

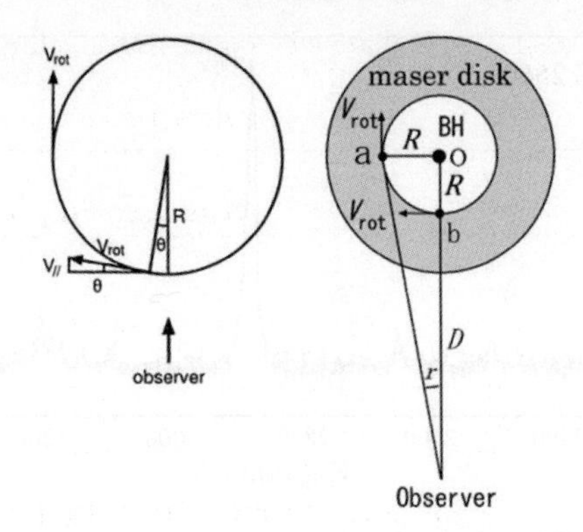

Figure 4. Geometry of the maser disk.

Equations (1) and (2) gives

$$V_{\rm rot} = \sqrt{\frac{GM_{\rm BH}}{r} + 2\pi G\Sigma_{\rm out}a_{\rm out}}. \tag{4}$$

Fitting the rotation curve (Figure 2c) with equation (4),

$$M_{\rm BH} = (4.14 \pm 0.17) \times 10^6 [M_{\rm o}]. \tag{5}$$

The disk mass within the outer edge of the maser disk is given by

$$M_{\rm disk} = \int_{a_{\rm in}}^{r_{\rm out}} \Sigma_{\rm disk}(r) 2\pi r dr = 2\pi\Sigma_{\rm out}a_{\rm out}\left(r_{\rm out} - a_{\rm in}\right) = (1.67 \pm 0.27) \times 10^6 [M_{\rm o}], \tag{6}$$

which is 40 % of the black hole mass.

3. Distance of IC2560

The linear radius of the maser disk is determined by

$$R = V_{\rm rot}^2/\alpha, \tag{7}$$

using the accelation of the systemic features, $\alpha = dV_{||}/dt = 2.593 \pm 0.027$ km s^{-1} yr^{-1} measured by Yamauchi *et al.* (2012) (Figure 4). The apparent radius, r, is directly determined by the rotation curve and the systemic features, as shown in Figure 5. Then the distance to the galaxy is determined to be $D = R/r = 44.5 \pm 6.1$ Mpc.

4. Hubble constant

The flat universe model ($K = 0$, $\Omega_{\rm m} + \Omega_\lambda = 1$) gives the angular-size distance,

$$D_{\rm A} \cong \frac{cz}{H_{\rm o}}\left(1 - \left(2 + \frac{3}{2}\Omega_{\rm m}\right)\frac{z}{2}\right), \tag{8}$$

where K is the curvature of the universe, $\Omega_{\rm m} = 0.315$ the matter density parameter and Ω_λ the dark energy density parameter. The velocity of IC2560 is $cz = 3160.0 \pm 4.7$ km s^{-1}, corrected the proper motion of our Galaxy to CMB. Figure 6 is the Hubble diagram of Pesce *et al.* (2020) adding IC2560 but removing NGC 4258 which is too close. Least square fitting gives the Hubble constant of $H_{\rm o} = 72.8 \pm 2.5$ km s^{-1} Mpc^{-1} which is different from the values obtained from observations of CMB, $H_{\rm o} = 69.32 \pm 0.80$ km s^{-1} Mpc^{-1} by Bennett *et al.* (2013) and $H_{\rm o} = 67.4 \pm 0.5$ km s^{-1} Mpc^{-1} by Planck collaboration (2020), confirming Hubble tension.

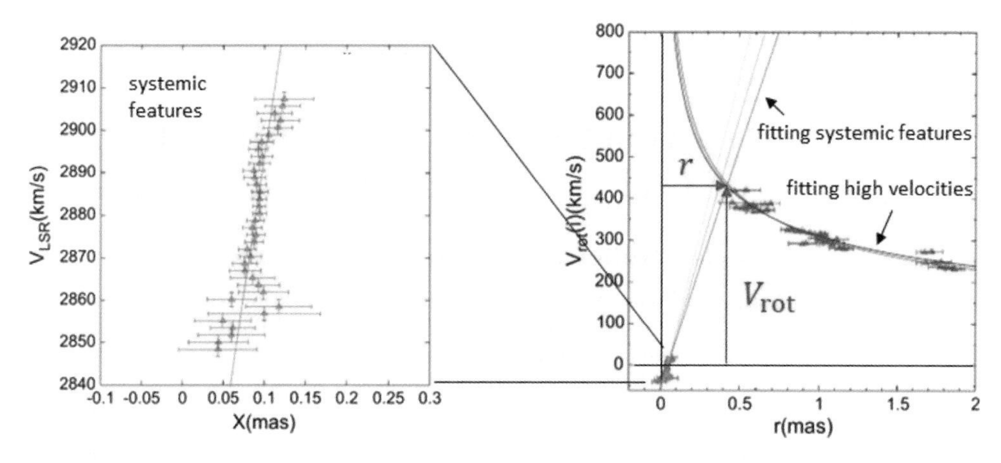

Figure 5. Velocities of the systemic and high-velocity features as a function of the distance from the center.

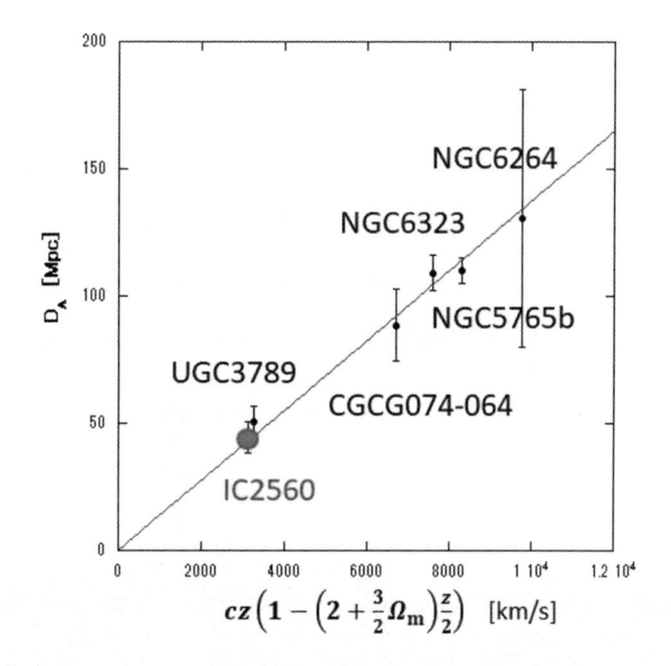

Figure 6. Relation between the recession velocity and the distance of megamasers.

References

Bennett, C. L., *et al.* 2013, *ApJS*, 208, 20

Braatz, J. A., Wilson, A. S., & Henkel, C., 1996, *ApJS*, 106, 51

Hure, J.-M., 2002, *A&A*, 395, L21

Hure, J.-M., & Hersant, F. 2007, *A&A*, 467, 907

Hure, J.-M., Hersant, F., Carreau, C., & Busset, J.-P. 2008, *A&A*, 490, 477

Ishihara, Y., Nakai, N., Iyomoto, N., Makishima, K., Diamond, P., & Hall, P. 2001, *PASJ*, 53, 215

Pesce, D. W., *et al.* 2020, *ApJ*, 891, L1

Planck collaboration., 2020, *A&A*, 641, A6

Yamauchi, A., Nakai, N., Ishihara, Y., Diamond, P., & Sato, N. 2001, *PASJ*, 64, 103

Cosmic Masers: Proper Motion toward the Next-Generation Large Projects
Proceedings IAU Symposium No. 380, 2024
T. Hirota, H. Imai, K. Menten, & Y. Pihlström, eds.
doi:10.1017/S1743921323002314

The Past, Present, and Groundbreaking Future of OH Megamaser Discoveries

Hayley Roberts and Jeremy Darling

Center for Astrophysics and Space Astronomy, Department of Astrophysical and Planetary Science, University of Colorado, 389 UCB, Boulder, CO 80309-0389, USA.
email: hayley.roberts@colorado.edu

Abstract. OH megamasers (OHMs) are luminous masers found in (ultra-)luminous infrared galaxies ([U]LIRGs). OHMs are signposts of major gas-rich mergers associated with some of the most extreme star forming regions in our universe. The dominant OH masing line, occurring at 1667 MHz, can spoof the 1420 MHz neutral hydrogen (H I) line in untargeted H I emission line surveys. While only ∼120 OHMs are currently known, H I surveys on next-generation radio telescopes, such as the Square Kilometre Array (SKA) and its precursors, will detect unprecedented numbers of OHMs. This surge in detections will not only fundamentally change what we know about the OHM population, but will also unlock our ability to implement OHMs as tracers of major mergers and extreme star formation on cosmic scales. Here we present predictions for the number of OHMs that will be detected by these surveys. We also present our novel methods for identifying these interlopers using a k-Nearest Neighbors machine learning algorithm. Preliminary data from H I surveys on precursor SKA telescopes is being used to vet and strengthen these methods as well as give us a first look at a new era in OHM science. From a detection of one of the most luminous OHMs to the discovery of a megamaser at a record-shattering redshift, these new sources are glimpses into how our understanding of the known OHM population will soon be expanding and shifting rapidly and how they will influence our understanding of galaxy evolution.

Keywords. OH megamasers, merging galaxies, starburst galaxies

---

1. Introduction

OH megamasers (OHMs) are rare, luminous 18 cm masers produced in the late stages of major galaxy mergers, generating isotropic line luminosities of $L_{\rm OH} \geq 10\,L_\odot$ with line widths ranging from 10 to 1000 $\rm km\,s^{-1}$. Often, OHMs with line luminosities of $L_{\rm OH} \geq 10^4\,L_\odot$ are referred to as OH *giga*masers but the distinction is arbitrary. This phenomena has only been discovered in ∼120 galaxies and predominantly at redshifts less than $z = 0.265$ (Roberts et al. in prep.). The longstanding record holder for highest redshift was IRAS 14070+0525, discovered by Baan et al. (1992), which held the record from 1992 until 2022.

OHMs are found in (ultra-)luminous infrared galaxies ([U]LIRGs) and are signposts of extreme star formation (Lockett & Elitzur 2008), high molecular gas density (Willett et al. 2011), and strong far-IR radiation (Baan et al. 1989). Though currently rare and known only at limited redshifts, understanding these sources will eventually allow us to study what role these sources play in galaxy evolution, particularly star formation feedback and black hole accretion. Further, as OHMs are found in late-stage major galaxy mergers, they can independently constrain the major merger rate (Roberts et al. 2021). They have also been used as in-situ magnetometers via observations of Zeeman

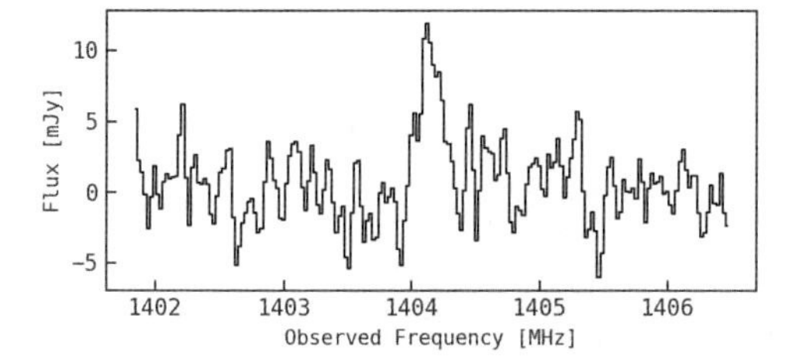

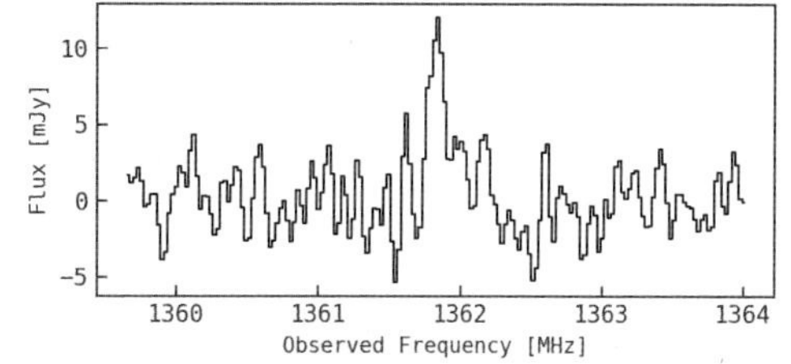

Figure 1. Example spectra from ALFALFA (Haynes et al. 2018) of OH and H I sources – the left panel shows an OH masing line from an OHM host ($\nu_{\mathrm{OH,rest}} = 1667$ MHz, $z = 0.188$) and the right panel shows an H I emission line from a non-masing starforming galaxy ($\nu_{\mathrm{HI,rest}} = 1420$ MHz, $z = 0.043$).

splitting (Robishaw et al. 2008; McBride *et al.* 2013). Their extraordinary properties and correlation with galaxy mergers make OHMs an invaluable tool for understanding galaxy evolution.

2. The Past: Previous OHM Searches

Despite their apparent utility for tracing major mergers and studying extreme star-forming galaxies, efficient methods for finding OHM host galaxies have eluded those searching for them (Roberts et al. in prep.; and references therein). In the 40 years since their discovery, significant efforts have been made to isolate the conditions or properties that can be used to identify OHM host galaxies. However, more has been understood about what is *not* associated with OHMs than with what is associated with them. While OH luminosity (L_{OH}) shows good correlation with IR luminosity (L_{IR}), ~80% of (U)LIRGs show no OHM activity (Darling & Giovanelli 2002; Lo 2005). While 50-90% of non-masing (U)LIRGs show evidence of active galactic nuclei (AGN), only 10-20% of OHMs indicate AGN activity (Willett et al. 2011). When it seemed that OHMs may be a distinct class of (U)LIRGs, no optical nuclear distinction could be found between masing and non-masing (U)LIRGs (Darling & Giovanelli 2006). Darling (2007) shows that the dense gas fraction may differentiate between masing and non-masing (U)LIRGs. However, this analysis is only done using eight OHM hosts at a limited redshift range and with unresolved dense gas measurements, requiring further investigation before this distinction can be confirmed. Existing OHMs have been extensively studied and there are still no clear markers of what conditions spur masing action nor what separates masing from non-masing (U)LIRGs.

The majority of past OHM searches have focused on targeting IR-luminous galaxies, which, in conjunction with the fact that only 20% of (U)LIRGs host OHMs, explains the low success rates in these searches. However, with no other properties of the host galaxies clearly distinguishing likely OHM hosts, it is impossible to formulate a more selective method for searching for OHMs. Further, until more are discovered, we lack the ability to entirely understand these sources and use them as tools in studying galaxy evolution.

3. The Present: Development of New OHM Finding Methods

First predicted by Briggs (1998), the 18 cm masing line from an OHM at z_{OH} can "spoof" the 21 cm neutral hydrogen (H I) emission line at a different redshift, z_{HI}, if $\nu_{\mathrm{HI}}/(1+z_{\mathrm{HI}}) = \nu_{\mathrm{OH}}/(1+z_{\mathrm{OH}})$ where $\nu_{\mathrm{HI}} = 1420.4$ MHz and $\nu_{\mathrm{OH}} = 1667.4$ MHz. The OH and H I emission lines have similar linewidths in their respective environments: H I in spiral galaxies and OH in major galaxy mergers, as shown in Figure 1. For many sources, distinguishing between these lines requires an independent measurement of the galaxy's spectroscopic redshift to determine the rest wavelength for an observed emission line.

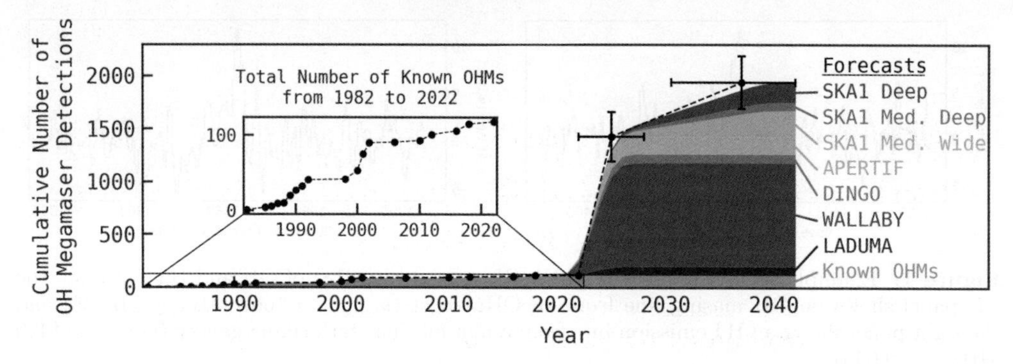

Figure 2. The cumulative number of OHMs detected up to 2023 and forecasts, broken down by current and upcoming H I survey contributions. The two markers with error bars are using projections from Roberts et al. (2021) and the errors in year are estimated using projected dates of survey completion considering typical telescope delays, particularly in the case of the SKA1, whose first light was recently predicted to be in 2027.

As a small demonstration of this, Morganti et al. (2006) fortuitously detected an OHM when searching for H I in galaxies. However, Suess et al. (2016) demonstrated the power of using H I surveys to find OHM hosts by identifying five previously unknown OHMs interloping as H I sources in the 40% data release of Arecibo Legacy Fast Arecibo L-Band Feed Array (ALFALFA; Haynes et al. 2011).

Roberts et al. (2021) presented new methods for flagging potential OHMs in H I surveys using machine learning algorithms and near- to mid-IR photometry. Using these methods, Roberts et al. (in prep.) demonstrates the ability of these methods to find new OHMs by finding five additional OHMs in the full ALFALFA data release (Haynes et al. 2018). These new OHM discoveries, however, are just the beginning.

4. The Groundbreaking Future: The OHM Renaissance in H I Surveys

While finding OHMs in H I surveys is a promising method for expanding the known population of OHM hosts, this method of OHM identification has suffered from limited H I surveys. However, we are currently on the cusp of a new era of H I science with the construction of the Square Kilometre Array (SKA) and its precursors. H I surveys are these telescopes will detect unprecedented numbers of H I sources and, subsequently, OHMs (Roberts et al. 2021). Recent detections from surveys on next-generation radio telescopes highlight this upcoming future. Hess et al. (2021) reports an OHM detection in APERTIF, an H I survey on the Westerbork Synthesis Radio Telescope. This detection presents one of the most luminous OHMs ever found and places an upper limit on the 1612 MHz OH satellite line, making it only the fourth OHM with such a measurement. In addition, the record for highest-redshift OHM was recently shattered with the discovery of an OHM at a redshift of $z = 0.52$ with LADUMA (Looking at the Distant Universe with the MeerKAT Array; Glowacki et al. 2022). However, these two detections are just a small glimpse as the future. Including surveys such as Widefield ASKAP L-band Legacy All-sky Blind surveY (WALLABY) and Deep Investigation of Neutral Gas Origins (DINGO) on the Australian SKA Pathfinder (ASKAP; Duffy et al. 2012), the OHM discovery space will be soon rapidly expanding.

Figure 2 shows projections of OHM detections for a number of H I surveys on SKA precursors as well as fiducial H I surveys of the first phase of the SKA, SKA1. Predictions are obtained from Roberts et al. (2021) which also contains detailed information for each survey. The inset plot shows how past OHM detections evolved in comparison. Even in the most conservative estimates, the next two decades of OHM detections will dominate those from the past four decades. In addition to the volume of detections, the quality and diversity of those discoveries will be unmatched, as demonstrated by the recent detections from APERTIF (Hess et al. 2021) and LADUMA (Glowacki et al. 2022). In addition, the tools necessary for identifying interloping OHMs, such as those presented in Roberts et al. (2021), are currently being tested and will be imperative for ushering in this exciting future of OHM science.

5. Conclusions

After forty years of limited OHM detections and discoveries, the era of next-generation H I surveys will be initiating a renaissance of OHM science. Using detections from these surveys in conjunction with novel OHM flagging methods, the number of known OHMs will expand by an order of magnitude, unveiling *thousands* of new sources and fundamentally altering the landscape of what we know about OHMs and their host galaxies. Using this newly expanded, diverse population of OHMs, we will be able to investigate what host galaxy properties are associated with OHMs and potentially finally isolate the mechanisms needed to identify OHM hosts. Further, these sources are science-rich for studying galaxy evolution, extreme star formation, and other host galaxy properties as discussed in these proceedings. This future of OHM detections will unlock the full potential of OHMs as tracers of some of the most extreme conditions in our universe.

References

Baan, W. A., Rhoads, J., Fisher, K., *et al.* 1992, *ApJL*, 396, L99

Baan, W. A., Haschick, A. D., & Henkel, C. 1989, *ApJ*, 346, 680

Briggs, F. H. 1998, *A&A*, 336, 815

Darling, J. & Giovanelli, R. 2006, *AJ*, 132, 2596

Darling, J. 2007, *ApJL*, 669, L9

Darling, J. & Giovanelli, R. 2002, *AJ*, 124, 100

Duffy, A. R., Meyer, M. J., Staveley-Smith, L., *et al.* 2012, *MNRAS*, 426, 3385

Glowacki, M., Collier, J. D., Kazemi-Moridani, A., *et al.* 2022, *ApJL*, 931, L7

Haynes, M. P., Giovanelli, R., Martin, A. M., *et al.* 2011, *AJ*, 142, 170

Haynes, M. P., Giovanelli, R., Kent, B. R., *et al.* 2018, *ApJ*, 861, 49

Hess, K. M., Roberts, H., Dénes, H., *et al.* 2021, *A&A*, 647, A193

Lo, K. Y. 2005, *ARA&A*, 43, 625

Lockett, P. & Elitzur, M. 2008, *ApJ*, 677, 985

McBride, J. & Heiles, C. 2013, *ApJ*, 763, 8

Morganti, R., de Zeeuw, P. T., Oosterloo, T. A., *et al.* 2006, *MNRAS*, 371, 157

Roberts, H., Darling, J., & Baker, A. J. 2021, *ApJ*, 911, 38

Roberts, H., Darling, J., Hess, K. M. & Baker, A. J. in prep., *ApJ*

Robishaw, T., Quataert, E., & Heiles, C. 2008, *ApJ*, 680, 981

Suess, K. A., Darling, J., Haynes, M. P., *et al.* 2016, *MNRAS*, 459, 220

Willett, K. W., Darling, J., Spoon, H. W. W., *et al.* 2011, *ApJ*, 730, 56

Top; Conference room in the Li-Ka Nangoku Hall. Photographs taken by Ka-Yiu Shum. Bottom; poster session. Taken by Kaito Kawakami.

Chapter 2
Black Hole Masses and the M-sigma Relation

Cosmic Masers: Proper Motion toward the Next-Generation Large Projects
Proceedings IAU Symposium No. 380, 2024
T. Hirota, H. Imai, K. Menten, & Y. Pihlström, eds.
doi:10.1017/S1743921323002478

Supermassive black hole mass growth in infrared-luminous gas-rich galaxy mergers and potential power of (sub)millimeter H_2O megamaser observations

Masatoshi Imanishi

National Astronomical Observatory of Japan, 2-21-1 Osawa Mitaka, Tokyo 181-8588, Japan.
email: masa.imanishi@nao.ac.jp

Abstract. We present our systematic infrared and (sub)millimeter spectroscopic observations of gas/dust-rich merging ultraluminous infrared galaxies (ULIRGs) to scrutinize deeply buried AGNs (mass-accreting supermassive black holes [SMBHs]). We have found signatures of optically elusive, but intrinsically luminous buried AGNs in a large fraction of nearby ($z < 0.3$) ULIRGs, suggesting that SMBH mass growth is ongoing in the ULIRG population. Using ALMA, we have detected compact (<100 pc), very luminous (>$10^4 L_\odot$), AGN-origin, 183 GHz (1.6 mm) H_2O megamaser emission in one merging ULIRG, demonstrating that the megamaser emission can be a very powerful tool to dynamically estimate SMBH masses, with the smallest modeling uncertainty of kpc-wide stellar and gas mass distribution, at dusty ULIRGs' nuclei, because of minimum extinction effects at millimeter. We present our current results and future prospect for the study of the SMBH mass growth in gas/dust-rich galaxy mergers, using (sub)millimeter AGN-origin H_2O megamaser emission lines.

Keywords. Supermassive black hole, Water megamaser, Ultraluminous infrared galaxies, Active galactic nuclei, Millimeter megamaser, Gas-rich galaxy mergers

---

1. Introduction: Co-evolution of stars and supermassive black holes in the universe

The apparent ubiquity of supermassive black holes (SMBHs) at the centers of present-day galaxies, and the correlation between the masses of SMBHs and spheroidal stars (e.g., Magorrian *et al.* 1998; Ferrarese & Merritt 2000; McConnell & Ma 2013) indicate that an active galactic nucleus (AGN; SMBH-driven activity) and a starburst (active star-formation; plausibly the progenitors of spheroids) are physically connected and have *co-evolved*. The widely accepted cold dark matter-based galaxy formation scenarios (e.g., White & Rees 1978) postulate that mergers of gas-rich galaxies with SMBHs at their centers are common throughout the history of the universe (Hopkins *et al.* 2008). Numerical simulations of such galaxy mergers predict that not only *many stars are formed* rapidly, but also *SMBHs grow in mass* through high mass accretion rates, at deeply embedded nuclear regions by gas and dust (Hopkins *et al.* 2006). Such gas/dust-rich galaxy mergers become infrared luminous and are usually observed as ultraluminous infrared galaxies (ULIRGs; infrared luminosity $L_{IR} > 10^{12} L_\odot$ †; Sanders & Mirabel (1996)). Thus, ULIRGs are believed to be an important phase for the growth of SMBH mass and the *co-evolution* of SMBHs and stars in the universe.

† Throughout this manuscript, we adopt the cosmological parameters $H_0 = 71$ km s^{-1} Mpc^{-1}, $\Omega_M = 0.27$, and $\Omega_\Lambda = 0.73$.

To discuss the SMBH mass growth in ULIRGs, it is crucial to estimate the energetic contributions from AGN activity (energetically dominated by a mass-accreting SMBH), by distinguishing from starburst activity which is powered by nuclear fusion inside stars. Unlike optically (0.3–1 μm) identifiable AGNs which are surrounded by toroidally distributed (torus-shaped) dusty medium (i.e., certain opening angle where ionizing UV photons can escape) (Baldwin *et al.* 1981; Veilleux & Osterbrock 1987; Kewley *et al.* 2001; Kauffmann *et al.* 2003), ULIRGs are major mergers of gas-rich galaxies and have large amounts of concentrated molecular gas and dust in nuclear regions (Sanders & Mirabel 1996). Putative compact AGNs at ULIRGs' nuclei can easily be obscured by gas and dust *in virtually all lines of sight* (i.e., almost no opening angle), and so optical detection of AGN signatures becomes extremely difficult. However, understanding the energetic importance of such optically elusive, but intrinsically luminous *buried* AGNs is vital to clarify the true nature of the ULIRG population (Hopkins *et al.* 2005) as well as the SMBH mass growth process during gas-rich galaxy mergers (Hopkins *et al.* 2008; DeBuhr *et al.* 2011), which must have happened very frequently in the early universe, if we are based on the widely accepted cold dark matter-based galaxy formation scenario (e.g., White & Rees 1978).

2. Observations at wavelengths of low extinction

It is well known from observations that nearby ($z < 0.3$) ULIRGs are energetically dominated by compact (<kpc) nuclear regions, rather than spatially extended (>a few kpc) star-formation in the host galaxies (e.g., Soifer *et al.* 2000; Diaz-Santos *et al.* 2010; Imanishi *et al.* 2011; Pereira-Santaella *et al.* 2021). It is of particular importance to distinguish between AGNs and compact nuclear starbursts, and estimate how they contribute energetically in nearby ULIRGs. Because AGNs (=mass-accreting SMBHs) are spatially more centrally concentrated than, and are surrounded by, compact nuclear starbursts in ULIRGs' nuclei, AGNs are generally more difficult to detect observationally. It is indispensable to observe at wavelengths of low dust extinction if we are to detect putative AGNs and properly estimate their energetic importance in dusty ULIRGs' nuclei. A buried AGN (mass-accreting SMBH) has two distinguished properties, when compared to a starburst (nuclear fusion inside stars). First, because the radiative energy generation efficiency of a mass-accreting SMBH (6–42% of $\dot{M}c^2$; Bardeen (1970); Thorne (1974)) is much higher than that of nuclear fusion inside stars (~0.7% of $\dot{M}c^2$), an AGN can produce very large luminosity from a compact region (i.e., high emission surface brightness) (Soifer *et al.* 2000; Thompson *et al.* 2005; Pereira-Santaella *et al.* 2021) and create a larger amount of hot (>100K) dust than a starburst can. Thus, an AGN can become luminous in the near- to mid-infrared (1–30 μm) because of this hot dust thermal radiation. Second, X-ray to UV luminosity ratio is substantially higher in an AGN than a starburst, because of strong X-ray emission in the vicinity of a mass-accreting SMBH through the inverse Compton upscattering of UV photons to X-rays (Ranalli *et al.* 2003; Shang *et al.* 2011). If we can detect at least one of these features characteristic to an AGN, then we can separate AGN and starburst activity, and estimate how an AGN energetically contributes to the bolometric luminosity of a ULIRG. From this information, we can convert to mass accretion rate onto a SMBH.

2.1. *X-rays*

Because an AGN is a much more luminous X-ray emitting source than a starburst, detection of intrinsically luminous X-ray emission can be the most straightforward way to argue for the presence of a luminous AGN. In particular, extinction effects become smaller at hard X-rays at >10 keV (e.g., Ryter 1996; Wilms *et al.* 2000). Thus, >10

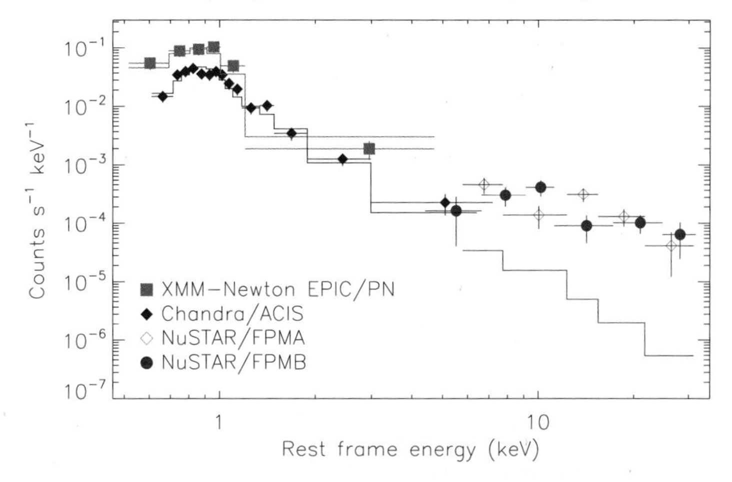

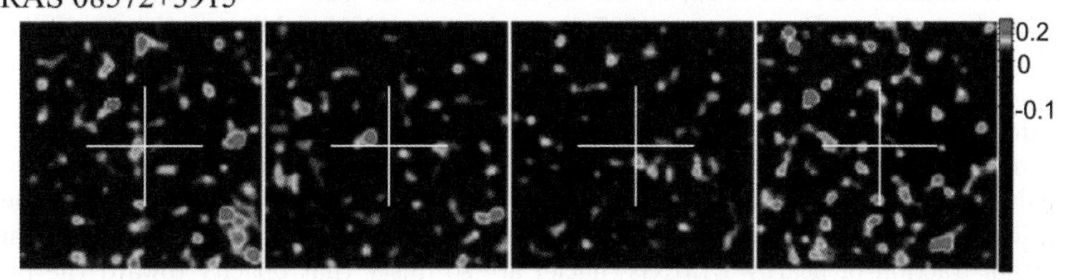

Figure 1. *(Top)*: X-ray spectrum of the LIRG NGC 6286 (L = $10^{11.4}L_\odot$; $z = 0.018$). Compared to the extrapolation from 1–5 keV X-ray emission which is dominated by starburst activity, there is an excess component at >5 keV. This excess emission is interpreted as originating from highly absorbed, but intrinsically luminous AGN activity (Ricci *et al.* 2016). *(Bottom)*: X-ray image of the ULIRG IRAS 08572+3915 (L = $10^{12.1}L_\odot$; $z = 0.058$) at *(Left)* 3–10 keV, *(Second left)* 10–20 keV, *(Second right)* 20–30 keV, and *(Right)* 30–79 keV (Teng *et al.* 2015). This ULIRG has no optical AGN signature, but is classified as buried AGN dominated in infrared spectroscopy (as we will explain later; see Figure 2 bottom middle). However, no significant X-ray emission is detected at >10 keV at the position of this ULIRG shown as the cross mark.

keV hard X-ray observations can be an effective tool to detect deeply buried AGNs by penetrating through a large column density of obscuring material at ULIRGs' nuclei. In fact, these hard X-ray observations have been successful to detect highly absorbed, but intrinsically luminous, X-ray emitting AGNs in a few luminous infrared galaxies (LIRGs; $L_{IR} = 10^{11-12}L_\odot$) and ULIRGs (e.g., Ricci *et al.* 2016, 2017; Oda *et al.* 2017; Yamada *et al.* 2021). An example is shown in Figure 1 (top). However, detection of >10 keV hard X-ray emission from ULIRGs has been unsuccessful in many cases, even for very nearby sources at $z < 0.1$ (e.g., Teng *et al.* 2015) (Figure 1 bottom). It is possible that these hard X-ray non-detected ULIRGs do not contain luminous AGNs and are starburst dominated. However, it is also possible that the putative buried AGNs in these ULIRGs suffer from heavily Compton thick X-ray absorption with hydrogen column density $N_H > 5 \times 10^{25}$ cm^{-2}, in which case detection of X-ray emission is extremely difficult even at >10 keV.

2.2. *Infrared*

Infrared 2.5–35 μm low-resolution spectroscopy is another powerful tool to detect deeply buried, but intrinsically luminous AGNs in nearby ($z < 0.3$) ULIRGs, because of much reduced dust extinction effects compared to the optical (0.3–1 μm). Furthermore, spectroscopy at this infrared wavelength range can distinguish between AGNs and starbursts, based on polycyclic aromatic hydrocarbon (PAH) emission and broad dust absorption features, as long as buried AGNs contribute significantly to observed infrared fluxes (Figure 2). This infrared spectroscopic energy diagnostic method has been applied to >100 nearby ULIRGs at $z < 0.3$ and been successful to detect luminous buried AGNs in roughly half of the observed nearby ULIRGs (e.g., Genzel *et al.* 1998; Imanishi *et al.* 2006a; Armus *et al.* 2007; Imanishi *et al.* 2007a, 2008; Nardini *et al.* 2008; Imanishi 2009; Veilleux *et al.* 2009; Nardini *et al.* 2009, 2010; Imanishi *et al.* 2010a,b). Most importantly, in this infrared spectroscopy, luminous buried AGN signatures have been found in many nearby ULIRGs with no AGN signatures in hard X-rays at 10–80 keV (Figure 1 bottom), due to heavily Compton thick X-ray absorption ($N_H > 5 \times 10^{25}$ cm^{-2}). This is because the ratios of X-ray absorption (N_H by gas and dust) to infrared dust extinction (A_V) toward obscured AGNs are known to be much higher than the Galactic value of $N_H/A_V \sim 2 \times 10^{21}$ cm^{-2} mag^{-1} (Predehl & Schmitt 1995), most likely because of the presence of large amount of dust-free X-ray absorbing gas inside the dust sublimation radius around AGNs (e.g., Alonso-Herrero *et al.* 1997; Burtscher *et al.* 2015; Ichikawa *et al.* 2019). Thus, infrared spectroscopy has turned out to be more effective than hard X-ray observations to detect deeply buried AGNs in nearby ULIRGs.

However, there remain many nearby ULIRGs which do not display buried AGN signatures in the infrared. It may be possible that these ULIRGs do not contain luminous AGNs. However, another scenario is that the putative buried AGNs can be even infrared elusive because of extremely high obscuration. In fact, there are arguments that a certain fraction of nearby ULIRGs' nuclei are so dusty that opacity toward the putative buried AGNs can become very large (optical depth $\tau > 1$) at infrared 2.5–35 μm (e.g., Downes & Eckart 2007; Sakamoto *et al.* 2008; Matsushita *et al.* 2009; Gonzalez-Alfonso *et al.* 2015; Scoville *et al.* 2017; Sakamoto *et al.* 2021; Pereira-Santaella *et al.* 2021). It is indispensable to observe at wavelengths of even lower extinction effects than 2.5–35 μm, if we are to distinguish between these two scenarios.

2.3. *(Sub)millimeter*

Extinction effects at (sub)millimeter 0.8–3.5 mm are a factor of >20 smaller than those at infrared 2.5–35 μm (Hildebrand 1983). Thus, (sub)millimeter observations can have a potential to detect infrared-elusive, extremely deeply buried AGNs in nearby ULIRGs. Then, how can we distinguish between an AGN and a starburst at (sub)millimeter? Because a luminous AGN can show much stronger X-ray and near- to mid-infrared (1–30 μm) hot (>100 K) dust emission than a starburst can, physical and chemical effects to surrounding molecular gas can be different depending on energy sources (e.g., Meijerink *et al.* 2007; Harada *et al.* 2010). Thus, molecular rotational J-transition emission line flux ratios can be different between an AGN and a starburst. If we find AGN-sensitive bright molecular J-transition lines in the (sub)millimeter wavelength range, we have a tool to detect infrared-elusive extremely deeply buried luminous AGNs in gas/dust-rich merging ULIRGs, by distinguishing from starbursts. Because ULIRG's nuclear regions are usually dominated by high density molecular gas, caused by merger-induced gas compression (Gao & Solomon 2004), molecules with large dipole moments, such as HCN, HCO^+, HNC ($\mu \gtrsim 3$ debye; C means ^{12}C throughout this manuscript), are better suited than widely used bright low-J ($J \lesssim 3$) lines of CO ($\mu = 0.1$ debye), to properly probe the physical and chemical properties of ULIRG's nuclear dense molecular gas. Pre-ALMA millimeter observations found a clear trend that optically identified luminous AGNs display higher HCN-to-HCO^+ flux ratios than starburst-dominated galaxies, at J=1–0 (rotational transition) at ~3.5 mm (e.g., Kohno 2005; Krips *et al.* 2008). Enhanced HCN

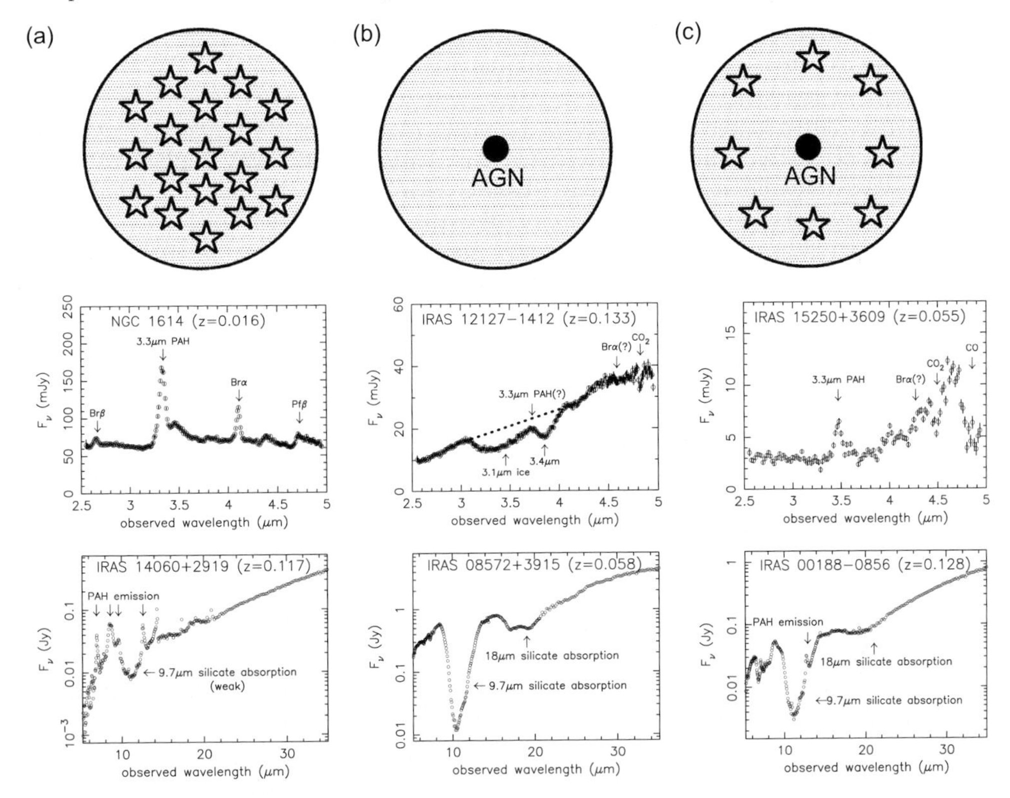

Figure 2. Infrared 2.5–35 μm low-resolution (R = 50–100) spectroscopic energy diagnostic method of galaxies, using polycyclic aromatic hydrocarbon (PAH) emission and dust absorption features (Imanishi *et al.* 2007a, 2008, 2010a). *(Top)*: Primary energy source and its geometry. *(Left)*: A pure starburst galaxy. Stellar energy sources and dust/gas are spatially well mixed. Prominent, large equivalent width, PAH emission features, excited by stellar far-UV photons in photo-dissociation regions, are found in infrared 2.5–35 μm spectra. The optical depths of dust absorption features are modest in this mixed dust/source geometry, because the observed flux largely comes from foreground, weakly obscured emission at a near side (which shows only weak dust absorption features), with small contributions from highly obscured emission at a far side (which can display deep dust absorption features). *(Middle)*: A pure buried AGN. The energy source (a mass-accreting SMBH) is more centrally concentrated than the surrounding dust. No PAH emission feature is seen, because PAH molecules are destroyed by strong AGN's X-ray radiation, in regions where PAH-exciting AGN's far-UV photons are available. PAH-free continuum emission from larger-sized hot (>100 K) dust heated by an AGN is dominant. The optical depths of dust absorption features can be very large in this centrally concentrated energy source geometry, because the so-called foreground screen dust model is applicable. *(Right)*: A buried AGN and starburst composite galaxy. PAH emission from starburst regions is seen, but its equivalent width is lower than a pure starburst-dominated galaxy, because of the dilution by AGN-originating PAH-free continuum emission. *(Second row)*: Infrared 2.5–5 μm spectra taken with the AKARI infrared satellite (Imanishi *et al.* 2008, 2010a). The 3.3 μm PAH emission feature is prominent in a starburst *(Left)*, but is weak in buried AGNs *(Middle and Right)*. The 3.1 μm and 3.4 μm absorption features come from ice-covered dust and bare carbonaceous dust, respectively. The red, steeply rising continuum flux with increasing wavelength, caused by hot dust thermal emission, also suggests the presence of a luminous buried AGN *(Middle and Right)*. Hydrogen recombination lines (Brα at 4.05 μm, Brβ at 2.63 μm, and Pfβ at 4.65 μm) can be strongly emitted in HII-regions in a starburst *(Left)*. Absorption features by CO_2 (at 4.26 μm) and/or CO (at 4.67 μm) are also seen for ULIRGs containing deeply buried AGNs *(Middle and Right)*. *(Bottom)*: Infrared 5–35 μm spectra taken with the Spitzer infrared satellite (Imanishi *et al.* 2007a). PAH emission features at 6.2 μm, 7.7 μm, 8.6 μm, and 11.3 μm, are prominent in a starburst *(Left)*, but are weak in buried AGNs *(Middle and Right)*. The 9.7 μm and 18 μm silicate dust absorption features can be strong in buried AGNs *(Middle and Right)*, but are weak in a starburst *(Left)*. All (L)LIRGs plotted in this figure do not display obvious AGN signatures in optical spectra (optically classified as HII-regions or LINERs).

abundance caused by AGN radiation and/or mechanical effects is regarded to be responsible for the elevated HCN emission in luminous AGNs (e.g., Kohno 2005; Izumi *et al.* 2013; Krips *et al.* 2008; Aladro *et al.* 2015; Saito *et al.* 2018; Nakajima *et al.* 2018; Takano *et al.* 2019; Kameno *et al.* 2020; Imanishi *et al.* 2020; Butterworth *et al.* 2022; Imanishi *et al.* 2023). This HCN-to-HCO^+ flux ratio method was applied to nearby ULIRGs, and it was demonstrated that ULIRGs with infrared-classified luminous buried AGNs tend to display higher HCN-to-HCO^+ flux ratios than known starburst-dominated galaxies at J=1–0 (e.g., Imanishi *et al.* 2006b, 2007b, 2009; Privon *et al.* 2015), at J=2–1 (e.g., Imanishi *et al.* 2022), at J=3–2 (e.g., Imanishi *et al.* 2016b, 2019) (Figure 3 bottom left), and at J=4–3 (e.g., Izumi *et al.* 2016; Imanishi *et al.* 2018) (Figure 3 top). In Figure 3 (bottom right), we plot the observed HCN-to-HCO^+ flux ratios at J=3–2 in the ordinate, as a function of infrared-estimated AGN's bolometric contributions (Nardini *et al.* 2010) in the abscissa. Not only infrared-identified luminous buried AGNs, but also infrared-classified starbursts, without luminous buried AGN signatures in the infrared, display high HCN-to-HCO^+ flux ratios with larger than unity, as expected for AGN-important galaxies. If the high HCN-to-HCO^+ flux ratios are a reliable AGN indicator, then this result may suggest that (i) almost all nearby ULIRGs contain luminous AGNs (i.e., mass-accreting SMBHs), and (ii) SMBH-mass growth is ubiquitous in the nearby ULIRG population. In this case, our next desire is to estimate SMBH mass and directly witness the SMBH mass growth in nearby gas/dust-rich merging ULIRGs.

3. H_2O megamaser emission and direct SMBH mass measurements in gas/dust-rich infrared luminous galaxy mergers

3.1. *Need for SMBH measurements at extinction free wavelengths*

It is *inferred* from host galaxies' stellar properties that SMBH masses in nearby ULIRGs can be as high as $M_{SMBH} = 10^{7-8} M_\odot$ (e.g., Genzel *et al.* 2001; Dasyra *et al.* 2006). It is crucial to verify this more directly. Investigating the dynamics of gas and/or stars at galaxy nuclear regions that are largely affected by the gravity of a SMBH, is one of the most direct and reliable methods to estimate the central SMBH mass (e.g., Miyoshi *et al.* 1995). Many dynamical studies of gas and/or stars at the nuclei of *less obscured galaxies* have been conducted to estimate the central SMBH masses, mostly in the optical (0.3–1 μm) and near-infrared (1–2.5 μm) wavelength ranges (e.g., Kormendy & Ho 2013). However, these conventional optical and near-infrared methods are not applicable to ULIRGs' nuclei, because of very large dust extinction. It is urgently needed to establish a solid method to probe the dynamics of material around mass-accreting SMBHs at dusty ULIRGs' nuclei, at almost-dust-extinction-free wavelengths.

As described before, dust extinction effects in the (sub)millimeter wavelength (0.8–3.5 mm) are much smaller than the optical and near-infrared, and even smaller than other widely explored wavelengths of small dust extinction, $\sim$10 keV hard X-rays and infrared $\sim$20 μm, by a factor of $>$20 (Hildebrand 1983). Furthermore, possible free-free absorption effects by ionized gas ($\propto \lambda^{2.1}$), which are sometimes significant in the centimeter wavelength (1–22 cm) at ULIRGs' nuclei (e.g., Leroy *et al.* 2011; Murphy *et al.* 2013), are usually negligible at (sub)millimeter. (Sub)millimeter observations of molecular rotational transition lines can thus potentially be the most powerful means to investigate the dynamics of gas around mass-accreting SMBHs at ULIRGs' nuclei, by penetrating through a large column density of gas and dust.

3.2. *H_2O megamaser emission*

H_2O has more complex rotational energy levels than simple molecules (e.g., CO, HCN, HNC) and it was theoretically predicted that population inversion can occur for a

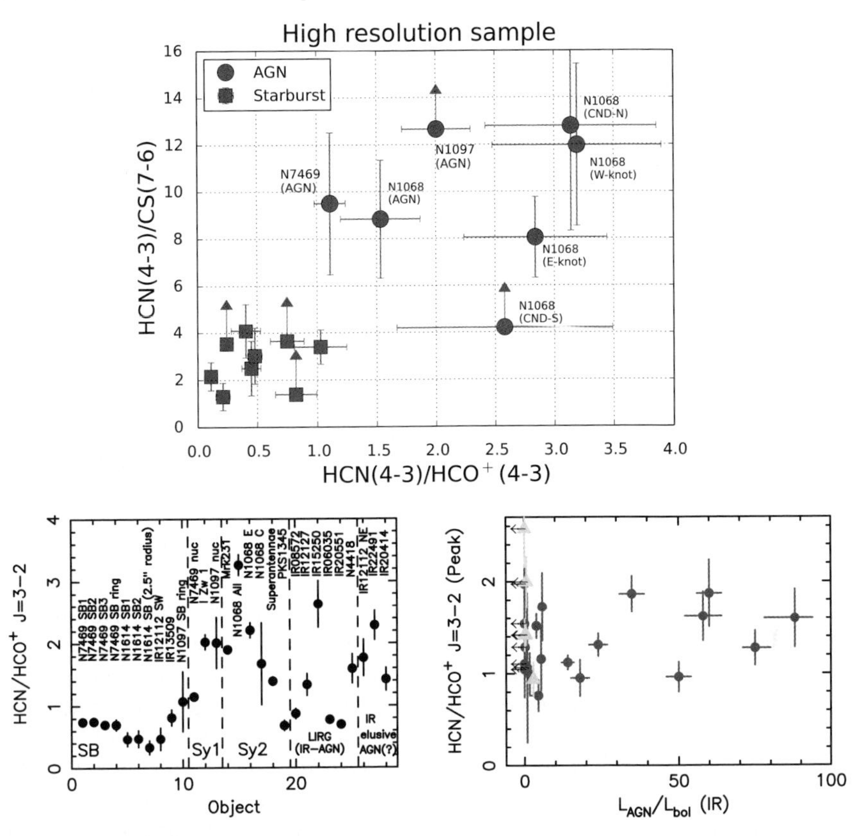

Figure 3. *(Top)*: Observed HCN-to-HCO^+ flux ratio at J=4–3 (abscissa) and HCN J=4–3 to CS J=7–6 flux ratio for galaxy nuclei observed with modestly high angular resolution (Izumi *et al.* 2016). AGN-important galaxies (red filled circles) tend to show higher HCN-to-HCO^+ flux ratios at J=4–3 than starburst-dominated galaxies (blue filled squares) in the abscissa. *(Bottom left)*: Observed HCN-to-HCO^+ flux ratios at J=3–2 (Imanishi *et al.* 2016b). "SB", "Sy1", "Sy2", "LIRG (IR-AGN)", and "IR elusive AGN(?)" mean starburst-dominated regions, optically identified Seyfert 1 type unobscured AGNs, optically identified Seyfert 2 type obscured AGNs, luminous infrared galaxies with optically elusive, but infrared-detected, buried AGN signatures, and candidates of infrared-elusive, but (sub)millimeter-detected, extremely deeply buried AGNs, respectively. Sources with AGN signatures tend to show elevated HCN-to-HCO^+ flux ratios, compared to starbursts. Some AGN-classified sources display non-high HCN-to-HCO^+ flux ratios, which can be explained by higher HCN flux attenuation by larger line opacity (not dust extinction), caused by enhanced HCN abundance in AGNs (Imanishi *et al.* 2016a,b). Optically thin ^{13}C isotopologue $H^{13}CN$-to-$H^{13}CO^+$ flux ratios are needed to remove the ambiguity coming from the possibly larger line opacity for HCN ($H^{12}CN$) than for HCO^+ ($H^{12}CO^+$). *(Bottom right)*: Comparison of infrared spectroscopically estimated AGN bolometric contribution (in %) by Nardini *et al.* (2010) (abscissa) and observed HCN-to-HCO^+ J=3–2 flux ratio at the nuclear continuum peak position within observed beam size (ordinate) in nearby ULIRGs (Imanishi *et al.* 2019). ULIRGs observed with small ($0.1''$–$0.2''$) beam sizes are plotted as red circles. Those observed with large ($0.5''$–$0.9''$) beam sizes are plotted as light blue triangles.

number of H_2O rotational transitions (e.g., 22 GHz $6_{1,6}$–$5_{2,3}$, 183 GHz $3_{1,3}$–$2_{2,0}$, and 321 GHz $10_{2,9}$–$9_{3,6}$ lines) through collisional excitation and/or infrared radiative pumping in AGN-illuminated warm and dense molecular gas (e.g., Neufeld *et al.* 1991; Yates *et al.* 1997; Maloney *et al.* 2002; Gonzalez-Alfonso *et al.* 2010; Gray *et al.* 2016). This population inversion can amplify background radiation and these *H_2O emission lines can be extremely bright* and high surface brightness *through maser phenomena.* In fact, H_2O

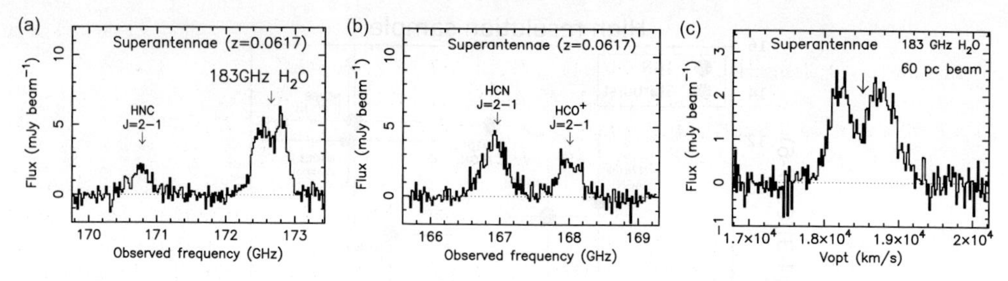

Figure 4. ALMA spectrum of the Superantennae for *(a)* HNC J=2–1 and 183 GHz H_2O lines and *(b)* HCN and HCO^+ J=2–1 lines, taken with ∼0.4″ (∼500 pc) beam (Imanishi *et al.* 2021). The abscissa is observed frequency (in GHz) and the ordinate is flux density (in mJy beam^{-1}). The *183 GHz H_2O emission line flux is even higher* than bright dense molecular gas tracers (HNC, HCN, and HCO^+ J=2–1 lines). *(c)*: Detailed velocity profile of the 183 GHz H_2O emission line in a newly taken ∼60 pc beam-sized spectrum (Imanishi *et al.* 2023b in preparation). The abscissa is optical LSR velocity (in km s^{-1}). The downward arrow indicates the systemic velocity (18510 km s^{-1}).

megamaser (>10L$_\odot$) emission was detected in many AGNs at 22 GHz (1.4 cm) through previously conducted centimeter observations (e.g., Braatz *et al.* 1997; Hagiwara *et al.* 2002; Greenhill *et al.* 2003a; Braatz *et al.* 2004; Henkel *et al.* 2005; Kondratko *et al.* 2006a,b; Yamauchi *et al.* 2017). The 22 GHz H_2O megamaser emission coming from AGN-illuminated innermost molecular gas has been used to probe gas dynamics in close proximity to central mass-accreting SMBHs and accurately measure the SMBH masses, based on follow-up centimeter VLBI (very long baseline interferometry) high-angular-resolution observations (e.g., Miyoshi *et al.* 1995; Greenhill *et al.* 1997, 2003b). Although this method is very powerful, it is applicable only to AGNs with extremely bright 22 GHz H_2O megamaser emission (e.g., Kuo *et al.* 2011, 2020a). SMBH mass in the ULIRG population has never been estimated in this method.

3.3. *(Sub)millimeter H_2O magamaser emission*

Imanishi *et al.* (2021) have recently detected a remarkably luminous (>10^{4}L$_\odot$) 183 GHz (1.6 mm) H_2O $3_{1,3}$–$2_{2,0}$ emission line in the merging ULIRG, the Superantennae (IRAS 19254−7245; $z = 0.0617$; 1″ = 1.2 kpc) (Figure 4). The detected 183 GHz H_2O emission line is unusually bright, even brighter than widely used major dense molecular gas tracers, HCN, HCO^+, and HNC J=2–1 lines (Figure 4a–b). Additionally, while emission of dense molecular gas tracers is spatially resolved with 500 pc beam-sized data (Figure 5a–b), the 183 GHz H_2O emission is not (Figure 5c), suggesting that the bright 183 GHz H_2O emission line comes from a very compact region. The most natural interpretation for the remarkably luminous, compact 183 GHz H_2O emission line is that we see 183 GHz H_2O *megamaser* emission in the AGN-illuminated innermost warm and dense molecular gas at the very center of the galaxy nucleus (Imanishi *et al.* 2021), as was detected in other very nearby (<20 Mpc) AGNs (Humphreys *et al.* 2005, 2016). No 22-GHz (1.4 cm) H_2O megamaser emission line was detected in the Superantennae (Greenhill *et al.* 2002).

Subsequent ALMA higher-spatial-resolution (∼60 pc beam) data revealed redshifted and blueshifted components at ∼100 pc scale for the 183 GHz H_2O emission line, suggesting that it is barely spatially resolved in the Superantennae (Figure 5d and 4c). Figure 6 shows a schematic diagram of the geometrical difference between the 183 GHz H_2O and dense molecular line emission in the Superantennae. Given that the 183 GHz H_2O megamaser emission selectively comes from the innermost region around a mass-accreting SMBH, this megamaser (non thermal) emission can be a more reliable tool to dynamically constrain the central SMBH mass than using spatially extended (∼kpc) thermal

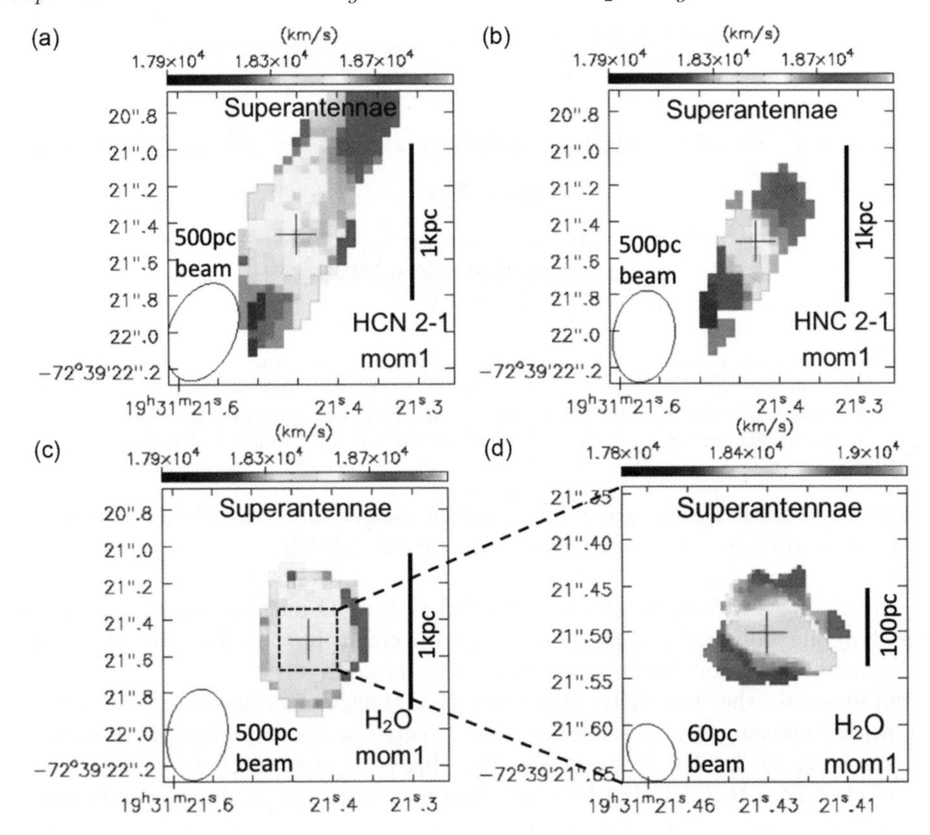

Figure 5. Intensity-weighted mean velocity (moment 1) map of *(a)* HCN J=2–1 line, *(b)* HNC J=2–1 line, *(c)* 183 GHz H_2O line of the Superantennae (Imanishi *et al.* 2021). The vertical bar in *(a)–(c)* corresponds to 1 kpc. Dense molecular gas tracers (HCN and HNC J=2–1 lines) display rotation pattern with north-western redshifted and south-eastern blueshifted components, because these lines are spatially resolved with ~500 pc beam. However, no such rotation signature is seen for the 183 GHz H_2O emission line, strongly suggesting that it is spatially unresolved (<220 pc). In *(d)*, a newly taken ~60 pc beam-sized moment 1 map of the 183 GHz H_2O line (Imanishi *et al.* 2023b in preparation) is displayed. The vertical bar in *(d)* corresponds to 100 pc. The compact 183 GHz H_2O emission is spatially resolved at ~100 pc scale, with north redshifted and south blueshifted components discernible.

(non maser) molecular line emission. From the redshifted and blueshifted components of the 183 GHz H_2O emission line (Figure 5d), enclosed mass is roughly estimated to be $\sim 10^8 M_\odot$, by assuming a simple Keplerian rotation. This is comparable to *inferred* SMBH masses in nearby ULIRGs ($M_{SMBH} = 10^{7-8} M_\odot$) from host galaxies' stellar properties (e.g., Genzel *et al.* 2001; Dasyra *et al.* 2006), although even smaller ALMA beam size is ultimately needed to better estimate the M_{SMBH} value. The *extremely bright, compact (<<1 kpc) AGN-origin 183 GHz H_2O megamaser emission* (Figure 6) *can provide the best achievable constraints on gas dynamics in the vicinity of mass-accreting SMBHs* in gas/dust-rich merging ULIRGs, with the smallest modeling uncertainty of spatially extended, >kpc-wide stellar and gas mass distribution (Boizelle *et al.* 2019).

3.4. *Future prospect*

To fully exploit the potential power of millimeter AGN-origin H_2O megamaser emission for the purpose of estimating SMBH mass in a more reliable manner and directly

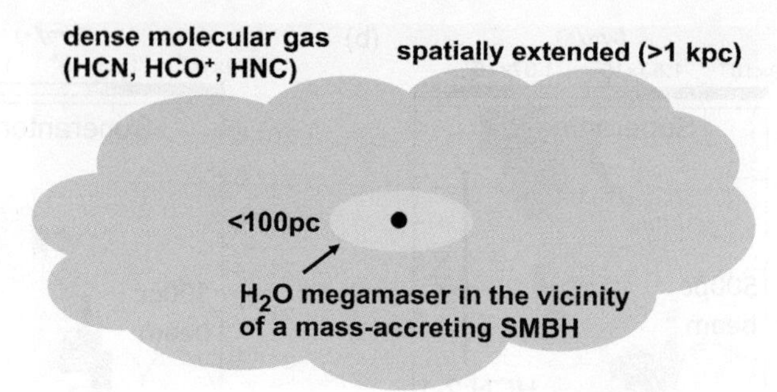

Figure 6. Schematic diagram of the H_2O megamaser emission and dense molecular gas. The AGN-origin H_2O megamaser emission comes from the very center of a galaxy nucleus (<<1 kpc), while dense molecular line emission comes from the entire nuclear region (kpc scale). Stars and low density molecular gas (probed with CO) distribute in a much wider area (>>1 kpc).

witnessing the SMBH mass growth in gas-rich galaxy mergers, we need even higher-spatial-resolution data. This is because not only the SMBH, but also gas and stars at $\lesssim$60 pc could contribute to the above estimated enclosed mass with $\sim 10^8 M_\odot$ for the Superantennae. ALMA observations of the 183 GHz H_2O megamaser emission with the longest baseline will provide a factor of $\sim$2 improvement in spatial resolution, compared to the available data for the Superantennae (Figure 5d).

In addition to the 183 GHz (1.6 mm) H_2O megamaser emission, the 321 GHz (0.9 mm) H_2O megamaser emission has been detected in nearby well-studied AGNs (e.g., Hagiwara *et al.* 2013, 2016; Pesce *et al.* 2016; Hagiwara *et al.* 2021; Kameno *et al.* 2023; Pesce *et al.* 2023). At 321 GHz (0.9 mm), even higher spatial resolution can be achieved with ALMA, because of shorter wavelength than at 183 GHz (1.6 mm). The smallest achievable angular resolution of ALMA at 321 GHz is $\lesssim 0.015''$ or as small as $\lesssim$18 pc for the Superantennae at $z \sim 0.06$. Because AGNs show bright H_2O megamaser emission at multiple lines, and not at only one line (Pesce *et al.* 2023), it is quite possible that bright AGN-origin 321 GHz H_2O megamaser emission is detected in the Superantennae. If we can detect bright, AGN-origin, compact 321 GHz H_2O megamaser emission in the Superantennae at $z \sim 0.06$, we should be able to constrain the central SMBH mass if it is $M_{SMBH} \gtrsim 5 \times 10^7 M_\odot$ (Davis 2014), given that the H_2O megamaser emitting disk in the Superantennae is highly edge-on as suggested from the spatially separated redshifted and blueshifted emission components (Figure 5d).

It is also very important to detect bright 183 GHz and/or 321 GHz H_2O megamaser emission lines in other moderately nearby ($z \lesssim 0.1$) ULIRGs as well, to verify the widely argued SMBH mass growth scenario in gas-rich galaxy mergers in more general. Previously conducted centimeter surveys of bright 22 GHz (1.4 cm) H_2O megamaser emission have shown that the megamaser detection rate can be as high as $\sim$50% if we target highly *obscured AGNs* (e.g., Panessa *et al.* 2020; Kuo *et al.* 2020a). It is quite possible that we will be able to detect bright 321 GHz (0.9 mm) and/or 183 GHz (1.6 mm) H_2O megamaser emission in multiple $z \lesssim 0.1$ ULIRGs hosting luminous obscured AGNs, if we can conduct the (sub)millimeter H_2O megamaser emission line survey. Once we can detect bright AGN-origin 321 GHz (0.9 mm) H_2O megamaser emission, then by using the highest achievable angular resolution with ALMA at 321 GHz ($\lesssim 0.015''$), we should be able to constrain SMBH mass in ULIRGs at $z \sim 0.1$, if the mass is $M_{SMBH} \gtrsim 6 \times 10^7 M_\odot$ (unless an H_2O megamaser emitting disk is highly face-on) (Davis 2014). High-spatial-resolution and high-sensitivity ALMA observations at (sub)millimeter wavelength will play a key role to most stringently test the SMBH mass growth scenario in gas-rich galaxy mergers in the universe.

4. Summary

We demonstrated from our infrared and (sub)millimeter spectroscopic observations that luminous buried AGNs are common and so SMBH mass growth is ongoing in nearby ($z < 0.3$) gas/dust-rich merging ULIRGs. It is awaited to constrain the central SMBH masses in the nearby ULIRG population at almost-extinction-free wavelengths, by penetrating through a large column density of nuclear obscuring material, to test the widely proposed SMBH mass growth scenario in gas-rich galaxy mergers. We argued that ALMA high-spatial-resolution and high-sensitivity observations of bright AGN-origin (sub)millimeter H_2O megamaser emission lines can be the most unique tool to achieve this goal, which is vital to understand the origin of the co-evolution of SMBHs and stars in our universe.

5. Acknowledgment

The author of this manuscript (M.I.) is supported by JP21K03632.

References

Aladro, R., Martin, S., Riquelme, D., *et al.* 2015, *A&A*, 579, A101
Alonso-Herrero, A., Ward, M. J., & Kotilainen, J. K. 1997, *MNRAS*, 288, 977
Armus, L., Charmandaris, V., Bernard-Salas, J., *et al.* 2007, *ApJ*, 656, 148
Baldwin, J. A., Phillips, M. M., & Terlevich, R. 1981, *PASP*, 93, 5
Bardeen, J. M. 1970, *Nature*, 226, 64
Boizelle, B. D., Barth, A. J., Walsh, J. L., *et al.* 2019, *ApJ*, 881, 10
Braatz, J. A., Henkel, C., Greenhill, L. J., Moran, J. M., & Wilson, A. S. 2004, *ApJ*, 617, L29
Braatz, J. A., Wilson, A. S., & Henkel, C. 1997, *ApJS*, 110, 321
Burtscher, L., Orban de Xivry, G., Davies, R. I., *et al.* 2015, *A&A*, 578, 47
Butterworth, J., Holdship, J., Viti, S., & Garcia-Burillo, S. 2022, *A&A*, 667, A131
Dasyra, K. M., Tacconi, L. J., Davies, R. I., *et al.* 2006, *ApJ*, 651, 835
Davis, T. A., 2014, *MNRAS*, 443, 911
DeBuhr, J., Quataert, E., & Ma, C-P. 2011, *MNRAS*, 412, 1341
Diaz-Santos, T., Charmandaris, V., Armus, L., *et al.* 2010, *ApJ*, 723, 993
Downes, D., & Eckart, A. 2007, *A&A*, 468, L57
Ferrarese, L., & Merritt, D. 2000, *ApJ*, 539, L9
Gao, Y., & Solomon, P. M. 2004, *ApJS*, 152, 63
Genzel, R., Lutz, D., Sturm, E., *et al.* 1998, *ApJ*, 498, 579
Genzel, R., Tacconi, L. J., Rigopoulou, D., *et al.* 2001, *ApJ*, 563, 527
Gonzalez-Alfonso, E., Fischer, J., Isaak, K., *et al.* 2010, *A&A*, 518, L43
Gonzalez-Alfonso, E., Fischer, J., Sturm, E., *et al.* 2015, *ApJ*, 800, 69
Gray, M. D., Baudry, A., Richards, A. M. S., *et al.* 2016, *MNRAS*, 456, 374
Greenhill, L. J., Booth, R. S., Ellingsen, S. P., *et al.* 2003a, *ApJ*, 590, 162
Greenhill, L. J., Ellingsen, S. P., Norris, R. P., *et al.* 2002, *ApJ*, 565, 836
Greenhill, L. J., Gwinn, C. R., Antonucci, R., & Barvainis, R. 1996, *ApJ*, 472, L21
Greenhill, L. J., Kondratko, P. T., Lovell, J. E. J., *et al.* 2003b, *ApJ*, 582, L11
Greenhill, L. J., Moran, J. M., & Herrnstein, J. R. 1997, *ApJ*, 481, L23
Hagiwara, Y., Diamond, P. J., & Miyoshi, M. 2002, *A&A*, 383, 65
Hagiwara, Y., Horiuchi, S., Doi, A., Miyoshi, M., & Edwards, P. G. 2016, *ApJ*, 827, 69,
Hagiwara, Y., Horiuchi, S., Imanishi, M., & Edwards, P. G. 2021, *ApJ*, 923, 251
Hagiwara, Y., Miyoshi, M., Doi, A., & Horiuchi, S. 2013, *ApJ*, 768, L38
Harada, N., Herbst, E., & Wakelam, V. 2010, *ApJ*, 721, 1570
Henkel, C., Peck, A. B., Tarchi, A., *et al.* 2005, *A&A*, 436, 75
Hildebrand, R. H. 1983, *QJRAS*, 24, 267
Hopkins, P. F., Hernquist, L., Cox, T. J., *et al.* 2005, *ApJ*, 630, 705
Hopkins, P. F., Hernquist, L., Cox, T. J., *et al.* 2006, *ApJS*, 163, 1
Hopkins, P. F., Hernquist, L., Cox, T. J., & Keres, D. 2008, *ApJS*, 175, 356

Humphreys, E. M. L. Greenhill, L. J., Reid, M. J., *et al.* 2005, *ApJ*, 634, L133
Humphreys, E. M. L., Vlemmings, W. H. T., Impellizzeri, C. M. V., *et al.* 2016, *A&A*, 592, L13
Ichikawa, K., Ricci, C., Ueda, Y., *et al.* 2019, *ApJ*, 870, 31
Imanishi, M. 2009, *ApJ*, 694, 751
Imanishi, M., Baba, S., Nakanishi, K., & Izumi, T. 2023, *ApJ*, 950, 75
Imanishi, M., Dudley, C. C., & Maloney, P. R. 2006a, *ApJ*, 637, 114
Imanishi, M., Dudley, C. C., Maiolino, R., *et al.* 2007a, *ApJS*, 171, 72
Imanishi, M., Hagiwara, Y., Horiuchi, S., Izumi, T., & Nakanishi, K. 2021, *MNRAS*, 502, L79
Imanishi, M., Imase, K., Oi, N., & Ichikawa, K. 2011, *AJ*, 141, 156
Imanishi, M., Maiolino, R., & Nakagawa, T. 2010b, *ApJ*, 709, 801
Imanishi, M., Nakagawa, T., Ohyama, Y., *et al.* 2008, *PASJ*, 60, S489
Imanishi, M., Nakagawa, T., Shirahata, M., Ohyama, Y., & Onaka, T. 2010a, *ApJ*, 721, 1233
Imanishi, M., Nakanishi, K., & Izumi, T. 2016a, *ApJ*, 825, 44
Imanishi, M., Nakanishi, K., & Izumi, T. 2016b, *AJ*, 152, 218
Imanishi, M., Nakanishi, K., & Izumi, T. 2018, *ApJ*, 856, 143
Imanishi, M., Nakanishi, K., & Izumi, T. 2019, *ApJS*, 241, 19
Imanishi, M., Nakanishi, K., Izumi, T., & Baba, S., 2022, *ApJ*, 926, 159
Imanishi, M., Nakanishi, K., & Kohno, K. 2006b, *AJ*, 131, 2888
Imanishi, M., Nakanishi, K., Tamura, Y., Oi, N., & Kohno, K. 2007b, *AJ*, 134, 2366
Imanishi, M., Nakanishi, K., Tamura, Y., & Peng, C. -H. 2009, *AJ*, 137, 3581
Imanishi, M., Nguyen, D. D., Wada, K., *et al.* 2020, *ApJ*, 902, 99
Izumi, T., Kohno, K., Aalto, S., *et al.* 2016, *ApJ*, 818, 42
Izumi, T., Kohno, K., Martin, S., *et al.* 2013, *PASJ*, 65, 100
Kameno, S., Harikane, Y., Sawada-Satoh, S., *et al.* 2023, *PASJ*, 75, L1
Kameno, S., Sawada-Satoh, S., Impellizzeri, C. M. V., *et al.* 2020, *ApJ*, 895, 73
Kauffmann, G., Heckman, T. M., Tremonti, C., *et al.* 2003, *MNRAS*, 346, 1055
Kewley, L. J., Heisler, C. A., Dopita, M. A., & Lumsden, S. 2001, *ApJS*, 132, 37
Kohno, K. 2005, in AIP Conf. Ser. 783, The Evolution of Starbursts, ed. S. Hüttemeister, E. Manthey, D. Bomans, & K. Weis (New York: AIP), 203 (astro-ph/0508420)
Kondratko, P. T., Greenhill, L. J., & Moran, J. M. 2006a, *ApJ*, 652, 136
Kondratko, P. T., Greenhill, L. J., Moran, J. M., *et al.* 2006b, *ApJ*, 638, 100
Kormendy, J., & Ho, L. C. 2013, *ARAA*, 51, 511
Krips, M., Neri, R., Garcia-Burillo, S., *et al.* 2008, *ApJ*, 677, 262
Kuo, C. Y., Braatz, J. A., Condon, J. J., *et al.* 2011, *ApJ*, 727, 20
Kuo, C. Y., Braatz, J. A., Impellizzeri, C. M. V., et al. 2020a, *MNRAS*, 498, 1609
Kuo, C. Y., Hsiang, J. Y., Chung, H. H., *et al.* 2020b, *ApJ*, 892, 18
Leroy, A., Evans, A., Momjian, E., *et al.* 2011, *ApJL*, 739, L25
Magorrian, J., *et al.* 1998, *ApJ*, 115, 2285
Maloney, P. R. 2002, *PASA*, 19, 401
Matsushita, S., Iono, D., Petitpas, G. R., *et al.* 2009, *ApJ*, 693, 56
McConnell, N. J. & Ma, C-P. 2013, *ApJ*, 764, 184
Meijerink, R., Spaans, M., & Israel, F. P. 2007, *A&A*, 461, 793
Miyoshi, M., Moran, J., & Herrnstein, J., *et al.* 1995, *Nature*, 373, 127
Murphy, E. J., Stierwalt, S., Armus, L., *et al.* 2013, *ApJ*, 768, 2
Nakajima, T., Takano, S., Kohno, K., Harada, N., & Herbst, E. 2018, *PASJ*, 70, 7
Nardini, E., Risaliti, G., Salvati, M., *et al.* 2008, *MNRAS*, 385, L130
Nardini, E., Risaliti, G., Salvati, M., *et al.* 2009, *MNRAS*, 399, 1373
Nardini, E., Risaliti, G., Watabe, Y., Salvati, M., & Sani, E. 2010, *MNRAS*, 405, 2505
Neufeld, D. A., & Melnick, G. J. 1991, *ApJ*, 368, 215
Oda, S., Tanimoto, A., Ueda, Y., *et al.* 2017, *ApJ*, 835, 179
Panessa, F., Castangia, P., Malizia, A., *et al.* 2020, *A&A*, 641, A162
Pereira-Santaella, M., Colina, L., Garcia-Burillo, S., *et al.* 2021, *A&A*, 651, A42
Pesce, D. W., Braatz, J. A., & Impellizzeri, C. M. V., 2016, *ApJ*, 827, 68

Pesce, D. W., Braatz, J. A., Henkel, C., *et al.* 2023, *ApJ*, submitted (arXiv:2302.02572)
Predehl, P., & Schmitt, J. H. M. M. 1995, *A&A*, 293, 889
Privon, G. C., Herrero-Illana, R., Evans, A. S., *et al.* 2015, *ApJ*, 814, 39
Ranalli, P., Comastri, A., & Setti, G. 2003, *A&A*, 399, 39
Ricci, C., Bauer, F. E., Treister, E., *et al.* 2016, *ApJ*, 819, 4
Ricci, C., Bauer, F. E., Treister, E., *et al.* 2017, *MNRAS*, 468, 1273
Ryter, C. E. 1996, *Ap&SS*, 236, 285
Saito, T., Iono, D., Ueda, J., *et al.* 2018, *MNRAS*, 475, L52
Sakamoto, K., Gonzalez-Alfonso, E., Martin, S., *et al.* 2021, *ApJ*, 923, 206
Sakamoto, K., Wang, J., Wiedner, M. C., *et al.* 2008, *ApJ*, 684, 957
Sanders, D. B., & Mirabel, I. F. 1996, *ARAA*, 34, 749
Scoville, N., Murchikova, L., Walter, F., *et al.* 2017, *ApJ*, 836, 66
Shang, Z., Brotherton, M. S., Wills, B. J., *et al.* 2011, *ApJ*, 196, 2
Soifer, B. T., Neugebauer, G., Matthews, K., *et al.* 2000, *AJ*, 119, 509
Takano, S., Nakajima, T., & Kohno, K. 2019, *PASJ*, 71, S20
Teng, S. H., Rigby, J. R., Stern, D., *et al.* 2015, *ApJ*, 814, 56
Thompson, T. A., Quataert, E., & Murray, N. 2005, *ApJ*, 630, 167
Thorne, K. S. 1974, *ApJ*, 191, 507
Veilleux, S., & Osterbrock, D. E. 1987, *ApJS*, 63, 295
Veilleux, S., Rupke, D. S. N., Kim, D.-C., *et al.* 2009, *ApJS*, 182, 628
Yamada, S., Ueda, Y., Tanimoto, A., *et al.* 2021, *ApJS*, 257, 61
Yamauchi, A., Miyamoto, Y., Nakai, N., *et al.* 2017, *PASJ*, 69, L6
Yates, J. A., Field, D., & Gray, M. D. 1997, *MNRAS*, 285, 303
White, S. D. M., & Rees, M. J. 1978, *MNRAS*, 183, 341
Wilms, J., Allen, A., & McCray, R. 2000, *ApJ*, 542, 914

Cosmic Masers: Proper Motion toward the Next-Generation Large Projects
Proceedings IAU Symposium No. 380, 2024
T. Hirota, H. Imai, K. Menten, & Y. Pihlström, eds.
doi:10.1017/S174392132300203X

A 4000 $M_\odot$ supermassive star as a possible source for the W1 kilomaser

Katarzyna Nowak and Martin G. H. Krause

Centre for Astrophysics Research, Department of Physics, Astronomy and Mathematics, University of Hertfordshire, College Lane, Hatfield AL10 9AB, UK. email: k.nowak@herts.ac.uk

Abstract. Supermassive stars have been proposed as the solution to a number of longstanding problems in globular cluster formation. The hypothetical stars have been suggested as potential polluters responsible for the observed chemical peculiarities within those clusters. In recent hydrodynamic simulations, we have demonstrated that accretion discs around such stars are stable even with large stellar accretion and flyby rates and produce H_2O kilomasers. We propose that the W1 kilomaser, associated with a super star cluster in the starburst galaxy NGC 253, may arise in an accretion disc around a supermassive star with a mass of around 4000 $M_\odot$.

Keywords. star clusters, masers, accretion discs, hydrodynamics

---

1. Globular clusters

Globular clusters are known as some of the oldest objects in the universe, with the oldest ones having an age of more than 13 Gyr. These ancient clusters exhibit unique chemical compositions, a characteristic that has puzzled astronomers for decades. Studies dating back to the 1960s revealed an interesting phenomenon known as the '2nd parameter problem', where globular clusters with the same metallicity display different horizontal branch morphologies in their colour-magnitude diagrams (Sandage & Wildey, 1967). Spectroscopic measurements conducted in the 2000s demonstrated that most Galactic and extra-galactic globular clusters exhibit multiple main sequences across the colour magnitude diagram (Anderson *et al.* 2002). This is considered to be a consequence of variations in helium abundance (Norris, 2004), providing evidence that globular clusters harbor multiple stellar populations (Bastian & Lardo, 2018). Globular clusters also display large variations in light elements, particularly Na-O, C-N and Mg-Al anticorrelations (Denissenkov *et al.* 1990), with the most prominent feature in most of them being the Na-O anticorrelation. A hot-hydrogen burning environment is needed to vary those abundances with the concurrent p-capturing reactions of the CNO-cycle ($\gtrsim$ 20 MK), NeNa ($\gtrsim$ 45 MK) and MgAl ($\gtrsim$ 70 MK) chains leading to the rise of those anticorrelations (Gratton *et al.* 2012; Prantzos *et al.* 2017). Most models, seeking to explain these anomalies in globular clusters invoke a process called self-enrichment, wherein certain stars within a cluster, known as polluters, enrich other stars within the same cluster.

2. Supermassive stars

Supermassive stars (SMS, >1000 $M_\odot$) are hypothetical stars, proposed as candidates for polluters responsible for the chemical peculiarities in those ancient clusters. These stars reach the essential central temperature required to activate the MgAl chain early in their evolution, when the abundance of helium is still low (Prantzos *et al.*

2017). During their early main sequence phase, the H-burning products of supermassive stars exhibit agreement with various observed anticorrelations in globular clusters (Denissenkov & Hartwick, 2014). It is assumed that supermassive stars are fully convective and release material through a radiatively driven wind at the beginning of their main sequence phase. The wind is quickly decelerated by interaction with dense gas in the embedded cluster. The ejecta would then mix with star-forming gas that either accretes on to protostars or collapses to form stars independently (Krause *et al.* 2020). The model proposed by Gieles *et al.* (2018), which suggests the concurrent formation of protoglobular clusters and supermassive stars, provides the correct chemical patterns through the "conveyor-belt" production of hot-H burning yields. This model also resolves the mass budget problem faced by other proposed polluter models, which typically require non-standard assumptions to produce the required mass of polluted gas. In this model, the supermassive star maintains its mass through accretion of protostars while losing mass via its wind. As a result, it cycles through a significantly larger amount of mass than its initial mass. However, a major drawback of this model is that no such objects have been observed to date (Renzini *et al.* 2022). The candidate forming massive clusters are located outside the Milky Way with very dense centers, where the hypothetical star would be obscured by gas and dust (Hollyhead *et al.* 2015). This makes direct observation challenging.

3. Kilomasers

An alternative method that could be used to detect those exotic objects is through kilomasers. 22.2 GHz water masers are commonly associated with massive star formation (Ellingsen *et al.* 2018). Krause *et al.* (2020) have suggested that the more luminous kilomasers could originate from the accretion disc around supermassive stars, similar to the even more luminous AGN megamasers. Recently, a very strong nuclear kilomaser, W1 (Fig. 1, right column, orange spectrum), has been found in a nearby galaxy (NGC 253) that is associated with a forming super star cluster (Gorski *et al.* 2019). Other kilomasers have also been observed in connection with intense star formation, e.g. in the Antennae galaxies (Darling *et al.* 2008), where spatial resolution has allowed for direct association with super star clusters (Brogan *et al.* 2010). W1 exhibits three distinct line systems: the prominent one at a systemic velocity of 116 km s^{-1} and two 'high velocity' features on either side of the systemic velocity at substantially lower flux. If a maser shows two or three of these corresponding lines (or line systems), it is referred to as a clean disc maser (Pesce *et al.* 2015), a characteristic typically observed in AGN megamasers. Although the spectrum of W1 resembles that of a disc maser, its luminosity is much weaker compared to typical AGN megamasers, being approximately two orders of magnitude lower. Some extragalactic kilomasers found in super star clusters have been compared to the Galactic high mass star forming region W49N, located roughly 11.1 kpc away (Gwinn *et al.* 1992). This region produces a large number of highly variable 22 GHz H_2O maser spots with the total luminosity of $\approx 1\ L_\odot$ (Zhang et al. 2013), making it also a kilomaser. The spectrum consists of 316 individual narrow lines (McGrath *et al.* 2004), but in all cases, the extragalactic kilomasers appear to have a more peaked and narrower spectrum. The spectrum of W1 is clearly different from the one of W49N.

4. Simulations and results

In a recent theoretical study (Nowak *et al.* 2022), we have verified with 2D hydrodynamic simulations that an accretion disc around a collisionally supported supermassive star would be able to survive and could produce collisionally pumped maser lines with fluxes and spectral shape similar to the high-velocity wings in W1 (we did not model

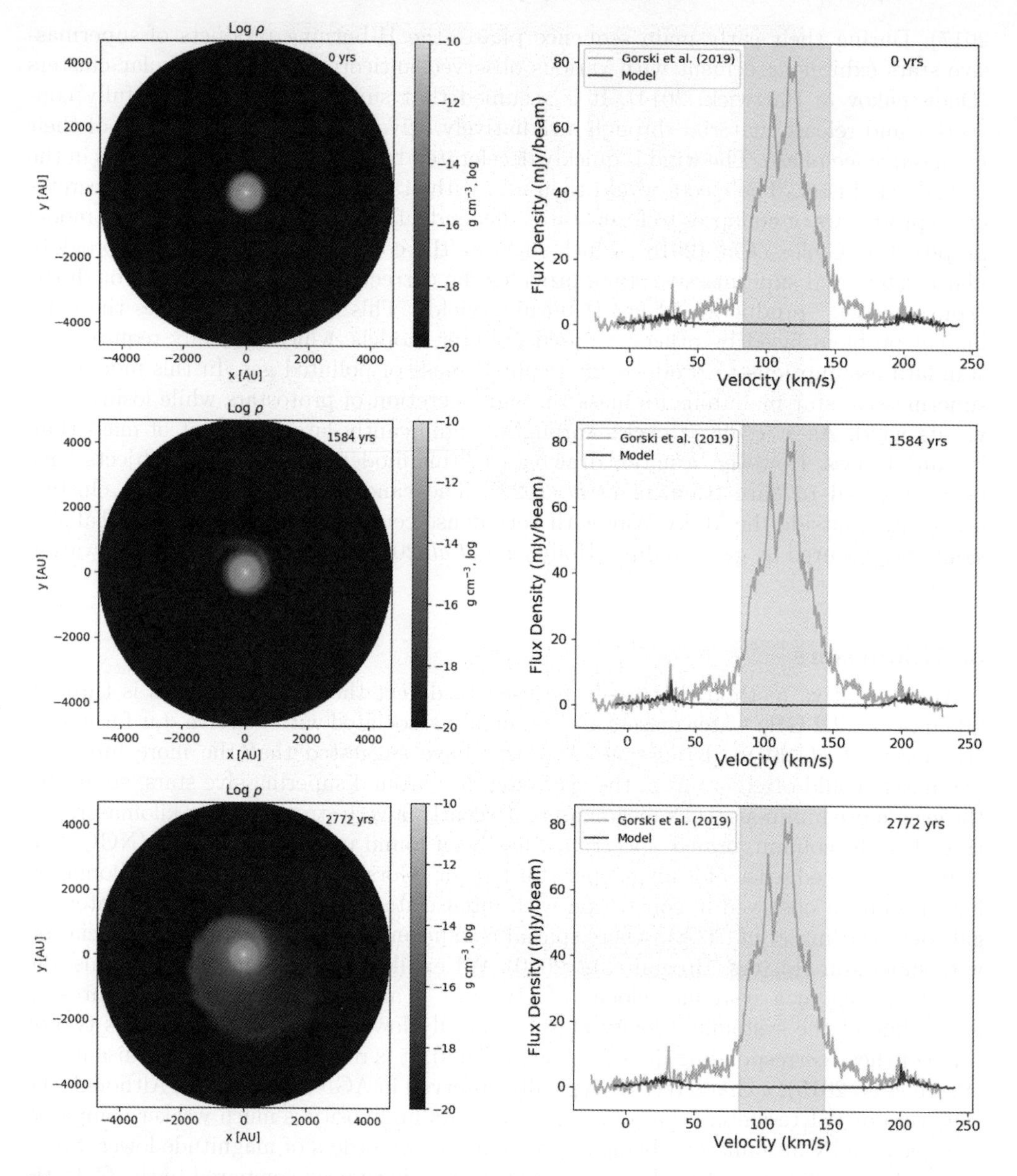

Figure 1. The time evolution of the reaction of an accretion disc around an SMS with a mass of 4000 $M_\odot$, a disc radius of 500 au, and a disc mass of 10 $M_\odot$, similar to Nowak *et al.* (2022), but using linear interpolation of perturbers' positions and improved modelling of long-range forces. The flyby rate considered in this hydrodynamics model is one stellar perturber per year. The density plots are presented in the left column, with the corresponding maser spectra on the right, for selected time steps. The model spectrum is shown in blue, whilst the W1 kilomaser from Gorski *et al.* (2019) for comparison is plotted in orange.

the radiatively pumped brighter central feature, indicated by shaded area, Fig. 1, right column, blue line). The maser spectrum, corresponding to a model with an SMS mass of 4000 $M_\odot$, exhibits high-velocity peaks that coincide with those observed in the W1 kilomaser. The density plots in Fig. 1 (left column), demonstrate the evolution of the disc as perturbers interact and create spiral arms within the disc. These perturbations in the disc structure have a significant impact on the peaks observed in the model maser, causing them to either move inward or outward, as well as increasing their flux. Based on our results, we propose that a supermassive star of this nature can provide an explanation for the observed W1 kilomaser.

5. Acknowledgements

The authors gratefully acknowledge the support by Science and Technology Facilities Council (ST/V506709/1). KN would like to thank International Astronomical Union and Royal Astronomical Society for generous travel grants.

References

Anderson J., in van Leeuwen F., Hughes J. D., Piotto G., 2002, *ASP-CS*, Vol. 265
Bastian N., Lardo C., 2018, *ARAA*, 56, 83
Brogan C., Johnson K., Darling J., 2010, *ApJ*, 716, L51
Darling J., Brogan C., Johnson K., 2008, *ApJ*, 685, L39
Denissenkov P. A., Denisenkova S. N., 1990, *Soviet Astron.*, 16, 275
Denissenkov P. A., Hartwick F. D. A., 2014, *MNRAS*, 437, L21
Ellingsen S. P., Voronkov M. A., Breen S. L., Caswell J. L., Sobolev A. M., 2018, *MNRAS*, 480, 4851
Gieles M. *et al.* 2018, *MNRAS*, 478, 2461
Gorski M. D. *et al.* 2019, *MNRAS*, 483, 5434
Gratton R. G., Carretta E., Bragaglia A., 2012, *A&AR*, 20, 50
Gwinn C. R., Moran J. M., Reid M. J., 1992, *ApJ*, 393, 149
Hollyhead K. *et al.* 2015, *MNRAS*, 449, 1106
Krause M. *et al.* 2020, *Space Sci. Revs*, 216, 64
McGrath E. J., Goss W. M., De Pree C. G., 2004, *ApJS*, 155, 577
Norris J. E., 2004, *ApJ*, 612, L25
Nowak K., Krause M. G H., Schaerer D., 2022, *MNRAS*, 516, 5507
Pesce D. W. *et al.* 2015, *ApJ*, 810, 65
Prantzos N., Charbonnel C., Iliadis C., 2017, *A&A*, 608, A28
Renzini A., Marino A. F., Milone A. P., 2022, *MNRAS*, 513, 2111
Sandage, A., Wildey, R. 1967, *ApJ*, 150, 469
Zhang B., Reid M. J., Menten K. M., Zheng X. W., Brunthaler A., Dame T. M., Xu Y., 2013, *ApJ*, 775, 79

Cosmic Masers: Proper Motion toward the Next-Generation Large Projects
Proceedings IAU Symposium No. 380, 2024
T. Hirota, H. Imai, K. Menten, & Y. Pihlström, eds.
doi:10.1017/S1743921323003307

OH megamaser emission in the outflow of the luminous infrared galaxy Zw049.057

Susanne Aalto[1], Boy Lankhaar[1,2], Clare Wethers[1], Javier Moldon[3] and Robert Beswick[4]

[1]Department of Space Earth and Environment, Chalmers University of Technology, SE-412 96 Gothenburg, Sweden. email: saalto@chalmers.se

[2]Leiden Observatory, Leiden University, 2300 RA, Leiden, The Netherlands

[3]Instituto de Astrofísica de Andalucía (IAA, CSIC), Glorieta de la Astronomía, s/n, E-18008 Granada, Spain

[4]Jodrell Bank Centre for Astrophysics, School of Physics and Astronomy, The University of Manchester, Alan Turing Building, Oxford Road, Manchester, M13 9PL, UK

Abstract. High resolution (0.”26 × 0.”13 (70 × 35 pc)) L-band (18 cm) OH megamaser (OHM) e-Merlin observations of the LIRG Zw049.057 show that the emission is emerging from a low velocity outflowing structure - which is foreground to a fast, dense and collimated molecular outflow detected by ALMA. The extremely dusty compact obscured nucleus (CON) of Zw049.057 has no (or only little) OHM emission associated with it - possibly because of too high number densities that quench the OHM. In contrast we detect 6 cm H_2CO emission primarily from the CON-region. We suggest that the OHM-region of Zw049.057 is not directly associated with star formation, but instead occurs in a wide-angle, slow outflow that surrounds the fast and dense outflow. The OHM is pumped by IR emission that likely stems from activities in the nucleus. We briefly discuss how OHM emission can be used as a probe of LIRG-CON galaxies.

Keywords. galaxies: evolution, galaxies: individual: Zw049.057, galaxies: active, galaxies: nuclei, ISM: molecules, ISM: jets and outflows

---

1. Introduction

Luminous and ultraluminous infrared galaxies (LIRG: $L_{\rm IR} \gtrsim 10^{11}$ $L_\odot$, ULIRG: $L_{\rm IR} \gtrsim 10^{12}$ $L_\odot$) are gas-rich galaxies that harbour intense dust-enshrouded activity in their centres - either in the form of a starburst or an AGN (Active Galactic Nucleus=accreting supermassive black hole). U/LIRGs are often interacting or merging systems and are key phenomena to our understanding of galaxy evolution. They drive powerful molecular outflows (Veilleux *et al.* 2020) that likely act as regulatory processes for star formation and AGN activity. Some U/LIRGs host so called Compact Obscured Nuclei (CONs) with extremely high inferred gas column densities ($N(H_2) > 10^{25}$ cm^{-2}) and dust opacities (e.g. Sakamoto *et al.* 2010; Aalto *et al.* 2015; Falstad *et al.* 2021). The nature of the buried activity in CONs is difficult to discern, but a luminous AGN and/or a starburst with a top-heavy initial mass function (IMF) are possibilities (e.g. Aalto *et al.* 2019). There is mounting evidence that CONs drive collimated dense molecular outflows (Barcos *et al.* 2018; Aalto *et al.* 2020; Yang *et al.* 2023) that are key to our understanding of the nature and growth of the embedded activity. U/LIRGs also often harbour luminous OH megamasers (OHM) (e.g. Mirabel & Sanders 1986) and we are investigating

OHMs as potential probes of CON evolution. Here we present some initial results on the LIRG-CON Zw049.057.

2. The LIRG-CON Zw049.057

Zw049.057 (from now on Zw049) is a nearby (D=56 Mpc (linear scale: ~270 pc arcsec^{-1} (Katgert *et al.* 1998))). Zw049 is a LIRG with an infrared luminosity of $L_{\rm IR} \sim 1.8 \times 10^{11}$ L$_\odot$ (Sanders *et al.* 2003) It has a highly inclined disk (Scoville *et al.* 2000) and is known to host a relatively low luminosity OHM (Baan *et al.* 1987; Martin *et al.* 1988). Its structure has not been imaged at high angular resolution before, neither has the exact nature of the nuclear power source been determined. The centre of Zw049 was first classified as a starburst based on optical spectroscopy (Baan *et al.* 1998), but has later been suggested to host an AGN due to its radio compactness and spectral index (Baan & Klöckner 2006). Its radio brightness temperature is however consistent with both a compact starburst and an AGN. McBride *et al.* (2013) suggest that Zw049 is in a transition phase between OH megamasers and galaxies with lower luminosity masers. The nucleus is highly dust obscured with luminous vibrationally excited emission from HCN and high surface brightness in the mm continuum (Falstad *et al.* 2015; Aalto *et al.* 2015; Falstad *et al.* 2021). Zw049 fulfills the criteria of a CON and is likely in a phase of rapid nuclear evolution. Falstad *et al.* (2018) find a fast outflow in CO J=6-5, J=2-1 with the Atacama Large mm and submm Array (ALMA), and the Submillimeter Array (SMA), and in cm-wave C-band transitions of OH with the Very Large Array (VLA)).

We have observed the 1665 and 1667 MHz OHM emission with e-Merlin and also the 6 cm formaldehyde H_2CO maser emission (FM). Furthermore we have observed multiple lines with ALMA, including the 265.89 GHz HCN J=3-2 line. *For details on line and continuum fluxes, the physical and chemical conditions, masses, proposed driving forces and structures of the collimated, and slow, outflows of Zw049 please see upcoming journal papers by Wethers et al. and Lankhaar et al. - currently in preparation.* Below are some brief notes on the results.

2.1. *OHM in the complex outflow*

Zw049 appears to have relatively convoluted outflow structure. Falstad *et al.* (2018) detected a fast ($v_{\rm projected}$>300 km s^{-1}) outflow in CO 6-5, 2-1 (in emission) and in cm-wave (4.66, 4.75, 4.77, 6.02, 6.03, 6.03 and 6.05 GHz) OH absorption with the VLA. With recent, very high resolution (0".027 × 0."024 (7 × 6 pc)) ALMA observations we detect luminous HCN J=3-2 emission from a dense ($n > 10^4$ cm^{-2}), collimated molecular outflow (Wethers *et al.* in prep., Lankhaar *et al.* in prep.) (Fig. 1). This is likely the same outflow seen at lower angular resolution in CO 6-5 by Falstad *et al.* (2018), but we find even higher velocities $v_{\rm projected}$=500 km s^{-1}. The dense outflow has an apparent, projected spatial extent of r~0."25 (~70 pc) and is collimated where the width is velocity-dependent with the most narrow structure at the highest velocities. Interestingly, using the ESO Multi Unit Spectroscopic Explorer (MUSE) instrument, Wethers *et al.* (in prep.) find an optical outflow on spatial scales at least an order of magnitude larger. The optical outflow also appears to be collimated and its potential relation to the molecular outflow requires further investigation. In addition, Falstad *et al.* (2018) show that there is a radio-jet-like structure extending to the north-west - i.e. not aligned with the molecular outflow. Instead the radio feature appear associated with a narrow dark feature in the HST J-band image (see Fig. 11 in Falstad *et al.* (2018)). This dark feature is either due to dust extinction or to a cloud in the nucleus that blocks radiation (Scoville *et al.* 2000) (if so then much of the minor axis light is scattered from the nucleus).

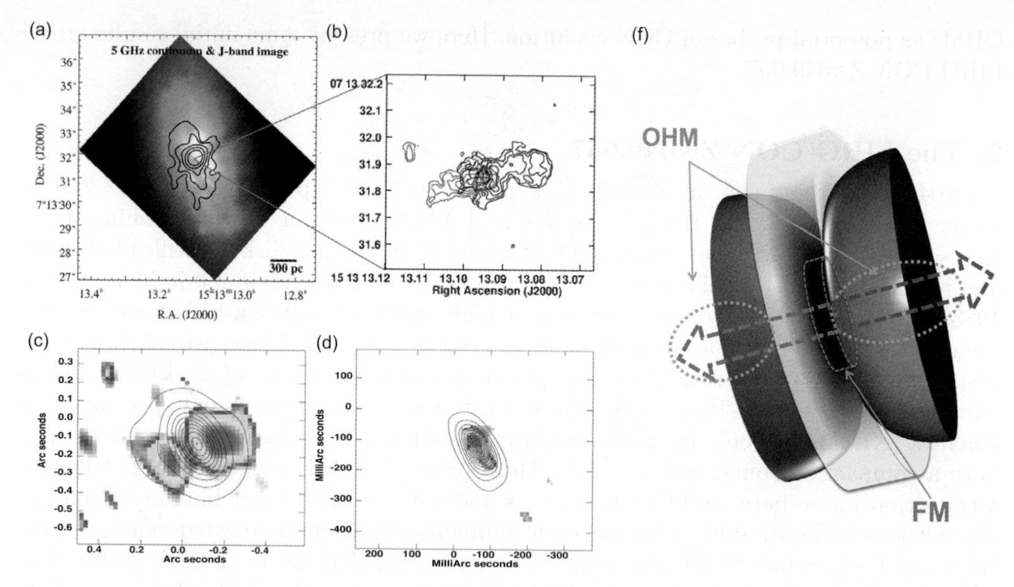

Figure 1. Top left panel **a)** Radio continuum contours overlaid on an HST J-band image (Falstad *et al.* 2018). The radio contours are at 1,4,16,64,256 $\times 15\mu$Jy beam^{-1}. Top centre panel **b)** Preliminary ALMA integrated intensity image of the dense fast and collimated outflow in HCN 3-2 (Lankhaar *et al.* and Wethers *et al.* in prep.). The image is constructed of emission at projected velocities >200 km s^{-1}. Lower left panels: preliminary integrated e-Merlin intensity images of **c)** the OHM line emission (left) and **d)** FM (right). Contours show L- and C-band continuum contours. Right panel: **f)** Preliminary schematic figure of the slow moving wind (in colour) and the fast collimated dense outflow (thick, dashed arrows) of the molecular outflow structure of Zw049. The suggested locations of OH megamaser (OHM) and formaldehyde (FM) megamaser emission are indicated in the figure. High resolution FM emission in Zw049 has previously been reported by Baan *et al.* (2017).

With e-Merlin (resolution 0."26 $\times$ 0."13 (70 $\times$ 35 pc))(Lankhaar *et al.* in prep.), we find that the distribution of the global OHM flux is distributed along the minor axis of Zw049 (Fig. 1). The emission has an east-west velocity gradient with the redshifted velocity to the east and blueshifted to the west with shifts from +20 km s^{-1} to -70 km s^{-1}. This is the same orientation, and size scale, as for the dense collimated molecular outflow, but the velocities are significantly smaller. There is no (or only little) OHM emission associated with the disk or the obscured nucleus.

We propose that the OHM emission is emerging from a larger scale, slower-moving foreground part of the Zw049 outflow. We present a simple cartoon in Fig. 1 of the proposed configuration. We suggest that the OH is pumped by IR emission from the nucleus and/or from the dense outflow. The gas in the CON nucleus is likely too dense for the OHM to operate while van der Walt & Mfulwane (2022) suggest that a FM can operate at $T_{\rm kin} > 100$ K and $n > 10^{4.5}$ cm^{-3}. If so, this explains the FM peak in parts of the CON - while OHM is suppressed in the same region. This opens the possibility of an evolutionary scheme where OHM emission is first suppressed in the LIRG-CON, but then emerges later in its evolution when a dense outflow has developed. ZW049 may therefore not be in transition *out of* an OHM stage, but instead on its way toward higher OHM luminosities when the gas densities in the nucleus have dropped below quenching levels.

3. OHM and an evolutionary sequence of LIRG-CONs

The LIRG-CONs are characterized by large $N(H_2)$ column densities, high average gas number densities ($n > 10^6$ cm^{-3}), high temperatures ($T > 100$ K,) one (or several) dust-embedded luminosity sources and high IR surface brightnesses (e.g. Aalto *et al.* 2015). The exact process behind the build-up of these large nuclear column densities is not clear. The central gas concentrations may be caused by large inflows of gas in galaxy interactions or mergers, or the return of low angular momentum gas in galactic fountains. The CON region itself is not expected to be a bright OHM source because of the quenching of the OH maser in dense environments. Therefore, in the early, pre-outflow phase of the CON, we do not expect OHM emission. As an outflow develops, OHM action is anticipated to be associated with foreground gas if IR photons can escape from the nucleus along the minor axis. Alternatively, IR emission is emerging from the inner dense outflow itself - providing background pumping emission for the OHM. In this latter scenario, the IR emission would emerge from star formation in the dense gas of the outflow. Such a scenario has yet not been confirmed to exist in outflowing gas (even though there have been suggestions).

The development of an outflow would help drive the evolution of the CON forward through removing gas from the CON-region - reducing the nuclear gas density and opacity. We therefore suggest that OHM is a good tracer of CON-evolution: 1) suppressed in the early pre-outflow stages, 2) then developing in foreground structures linked to the dense outflow, 3) peaking when OHM emission is emerging from both the post-CON nucleus and in structures in the outflow. OHM activity then subsides when gas and dust is dispersed from the central region.

High resolution OHM observations in a sample of U/LIRG CONs, pre- and post-CONs are necessary to test this scenario. It is important to also combine these studies with FM observations, and high resolution mm/submm observations of dense outflows. High resolution OHM observations will also reveal if the CON is surrounded by extended dusty star formation regions that produce OHM emission unrelated to the CON activity. We conclude that it is precarious to generally assume that the OHM emission is associated with disk star formation and that high-resolution OHM observations are key to determine its origin.

4. Acknowledgements

S.A, CW gratefully acknowledge support from the ERC Advanced Grant 789410.

References

Aalto, S., Martín, S., Costagliola, F., *et al.* 2015, *A&A*, 584, 42
Aalto, S., Muller, S., König, F., *et al.* 2019, *A&A*, 627, 147
Aalto, S., Falstad, N., Muller, S., *et al.* 2019, *A&A*, 627, 147
Barcos-Muñoz, L., Aalto, S., Thompson, T. A., *et al.* 2019, *ApJ*, 853, 28
Baan, W. A., Henkel, C., Haschick, A. D., 1987, *ApJ*, 320, 154
Baan, W. A., Salzer, J.J., LeWinter, R. D., 1998, *ApJ*, 509, 633
Baan, W. A., & Klöckner, H. -R., 2006, *A&A*, 449, 559
Baan, W. A., An, T., Klöckner, H. -R., Thomasson, P., 2017, *MNRAS*, 469, 916
Falstad, N., Gonzalez-Alfonso, E., Aalto, S., *et al.* 2015, *A&A*, 580, 52
Falstad, N., Aalto, S., Mangum, J. G., *et al.* 2018, *A&A*, 609, 75
Falstad, N., Aalto, S., König, S., *et al.* 2021, *A&A*, 649, 105
Katgert, P., Mazure, A., den Hartog, R., *et al.* 2015, *A&AS*, 129, 399
Martin, J. M., Bottinelli, L., Dennefeld, C., *et al.* 1988, *A&A*,195, 71
McBride, J., Heiles, C., Elitzur, M. 2013, *ApJ*, 774, 35

Mirabel, I.F., & Sanders, D. B., 1986, *ApJ*, 322, 688
Sakamoto, K., Aalto, S., Evans, A., *et al.* 2010, *ApJ*, 725, 228
Sanders, D., B., Mazzarella, J. M., Kim, D.-C., *et al.* 2003, *AJ*, 126, 1607
Scoville, N. Z., Evans, A. S., Thompson, R, *et al.* 2000, *AJ*, 119, 991
van der Walt, D. J., & Mfulwane, L. L., 2022, *A&A*, 657, 63
Veilleux, S., Maiolino, R., Bollatto A., Aalto, S., 2020, *A&AR*, 28, 1
Yang, C., Aalto, S., König, S., *et al.* 2023, in *arXiv e-prints*, arXiv:2307.07641

Cosmic Masers: Proper Motion toward the Next-Generation Large Projects
Proceedings IAU Symposium No. 380, 2024
T. Hirota, H. Imai, K. Menten, & Y. Pihlström, eds.
doi:10.1017/S174392132300248X

IC 485 : A candidate for a new disk-maser galaxy

Elisabetta Ladu[1,2], Andrea Tarchi[2], Paola Castangia[2], Gabriele Surcis[2], James A. Braatz[3], Francesca Panessa[4] and Dominic Pesce[5]

[1]Dipartimento di Fisica, Universitá degli Studi di Cagliari, S.P.Monserrato-Sestu km 0,700, I-09042 Monserrato (CA), Italy. elisabetta.ladu@inaf.it

[2]INAF-Osservatorio Astronomico di Cagliari, via della Scienza 5, 09047, Selargius (CA), Italy

[3]National Radio Astronomy Observatory, 520 Edgemont Road, Charlottesville, VA 22903, USA

[4]INAF - Istituto di Astrofisica e Planetologia Spaziali, via Fosso del Cavaliere 100, I-00133 Roma, Italy

[5]Harvard-Smithsonian Center for Astrophysics, 60 Garden Street, Cambridge, MA 02138, USA

Abstract. Fundamental physical quantities of the nuclear regions of the Active Galactic Nuclei can be obtained using megamaser studies. In particular, disk-masers associated with accretion disks around the supermassive black holes are used, through high angular resolution measurements, to trace the disk geometry, to estimate the BH mass and to measure accurate distances to their host galaxies. In this contribution, we present the first results in continuum and spectral-line mode of a high-sensitivity, multi-epoch VLBI study of the nuclear region of the megamaser LINER galaxy IC 485.

Keywords. masers, technique:high angular resolution, galaxy:active, galaxy:individual (IC 485)

1. Introduction

Galaxies with intense activity in their nuclear regions are defined Active Galactic Nuclei (AGNs). They are classified according to the Unified Model by Urry & Padovani (1995) and according to their radio activity. In the framework of megamaser studies, the most relevant are the "radio-quiet" AGNs: Seyfert 2 and LINERs (e.g. Lal *et al.* (2011), Marquez *et al.* (2017)) where most of the known megamasers are observed. Three AGN components are associated with the activity of H_2O maser emission: disk, jets and outflows (e.g. Tarchi (2012)). In particular, disk masers are associated with the nuclear accretion disk and they permit to trace disk geometry, to measure the rotation velocity and the mass of the supermassive black hole (SMBH) (e.g. Gao *et al.* (2017), Pesce *et al.* (2020)) to estimate distances to the parent galaxies and cosmological measurement of H_0, (e.g. Reid *et al.* (2013), Braatz *et al.* (2013)). These studies are possible thanks to the high angular resolution of the Very Long Baselines Interferometry (VLBI) technique which also may allow to discern the observational signatures of a new peculiar typology of maser: "inclined maser disk" where the maser detection is possible thanks the gravitational lensing (Darling 2017).

Table 1. Observational details of the six epochs analyzed. [a]The rms has been calculated with the task IMEAN. [b]The noise in the cube map, indicated in boldface, is calculated in a range of 500 channels where there is not maser emission.

Project name	Epoch	Band	Array	# IF	Band-width	Numb. Channels	Integr. time (h)	rms[a] ($\frac{mJy}{beam}$)
BT142_A2	2018.11	L	VLBA	4	64 MHz	128	2.5	0.029
BT142_A1	2018.16	K	VLBA	1	64 MHz	4096	2.5	**3.02**[b] ; 0.064
ET038_A	2018.39	C	EVN	8	16 MHz	32	1.4	0.028
ET038_B	2018.40	L	EVN	8	16 MHz	32	2.3	0.018
BT142_A2	2018.59	L	VLBA	4	16 MHz	128	2.5	0.05
BT145_A1	2018.83	K	VLBA	2	16 MHz	4096	3.5	**5.87** ; 0.066

2. Outline of the project and target details

The present study is part of an ongoing Ph.D. project focused on a detailed study of H_2O (and OH) megamasers at VLBI scale in radio quiet AGN galaxies (Seyfert and/or LINERs). This will permit i) to increase the physical and geometric information of the maser phenomenon; ii) to expand the number of cases analyzed and thus collect evidence for (or against) the Unified Model. Indeed, the galaxy IC 485 is one of the targets included in the sample of this project. IC 485, with systemic velocity $V_{sys}^{hel,opt,} = 8338\,\mathrm{km\,s^{-1}}$, has been optically classified as a Sa spiral galaxy located at the distance of 122.0 ± 8.5 Mpc, assuming $H_0 = 70\,\mathrm{km\,s^{-1}\,Mpc^{-1}}$ (Kamali *et al.* 2017). The spectroscopic classification of the galaxy is unclear yet. The galaxy is presented as a LINER and is included in the sample of “inclined maser disk” candidate reported by Darling (2017). On the other hand, a classification as a Seyfert 2 was proposed by Kamali *et al.* (2017). Darling in his Very Large Array (VLA) observations reports a broad multi-component maser with a peak of 78 ± 2 mJy with a high luminosity ($L_{\rm iso} = 868 \pm 46\, L_\odot$). Unresolved and faint radio continuum emission was observed with the VLA at 1.4 GHz in the NVSS (Condon *et al.* 2002) and FIRST. Detections at 20 GHz and at 33 GHz although with a low significance or as tentative were also reported by Darling (2017) and Kamali *et al.* (2017), respectively.

3. Observations and data reduction

IC 485 was observed with the Very Long Baseline Array (VLBA†) and with the European VLBI Network (EVN‡) in six epochs during 2018. In order to measure absolute position and to correct phase variation, we observed in phase-referencing mode using J0802 + 2509 at 1.77° from the target. The configuration at K-band was chosen in order to cover the main broad maser emission line with a velocity resolution of $0.2\,\mathrm{km\,s^{-1}}$ and to leave enough line-free channels for continuum subtraction and to produce a K-band continuum image. We reduced and analyzed the data utilizing the NRAO Astronomical Image Processing System (AIPS§) with standard procedure of calibration and the spectra were analyzed with the software CLASS¶. Details of the observations are reported in Table 1.

4. Results and discussion

At the L- and C- band, no continuum source was detected above the 5σ noise level (the rms are reported in Table 1). At K-band, 9 unresolved continuum sources were detected with a SNR$\geq 5\sigma$ (Fig. 1, upper pannel). The results suggested that the diffuse and continuum emission reported in Darling (2017) with VLA was resolved-out by

† The National Radio Astronomy Observatory is a facility of the National Science Foundation operated under cooperative agreement by Associated Universities, Inc.

‡ The European VLBI Network is a joint facility of independent European, African, Asian, and North American radio astronomy institutes.

§ www.aips.nrao.edu/

¶ www.iram.fr/IRAMFR/GILDAS

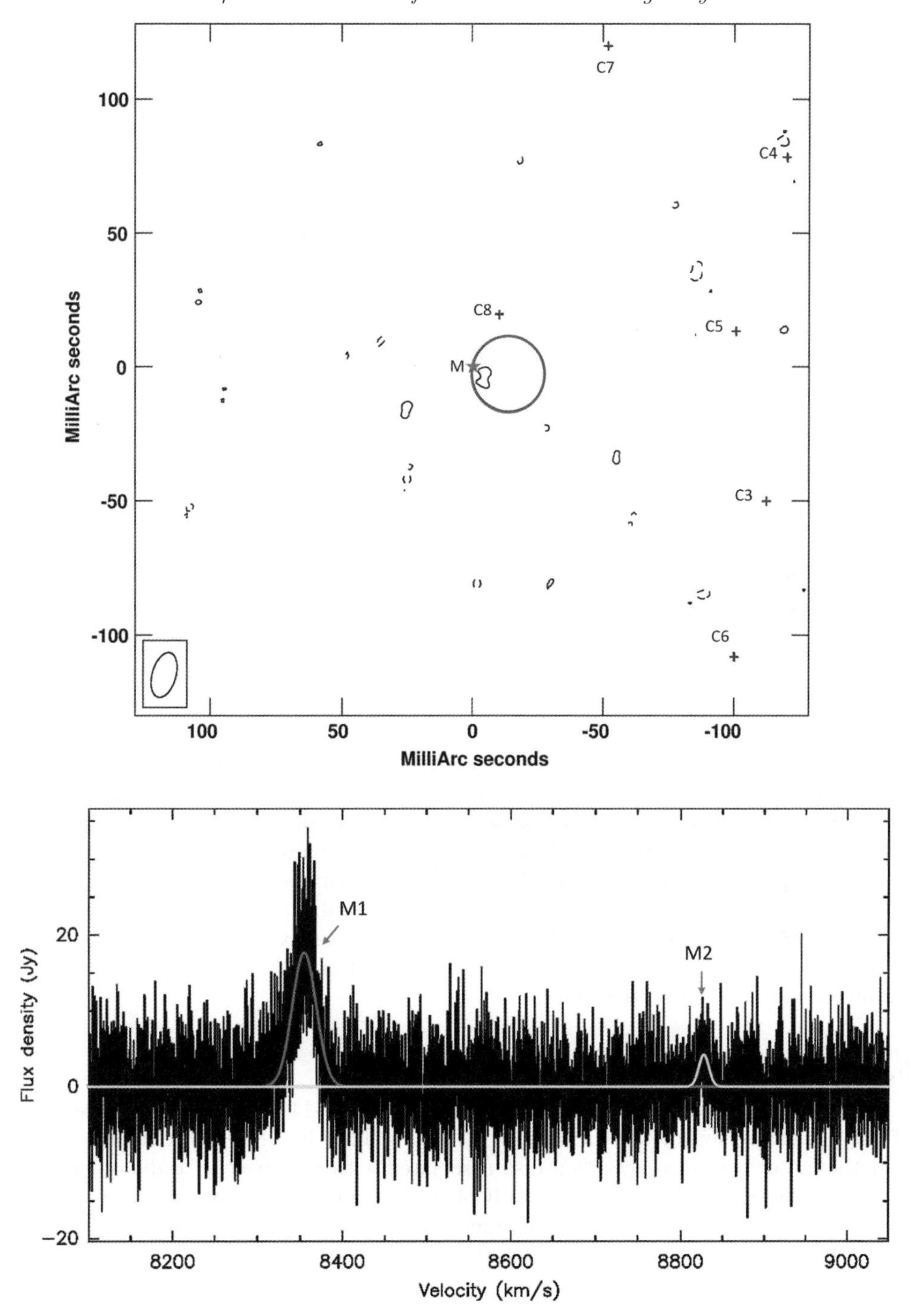

Figure 1. *Top panel*: The L-band EVN radio continuum map of the nuclear region of IC 485. Contour levels are $(-3,\ 3,\ 4,\ 5) \times 18\,\mu\text{Jy beam}^{-1}$. The blue crosses pinpoint the continuum sources detected at K-band. The red star indicate the position of the main maser feature. The purple circle represents the position of the VLA continuum source detected by Darling (2017). *Botton panel*: The spectrum of the main features observed with the VLBA in the epoch 2018.83.

Table 2. Parameters of the maser features. The columns indicate the epoch of detection, the name of the maser features, right ascension, declination, the peak velocity and the isotropic luminosity.

Epoch (VLBA)	**Maser**	**RA** (α_{2000}) [$^{h}:^{m}:^{s}$]	**Dec** (δ_{2000}) [$^{\circ}:':''$]	**Peak velocity** [$\mathrm{km\,s^{-1}}$]	L_{iso} ($L_{\odot}$)
2018.16	M1	08:00:19.75253	26:42:05.0523	8353.6±0.1	524±56
	M1	08:00:19.75252	26:42:05.0525	8354.8±0.5	239±38
2018.83	M1B	08:00:19.75247	26:42:05.0520	8355±2	76±18
	M2	08:00:19.75252	26:42:05.0528	8827±1	24±16

VLBA and EVN interferometer. The AGN in IC 485 may be considered radio silent or that its contribution is faint in radio band. Another explanation is that the continuum radio emission is dominated by the star forming regions in the host galaxy. In the spectral line mode, we report maser emission from three features. Two features are observed close to systemic velocity: a main bright maser component, labeled M1; a weaker tentative feature, M1B; and a third feature, M2, is observed at a velocity shifted w.r.t. the systemic velocity (Fig. 1, botton pannel). The main maser feature M1 is observed on both epochs. The details of the features detected are reported in Table 2. The position and the linear distribution of the two components, M1 and M2, allow us to speculate a disk-maser nature with a radius of 0.2 pc. Assuming a Keplerian rotation, the black hole mass in the center of the nuclear region estimated is $M_{BH} \approx 10^7 M_{\odot}$. This value is consistent with the one expected for a SMBH in a Seyfert or a LINER galaxy (e.g. Kuo *et al.* (2010)). Our hypothesis is built on the high sensitivity observations reported in Pesce *et al.* (2015), where the three components (although the blue component was reported as tentative) associable with a disk-maser are observable in the spectrum of IC 485. Furthermore, a recent survey of the H_2O emission at 183 GHz in AGNs already known for the presence of 22 GHz maser-disk reveals how the sub-mm transition can similarly trace the same accretion disk as that at cm wavelength (Pesce *et al.* 2023). The detection of the two transitions of masers in these galaxies and in our target, IC 485, seem to support and encourage our hypothesis of disk-maser. Interestingly, the feature M1B is spatially distinct from M1 and detached by the disk, suggesting a distinct origin, possibly associated with a jet or an outflow maser. This may hint at a composite nature of the maser in IC 485, already known to be observed in other megamaser galaxies (e.g. Gallimore *et al.* 1996 for NCG1068).

5. Summary

Here, we report the main results of six-epochs VLBA and EVN multi-band observations of the nuclear region of the megamaser LINER galaxy IC 485. The outcome suggests that the maser traces an accretion disk associated with a relatively "radio-quiet" AGN. Therefore the scenario for the maser in IC 485 being a candidate for an "inclined maser" seems to be ruled out. Further measurements are presently ongoing aimed at confirming the disk nature and better constraining the nuclear components' parameters.

References

Braatz, J., Reid, M., and Kuo, C. Y, Impellizzeri, V., Condon, J. and Henkel, C., Lo, K. Y. and Greene, J. Gao, F. Zhao, W. 2013, *IAUS*, 289, 225

Condon, J. J. Cotton, W. D. Broderick, J. J. 2002, *AJ*, 124, 675

Darling, J. 2017, *ApJ*, 837, 100

Gallimore, J. F., Baum, S. A., O'Dea, C. P. 1996, *American Astronomical Society Meeting Abstracts*, 189, 109.05

Gao, F., Braatz, J. A., Reid, M. J., Condon, J. J., Greene, J. E., and Henkel, C., Impellizzeri, C. M. V., Lo, K. Y., Kuo, C. Y., Pesce, D. W., Wagner, J., Zhao, W. 2017, *ApJ*, 834, 52

Kamali, F., Henkel, C., Brunthaler, A., Impellizzeri, C. M. V., Menten, K. M., Braatz, J. A., Greene, J. E., Reid, M. J., Condon, J. J., Lo, K. Y., Kuo, C. Y., Litzinger, E., Kadler, M. 2017, *A&A*, 605, A84

Kuo, C. Y., Braatz, J. A., Condon, J. J., Impellizzeri, C. M. V., Lo, K. Y., Zaw, I., Schenker, M., Henkel, C., Reid, M. J., Greene, J. E. 2010, *ApJ*, 727, 20

Lal, Dharam V., Shastri, Prajval, Gabuzda, Denise C. 2011, *ApJ*, 731, 68

Márquez, I., Masegosa, J., González-Martin, O. Hernández-Garcia, L., Pović, M., Netzer, H., Cazzoli, S. del Olmo, A. 2017, *Frontiers in Astronomy and Space Sciences*, 4, 34

Pesce, D. W., Braatz, J. A., Condon, J. J., and Gao, F. Henkel, C., Litzinger, E., Lo, K. Y., Reid, M. J. 2015 *ApJ*, 810, 65

Pesce, D. W., Braatz, J. A., Reid, M. J., Condon, J. J., Gao, F., Henkel, C., Kuo, C. Y., Lo, K. Y., Zhao, W. 2020, *ApJ*, 890, 118

Pesce, D. W., Braatz, J. A., Henkel, C., Humphreys, E. M. L. and Impellizzeri, C. M. V., Kuo, C Y. 2023, *ApJ*, 948, 134

Reid, M. J., Braatz, J. A. Condon, J. J., Lo, K. Y. Kuo, C. Y., Impellizzeri, C. M. V., and Henkel, C. 2013, *ApJ*, 767, 154

Tarchi, A. 2012, *IAUS*, 287, 323

Urry, C. M., and Padovani, P. 1995 *PASP*, 107, 803

Cosmic Masers: Proper Motion toward the Next-Generation Large Projects
Proceedings IAU Symposium No. 380, 2024
T. Hirota, H. Imai, K. Menten, & Y. Pihlström, eds.
doi:10.1017/S1743921323002077

What's behind the corner: Maser emission in nearby and distant galaxies with the new radio facilities

Andrea Tarchi[1], Paola Castangia[1], Gabriele Surcis[1], Elisabetta Ladu[1,2] and Elena Yu Bannikova[3,4]

[1]INAF-Osservatorio Astronomico di Cagliari, Via della Scienza 5, 09047, Selargius (CA), Italy.
email: andrea.tarchi@inaf.it

[2]Department of Physics, University of Cagliari, S.P.Monserrato-Sestu km 0,700, I-09042 Monserrato (CA), Italy

[3]INAF - Astronomical Observatory of Capodimonte, Salita Moiariello 16, Naples I-80131, Italy

[4]Institute of Radio Astronomy, National Academy of Sciences of Ukraine, Mystetstv 4, Kharkiv UA-61002, Ukraine

Abstract. Extragalactic maser sources are unique tools to derive fundamental physical quantities of the host galaxies, e.g, geometry of accretion disks around super-massive black holes and precise black hole masses, and study in detail the interaction region of nuclear jets/outflows with the interstellar medium, in nearby and distant Active Galactic Nuclei. So far, however, extragalactic maser searches have yielded detection of few percent, and only relatively few maser sources have been found. Because of their unprecedented sensitivity, new upcoming facilities, like the SKA and the ngVLA, will allow to significantly increase the number of known (water) maser sources. This will lead to the chance of performing statistically-relevant studies of the maser phenomenon (and its occurrence), derive extragalactic masers luminosity functions, and ultimately (in particular, through the aid of longer-baselines arrays options) to perform the studies described above for larger samples and up to cosmological distances.

Keywords. masers, galaxy:active, galaxy:individual (TXS2226−184)

1. Introduction

H_2O masers associated with active galactic nuclei (AGN, the 'megamasers') have been related with three distinct phenomena: (i) nuclear accretion disks, where they can be used to derive the disk geometry, enclosed nuclear mass, and distance to the host galaxy (see, e.g., Braatz *et al.* 2010 for UGC 3789), or sometimes with the inner boundary of a dusty torus in the region of the interaction with the outflows (e.g.,Bannikova *et al.* 2023, for NGC1068); (ii) radio jets, where can provide important information about the evolution of jets and their hotspots (e. g. Peck *et al.* 2003 for Mrk 348, and, more recently, Castangia *et al.* 2019, for IRAS 15480); (iii) nuclear outflows, tracing the velocity and geometry of nuclear winds at < 1 pc from the nucleus, as in the case of Circinus (Greenhill *et al.* 2003) and NGC 3079 (Kondratko *et al.* 2005), where they offer a promising means to probe the structure and motion of the clouds in the Toroidal Obscuring Region (TOR) predicted by clumpy torus models (e.g., Nenkova *et al.* 2008). In all these cases, water megamasers provide vital information on the structures, dynamics and composition of the molecular gas at close proximity to the AGN.

2. Recent (VLBI) studies by our team

Our group is presently leading a number of (VLBI) studies focused on a few promising extragalactic maser sources in order to exploit the information obtained in the framework of a better understanding of the unified scheme for AGN. In particular, the main targets are:

• the jet-maser in the nucleus of the Seyfert 2 galaxy IRAS15480−0344, detected in a sample that has provided one of the highest maser detection rate (50%), and followed-up in a recent survey with the SRT (for details, see P. Castangia's contribution, this volume);

• the luminous maser in the LINER galaxy IC 485, for which our recent EVN and VLBA observations hint at a disk maser nature, possibly accompanied by a jet/outflow maser component (for details, see Elisabetta Ladu's contribution, this volume);

• the nuclear gigamaser in TXS 2226-184 where, for the first time, the location of the H_2O masers w.r.t. the continuum emission has been obtained (see following sections)

2.1. *The gigamaser in TXS2226−184*

TXS2226−184 has been optically identified as an elliptical/S0 galaxy (Koekemoer *et al.* 1995), even though Falcke *et al.* (2000) proposed a different classification as a possible later-type galaxy. Furthermore, it is spectroscopically classified as a LINER (Bennert *et al.* 2004). TXS 2226-184 is located at a distance of 107.1 Mpc (Kuo *et al.* 2018) and, in this galaxy, a water megamaser was first detected in 1995, whose emission was so bright (6100 solar luminosities) that the source was labeled as gigamaser (Koekemoer *et al.* 1995). High angular resolution measurements were reported only in a conference proceeding on 2005 (Ball *et al.* 2005), although no absolute position was measured for the maser spots. More recently, Surcis *et al.* (2020) reported that the maser emission is resolved into 6 maser features forming a linear/arc structure without a regular velocity gradient (Figure 1). Neither continuum nor polarized emission at 22 GHz is detected down to a level of 0.2 mJy/beam and 15%, respectively. The maser features are associated with the most luminous radio continuum knot reported by Taylor *et al.* (2004) at 1.4 GHz. In order to draw a definite conclusion on the maser nature, however, a key question has to be addressed: where is the actual location of the nucleus/SMBH?

2.2. *Where is the nucleus in TXS2226−184?*

On February and March 2021, we observed the nuclear region of TXS2226−184 with the EVN+eMERLIN array at L, C, and K band. So far, only the data of the two lowest frequencies were calibrated, and a preliminary analysis performed.

The new maps (Figure 1) makes evident that the spectral index is highly inverted at the location of the maser and where the optical depth (as derived by Taylor *et al.* 2004) is the largest. This may indicate that the nucleus (taken as the SMBH or base of the jet) is actually coincident with the maser emitting region. If this is the case, the most likely option for the maser nature is related to the base of the jet, to a peculiar accretion disk, or to the inner region of the dusty torus (Tarchi et al. in prep).

3. Great expectations for extragalactic maser studies

Despite each megamaser source represents a goldmine of information on the nuclear ejection/accretion activity of the host galaxy, overall maser detection rates are of only a few percent. Indeed, out of about 3500 galaxies surveyed, only 180 water maser sources have been found, mostly in radio quiet AGN, mostly classified as Seyfert 2s or LINERs, in the local Universe ($z < 0.05$; Braatz *et al.* 2018). Therefore, more sources are needed for detailed studies, maser luminosity functions, and statistical considerations.

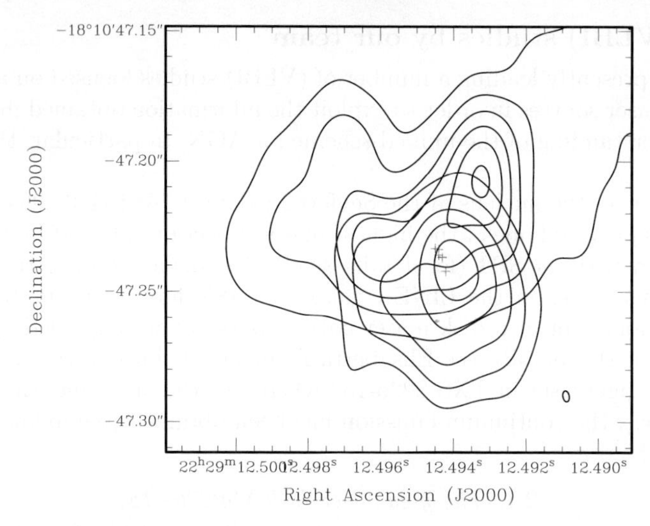

Figure 1. L and C bands EVN contour maps of the nuclear radio continuum emission of TXS2226−184. The C-band data have been convolved to the same resolution of the L-band data (HPBW: 30×15 mas). Contours at 1.6 GHz (in black) are drawn at 0.2, 0.4. 0.6. 0.8, ..., 1.6 mJy/beam. Contours at 5 GHz (in brown) are drawn at 1.5, 2.0, 2.5, 3.0, 3.5, 4.0 mJy/beam. Red and blue crosses indicate the maser positions as derived by Surcis *et al.* (2020).

3.1. *Targetted searches*

Taking profit of their enhanced sensitivity w.r.t. existing radio telescopes, the SKA and the ngVLA will be able to deliver relevant results by performing targeted searches of extremely large samples of galaxies. The spectral-line sensitivity (a few mJy in an 1-km/s wide channel) typically sought for with a 100-m class single-dish (i.e., the antennas that are typically used in water maser searches) will be reached by the aforementioned facilities in a factor 10 less time, thus allowing to search (with comparable sensitivity) much larger samples. This will yield the detection (and, possibly, the simultaneous sub-arcsecond position, useful for VLBI studies) of many more objects locally and, also when considering that the water maser luminosity function (LF) is a relatively steep power-law, of a still significant number of maser sources also at larger distances (up to z ∼0.5) and, seemingly, in classes of galaxies where, so far, no maser have been found.

3.2. *Blind searches*

When approaching cosmological distances $z \geq 1$, blind surveys of water maser sources with the SKA and the ngVLA may also become extremely profitable, particularly if, as reported in the literature, the LF evolves with redshift by $(1+z)^m$ (with $m = 4$, or 8). Under this latter assumption, as shown in Figure 2, also blind searches with these new facilities would already yield, at $z \geq 1$, a significant number of expected masers, when areas of the order of a square degree is attained. This is surely feasible thanks to the enhanced sensitivity and the large field of view of the aforementioned arrays, and together with the large instantaneous bandwidths (∼ 2.5 GHz) available, that allow one to cover all at once relevant redshift ranges.

Needless to say, in order to maximize the return of any new maser detection and, more generally, to exploit masers as tools to derive fundamental information on AGN (and not only), as those described in the previous section, high-angular resolution studies (mainly with VLBI) are mandatory. In this framework, the option offered for the SKA

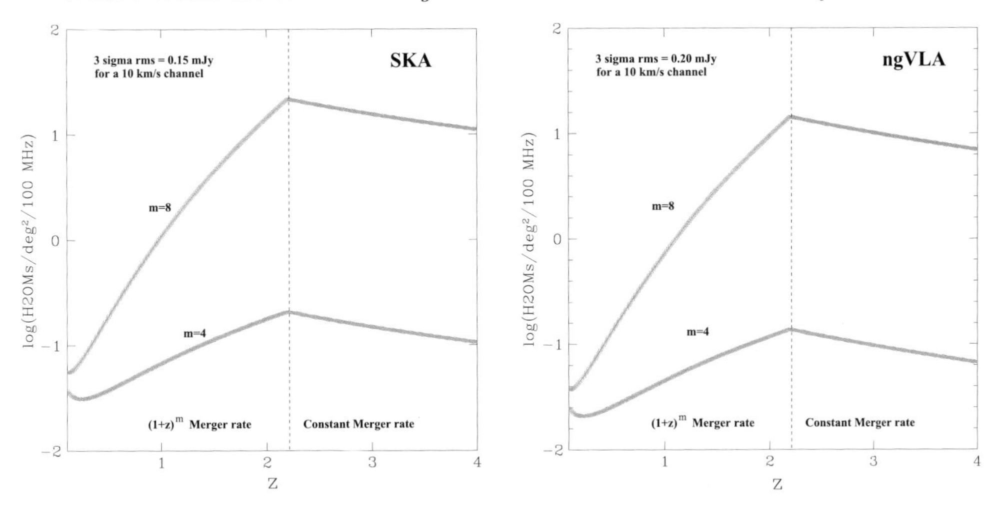

Figure 2. Density of detectable water masers as a function of redshift (0.1$\leq z \leq$4) with the SKA-1 MID (left) and ngVLA (right) arrays, assuming a LF evolution, increasing with redshift by $(1+z)^m$, with m equal to 4 or 8 up to z =2.2, and then becoming constant. The values are computed considering a survey covering an area of a square degree and an integration time of 15 hours per pointing (Tarchi et al. in prep.).

and the ngVLA to access baselines as long as, at least, those presently provided by the main existing VLBI networks becomes indispensable, especially given that an increased sensitivity will be offered by the arrays as a whole.

References

Ball, G.H., Greenhill, L.J., Moran, J.M., Zaw, I., Henkel, C. 2005, *ASP Conf. Ser.*, 340, 235
Bannikova, E. Yu, Akerman, N., Capaccioli, M., *et al.* 2023, *MNRAS*, 518, 742
N. Bennert, H. Schulz, & C. Henkel, 2004, *A&A*, 419, 127
Braatz, J.A., Reid, M.J., Humphreys, E.M.L, *et al.* 2010, *ApJ*, 718, 657
Braatz, J.A.; Condon, J., Henkel, C., *et al.* 2018, *IAUS*, 336, 86
Castangia, P., Surcis, G., Tarchi, A., *et al.* 2019, *A&A*, 629, 25
Falcke, H., Wilson, A.S., Henkel, C., Brunthaler, A., Braatz J.A. 2000, *ApJ*, 530 L13
Greenhill, L.J., Booth, R.S., Ellingsen, S.P., *et al.* 2003, *ApJ*, 590, 162
Koekemoer, A.M., Henkel, C., Greenhill, L.J., *et al.* 1995, *Nature*, 378, 697
Kondratko, P.T., Greenhill, L.J., & Moran, J.M. 2005, *ApJ*, 618, 618
Kuo, C.Y., Constantin, A., Braatz, J.A., *et al.* 2018, *ApJ* 860, 169
Nenkova, M., Sirocky, M.M., Nikutta, R., Ivezić, Z., Elitzur, M. 2008, *ApJ*, 685, 160
Peck, A. B., Henkel, C., Ulvestad, J.S., *et al.* 2003, *ApJ*, 590, 149
Surcis, G., Tarchi, A., & Castangia, P. 2020, *A&A*, 637, 57
G.B. Taylor, A.B. Peck, J.S. Ulvestad, C.P. ODea 2004, *ApJ*, 612, 780

Cosmic Masers: Proper Motion toward the Next-Generation Large Projects
Proceedings IAU Symposium No. 380, 2024
T. Hirota, H. Imai, K. Menten, & Y. Pihlström, eds.
doi:10.1017/S1743921323002727

Water megamaser emission in hard X-ray selected, highly obscured AGNs

Paola Castangia[1], Andrea Tarchi[1], Roberto Della Ceca[2], Alessandro Caccianiga[2], Paola Severgnini[2], Gabriele Surcis[1], Andrea Melis[1], Francesca Panessa[3], Angela Malizia[4] and Loredana Bassani[4]

[1]INAF-Osservatorio Astronomico di Cagliari, Via della Scienza 5, 09047, Selargius (CA), Italy.
email: paola.castangia@inaf.it

[2]INAF-Osservatorio Astronomico di Brera, via Brera 28, 20121 Milan, Italy

[3]INAF-Istituto di Astrofisica e Planetologia Spaziali di Roma (IAPS), Via del Fosso del Cavaliere 100, 00133 Roma, Italy

[4]INAF-Osservatorio di Astrofisica e Scienza dello Spazio, via P. Gobetti 101, I-40129 Bologna, Italy

Abstract. We took profit of the availability of large catalogs of active galact nuclei (AGNs) selected in the hard X-ray from satellite missions (e. g., INTEGRAL, Swift/BAT) to investigate the relation between the occurrence of water maser emission and the X-ray properties of the nuclei on a statistically meaningful basis. Our studies demonstrate that the hard X-ray selection may significantly enhance the maser detection rate over comparably large optical or infrared surveys. Here, we report on a recent survey to search for water maser emission with the Sardinia Radio Telescope (SRT) in a sample of heavily absorbed AGN taken from the 70 months Swift/BAT catalog.

Keywords. masers, galaxy: active, galaxy: nuclei

---

1. A search for H_2O megamasers with the SRT

Despite the important information that can be derived for AGNs and cosmology, water megamasers are rare, with detection rates of a few percent in optically-selected samples of AGNs. Statistical studies on the X-ray properties of H_2O maser galaxies reveal that objects with the higher X-ray luminosity and/or the higher column density more likely host masers. Studying a sample of 36 heavily absorbed AGNs ($N_H > 10^{23}$ cm^{-2}) including Compton-thick (CT) sources, we obtained a remarkable maser fraction of (50±12)%, one of the highest ever found in extragalactic maser surveys (Castangia *et al.* 2019). The sample was selected in the local Universe through a combination of mid-IR (IRAS) and X-ray (XMM-Newton) data (for details see Severgnini *et al.* 2012). In order to confirm the indications provided by previous studies on a firm statistical basis, we also selected a large sample of 380 hard X-ray AGNs detected above 20 keV with INTEGRAL/IBIS and searched for water megamaser emission among them in the literature and through new dedicated observations (Panessa *et al.* 2020). We also considered a sub-sample of 87 sources, limited in volume and statistically complete. We found detection fractions of (15±3)% and (19±5)% for the large and the complete samples, respectively. The fraction of detected masers increases in type 2 Seyfert galaxies (22±5)% and in CT AGNs, (56±18)% of which host water maser discs, in agreement with the results of

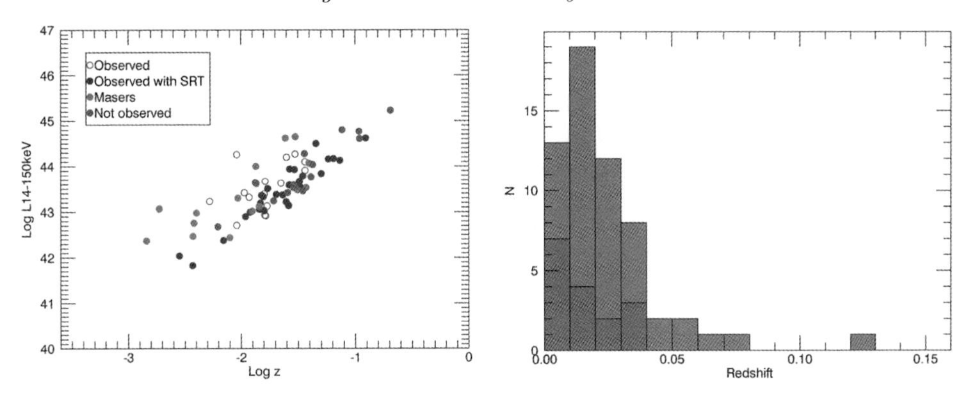

Figure 1. *Left*: Hard X-ray luminosity vs. redshift for the Ricci *et al.* (2017) sample. *Right*: Redshift distribution of the observed galaxies (blue) and of the known water masers (red) in the same sample.

Castangia *et al.* (2019). Overall, we conclude that hard X-ray samples of AGNs provide the opportunity of significantly increasing the maser detection efficiency over comparably large optical or infrared surveys (Panessa *et al.* 2020).

With the aim of detecting new luminous maser sources and comparing the results obtained with those of Castangia *et al.* (2019), we have observed a sample of 28 heavily obscured, hard X-ray selected AGNs, with the SRT (Prandoni *et al.* 2017). The targets were selected from the 75 Swift/BAT AGNs with $N_H > 10^{23}$ cm^{-2} reported in Ricci *et al.* (2017) (see the left panel of Fig. 1), 46 of which have already been observed in previous surveys (e. g. Megamaser Cosmology Project, Kondratko *et al.* 2006). Water maser emission were detected in 16 of them. From the 75 AGNs, we have selected the objects which can be observed with the SRT (Dec. $> -40°$) and whose maximum elevation at the SRT site is $\geq 30°$. Observations have been performed between December 2018 and May 2019. We employed two of the seven beams of the multi-feed K-band receiver in nodding mode and the SARDARA backend (Melis *et al.* 2018), with a bandwidth of 420 MHz and 16384 channels. This setup yielded a velocity coverage (at 22 GHz) of 3000 km s^{-1} and a channel spacing of 26 kHz ($\sim$0.35 km s^{-1}). We reached root mean square (rms) noise levels between 10 and 30 mJy per channel, slightly higher than those reported for the other samples (Castangia *et al.* 2019; Panessa *et al.* 2020). The survey yielded no confident detection, however, we obtained a number of tentative (signal-to-noise ratio $<$5 and/or presence of radio fequency interferences, RFI, in the spectra) water maser lines. The most promising of these features were observed again in April 2021, but not confirmed. This indicates that they are most likely produced by RFI, which affect about 50% of the spectra, or backend artifacts. Nevertheless, we cannot completely exclude the possibility that they are extremely variable water maser lines.

A preliminary statistical analysis of the Ricci *et al.* (2017) sample reveals an overall maser detection rate of (27±7)% (16/59), which is slightly higher than that obtained for the INTEGRAL complete sample, but significantly lower than the maser fraction extrapolated from the Severgnini *et al.* (2012) sample. While the higher fraction with respect to the INTEGRAL complete sample was expected due to the higher level of X-ray obscuration, the lower maser detection rate compared to the results of Castangia *et al.* (2019) is somewhat surprising. Indeed, the two samples have comparable redshift distributions (see the right panel of Fig. 1) and similar amounts of X-ray absorption. These results suggest that a key role in increasing maser detection efficiency might be played by a selection criterion that involves the mid-IR emission and deserves further investigations.

References

Castangia, P., Tarchi, A., Caccianiga, A., Severgnini, P., Della Ceca, R. 2019, *A&A*, 629A, 25C
Kondratko, P. T., Greenhill, L. J., & Moran, J. M. 2006, *ApJ*, 652, 136
Melis, A., Concu, R., Trois, A., *et al.* 2018, *JAI*, 850004M
Panessa, F., Castangia, P., Malizia, A., Bassani, L., Tarchi, A., *et al.* 2020, *A&A*, 641, 162
Prandoni, I., Murgia, M., Tarchi, A., *et al.* 2017, *A&A*, 608, 40
Ricci, C., Trakhtenbrot, B., Koss, M. J., *et al.* 2017, *ApJS*, 233, 17
Severgnini, P., Caccianiga, A., & Della Ceca R. 2012, *A&A*, 542, 46

Cosmic Masers: Proper Motion toward the Next-Generation Large Projects
Proceedings IAU Symposium No. 380, 2024
T. Hirota, H. Imai, K. Menten, & Y. Pihlström, eds.
doi:10.1017/S1743921323002193

A Holistic Search for Megamaser Disks and their Role in Feeding Supermassive Black Holes

Anca Constantin[1], **Cameron Kelahan**[1,2], **C. Y. Kuo**[3] and **J. A. Braatz**[4]

[1]Department of Physics and Astronomy, James Madison University, Harrisonburg, VA 22807, USA. email: constaax@jmu.edu

[2]Southeastern Universities Research Association, NASA Goddard Space Flight Center

[3]Physics Department, National Sun Yat-Sen University, No. 70, Lien-Hai Rd, Kaosiung City 80424, Taiwan, R.O.C

[4]National Radio Astronomy Observatory, 520 Edgemont Road, Charlottesville, VA 22903, USA

Abstract. If water megamaser disk activity is intimately related to the circumnuclear activity from accreting supermassive black holes, a thorough understanding of the co-evolution of galaxies with their central black holes should consider the degree to which the maser production correlates with traits of their host galaxies. This contribution presents an investigation of multiwavelength nuclear and host properties of galaxies with and without water megamasers, that reveals a rather narrow multi-dimensional parameter space associated with the megamaser emission. This "goldilocks" region embodies the availability of gas, the degree of dusty obscuration and reprocessing of the central emission, the black hole mass, and the accretion rate, suggesting that the disk megamaser emission in particular is linked to a short-lived phase in the intermediate-mass galaxy evolution, providing new tools for both 1) further constraining the growth process of the incumbent AGN and its host galaxy, and 2) significantly boosting the maser disk detection by efficiently confining the 22 GHz survey parameters.

Keywords. masers, galaxies: active

---

1. The Hunt for Water Megamasers: the current state of affairs

Current discovery surveys for water megamasers ($L_{\mathrm{H_2O}} \geq 10L_{\mathrm{sun}}$) do not appear to improve the success rate; searches remain rather blind and expensive. Of more than 6000 galaxies surveyed for 22 GHz emission in their centers, only 180 (or $\lesssim 3\%$) of them have been found to host water maser emission; among these, only $\sim$30% show megamaser emission in a disk-like configuration†. An efficient scrutiny for new such systems requires a good understanding of the special physical characteristics that nurture them in galaxy centers which, to date, remains ambiguous. While there is some evidence that megamasers may be associated with the molecular disk or torus that surrounds and (partially) obscures an actively accreting massive black hole harbored by a galactic nucleus (AGN), the true connection between megamasers and AGN activity, or other

† i.e., based on continuously updated results of all 22 GHz surveys of galaxies, via the Megamaser Cosmology Project, or MCP; http://wiki.gb.nrao.edu/bin/view/Main/MegamaserCosmologyProject

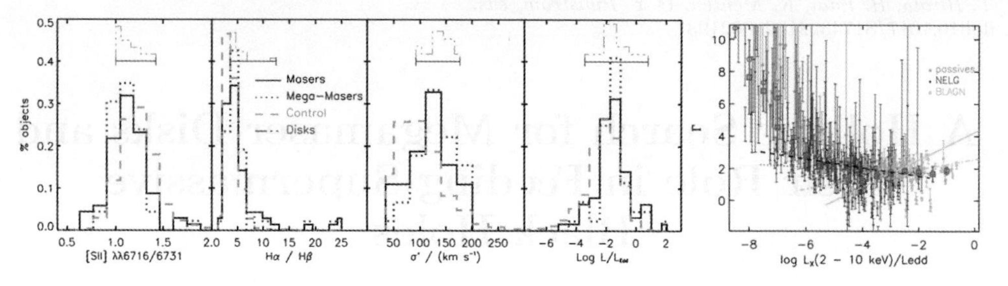

Figure 1. *Left:* Distributions of [S II]$\lambda\lambda$6716/6731 line flux ratios (as proxy for gas densities), Balmer Decrements H_α/H_β, stellar velocity dispersions σ^*, and Eddington ratios ($L/L_{\rm Edd}$) for masers (black), megamasers (dotted blue), disks (dot-dashed red), and non-masers (as control sample; gray dashed). *Right:* Inflection in the $\Gamma - L_X/L_{\rm Edd}$ relation at $L/L_{\rm Edd} \sim 1\%$ illustrated with a sample of ~600 nearby galaxy nuclei from the Chandra Source Catalog with SDSS spectra (Constantin et al. 2023, in prep)

nuclear galactic properties, is still an open question. AGN identification is truly a multi-scale, multi-component, multi-wavelength challenge as for most of the active galaxies, a mix of emission processes (including star-forming regions, shocks, turbulence, etc.), nuclear obscuration, as well as host galaxy starlight, obfuscate their true classification. Searching for maser disks will therefore need to match in sophistication.

2. A "goldilocks" region of nuclear and host galaxy traits for maser disks

Kuo et al. (2018, 2020) have identified some special characteristics that nurture the maser emission in galaxy centers: the megamaser (disk maser) detection rate gets boosted abruptly from $\lesssim$3%($\lesssim$2%) to ~12%(~5%) for a certain wedge in the $w_1 - w_4$ mid-IR (*WISE*) colors and 12 μm AGN luminosities $L_{12\mu m}^{AGN} > 10^{42}$ erg s^{-1}, and to 15%(7%) and 20%(9%) if one select galaxies with $N_{\rm H} \geq 10^{23}$ cm^{-2} and $N_{\rm H} \geq 10^{24}$ cm^{-2}, respectively. These results identify efficient ways of targeting galaxies with megamaser disk emission and strengthen evidence that megamasers are associated with the molecular disk or torus that surrounds and (at least partially) obscures the central AGN. Nevertheless, the completeness rates remain compromised, as a fraction of maser disks would remain undiscovered under such selection criteria.

Interestingly, when adding optical properties (e.g., Constantin 2012), the maser disk galaxies associate with a narrow range of properties, including: the density of gas ($n_e \sim$ 100's cm^{-3}, as measured by the [S II]$\lambda\lambda$6716/6731 line flux ratios), the presence and geometric distribution of obscuration (e.g., Balmer Decrements H_α/H_β), the accretion rate (Eddington ratios; $L/L_{\rm Edd} \sim 10^{-2}$), and the black hole mass ($M_{\rm BH} \sim 10^7 M_{\rm sun}$, via stellar velocity dispersions σ^*), which also corresponds to the location of an inflection point in the X-ray spectral index Γ vs. $L/L_{\rm Edd}$ trend (Figure 1); this inflection is present for a wide range of galactic properties including the AGN fraction, the BH and host mass, the optical spectral classification, presence or absence of the broad line region, and even the host morphology.

These findings suggest that the disk megamaser emission in particular is linked to a short-lived phase in the intermediate-mass galaxy evolution, characterized by an apparent transition in both the availability of the intrinsic absorption (being blown away for higher accretion rates) and the accretion rate of the central engine (from inefficient, or ADAF to the standard quasar-like, Shakura-Sunyaev mode); the inflection location at $L/L_{\rm Edd} \sim 10^{-2}$ seems to also be where the "changing-look" AGN fluctuate from type 1 to type 1.8(2?) Seyferts, and sharp transitions in the column densities N_H have been detected (e.g., Noda & Done 2018; Ricci & Trakhtenbrot 2022). The accretion mode change might thus influence the maser production, also linked with the (just right!) amount of obscuring material.

References

Constantin, A. 2012, *JPhCS*, 372, 012047
Kuo, C. Y., Constantin, A., Braatz, J.A., *et al.* 2018, *ApJ*, 860, 169
Kuo, C. Y., Hsiang, J. Y., Chung, H. H., Constantin, A., *et al.* 2020, *ApJ*, 892, 18
Noda H. & Done C. 2018, *MNRAS*, 480, 3898
Ricci C. & Trakhtenbrot B. 2022, *Nature Astronomy*, arXiv:2211.05132

Cosmic Masers: Proper Motion toward the Next-Generation Large Projects
Proceedings IAU Symposium No. 380, 2024
T. Hirota, H. Imai, K. Menten, & Y. Pihlström, eds.
doi:10.1017/S1743921323002491

Study of Active Galactic Nuclei using the Water Vapour Masers

Deepshikha[1], **Nakai Naomasa**[1], **Yamazaki Madoka**[2] **and Yamauchi Aya**[3]

[1]Kwansei Gakuin University. email: isr24577@kwansei.ac.jp

[2]University of Tsukuba

[3]National Astronomical Observatory of Japan

Abstract. The mass of the black hole separated from the mass of the maser disk is calculated using the mega-maser technique for 15 maser-galaxies and the corresponding Magorrian relationship ($M_{BH} - \sigma$) is analysed.

Keywords. Active Galactic Nuclei, Maser disk, Black hole

---

1. Introduction

In the surroundings of some Active Galactic Nuclei (AGN), the hot (500K) dense molecular gas (10^{7-10} cm^{-3}) that probes the region closer to the central engine i.e. a black hole (BH), radiates extremely luminous mega-maser emission. They have a characteristic spectral profile usually with two distinct groups: one centred near the systemic recession velocity (V_{sys}) of the galaxy while the other includes the high-velocity features, blue-shifted or red-shifted, whose Very Long Baseline Interferometry (VLBI) observations are used to plot the rotational velocity (V_{rot}) curves.

2. Methods

Using the VLBI images of the distribution of maser spots, the position angles of the mega-maser disks are calculated assuming the disks to be edge-on and the rotation curves (Figure 1) are plotted using $V_{rot} = V_{LSR} - V_{sys}$, where V_{LSR} is the line-of-sight velocity with respect to the Local Standard of Reference. Since the sub-parsec disks used for analysis are in gravitational potentials near the BH, special and general relativistic corrections are done in the data to increase the accuracy (although the effects are minimal). Moreover, in order to see if the disk is following the Keplerian rotation or not, a position-velocity relationship (1) is fitted in Figure 1 with α as one of the free parameters. If $|\alpha| = 0.5$, the orbit is Keplerian (point mass, BH) whereas if $|\alpha| < 0.5$, it is Non-Keplerian (extended/non-negligible mass, BH+maser disk)

$$V_{LSR} = A(X - X_0)^{\alpha} + V_{sys} \tag{1}$$

where A and α are constants, X is a coordinate along the edge-on disk, and X_0 is the position of BH.

2.1. *Mass Model*

The Non-Keplerian rotation indicates that the mass causing the gravitational field is not reducible to a point but is extended. The maser disk is assumed to be in a plane that

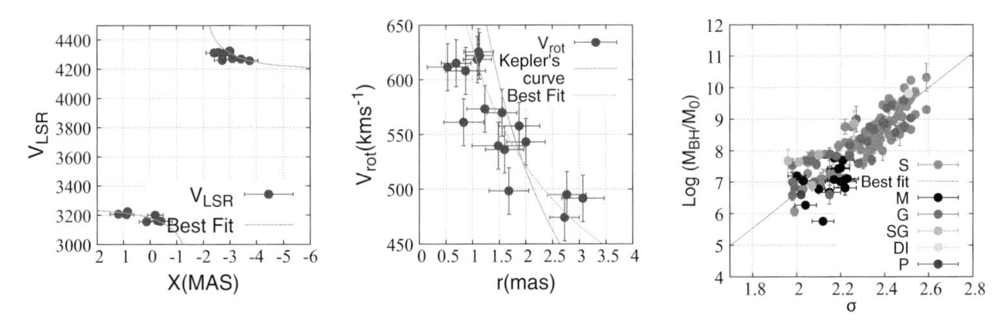

Figure 1. (left) Plot for line-of-sight velocities for blue-shifted and red-shifted maser features of NGC3393 (as an example), (middle) Keplerian and Non-Keplerian rotation curves for NGC3393, and (right) The Maggorian Relationship due to the stellar dynamical modelling [S], megamaser technique [M], gas dynamical modelling [G],stellar gas dynamical modelling [SG], direct imaging [DI], and due to the proper motion [P]

contains the accretion disk and an outer gaseous disk surrounding the fixed BH (Huré 2002). To calculate M_{BH}, a position-velocity curve is fitted using the equation (2) (Huré 2002; Huré *et al.* 2007, 2008; Mamyoda *et al.* 2009) where the disk is assumed to be a Mestel disk.

$$V_{rot} = |V_{LSR} - V_{sys}| = \sqrt{\frac{GM_{BH}}{X - X_0} + 2\pi G \Sigma_{out} a_{out}}, \tag{2}$$

where a_{out} is the outer edge of the gaseous disk and Σ_{out} is the surface density of the gas at a_{out}.

3. Results

The masses calculated by the mega-maser technique using the existing maser data (Gao *et al.* 2017; Kuo *et al.* 2011; Kondratko *et al.* 2008; Gao *et al.* 2016; Zhao *et al.* 2018) are plotted in the $M_{BH} - \sigma$ plot (Figure 1 where σ is the velocity dispersion of the stars in the bulge of the galaxy) but they are not included in the linear fit, showing that these are usually lighter than those calculated using other techniques.

4. Conclusion

(1) The orbits were checked to see whether they are Keplerian or not and if the mass distribution is pointed or extended.

(2) M_{BH} separated from the M_{disk} was calculated while adapting the disk-mass model using the existing maser data of 15 mega-maser galaxies assuming the disk to be Mestel.

(3) $M_{BH} - \sigma$ relationship was plotted using the data from Nandini *et al.* (2019) that includes the non-maser techniques. The calculated masses were plotted in the same graph for comparison. It may be concluded that since maser discs are extremely close to the central engines and cover lesser radii, this technique gives more precise M_{BH} separated from the M_{disk}. Hence, most of the M_{BH} of mega-maser galaxies are much lower than the mass expected by the $M_{BH} - \sigma$ relationship of the non-maser techniques.

Supplementary material

To view supplementary material for this article, please visit http://dx.doi.org/10.1017/S1743921323002491

References

Gao, F., Braatz, J., A., Reid, M., J., Lo, K., Y., Condon, J., J., Henkel, C., Kuo, C., Y., Impellizzeri, C., M., V., Pesce, D., W., Zhao, W. 2016, *ApJ*, 817, 128

Gao, F., Braatz, J., A., Reid, M.,J., Condon, J., J., Greene, J., E., Henkel, C., Impellizzeri, C., M., V., Lo, K., Y., Kuo, C., Y., Pesce, D., W. 2017, *ApJ*, 834, 52

Huré, J.,-M. 2002, *A&A*, 395, L21

Huré, J.,-M., & Hersant, F. 2007, *A&A*, 467, 907

Huré, J.,-M., Hersant, F., Carreau, C., Busset, J., -P. 2008, *A&A*, 490, 477

Kondratko, P., T., Greenhill, L., J., Moran, J., M. 2008, *ApJ*, 678, 87

Kuo, C. Y., Braatz, J., A., Condon, J., J., Impellizzeri, C., M., V., Lo, K., Y., Zaw, I., Schenker, M., Henkel, C., Reid, M., J., Greene, J., E. 2011, *ApJ*, 727, 20

Mamyoda, K., Nakai, N., Yamauchi, A., Diamond, P., Huré, J., -M. 2009, *PASJ*, 61, 1143

Nandini, S., Alister, W., G., Benjamin, L., D. 2019, *ApJ*, 887, 10

Reid, M., J., Braatz, J., A., Condon, J., J., Greenhill, L., J., Henkel, C., Lo, K., Y. 2009, *ApJ*, 695, 287

Zhao, W., Braatz, J., A., Condon, J., J., Lo, K., Y., Reid, M., J., Henkel, C., Pesce, D., W., Greene, J., E., Gao, F., Kuo, C., Y. 2018, *ApJ*, 854, 124Z

Cosmic Masers: Proper Motion toward the Next-Generation Large Projects
Proceedings IAU Symposium No. 380, 2024
T. Hirota, H. Imai, K. Menten, & Y. Pihlström, eds.
doi:10.1017/S1743921323002119

Water masers associated with AGN in radio galaxies

Satoko Sawada-Satoh

Osaka Metropolitan University, Sakai-shi, Osaka 599-8531, Japan. email: swdsth@gmail.com

Abstract. We present dual-frequency VLBI observations of a nearby radio galaxy NGC 4261 at 22 and 43 GHz using the East Asia VLBI Network. In particular, the first sub-pc scale image of the 22 GHz water megamaser line in the circumnuclear region of NGC 4261 is shown. Our results suggest that the megamaser emission in NGC 4261 can be associated with the inner radius of the obscuring disk, as it is proposed for the nearest radio-loud megamaser source NGC 1052. An alternative hypothesis on the megamaser association is the shock region of the interaction between the jet and ambient molecular clouds.

Keywords. galaxies: active, galaxies: nuclei, radio lines: galaxies

1. Introduction

22 GHz water megamasers have been explained as a possible signature of AGN phenomena such as acccretion disk (e.g. NGC 4258) and nuclear outflow (e.g. Circinus). They are mostly associated with radio-quiet AGNs. Water megamasers have been found in a limited number of radio-loud AGNs such as NGC 1052 (Braatz *et al.* 1994; Claussen *et al.* 1998), TXS 2226-184 (Koekemoer *et al.* 1995; Surcis *et al.* 2020), 3C 403 (Tarchi *et al.* 2003, 2007), Mrk 348 (Peck *et al.* 2003) and NGC 4261 (Wagner 2013). Past multi-frequency VLBI observations of the nearest radio-loud water megamaser NGC 1052 have proposed a model that maser clouds in the radio galaxy NGC 1052 lie in a circumnuclear torus, and amplify the continuum seed emission from the jet knots in the background (Sawada-Satoh *et al.* 2008). However, the origin and excitation mechanism of water megamasers in radio-loud AGN still remains unclear to date. Therefore, determining the location and kinematics of water maser gas in the second-nearest radio-loud water megamaser source NGC 4261 is essential.

2. Water megamaser emission in NGC 4261

NGC 4261 is a nearby radio galaxy with a symmetric two-sided radio jet along the east-west direction. The western and eastern jet approach and recede from the observers, respectively (Haga *et al.* 2015). This galaxy is known to have a pc-scale obscuring disk or torus traced by ionized gas (Jones & Wehrle 1997; Jones *et al.* 2000) and neutral atomic hydrogen (van Langevelde *et al.* 2000) surrounding the central source.

Our VLBI observations marginally detected water megamaser line with a peak flux density of 12.9 mJy at 2289 km s^{-1}, slightly redshifted relative to the V_{sys} (Sawada-Satoh *et al.* 2023). Imaged velocity-integrated intensities (moment 0) maps reveal a prominent elongated structure, just east of the continuum peak position at 22 GHz (figure 1ac), where the free-free absorption opacity due to the ionized gas was high on the eastern receding jet. This suggests that the water maser gas spatially coincides with the ionized gas. The inner surface of the disk is directly illuminated with by

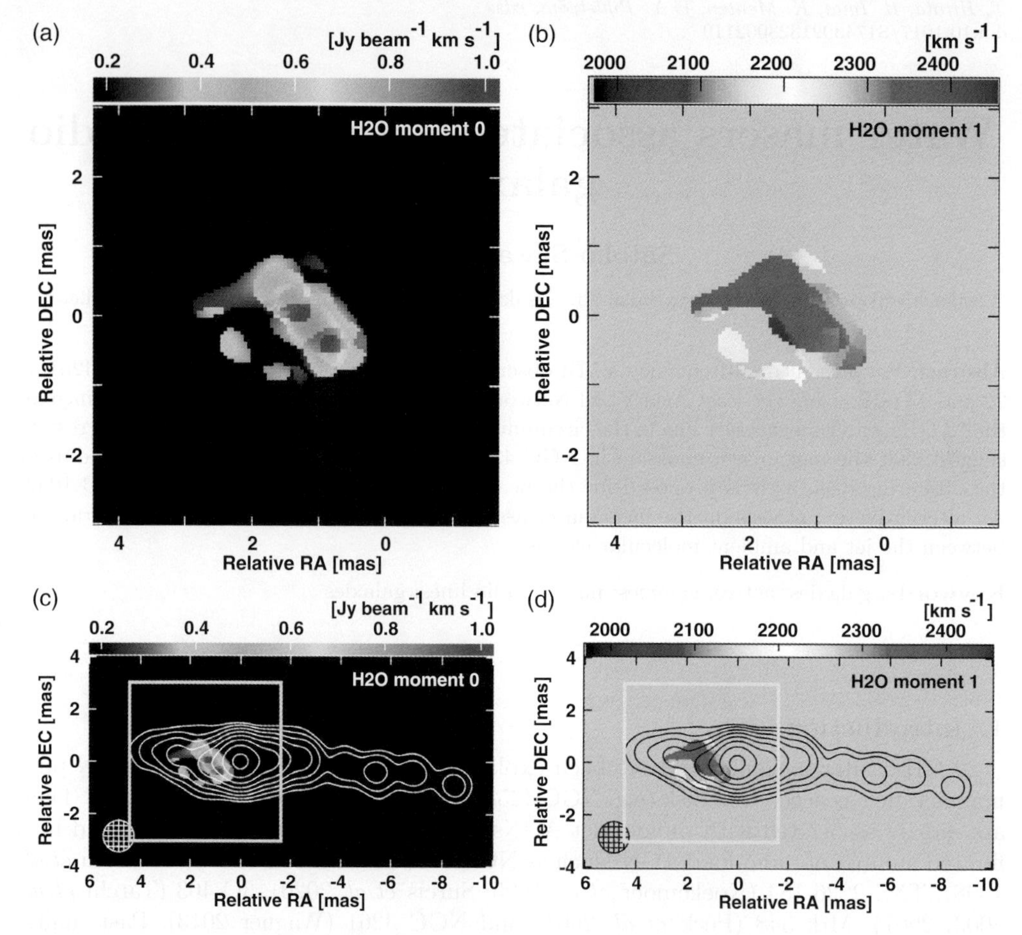

Figure 1. Close-up view of the (a) moment 0 and (b) moment 1 maps of the water megamaser. Relative distributions of (c) the moment-0 and (d) moment-1 maps with respect to the 22 GHz continuum image (Sawada-Satoh *et al.* 2023).

X-ray radiation from the central source, and an ionized gas layer is formed on the surface. Excited water molecular gas inside the near side of disk amplify the background continuum seed emission from the receding jet, and produce the luminous water megamaser emission. Intensity-weighted velocity (moment 1) maps show that the redshifted emission arises from the elongated structure (figure 1bd). The redshifted velocity could be ongoing infall motion from the disk toward the center. This is analogous to the multi-phase circumnuclear torus model in the nearest radio-loud water megamaser source NGC 1052 (Sawada-Satoh *et al.* 2008). Further high-sensitivity VLBI imaging would be helpful in better understanding the complex gas kinematics in the immediate vicinity of the SMBH.

References

Braatz, J. A., Wilson, A. S., & Henkel, C. 1994, *ApJL*, 437, L99.
Claussen, M. J., Diamond, P. J., Braatz, J. A., *et al.* 1998, *ApJL*, 500, L129.
Haga, T., Doi, A., Murata, Y., *et al.* 2015, *ApJ*, 807, 15.
Jones, D. L. & Wehrle, A. E. 1997, *ApJ*, 484, 186.
Jones, D. L., Wehrle, A. E., Meier, D. L., *et al.* 2000, *ApJ*, 534, 165.
Koekemoer, A. M., Henkel, C., Greenhill, L. J., *et al.* 1995, *Nature*, 378, 697.
Peck, A. B., Henkel, C., Ulvestad, J. S., *et al.* 2003, *ApJ*, 590, 149.

Sawada-Satoh, S., Kameno, S., Nakamura, K., *et al.* 2008, *ApJ*, 680, 191.
Sawada-Satoh, S., Kawakatu, N., Niinuma, K., *et al.* 2023, *PASJ*, in press.
Surcis, G., Tarchi, A., & Castangia, P. 2020, *A&A*, 637, A57.
Tarchi, A., Henkel, C., Chiaberge, M., *et al.* 2003, *A&A*, 407, L33.
Tarchi, A., Brunthaler, A., Henkel, C., *et al.* 2007, *A&A*, 475, 497.
van Langevelde, H. J., Pihlström, Y. M., Conway, J. E., *et al.* 2000, *A&A*, 354, L45
Wagner, J. 2013, *A&A*, 560, A12.

Review talk in the session Black Hole Masses and the M-sigma Relation by Masatoshi Imanishi. Taken by Ka-Yiu Shum.

Chapter 3
Structure of the Milky Way

Cosmic Masers: Proper Motion toward the Next-Generation Large Projects
Proceedings IAU Symposium No. 380, 2024
T. Hirota, H. Imai, K. Menten, & Y. Pihlström, eds.
doi:10.1017/S1743921323002545

Galactic Astrometry with VLBI

Kazi L. J. Rygl

INAF-Istituto di Radioastronomia, Via P. Gobetti 101, I-40129 Bologna, Italy.
email: kazi.rygl@inaf.it

Abstract. Astrometric very long baseline interferometry (VLBI) observations of stellar masers are an excellent method to determine distances and proper motions in our Galaxy. Large maser astrometry surveys, the Bar and Spiral Structure Legacy survey and the VLBI Exploration for Radio Astrometry, allowed astronomers to determine fundamental Galactic parameters, such as the rotation curve and the distance to the Galactic centre, as well as to trace the spiral arms. In this review, the results of these surveys will be summarised and compared with astrometric measurements using other methods.

Keywords. Masers, astrometry, VLBI, Milky Way

---

1. Introduction

Hubble Space Telescope images of nearby, face-on barred spiral galaxies beautifully show their spiral arm structures (Fig. 1). These are traced by bright early-type stars and dark dust lanes. Understanding the spiral structure of our own Galaxy is much harder due to the position of the Sun in the Galactic disk. For us, the Milky Way appears as a starry and dusty lane on the sky (Fig. 2), where at each Galactic longitude we see the emission of the spiral arms in our line of sight, located at different distances from us, superimposed. The high dust extinction in the spiral arms further limits optical wavelengths to detect emission from far away sources. Precise locations of spiral arms in the Galactic plane, their number and structure are therefore much more difficult to measure for the Milky Way than for the nearby face-on galaxies from Fig. 1.

Spectral line Galaxy-wide surveys of HI or CO have the potential to unravel the spiral arm structure in (Galactic) longitude-velocity $(l - v)$ diagrams through Galactic rotation patterns. The $(l - v)$ diagram of the CO (1–0) line survey of the Galaxy from Dame *et al.* (2001) shows clearly the spiral arms signatures as coherent emission features, whilst the detection of the ellipsoidal brightness distribution at 2.4 μm have shown that the Milky Way is a barred galaxy (Matsumoto *et al.* 1982). However to pinpoint *where* those spiral arms are in three dimensions requires to measure distances on Galactic size scales of about 10 kpc or more at a wavelength that is not obscured by dust.

Using very long baseline astrometry (VLBI) at radio frequencies, which are not affected by dust extinction, astronomers can achieve parallax uncertainties of 10 μarcseconds (e.g., Reid and Honma 2014). Masers around high-mass star-forming regions (HMSFRs) and evolved stars provide bright, point-like targets that can be used for VLBI measurements. In particular, astrometry of HMSFRs, which are known to be located mainly in the spiral arms, is a very powerful method for measuring the structure of the spiral arms and fundamental parameters of our Milky Way, such the Galactic rotation and the distance to the Galactic Centre. In the next sections, I will discuss Galactic astrometry using radio VLBI and specifically radio VLBI of masers, how the two main maser astrometry surveys were conducted, and some of the most important results regarding the Galactic structure obtained by these surveys. After that, I describe the present, new maser survey

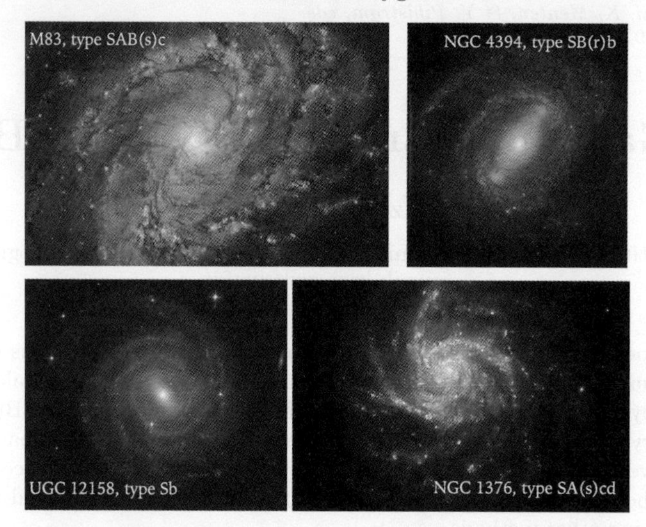

Figure 1. Hubble Space Telescope images of four nearby face-on barred spiral galaxies. Image credits: NASA, ESA, Hubble Heritage Team (STScI/AURA), W. Blair, J. Schmidt and R. Thompson.

Figure 2. The Milky Way as seen from the Atacama Large Millimeter/submillimeter Array site. Image credit: D. Kordan/ESO.

in the Southern hemisphere and improvements on the astrometric accuracy obtained by new calibration methods.

A number of reviews on Galactic maser astrometry with VLBI and comparisons with *Gaia* are readily available, these are Reid and Honma (2014); Xu *et al.* (2018); Immer and Rygl (2022).

2. Astrometry at radio wavelengths using VLBI

2.1. *VLBI arrays for astrometry*

Table 1 lists the VLBI arrays used for maser astrometry.

In Asia, there is the VLBI Experiment for Astrometry† (VERA) array which contains four 20m-antennas located in Japan. Most of the maser astrometry performed with VERA uses the 22 GHz water and SiO masers at ∼43 GHz, however also 6.7 GHz methanol maser astrometry has been done (Matsumoto *et al.* 2011). The VERA antennas are

† The VERA website is available at `https://www.miz.nao.ac.jp/veraserver/index.html`

Table 1. VLBI arrays for maser astrometry.

Array	Max baseline (km)	Angular resolution @6.7 GHz (mas)	Maser species
VERA	2,300	∼4	CH_3OH, H_2O, SiO
VLBA	8,600	∼1	OH, CH_3OH, H_2O, SiO
EVN	10,200	∼1	OH, CH_3OH, H_2O, SiO
LBA	9,800	∼1	OH, CH_3OH, H_2O
ASCI	3,500	∼3	CH_3OH

equipped with a dual-beam receiver system that permits to observe the maser target and the phase calibrator simultaneously (Honma *et al.* 2008).

The Very Long Baseline Array† (VLBA) is a VLBI network in the USA consisting of ten identical 25-m antennas. Its largest baseline is comprised by the station on the Virgin Islands and the station in Hawaii. The VLBA can observe a wide range of masers for astrometry: hydroxyl (OH) around 1.6 GHz; methanol at 6.7 and 12.2 GHz; water at 22 GHz, and the SiO masers both around 43 and 86 GHz. The VLBA was the first VLBI array to perform maser astrometry observations.

The European VLBI Network‡ (EVN) is a network of (inhomogenous) antennas from Europe, China, Russia, South Africa and Puerto Rico making it the cm VLBI network with the largest baseline separation. Thanks to a number of large dishes, such as the 100m Effelsberg, 65m Tianma and 64m Sardinia Radio Telescope, the EVN is the most sensitive VLBI array on Earth. The EVN can observe all masers from OH masers around 1.6 GHz to SiO masers around 43 GHz. It is the first VLBI network to have measured 6.7 GHz methanol maser parallaxes (Rygl *et al.* 2010). The EVN also observes together with the VLBA, VLA and Green Bank Observatory; these are called global VLBI observations.

In Oceania there are two VLBI arrays, the Long Baseline Array§ (LBA) and the AuScope-Ceduna-Interferometer (ASCI). These are the only two Southern hemisphere VLBI arrays, which are essential for astrometry in the Galactic Centre region and the fourth Galactic quadrant. The LBA consist of 7 antennas of which 70m Tidbinbilla and 64m Parkes are the largest antennas. The LBA can observe the OH maser around 1.6 GHz, the methanol masers at 6.7 and 12.2 GHz and the 22 GHz water masers. Krishnan *et al.* (2015) performed the first LBA maser astrometry using the 6.7 GHz methanol masers. The ASCI consists of five (inhomogenous) antennas, of which the largest are the 30m Ceduna and the 27m Hobart antennas (Hyland *et al.* 2018). ASCI has receivers in the range of 2.3 - 14 GHz, which include the methanol maser transitions at 6.7 and 12.2 GHz.

2.2. *Targets for VLBI astrometry*

There are number of compact, non-thermal radio-emitting objects that are interesting for Galactic VLBI astrometry. A few of these are described below.

Non-thermal radio continuum emission in low-mass young stellar objects (YSOs) arises primarily from their magnetosphere. The Goulds Belt Distances Survey (GOBELINS) is a large VLBA programme to measure the parallax and proper motions of YSOs in the nearby star-forming clouds to obtain cloud distances and the 3-dimensional structure of the cloud (e.g., Ortiz-León *et al.* 2017). Toward some of these radio YSOs there is relatively (with respect to high-mass YSOs) little exctinction and a comparison between optical *Gaia* and radio astrometry can be made. For example, Ortiz-León *et al.* (2018)

† The VLBA website is available at `https://science.nrao.edu/facilities/vlba`
‡ The EVN website is available at `https://www.evlbi.org/`
§ The LBA website is available at `https://www.atnf.csiro.au/vlbi/overview/index.html`

have shown there is a good agreement between radio VLBI and *Gaia* astrometry for the YSOs in the Ophiuchos cloud.

PSRπ is a large VLBA project to measure radio pulsar distances and proper motions (Deller *et al.* 2016, 2019) which, when combined with their dispersion measures, can help to improve our knowledge of the Galactic electron distribution. The project has obtained parallaxes for 57 pulsars with a median uncertainty of 40 μas.

Black hole X-ray binaries (BHXBs) are compact radio sources. Black hole formation through a supernova explosion is expected to give these systems a larger natal kick than the direct collapse mechanism. Thus, by measuring the proper motion of BHXBs one try to model the natal kick velocity of that system and learn about the formation of stellar-mass black holes. For example, Atri *et al.* (2019) used VLBI astrometry of the LBA, VLBA and EVN to measure the parallax and proper motion of three BHXB from which they modelled the natal kick velocity distributions.

Maser astrometry and polarization studies can reveal structure, kinematics and magnetic field properties of the masing gas on tens to thousands au-scales around YSOs as shown by Sanna *et al.* (2010, 2015) for HMSFR G023.01-00.41. Recently Moscadelli *et al.* (2022, 2023) used global VLBI observations of 22 GHz water masers to provide the first observational evidence of a disk wind in the high-mass YSO IRAS 21078+5211. Maser astrometry can also be used to study the structure of star-forming complexes. For example, Rygl *et al.* (2012) measured maser astrometry towards five HMSFRs toward the Cygnus X region, determining the membership of the Cygnus X complex, its distance and its structure and velocity distribution. On larger scales, using a number of HMSFRs maser astrometry can reveal the structure and kinematics of spiral arms. Sakai *et al.* (2019) found that HMSFRs in the inner part of the Perseus spiral arm have a significant radial inward motion which agrees with models of star-formation through gas shocks when encountering the slower-rotation spiral arm (Roberts 1969, 1972). Zooming out even more, to Galactic scales, one can use maser astrometry to measure the spiral arm positions and determine fundamental parameters of the Milky Way as discussed in the next sections.

3. Measuring Galactic structure through maser VLBI astrometry

High-mass (OB-type) stars form in dense, molecular clouds that commonly found in the spiral arms. Figure 1 shows how the spiral arms are the locations of extinction, the 'dark lanes' and the bright OB-type stars (for an overview of spiral arm tracers see Xu *et al.* 2018). Time scales for high-mass star formation are of the order of 10^5 years or less (see Motte *et al.* 2018, and references therein) and lifetimes of OB type stars are of the order of few 10^5 years (Weidner and Vink 2010). These very short lifetimes imply that these objects could not have moved far from their birth sites and are likely still carrying the kinematic imprint of the gas from which they formed. This is why HMSFRs and OB-type stars are excellent tracers of spiral arm structure and Galactic rotation. The advantage of VLBI maser astrometry of HMSFRs is the lack of extinction at radio wavelengths combined with milliarcsecond angular resolution that allows astronomers to measure parallaxes and proper motions out to even 20 kpc (Sanna *et al.* 2017).

3.1. *Masers in high-mass star-forming regions and evolved stars*

The most common masers found towards HMSFRs are the 22 GHz water masers, which may be located in the disk around the YSO, in the envelope, in the disk-wind, but also in the outflows, as is the case for the water masers in the Turner-Welch object in W 3 (Hachisuka *et al.* 2015). For methanol masers two transitions, at 6.7 and 12.2 GHz, are used for astrometry. These maser species are more often found in the disk or envelope around the YSO. Bartkiewicz *et al.* (2020) found that the 6.7 GHz methanol masers in high-mass YSO G23.657-00.127 are expanding, and are likely tracing a spherical outflow or a wide-angle wind at the base of the proto-stellar jet. Most methanol astrometry studies use the 6.7 GHz transition because it is brighter, ubiquitous, and so far found exclusively in HMSFRs Menten (1991). A few SFRs may also exhibit SiO masers in their outflows; for example, Kim *et al.* (2008) used 43 GHz SiO masers in Source I to derive a parallax to Orion. Their result was in excellent agreement with the parallax of non-thermal emission of low-mass stars in Orion from Menten *et al.* (2007) and Sandstrom *et al.* (2007).

Evolved star envelopes are rich in maser emission (see Matthews 2023). They mainly emit in 1.6 GHz OH, 22 GHz water and 43 (and 86) GHz SiO masers. For example, Matsuno *et al.* (2020) performed astrometric 22 GHz water maser observations of the asymptotic giant branch (AGB) star BX Cam. In addition to the parallax and proper motions, they measured that the maser features have an expansion velocity of about 15 km s^{-1}. The evolved stars with maser emission are typically AGB stars or red super giants (RSGs). RSGs evolve from high-mass stars and have a lifetime of $\sim$ tens of Myr. AGB stars evolve from low-mass stars and have therefore much larger ages. This implies that both these kinds of evolved stars, but especially the AGB stars, are not very good tracers of spiral structure as they have moved away from their natal sites. However, on the contrary to the high extinction towards HMSFRs which makes any optical measurements impossible, many evolved stars have optical parallaxes measured by *Gaia* and a comparison between radio and optical astrometry is possible. In particular, for the validation of the *Gaia* reference frame of bright stars Lindegren (2020) compared the *Gaia* astrometry with VLBI astrometry of continuum emitting of low-mass YSOs and masing evolved stars.

3.2. *Large astrometric surveys*

Recently two large VLBI maser parallax programmes have published their collective results. The two programmes are the Bar and Spiral Structure Legacy (BeSSeL) survey (Brunthaler *et al.* 2011; Reid *et al.* 2019) and the VERA collaboration (VERA Collaboration *et al.* 2020; Honma 2023).

The BeSSeL survey is a key science project of the VLBA and used 6.7 and 12.2 GHz methanol and 22 GHz water masers. The VERA survey used 22 GHz water and SiO masers around 43 GHz.

3.3. *Parallax and proper motions from relative astrometry*

Both BeSSeL and VERA surveys measure relative positions of the maser spots with respect to one or more position reference sources (compact quasars) over the course of one year. This allows to measure and separate the sinusoidal parallax signature and the linear proper motion of the target in each coordinate.

The number of epochs used for astrometric observations depends mainly on the maser species. The less variable methanol masers were observed in four epochs, sampling the minima and maxima of the Right Ascension parallax signal. Water masers, which are more variable have typically 6 epochs spread over one year in the BeSSeL survey, but sometimes more epochs over a longer duration were observed. VERA water maser and SiO astrometry typically observed every one or two months for a year, but some targets were tracked for more than 2 years.

Both these surveys rely on phase referencing observations, where the atmospheric delay is calibrated on a nearby position reference source (typically a quasar) that is assumed to have the same delay as the target. This method allows to achieve tens of microarcsecond *relative* position accuracy over baselines of 8,000 km for a 1° separation between target

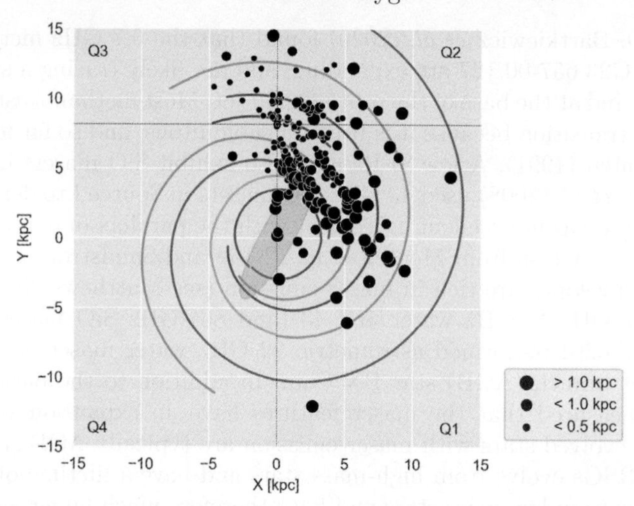

Figure 3. HMSFR distribution in the Milky Way based on maser parallaxes, shown in a plan-view of the Milky Way seen from the north Galactic pole. Data used are from Reid *et al.* (2019); VERA Collaboration *et al.* (2020); Xu *et al.* (2021); Hyland *et al.* (2022b), and references therein. Adapted figure from Immer and Rygl (2022).

and position reference source. This is the accuracy required to reach the ~ 10 microarcsecond parallax uncertainties necessary for Galactic-scale distances. In the case of maser astrometry, inverse-phase referencing is done, calibrating on a bright maser spot as these have a much higher signal-to-noise ratio than the position reference sources. More details can be found in the review by Reid and Honma (2014).

4. Structure and fundamental parameters of our Milky Way: results from maser astrometry

Figure 3 shows a plan-view of the Milky Way with the positions of 233 HMSFRs and RSG with maser parallax astrometry taken from Reid *et al.* (2019); VERA Collaboration *et al.* (2020); Xu *et al.* (2021); Hyland *et al.* (2022b). Around the Sun's position in the plane, the Milky Way is divided in four quadrants. It is clear that that the first and second quadrants are rather well covered by the maser astrometry surveys, the third only in part and the fourth hardly. The more filled quadrants correspond to the part of the sky that can be observed well from the Northern Hemisphere were most of the VLBI arrays, that have been used for maser astrometry up to now, are located.

Furthermore, there are notably few targets on the far side of the Galaxy (beyond the Galactic Centre), in the Galactic Centre region and in the outer regions of the Milky Way (at large Galactocentric radii). It is more difficult to measure smaller parallax signatures, and Fig. 3 also shows that the farther targets have typically larger uncertainties. The other reason is the scarceness of maser astrometry targets where there is fewer star formation activity. The Galactic Centre region has a lower star formation rate than the disk and thus fewer maser targets (Barnes *et al.* 2017). Any new maser astrometry result in this region is very important for our understanding of the structure and dynamics of the Galactic bar, see also Kumar *et al.* (2023). At large Galactocentric radii the star formation activity is notably less than in the inner Galaxy. This is visible in Fig. 3 by fewer astrometric data available for these regions.

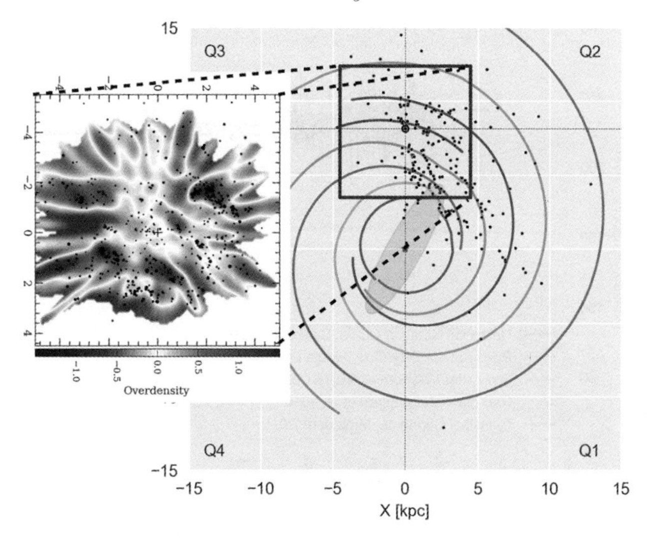

Figure 4. Same as Figure 3, but now showing the area for which *Gaia* DR3 astrometry of OB-type stars is available. The inset shows the over density of the OB-type stars from Gaia Collaboration *et al.* (2022a) rotated to follow the same orientation as the underlying figure.

4.1. *Spiral arm structure*

Combining the parallax and proper motions with the sky positions and a local-standard-of-rest (LSR) velocity, one can obtain a 3-dimensional position and velocity data point for each maser astrometry target.

To derive the spiral arm structure, Reid *et al.* (2019) assigned spiral arm membership to each maser astrometry target as follows. Where possible, spiral arm designations were assigned by associating their Galactic longitude and LSR velocity to molecular clouds of a spiral arm in the $(l - v)$ diagram of CO or HI data (Weaver 1974; Cohen *et al.* 1980; Dame *et al.* 2001). For objects with unclear arm associations through this method, also their parallax, proper motion, and Galactic latitude were used through the Bayesian distance calculator (Reid *et al.* 2016, 2019) to find the best matching spiral arm.

Reid *et al.* (2019) then fitted log-periodic spirals to HMSFRs with spiral arm memberships to derive the parameters for the spiral arms. These derived spiral arm models are shown in color in Figures 3, 4.

Other tracers of spiral arms are OB-type stars for which the *Gaia* satellite can measure parallaxes and proper motions. Due to the dust extinction in the Galactic plane, the latest *Gaia* DR3 data have still a rather limited range of ~3 kpc distance from the Sun where their parallax uncertainties are 10% or better. Nevertheless, within this range, Fig. 4 shows that the OB-type star traced arms Gaia Collaboration *et al.* (2022a) match well with those from maser VLBI astrometry, as expected.

4.2. *Galactic rotation curve*

The motion of each maser astrometry target is a combination of its Galactic rotation at its Galactocentric radius, its peculiar motion, and the Solar motion. Therefore, Galactic rotation curves can be fitted to the three dimensional space velocities of a large number of HMSFRs (and a few super giants) spread over the Galactic plane (as shown in Fig. 3). Both the BeSSeL and the VERA teams used Bayesian Markov Monte Carlo Chain fitting routines, but with different parametrisations of the Galactic rotation curve, a (slightly)

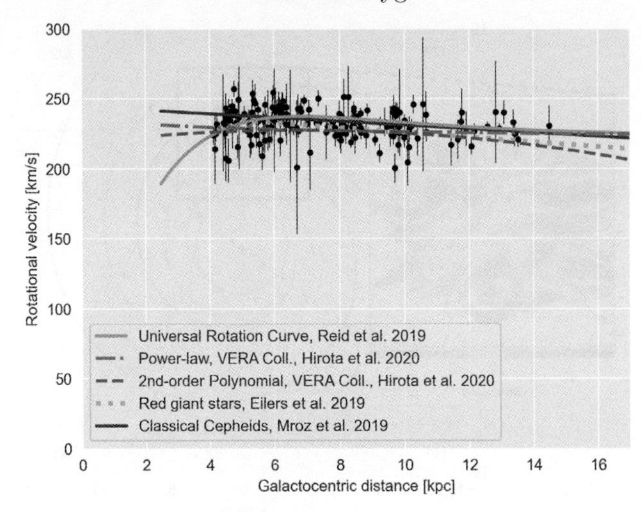

Figure 5. Galactic rotation curves from VLBI maser astrometry (Reid *et al.* 2019; VERA Collaboration *et al.* 2020), from RGSs (Eilers *et al.* 2019) and Classical Cepheids (Mróz *et al.* 2019). The rotational velocity of maser sources (Reid *et al.* 2019; VERA Collaboration *et al.* 2020) is shown by black dots. Figure taken from Immer and Rygl (2022).

different number of maser targets and different priors (for example the Solar motion prior). Details of the fitting can be found in Reid *et al.* (2019) for the BeSSeL survey and in VERA Collaboration *et al.* (2020) for the VERA survey.

In Figure 5, the rotation curves derived by maser astrometry (Reid *et al.* 2019; VERA Collaboration *et al.* 2020) are shown and compared to those from RSGs (Eilers *et al.* 2019) and classical Cepheids (Mróz *et al.* 2019). The maser-based rotation curves were derived from data at Galactocentric radii $\geq$4 kpc since the motions in the inner Galaxy are expected to be influenced by the Galactic bar. While the shape of each of these curves is a bit different, within the region covered by maser data (from 4 to almost 15 kpc, but getting scarcer with larger radii) the rotation curves are quite similar and nearly flat. Note that the RSGs and classical Cepheids rotation curves are derived from Galactocentric radii out to 20 or even 25 kpc. Nevertheless, they agree rather well with the maser-based rotation curves. Figure 5 shows that that the rotational velocity is expected to decrease slowly with radius for all curves. However, the RGSs curve (Eilers *et al.* 2019) is decaying faster than that of the classical Cepheids (Mróz *et al.* 2019). To investigate this decrease with maser astrometry more data in the outer Galaxy are needed. Unfortunately, the star-formation activity in the Outer spiral arm is rather low, and finding HMSFRs with bright enough masers for astrometry is not an easy task.

4.3. *Distance to the Galactic Centre*

By fitting the Galactic rotation curve to maser astrometry data also a distance to the Galactic Centre (R_0) is obtained. Figure 6 shows on the top the two measurements from maser astrometry: $R_0 = 8.15 \pm 0.15$ kpc (Reid *et al.* 2019) and $R_0 = 7.92 \pm 0.16_{\rm stat} \pm 0.3_{\rm sys}$ kpc (VERA Collaboration *et al.* 2020). These two measurements agree well within their uncertainties. It is interesting to see how good these measurements agree with Galactic distances from other methods. The maser astrometry values agree well with the R_0 measurements from other methods as shown in Fig. 6: kinematics of Galactic bar stars (8.23 ± 0.12 kpc, Leung *et al.* 2023), radial velocities of stars (8.27 ± 0.29 kpc, Schönrich 2012), *Gaia* DR2 astrometry of OB-type stars

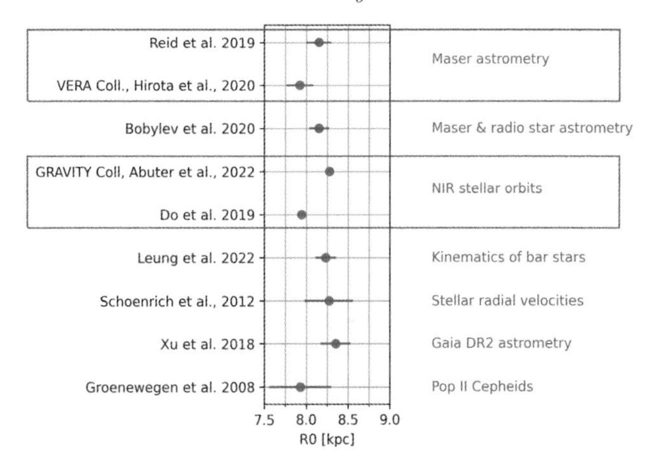

Figure 6. Distance to the Galactic Centre (R_0) as measured by various methods.

(8.35 ± 0.18 kpc, Xu *et al.* 2018), maser and radio star astrometry ($8.15^{+0.04}_{-0.20}$ kpc, Bobylev *et al.* 2020) and Pop II Cepheids (7.93 ± 0.37 kpc, Groenewegen *et al.* 2008). Above all, the two most precise measurements of R_0, based on the measurements of stellar orbits around the Galactic Centre supermassive black hole, obtained $R_0 = 7.946 \pm 0.050_{\rm stat} \pm 0.032_{\rm sys}$ kpc using adaptive optics techniques with the Keck telescope (Do *et al.* 2019), and $8.277 \pm 0.009_{\rm stat} \pm 0.033_{\rm sys}$ kpc with the Very Large Telescope Interferometer (Gravity Collaboration *et al.* 2022). While these values are in a slight tension with each other, both of them agree well with the maser astrometry values, which have larger uncertainties.

4.4. *Sgr A* as the dynamic centre of the Milky Way*

The full angular velocity of the Sun, $\Omega_\odot$, obtained from maser astrometry is $\Omega_\odot = 30.32 \pm 0.27$ km s^{-1} kpc^{-1} (Reid *et al.* 2019) and $\Omega_\odot = 30.17 \pm 0.27$ km s^{-1} kpc^{-1} (VERA Collaboration *et al.* 2020). These two values, which are in good agreement with each other, can then be compared to the apparent proper motion of Sgr A* in the direction of Galactic longitude, -6.411 ± 0.008 mas yr^{-1}, measured by Reid and Brunthaler (2020). Assuming that Sgr A* is stationary, this motion can be interpreted as the reflex motion of the Sun in the direction of Galactic rotation. This would yield 30.39 ± 0.04 km s^{-1} kpc^{-}1 as angular velocity at the Sun, which is indeed in excellent agreement.

This agreement underlines the validity of the assumption of Sgr A* being stationary in the rotating Galactic disk, making it the dynamic centre of the Milky Way. Notably, Galactic-wide velocity fields mapped by HMSFRs over ~ 10 kpc radii and the stellar orbits at 1000s au around Sgr A* yield a consistent distance to the Galactic Centre.

4.5. *Galactic mid plane and warp*

Reid *et al.* (2019) have shown that HMSFRs with Galacto-centric distances ≤ 7 kpc, thus not affected by the Galactic warp, have a very narrow distribution in Galactic latitude. The peak of their distribution is at negative Galactic latitudes and not at 0° which would be expected if the Sun would be in the midplane of the disk. These authors used the latitude distribution of maser-bearing HMSFRs to derive the height of the Sun with respect to the midplane finding $Z_\odot = 5.5 \pm 5.8$ pc. Using this value to correct the IAU Z-plane, shifts the position of Sgr A* (which has Z_{IAU}=-6.5 pc when using d=8.15 kpc) into to the Galactic midplane where one would expect it to be if Sgr A* is our dynamical centre.

In the outer parts of the Milky Way, maser astrometry can be used to trace the Galactic warp. Sakai *et al.* (2020) performed maser astrometry of Outer arm sources in the second quadrant. They found that between Galactic azimuth† of 20 and 50 degrees the scale height of the HMSFRs increases in a sinusoidal wave fashion from Galactocentric distances of 10 kpc onwards reaching heights above the midplane of Z=200-600 pc. In addition to positions, Sakai *et al.* (2020) also investigated the W-motion (in direction of the north Galactic pole) finding it could be fitted well with a sinusoidal curve of about 10 km s^{-1} which reaches 0 km s^{-1} at the Galactocentric radius with the highest Z. This wave-like distribution is also seen in the classical Cepheids distribution of Skowron *et al.* (2019) in that part of the Milky Way. Classical Cepheids with their ages of $\leq$ 400 Myr are considered still relatively good tracers of the Galactic disk. The advantage of Cepheid studies is that there are many more Cepheid stars available and that these reach out to larger Galactocentric radii than maser-bearing stars.

4.6. *High proper motions in the Scutum arm: evidence of the Galactic bar?*

Aside of measuring locations and Galactic rotation velocity components, one obtains the three-dimensional peculiar motions of the maser-bearing SFR. When measuring the parallaxes and proper motions of the Scutum spiral arm, Immer *et al.* (2019) found that there were a number of HMSFRs in this spiral arm, at azimuths where the arm is in the vicinity of the tip of the Galactic bar, that showed exceptionally large ($>$ 20 km s^{-1}) peculiar motions. The authors put forward a combined gravitational attraction of the Galactic bar and the spiral arm potential as the most likely explanation. This fascinating kinematic signature of star-forming gas near the tip of the Galactic Bar was modelled by Li *et al.* (2022) using hydrodynamical simulations of gas in a Milky-Way-like barred spiral galaxy. These simulations reproduced very well the velocity field at the positions of the HMSFRs. In the residual (after subtraction the Galactic rotation) radial velocity component of the gas, the rotation of the gas around the Galactic bar is very evident. *Gaia* astrometry of RGB stars found the same distribution in radial velocities as in the simulations. The combination of maser and stellar astrometry with simulations testify to the strong influence of the Galactic bar on the dynamics of the Milky Way.

5. The future: next generation large projects, new calibration techniques and better comparisons with *Gaia*

5.1. *Southern hemisphere VLBI maser astrometry*

Figure 3 shows the VLBI maser astrometry coverage of the Galactic plane with the rather empty fourth and third quadrants. Southern Hemisphere VLBI maser astrometry is necessary to study this part of the Milky Way, and also the Galactic Centre which is particularly interesting because of the bar. The new, Southern Hemisphere Parallax Interferometric Radio Astrometry Legacy Survey (SπRALS) is a 6.7 GHz methanol maser astrometry survey using the ASCI array. Hyland *et al.* (2022b) published the first astrometric results of this survey, showing to be able to obtain consistent results for HMSFR G232.62+00.99 to those from 12.2 methanol maser astrometry (Reid *et al.* 2009) and delivering maser astrometry for one new target (see also Hyland *et al.* 2023). Furthermore, Kumar *et al.* (2023) have been using ASCI 6.7 GHz data of the Galactic Centre region to study the structure and dynamics of the bar. To improve the Galactic parameters

† Galactic azimuth has the origin in the Galactic Centre and is defined as having zero degree in the direction to the Sun, increasing clock-wise.

derived from maser astrometry, it is crucial to have a good coverage the Galactic velocity field, so results of the SπRALS project are eagerly awaited.

5.2. *New phase referencing techniques*

When doing 6.7 GHz methanol maser astrometry, it was noted that uncompensated dispersive delays due to the ionosphere can cause positional shifts. Reid *et al.* (2017); Wu *et al.* (2019) then devised an approach to estimate and correct for these position shifts, following the MultiView method (Rioja *et al.* 2017; Rioja 2023) but in the image plane. To improve the ionospheric calibration, the SπRALS observations use the MultiView method (Rioja *et al.* 2017), that employs multiple position reference sources at different position angles of the maser target to interpolate the ionospheric delay at the target position. In fact, as the bright maser emission is used for phase-referencing, this method is called inverse MultiView. Hyland *et al.* (2022a) showed that for their two HMSFRs their parallax uncertainties using inverse MultiView were around 10 microarcseconds while when using inverse phase referencing to the various position reference sources the uncertainties were much bigger (in the range of 25 to 60 microarcseconds) and increasing with the separation between the maser and position reference target. A review of the technical improvements including the MultiView technique is discussed in Rioja and Dodson (2020).

5.3. Gaia *comparisons with maser astrometry*

In the latest DR3, *Gaia* has released the astrometry of almost 1.5 billion stars (Gaia Collaboration *et al.* 2022b; Jordi *et al.* 2023). For the AGB star, BX Cam, discussed earlier in Section 3.1, Matsuno *et al.* (2020) measured a 22 GHz water maser parallax of 1.73 ± 0.03 mas. The DR2 *Gaia* parallax of this star, 4.134 ± 0.25 mas, was quite problematic. However, with the improved astrometric accuracy of DR3 the newest *Gaia* DR3 parallax of 1.764 ± 0.10 mas for BX Cam is in very good agreement. This opens up many venues of interesting comparisons between VLBI maser astrometry and optical *Gaia* astrometry for stellar maser targets.

6. Summary

The large VLBI maser astrometry surveys BeSSeL and VERA have measured astrometry to more than 200 HMSFRs and evolved stars. With these data spiral arms have been traced in the Galactic plane. The measured 3-D velocity field of these maser stars permitted an accurate determination of the distance to the Galactic centre and the Galactic rotation. Evidence of the Galactic warp, both in position and velocity, has been measured from maser parallaxes. In the inner Galaxy, the peculiar velocities of HMSFRs are a testimony to the large influence of the bar on Galactic dynamics. With the new maser astrometric survey SπRALS in the Southern Hemisphere exciting new astrometric results will map out the uncharted territory in the forth and third quadrant of Milky Way, which will allow astronomers to improve determination of the Galactic parameters in the future.

References

Atri, P., Miller-Jones, J. C. A., Bahramian, A., *et al.* 2019, *MNRAS*, 489(3), 3116
Barnes, A. T., Longmore, S. N., Battersby, C., *et al.* 2017, *MNRAS*, 469(2), 2263
Bartkiewicz, A., Sanna, A., Szymczak, M., Moscadelli, L., *et al.* 2020, *A&A*, 637, A15
Bobylev, V. V., Krisanova, O. I., & Bajkova, A. T. 2020, *Astronomy Letters*, 46(7), 439
Brunthaler, A., Reid, M. J., Menten, K. M., *et al.* 2011, *Astronomische Nachrichten*, 332(5), 461
Cohen, R. S., Cong, H., Dame, T. M., & Thaddeus, P. 1980, *ApJ*, 239, L53

Dame, T. M., Hartmann, D., & Thaddeus, P. 2001, *ApJ*, 547(2), 792
Deller, A. T., Goss, W. M., Brisken, W. F., *et al.* 2019, *ApJ*, 875(2), 100
Deller, A. T., Vigeland, S. J., Kaplan, D. L., *et al.* 2016, *ApJ*, 828(1), 8
Do, T., Hees, A., Ghez, A., *et al.* 2019, *Science*, 365(6454), 664
Eilers, A.-C., Hogg, D. W., Rix, H.-W., & Ness, M. K. 2019, *ApJ*, 871(1), 120
Gaia Collaboration, Drimmel, R., Romero-Gomez, M., Chemin, L., *et al.* 2022a, *A&A*, 674, A37
Gaia Collaboration, Vallenari, A., Brown, A. G. A., Prusti, T., *et al.* 2022b, *A&A*, 674, A1
Gravity Collaboration, Abuter, R., Aimar, N., Amorim, A., *et al.* 2022, *A&A*, 657, L12
Groenewegen, M. A. T., Udalski, A., & Bono, G. 2008, *A&A*, 481(2), 441
Hachisuka, K., Choi, Y. K., Reid, M. J., *et al.* 2015, *ApJ*, 800(1), 2
Honma, M., Kijima, M., Suda, H., *et al.* 2008, *PASJ*, 60(5), 935
Honma, M. 2023, in this volume
Hyland, L. J., Ellingsen, S. P., & Reid, M. J. 2018, in A. Tarchi, M. J. Reid, M. J. & P. Castangia (eds.), *Astrophysical Masers: Unlocking the Mysteries of the Universe*, Proc. IAU Symposium, 336, p. 154
Hyland, L. J., Reid, M. J., Ellingsen, S. P., *et al.* 2022a, *ApJ*, 932 (1), 52
Hyland, L. J., Reid, M. J., Orosz, G., *et al.* 2022b, *ApJ*, 953(1), 21
Hyland, L. J., *et al.* 2023, in this volume
Immer, K., Li, J., Quiroga-Nuñez, L. H., *et al.* 2019, *A&A*, 632, A123
Immer, K. & Rygl, K. L. J. 2022, *Universe*, 8(8), 390
Jordi, C., *et al.* 2023, in this volume
Kim, M. K., Hirota, T., Honma, M., *et al.* 2008, *PASJ*, 60, 991
Krishnan, V., Ellingsen, S. P., Reid, M. J., *et al.* 2015, *ApJ*, 805(2), 129
Kumar, J., *et al.* 2023, in this volume
Leung, H. W., Bovy, J., Mackereth, J. T., *et al.* 2023, *MNRAS*, 519(1), 948
Li, Z., Shen, J., Gerhard, O., & Clarke, J. P. 2022, *ApJ*, 925(1), 71
Lindegren, L. 2020, *A&A*, 633, A1
Matthews, L. 2023, in this volume
Matsumoto, N., Honma, M., Isono, Y., *et al.* 2011, *PASJ*, 63, 1345
Matsumoto, T., Hayakawa, S., Koizumi, H., *et al.* 1982, in G. R. Riegler & R. D. Blandford (eds.), *The Galactic Center*, AIP Conference Series, 83, p. 48
Matsuno, M., Nakagawa, A., Morita, A., *et al.* 2020, *PASJ*, 72(4), 56
Menten, K. M. 1991, *ApJ*, 380, L75
Menten, K. M., Reid, M. J., Forbrich, J., & Brunthaler, A. 2007, *A&A*, 474(2), 515
Moscadelli, L., Sanna, A., Beuther, H., Oliva, A., & Kuiper, R. 2022, *Nature Astronomy*, 6, 1068
Moscadelli, L., *et al.* 2023, in this volume
Motte, F., Bontemps, S., & Louvet, F. 2018, *ARA&A*, 56, 41
Mróz, P., Udalski, A., Skowron, D. M., *et al.* 2019, *ApJ*, 870(1), L10
Ortiz-León, G. N., Loinard, L., Dzib, S. A., *et al.* 2018, *ApJ*, 865(1), 73
Ortiz-León, G. N., Loinard, L., Kounkel, M. A., *et al.* 2017, *ApJ*, 834(2), 141
Reid, M. J. & Brunthaler, A. 2020, *ApJ*, 892(1), 39
Reid, M. J., Brunthaler, A., Menten, K. M., *et al.* 2017, *AJ*, 154(2), 63
Reid, M. J., Dame, T. M., Menten, K. M., & Brunthaler, A. 2016, *ApJ*, 823(2), 77
Reid, M. J. & Honma, M. 2014, *ARA&A*, 52, 339
Reid, M. J., Menten, K. M., Brunthaler, A., *et al.* 2019, *ApJ*, 885(2), 131
Reid, M. J., Menten, K. M., Brunthaler, A., *et al.* 2009, *ApJ*, 693(1), 397
Rioja, M. J., Dodson, R., Orosz, G., Imai, H., & Frey, S. 2017, *AJ*, 153(3), 105
Rioja, M. J. & Dodson, R. 2020, *A&A Rev.*, 28(1), 6
Rioja, M. J. 2023, in this volume
Roberts, W. W. 1972, *ApJ*, 173, 259
Roberts, W. W. 1969, *ApJ*, 158, 123
Rygl, K. L. J., Brunthaler, A., Reid, M. J., *et al.* 2010, *A&A*, 511, A2
Rygl, K. L. J., Brunthaler, A., Sanna, A., *et al.* 2012, *A&A*, 539, A79
Sakai, N., Nagayama, T., Nakanishi, H., *et al.* 2020, *PASJ*, 72(4), 53

Sakai, N., Reid, M. J., Menten, K. M., Brunthaler, A., & Dame, T. M. 2019, *ApJ*, 876(1), 30
Sandstrom, K. M., Peek, J. E. G., Bower, G. C., Bolatto, A. D., & Plambeck, R. L. 2007, *ApJ*, 667(2), 1161
Sanna, A., Moscadelli, L., Cesaroni, R., *et al.* 2010, *A&A*, 517, A78
Sanna, A., Moscadelli, L., Surcis, G., *et al.* 2017, *A&A*, 603, A94
Sanna, A., Surcis, G., Moscadelli, L., *et al.* 2015, *A&A*, 583, L3
Schönrich, R. 2012, *MNRAS*, 427(1), 274
Skowron, D. M., Skowron, J., Mróz, P., *et al.* 2019, *Science*, 365(6452), 478
VERA Collaboration, Hirota, T., Nagayama, T., Honma, M., *et al.* 2020, *PASJ*, 72(4), 50
Weaver, H. 1974, in F. J. Kerr & S. C. Simonson (eds.), *Galactic Radio Astronomy*, 60, 573
Weidner, C. & Vink, J. S. 2010, *A&A*, 524, A98
Wu, Y. W., Reid, M. J., Sakai, N., *et al.* 2019, *ApJ*, 874(1), 94
Xu, Y., Bian, S. B., Reid, M. J., *et al.* 2021, *ApJS*, 253(1), 1
Xu, Y., Hou, L.-G., & Wu, Y.-W. 2018, *Research in Astronomy and Astrophysics*, 18(12), 146

Cosmic Masers: Proper Motion toward the Next-Generation Large Projects
Proceedings IAU Symposium No. 380, 2024
T. Hirota, H. Imai, K. Menten, & Y. Pihlström, eds.
doi:10.1017/S174392132300306X

Galactic Maser Astrometry with VERA

Mareki Honma[1,2,3], Tomoya Hirota[1,3], Kazuya Hachisuka[1], Hiroshi Imai[4], Hideyuki Kobayashi[1,3], Takaaki Jike[1], Akiharu Nakagawa[4], Tomoaki Oyama[1], Kazuyshi Sunada[1], Daisuke Sakai[1], Nobuyuki Sakai[5] and Aya Yamauchi[1]

[1]National Astronomical Observatory of Japan, Oshu, Iwate 023-0861, Japan.
email: mareki.honma@nao.ac.jp

[2]University of Tokyo, Bunkyo, Tokyo 113-0033, Japan

[3]SOKENDA, Mitaka, Tokyo 181-8588, Japan

[4]Kagoshima University, Kagoshima, Kagoshima 890-0065, Japan

[5]National Astronomical Research Institute of Thailand, Chiangmai 50180, Thailand

Abstract. VERA has been regularly conducting astrometry of Galactic maser sources for ∼ 20 years, producing more than 100 measurements of parallaxes and proper motions of star-forming regions as well as AGB stars. By combining the observational results obtained by VLBA BeSSeL, EVN, and LBA, maser astrometry provides a unique opportunity to explore the fundamental structure of the Galaxy. Here we present the view of the Galaxy revealed by the maser astrometry, and also discuss the importance of maser astrometry in the era of GAIA by comparing the results obtained by VLBI and GAIA. We also present our view of "proper motions toward the future" of the relevant field, expected in the next decade based on global collaborations.

Keywords. VLBI, maser, Milky Way Galaxy

1. Introduction

Maser emissions are one of most powerful tools for astrometric measurements, as their compactness and high intensities allow us to conduct VLBI observations with high-angular resolution. In fact, all the major VLBI arrays in the world, including VERA (VLBI Exploration of Radio Astrometry), VLBA (Very Long Baseline Array), EVN (European VLBI Network) in the northern hemisphere and LBA (Long Baseline Array) in the southern hemisphere, are actively conducting astrometric observations of maser sources (for reviews, see e.g., Reid & Honma 2014, Immer & Rygl 2022). Among them, VERA is a dedicated array for maser astrometry, and has its unique feature of dual-beam observing system to effectively compensate for tropospheric fluctuations, the main source of astrometric errors above 10 GHz.

The construction of the VERA array was completed in 2002. One of the four stations is located in the Kagoshima prefecture, where this maser symposium is held. Since its start of science operations, NAOJ and the radio astronomy group in Kagoshima University have been working closely in observations and producing science results from VERA. For these reasons, we are very happy to host one of the series of "Cosmic Maser" symposiums in Kagoshima, with the Local Organizing Committee mainly organized by Kagoshima University. We would like to express our deepest gratitude to those who contributed to the success of the symposium, including the LOC, SOC as well as in-person and on-line participants.

The power of VLBI in astrometry has been demonstrated well during the last 15 years, since the initial astrometric results for kpc-scale distances were reported around 2006–2007 (Xu *et al.* 2006, Hachisuka *et al.* 2006, Hirota *et al.* 2007, Honma *et al.* 2007). These studies have proved that phase-referencing VLBI can achieve an astrometric accuracy of 10 μas, which is high enough to measure a distance of 10 kpc. To date, VLBI maser astrometry has already provided parallaxes and proper motions for more than 200 sources (Reid et al. 2019, Hirota *et al.* 2020).

In the meantime, at optical bands, the astrometric satellite GAIA has been producing an enormous number of astrometric results for the stars in the Milky Way Galaxy. Although the target astrometric accuracy is similar for GAIA and maser astrometry with VLBI, there is an essential difference between the two approaches: while GAIA observes stars in the optical bands, VLBI astrometry traces radio emissions from maser sources, which are mainly associated with star-forming regions or sometimes other populations like Mira variables and super-giant stars.

Thus, the two approaches, GAIA and VLBI, trace the two different aspects of the Galactic sources, and the measurements are technically independent as well. Also notably different is the location of the sources in the Galaxy: while radio observations can penetrate through the middle of the Galaxy's disk, GAIA observations suffer from large extinction toward the Galactic plane. The two are thus supplementary and complementary, and their comparison is more important than ever before for understanding the detailed structure of the Milky Way Galaxy. In this proceeding paper, we would like to summarize what we have obtained so far using VERA and other VLBI observations, and in particular to compare to the astrometric results from GAIA.

2. Current view of the Milky Way Galaxy seen with Maser Astrometry

Astrometric observations have been regularly conducted with VERA, VLBA, EVN, LBA and other arrays for the last 15 years. During this period, the Galactic structure, including fundamental parameters, the shape of spiral arms and rotation curve, have been refined with the increase of available data of maser astrometry. The pioneering work was conducted with 18 sources by Reid *et al.* (2009), and then later studies evolved with ~50 sources in Honma *et al.* (2012), and ~100 sources in Reid *et al.* (2014). To date, astrometric measurements of maser sources have been obtained for more than 200 objects (Reid et al. 2019, Hirota *et al.* 2020). Figure 1 shows a plan view of the Milky Way Galaxy traced by the maser astrometry compiled in Hirota *et al.* (2020), super-imposed on an artistic-view of the spiral structures of the Milky Way Galaxy.

Figure 1 clearly demonstrates that the Galactic spiral arms are traced well up-to 10 kpc scale, particularly in the first ($0 < l < 90$ deg) and the second ($90 < l < 180$ deg) Galactic quadrant. One can trace the local arm as well as the major spiral arms of the Milky Way Galaxy including Scutum, Sagittarius, Perseus, and Outer arms. Remarkably contrasting is the lack of star-forming regions along the Galactic bar. Note that this is a real feature as long as the first quadrant is concerned, where there is little observational bias (though it exists for the fourth quadrant). This tendency is probably related with the star formation deficiency in the bar region, as is expected from simulation works of the Milky Way Galaxy.

The other remarkable feature in Figure 1 is the lack of maser sources in the third and fourth quadrant, which is a fully observational effect due to the geographic locations of VLBI arrays biased toward the northern hemisphere. Further exploration of the southern sky is essential to obtain the global view of the Milky Way Galaxy, and hopefully forthcoming observations with LBA, AuScope and SKA in future will be able to fill the gap in the southern sky. We would like to emphasize that there are only few sources in

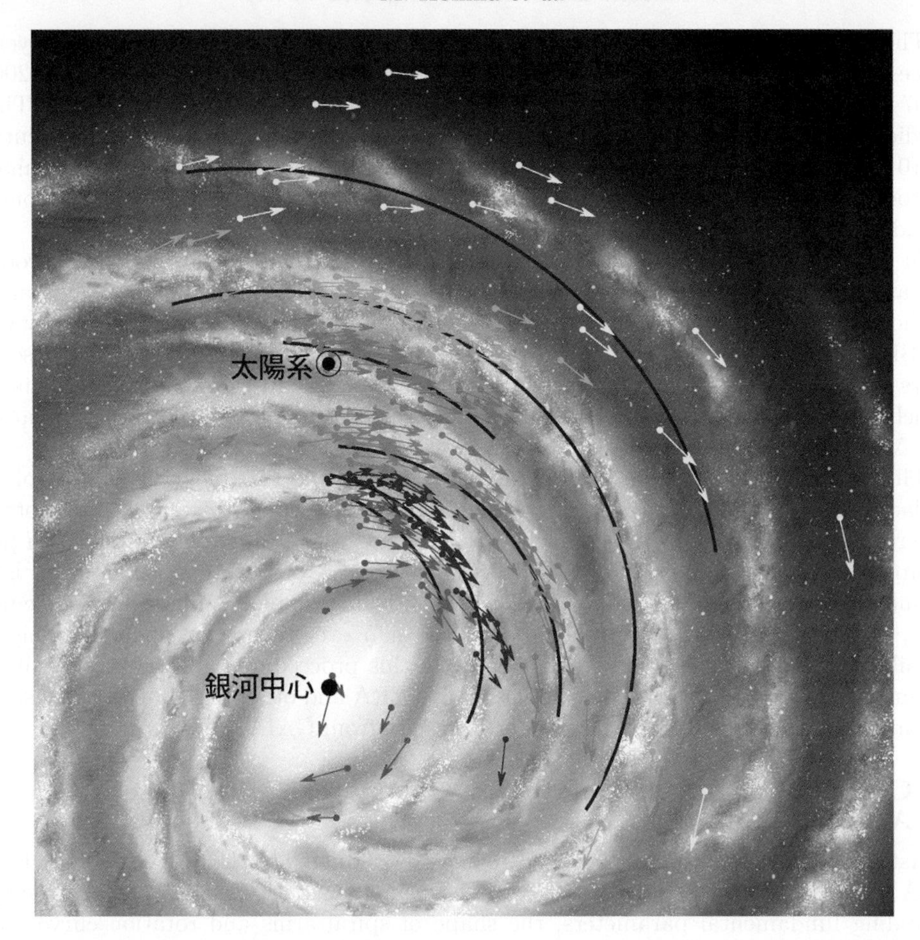

Figure 1. Plan view of the Milky Way Galaxy traced with maser astrometry, superposed on an artistic view of the Milky Way. Individual points show the location of star-forming regions with masers, and arrows show the proper motions of the maser source.

the Galactic disk beyond the Galactic Center (e.g., distance from the Sun larger than 10 kpc). This is mainly due to the observational limit in both sensitivity and astrometric accuracy, since the sources there have smaller maser fluxes as well as parallaxes. How to overcome this issue will be certainly one of the fundamental steps toward the next decades, and the use of 3-d kinematic distance (the combination of proper motions and $V_{\rm LSR}$) seems a promising tool for solving the problem (Yamauchi *et al.* 2016, Reid 2022, Sakai *et al.* 2023), in addition to improvement of astrometric accuracy.

Figure 1 also shows the proper motions of maser sources, which basically follow the Galactic rotation (clock-wise rotation in the figure). Galaxy-scale directional changes of the proper motions is evident, and based on the statistics of these proper motions, one can determine the location of the dynamical center of the Galaxy as well as the mean circular velocity. Hence one can accurately measure the fundamental parameters of the Milky Way Galaxy, namely the distance from the Sun to the Galaxy center, R_0, and the rotation velocity of the Local Standard of Rest, Θ_0. Table 1 summarizes the history of fundamental parameter determinations with maser astrometry, from Reid *et al.* (2009) to more recent one by Hirota *et al.* (2020). As one can see from the table, both R_0 and Θ_0 are converging to certain values with decreasing error bars. According to the most recent results (Reid et al. 2019, Hirota *et al.* 2020), R_0 is around 8.0 kpc with an error bar of

Table 1. Comparison of Galactic fundamental constant determinations with maser astrometry.

Reference	N_{src}	R_0 (kpc)	Θ_0 (km s^{-1})
Reid *et al.* 2009	18	8.4 ± 0.6	254 ± 16
Honma *et al.* 2012	52	8.05 ± 0.45	238 ± 14
Reid & Honma 2014	103	8.34 ± 0.16	240 ± 8
Reid et al. 2019	~200	8.15 ± 0.15	236 ± 7
Hirota *et al.* 2020	224	7.92 ± 0.16	227 ± 7

$\pm 2\%$, and Θ_0 230 km/s with an error of $\pm 3\%$. R_0 is in good agreement with independent measurements based on the stellar motions around Sgr A*, the supermassive black hole at the center of the Galaxy. Do *et al.* (2019) provided R_0 of 7.946 ± 0.050(stat.) ± 0.032(sys.) kpc, and Gravity Collaboration (2019) obtained R_0 of 8.178 ± 0.013(stat.) ± 0.022(sys.) kpc. The consistency between the maser astrometry and the stellar motions around Sgr A* ensures that the Sgr A* is indeed located at the dynamical center of the Galaxy.

The rotation curve of the Milky Way Galaxy is accurately determined with maser astrometry of ~200 sources (Reid et al. 2019, Hirota *et al.* 2020). Basically, the rotation curve shows a flat feature between 5 kpc and 15 kpc, being similar to the rotation curves of other spiral galaxies in a similar size. The flatness of the rotation curve confirms that the Galaxy contains plenty of dark matter particularly in the outer regions, as is expected to be in the form of a dark halo surrounding the Galactic disk. On the other hand, the rotation curve toward the inner region shows a clear deviation from the flat rotation: the overall shape shows a positive gradient of the rotation velocity (i.e., the rotation velocity increases with the Galacto-centric radius), with larger scatter of individual maser sources compared to that of the outer regions. This feature of the rotation curve is consistent with the existence of the Galactic bar, which causes asymmetric and rather complex orbits of the gas around the bar.

Another interesting feature that is currently being partially traced is the warp of the Galactic disk. Studies such as Sakai *et al.* (2020) and Immer & Rygl (2022) demonstrated that the disk warp toward the north is seen in the first Galactic quadrant, being consistent with the HI and stellar disk. However, to obtain the complete view of the disk warp, it is certainly necessary to extend the disk region covered by the maser observations, both in the southern hemisphere and in the region behind the Galactic center.

3. Comparison of GAIA and maser astrometry

The GAIA satellite, launched by ESA, has been producing huge amount of astrometric data of stars in the Galaxy seen at optical bands, and it is worth conducting direct comparison of the Galaxy's views obtained by GAIA and maser astrometry as well as individual sources observable by both GAIA and VLBI. The first step is to compare the common populations that can be observed with both optical and radio, namely, young stars (note that OB stars at optical bands and star-forming regions seen at radio are close in ages, with age difference of an order of 10^{5-6} years, that is much shorter than the time scale of Galaxy rotation, being 10^9 yr).

Drimmel *et al.* (2023) created a spiral arm map within 5 kpc from the Solar system using OB stars traced by GAIA. The population of the young stars successfully traces the Sagittarius Arm, Local Arm and Perseus Arm, being consistent with the arms located by maser astrometry. To trace more distant regions, Drimmel *et al.* (2023) used AGB stars (brighter but older populations) as well to map the region within 10 kpc, but the spiral structures are less prominent with such an old population, being in contrast to the map traced by masers showing spiral structures even at a distance of 10 kpc or more. This comparison clearly demonstrates the importance of observing multiple tracers (both young and old populations, with optical and radio observations) to understand the Galaxy structure, since these different populations are complementary to each other in terms of population ages as well as observing technique. The latter is particularly important for exploring the distant regions along the Galactic plane, since the optical observations are severely hampered by the dust obscuration.

GAIA and maser astrometry provide good opportunities of direct comparison of astrometric results for some types of stars that are observable at both optical and radio. Among such populations are AGB stars with strong maser emissions in H_2O and/or SiO, and young low-mass stars emitting continuum emissions are another type of targets, although they are not maser sources. Xu et al. (2019) conducted a direct comparison of parallaxes for these stars, showing that optical and radio astrometry are broadly consistent with each other. Nakagawa *et al.* (2019) conducted a similar study for AGB stars, confirming the two methods generally provide consistent results. However, these studies also revealed that there are some cases of discrepancy in parallax. Most notable examples were SV Peg (Sudoh *et al.* 2019) and BX Cam (Matsuno *et al.* 2020), for which GAIA DR2 and maser astrometry showed parallax discrepancy at more than 5-σ level. However, according to most updated results (GAIA DR3, Vallenari *et al.* 2023), the parallaxes are more consistent, with difference less than 2-σ level. Such a comparison demonstrates that confirmation and cross-check of parallaxes between optical and radio astrometry are essential for providing a firm basis for future astrometry.

4. Proper motion toward futures

As seen in figure 1, maser astrometry successfully traces the Galaxy scale structure, particularly well in the second Galactic quadrant. In the meantime, there is a lack of sources in the distant region (i.e., 10 kpc or more from the Sun) and in the southern hemisphere. The "proper motion toward the future" in the field of Galaxy-scale astrometry is to cover the regions currently unexplored. Since the parallax becomes small there (i.e., less than 100 μas), it is necessary to improve the accuracy of VLBI astrometry. In Asia, we have been developing East Asian VLBI Network (EAVN, Tao *et al.* 2018, Akiyama et al. 2022), which is a combined array in the collaboration with China, Japan and Korea. EAVN will by far improve the array sensitivity and the parallax accuracy compared to VERA, with the maximum baseline doubled and the total aperture area increased by an order of magnitude.

Also interesting for future astrometry is the use of proper motion and radial velocity to determine the 3-d kinematic distance. A good demonstration of such a method was done by Yamauchi *et al.* (2016), which located a maser source at 19 kpc based only on kinematics without parallax, and later confirmed by parallax measurement by VLBA (Sanna *et al.* 2017). Recently Reid (2022) also extensively discussed the power of 3-d kinematic distance, concluding that the methods provide fairly accurate distances except Galactic bar regions, as long as non-circular motion is substantially small. Sakai *et al.* (2023) recently reported another case of distance measurements with 3-d kinematics, locating a star forming region G034.84-00.95 at a distance of 19$\pm$1 kpc. Note that this technique is potentially applicable to non-VLBI observations, for instance, such as high-resolution ALMA observations that are able to measure proper motions of cores in star-forming regions.

Finally, in order to explore the southern part of the Galaxy, we definitely need the VLBI arrays in the southern hemisphere. Currently LBA is operational in VLBI astrometry (e.g., Krisnan *et al.* 2015, Krisnan *et al.* 2017), and more recently, AuScope started producing astrometric results by utilizing a sophisticated phase-referencing technique called Multi-View (Rioja & Dodson 2020, Hyland *et al.* 2022). In the next years, hopefully these existing arrays will produce more results on astrometry in the southern maser parallaxes, and then in the next decade, the advent of SKA will lead to a major breakthrough in the exploration of the southern sky, when it is combined with existing radio telescopes to build up the most powerful VLBI array in the southern hemisphere based on global collaborations.

References

Akiyama, K., *et al.* 2023, *Galaxies*, 10, 113
Do, T., *et al.* 2019, *Science*, 365, 664
GAIA collaboration, Drimmel, R., *et al.* 2023, *A&A*, 674, 37
GAIA collaboration, Vallenari, A., *et al.* 2023, *A&A*, 674, 1
Gravity Collaboration 2019, *A&A*, 625, L10
Hachisuka, K., *et al.* 2006, *ApJ*, 645, 337
Hirota, T., *et al.* 2007, *PASJ*, 59, 897
Hirota, T., *et al.* 2020, *PASJ*, 72, 50
Honma, M., *et al.* 2007, *PASJ*, 59, 889
Honma, M., *et al.* 2012, *PASJ*, 64, 136
Hyland, L., *et al.* 2022, *ApJ*, in press
Immer, K. & Rygl, K. 2022, *Universe review*, 8, 390
Krishnan, V., *et al.* 2015, *ApJ*, 805, 129
Krishnan, V., *et al.* 2017, *MNRAS*, 465, 1095
Matsuno, M., *et al.* 2020, *PASJ*, 72, 56
Nakagawa, A., *et al.* 2019, *Proceedings of the IAU symposium*, 343, 476
Rioja, M. & Dodson, R. 2020, *A&AR*, 28, 6
Reid, M. J. & Honma, M. 2014, *ARAA*, 52, 339
Reid, M. J., *et al.* 2009, *ApJ*, 700, 137
Reid, M. J., *et al.* 2014, *ApJ*, 783, 130
Reid, M. J., *et al.* 2019, *ApJ*, 885, 131
Reid, M. J. 2022, *AJ*, 164, 133
Sanna, A., *et al.* 2017, *Science*, 358, 227
Sakai, N., *et al.* 2020, *PASJ*, 72, 53
Sakai, N., *et al.* 2023, *PASJ*, 75, 208
Sudoh, H., *et al.* 2019, *PASJ*, 71, 16
Tao, A., Sohn, B. W. & Imai, H. 2018, *Nature Astronomy*, 2, 118
Xu, Y., *et al.* 2006, *Science*, 311, 54
Xu, S., *et al.* 2019, *ApJ*, 875, 114
Yamauchi, A., *et al.* 2016, *PASJ*, 68, 80

Cosmic Masers: Proper Motion toward the Next-Generation Large Projects
Proceedings IAU Symposium No. 380, 2024
T. Hirota, H. Imai, K. Menten, & Y. Pihlström, eds.
doi:10.1017/S1743921323002028

Galactic astrometry with Gaia

Carme Jordi[1,2,3,4]

[1]Institut de Ciències del Cosmos (ICCUB), Universitat de Barcelona (UB), Martí i Franquès 1, E-08028 Barcelona, Spain. email: carme@fqa.ub.edu

[2]Departament de Física Quàntica i Astrofísica (FQA), Universitat de Barcelona (UB), Martí i Franquès 1, E-08028 Barcelona, Spain

[3]Institut d'Estudis Espacials de Catalunya (IEEC), c. Gran Capità, 2-4, E-08034 Barcelona, Spain

[4]Institut dEstudis Catalans, c. Carme 47, E-08001, Barcelona, Spain

Abstract. *Gaia* space mission of the European Space Agency was launched at the end of 2013 and will continue operations until 2025. The published data releases revolutionize the view of the Milky Way galaxy and beyond thanks to its unprecedented astrometry, photometry and spectroscopy. The paper reviews the products of the last data release of the Gaia mission and some of the scientific impacts of the data. We also discuss the future perspectives of Galaxy astrometry from space.

Keywords. Galaxy, astrometry, photometry, spectroscopy

---

1. Introduction

Gaia space mission (Gaia Collaboration, Prusti *et al.* 2016) by the European Space Agency was launched to space in December 2013. Routine operations are going on smoothly since July 2014 at an average rate of 70 million source observations per day. The nominal mission of 5 years ended in July 2019 and the mission is currently in an extended operation period that will last until the exhaustion of cold gas around the first quarter of 2025.

The satellite is in orbit around the Earth-Sun Lagrange point L2 at 1.5 million kilometers from Earth. It holds two telescopes each one pointing to a different field-of-view. Both field-of-views are separated a large basic angle of 106.5 degrees. The rotation around the own spin axis, the precession of this spin axis around the direction Earth-Sun and the yearly rotation around the Sun makes that the two telescopes scan the full-sky about every six months.

The focal plane is equipped with 106 CCD cameras operating in Time-Delay-Integration mode with about 1 Giga pixels in total, the largest camera sent to space ever. These cameras serve different purposes and instruments: on board detection of point like sources (mainly stars, but also small bodies of our Solar System, far galaxies and quasars), astrometry, photometry, low and medium spectroscopy, monitoring of the basic angle and wave front sensors.

The repeated observations over time of the same objects allow the determination of positions, brightness, spectral energy distribution and radial velocity as well as their variations thus proper motions, parallaxes, changes in brightness, colours and radial velocities. The unprecedented precision of the data and the volume of the data makes the mission *Gaia* a revolution for many fields of astrophysics, from the Milky Way Galaxy itself, to stellar physics, exoplanetary science, extragalactic astronomy and general relativity.

So far, the *Gaia* Data Processing and Analysis Consortium (DPAC) has published three releases: DR1 (Gaia Collaboration, Brown *et al.* 2016), DR2 (Gaia Collaboration, Brown *et al.* 2018) and DR3 (Gaia Collaboration, Vallenari *et al.* 2023), which includes its early release EDR3 (Gaia Collaboration, Brown *et al.* 2021). This paper reviews the content of the last data release and highlights some of the main results.

After the present introduction, Sec. 2 describes the main products of the third *Gaia* release, Sec. 3 describes some results of performance verification papers and from community at large, and Sec. 4 describes the steps forward the future, including the next generation of future space astrometry.

2. (E)DR3 content

Gaia DR3 (Gaia Collaboration, Vallenari *et al.* 2023) published in June 2022 includes data for about 1.8 billion sources, mainly stars in our Milky Way galaxy. Astrometry (positions, proper motions and parallaxes) and photometry (apparent brightness in white band G and colour blue-red $G_{\mathrm{BP}} - G_{\mathrm{RP}}$) for about 1.5 million sources were advanced in the early release in December 2020 (Gaia Collaboration, Brown *et al.* 2021). Therefore, DR3 extends DR2 in terms of number of variable stars, radial velocities, astrophysical parameters and Solar System objects and completes EDR3 with novel products published for the first time like rotational velocities, chemical abundances, multiple systems and galaxies and quasars. The release is also complemented by the publication light curves for all sources in a field-of-view of one degree centered on Andromeda's galaxy. All data is accessible at the *Gaia* archive site https://gea.esac.esa.int/archive/.

2.1. *Astrometry*

The astrometric content is fully described in Lindegren *et al.* (2021) and the summary can be found in the EDR3 publication (Gaia Collaboration, Brown *et al.* 2021). The uncertainties of positions, proper motions and parallaxes have decreased both by the longer time baseline of the observations (from 22 months in DR2 to 34 months in DR3) and for the improvements introduced in all steps of the data processing pipeline by DPAC. Systematic errors still dominate the bright regime (to about $G = 13$), while the faint regime is dominated by the photon-noise statistics (see Fig. 7 in Lindegren *et al.* (2021)).

Worth to notice that systematic errors as mapped through quasars parallaxes and proper motions are much smaller than those in DR2, and both at large and small angular scales as shown in Fig. 1. The later have been derived by analysing the parallaxes and proper motions of stars in the Large Magellanic Cloud area. The global parallax zero point is -17 μas and the rms angular variations are 26 μas and 33 μas yr^{-1} for parallaxes and proper motions, respectively. The dependencies on magnitude, colour and sky position can be calculated using the code in https://gitlab.com/icc-ub/public/gaiadr3_zeropoint.

2.2. *Medium resolution spectroscopy*

One of jewels of DR3 is its spectroscopic content. The medium resolution spectrograph (RVS) has provided the third component of the velocity, the radial velocity, for 33 million stars, the largest catalogue of radial velocities ever (Katz *et al.* 2023). This is complemented with 3.5 millions of rotational velocities, 2.5 millions of chemical abundance determinations and with 1 million spectra, more spectra than those collected over

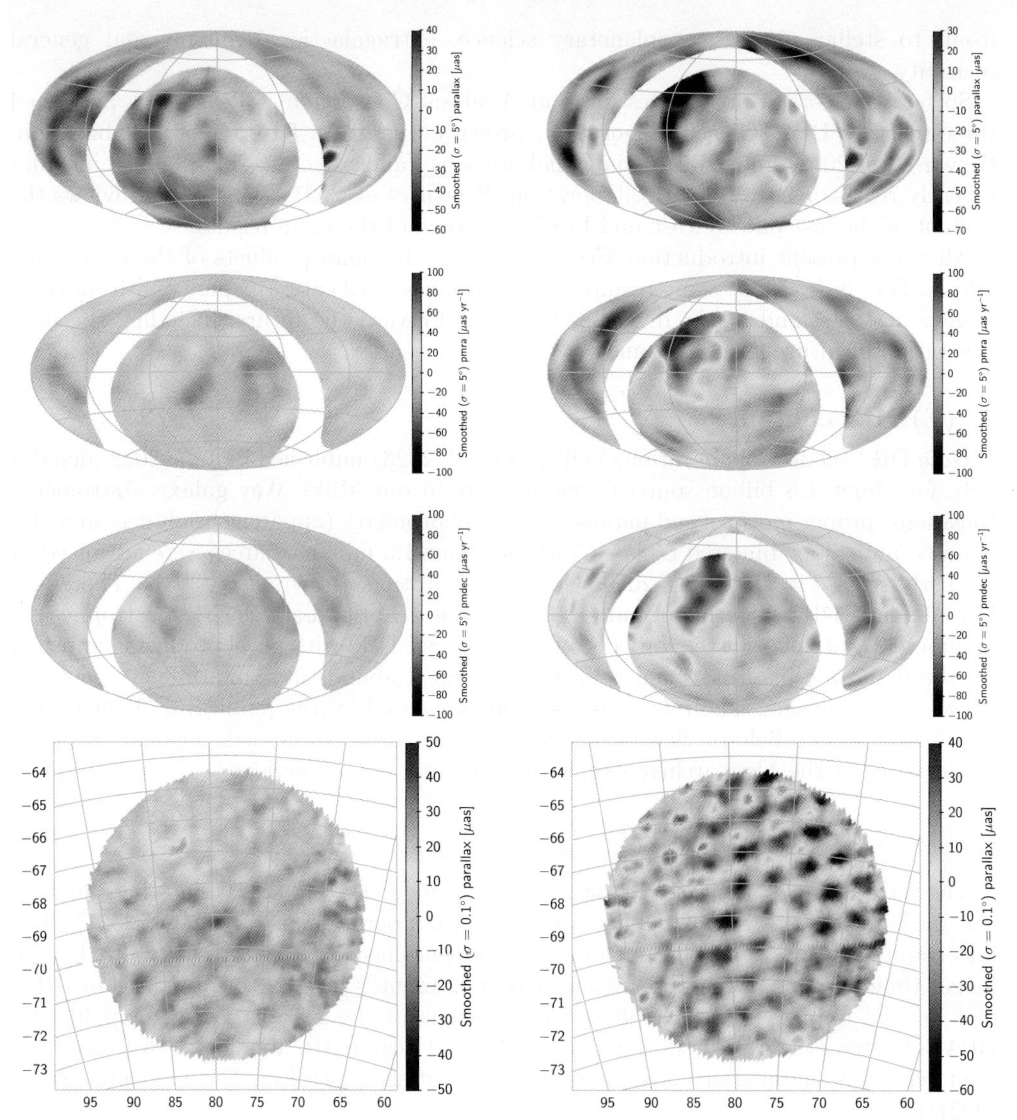

Figure 1. Figures 13 and 14 of Lindegren *et al.* (2021). The three top panels show the all sky smoothed maps of quasars parallaxes and proper motions, while the low panels show the smoothed maps of parallaxes in the Large Magellanic Cloud area. Left panels are for EDR3/DR3 and right panels for DR2.

all centuries. The sample of stars cover the range of $T_{\rm eff} \in [3100,\ 14500]$ K for the bright stars ($G_{\rm rvs} \leq 12$) and [3100, 6750] K for the fainter stars, reaching a few kilo-parsecs beyond the Galactic centre in the disc and up to about $10 - 15$ kpc vertically into the inner halo. The median uncertainties of the radial velocities are of 1.3 km s^{-1} at $G_{\rm rvs} = 12$ and 6.4 km s^{-1} at $G_{\rm rvs} = 14$. The radial velocity scale is in satisfactory agreement with APOGEE, GALAH, GES and RAVE, with systematic differences that mostly do not exceed a few hundreds m s^{-1}. The velocity zero point has a small systematic shift starting around $G_{\rm rvs} = 11$ and reaching about 400 m s^{-1} at $G_{\rm rvs} = 14$. From the RVS spectra rotational velocities, astrophysical parameters and chemical abundances have also been derived (see Sec. 3), as well as radial velocity variables have been identified.

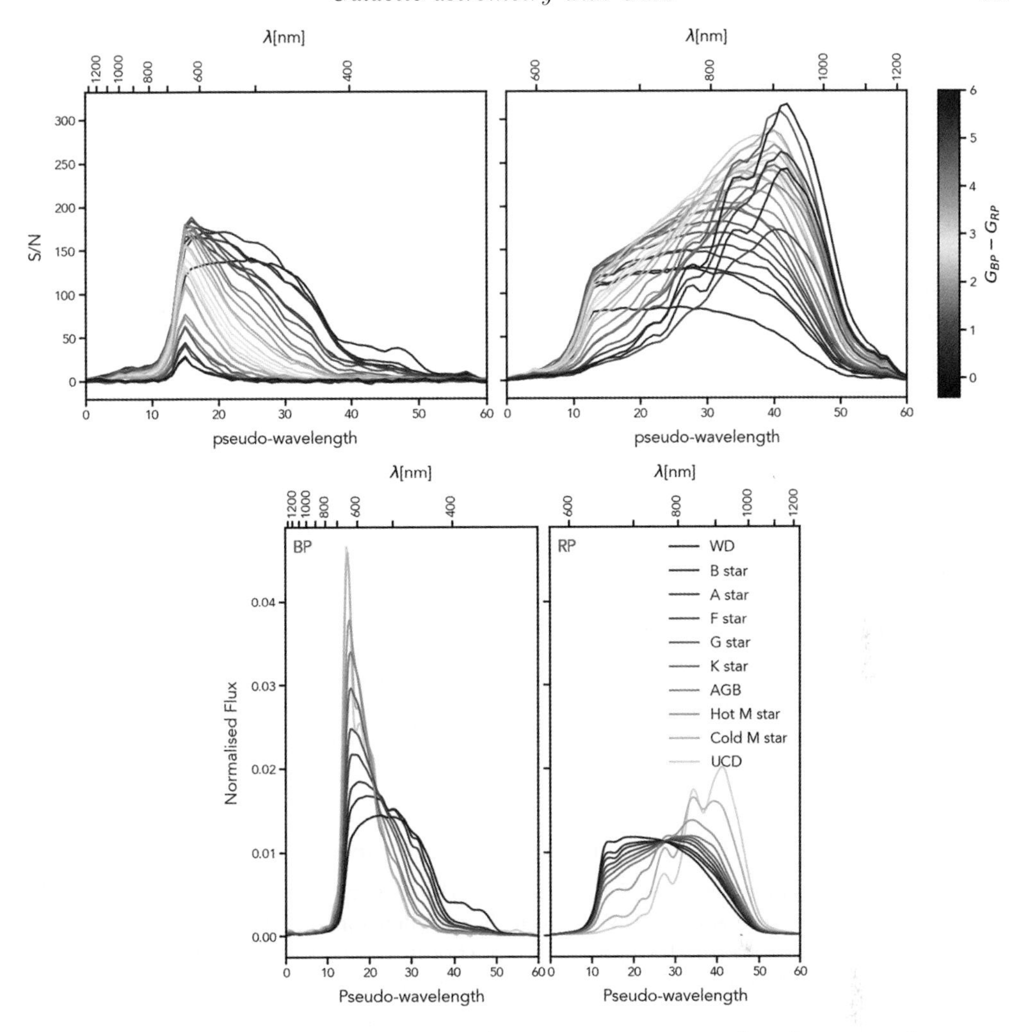

Figure 2. Figures 24 (top) and 26 (bottom) from De Angeli *et al.* (2023) showing the signal-to-noise ratio for stars of $G = 16$ and different colors $G_{\rm BP} - G_{\rm RP}$ and the normalised internal mean spectra for different types of stars. BP are shown in the left and RP in the right.

2.3. *Low resolution spectroscopy*

Low-resolution spectroscopy is acquired for every source transiting the *Gaia* focal plane. Spectral energy distribution (SED) are obtained in the blue (BP) and red (RP) globally covering the wavelength range [330, 1050] nm. These SEDs are the basis of the object classification into stars, quasars, galaxies and so on, and the determination of the astrophysical parameters (see Sec. 3). Bottom panel of Fig. 2 shows the normalized SEDs with different shapes for different types of stars. Top panel of Fig. 2 shows the signal-to-noise ratio for stars of $G = 16$ and different colors $G_{\rm BP} - G_{\rm RP}$. As expected, this signal-to-noise ratio depends on the wavelength, the magnitude and the colour of the objects, with sources around $G = 15$ having signal-to-noise above 100 in some wavelength ranges and reaching 1000 in the central part of the RP wavelength range for sources with magnitudes $9 < G < 12$ (De Angeli *et al.* 2023). The SEDs are provided internally and externally calibrated using a set of spectrophotometric standard stars. The spectra are represented by a set of coefficients and basis functions, in such a way that the analysis of the coefficients suffices for the classification and the detection of spectral features. In addition, the tool GaiaXPy is also provided to represent the spectra in traditional form or to compute synthetic photometry from the SEDs. DR3 includes about 220 SEDs most of them for objects brighter than $G = 17.65$.

2.4. *Derived products*

All the above data, astrometry, photometry and spectroscopy, are the basis for the multiple derived products included in DR3: (a) the classification of 1.5 billion sources into stars, galaxies, quasars and so on, (b) physical parameter determination for 470 million stars (temperatures, surface gravity, absorption, luminosities and radii and masses and ages for 128 million stars, (c) the analysis of 10 million light curves including more variability types than in DR2, (d) properties of 813 000 binary systems including binaries with compact objects, detection of 297 exoplanets, of which 114 are new candidates, (e) astrophysical parameters for 470 millions of stars from low resolution spectroscopy and for 5.6 million stars from medium resolution spectroscopy, (f) chemical abundances for 2.5 million stars from medium resolution spectroscopy, (g) data for 3 million galaxies and 2 million quasars, (h) orbits for 156 000 asteroids, BP/RP reflectance spectra of 60 000 asteroids. See all papers in https://www.cosmos.esa.int/web/gaia/dr3-papers.

3. Results

The richness of results based on *Gaia* data releases is enormous and impossible to review here. Only some of the results of the performance verification papers and some of the results related with archaeology of the Galaxy are described in the following. The selection is incomplete and biased by the author.

3.1. *Milky Way structure and dynamics*

The dynamical perturbations of the thin and thick discs due to accretion of galaxies by the Milky Way were soon realized after the publication of DR2. The perturbations of the orbits of the stars in the solar neighbourhood due to the pass of the Sagittarius dwarf galaxy are still noticeable in the present Antoja *et al.* (2018), and a set of stars in retrograde orbits confirm their extragalactic origin and has been called Gaia-Enceladus galaxy Helmi *et al.* (2018). The triggered star formation in bursts due to the subsequent passes of the Sagittarius dwarf galaxy was also detected through the analysis of stellar populations Ruiz-Lara *et al.* (2020). EDR3 provided new insights of these and other mergers in terms of dynamical perturbations and stellar populations (Gaia Collaboration, Antoja *et al.* 2021; Ramos *et al.* 2022).

The map of radial velocities (Fig. 3 left) shows the pattern of approaching and receding stars with respect to the Solar System due to the Galactic rotation and Solar System motion with respect to the local standard of rest. The 3D position and 3D motion information has been used to map the spiral structure associated with star formation $4 - 5$ kpc from the Sun by using 2800 Classical Cepheids younger than 200 million years, which show spiral features extending as far as 10 kpc from the Sun in the outer disc (Gaia Collaboration, Drimmel *et al.* 2021). The velocity field has been mapped through red giant branch stars with radial velocities as far as 8 kpc from the Sun, including the inner disc. The spiral structure revealed by the young populations is consistent with recent results using Gaia EDR3 astrometry and source lists based on near infrared photometry, showing the Local (Orion) arm to be at least 8 kpc long, and an outer arm consistent with what is seen in HI surveys, which seems to be a continuation of the Perseus arm into the third quadrant. Meanwhile, the subset of RGB stars with velocities clearly reveals the large scale kinematic signature of the bar in the inner disc, as well as evidence of

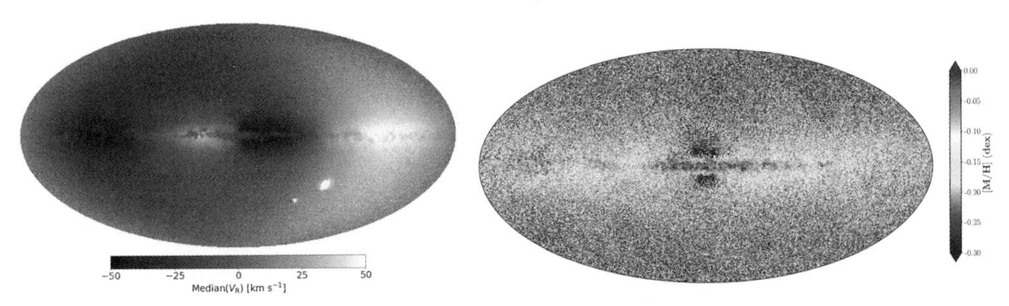

Figure 3. *Left*: Figure 5 from Katz *et al.* (2023) showing the sky map of the median radial velocity in each pixel with respect to the barycenter of the Solar system. The pattern of receding and approaching stars is the combination of the motion of the Sun and the galactic rotation. *Right*: Figure 2 from Gaia Collaboration, Recio-Blanco *et al.* (2023) showing the median of the stellar metallicity [M/H] in each pixel.

streaming motions in the outer disc that might be associated with spiral arms or bar resonances.

The all-sky Gaia chemical cartography (Fig. 3 right) allows a powerful and precise chemo-dynamical view of the Milky Way with unprecedented spatial coverage and statistical robustness (Gaia Collaboration, Recio-Blanco *et al.* 2023). The abundances have the required coverage and precision to unveil galaxy accretion debris and heated disc stars on halo orbits, and to allow the study of the chemo-dynamical properties of globular clusters. The abundance maps reveal the strong vertical symmetry of the Galaxy and the flared structure of the disc. There is a strong correlation of the kinematic disturbances of the disc with chemical patterns, both for young objects that trace the spiral arms and for older populations. Abundance elements (mainly iron-peak and α elements) trace the thin and thick disc properties in the solar cylinder. The abundances of open clusters, the largest sample analysed so far, show a steepening of the radial metallicity gradient with age, which is also observed in the young field population.

3.2. *Open clusters*

Simultaneous analysis of proper motions, parallaxes and photometry has allowed deep understanding of open clusters. The pre-*Gaia* samples (for instance Dias *et al.* (2002)) were revisited and enlarged. Hundreds of new open clusters have been identified in DR2 and EDR3 (for instance Castro-Ginard *et al.* (2022)). Young clusters and stellar associations show rich substructures with subpopulations of slightly different ages and kinematics unveiling the sequential process of formation line in Vela-Puppis (Cantat-Gaudin *et al.* 2019). Disruption, evaporation and streams have been also detected in many evolved clusters (Tarricq *et al.* 2022; Ratzenböck *et al.* 2020). All old clusters towards the center of the Galaxy reported in the pre-*Gaia* era, which should have been destroyed long time ago, show that they stars have incoherent proper motions and parallaxes, thus they are simply fortuitous projections on the sky and not physical groups (Cantat-Gaudin & Anders 2020). Young open clusters have been used to trace the spiral arms as seen in Fig. 4. They are in good agreement with other tracers like Cepheids, OB stars or masers.

3.3. *Acceleration of the Solar System*

The apparent proper motions of 1.2 million compact (QSO-like) extragalactic sources reveal a systematic pattern due to the acceleration of the Solar System barycentre with respect to the rest frame of the Universe (Gaia Collaboration, Klioner *et al.* 2021). The

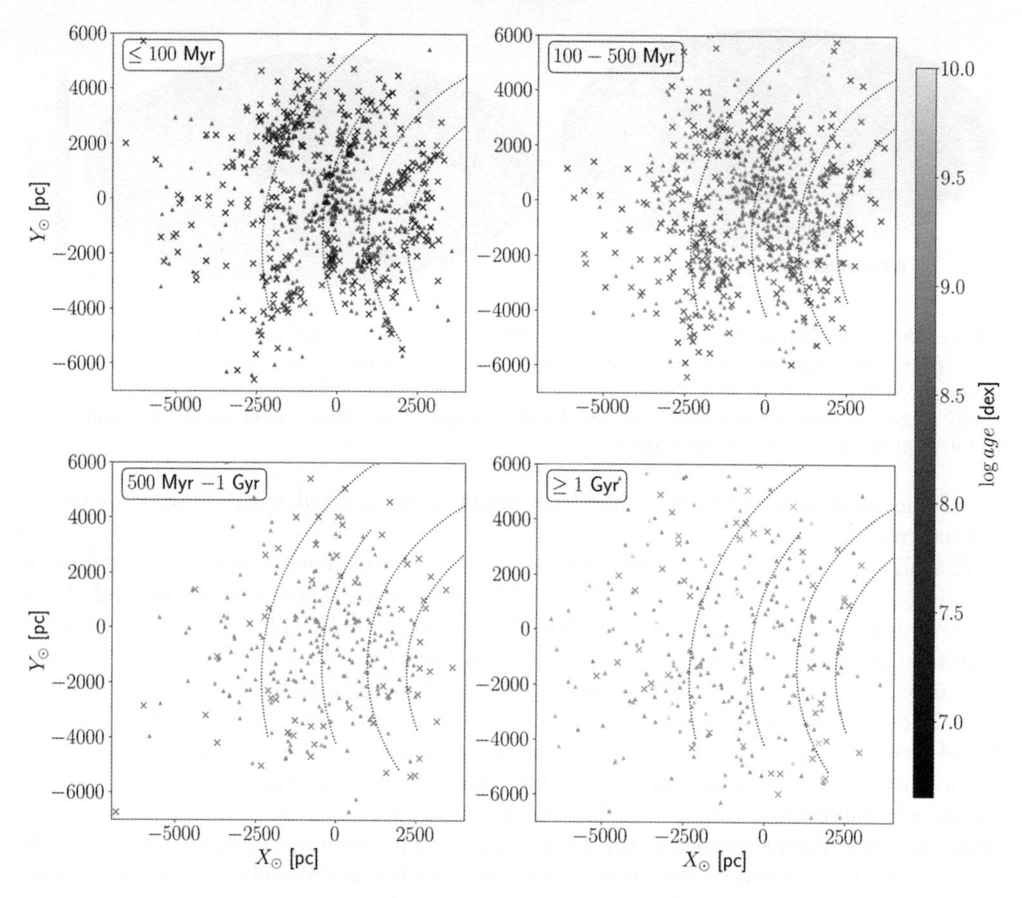

Figure 4. Figure 7 from Castro-Ginard *et al.* (2022). It shows the distribution of the new (crosses) and known (triangles) open clusters projected onto the Galactic plane for different age bins. The dotted lines show the spiral arms.

result is a proper motion amplitude of 5.05 ± 0.35 μas yr^{-1} that means an acceleration of $2.32 \pm 0.16 \cdot 10^{-10}$ m s^{-2}, in good agreement with the acceleration expected from current models of the Galactic gravitational potential. The accelerations point towards $\alpha = (269.1 \pm 5.4)^{\text{o}}$, $\delta = (-31.6 \pm 4.1)^{\text{o}}$, probably due to the presence of the Magellanic Clouds. We expect that future Gaia data releases will provide estimates of the acceleration with uncertainties substantially below 0.1μas yr^{-1}.

3.4. *Magellanic Clouds*

In spite that the *Gaia* proper motions and parallaxes for the Magellanic Clouds are at the limit of usability, radial and tangential velocity maps and global profiles have been derived (Gaia Collaboration, Luri *et al.* 2021) together with the spatial structure and motions in the central region, the bar, and the disc, providing new insights into features and kinematics. For the first time, the two planar components of the ordered and random motions are derived for multiple stellar evolutionary phases in a galactic disc outside the Milky Way, showing the differences between younger and older phases. Finally, *Gaia* data has allowed to resolve the Magellanic Bridge, and to trace the density and velocity flow of the stars from the SMC towards the LMC not only globally, but also separately for young and evolved populations.

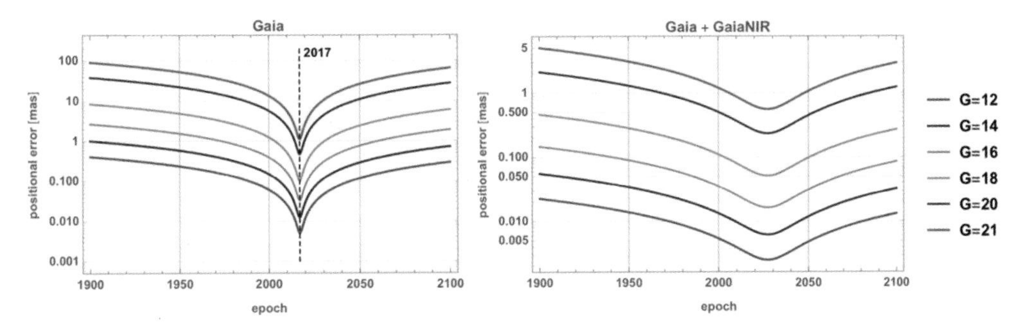

Figure 5. Figure 3 from Hobbs *et al.* (2023) showing the degradation over time of positions due to uncertainties on proper motions for *Gaia* (left) and a combination of *Gaia* and the future mission (right).

4. Near and far future

The satellite continues operating smoothly and ESA has approved the third extension of operations until 2025 where the cold gas to control the attitude will exhaust. We will have collected 10 years of science data that will overpass the initial expectations allowing better determination of all parameters and, more importantly, to detect of accelerations.

For the last quarter of 2023, DPAC foresees the publication of the Focused Product Release. It will include updated astrometry for Solar System objects, astrometry and photometry from engineering images for Omega Cen, the first results of quasars' environment analysis for gravitational lenses search, extended radial velocity epoch data for Long Period Variables, and diffuse interstellar bands from aggregated RVS spectra.

The results of the 5-yrs nominal mission duration, including both mean and epoch data, will constitute the fourth data release (DR4), and the complete data for the whole duration of the mission the fifth and final one (DR5), see https://www.cosmos.esa.int/web/gaia/release. DPAC is intensively working on the production of DR4 that is foreseen not before the end of 2025. DR5 is expected to be ready 5 years after the end of operations.

In spite of the revolution of *Gaia*, the optical regime limits the study of the Galaxy: the obscured regions by the dust are not directly accessible, and the observations of cold white dwarfs and brown dwarfs are limited, for instance. Current technology developments seem to promise competitive devices in the NIR for high precision astrometry. Therefore, the scientific community around *Gaia* has started to move towards the acquisition of all-sky optical and near infrared space astrometry in the future. This not only has a set of important science cases but also another key aspect: the combination of current positions obtained with *Gaia* combined with positions of a future mission will overcome the degradation of the positions due to the proper motions uncertainties, which would allow to maintain the realisation of the Celestial Reference Frame (see Fig. 5). Such a mission proposal (Hobbs *et al.* 2023) was submitted to ESA's call Voyage 2050, and the outcome of the evaluation is that space astrometry is one of the large themes in the Voyage 2050 framework, therefore providing excellent perspectives for the future.

Acknowledgements

This work has made use of results from the European Space Agency (ESA) space mission Gaia, the data from which were processed by the Gaia Data Processing and Analysis Consortium (DPAC). Funding for the DPAC has been provided by national institutions, in particular the institutions participating in the Gaia Multilateral Agreement. The Gaia mission website is http://www.cosmos.esa.int/gaia. The author is current member of the ESA Gaia mission team and of the Gaia DPAC. This work was (partially) supported by the Spanish MICIN/AEI/10.13039/501100011033 and by "ERDF A way of making Europe" by the "European Union" through grant PID2021-122842OB-C21, and the Institute of Cosmos Sciences University of Barcelona (ICCUB, Unidad de Excelencia 'María de Maeztu') through grant CEX2019-000918-M.

References

Antoja, T., Helmi, A., Romero-Gómez, M. *et al.* 2018, *Nature*, 561, 360
Cantat-Gaudin, T., Jordi, C., Wright, N. J. *et al.* 2019, *A&A*, 626, A17
Cantat-Gaudin, T., JAnders, F., 2020, *A&A*, 633, A99
Castro-Ginard, A., Jordi, C., Luri, X. *et al.* 2022, *A&A*, 661, A118
De Angeli, F., Weiler, M., Montegriffo, P., *et al.* 2023, *A&A*, in press, arXiv:2206.06143
Dias, W. S., Alessi, B. S., Moitinho, A., Lépine, J. R. D. 2002, *A&A*, 389, 871
Gaia Collaboration, Antoja, T., McMillan, P. J., *et al.* 2021, *A&A*, 649, A8
Gaia Collaboration, Brown, A. G. A., Vallenari, A., *et al.* 2016, *A&A*, 595, A2
Gaia Collaboration, Brown, A. G. A., Vallenari, A., *et al.* 2018, *A&A*, 616, A1
Gaia Collaboration, Brown, A. G. A., Vallenari, A., *et al.* 2021, *A&A*, 649, A1
Gaia Collaboration, Drimmel, R., Romero-Gómez, M., *et al.* 2023, *A&A*, in press, arXiv:2206.06207
Gaia Collaboration, Klioner, S. A., Mignard, F., *et al.* 2021, *A&A*, 649, A9
Gaia Collaboration, Luri, X., Chemin, L., *et al.* 2021, *A&A*, 649, A7
Gaia Collaboration, Prusti, T., de Bruijne, J. H. J., *et al.* 2016, *A&A*, 595, A1
Gaia Collaboration, Recio-Blanco, A., Kordopatis, G., *et al.* 2023, *A&A*, in press, arXiv:2206.05534
Gaia Collaboration, Vallenari, A., Brown, A. G. A., *et al.* 2023, *A&A*, in press, arXiv:2208.00211
Helmi, A., Babusiaux, C., Koppelman, H. H., *et al.* 2018, *Nature*, 563, 85
Hobbs, D., Brown, A., Høeg, E., *et al.* 2021, *Experimental Astronomy*, 51, 783
Katz, D., Sartoretti, P., Guerrier, A., *et al.* 2023, *A&A*, in press, arXiv:2206.05902
Lindegren, L., Klioner, S. A., Hernández, J., *et al.* 2021, *A&A*, 649, A2
Ratzenböck, S., Meingast, S., Alves, J., Möller, T., Bomze, I., 2020, *A&A*, 639, A64
Ramos, P., Antoja, T., Yuan, Z., *et al.* 2022, *A&A*, 666, A64
Ruiz-Lara, T., Gallart, C., Bernard, E. J., Cassisi, S., 2020, *Nature Astronomy*, 4, 965
Tarricq, Y., Soubiran, C., Casamiquela, L., *et al.* 2021, *A&A*, 659, A59

Cosmic Masers: Proper Motion toward the Next-Generation Large Projects
Proceedings IAU Symposium No. 380, 2024
T. Hirota, H. Imai, K. Menten, & Y. Pihlström, eds.
doi:10.1017/S1743921323002430

The origin of the Perseus-arm gap revealed with VLBI astrometry

Nobuyuki Sakai[1], **Hiroyuki Nakanishi**[2], **Kohei Kurahara**[3], **Daisuke Sakai**[3,4], **Kazuya Hachisuka**[3], **Jeong-Sook Kim**[5] **and Osamu Kameya**[3,6]

[1]National Astronomical Research Institute of Thailand (Public Organization).
email: nobuyuki@narit.or.th

[2]Graduate School of Science and Engineering, Kagoshima University

[3]National Astronomical Observatory of Japan, Oshu-shi, Iwate 023-0861, Japan

[4]The Iwate Nippo Co., Ltd.

[5]Ulsan National Institute of Science and Technology / Chungbuk National University

[6]Oshu Space & Astronomy Museum

Abstract. The Perseus arm has a gap in Galactic longitudes (l) between 50° and 80° where the arm has little star formation activity. To understand the gap, we conducted VERA (VLBI Exploration of Radio Astrometry) astrometry and analyzed archival H I data. We report on parallax and proper motion results from four star-forming regions, of which G050.2800.39 and G070.33+01.59 are associated with the gap. Perseus-arm sources G049.41+00.32 and G050.2800.39 lag relative to a Galactic rotation by 77 $\pm$ 17 km s^{-1} and 31 $\pm$ 10 km s^{-1}, respectively. The noncircular motion of G049.41+00.32 cannot be explained by the gravitational potential of the arm. We discovered rectangular holes with integrated brightness temperatures less than 30 K arcdeg in l vs. $V_{\rm LSR}$ of the H I data. Also, we found extended H I emission on one side of the Galactic plane when integrating the H I data over the velocity range covering the hole. G049.41+00.32 and G050.2800.39 are moving toward the emission. The Galactic H I disk at the same velocity range showed an arc structure, indicating that the disk was pushed from the lower side of the disk. All the observational results might be explained by a cloud collision with the Galactic disk.

Keywords. Galaxy:disk, Galaxy:kinematics and dynamics, masers, instrumentation:interferometers, parallaxes

---

1. Introduction

The Perseus arm is one of the two (e.g., Drimmel 2000; Churchwell et al. 2009) or four (e.g., Georgelin & Georgelin 1976; Russeil 2003) main arms of the Milky Way, based on radio, infrared and optical observations. Using the precise locations of very young objects, Xu et al. (2023) revealed that the inner part of the Milky Way shows a two-arm symmetry (the Norma and Perseus arms) and the two arms extend to the outer part where there are several long irregular arms (the Centaurus, Sagittarius, Carina, Outer, and Local arms). However, the Perseus arm shows a "gap structure" in the Galactic-longitude range between l = ~50° and ~80° where the molecular line emission of CO (J=1-0) is faint compared to the other part of the arm (Zhang et al. 2013). The same tendency has been confirmed in distributions of Massive Young Stellar Objects (MYSOs) via the MSX source survey of Urquhart et al. 2014) and molecular cores (via 1.1-mm dust continuum data by Shirley et al. 2013).

We aim to understand physical mechanisms of the creation of the Perseus-arm gap based on VLBI astrometry observations. The observations combined with spectroscopic data allow us to reveal the accurate location and 3D motion of the gap structure via measurements of trigonometric parallaxes, proper motion components, and line-of-sight velocities.

2. Observations

Based on the position-velocity (i.e., Galactic longitude l vs. LSR velocity $V_{\rm LSR}$) diagram of CO (Dame et al. 2001), we selected four MYSOs which are associated with the Perseus-arm gap. Using VERA (VLBI Exploration of Radio Astrometry), we conducted 39-epochs VLBI astrometry for 22 GHz H_2O maser emissions from the four MYSOs between the December of 2015 and the June of 2020.

The observing system of VERA including VLBI back-end system is summarized in (Oyama et al. 2016). Left-handed circular polarization data were recorded at 1,024 Mbps or 2 Gbps with 2-bit quantization and were correlated with Mizusawa software correlator†. A frequency spacing of 15.625 or 31.25 kHz was applied for 22 GHz H_2O masers, which corresponds to a velocity spacing of 0.21 or 0.42 km s^{-1}.

3. Data reduction

Data reduction was performed with the Astronomical Image Processing System (AIPS; van Moorsel et al. 1996). Basic amplitude and phase calibrations were applied for the data by referring to previous VERA astrometry papers (e.g., figure 11 of Kurayama et al. 2011).

Obtained masers' positions relative to a background continuum source were modeled using the following equations (see Kamezaki et al. 2012 for details):

$$\Delta\alpha\cos\delta = \pi(-\sin\alpha\cos\lambda_\odot + \cos\epsilon\cos\alpha\sin\lambda_\odot) + (\mu_\alpha\cos\delta)t + \alpha_0\cos\delta, \quad (1)$$

$$\Delta\delta = \pi(\sin\epsilon\cos\delta\sin\lambda_\odot - \cos\alpha\sin\delta\cos\lambda_\odot - \cos\epsilon\sin\alpha\sin\delta\sin\lambda_\odot) + \mu_\delta t + \delta_0 \quad (2)$$

where ($\Delta\alpha\cos\delta$, $\Delta\delta$) are the displacements of the observed maser spot, π the trigonometric parallax, (α, δ) are the right ascension and declination of the source, $\lambda_\odot$ is the ecliptic longitude of the Sun, ϵ is the obliquity of the ecliptic, ($\mu_\alpha\cos\delta$, μ_δ) are proper motion components in right ascension and declination directions, respectively, and ($\alpha_0\cos\delta$, δ_0) are right ascension and declination when $t = 0$.

4. Results and Discussions

4.1. *New parallax and proper-motion results*

We obtained new parallax and proper motion results for four MYSOs (see Table 1), of which G070.33+01.59 shows a fractional parallax error of 50%. Although the distance estimation via the parallax result of G070.33+01.59 is biased due to the large error (see Bailer-Jones 2015), we succeeded to estimate a source distance of 7.7±1.0 kpc based on the measured proper motion and the LSR velocity determined by a molecular line observation. The above distance is called "2D kinematic distance", which is less affected by noncircular motion (i.e., peculiar motion) compared to the conventional 1D kinematic distance based on the LSR velocity and a model of Galactic rotation curve. In Table 1, spiral-arm assignments of the four MYSOs are conducted using the Bayesian distance calculator‡ of Reid et al. (2016) where previous parallax results are used as prior information. Also, we input the information of measured proper motions into the calculator. As a result, G050.28−00.39 and G070.33+01.59 are classified in the Perseus arm.

† https://www.miz.nao.ac.jp/veraserver/system/fxcorr-e.html
‡ http://bessel.vlbi-astrometry.org/node/378

Table 1. New parallax and proper motion results.

Target	Parallax (π) (mas)	Distance (kpc)	$\mu_\alpha \cos\delta$ (mas yr^{-1})	μ_δ (mas yr^{-1})	$V_{\rm LSR}$ (km s^{-1})	Spiral arm	Ref.
G050.28-00.39	0.140±0.018	$7.1^{+1.1}_{-0.8}$	−3.29±0.56	−5.52±0.37	17±3	Perseus	a
G053.14+00.07	0.726±0.038	1.4±0.1	−1.27±1.08	−7.15±1.07	22±4	–	a
G070.33+01.59	0.074±0.037*	–	−2.82±0.29	−4.68±0.28	−23±5	Perseus	b
G079.08+01.33	0.118±0.035	$8.5^{+3.6}_{-1.9}$	−2.49±0.14	−3.36±0.24	−64±1†	Outer	c

Column 1: 22 GHz H_2O maser source; Columns 2-3: parallax and corresponding distance; Columns 4-5; proper motion components in right ascension and declination directions, respectively; Column 6: LSR velocity; Column 7: The spiral-arm assignment based on the Bayesian distance calculator of Reid et al. (2016). The symbol hyphen, –, indicates *unknown* result; Column 8: the reference of the LSR velocity estimated from a molecular line observation.
References: (a) Shirley et al. (2013); (b) Anglada et al. (1996); (c) Miville-Deschênes et al. (2017).
*Since the fractional parallax error is 50%, distance estimation by simply inverting the parallax results in a significant bias (see Bailer-Jones 2015).
†Different LSR velocities $V_{\rm LSR} = -18\pm5$ km s^{-1} (Yang et al. 2002) and -64 ± 1 km s^{-1} have been assigned for the source based on ^{12}CO (J=1−0) observations. The latter LSR velocity is applied in this paper.

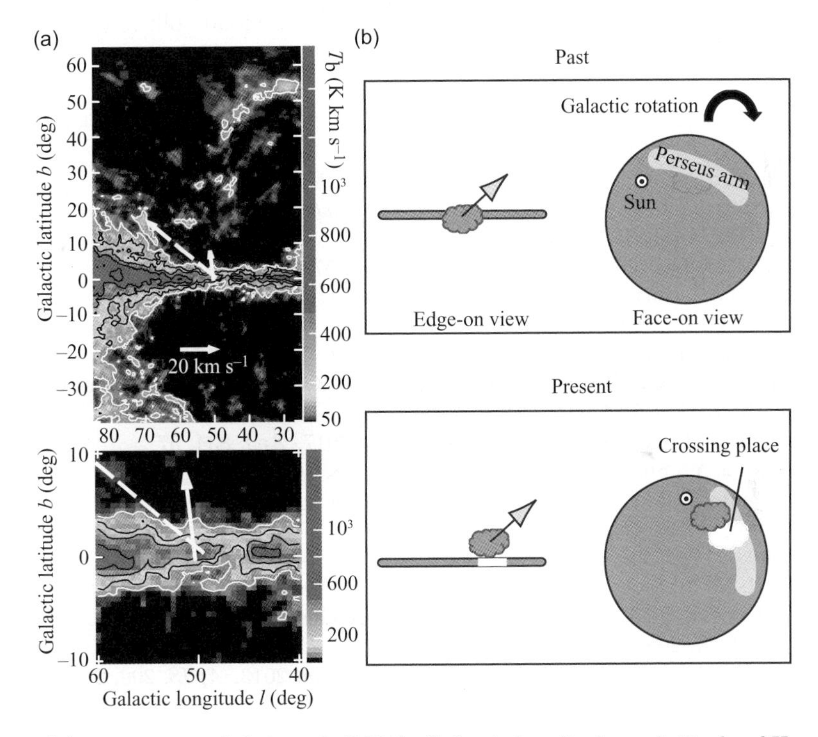

Figure 1. **(a)** Referring to Sakai et al. (2022). Galactic longitude vs. latitude of H I, obtained by integrating the LAB data (Kalberla et al. 2005) over the range of LSR velocity [−25, −5] km s^{-1}. The color shows the integrated brightness temperature of the H I emission. Arrows display noncircular motions of MYSOs G049.41+00.32 and G050.28−00.39. Both the sources are associated with an end-point of the Perseus-arm gap (at Galactic longitude ~50°). Bottom panel is an enlarged view of top panel. **(b)** Cartoon proposing a scenario how the Perseus-arm gap is created. Top panel indicates that the Galactic disk was pushed from the negative Galactic latitude side previously (i.e., a cloud collision with the Galactic disk) while bottom shows the current picture of the Milky Way reflecting observational results of H I and VLBI astrometry results.

4.2. *A possible physical mechanism on the creation of the Perseus-arm gap*

Using the new and previous VLBI astrometry results of the Perseus arm, we discuss kinematics especially for the noncircular motion of the arm. As shown by Fig. 4

of Sakai et al. (2022), the maximum and systematic noncircular motion of the arm is confirmed at an end-point of the Perseus-arm gap (at Galactic longitude $l\sim50°$; see white arrows in Fig. 1). Especially, one MYSO G049.41+00.32 lags relative to a universal rotation curve of (Reid et al. 2019; i.e., A5 model) by 77±17 km s^{-1}. Such a large noncircular motion cannot be explained by the gravitational potential of a spiral arm (e.g., see Sakai et al. 2019).

Figure 1 shows LAB H I data (Kalberla et al. 2005) integrated over the LSR velocity range [−25, −5] km s^{-1} covering the LSR velocity of G049.41+00.32, −21±1 km s^{-1} (Svoboda et al. 2016). The figure indicates the existence of H I stream above the Galactic plane (at $b > 0°$). The H I mass of the stream is $\sim8\times10^6$ $M_\odot$ (solar mass) at the distance of G049.41+00.32, $7.6^{+2.3}_{-1.4}$ kpc (Reid et al. 2019). Also, we can see arc-shape structure in the Galactic disk, which is offset by ∼0.4° from G049.41+00.32.

To explain all the observational results as well as the physical origin of the Perseus-arm gap, we propose a cloud collision with the Galactic disk (see the cartoon of Fig. 1). To confirm the hypothesis, further studies should be conducted in the future (e.g., the distance and metallicity of the H I stream; the time of a cloud collision).

References

Anglada, G., Estalella, R., Pastor, J., Rodriguez, L. F., & Haschick, A. D. 1996, *ApJ*, 463, 205
Bailer-Jones, C. A. L. 2015, *PASP*, 127, 994
Churchwell, E., Babler, B. L., Maede, M. R., *et al.* 2009, *PASP*, 121, 213
Dame, T. M, Hartmann, D., & Thaddeus, P. 2001, *ApJ*, 547, 792
Drimmel, R. 2000, *A&A*, 358, L13
Georgelin, Y.-M., & Georgelin, Y.-P. 1976, *A&A*, 49, 57
Kalberla, P. M. W., Burton, W. B., Hartmann, D., Arnal, E. M., Bajaja, E., Morras, R., & Pöppel, W. G. L. 2005, *A&A*, 440, 775
Kamezaki, T., Nakagawa, A., Omodaka, T., *et al.* 2012, *PASJ*, 64, 7
Kurayama, T., Nakagawa, A., Sawada-Satoh, S., Sato, K., Honma, M., Sunada, K., Hirota, T., & Imai, H. 2011, *PASJ*, 63, 513
Miville-Deschênes, M. -A., Murray, N., & Lee, E. J. 2017, *ApJ*, 834, 57
Oyama, T., Kono, Y., Suzuki, S., *et al.* 2016, *PASJ*, 68, 105
Reid, M. J., Dame, T. M., Menten, K. M., & Brunthaler, A. 2016, *ApJ*, 823, 77
Reid, M. J., *et al.* 2019, *ApJ*, 885, 131
Russeil, D. 2003, *A&A*, 397, 133
Sakai, N., Reid, M. J., Menten, K. M., Brunthaler, A., Dame, T. 2019, *ApJ*, 876, 30
Sakai, N., Nakanishi, H., Kurahara, K., Sakai, D., Hachisuka, K., Kim, J. -S., & Kameya, O. 2022, *PASJ*, 74, 209
Shirley, Y. L., Ellsworth-Bowers, T. P., Svoboda, B., *et al.* 2013, *ApJS*, 209, 2
Svoboda, B. E., *et al.* 2016, *ApJ*, 822, 59
Urquhart, J. S., *et al.* 2007, *A&A*, 474, 891
Urquhart, J. S., Figura, C. C., Moore, T. J., *et al.* 2014, *MNRAS*, 437, 1791
van Moorsel, G., Kemball, A., & Greisen, E. 1996, in ASP Conf. Ser., 101, Astronomical Data Analysis Software and Systems V, ed. G. H. Jacoby & J. Barnes (San Francisco: ASP), 37
Xu, Y., Hao, C. J., Liu, D. J., *et al.* 2023, *ApJ*, 947, 54
Yang, J., Jiang, Z., Wang, M., Ju, B., & Wang, H. 2002, *ApJS*, 141, 157
Zhang, B., Reid, M. J., Menten, K. M., *et al.* 2013, *ApJ*, 775, 79

Cosmic Masers: Proper Motion toward the Next-Generation Large Projects
Proceedings IAU Symposium No. 380, 2024
T. Hirota, H. Imai, K. Menten, & Y. Pihlström, eds.
doi:10.1017/S1743921323002272

Kinematics in the Galactic Center with SiO masers

Jennie Paine and Jeremy Darling

Center for Astrophysics and Space Astronomy, Department of Astrophysical and Planetary Sciences University of Colorado, 389 UCB, Boulder, CO 80309-0389, USA.
email: jennie.paine@colorado.edu

Abstract. Stellar SiO masers are found in the atmospheres of asymptotic giant branch (AGB) stars with several maser transitions observed around 43 and 86 GHz. At least 28 SiO maser stars have been detected within ~2 pc projected distance from Sgr A* by the Very Large Array (VLA) and Atacama Millimeter/submillimeter Array (ALMA). A subset of these masers have been studied for several decades and form the basis of the radio reference frame that anchors the reference frame for infrared stars in the Galactic Center (GC). We present new observations of the GC masers from VLA and ALMA. These new data combined with extant maser astrometry provide 3D positions, velocities, and acceleration limits. The proper motions and Doppler velocities are measured with unprecedented precision for these masers. We further demonstrate how these measurements may be used to trace the stellar and dark matter mass distributions within a few pc of Sgr A*.

Keywords. Silicon monoxide masers, stellar masers, Galactic Center, stellar kinematics

1. Introduction

Stellar kinematics probe the gravity profile in which the stars reside, which in the centers of galaxies will be dominated by a supermassive black hole, stars, and dark matter. The dark matter profiles in galactic centers are particularly of interest due to the discrepancy between simulations, which predict steep profiles (cusps), and observations, which find flat profiles (cores). This difference is known as the core-cusp problem (see de Blok 2010 for a review). Dark matter profiles on scales of several parsec and smaller are additionally poorly constrained by both simulations and observations.

The question of the distribution of dark matter in galactic centers is also tied to the presence and growth of the central black hole. If the supermassive black hole grows adiabatically, then the dark matter profile will steepen within the sphere of influence of the black hole (Gondolo & Silk 1999), known as a dark matter spike, and stars respond similarly to the presence of a black hole. The stellar cusp in our galaxy is shallower than predicted by theory (Schödel *et al.* 2018; Habibi *et al.* 2019), suggesting the absence of a dark matter spike, but the two profiles are not necessarily coupled if the stellar cluster formed after the growth of the black hole.

Determining the distribution and kinematics of stars in the Galactic Center (GC) is challenging due to high extinction and stellar crowding, but radio observations are unaffected by these issues. Thus, one can use stellar maser emission detected at radio frequencies to uniquely measure stellar kinematics with high resolution within a several parsec range around the central galactic black hole Sgr A*. We present observations of the 43 and 86 GHz masers in the GC from the Very Large Array (VLA) and the Atacama Large Millimeter/submillimeter Array (ALMA), and their resulting kinematics over several decades of observations. We show that the velocities and accelerations of these stars may be used to constrain the mass distribution around Sgr A*, though the measurements are complicated by the intrinsic variability of stellar SiO masers. Finally, we discuss the results of recent GC maser observations.

2. Stellar SiO masers in the Galactic Center

Sgr A* is the supermassive black hole at the center of our galaxy and a bright, compact radio source. The first strong evidence for the existence of the black hole was provided by near-infrared (near-IR) observations of stars on short period orbits around Sgr A* (Ghez *et al.* 2000, 2005; Schödel *et al.* 2002; Eisenhauer *et al.* 2005). The orbits of these "S" stars suggest a black hole mass of $4.15 \times 10^6\ M_\odot$ and a distance to the GC of 8.2 kpc Ghez *et al.* (2008); Genzel *et al.* (2010). Additionally, following the pericenter passage of S2 — a star on a highly eccentric orbit with pericenter at 120 AU from Sgr A* (approximately 1400 times the black hole's event horizon) — IR stellar orbits were able to test General Relativity (e.g. gravitational redshift and Schwarzschild precession; GRAVITY Collaboration *et al.* 2018; Do *et al.* 2019; GRAVITY Collaboration *et al.* 2020).

Since there is no obvious, bright near-IR counterpart source to Sgr A* from which to reference the positions of S stars, the IR stellar orbit measurements rely on the definition of an astrometric reference frame. Recently, the reference frame has been defined using the proper motions of SiO maser-emitting stars in the vicinity of Sgr A*, since the masers are observed in radio frequencies where the Sgr A* radio source is also detected. Observations of the GC over the past several decades have identified at least 28 stellar SiO masers within a few parsecs of Sgr A* (e.g. Menten *et al.* 1997; Reid *et al.* 2007; Li *et al.* 2010; Borkar *et al.* 2020; Paine & Darling 2022), and a subset of these maser-emitting stars are bright IR sources used for the reference frame. Menten *et al.* (1997) first proposed measuring the SiO proper motions relative to the Sgr A* radio continuum and matching the radio positions to the IR counterparts to determine the location of Sgr A* in IR images and establish a reference frame where Sgr A* is at rest. Since then, several iterations of the reference frame have been determined using SiO maser proper motions (e.g. Menten *et al.* 1997; Reid *et al.* 2007; Yelda *et al.* 2010; Sakai *et al.* 2019). The advantage of the radio reference frame is that as proper motion uncertainties decrease in time by $t^{-3/2}$, so does the reference frame improve. However, projecting the maser proper motions forward in time is a large source of uncertainty in the reference frame and resulting IR orbital measurement, so continuous monitoring of the SiO masers is required to improve the precision of IR measurements.

Stellar masers typically occur in the extended atmospheres and circumstellar envelopes of red giants and asymptotic giant branch (AGB) stars. SiO emission is found closest to the star compared to other maser lines, at radii of $\sim$ a few AU, interior to the dust formation point which drives the stellar wind. The observed SiO maser lines are rotational transitions typically in excited vibrational states. Several SiO lines are observed at frequencies around 43 GHz ($J = 1-0$) and 86 GHz ($J = 2-1$). VLBI observations of stellar SiO masers typically resolve the emission into discrete spots in a ring-like structure around the star which may be variable over relatively short timescales (e.g. Gonidakis *et al.* 2013). Individual maser spots track the local motion of material around the star, which is not always symmetric, and components may turn on or off between observations. The variability of the maser components may be caused by changes in the dominant pumping mechanism or acceleration of the maser medium with respect to the star disrupting the requirement of velocity coherence.

The result for lower resolution observations which detect the aggregate maser emission of many components is that SiO maser spectra often show multiple peaks distributed over a velocity range of $\sim 10\ \mathrm{km\ s^{-1}}$ relative to the systemic stellar velocity (Jewell *et al.* 1991), and which may vary in time. Different maser transitions observed simultaneously may not match in the location of spectral peaks since the transitions are not necessarily coincident around the star. Maser variability will also introduce intrinsic scatter in the astrometry over time as the position centroid may move around the area of maser ring.

3. Current maser kinematic measurements

In Paine & Darling (2022), we presented five epochs of observations of stellar SiO masers in the GC using VLA and ALMA. Interferometers like VLA and ALMA do not measure images of the sky directly, but instead sample the Fourier transform of the sky brightness, known as the visibilities. In cases where the observed source can be represented with a simple model (or a linear combination of simple models as in the case of a sample of point sources), one may forgo the creation of images and instead fit to the visibilities (e.g. Martí-Vidal *et al.* 2014; see also Pearson 1995 for a discussion of visibility model fitting and its applications). We use visibility fitting of the masers' positions and spectra to maximize the resolution of the observations — the smallest resolvable angular size is smaller than the diffraction limit of the interferometer for high signal-to-noise sources — and to minimize systematic differences from comparing data between different epochs and telescopes.

Combined with extant maser astrometry, we found proper motions and proper accelerations for the GC maser sample. The corresponding 2D stellar maser kinematics are measured with 0.5 km s^{-1} and 0.04 km s^{-1} yr^{-1} precision for velocities and accelerations, respectively. The radial velocities and accelerations are measured with 0.5 km s^{-1} and 0.1 km s^{-1} yr^{-1} precision, respectively. These measurements are the current benchmark for the published kinematics of these stars, though the precision and accuracy of the kinematics are heavily impacted by the intrinsic variability of the SiO maser sources, as described above.

By modeling the expected kinematics of stars in circular orbits about the GC, we identified several stars with anomalous velocity or acceleration measurements which are significantly higher than anticipated (discussed below). Figure 1 shows the 3D acceleration measurements and velocity-based enclosed mass limits for our sample of stars as a function of projected distance from Sgr A*. We compare these to models of the total mass profile, both with and without a theoretical dark matter spike, to demonstrate how the maser kinematics may differentiate between dark matter models in the GC with improved precision from more observations.

4. Implications from continued maser monitoring

Since the publication of Paine & Darling (2022), we have observed one additional epoch from VLA and one from ALMA, which generally improve the precision of the kinematic measurements. Several anomalous velocities or accelerations are also confirmed using the new observations. For example, the stars SiO-16, SiO-21, and SiO-25 (the outlying stars in the upper right of the left figure of Figure 1) have proper motion-based velocities > 1000 km s^{-1}, which are about an order of magnitude larger than anticipated for their locations in the GC. These large velocities may be the product of measurement error or incorrect distance estimates if the stars are actually located in the foreground rather than the GC. However, if accurate, these velocities would suggest that the stars are not bound to the GC.

Additionally, several stars, such as SiO-14 and IRS 15NE, show significant or marginally significant acceleration in the velocity centroid of the maser spectra over time which cannot be explained by acceleration of the star about the GC alone. As noted in Section 2, SiO maser spectra are highly variable about the systemic stellar velocity and

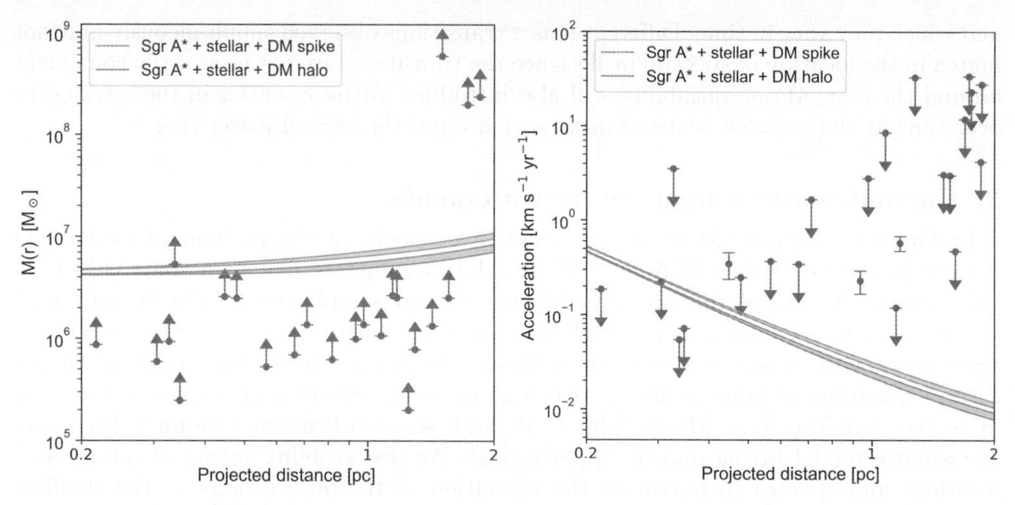

Figure 1. Figures adapted from Paine & Darling (2022). **Left:** Enclosed mass lower limits calculated from the stellar 3D velocities as a function of projected distance. The dark matter halo model (green region) is a generalized NFW profile (McMillan 2017), and the maximal dark matter spike model (orange region) includes an additional sharp spike in the dark matter density profile (Lacroix 2018). **Right:** 3D stellar acceleration magnitude upper limits as a function of projected distance. Shaded regions show expected accelerations due to models of the interior stellar and dark matter mass. Error bars indicate stars with at least 3σ measurement of the total acceleration.

therefore acceleration of the aggregate maser source may not reflect real acceleration of the star. However, one would expect to observe this as scatter about the trend, so the large, approximately secular changes in velocity observed for these stars requires further investigation.

References

Borkar, A., Eckart, A., Straubmeier, C., *et al.* 2020, *Multifrequency Behaviour of High Energy Cosmic Sources - XIII*, 33

de Blok, W. J. G. 2010, *Advances in Astronomy*, 2010, 789293

Do, T., Hees, A., Ghez, A., *et al.* 2019, *Science*, 365, 664

Eisenhauer, F., Genzel, R., Alexander, T., *et al.* 2005, *ApJ*, 628, 246

Genzel, R., Eisenhauer, F., & Gillessen, S. 2010, *Reviews of Modern Physics*, 82, 3121

Ghez, A. M., Morris, M., Becklin, E. E., *et al.* 2000, *Nature*, 407, 349

Ghez, A. M., Salim, S., Hornstein, S. D., *et al.* 2005, *ApJ*, 620, 744

Ghez, A. M., Salim, S., Weinberg, N. N., *et al.* 2008, *ApJ*, 689, 1044

Gondolo, P. & Silk, J. 1999, *Phys. Rev. Lett.*, 83, 1719

Gonidakis, I., Diamond, P. J., & Kemball, A. J. 2013, *MNRAS*, 433, 3133

GRAVITY Collaboration, Abuter, R., Amorim, A., *et al.* 2018, *A&A*, 615, L15

GRAVITY Collaboration, Abuter, R., Amorim, A., *et al.* 2020, *A&A*, 636, L5

Habibi, M., Gillessen, S., Pfuhl, O., *et al.* 2019, *ApJ* (Letters), 872, L15

Jewell, P. R., Snyder, L. E., Walmsley, C. M., *et al.* 1991, *A&A*, 242, 211

Lacroix, T. 2018, *A&A*, 619, A46

Li, J., An, T., Shen, Z.-Q., *et al.* 2010, *ApJ* (Letters), 720, L56

Martí-Vidal, I., Vlemmings, W. H. T., Muller, S., *et al.* 2014, *A&A*, 563, A136

McMillan, P. J. 2017, *MNRAS*, 465, 76

Menten, K. M., Reid, M. J., Eckart, A., *et al.* 1997, *ApJ* (Letters), 475, L111

Paine, J. & Darling, J. 2022, *ApJ*, 927, 181

Pearson, T. J. 1995, *Very Long Baseline Interferometry and the VLBA NRAO Workshop No. 22, Astronomical Society of the Pacific Conference Series*, 82, 267

Reid, M. J., Menten, K. M., Trippe, S., *et al.* 2007, *ApJ*, 659, 378
Sakai, S., Lu, J. R., Ghez, A., *et al.* 2019, *ApJ*, 873, 65
Schödel, R., Ott, T., Genzel, R., *et al.* 2002, *Nature*, 419, 694
Schödel, R., Gallego-Cano, E., Dong, H., *et al.* 2018, *A&A*, 609, A27
Yelda, S., Lu, J. R., Ghez, A. M., et al. 2010, *ApJ*, 725, 331

Cosmic Masers: Proper Motion toward the Next-Generation Large Projects
Proceedings IAU Symposium No. 380, 2024
T. Hirota, H. Imai, K. Menten, & Y. Pihlström, eds.
doi:10.1017/S1743921323002855

Trigonometric parallax, proper motion, and structure of three southern hemisphere methanol masers

Lucas J. Hyland[1], **Simon P. Ellingsen**[1], **Mark J. Reid**[2], **Jayender Kumar**[1,3] **and Gabor Orosz**[1,4]

[1]School of Natural Sciences, University of Tasmania, Private Bag 37, Hobart, Tasmania 7001, Australia. email: lucas.hyland@utas.edu.au

[2]Center for Astrophysics | Harvard & Smithsonian, Cambridge, MA 02138, USA

[3]CSIRO, Space and Astronomy, PO Box 76, Epping, NSW 1710, Australia

[4]Joint Institute for VLBI ERIC, Oude Hoogeveensedijk 4, 7991PD Dwingeloo, Netherlands

Abstract. We present the trigonometric parallax, proper motion, and structure of three methanol masers from the Southern Hemisphere Parallax Interferometric Radio Astrometry Legacy Project (SπRALS). All three masers have better than 5% parallax accuracy, which we attribute to the new inverse MultiView calibration technique.

Keywords. astrometry - proper motions, parallaxes; masers - methanol; techniques - Very Long Baseline Interferometry

1. Introduction

The spiral structure of the 4th quadrant of the Milky Way; $180^\circ < l < 359^\circ$ and visible almost exclusively to the southern hemisphere; has long remained shrouded in uncertainty. Until very recently, accurate trigonometric parallax measurements to masers in high mass star forming regions have been unattainable. The best maser species candidates for trigonometric parallax measurements are 22 GHz water and 6.7 GHz methanol; the former species is structurally variable and should be observed within a 1 yr span before maser spots may disappear. Overall, at least 8 observational epochs are ideal for parallax measurements of both species. In the past, scheduling for strict time constraints posed a challenge for the Australian Long Baseline Array (LBA), the sole Very Long Baseline Interferometric (VLBI) array in the southern hemisphere. The species at 6.7 GHz is much more stable and theoretically can be observed over many years. However at this frequency, uncorrected dispersive delays caused by the ionosphere can significantly and adversely affect astrometric observations. (e.g., Krishnan *et al.* 2015; Reid *et al.* 2017).

In order to address the issue of scheduling, the University of Tasmania outfitted two of their 12 m geodetic antennas (Hobart 12 m and Katherine 12 m; Lovell *et al.* 2013) with 6.7 GHz receivers capable to make an array of four antennas with Ceduna 30 m

Table 1. Methanol maser names (1), correlated positions (2-3), measured parallaxes (4), proper motions in the north-south (5) and east-west (6) directions, and reference (7). Units are given on the second row where applicable.

Name	R.A (*h m s*)	Decl. (° ′ ″)	π (mas)	μ_x (mas yr^{-1})	μ_y (mas yr^{-1})	ref.
G309.92+00.47	13 50 41.78	−61 35 10.2	0.291 ± 0.011	-5.560 ± 0.025	-1.561 ± 0.070	-
G323.74−00.26	15 31 45.45	−56 30 50.1	0.364 ± 0.009	-3.239 ± 0.025	-3.976 ± 0.039	1
G328.80+00.63	15 55 48.45	−52 43 06.6	0.381 ± 0.016	-3.492 ± 0.044	-3.276 ± 0.116	2

Notes: 1 - Hyland *et al.* (2022); 2 - Kumar (2023)

(McCulloch *et al.* 2005) in South Australia and Warkworth 30 m (Woodburn *et al.* 2015) in Auckland, New Zealand (operated at the time by the Auckland University of Technology). This array has the flexibility to dedicate hundreds of hours per year exclusively to the astrometric observations of 6.7 GHz methanol masers (Hyland *et al.* 2018; Hyland *et al.* 2021).

The final piece of the puzzle was the development of inverse MultiView (Hyland *et al.* 2021, 2022). This calibration technique facilitates the accurate removal of tropospheric and ionospheric residual delays and allows astrometric positioning down to the uncertainty caused by the thermal noise. In addition, it allows the use of quasar calibrators much further away than traditionally accepted (i.e., 7° at 8.4 GHz).

And so the Southern Hemisphere Parallax Interferometric Radio Astrometry Legacy Survey (SπRALS; Hyland *et al.* 2021) was started, with the aim to measure the spiral structure of the 4th quadrant of the Milky Way using trigonometric parallaxes of 6.7 GHz methanol masers. Here we present our first three novel maser parallaxes, their proper motions, and structure.

2. Results

Table 1 contains the measured trigonometric parallaxes and proper motions for the southern hemisphere methanol masers G309.92+00.47, G323.74−00.26, and G328.80+00.63. The parallax and proper motion fits to the astrometric data for each maser are shown in Figure 1. Two of the masers have parallax accuracy ~ 10 μas, the current gold standard for relative astrometry. Even for parallax measurements of 22 GHz water masers on the Very Long Baseline Array (VLBA), parallax accuracy at this level is uncommon.

The structures of all three masers are shown in Figure 2. All three masers have a multitude of emission regions, and show a diverse range of overall structures. G323.74−00.26 appears to trace a shock, G309.92+00.47 has a linear structure with a clear velocity gradient, and G328.80+0.63 has a 'paired' structure.

The data for G323.74−00.26 is published in Hyland *et al.* (2022) and for G328.80+00.63 is presented in Kumar (2023). The analysis of G309.92+00.47 is still in the preliminary stage and the final results will be fully presented in an upcoming publication.

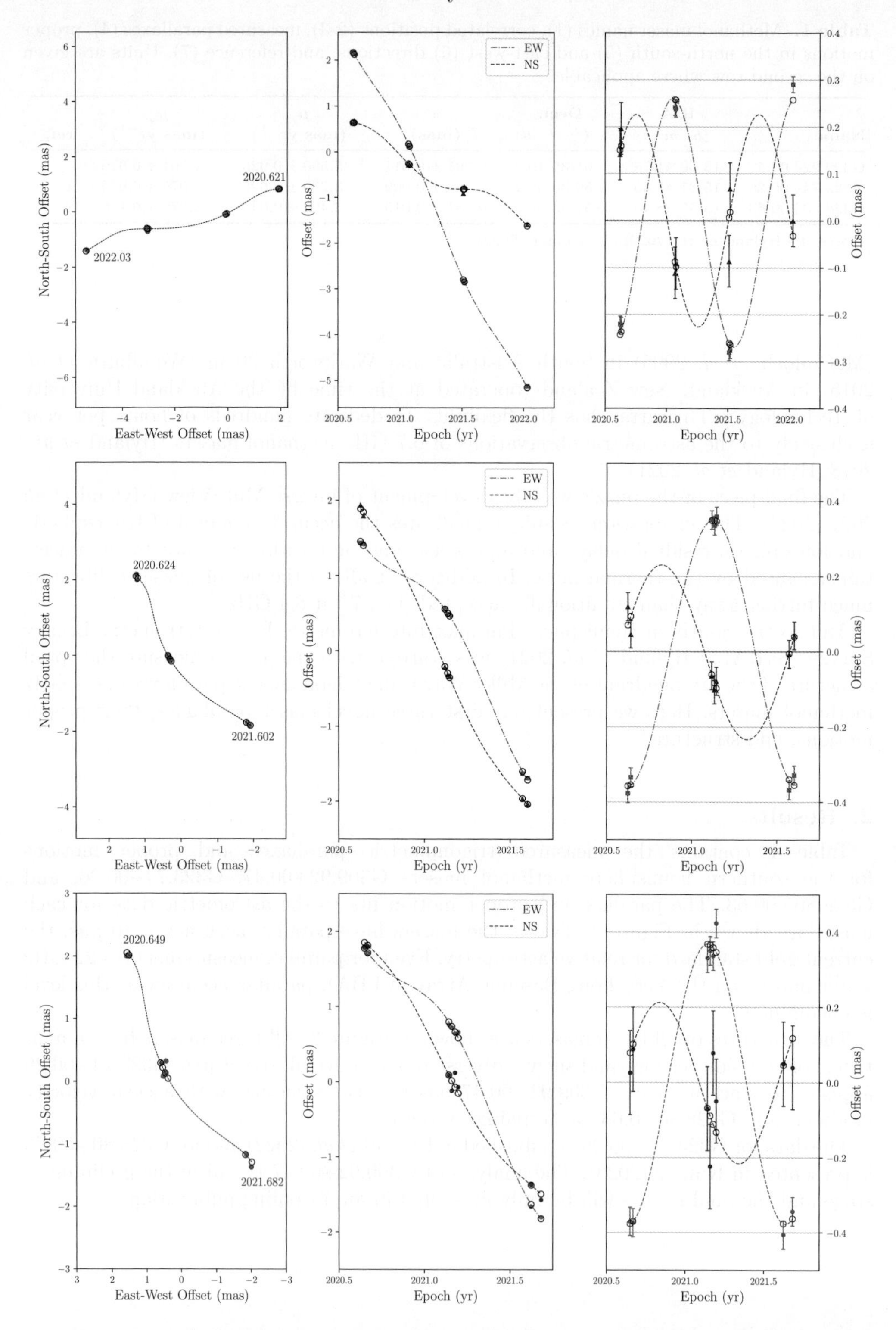

Figure 1. Trigonometric parallax and proper motion fits to astrometric data for all three masers. Top to bottom: Fits for G309.92+00.47, G323.74−00.26, and G328.80+00.63. Left to right: Total sky motions relative to reference positions, sky positions decomposed into EW (green dot-dashed) and NW (blue dashed) positions over time, and proper motion-subtracted position over time showing the parallax motion.

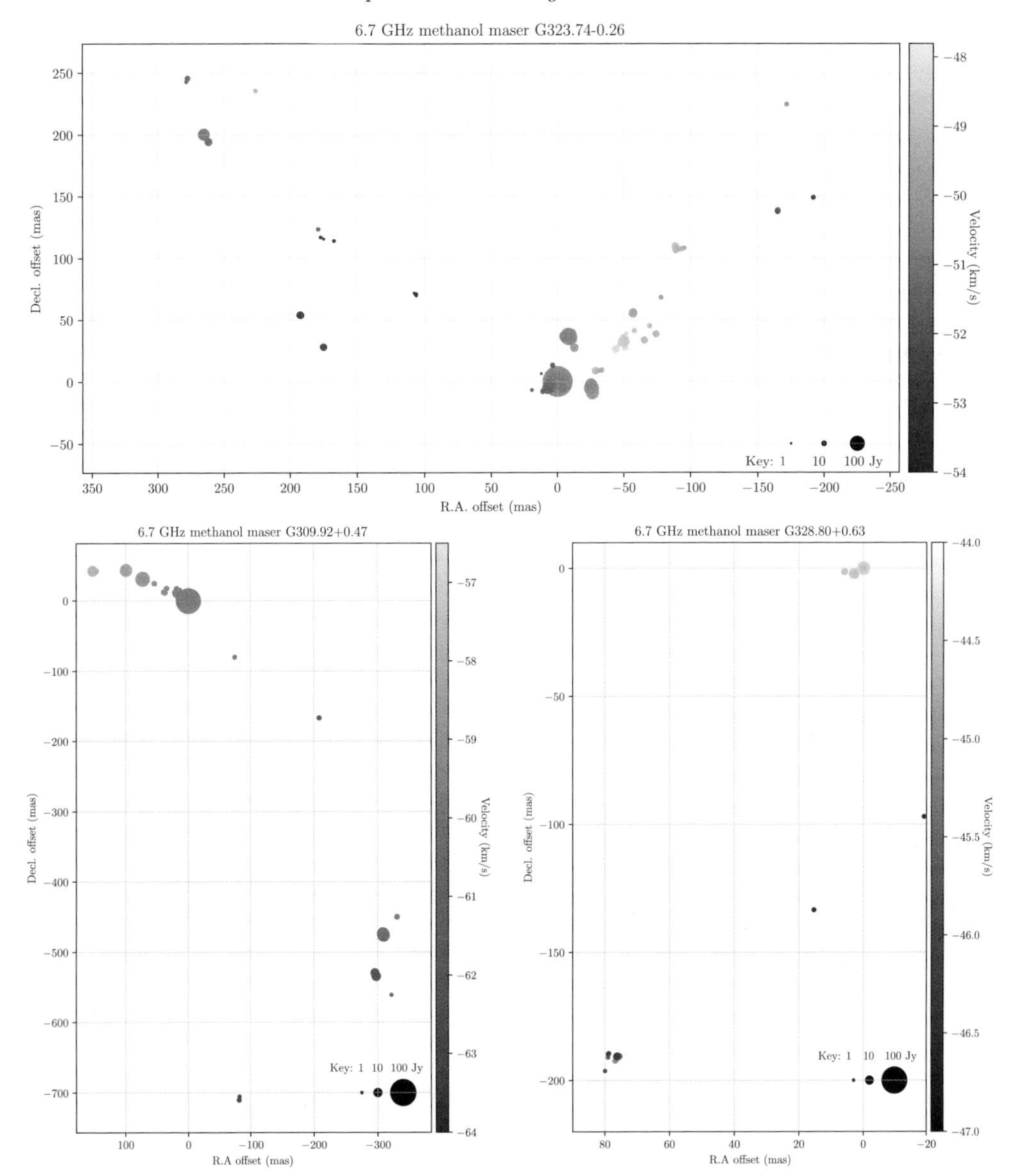

Figure 2. Spot maps displaying the structures of the three masers, giving each spot position relative to a reference feature. The size of the spot indicates the flux density (Jy), with a key given at the bottom right of each map. A color bar on each map indicates the LSR velocity of the spot.

References

Hyland, L. J., Ellingsen, S. P., & Reid, M. J. 2018, in: *Unlocking the Mysteries of the Universe*, Proc IAU Symposium, 336, 154

Hyland, L. J. 2021, *SPIRALS - Southern Hemisphere Parallax Interferometric Radio Astrometry Legacy Survey*, PhD thesis, School of Natural Sciences, University of Tasmania

Hyland, L. J., Reid, M. J., Ellingsen, S. P., Rioja, M. J., Dodson, R., Orosz, G., Masson, C. R., & McCallum, J. M. 2022, *ApJ*, 932, 52

Hyland, L. J., Reid, M. J., Orosz, G., Ellingsen, S. P., Weston, S. D., Kumar, J., Dodson, R., Rioja, M. J., Hankey, W. J., Yates-Jones, P. M., *et al.* 2022, arXiv e-prints, arXiv:2212.03555

Kumar, J. 2023, *The Structure and 3D Kinematics within 4 kpc of the Galactic Centre*, PhD thesis, School of Natural Sciences, University of Tasmania

Krishnan, V., Ellingsen, S. P., Reid, M. J., Brunthaler, A., Sanna, A., McCallum, J., Reynolds, C., Bignall, H. E., Phillips, C. J., *et al.* 2015, *ApJ*, 805, 129

Lovell, J. E. J., McCallum, J. N., Reid, P. B., McCulloch, P. M., Baynes, B. E., Dickey, J. M., Shabala, S. S., Watson, C. S., Titov, O., Ruddick, R., *et al.* 2013, *J. Geod.*, 87, 527

McCulloch, P. M., Ellingsen, S. P., Jauncey, D. L., Carter, S. J. B., Cimò, G., Lovell, J. E. J., & Dodson, R. G. 2005, *AJ*, 129, 2034

Reid, M. J., Brunthaler, A., Menten, K. M., Sanna, A., Xu, Y., Li, J. J., Wu, Y., Hu, B., Zheng, X. W., *et al.* 2017, *AJ*, 154, 63

Woodburn, L., Natusch, T., Weston, S., Thomasson, P., Godwin, M., Granet, C., & Gulyaev, S. 2015, *PASA*, 32, 17

Cosmic Masers: Proper Motion toward the Next-Generation Large Projects
Proceedings IAU Symposium No. 380, 2024
T. Hirota, H. Imai, K. Menten, & Y. Pihlström, eds.
doi:10.1017/S1743921323002211

Mapping the Far Side of the Milky Way

Mark J. Reid

Center for Astrophysics | Harvard & Smithsonian. email: mreid@cfa.harvard.edu

Abstract. Great progress has been made using VLBI techniques to measure trigonometric parallaxes to masers associated with young, high-mass stars, in order to map the spiral structure of the Milky Way. However, large numbers of parallax distance have only been obtained over about half of the Galaxy. Here I discuss the use of 3-dimensional kinematic distances for completing the map with many sources well past the Galactic Center.

Keywords. Milk Way, Spiral Structure, Distances, VLBI

1. Introduction

Over the last decade, the Bar and Spiral Structure Legacy (BeSSeL) Survey and the VLBI Exploration of Radio Astrometry (VERA) have been measuring trigonometric parallaxes for large numbers of maser sources associated with high-mass star forming regions. These masers are extremely young, < 1 Myr, and are excellent tracers of spiral arms. A major goal of these projects is to map the spiral structure of the Milky Way and to date they have done this across a large portion of the Galaxy, including the 1st quadrant out to distances from the Sun of ~ 10 kpc, and the 2nd and a portion of the 3rd quadrants. What remains to be mapped in detail is the entire 4th quadrant and the 1st quadrant beyond the Galactic Center.

The near side of the 4th quadrant is currently being surveyed by the Southern Hemisphere Parallax Interferometric Radio Astronometry Legacy Survey (SπRALS) using four antennas spanning the Australian continent and a fifth antenna in New Zealand. However, measuring parallaxes for sources well past the Galactic Center in both the 1st and 4th quadrants would require tens of thousands of hours of time on VLBI arrays to achieve distant accuracies of better than $\pm 10\%$. For a source at 20 kpc distance (parallax $= 0.05$ mas), this would require a ± 0.005 mas uncertainty which is difficult to achieve. And, even if achieved, this would correspond to ± 2 kpc distance uncertainty, which is roughly the separation of spiral arms and marginal for mapping spiral structure.

In this talk, I discuss a novel method for estimating distances to sources well past the Galactic Center, which promises to finish the mapping of the spiral structure of the Milky Way. The method uses 3-dimensional (3-D) kinematic distance estimates and only requires measurements of Doppler velocities and proper motions. The first application of this method was by Yamauchi *et al.* (2016) using the VERA array. They obtained a distance to a water maser in the massive star forming region G007.47+0.06 of 20 ± 2 kpc, which places this source 12 kpc past the Galactic Center. This result was confirmed by Sanna *et al.* (2017) with an extremely accurate trigonometric parallax measurement (uncertainty of ± 0.006 mas). Reid (2022) presented a detailed evaluation of the accuracies and limitations of 3-D kinematic distances, which I summarize here.

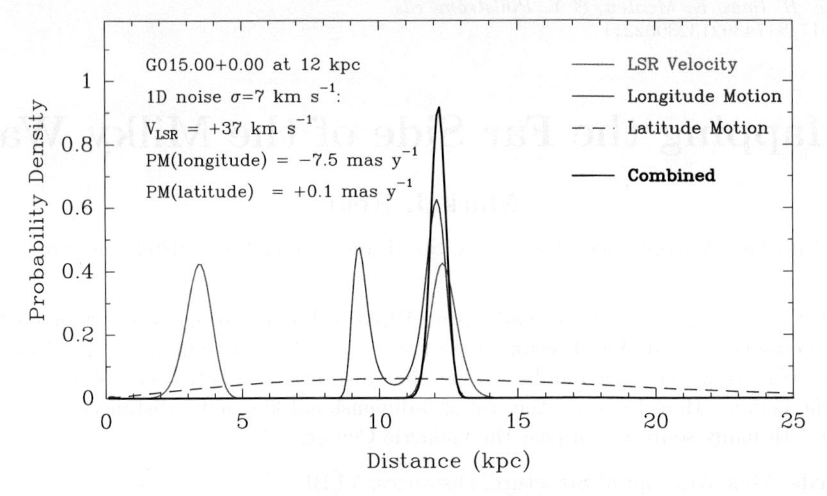

Figure 1. 3-D kinematic distance simulation after Reid (2022). Plotted are distance probability densities for each of three components of motion as listed in the legend. The simulated measurements are for a source at 12 kpc distance and Galactic longitude of 15° experiencing a near circular Galactic orbit with Gaussian random noise of 7 km s^{-1} added.

2. 3-D Kinematic Distance Method

For over a half-century, distances to many sources, including atomic and molecular clouds of gas, ionized hydrogen regions, and stars have been estimated from their Galactic coordinates and line-of-sight component of velocity (V_{lsr}), and called kinematic distances. These use 1-dimension of a source's velocity vector in conjunction with a model of Galactic rotation. 3-D kinematic distances expand on kinematic distances and offer not only improved accuracy, but also greater robustness.

A simulation of a 3-D kinematic distance estimate is shown in Fig. 1. The red line gives the probability density as a function of distance for a source at Galactic longitude 15°, based on its $V_{lsr} = 37$ km s^{-1} (i.e. standard kinematic distance). There are two probability peaks, one at 3 kpc and one at 12 kpc, demonstrating the well-known distance ambiguity problem for sources toward the inner Galaxy. While there are ways to decide between "near" and "far" kinematic distances, for example using hydrogen emission along the same line-of-sight, these are not always robust. In the figure, the solid blue line gives the probability density for the proper motion component in Galactic longitude, which peaks at 9 and 12 kpc distances. This distance information comes from proper motion measurement and is independent of, and complementary to, that from V_{lsr}. Finally, the dashed blue line gives the probability density for the proper motion component in Galactic latitude. While usually not strongly constraining, this extra distance information comes from the fact that, for a fixed velocity dispersion perpendicular to the Galactic plane, one expects a smaller angular motion the more distant a sources is.

In this example, the posterior likelihood for distance, which comes from the product of the three probability density functions, gives a distance estimate of $12.2 \pm 0.3 \pm 0.1$ kpc (where the first uncertainty is the formal statistical value and the second uncertainty is an estimate of systematic error. The formal statistical uncertainty is derived from the width of the combined posterior likelihood function, and the systematic error is estimated as half of the separation between the V_{lsr} and longitude motion peaks nearest the combined value. This provides a way to evaluate possible systematic error owing, for example, to a large peculiar (non-circular) motion or a significant error in the assumed rotation curve.

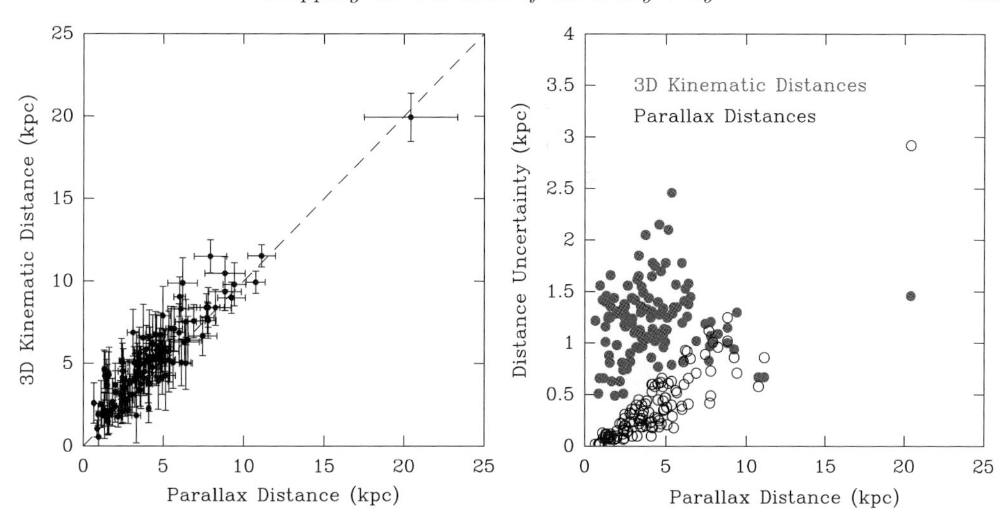

Figure 2. Comparison of trigonometric parallax and 3-D kinematic distances after Reid (2022). Plotted are data from Table 1 of Reid *et al.* (2019), collated from BeSSeL Survey and VERA project results and filtered to exclude sources with fractional parallax uncertainty greater than 15%. Note the general good agreement between the two methods and especially for G007.47+0.06 at near 20 kpc distance. The left panel plots 3-D kinematic distances vs. parallax distances. The right panel plots the distance uncertainties from the two methods: black circles for parallaxes and red dots for 3-D kinemtic distances.

3. The Accuracy of 3-D Kinematic Distances

Given the more than 200 sources with published parallaxes and proper motions from the BeSSeL Survey and the VERA project, we can empirically evaluate the accuracy and robustness of 3-D kinematic distances against their trigonometic parallax distances. This is shown in Fig. 2. The left panel directly compares the two distance estimates on a source-by-source basis. As can be seen, the 3-D kinematic distances generally agree well with the direct parallax estimates. For sources with (parallax) distance less than about 5 kpc, there is a tendency for the 3-D kinematic distances to be ~ 1 kpc too large. This bias can be traced to astrophysical noise, owing to non-circular motions of ≈ 7 km s^{-1} (Gaussian 1σ) per motion component, which leads to an asymmetric probability density function for Galactic longitude when dividing velocity by distance to get angular motion. This bias increases as distance decreases. (There are some other regions which have 3-D kinematic distance biases; see Reid (2022) for details.)

The right panel of Fig. 2 plots the uncertainties in the two methods. For sources with (parallax) distances less than about 8 kpc parallax distances are generally superior to 3-D kinematic distances. However, for greater distances the situtation is reversed, with 3-D kinematic distances generally yielding better results. Indeed, the 3-D distances tend to have uncertainties of ≈ 1.5 kpc, with no obvious increase in uncertainty with distance. This is in contrast to a parallax with uncertainty, σ_π, when converted to distance uncertainty, σ_d, formally scales as $\sigma_d = d^2\sigma_\pi$. (Note, the parallax-distance uncertainties plotted in Fig. 2, seem to scale roughly as d^1, which may be partially explained by our selection criterion.)

Why are 3-D kinematic distances so precise at great distances? Firstly, it is worth noting that proper motion uncertainty, σ_{pm}, scales with time spanned, Δt, as $\sigma_{pm} \propto \Delta t^{-1} \times \Delta t^{-1/2}$ assuming uniform measurement sampling in time. The first term comes from fitting a straight line to position vs time data and the longer the time spanned the lower the uncertainty in the slope. The second term comes from $1/\sqrt{N}$ statistics for N

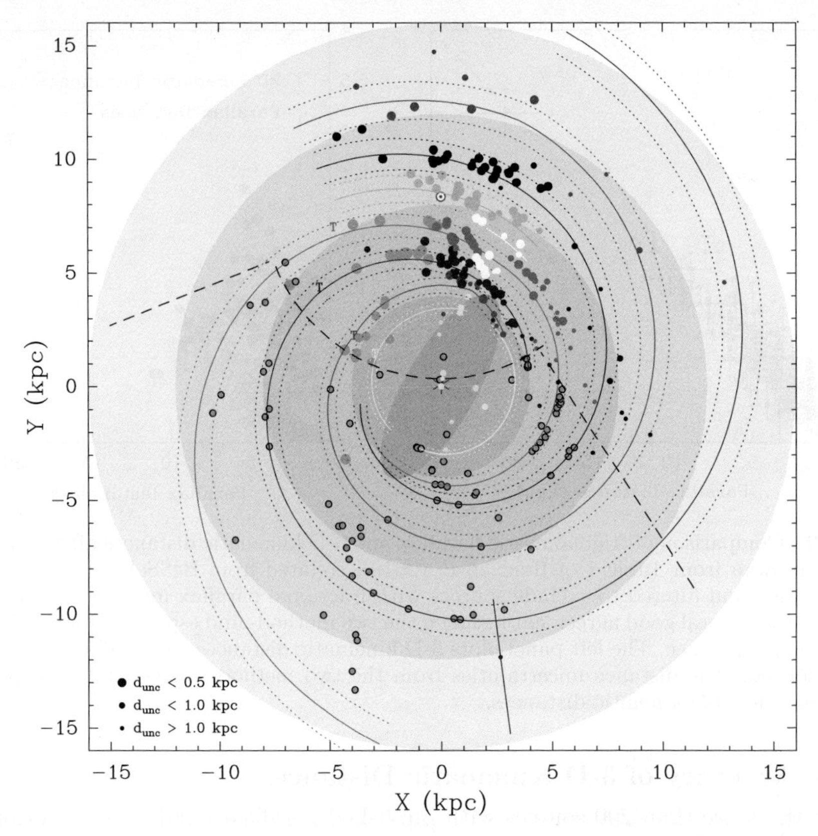

Figure 3. Plan view of the Milky Way showing high-mass stars with trigonometric parallaxes after Reid *et al.* (2019). Different colors indicate spiral arms/spurs from previous and on-going parallax measurements. Dashed black lines enclose candidate sources *(green with black circles)* for 3D kinematic distances. Note the locations of current and proposed sources are from 1D velocities and are quite uncertain.

measurements. Thus, proper motion uncertainty improves as $\Delta t^{-3/2}$, much faster than parallax uncertainty which scales as $\Delta t^{-1/2}$.

Secondly, the proper motion "signal" for a source on the far side of the Galaxy at the same distance from the Galactic center as the Sun (i.e., about 16 kpc from the Sun), has a (heliocentric) proper motion of twice the circular speed, $\Theta_0 \approx 236$ km s^{-1}, since the Sun and the source are moving in opposite directions. Thus, the proper motion "signal" is about $2 \times 236/16$ km s^{-1} kpc^{-1}, which is roughly 6 mas y^{-1}. Compared the parallax at 16 kpc of 0.06 mas, the proper motion is about 100-times larger and much more easily measured. Even if one admits a rather large ± 20 km s^{-1} uncertainty in either the speed of the source or the rotation curve of Galaxy, this is only a 4% effect. In conclusion, 3-D kinematic distances should be the method of choice for sources on the far side of the Milky Way, provided one avoids the Galactic bar region where very large non-circular motions are expected.

4. Masers in High-Mass Star Forming Regions

Fig. 3 presents a map of the Milky Way. Maser sources with parallax measurements are shown with dots, which are color coded by spiral arm mostly based on $l - V$ plots of CO emission. Green dots circled in black are methanol masers, placed using standard (inaccurate) kinematic distances, which are candidates for 3-D kinematic distances. Efforts are currently underway to measure proper motions to these methanol and other water masers (not shown) in order to better constain the spiral arm model of the Milky Way.

References

Reid, M. J., Menten, K. M., Burnthaler, A. *et al.* 2019, *ApJ*, 885, 131
Reid, M. J. 2022, *AJ*, 164, 133
Sanna, A., Reid, M. J., Dame, T. M., Menten, K. M. & Brunthaler, A. 2017, *Science*, 358, 227
Yamauchi, A., Yamashita, K., Honma, M. *et al.* 2016, *PASJ*, 68, 60

Cosmic Masers: Proper Motion toward the Next-Generation Large Projects
Proceedings IAU Symposium No. 380, 2024
T. Hirota, H. Imai, K. Menten, & Y. Pihlström, eds.
doi:10.1017/S1743921323002648

Estimating distances to AGB stars using IR data

Rajorshi Bhattacharya[1], Ylva M. Pihlström[1] and Loránt O. Sjouwerman[2]

[1]University of New Mexico, Albuquerque, NM, USA. email: rbhattacharya1995@unm.edu

[2]National Radio Astronomy Observatory, Socorro, NM, USA

Abstract. We present a method to estimate distances to AGB stars, utilizing the rich infrared data sets available for these infrared-bright targets. The method is based on the assumption that stars with intrinsically similar properties (metallicity, initial mass, etc.) produce similar spectral energy distributions (SEDs) and similar luminosities. We here discuss the results for AGB stars belonging to the BAaDE survey sample whose distances were calibrated using the template SEDs of stars with their VLBI parallaxes. As VLBI parallaxes are only known for a handful of sources, the resulting templates only cover a small subset of the BAaDE sample. Additional methods to derive suitable templates will therefore also be required. The work on expanding the template set is promising, although more fine tuning is still needed.

Keywords. Asymptotic giant branch stars, Milky Way, infrared, spectral energy distributions

---

1. Motivation

The Bulge Asymmetries and Dynamical Evolution (BAaDE) project has delivered line of sight velocities of more than 10,000 AGB stars in the Milky Way, primarily in the bulge region. To incorporate these velocities into dynamical models and to distinguish between different AGB populations, it is crucial to estimate the distances and 3D positions of these stars. Distance estimates would also enable luminosities and mass-loss rates to be evaluated. The vast majority of BAaDE AGB stars lack reliable Gaia parallaxes, hence we are exploring alternative methods to determine distances to AGB stars. The proposed method is advantageous as it builds on utilizing existing infrared catalogs, and can be used for AGB stars throughout the Galaxy.

2. Methodology

The initial assumption is that stars with similar intrinsic properties produce similar SEDs and are of similar luminosity. To test whether the spectral energy distributions (SED) of a template and a target star has similar characteristics, we use three different colors: 2MASS [J]-[K], MSX [A]-[D], and [K]-[A]. 2MASS J and K bands have wavelengths of 1.235 μm and 2.159 μm, and 8.28 μm and 14.65 μm in MSX A and D bands, respectively. After extinction correction, a distance estimate is extracted by scaling the individual target SED fluxes with a distance-calibrated template SED.

3. Comparison to VLBI and Gaia parallax data

To test the method, we selected 26 sources with known VLBI parallaxes Xu *et al.* (2019). A single VLBI source (R Cnc) was used as a template to calculate SED distances

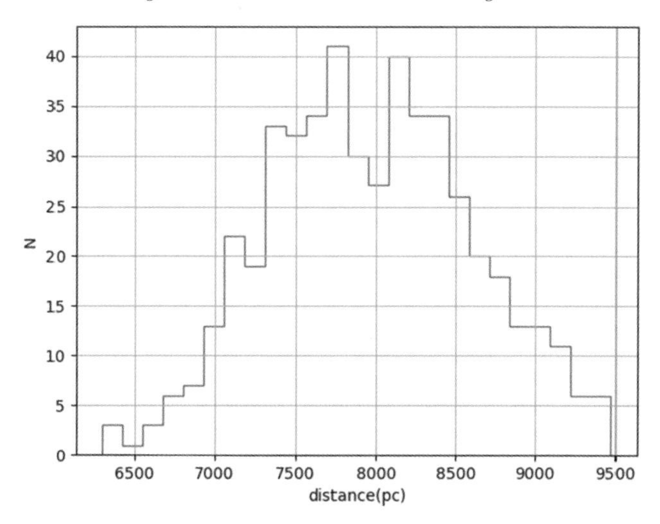

Figure 1. Histogram of obtained SED distances for the BAaDE sources with $|l| < 1°$ and $|b| < 1°$.

to the other VLBI sources. A good agreement was achieved, with all SED distance estimates ending up within 2σ of the VLBI parallax distances. In addition to R Cnc, several other sources were used as templates, consistently reproducing a close correlation between the VLBI and SED distance estimates.

As we obtained good agreement with VLBI parallax distances, we continued testing with the subset of BAaDE sources that had Gaia parallaxes with parallax errors $< 20\%$ ($\sim$ 1,000 objects). Initially this resulted in no obvious correlation. However, Andriantsaralaza *et al.* (2022) points out difficulties for Gaia in providing reliable parallaxes for the obscured AGB stars, even for sources with parallax errors $< 20\%$. They also point out that for brighter Gaia sources the errors are underestimated, and the relative parallax errors therefore must be corrected (for a source having $8 < G < 12$, the error gets underestimated by a factor of 2.5 whereas for $G < 8$, the error gets underestimated by a factor of 5). Following their recommendation resulted in a much smaller sample, but now a clear correlation between the Gaia parallax and SED distances was obtained.

4. Expanding template set with Galactic Center sources

As there are only a small set of AGBs with VLBI and/or reliable Gaia parallaxes, a larger set of templates must be defined to cover the full property range of the BAaDE sample. Sources with $|l| < 1°$ and $|b| < 1°$, and with absolute line-of-sight velocities > 100 km s^{-1} were selected to form a sample most likely to be located close to the Galactic Center, in the bulge region. We adopted a distance of 8.178 kpc (Abuter *et al.* 2019) for a single source used as the template, and then obtained distances to the other sources. The distribution of source distances can be seen in Figure 1, showing that the targets are all within the bulge region as expected.

Supplementary material

To view supplementary material for this article, please visit http://dx.doi.org/10.1017/S1743921323002648

References

Abuter, R., Amorim, A., Bauböck, M., Berger, J., Bonnet, H., Brandner, W., Clénet, Y., Du Foresto, V. C., De Zeeuw, P., Dexter, J., *et al.* 2019, *A&A*, 625, L10.

Andriantsaralaza, M., Ramstedt, S., Vlemmings, W. T., & De Beck, E. 2022, *arXiv preprint arXiv:2209.03906*,.

Xu, S., Zhang, B., Reid, M. J., Zheng, X., & Wang, G. 2019, *ApJ*, 875, 114.

Cosmic Masers: Proper Motion toward the Next-Generation Large Projects
Proceedings IAU Symposium No. 380, 2024
T. Hirota, H. Imai, K. Menten, & Y. Pihlström, eds.
doi:10.1017/S1743921323003010

Astrometry of Water Maser sources in the Outer Galaxy with VERA

Hiroyuki Nakanishi[1], **Nobuyuki Sakai**[2], **Kohei Kurahara**[3] **and VERA Outer Rotation Curve project members**

[1]Graduate School of Science and Engineering, Kagoshima University.
email: hnakanis@sci.kagoshima-u.ac.jp

[2]National Astronomical Research Institute of Thailand

[3]National Astronomical Observatory of Japan

Abstract. While the rotation curve of the inner Galactic disk is well determined, study of the outer rotation curve requires observational measurements of distances and proper motions of individual sources in the Outer Galaxy. We report astrometric observation for water maser sources in the Outer Galactic disk conducted with VERA, aiming to measure the Outer Rotation Curve. We have measured annual parallaxes and proper motions for these objects. Our result was consistent with recent other works based on astrometry and classical Cepheid observations. Epicyclic frequency seems to suggest that 2 and 4 spiral mode are dominant in the inner and outer Galaxy, respectively.

Keywords. VERA, Water Maser, Milky Way Galaxy, Rotation Curve

---

1. Introduction

Rotation curve is one of the essential parameters of the Galaxy. Inner rotation curve can be well determined by measuring the terminal velocities of the HI and CO data. However, outer rotation curve needs to be derived by measuring both distance and velocity of each object in the outer Galaxy. Period-Luminosity relation of Cepheids is one of the tools to measure the outer rotation curve (e.g., Mróz *et al.* 2019). While recently Gaia Collaboration *et al.* (2018) provided a velocity field of the Milky Way Galaxy using 3.2 million stars, there are some reports that Gaia results are sometimes contradictory to VLBI results (e.g., Matsuno *et al.* 2020). Hence, it is necessary to check the result independently with VLBI observational work.

2. VERA ORC Project

We aim to measure annual parallaxes and proper motions for H_2O maser (22 GHz) sources in the Outer Galaxy with VERA in order to study Outer Rotation Curve (ORC).

We selected 76 objects in the Galactic longitude and latitude ranges of $90° < l < 240°$, $-10° < b < 10°$, respectively. Through monitoring observations, we had carried out VLBI astrometric observations for 35 bright maser sources because a maser's flux generally varies.

Data reductions were conducted by applying phase-referencing technique to increase Signal-to-Noise ratio and to improve positional accuracy with a software AIPS.

Parallaxes and proper motions for 24 objects have been measured and a part of them have been published (Sakai *et al.* 2012; Nakanishi *et al.* 2015; Sakai *et al.* 2015; Koide *et al.*

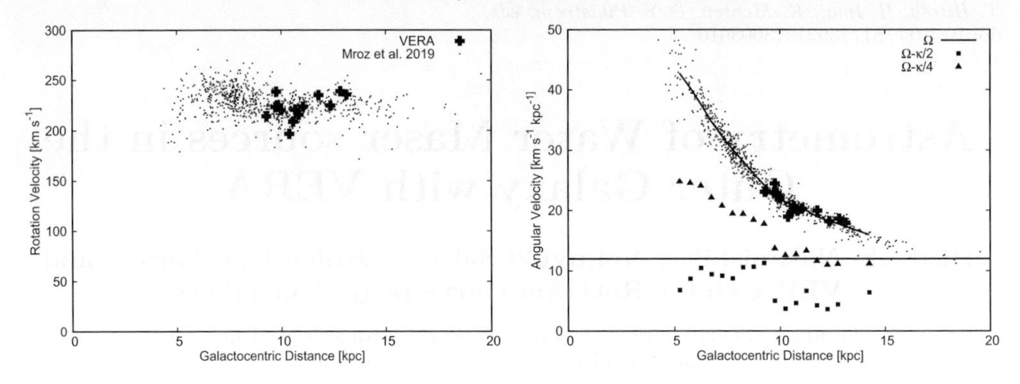

Figure 1. Rotation curve (left) and angular velocity (right) as functions of the Galactocentric distance.

2019; VERA Collaboration *et al.* 2020) so far. Details of data reductions are described in these references.

3. Result and Discussion

We have obtained parallaxes and proper motions and calculated three-dimensional position (R, θ, z) and velocity $(V_{\mathrm{R}}, V_\theta, V_z)$ for 24 objects. As shown in Figure 1, almost flat outer rotation curve was obtained and consistent with recent works such as Reid *et al.* (2019) and Mróz *et al.* (2019).

Angular velocity (Ω) and epicyclic frequency (κ) were calculated. Plotting $\Omega - \kappa/2$ and $\Omega - \kappa/4$ as shown in the right panel of Figure 1, we found that the former and latter are almost constant at about 10 km/s/kpc in the inner and outer Galaxy, respectively. These values of the former and latter are pattern speed of 2- and 4-arm spiral. If pattern speed of 12 km/s/kpc (Eilers *et al.* 2020), 2- and 4-arm spirals seem to stably exist in the inner and outer Galaxy, respectively. According to some former works on spiral mode such as Fuchs & Möllenhoff (1999) and Bottema (2003), it is suggested that number of mode increases with the Galactocentric distance. Our result is consistent with these former works.

The swing amplification theory (e.g., Fujii *et al.* 2011; Baba, *et al.* 2013) suggests that the dominating number of spiral arms, m, is estimated as $m \sim \kappa^2 R/(4\pi G\Sigma)$ using epicyclic frequency (κ), Galactocentrid distance (R), Gravitational constant (G), and surface mass density (Σ). If the disk is an exponential $(\Sigma \propto e^{-R/h})$, where h is scale length, the number of spiral number (m) tends to increase with the Galactocentric distance. This is also consistent with plots of $\Omega - \kappa/2$ and $\Omega - \kappa/4$.

Furthermore, recent study on the Galactic structure by Xu *et al.* (2023) suggests that the Milky Way Galaxy has 2-arm spiral in the inner region and 4-arm spiral in the outer region and seems to support our results.

References

Baba, J., Saitoh, T. R., & Wada, K. 2013, *ApJ*, 763, 46
Bottema, R. 2003, *MNRAS*, 344, 358
Eilers, A.-C., Hogg, D. W., Rix, H.-W., *et al.* 2020, *ApJ*, 900, 186
Fuchs, B. & Möllenhoff, C. 1999, *A&A*, 352, L36
Fujii, M. S., Baba, J., Saitoh, T. R., *et al.* 2011, *ApJ*, 730, 109
Gaia Collaboration, Katz, D., Antoja, T., *et al.* 2018, *A&A*, 616, A11
Koide, N., Nakanishi, H., Sakai, N., *et al.* 2019, *PASJ*, 71, 113
Matsuno, M., Nakagawa, A., Morita, A., *et al.* 2020, *PASJ*, 72, 56
Mróz, P., Udalski, A., Skowron, D. M., *et al.* 2019, *ApJL*, 870, L10

Nakanishi, H., Sakai, N., Kurayama, T., *et al.* 2015, *PASJ*, 67, 68
Reid, M. J., Menten, K. M., Brunthaler, A., *et al.* 2019, *ApJ*, 885, 131
Sakai, N., Honma, M., Nakanishi, H., *et al.* 2012, *PASJ*, 64, 108
Sakai, N., Nakanishi, H., Matsuo, M., *et al.* 2015, *PASJ*, 67, 69
VERA Collaboration, Hirota, T., Nagayama, T., *et al.* 2020, *PASJ*, 72, 50
Xu, Y., Hao, C. J., Liu, D. J., *et al.* 2023, *ApJ*, 947, 54

Cosmic Masers: Proper Motion toward the Next-Generation Large Projects
Proceedings IAU Symposium No. 380, 2024
T. Hirota, H. Imai, K. Menten, & Y. Pihlström, eds.
doi:10.1017/S1743921323003113

Astrometric observations of water maser sources toward the Galactic Center with VLBI

Daisuke Sakai[1], **Tomoaki Oyama**[1], **Hideyuki Kobayashi**[1,2] **and Mareki Honma**[1,3]

[1]National Astronomical Observatory of Japan, Oshu-shi, Iwate 023-0861, Japan.
email: daisuke.sakai@nao.ac.jp

[2]National Astronomical Observatory of Japan, Mitaka, Tokyo 181-8588, Japan

[3]Department of Astronomy, Graduate School of Science, The University of Tokyo, Tokyo, 181-8588, Japan

Abstract. The Central Molecular Zone (CMZ) in the Galactic Center region shows outstanding non-circular motion unlike the Galactic disk. While several models describing this non-circular motion have been proposed, a uniform kinematic model of the CMZ orbit has not yet emerged. To uncover the dynamics of the Galactic center region, we conducted VLBI astrometric observations of 22 GHz water maser sources towards the Galactic center using VERA. By measuring parallaxes and proper motions, we can determine whether each source is actually located in the CMZ or not, and identify the three-dimensional positions and velocities in the non-circular orbit if the source is indeed located in the CMZ. We present the results of our astrometric study for several maser sources associated with molecular clouds towards the Galactic center. The astrometric observations toward Sgr B2(M) indicated that Sgr B2 complex is moving toward the positive Galactic longitude relative to Sgr A*.

Keywords. Galaxy:kinematics and dynamics, masers, instrumentation:interferometers, parallaxes

---

1. Introduction

The Central Molecular Zone (CMZ) of our Galaxy, which ranges from −2 to +2 degrees of the Galactic longitude, has very peculiar properties. Star formation activity in this region is quite inactive despite high molecular gas density ($n = 10^3$–$10^5\ \mathrm{cm}^{-3}$) and wide line width ($\Delta V = 20$–$50\ \mathrm{km\,s}^{-1}$) (Morris & Serabyn 1996). Two representative models are proposed to explain these properties in the CMZ. First model is closed orbit model proposed by Molinari *et al.* (2011). In this mode, the orbital velocity is constant, and the star formation is triggered by cloud-cloud collision in the intersection between different orbits. Another model is open orbit model proposed by Kruijssen *et al.* (2015). In open orbit model, the tidal compression in the orbit triggers the star formation. Understanding the dynamical model of the CMZ is important to study not only whole property of the CMZ but also star-formation of individual molecular cloud in the CMZ. Three-dimensional velocity information will be critical for constraining the orbital models of the CMZ because most proposed models are devised to reproduce the line-of-sight velocity profiles of the molecular clouds in this region.

Table 1. Parallax and proper motions toward the Galactic center.

Target	Parallax (mas)	Distance (kpc)	$\mu_\alpha \cos\delta$ (mas yr^{-1})	μ_δ (mas yr^{-1})	$V_{\rm LSR}$ (km s^{-1})
Sgr B2(M)	0.133±0.0038	$7.5^{+3.0}_{-1.7}$	-2.17 ± 0.03	-2.63 ± 0.06	+60
G000.16-00.44	–	–	0.35 ± 0.52	-1.46 ± 0.57	+10
G359.94-00.14	–	–	-1.69 ± 0.26	-0.37 ± 0.30	−35
G000.21-00.00	–	–	-0.93 ± 0.89	0.35 ± 0.50	+47

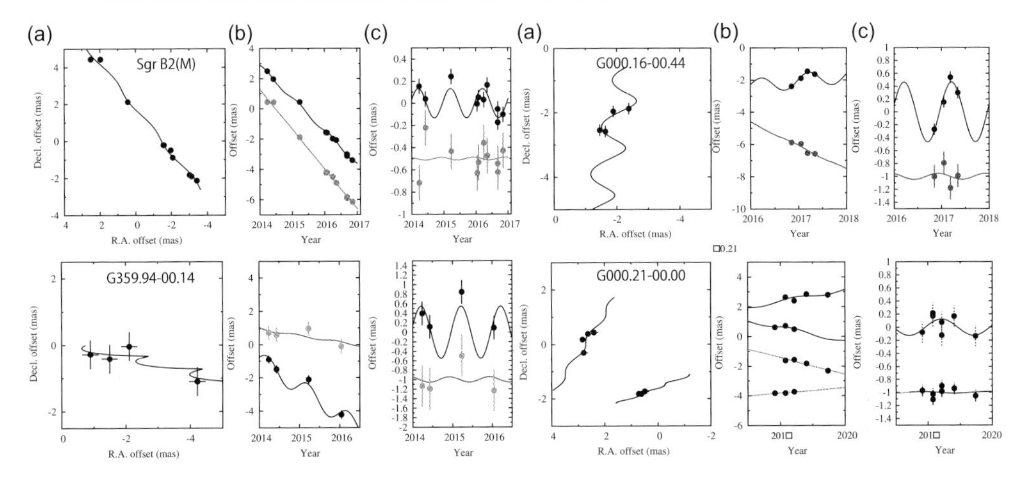

Figure 1. Astrometric results for four target sources. (a) Absolute proper motions of the maser source. The filled circles show the observed points from phase referencing. (b) Motions toward right ascension and declination as a function of time. The black circles show the motion in the right ascension direction, and the gray circles show the motion in the declination direction. (c) Results of parallax fitting. Error bars are evaluated so that a χ^2 value in the model fitting become unity.

2. Observations

We conducted VLBI monitoring observations of four 22 GHz H_2O maser sources using VERA (VLBI Exploration of Radio Astrometry). Using the dual-beam system of VERA, the target sources and a position reference source, i.e., J1745-2820, were simultaneously observed. Target sources and the number of observations for each source are listed in Table 1.

3. Results and Discussion

We succeeded to measure parallaxes and proper motions for Sgr B2(M), and to measure proper motions for G000.16-00.44, G359.94-00.14, and G000.21-00.00. Figure 1 shows astrometric results for each target source. From parallax and proper motion, Sgr B2(M) is located at the Galactic center distance, and moving toward positive Galactic longitude. G359.94-00.14 and G000.21-00.00 show proper motions moving toward positive Galactic longitude, suggesting that these sources are located at closer side of CMZ. Note that it is necessary to conduct a more detailed investigation, incorporating information such as annual parallax to conclude whether these sources reside at CMZ.

The results from proper motion measurements suggest a preference for the open orbit model among the dynamic models of the CMZ, yet a definitive distinction requires observations of a greater number of sources. In order to increase the number of sources available for astrometric observation, we are considering investigations of other types of masers, such as SiO masers, and observations with larger arrays, like EAVN (East Asian VLBI Network).

References

Kruijssen, J. M. D., Dale, J. E., & Longmore, S. N. 2015, *MNRAS*, 447, 1059
Molinari, S., Bally, J., Noriega-Crespo, A., *et al.* 2011, *ApJL*, 735, L33
Morris, M. & Serabyn, E. 1996, *ARAA*, 34, 645

Cosmic Masers: Proper Motion toward the Next-Generation Large Projects
Proceedings IAU Symposium No. 380, 2024
T. Hirota, H. Imai, K. Menten, & Y. Pihlström, eds.
doi:10.1017/S1743921323001977

Water Masers in the Galactic Center

Dylan Ward[1], Jürgen Ott[1,2] and David S. Meier[1,2]

[1]New Mexico Institute of Mining and Technology, Socorro, NM, USA.
email: dylan.ward@student.nmt.edu

[2]National Radio Astronomy Observatory, Socorro, NM, USA

Abstract. The Central Molecular Zone (CMZ) makes up roughly the inner 500 pc of the Milky Way and has a large amount of dense hot gas, strong magnetic fields, and highly energetic particles. The Survey of Water and Ammonia in the Galactic Center (SWAG) is a major imaging line survey using the Australia Telescope Compact Array with the goal to map out the molecular content in the entire CMZ. SWAG data includes the 22 GHz H_2O maser transition which is typically used as a tracer for phases of star formation, including both young stellar objects (YSOs) and evolved stars, such as asymptotic giant branch (AGB) stars. The SWAG H_2O survey is significantly deeper with better resolution than existing surveys that cover the entire CMZ. The goal is to create a robust catalog of the maser positions, spectral properties, and the sources they trace. The H_2O maser catalog shows 703 H_2O masers which increases the amount of detected H_2O masers in the CMZ by more than an order of magnitude. The H_2O masers have a more symmetric distribution in the Galactic center than that of the gas. Cross-correlation with other observations and catalogs will provide information relating maser properties to YSOs and AGB stars, for which multiple maser components will provide outflow properties. We will also connect the surrounding molecular gas to the YSO maser velocities.

Keywords. maser, Galaxy: center, stars

---

1. Introduction

The inner 500 pc of the Galactic center is known as the Central Molecular Zone (CMZ). The CMZ contains a large amount of dense molecular gas, shows an asymmetric distribution of gas, has higher temperatures and pressures compared to the disk, and is believed to undergo star formation in an episodic nature (Bryant 2021). Thus, the CMZ is a favorable region to study H_2O masers. The 22 GHz H_2O maser traces young and evolved stars and is unaffected by optical extinction, so it can be observed in the Galactic center. The observations were carried out with the Australia Telescope Compact Array in a mosaic of ~ 6500 pointings over ~ 600 hours spread over 3 years.

2. Results

The maser positions were gathered using the astrodendro python package as a clump finding tool with a minimum detection limit of 5σ, which is equivalent to ~ 0.45 K. The H_2O maser positions can be seen as orange circles plotted over an integrated intensity map of ammonia in Figure 1. For the first step of the data analysis, the luminosity of each maser was calculated using 8.178 kpc for the distance to Galactic center, determined by GRAVITY (2019). From this, Figures 2(a) and 2(b) were created. Figure 2(a) shows the maser positions plotted over the field of view where the larger the circles represent larger luminosities. Figure 2(b) is a histogram of the number of sources in a certain luminosity range, where the green line represents a maser with a mean 5σ luminosity above the noise. The amount of masers drops sharply at $L > 10^{-6} L_\odot$ which is similar

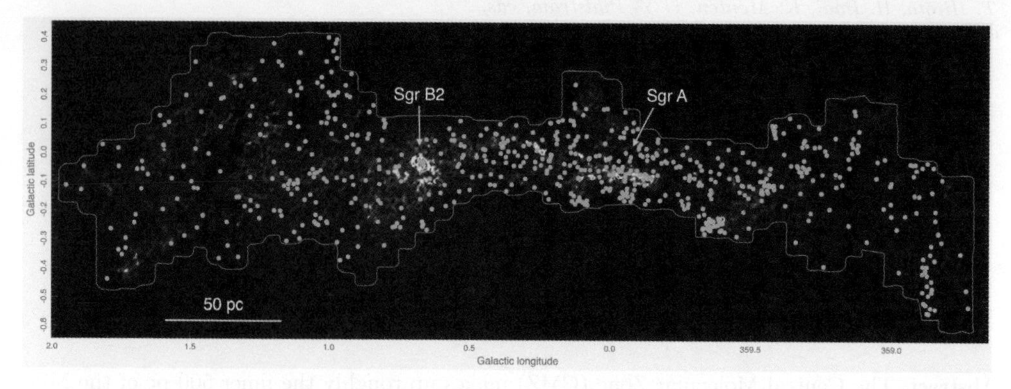

Figure 1. The orange outline represents the survey's field of view. H_2O maser locations are represented as orange circles plotted over an integrated intensity NH_3 (3,3) map (in blue).

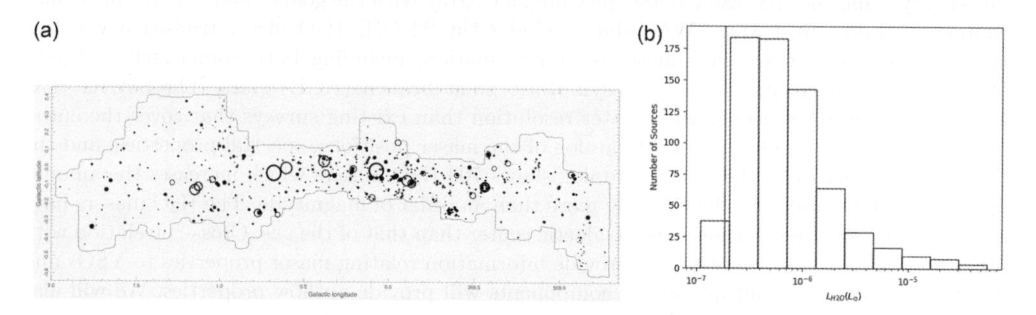

Figure 2. a) Shows maser positions vs luminosity. Larger circles represent higher luminosities. b) Histogram of luminosity of the masers. The green line represents a maser with a mean 5σ luminosity.

to H_2O luminosity for evolved stars from Palagi (1993). However, if the sources are in the disk, then the luminosities will be off by a factor of 4. This analysis was done before source identification has been completed.

3. Future Work

Once the sources have been identified by cross-correlation with other catalogs, the luminosity histogram will be split into 2 separate histograms: one for young stars and one for evolved stars; this will provide a much better statistical study on H_2O masers in the Galactic center than Palagi (1993). After the luminosity analysis has been completed, a comparison to the spectral properties of CH_3OH and SiO masers from other catalogs at the same location of the H_2O masers will be done.

Acknowledgements

Support for this work was provided by the NSF through the Grote Reber Fellowship Program administered by Associated Universities, Inc./National Radio Astronomy Observatory. This research made use of astrodendro, a Python package to compute dendrograms of Astronomical data (www.dendrograms.org/)

Supplementary material

To view supplementary material for this article, please visit http://dx.doi.org/10.1017/S1743921323001977

References

Bryant, Aaron and Krabbe, Alfred 2021, *New Astron. Revs*, 93, 101630
GRAVITY Collaboration, *et al.* 2019, *A&A*, 625, 10
Palagi, F., *et al.* 1993, *A&A*, 101, 153

Cosmic Masers: Proper Motion toward the Next-Generation Large Projects
Proceedings IAU Symposium No. 380, 2024
T. Hirota, H. Imai, K. Menten, & Y. Pihlström, eds.
doi:10.1017/S1743921323002387

Searching masers from the Sagittarius stellar stream

Yuanwei Wu[1], Bo Zhang[2], Yan Gong[3], Wenjin Yang[3] and Nicolas Mauron[4]

[1]National Time Service Center, Chinese Academy of Sciences, Xi'an 710600, China.
email: yuanwei.wu@ntsc.ac.cn

[2]Shanghai Astronomical Observatory, Chinese Adademy of Sciences, Shanghai 200030, China

[3]Max-Plank-Institut für Radioastronomie, auf dem Hügel 69, 53121 Bonn, Germany

[4]Université de Montpellier, Laboratoire Univers et Particules de Montpellier CNRS/UM, Place Batailon, 34095 Montpellier, France

Abstract. Large scale optical and infrared surveys have revealed numbers of accretion-derived stellar features within the halo of the Galaxy. These coherent tail-like features are produced by encounters with satellite dwarf galaxies. We conducted an SiO and H_2O maser survey towards O-rich AGBs towards the orbital plane of the Sgr Stellar Stream from 2016. Up to now, maser emissions have been found from 60 sources, most of which are detected for the first time. However, their distances and kinematics suggest they are still disk stars.

Keywords. Masers, stars: AGB and post-AGB, Galaxy: structure

1. Motivation

It is well known that our milky Way is a barred spiral galaxy. Within the past decade, the spiral strcuture and kinematics of the Milky Way have been well studied by measuring parallaxes and proper motions of interstellar masers (Reid et al. 2019). Apart from spiral features in the disk, giant stellar features also exists in the halo of the Milky Way.

According to the hierarchical galaxy formation theory, the Milky Way was assembled through the accretion of smaller systems (Blumenthal *et al.* 1984). The interaction between the Milky Way and its satellite galaxies can produce large-scale tidal streams in the halo region with galactocentric distances ranging from ~10 kpc to more than 100 kpc (Lynden-Bell & Lynden-Bell 1995). During the last two decades, various optical and infrared surveys, as well as recent Gaia data, have identified more than 90 stellar streams in the Milky way (Mateu 2023).

Once maser emissions were found in a stellar stream, it will be possible to use masers to study kinematics of the stream. Therefore, we launched a maser survey towards the Sagittarius stellar stream (hereafter Sgr stream), the most prominent and well-studied stream in the Milky way halo.

2. Sample and Survey Results

Up to now, we have searched around 400 sources selected from two samples. The 1st sample was selected from the WISE all-sky point source catalog by their infrared color and magnitude. Details of this WISE selected sample can be found in Wu *et al.* (2018). The 2nd sample was selected from Mauron *et al.* (2019)'s O-rich AGB star catalog. Figure 1 shows the infrared color-color and color-magnitude diagram of these two samples.

Table 1. Summary of Sgr stream Maser Survey

Telescope	Maser	Frequency	Observed	Detected	Obs. Dates	Ref.
Nobeyama 45m	H_2O	22.235 GHz	49	42	May 2016	Wu *et al.* (2022)
Tidbinbilla 70m	H_2O	22.235 GHz	127	7	Nov. 2016, Mar. 2017	Wu *et al.* (2022)
Nobeyama 45m	SiO	42.820, 43.122 GHz	221	44	Apr., May 2016	Wu *et al.* (2018)
Effelsberg 100m	SiO	42.820, 43.122 GHz	52	8	Sep. 2022 ~ Feb. 2023	Yang *et al.* (2023)
Tianma 65m	SiO	42.820, 43.122 GHz	50	0	Jan. 2023	Yang *et al.* (2023)

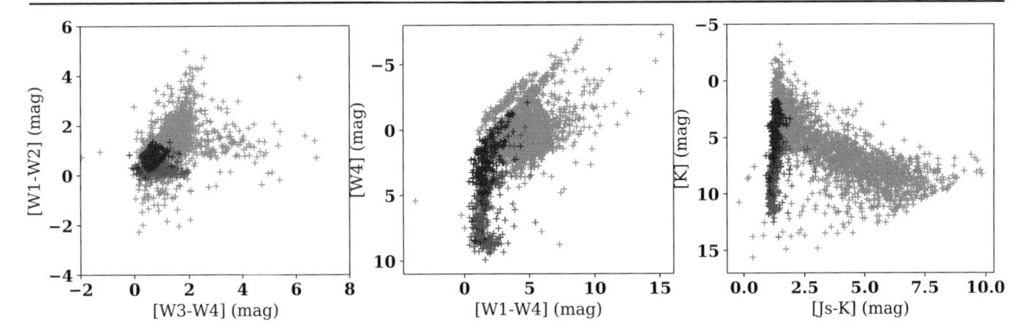

Figure 1. Infrared color-color and color-magnitude diagrams of O-rich AGB stars.The grey dots show the Galactic disk O-rich AGBs associated with SiO masers. The blue plus signs are the 1st sample selected from WISE all-sky catalog (Wu *et al.* 2018). The red plus signs are the 2nd sample selected from Mauron *et al.* (2019)'s O-rich stars in the Galactic halo and Sgr stream. The grey plus signs are disk SiO masers.

In Table 1, we summarize the survey results currently finished. In year 2016 and 2017, we searched SiO masers towards 221 sources and H_2O masers towards 176 sources. In the survey, 52 sources are found to be associated with at least one transitions of SiO or H_2O masers, including 43 SiO maser emissions and 21 H_2O maser emissions (Wu *et al.* 2018, 2022). In year 2022 and 2023, we searched SiO masers towards 102 sources selected from Mauron *et al.* (2019), and found maser emissions from 8 sources (Yang *et al.* 2023).

3. Discussions

We investigate the Galactic location and kinematics of these sources. All these sources are within 8 kpc of the Sun, including the newest 8 detections found by Effelsberg 100m (Yang *et al.* 2023). The 3D velocities of these sources are more aligned with the Milky Way plane rather than the Sgr stellar stream orbital plane, suggesting these sources are actually still disk stars rather than Sgr stream debris (Wu *et al.* 2022; Yang *et al.* 2023). The faintest SiO maser has a flux density of 0.16 Jy, which is detected above 5σ at a noise level of 0.02 Jy. Frankly speaking, detecting masers beyond 10 or even 20 kpc towards the Sun can be very challenging task, but still possible.

Kinematically, these maser traced O-rich stars are nearly all move away from the Galactic center, which is also founded for solar neighbourhood Miras (within 2 kpc) (Feast & Whitelock 2000). On the other hand, the lag ($\sim$100 km s^{-1}) of rotational speed of nearby Miras reported by (Feast & Whitelock 2000) is not seen for the maser-traced AGB stars found in this survey.

References

Blumenthal, G. R., Faber, S. M., Primack, J. R., & Rees, M. J. 1984, *Nature*, 311, 517
Feast, M. W., & Whitelock, P. A. 2000, *MNRAS*, 317, 460

Lynden-Bell, D., & Lynden-Bell, R. M. 1995, *MNRAS*, 275, 429
Mateu, C. 2023, *MNRAS*, 520, 5225
Mauron, N., Maurin, L. P. A., & Kendall, T. R. 2019, *A&A*, 626, A112
Reid, M. J., Menten, K. M., Brunthaler, A., *et al.* 2019, *ApJ*, 885, 131
Wu, Y., Zhang, B., Li, J., & Zheng, X.-W. 2022, *MNRAS*, 516, 1881
Wu, Y. W., Matsunaga, N., Burns, R. A., & Zhang, B. 2018, *MNRAS*, 473, 3325
Yang, W. J. Wu, Y. W., Gong, Y., Zhang, B., & Mauron, N. 2023, in preparation

Review talk in the session Structure of the Milky Way by Kazi L. J. Rygl. Taken by Ka-Yiu Shum.

Chapter 4
Dynamics of Formation of Massive Stars

Cosmic Masers: Proper Motion toward the Next-Generation Large Projects
Proceedings IAU Symposium No. 380, 2024
T. Hirota, H. Imai, K. Menten, & Y. Pihlström, eds.
doi:10.1017/S1743921323002326

Evolutionary Trends in Star Formation

J. S. Urquhart

Centre for Astrophysics and Planetary Science, University of Kent, Canterbury, CT2 7NH, UK. email: j.s.urquhart@kent.ac.uk

Abstract. Over the past 20 years, the Galactic plane has been surveyed at high resolution at wavelengths from 1 micron through to 20 cm. The combination of these surveys has produced large samples of deeply embedded young stars located across the Galactic disc. These continuum surveys are complemented by spectral line surveys of thermal, radio recombination, and molecular maser (OH, H_2O, CH_3OH) lines. The identified sources cover the whole range of evolutionary stages in the star formation process, allowing the physical properties of these stages to be measured. This information has been used to calculate the star formation efficiency and star formation rate of the Milky Way and to evaluate the impact of environment and location within the disc. This review provides an overview of some of the most significant studies in recent years and discusses how the evolutionary sequence has been used to investigate the correlation of other star formation tracers and maser associations.

Keywords. Masers, star: formation, ISM: molecules, ISM: surveys

---

1. Introduction

Massive stars ($> 8\,M_\odot$ and $10^3\,L_\odot$) play an important role in many astrophysical processes and in shaping the morphological, dynamical and chemical structure of their host galaxies (Kennicutt & Evans 2012). These stars have a profound impact on their local environment through powerful outflows, strong stellar winds and copious amounts of optical/far-UV radiation, which shape the interstellar medium (ISM) and regulate star formation, and ultimately governs the evolution of their host galaxy (McKee & Ostriker 2007). During their lives they reprocess huge amounts of material and are responsible for most of the heavy elements in the universe, which are returned to the ISM through stellar winds, and at the ends of their lives, in supernovae explosions. Furthermore, emission from massive stars dominates the light seen from distant galaxies, and therefore, galaxy models are critically dependent on various assumptions that are made about high-mass star formation, such as the star formation rate (SFR; Davies *et al.* 2011) or the universality of the initial mass function (IMF; Kroupa & Weidner 2003), if we are to understand how they evolved over cosmological timescales. Understanding the formation and early evolution of massive stars is therefore a fundamental goal for modern astrophysics.

Studies of external galaxies are generally restricted to studies of global star formation properties integrated over entire complexes (resolution ~90 pc; Rosolowsky *et al.* 2021) or even entire galaxies (Gao & Solomon 2004). Studies of star formation within our own Galaxy, however, are able to probe star forming regions in far greater detail and the large range of environments available have extragalactic analogues (Kruijssen & Longmore 2013), e.g., the Galactic centre with its extreme UV-radiation and cosmic ray fluxes, intense star formation regions found in the disc (e.g., W43 and W51; often described as "mini-starbursts") and low-metallicity environment found in the outer Galaxy. The Milky Way, therefore, provides our best opportunity to understand the processes involved in massive star formation in both Galactic and extragalactic environments.

Despite their importance our understanding of the initial conditions required, and processes involved, in the formation and early evolution of massive stars, is still rather poor. There are a number of reasons for this: massive stars are rare and relatively few are located closer than a few kpc from the Sun; they form almost exclusively in clusters, making it hard to distinguish between the properties of the cluster and individual members; and they evolve rapidly, reaching the main sequence while still deeply embedded in their natal environment, and consequently, the earliest stages can only be probed at far-infrared and (sub)millimetre wavelengths (see Motte *et al.* 2018 for a review). To overcome these difficulties requires a combination of unbiased multi-wavelength Galactic plane surveys to identify large samples of embedded massive stars and high resolution to study their properties in detail.

Our ability to make significant progress in this field has been dramatically enhanced in recent years with the completion of a large number of Galactic plane surveys that cover the whole wavelength range from the near-infrared to the radio, e.g., UKIDSS (Lucas *et al.* 2008), GLIMPSE (Churchwell *et al.* 2009), Hi-GAL (Molinari *et al.* 2010), ATLASGAL (Schuller *et al.* 2009) and CORNISH (Hoare *et al.* 2012) (see Table 1 for a more comprehensive list). These unbiased surveys provide the large spatial volumes required to address the major problem in studying Galactic massive star formation, its intrinsic rarity. These continuum surveys are complemented by a number of unbiased and targeted spectral line surveys that can constrain the macroscopic properties of star forming environments, e.g., COHRS (Dempsey *et al.* 2013), SEDIGISM (Schuller *et al.* 2021) and MALT90 (Jackson *et al.* 2013).

Combined, these surveys provide a global view of massive star formation and enable the identification and characterisation of statistically significant samples over the full range of evolutionary stages from pre-stellar through to the post-compact HII region stage when the star emerges from its natal clump (see Fig. 1). The availability of these surveys and upgrades of facilities like the Jansky Very Large Array presents an exciting opportunity to make significant progress in our understanding of the formation and evolution of massive stars and their role in driving galaxy evolution.

In this review we will look at the progress that has been made in identifying large samples of embedded high-mass star forming environments over the past 20 years and give an overview of the current state of the art.

2. The Evolutionary Sequence

Most stars form in high-mass star-forming clusters found within the densest parts of giant molecular clouds (Williams & McKee 1997); these are often referred to as clumps. These have volume densities of 10^{4-5} cm^{-3}, sizes of $\sim$1 pc and masses of 500–1000 $M_\odot$ (Urquhart *et al.* 2014b). These dense clumps are initially in a quiescent state (starless) but are gravitationally bound and continue to accrete material from their surroundings that flows in along filamentary networks. These clumps become gravitationally unstable and begin to collapse, resulting in the formation of a protostar, which will become visible first at far-infrared wavelengths (e.g., 70 μm), then at mid-infrared and near-infrared wavelengths as it accretes more material and becomes hotter, resulting in the peak of its spectral energy distribution (SED) shifting to shorter wavelengths. This mid-infrared stage is often referred to as the massive young stellar object (MYSO) stage and precedes the stars arriving on the main sequence stars, which is indicated by the formation of an HII region.

Figure 1 presents a schematic that shows the main evolutionary stages and the wavelength ranges that the various stages are detectable. We use this to identify 4 observationally distinct evolutionary stages: 1) the quiescent stage is far-infrared quiet and

Table 1. Summary of Galactic plane surveys.

Survey	Wavelength	Beam ($''$)	ℓ Coverage ($^\circ$)	b Coverage ($^\circ$)	Probe	Reference
IPHAS	Hα	1.7	$30^\circ < l < 210^\circ$	$\|b\| < 5^\circ$	Nebulae & stars	Drew *et al.* (2005)
UKIDSS	JHK	0.8	$-2^\circ < l < 230^\circ$	$\|b\| < 1^\circ$	Stars, Nebulae	Lucas *et al.* (2008)
VVV	ZYJHK	0.8	$-65^\circ < l < 10^\circ$	$\|b\| < 2^\circ$	Nebulae Nebulae	Minniti *et al.* (2010)
GLIMPSE	4-8	2	$-65^\circ < l < 65^\circ$	$\|b\| < 1^\circ$	Stars, Hot Dust	Churchwell *et al.* (2009)
MSX	8-21	18	All	$\|b\| < 5^\circ$	Warm Dust	Price *et al.* (2001)
MIPSGAL	24,70	6, 20	$-65^\circ < l < 65^\circ$	$\|b\| < 1^\circ$	Warm Dust	Carey *et al.* (2009)
AKARI	50-200	30-50	All sky		Cool Dust	White *et al.* (2009)
Hi-GAL	70-500	10-34	All	$\|b\| < 1^\circ$a	Cool Dust	Molinari *et al.* (2010)
JPS	450,850	8-14	$10^\circ < l < 60^\circ$	$\|b\| < 1^\circ$	Cool Dust	Moore *et al.* (2015)
ATLASGAL	850	19	$-60^\circ < l < 60^\circ$	$\|b\| < 1.5^\circ$	Cool Dust	Schuller *et al.* (2009)
BOLOCAM	1100	33	$-10^\circ < l < 90^\circ$	$\|b\| < 0.5^\circ$	Cool Dust	Aguirre *et al.* (2011)
GRS	^{13}CO 1-0	46	$18^\circ < l < 56^\circ$	$\|b\| < 1^\circ$	Molecular Gas	Jackson *et al.* (2006)
MMB	6.7 GHz	192b	$-180^\circ < l < 60^\circ$	$\|b\| < 2^\circ$	Methanol Masers	Green (2009)
HOPS	22 GHz	132b	$-180^\circ < l < 60^\circ$	$\|b\| < 2^\circ$	Water Masers	Walsh *et al.* (2011)
OH	1.6 GHz	∼10	$-45^\circ < l < 45^\circ$	$\|b\| < 3^\circ$	Hydroxyl Masers	Sevenster *et al.* (2001)
CORNISH -North	6 cm	1.5	$10^\circ < l < 65^\circ$	$\|b\| < 1^\circ$	Compact Ionized Gas	Hoare *et al.* (2012)
CORNISH -South	6 cm	3	$300^\circ < l < 350^\circ$	$\|b\| < 1^\circ$	Ionized Ionized Gas	Irabor *et al.* (2023)
GLOSTAR	6 cm	1.5-20	$-2^\circ < l < 60^\circ$	$\|b\| < 1^\circ$	Compact Ionized Gas	Brunthaler *et al.* (2021)
THOR	20 cm	20	$14.4^\circ < l < 67.4^\circ$	$\|b\| < 1.25^\circ$	Atomic and Ionized Gas	Beuther *et al.* (2016)
S/V/CGPS	21 cm	60	$-107^\circ < l < 147^\circ$	$\|b\| < 1.3^\circ$	Gas Gas	Stil *et al.* (2006)
MAGPIS	20 cm	5	$5^\circ < l < 48^\circ$	$\|b\| < 0.8^\circ$	Diffuse Ionized Gas	Helfand *et al.* (2006)
MGPS-2	35 cm	45	$-115^\circ < l < 0^\circ$	$\|b\| < 10^\circ$	Diffuse Ionized Gas	Murphy *et al.* (2007)

Notes: This table is a modified version of the table 1 of Hoare *et al.* 2012.

so precedes the formation of any protostellar objects; 2) the protostellar stage is far-infrared bright, indicating the formation of a protostellar object, but that is not yet evolved enough to be mid-infrared bright; 3) the MYSO stage, where the embedded protostellar object is luminous ($> 10^3\,L_\odot$; Wynn-Williams 1982) and hot enough to be detected at mid-infrared wavelengths but has not yet started to form a detectable H II region; and finally 4) the H II region stage when the star has joined the main sequence and begun to ionize its surrounding natal gas and can now be detected at radio wavelengths (See Figure 2 for spectral energy distributions of the four evolutionary stages).

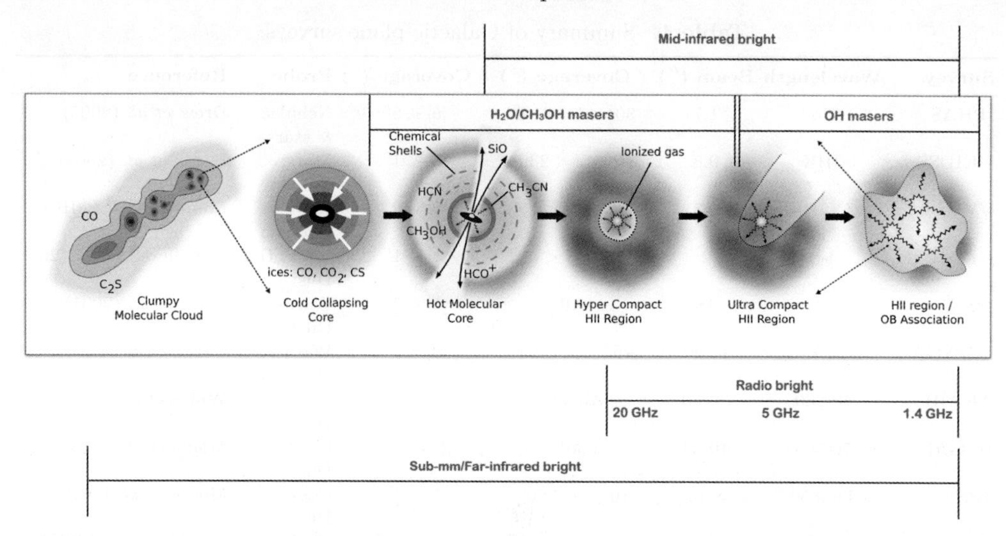

Figure 1. Schematic for the evolutionary sequence of massive star formation. Infrared and radio surveys are sensitive to the more evolved stages shown to the right while submillimetre surveys are sensitive to all evolutionary stages. The greyscale indicates the distribution of cold gas and dust with the darker areas corresponding to denser regions. The protostellar objects are indicated by the yellow circles, warmer gas by the orange colouring and the hotter ionized gas is shown in red. Molecular lines used to trace the properties of the molecular gas in different stages are also noted. The red labels and associated horizontal lines show the wavelengths and time ranges that different stages are observable, and the green labels indicate the maser associations. Image Credit: Cormac Purcell.

3. Legacy of the IRAS

Given that high-mass star formation occurs deep inside dense clumps, behind hundreds of magnitudes of visual extinction, all of the earliest stages are inaccessible from ground based telescopes. Progress in this area only became possible with the launch of the Infrared Astronomical Satellite (IRAS) in 1983. IRAS surveyed 96% of the sky in 4 mid- and far-infrared photometric bands (12, 25, 60 and 100 μm) detecting more than 250,000 sources (IRAS Point Source Catalogue or PSC Version 2; Beichman *et al.* 1988) that could be used to develop colour selection criteria to identify embedded high-mass young stellar objects (e.g., Campbell *et al.* 1989; Chan *et al.* 1996). These colour selected samples are not able to distinguish between MYSO and the more evolved HII region stages, however, the latter stage is bright at radio wavelengths. Work by Wood & Churchwell (1989) and Kurtz *et al.* (1994) with the VLA was able to separate these two stages, resulting in the identification of ∼100 ultracompact (UC) HII regions and allowing their physical properties to be characterised.

Molinari *et al.* (1996) applied colour cuts to the IRAS PSC and identified 260 objects with properties consistent with being embedded high-mass protostars. Approximately half of which have colours consistent with being UC HII regions (according to the work of Wood & Churchwell 1989); they refer to these as the 'High' group with the remaining sources referred to as the 'Low' group, which have been the starting point for many follow-up studies (e.g., Wu *et al.* 2006). Sridharan *et al.* (2002) used data from a higher spatial resolution imaging of the original IRAS data (HIRES; Aumann *et al.* 1990) to identify a sample of high-mass protostellar objects and used molecular line and radio continuum data to characterise the sample (Beuther *et al.* 2002) and to exclude more evolved HII regions.

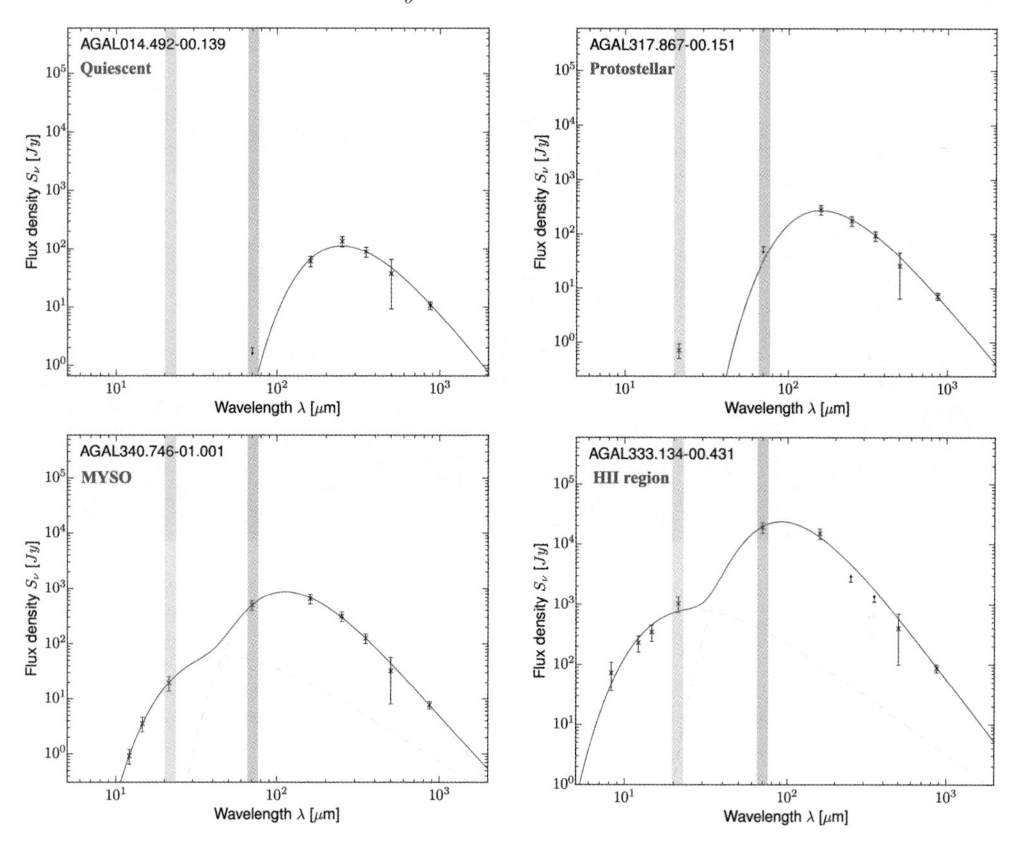

Figure 2. Examples of spectral energy distributions (SEDs) from the four evolutionary stages described in the text. These are taken from the ATLASGAL survey and have been adapted from figure 7 of König *et al.* (2017). The source names and evolutionary stage are given in the upper left corner of each plot. The mid-infrared flux values have been drawn from the MSX, WISE or MPISGAL surveys while the far-infrared and submillimetre fluxes (70-500 μm) have been drawn from HiGAL and the 870 μm comes from ATLASGAL. The blue curve shows the results of greybody fits to the photometric points (a single component for the quiescent and protostellar clumps and two components to the MYSO and HII region associated clumps). The red and green vertical lines indicate wavelengths that are used to distinguish between quiescent and protostellar clumps, and protostellar and more evolved clumps (i.e., those hosting MYSOs and HII regions). Image Credit: Carsten König.

These early studies were limited by the IRAS angular resolution ($\sim 2'$), so focus on bright isolated sources and tended to be biased away from intense star formation regions and from the Galactic mid-plane where source confusion is high. In the next section we will discuss a survey that took advantage of the higher-resolution mid-infrared photometry provided by the Midcourse Space Experiment (MSX) to produce a large and well-selected catalogue of MYSOs and UC HII regions.

4. The Red MSX Source (RMS) Survey

The Red MSX Source (RMS) survey is a systematic search of the entire Galaxy mid-plane for MYSOs and HII regions. This survey makes use of the higher resolution provided by the MSX satellite (Price *et al.* 2001) to conduct a large comprehensive search for young embedded high-mass stars. This reduces the biases affecting the early studies based on IRAS data.

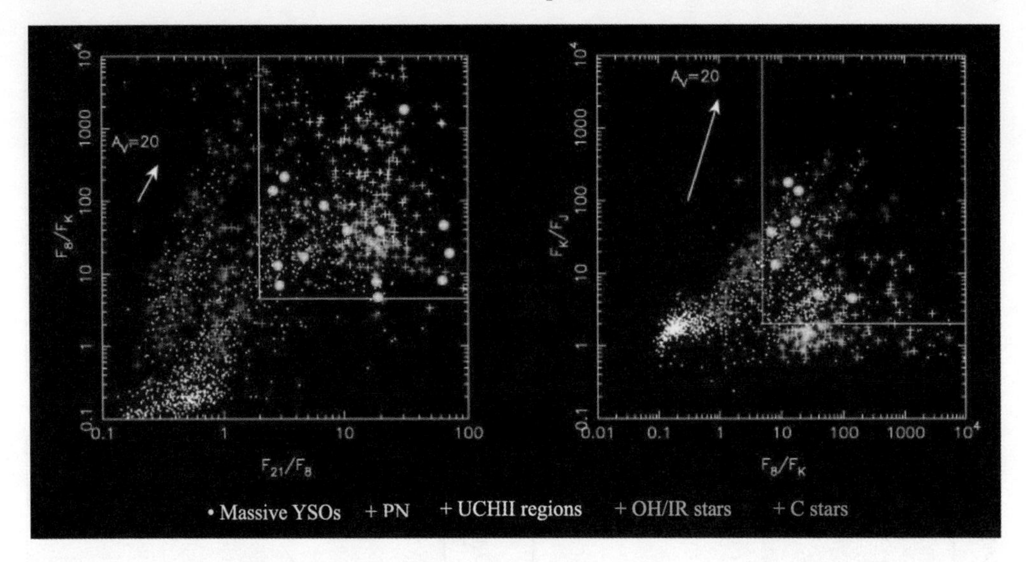

Figure 3. Colour-colour plots showing the distribution of different kinds of objects encountered in the search for embedded high-mass protostellar objects. The colours indicate the ratios of the fluxes at different wavelengths where 8 and 21 refer to the MSX fluxes at 8 and 21 μm and the J, H, K refer to the 2MASS magnitudes. The regions enclosed by the white boxes are the regions of the colour-colour space where the previously identified MYSOs are located and the selection criteria used for identifying the RMS sample of MYSO candidates. Image Credit: Stuart Lumsden.

The MSX was launched in 1996 and surveyed the entire Galactic plane ($|b| < 5°$) in four mid-infrared spectral bands between 6 and 25 μm at a spatial resolution of ~18″ (Price *et al.* 2001). The resulting point source catalogue contains 440,483 sources (Egan *et al.* 2003). The RMS team started by selecting MYSOs that have similar mid-infrared colours as embedded high-mass protostars identified in the previous IRAS studies. To this they added near-infrared photometry from the 2MASS survey (Skrutskie *et al.* 2006) to exclude more evolved sources. These colour cuts and the visual inspection of the images to eliminate extended sources resulted in the identification of ~2000 MYSO candidates (Lumsden *et al.* 2002).

Figure 3 shows the colour-colour plots used to identify the MYSO candidates by the RMS team. These plots demonstrate how effective these colour criteria are at excluding evolved stars from the sample, however, they are not able to remove the contaminating sources completely or to differentiate between MYSOs or H II regions. To accomplish this required a comprehensive multi-wavelength campaign of follow-up observations including near-infrared spectroscope (Clarke *et al.* 2006; Wheelwright *et al.* 2010; Ilee *et al.* 2013; Cooper *et al.* 2013) mid-infrared continuum (Mottram *et al.* 2007), molecular line (Urquhart *et al.* 2007b, 2008, 2011), and radio continuum (Urquhart *et al.* 2007a, 2009).

These efforts have resulted in the production of a sample of approximately 600 YSOs, 110 of which have luminosities above 20,000 $L_\odot$ and are therefore considered genuine MYSOs, and a similar number of UC H II regions (Lumsden *et al.* 2013; Urquhart *et al.* 2014a). This is the largest and most complete sample of MYSOs and UC H II regions produced to date.

5. Radio Continuum Surveys

Radio continuum surveys are an indispensable part of defining an evolutionary sequence for high-mass star formation. Arguably, this is not a formation stage given that

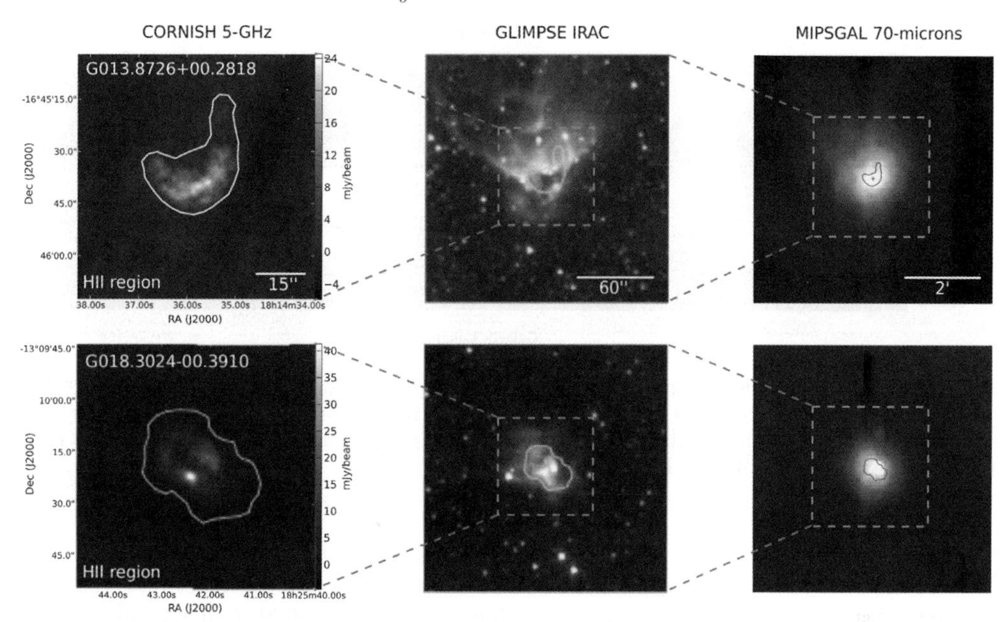

Figure 4. Examples of two UC HII regions identified in the CONRISH-North survey. This has been adapted from figure 22 from Purcell *et al.* (2013). The left panels show the radio emission and the green polygon the area over which the flux is integrated. The middle panels are three colour composite images showing the mid-infrared environment and the right panels show the far-infrared emission from MIPSGAL. Image Credit: Cormac Purcell.

accretion has been terminated and the star has arrived on the main sequence. However, the detection of the radio emission from the ionized gas surrounding the newly formed OB star allow MYSOs and more evolved UC HII regions to be separated, which is not possible from the infrared emission alone. For this reason, it is worth reviewing the contribution of radio surveys have made to the development of the evolutionary sequence.

There are several HII region stages that have been identified. The hypercompact (HC) HII regions are the most compact with diameter < 0.03 pc and densities $n_e > 10^6$ cm^{-3} (Kurtz & Hofner 2005) and are optically thick below 10 GHz (where most radio surveys have been conducted). These expand into the ultracompact (UC) HII region stage defined as having diameters of < 0.1 pc and densities $n_e > 10^4$ cm^{-3} (Wood & Churchwell 1989). The expansion of the HII region continues and the electron density continues to decrease through the compact HII region stage ($n_e > 10^4$ cm^{-3} and diameter > 0.1 pc; Wood & Churchwell 1989) until it breaks out of its natal cloud and becomes optically visible, at which point it is described as a classical HII region. All of the HII region stages are unlikely to be distinct stages but rather observationally determined size scales on a continuum (Hoare *et al.* 2007).

The HII region most commonly associated with the early stages of high-mass star formation is the UC HII region stage. At this point in the HII regions evolution, its electron density has dropped to a point that the nebula has become optically thin at 5 GHz and this frequency is where most radio surveys have been targeted (e.g., CORNISH. Hoare *et al.* 2012; GLOSTAR. Brunthaler *et al.* 2021). The HC HII region stage is the earliest manifestation of a HII region, however, this is thought to be very short-lived (only 21 have been detected; Yang *et al.* 2021) and the higher frequency (>10-20 GHz) and small beam makes the large systematic survey needed to search for these unfeasible with current telescopes.

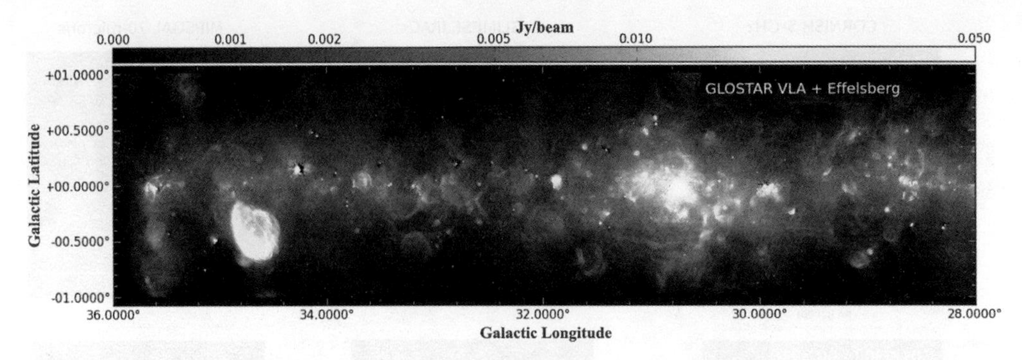

Figure 5. Radio continuum image of the GLOSTAR pilot region in the range $28° < \ell < 36°$. This image has been produced by combining of the VLA D configuration and the Effelsberg single-dish continuum data. This image has been adapted from figure 5 of Brunthaler *et al.* (2021). Image Credit: Andreas Brunthaler.

The CORNISH survey consists of two parts: CORNISH-North conducted with the Very Large Array (VLA) and CORNISH-South conducted with the Australia Telescope Compact Array (ATCA). These two surveys have identified ~7,000 compact radio sources (Purcell *et al.* 2013; Irabor *et al.* 2023) of which ~800 have been classified as compact H II regions including 494 that have been classified as UC H II regions (Irabor *et al.* 2023; Kalcheva *et al.* 2018). Figure 4 shows two examples of H II regions detected in the CORNISH-North survey region (Purcell *et al.* 2013).

The GLOSTAR survey is a new VLA survey that is covering most of the 1st Galactic quadrant ($-2° < \ell < 60°$ and $|b| < 1°$) and Cygnus X star formation region ($76° < \ell < 83°$ and $-1° < b < 2°$) in the frequency range 4-8 GHz and the Effelsberg 100-m telescope (Brunthaler *et al.* 2021). The image resulting from the combination of the VLA D-array and Effelsberg data of the pilot region is shown in Fig. 5. The setup includes spectral windows covering the 6.7 GHz methanol maser transition (Ortiz-León *et al.* 2021; Nguyen *et al.* 2022), formaldehyde and radio recombination lines. Analysis of these data is still in the early stages but results are already starting to emerge (e.g., Medina *et al.* 2019; Nguyen *et al.* 2021; Dokara *et al.* 2021; Dzib *et al.* 2023) that demonstrate the impact this survey will have in the future.

6. Dust Continuum Surveys

The radio and infrared surveys discussed so far have been very successful in identifying large well-selected samples of MYSOs and UC H II regions across the Galactic mid-plane. However, these only provide details on already quite evolved stages in the high-mass star formation process and completely miss the very earliest pre-stellar and protostellar stages. Early attempts to probe these younger evolutionary stages was led by SIMBA on the SEST and SCUBA on the JCMT (e.g., Faúndez *et al.* 2004; Hill *et al.* 2005; Thompson *et al.* 2006). These resulted in the detection of hundreds of dense clumps including many starless clumps, however, they were often made towards MYSOs and UC H II regions identified from studies of IRAS data and so suffer from the same problems described earlier.

ATLASGAL traces dust emission at 870 μm providing the first unbiased survey of the inner Galactic disk (Schuller *et al.* 2009; see Fig. 6 for survey coverage and mass sensitivity). The survey has an angular resolution of $\sim 18''$ and sensitivity of 60 mJy beam^{-1}. The dust emission is optically thin at this frequency and so ATLASGAL is able to probe all evolutionary stages associated with high-mass star formation (see Figure 1).

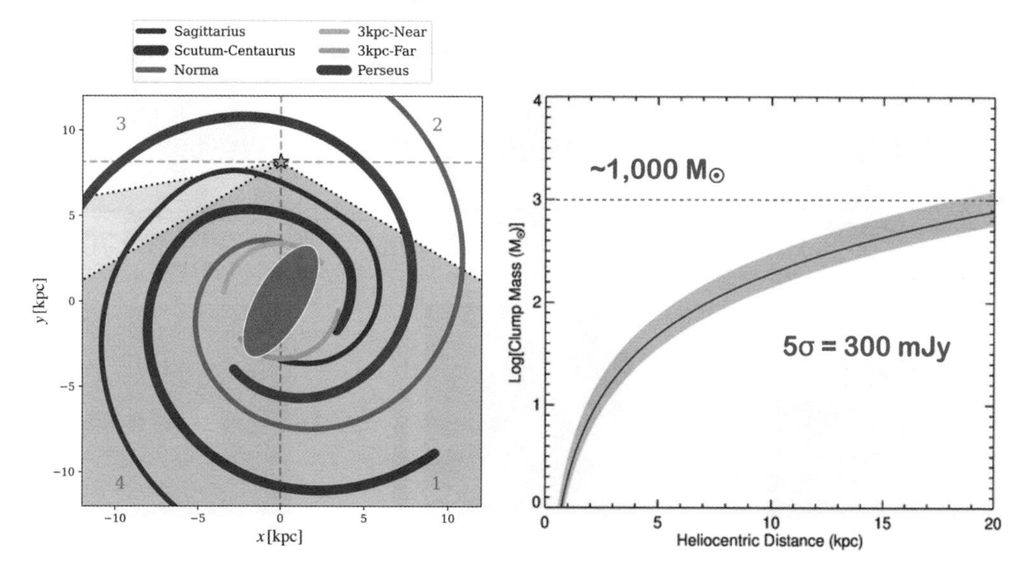

Figure 6. Left: ATLASGAL coverage of the inner Galactic plane. The dark grey shows the initial survey region ($-1.5° < b < 1.5°$) and the light grey the outer Galaxy extension ($-2° < b < 1°$). The spiral arms and the positon and orientation of the Galactic bar are included to show the structures covered in the survey. The position of the Sun is indicated by the yellow star symbol and the numbers in the corners show the Galactic quadrants. Right: Corresponding mass sensitivity as a function of heliocentric distance. The black curve shows the 5σ mass sensitivity as a function of heliocentric distance assuming a dust temperature of 20 K, the grey shading on either side corresponds to an temperature uncertainty of ± 5 K. The dashed horizontal line shows the clump mass completeness limit for ATLASGAL.

A catalogue of $\sim$10,000 dense clumps (Contreras *et al.* 2013; Urquhart *et al.* 2014c; Csengeri *et al.* 2014) has been identified from the survey maps and their properties have been fully characterised. Their distances are taken from the literature where available, and radial velocities obtained from the follow-up molecular line observation (Wienen *et al.* 2012; Giannetti *et al.* 2013; Csengeri *et al.* 2016; Kim *et al.* 2017; Wienen *et al.* 2018) have been used to determine kinematic distances (Wienen *et al.* 2015; Urquhart *et al.* 2018) for the rest of the clumps. The SED fits to the photometry are used to determine the luminosities and dust temperatures (König *et al.* 2017; see Fig. 2), which is in turn used to determine the clump masses, column and volume densities (Urquhart *et al.* 2022).

As the first step towards defining an evolutionary sample, the ATLASGAL catalogue was cross-matched with the CORNISH-North catalogue (Purcell *et al.* 2013), the Methanol Multibeam (MMB) survey (Green 2009)† and the RMS survey (Lumsden *et al.* 2013) to identify clumps associated with previously identified high-mass protostars (MYSOs and UC HII regions; Urquhart *et al.* 2013b,a, 2014b). These comparisons identified a sample of $\sim$1400 high-mass star-forming clumps, however, the earliest stages are still missing.

To identify the protostellar and quiescent clumps the 8, 24 and 70 μm images towards the remaining clumps were visually inspected. In Figure 7 an example of these three wavelength images for each of the four evolutionary stages described in Section 2 are presented. A brief summary of the classification scheme used is: clumps that are dark at all wavelengths are classified as quiescent, clumps associated with a 70 μm source but dark

† Methanol maser emission at 6.7 GHz is considered to be exclusively associated with embedded high-mass young stars (Breen *et al.* 2013).

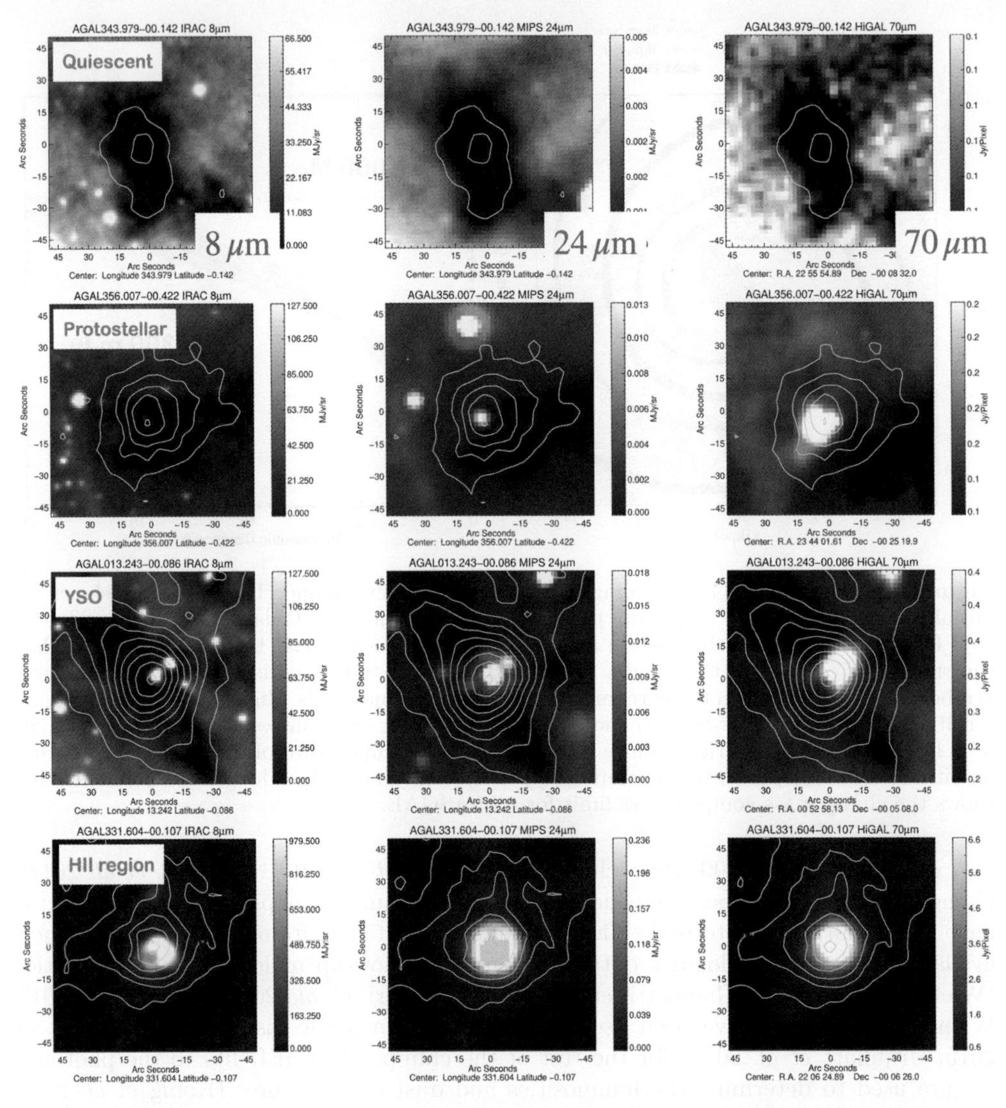

Figure 7. Examples of the images used to classify ATLASGAL sources into one of the four observationally defined evolutionary types. The contours show the distribution of the 870-μm dust emission traced by ATLASGAL. This has been adapted from figure 4 of Urquhart *et al.* (2022).

or weak at 24 μm are classified as protostellar, clumps associated with compact bright point sources at all three wavelengths and not associated with compact radio emission are classified as YSOs, and clumps associated with extended emission in all three images or compact point sources in all three images and radio emission are classified as HII regions.

In total, ~8,500 clumps have been classified with ~5,000 clumps being placed in one of the four evolutionary stages (Urquhart *et al.* 2018, 2022). This results in roughly equal number of clumps in each category, indicating that the lifetimes of these stages are also approximately equal. In the next section, we will describe how this evolutionary sample has been used to investigate how the physical properties change during the star formation process.

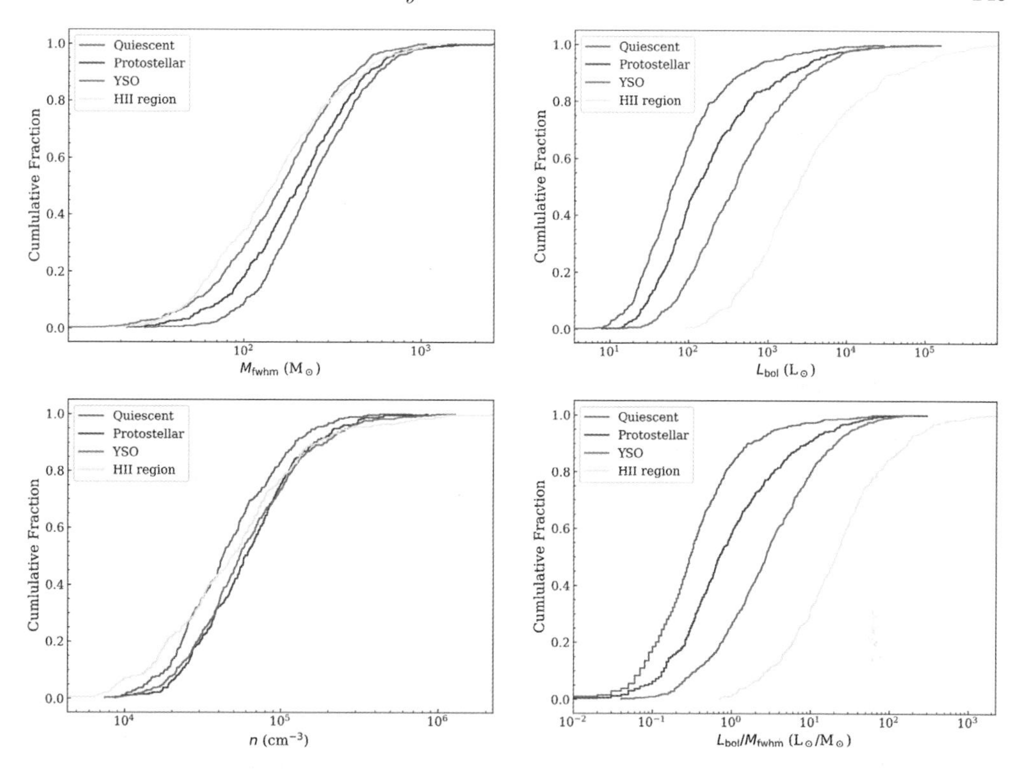

Figure 8. Cumulative distribution functions for the four main evolutionary stages. In the left panels the clump mass and volume densities are shown; these are similar for all stages. In the right panels the bolometric luminosity and the luminosity-to-mass ($L_{\rm bol}/M_{\rm fwhm}$)-ratio are shown; the evolutionary stages are well separated for these parameters making them good evolutionary diagnostics. For all of these plots a distance-limited sample of 2-4 kpc is used to avoid any possible distance bias. This figure has been adapted from figure 7 of Urquhart *et al.* (2022).

7. Towards an Evolutionary Sequence

Figure 8 presents plots comparing the clump mass, density, luminosity and luminosity-to-mass ($L_{\rm bol}/M_{\rm fwhm}$) ratio of a distance-limited sample (2-4 kpc) for the four evolutionary samples discussed in the previous section. The mass is calculated from the 870 μm emission above the half-maximum intensity to avoid a temperature bias due to evolution (see Urquhart *et al.* 2022 for details). No significant difference is revealed for the evolutionary samples in the mass and volume density distributions, however, significant differences are found for the luminosity and $L_{\rm bol}/M_{\rm fwhm}$-ratio, with both of these distributions showing a steady progression to higher values as a function of evolution. The luminosity is expected to increase strongly as the embedded protostar evolves, however, it is also linked to the mass of the protostar so that an evolved low-mass protostar can have a luminosity similar to that of a less-evolved high-mass protostar. The $L_{\rm bol}/M_{\rm fwhm}$-ratio avoids this degeneracy and is therefore considered to be a better indicator of the evolutionary state of clumps (Molinari *et al.* 2008; Urquhart *et al.* 2022).

In left panel of Fig. 9 the $L_{\rm bol}/M_{\rm fwhm}$-ratio for all ATLASGAL clumps and the mean values for the four evolutionary stages is shown. If the observationally identified evolutionary stages represent distinct stages in the star formation process one might expect to see points of inflection and/or jumps in the cumulative distribution function at the transition points of a clump's evolution. However, the distribution is smooth and devoid

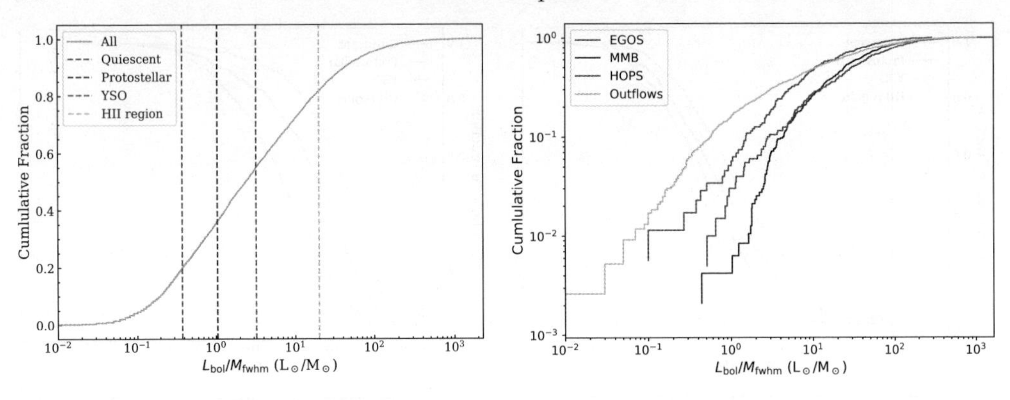

Figure 9. Left: Cumulative distribution function for the $L_{\rm bol}/M_{\rm fwhm}$-ratio of the combined star formation sample (rose curve). The vertical lines indicate the mean of the lognormal distribution of the four evolutionary stages taken from the lower right panel of Figure 8. Right: Cumulative distribution function for the $L_{\rm bol}/M_{\rm fwhm}$-ratio as a function of different star formation tracers (see text for details).

of any features that are coincident with the mean values of $L_{\rm bol}/M_{\rm fwhm}$-ratios for the four evolutionary stages. This lack of any abrupt changes suggests that star formation is a smooth and continuous process. These stages are useful in identifying groups of protostellar objects with similar properties and/or ages, but do not themselves represent fundamentally changes in the physical mechanisms involved.

The ATLASGAL evolutionary sample has been cross-matched with four other star formation tracers to test the validity and to put these other tracers into context. The catalogue of ~300 extended green objects (EGOs; Cyganowski *et al.* 2008), catalogues of molecular outflows produced by Maud *et al.* (2015); de Villiers *et al.* (2014); Yang *et al.* (2018, 2021), the methanol multibeam (MMB) survey (Green 2009) and water masers identified by HOPS (Walsh *et al.* 2011, 2014) were used. EGOs are identified by their enhanced emission in the 4.5-μm band that contains the rotationally excited H_2 ($v = 0 - 0$, S(9, 10, 11)) and CO ($v = 1 - 0$) band-head lines, which are indicative of outflow activity and active accretion (Cyganowski *et al.* 2008). Molecular outflows are intimately associated with accretion disks and are therefore a strong indication that star formation is taking place in a clump. Class II Methanol masers (Menten 1991) are thermally pumped and require the presence of high densities and a strong mid-infrared radiation source and so are thought to be associated exclusively with high-mass protostellar objects (Minier *et al.* 2003; Breen *et al.* 2013) and dense gas (Urquhart *et al.* 2015; Billington *et al.* 2019). Water masers are collisionally pumped and, when encountered in star formation environments, are thought to be excited in the cavities of molecular outflows.

The right panel of Figure 9 shows the cumulative distribution functions for the $L_{\rm bol}/M_{\rm fwhm}$-ratio of clumps associated with each of the four star formation tracers. Inspection of these curves reveals that the association rates increase as a function of evolution, and this trend holds even up to the HII region stage for all but the EGOs, where the association rate peaks at the YSO stage and then begins to decrease. The association rates for EGOs, water and methanol masers with clumps classified as quiescent are zero, which is consistent with their classification as starless (as suggested by the lack of a 70-μm point source). However, there is a significant fraction associated with molecular outflows, which is a strong indication that star formation is already underway in them.

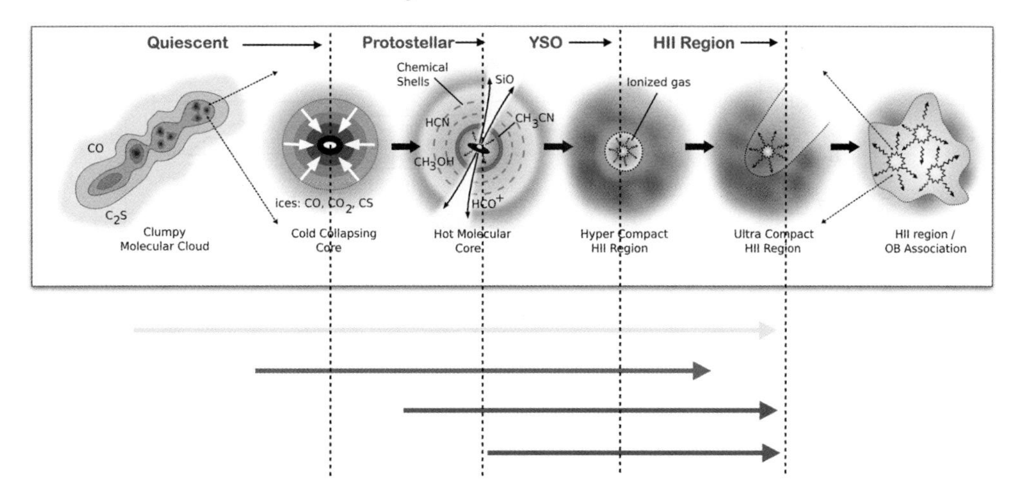

Figure 10. Schematic is the same as shown in Figure 1 but has been adapted to show approximately when the star formation tracers discussed in the text first appear and when they start to decline in frequency. The arrows correspond to outflows (top), EGOs (upper middle), water masers (lower middle) and methanol masers (bottom).

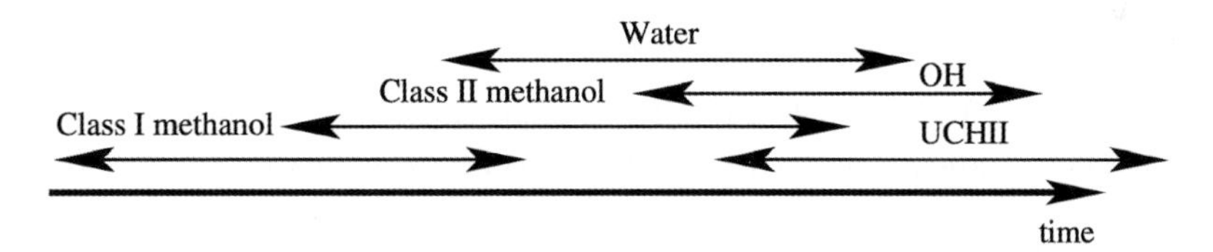

Figure 11. Model for an evolutionary sequence high-mass star formation regions constructed from maser emission. Image credit: This has been adapted from figure 2 of Ellingsen *et al.* (2007). Image Credit: Simon Ellingsen.

This analysis reveals that molecular outflows are the earliest signpost for star formation, followed by EGOs, and then by water and methanol masers (for more detailed analysis of methanol and water masers and star formation evolutionary sequences, see Breen *et al.* 2018; Billington *et al.* 2020; Ladeyschikov *et al.* 2020). Figure 10 presents a new version of the evolutionary schematic presented in Figure 1 that has been modified to indicate the approximate lifetimes that corresponds to the evolutionary stages of the four star formation tracers discussed in this section.

8. Masers as a Tool for Understanding Star formation

Another potentially useful probe to investigate the evolutionary sequence for high-mass stars is emission from maser transitions. Maser emission is ubiquitous in the Galaxy with different species being found to be attributed to specific physical processes, that in turn are associated with particular celestial objects (e.g., late-type stars, HII regions, star-forming regions; Elitzur 1992).

The first model for using a combination of maser transitions to construct an evolution sequence for high-mass star formation was proposed by Ellingsen *et al.* (2007). This was produced from the detection statistics obtained from OH, water and class I and class II methanol maser surveys (see Fig. 11 for schematic). This model was extended by Breen *et al.* (2010) with the incorporation of 12.2 GHz class II methanol maser transition and addition of relative lifetimes of the masers.

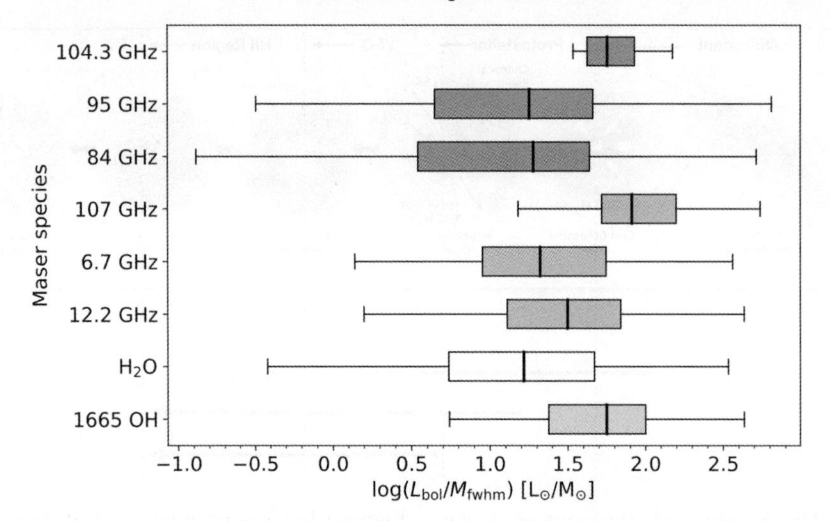

Figure 12. Box plot showing the distribution of $L_{\rm bol}/M_{\rm fwhm}$-ratios for ATLASGAL clumps associated with a wide range of masers species. This plot has been produced using the results from Ladeyschikov *et al.* (2022) and Yang *et al.* (2023) and appears in Yang *et al.* 2023 (this proceeding) and reproduced here with their permission. Image Credit: Wenjin Yang.

These early efforts to establish a reliable model using masers has been built upon in a series of recent studies that have combined results from maser surveys (e.g., HOPS Walsh *et al.* 2011, 2014, MMB Green 2009) and the $L_{\rm bol}/M_{\rm fwhm}$-ratio from ATLASGAL (Urquhart *et al.* 2018, 2022) as a measure of evolution. The first of these combined water, class II 6.7 and 12.2 GHz methanol and OH masers (Billington *et al.* 2020). This was followed by studies of 44 and 95 GHz class I methanol maser (Ladeyschikov *et al.* 2020) and water masers (Ladeyschikov *et al.* 2022). The range of maser species has recently been expanded again with the inclusion of the 84 and 104 GHz class I methanol masers and 107 GHz class II methanol maser (Yang *et al.* 2023).

Figure 12 presents a box plot that summarises all of these recent works. This shows there to be a lot of overlap between the different maser species but there are also significant differences with some maser species predominately associated with earlier evolutionary stages (e.g., 84 and 95 GHz class I methanol masers) while others are associated with more evolved stages (1665 MHz OH, 107 GHz class II methanol and 104.3 GHz class I methanol masers). Further work is needed to develop these results into a predictive model but the current work is promising.

9. Summary

The huge international effort to survey the Galactic plane at infrared, submillimetre and radio wavelengths have provided the large coverage necessary to identify large well-selected and representative samples of embedded high-mass star forming region, and start to develop an evolutionary sequence for high-mass star formation. In this review, we have given an overview of the progress that has been made over the past 20 years and how access to high-resolution far-infrared and submillimetre has allowed earlier evolutionary stages to be identified and their properties characterised.

The ATLASGAL has been used to produce a catalogue of dense high-mass clumps that includes examples of all embedded stages associated with high-mass star formation. This catalogue has been cross-matched with mid-infrared and radio surveys to produce a sample of ∼5000 clumps classified into one of four evolutionary stages; these being quiescent, protostellar, YSO and HII region. Comparing the luminosities and $L_{\rm bol}/M_{\rm fwhm}$-ratios for these four stages shows a clear trend with both parameters increasing with evolution, however, the cumulative distribution of these parameters is smooth and does not reveal any sharp jumps or discontinuities between the different evolutionary type. This suggests that star formation is a smooth and continuous process and that the evolutionary stages identified themselves do not represent fundamentally different stages or changes in the physical mechanisms involved.

The ATLASGAL $L_{\rm bol}/M_{\rm fwhm}$-ratio has been used as a measure of clump evolution to determine when other commonly used star formation tracers first appear and how long they persist for. This analysis has revealed that the appearance of a molecular outflow is the earliest manifestation of the star formation process, appearing before the formation of a detectable 70 μm point source, and persisting until after the formation of the HII region stage. Next to appear are the EGOs, which appear after the molecular outflow but still before the protostar becomes viable. The water and methanol masers are the last of the star formation tracers examined here to appear with the water masers appearing shortly before the methanol masers at approximate midway through the protostellar stage persisting until the appearance of the HII region. The $L_{\rm bol}/M_{\rm fwhm}$-ratio has also been used as a measure of evolution in a number of studies that focused on developing an evolutionary model based on different maser species that are associated with high-mass star formation. However, there is significant overlap between in the $L_{\rm bol}/M_{\rm fwhm}$-ratio distributions and although there is evidence that some maser species are more prevalent in the earlier stages and some in the more evolved stages.

Overall, significant progress has been made in recent years in constructing an evolutionary sequence and with surveys currently underway and new facilities coming on line the situation will undoubtedly improve in the coming years.

References

Aguirre J. E., Ginsburg A. G., Dunham M. K., *et al.* 2011, *ApJS*, 192, 4
Aumann H. H., Fowler J. W., and Melnyk M., 1990, *AJ*, 99, 1674
Beichman, C. A., Neugebauer, G., Habing, H. J., *et al.* 1988, Infrared astronomical satellite (IRAS) catalogs and atlases. Volume 1: Explanatory supplement, 1
Beuther H., Walsh A., Schilke P., *et al.* 2002, *A&A*, 390, 289
Beuther H., Bihr S., Rugel M., *et al.* 2016, *A&A*, 595, A32
Billington S. J., Urquhart J. S., König C., *et al.* 2019, *MNRAS*, 490, 2779
Billington S. J., Urquhart J. S., König C., *et al.* 2020, *MNRAS*, 499, 2744
Breen S. L., Ellingsen S. P., Caswell J. L., and Lewis B. E., 2010, *MNRAS*, 401, 2219
Breen S. L., Ellingsen S. P., Contreras Y., *et al.* 2013, *MNRAS*, 435, 524
Breen S. L., Contreras Y., Ellingsen S. P., *et al.* 2018, *MNRAS*, 474, 3898
Brunthaler A., Menten K. M., Dzib S. A., *et al.* 2021, *A&A*, 651, A85
Campbell B., Persson S. E., and Matthews K., 1989, *AJ*, 98, 643
Carey S. J., Noriega-Crespo A., Mizuno D. R., *et al.* 2009, *PASP*, 121, 76
Chan S. J., Henning T., and Schreyer K., 1996, A&AS, 115, 285
Churchwell E., Babler B. L., Meade M. R., *et al.* 2009, *PASP*, 121, 213
Clarke A. J., Lumsden S. L., Oudmaijer R. D., *et al.* 2006, *A&A*, 457, 183
Contreras Y., Schuller F., Urquhart J. S., *et al.* 2013, *A&A*, 549, A45
Cooper H. D. B., Lumsden S. L., Oudmaijer R. D., *et al.* 2013, *MNRAS*, 430, 1125
Csengeri T., Urquhart J. S., Schuller F., *et al.* 2014, *A&A*, 565, A75
Csengeri T., Leurini S., Wyrowski F., *et al.* 2016, *A&A*, 586, A149
Cyganowski C. J., Whitney B. A., Holden E., *et al.* 2008, *AJ*, 136, 2391
Davies B., Hoare M. G., Lumsden S. L., *et al.* 2011, *MNRAS*, 416, 972
de Villiers H. M., Chrysostomou A., Thompson M. A., *et al.* 2014, *MNRAS*, 444, 566
Dempsey J. T., Thomas H. S., and Currie M. J., 2013, *ApJS*, 209, 8

Dokara R., Brunthaler A., Menten K. M., *et al.* 2021, *A&A*, 651, A86
Drew J. E., Greimel R., Irwin M. J., *et al.* 2005, *MNRAS*, 362, 753
Dzib S. A., Yang A. Y., Urquhart J. S., *et al.* 2023, *A&A*, 670, A9
Egan M. P., Price S. D., Kraemer K. E., *et al.* 2003, VizieR Online Data Catalog, 5114, 0
Elitzur M., 1992, *ARA&A*, 30, 75
Ellingsen S. P., Voronkov M. A., Cragg D. M., *et al.* 2007, in Chapman J. M., and Baan W. A., eds, Vol. 242, Astrophysical Masers and their Environments. pp 213–217 (arXiv:0705.2906), doi:10.1017/S1743921307012999
Faúndez S., Bronfman L., Garay G., *et al.* 2004, *A&A*, 426, 97
Gao Y., and Solomon P. M., 2004, *ApJS*, 152, 63
Giannetti A., Brand J., Sánchez-Monge Á., *et al.* 2013, *A&A*, 556, A16
Green J. A. a. a., 2009, *MNRAS*, 392, 783
Helfand D. J., Becker R. H., White R. L., *et al.* 2006, *AJ*, 131, 2525
Hill T., Burton M. G., Minier V., *et al.* 2005, *MNRAS*, 363, 405
Hoare M. G., Kurtz S. E., Lizano S., *et al.* 2007, Protostars and Planets V, pp 181–196
Hoare M. G., Purcell C. R., Churchwell E. B., *et al.* 2012, *PASP*, 124, 939
Ilee J. D., Wheelwright H. E., Oudmaijer R. D., *et al.* 2013, *MNRAS*, 429, 2960
Irabor T., Hoare M. G., Burton M., *et al.* 2023, *MNRAS*, 520, 1073
Jackson J. M., Rathborne J. M., Shah R. Y., *et al.* 2006, *ApJS*, 163, 145
Jackson J. M., Rathborne J. M., Foster J. B., *et al.* 2013, *PASA*, 30, 57
Kalcheva I. E., Hoare M. G., Urquhart J. S., *et al.* 2018, *A&A*, 615, A103
Kennicutt R. C., and Evans N. J., 2012, *ARA&A*, 50, 531
Kim W.-J., Wyrowski F., Urquhart J. S., *et al.* 2017, preprint, (arXiv:1702.02062)
König C., Urquhart J. S., Csengeri T., *et al.* 2017, *A&A*, 599, A139
Kroupa P., and Weidner C., 2003, *ApJ*, 598, 1076
Kruijssen J. M. D., and Longmore S. N., 2013, *MNRAS*,
Kurtz S., and Hofner P., 2005, *AJ*, 130, 711
Kurtz S., Churchwell E., and Wood D. O. S., 1994, *ApJS*, 91, 659
Ladeyschikov D. A., Urquhart J. S., Sobolev A. M., *et al.* 2020, *AJ*, 160, 213
Ladeyschikov D. A., Gong Y., Sobolev A. M., *et al.* 2022, *ApJS*, 261, 14
Lucas P. W., Hoare M. G., Longmore A., *et al.* 2008, *MNRAS*, 391, 136
Lumsden S. L., Hoare M. G., Oudmaijer R. D., and Richards D., 2002, *MNRAS*, 336, 621
Lumsden S. L., Hoare M. G., Urquhart J. S., *et al.* 2013, *ApJS*, 208, 11
Maud L. T., Moore T. J. T., Lumsden S. L., *et al.* 2015, *MNRAS*, 453, 645
McKee C. F., and Ostriker E. C., 2007, *ARA&A*, 45, 565
Medina S. N. X., Urquhart J. S., Dzib S. A., *et al.* 2019, *A&A*, 627, A175
Menten K. M., 1991, *ApJ*, 380, L75
Minier V., Ellingsen S. P., Norris R. P., and Booth R. S., 2003, *A&A*, 403, 1095
Minniti D., Lucas P. W., Emerson J. P., *et al.* 2010, New A, 15, 433
Molinari S., Brand J., Cesaroni R., and Palla F., 1996, *A&A* 308, 573
Molinari S., Pezzuto S., Cesaroni R., *et al.* 2008, *A&A*, 481, 345
Molinari S., Swinyard B., Bally J., *et al.* 2010, *A&A*, 518, L100
Moore T. J. T., Plume R., Thompson M. A., *et al.* 2015, *MNRAS*, 453, 4264
Motte F., Bontemps S., and Louvet F., 2018, *ARA&A*, 56, 41
Mottram J. C., Hoare M. G., Lumsden S. L., *et al.* 2007, *A&A*, 476, 1019
Murphy T., Mauch T., Green A., *et al.* 2007, *MNRAS*, 382, 382
Nguyen H., Rugel M. R., Menten K. M., *et al.* 2021, *A&A*, 651, A88
Nguyen H., Rugel M. R., Murugeshan C., *et al.* 2022, *A&A*, 666, A59
Ortiz-León G. N., Menten K. M., Brunthaler A., *et al.* 2021, *A&A*, 651, A87
Price S. D., Egan M. P., Carey S. J., *et al.* 2001, *AJ*, 121, 2819
Purcell C. R., Hoare M. G., Cotton W. D., *et al.* 2013, *ApJS*, 205, 1
Rosolowsky E., Hughes A., Leroy A. K., *et al.* 2021, *MNRAS*, 502, 1218
Schuller F., Menten K. M., Contreras Y., *et al.* 2009, *A&A*, 504, 415
Schuller F., Urquhart J. S., Csengeri T., *et al.* 2021, *MNRAS*, 500, 3064

Sevenster M. N., van Langevelde H. J., Moody R. A., *et al.* 2001, *A&A*, 366, 481
Skrutskie M. F., Cutri R. M., Stiening R., *et al.* 2006, *AJ*, 131, 1163
Sridharan T. K., Beuther H., Schilke P., *et al.* 2002, *ApJ*, 566, 931
Stil J. M., Taylor A. R., Dickey J. M., *et al.* 2006, AJ, 132, 1158
Thompson M. A., Hatchell J., Walsh A. J., *et al.* 2006, *A&A*, 453, 1003
Urquhart J. S., Busfield A. L., Hoare M. G., *et al.* 2007a, *A&A*, 461, 11
Urquhart J. S., Busfield A. L., Hoare M. G., *et al.* 2007b, *A&A*, 474, 891
Urquhart J. S., Busfield A. L., Hoare M. G., *et al.* 2008, *A&A*, 487, 253
Urquhart J. S., Hoare M. G., Purcell C. R., *et al.* 2009, *A&A*, 501, 539
Urquhart J. S., Morgan L. K., Figura C. C., *et al.* 2011, *MNRAS*, 418, 1689
Urquhart J. S., Moore T. J. T., Schuller F., *et al.* 2013a, *MNRAS*, 431, 1752
Urquhart J. S., Thompson M. A., Moore T. J. T., *et al.* 2013b, *MNRAS*, 435, 400
Urquhart J. S., Figura C. C., Moore T. J. T., *et al.* 2014a, *MNRAS*, 437, 1791
Urquhart J. S., Moore T. J. T., Csengeri T., *et al.* 2014b, *MNRAS*, 443, 1555
Urquhart J. S., Csengeri T., Wyrowski F., *et al.* 2014c, *A&A*, 568, A41
Urquhart J. S., Moore T. J. T., Menten K. M., *et al.* 2015, *MNRAS*, 446, 3461
Urquhart J. S., König C., Giannetti A., *et al.* 2018, *MNRAS*, 473, 1059
Urquhart J. S., Wells M. R. A., Pillai T., *et al.* 2022, *MNRAS*, 510, 3389
Walsh A. J., Breen S. L., Britton T., *et al.* 2011, *MNRAS*, 416, 1764
Walsh A. J., Purcell C. R., Longmore S. N., *et al.* 2014, *MNRAS*, 442, 2240
Wheelwright H. E., Oudmaijer R. D., de Wit W. J., *et al.* 2010, *MNRAS*, 408, 1840
White G. J., Etxaluze M., Doi Y., *et al.* 2009, in Onaka T., White G. J., Nakagawa T., and Yamamura I., eds, Astronomical Society of the Pacific Conference Series Vol. 418, AKARI, a Light to Illuminate the Misty Universe. p. 67
Wienen M., Wyrowski F., Schuller F., *et al.* 2012, *A&A*, 544, A146
Wienen M., Wyrowski F., Menten K. M., *et al.* 2015, *A&A*, 579, A91
Wienen M., Wyrowski F., Menten K. M., *et al.* 2018, *A&A*, 609, A125
Williams J. P., and McKee C. F., 1997, *ApJ*, 476, 166
Wood D. O. S., and Churchwell E., 1989, *ApJS* 69, 831
Wu Y., Zhang Q., Yu W., *et al.* 2006, *A&A*, 450, 607
Wynn-Williams C. G., 1982, *ARA&A*, 20, 587
Yang A. Y., Thompson M. A., Urquhart J. S., and Tian W. W., 2018, *ApJS*, 235, 3
Yang A. Y., Urquhart J. S., Thompson M. A., *et al.* 2021, *A&A*, 645, A110
Yang W., Gong Y., Menten K. M., *et al.* 2023, arXiv e-prints, p. arXiv:2305.04264

Cosmic Masers: Proper Motion toward the Next-Generation Large Projects
Proceedings IAU Symposium No. 380, 2024
T. Hirota, H. Imai, K. Menten, & Y. Pihlström, eds.
doi:10.1017/S1743921323002508

Masers in accretion burst sources

Olga Bayandina[1] and the M2O collaboration:
Agnieszka Kobak, Alessio Caratti o Garatti, Alexander Tolmachev, Alexandr Volvach, Alexei Alakoz, Alwyn Wootten, Anastasia Bisyarina, Andrews Dzodzomenyo, Andrey Sobolev, Anna Bartkiewicz, Artis Aberfelds, Bringfried Stecklum, Busaba Kramer, Callum Macdonald, Claudia Cyganowski, Fransisco Colomer, Cristina Garcia Miro, Crystal Brogan, Dalei Li, Derck Smits, Dieter Engels, Dmitry Ladeyschikov, Doug Johnstone, Elena Popova, Emmanuel Proven-Adzri, Fanie van den Heever, Gabor Orosz, Gabriele Surcis, Gang Wu, Gordon MacLeod, Hendrik Linz, Hiroshi Imai, Huib van Langevelde, Irina Val'tts, Ivar Shmeld, James O. Chibueze, Jan Brand, Jayender Kumar, Jimi Green, Job Vorster, Jochen Eislöffel, Jungha Kim, Koichiro Sugiyama, Karl Menten, Katharina Immer, Kazi Rygl, Kazuyoshi Sunada, Kee-Tae Kim, Larisa Volvach, Luca Moscadelli, Lucas Jordan, Lucero Uscanga, Malcolm Gray, Marian Szymczak, Mateusz Olech, Melvin Hoare, Michał Durjasz, Mizuho Uchiyama, Nadya Shakhvorostova, Pawel Wolak, Sergei Gulyaev, Sergey Khaibrakhmanov, Shari Breen, Sharmila Goedhart, Silvia Casu, Simon Ellingsen, Stan Kurtz, Stuart Weston, Tanabe Yoshihiro, Tim Natusc, Todd Hunter, Tomoya Hirota, Willem Baan, Wouter Vlemmings, Xi Chen, Yan Gong, Yoshinori Yonekura, Zsófia Marianna Szabó, Zulema Abraham

[1]INAF - Osservatorio Astrofisico di Arcetri, Largo E. Fermi 5, 50125 Firenze, Italy.
email: olga.bayandina@inaf.it

Abstract. Recently, remarkable progress has been made in understanding the formation of high mass stars. Observations provided direct evidence that massive young stellar objects (MYSOs), analogously to low-mass ones, form via disk-mediated accretion accompanied by episodic accretion bursts, possibly caused by disk fragmentation. In the case of MYSOs, the mechanism theoretically provides a means to overcome radiation pressure, but in practice it is poorly studied - only three accretion bursts in MYSOs have been caught in action to date. A significant contribution to the development of the theory has been made with the study of masers, which have proven to be a powerful tool for locating "bursting" MYSOs. This overview focuses on the exceptional role that masers play in the search and study of accretion bursts in massive protostars.

Keywords. Stars: formation, Masers

1. Introduction

Massive stars play a critical role in, both physical and chemical, formation of galaxies (e.g. Greif 2015). However, the question of how massive stars themselves form remains debatable.

It seems reasonable to assume that massive young stellar objects (MYSOs) should follow the same mechanisms that we see in low-mass protostars, and thus the formation of massive stars should be a scaled-up version of low-mass star formation. Nevertheless, as a protostar accumulates mass, it also emits more and more energy. At some point, the growing radiation of the protostar inevitably begins to push away the surrounding matter, thus cutting itself off from the mass supply (e.g. Hosokawa *et al.* 2012). The situation only worsens if radiation pressure is trapped in a dense gas and dust shell. And this applies to massive protostars, which evolve much faster than low-mass stars and remain deeply embedded in their parental envelopes throughout the early stages of the evolution (e.g. Zinnecker & Yorke 2007).

Although radiation pressure plays an important role in the formation of MYSOs, it can fatally stop the accumulation of mass only under the assumption that stellar evolution is laminar and continuous. In reality, nature has come up with a few ways to get around the issue. Instead of pushing matter uniformly in all directions, the radiation pressure finds an outlet in certain, physically favorable directions, thus forming collimated flows without destroying the accretion of matter onto a star (Moscadelli *et al.* 2020). Another critical point is the fact that the rate of accretion onto a protostar is not constant, but episodic, with episodes of rapid accretion (bursts). The prevailing idea at the moment is that MYSOs form via disk-mediated accretion, accompanied by episodic accretion bursts, possibly caused by disk fragmentation (e.g. Meyer *et al.* 2019).

The problem is that, despite the general theoretical understanding of the process, we struggle to gather a statistically significant sample of massive protostars going through these critical burst accretion episodes. When we discover a new object or event, the question arises whether that thing is so unique that we have never been lucky enough to catch it in action before, or whether our observation methods and samples were missing the point. The way to test these two possibilities is to conduct extensive observations. A good example of this approach are fast radio bursts (FRBs), which have gone from the first discovery in 2007 to hundreds of events detected to date (CHIME/FRB Collaboration *et al.* 2021). Massive stars per se are a challenging observational target. They are rare and located at far distances (e.g. Zinnecker & Yorke 2007), making any statistical or high-resolution study taxing. As mentioned above, massive stars evolve rapidly, remaining hidden in the natal clouds for most of their evolution. Dense envelopes obscure MYSOs from direct observations, leaving only the window of sub-millimeter, IR, and radio wavelengths open. Additionally, the rapid evolution means that any particular evolutionary stages are short and demand quick-response facilities and observational techniques. Models suggest that accretion bursts constitute only $\sim$1.7% of the formative first 60 kyr of massive stars (Meyer *et al.* 2019). No accretion burst in MYSO were found up until 2016, when the first such event was discovered in IR observations (Caratti o Garatti *et al.* 2017). Since then, only two more accretion bursts have been detected "in action", e.g. during the burst epoch (Hunter *et al.* 2017; Stecklum *et al.* 2021), but this is a promising start, because these detections were made possible by a change in the observational strategy.

Accretion bursts traced by continuum emission require "pre-burst" data and confirmation via multi-epoch observations with mm-interferometers such as ALMA, NOEMA, or the SMA - thus limiting the target sample to well-studied MYSOs. Moreover, to our knowledge, there is no monitoring program for (sub-)mm continuum emission with a cadence that would detect bursting sources during a burst onset. A turning point in the search for accretion bursts occurred when it was noticed that large accretion events in MYSO are accompanied by maser flares (e.g. Hunter *et al.* 2017; Szymczak *et al.* 2018b; MacLeod *et al.* 2018). Bright and compact, masers trace physical conditions and dynamics of the environment around MYSOs, yet are easily accessible to observation using a variety of instruments, from single-dish telescopes (e.g. Szymczak *et al.* 2018a) to space-VLBI (e.g. Sobolev *et al.* 2018). In contrast to continuum sources, pre-burst data are available for most maser sources thanks to maser surveys conducted using single-dish telescopes and interferometers (see Maserdb, the database of astrophysical masers, Ladeyschikov et al. 2019).

Masers are numerous, providing a rich selection of possible targets, but also fragile, being finely tuned to local physical conditions (e.g. Ellingsen *et al.* 2007 and Breen *et al.* 2019a). Changes in temperature, density, or velocity field of masering region can result in brightening (flare) or fading of maser emission. Spatial distribution of masers can reveal the temperature, density, and radiation enhancements in the region, while the kinematics of the maser spots can indicate gas motions. Hence the study of maser emission can provide us with the record of the evolution of a particular star, with maser flares highlighting accretion bursts, as no other tracer can.

2. Sample of the known accretion burst sources

To date only three definite accretion bursts, observed during the accretion events and confirmed both in IR and maser observations, have been found: S255IR (Caratti o Garatti *et al.* 2017), NGC6334I (Hunter *et al.* 2017), and G358.93-0.03 (Stecklum *et al.* 2021). The first two sources, S255IR and NGC6334I, exercised accretion bursts in 2015 (Caratti o Garatti *et al.* 2017; Hunter *et al.* 2017) and set the scene for the follow-up search of the sources of this kind. The discovery of the burst in the latest source, G358.93-0.03, happened in 2019 (Stecklum *et al.* 2021) and was the product of such a dedicated quest.

In all three mentioned cases of accretion bursts, outstanding change in flux density and structure of sub-millimeter, IR, and radio emission from the sources has been noted (Caratti o Garatti *et al.* 2017; Hunter *et al.* 2017; Stecklum *et al.* 2021). IR-brightening of the central source is one of the defining features of accretion bursts in massive star formation, however even such a small sample of the sources showed a surprising variety of the IR emission parameters, with G358.93-0.03 being the first NIR- and (sub)mm-dark but FIR-loud accretion burst (citealpStecklum2021).

According to the analysis of archival data, in addition to the core sample, at least three more sources are thought to experience accretion bursts. The source M17 MIR showed correlated variations in MIR and 22 GHz water maser fluxes, indicating two accretion bursts separated by a six years long quiescent phase (Chen *et al.* 2021). The object is extremely young, and minor accretion bursts are expected to be frequent in the very early stages of massive star formation (e.g. Meyer *et al.* 2019). An accretion burst in another source, V723 Car, was suspected on basis of NIR images analysis, but since the burst was found post factum, no information on the accretion luminosity is available (Tapia *et al.* 2015). This source also has no associated masers. In contrast, G323.46-0.08 is thought to go through an accretion burst solely based on 6.7 GHz maser data without IR confirmation (Proven-Adzri *et al.* 2019). The periodic 6.7 GHz methanol maser in the source showed a flare and appearance of new maser features (Proven-Adzri *et al.* 2019).

A few more sources showed features that can potentially be interpreted as signs of accretion bursts and be added to the sample. Of particular interest are sources housing periodic masers, such as G351.78-0.54 (MacLeod, G. C. & Gaylard, M. J. 1996), G107.298+5.639 (Stecklum *et al.* 2018; Olechi *et al.* 2020) and G323.46-0.08 (Proven-Adzri *et al.* 2019). The periodic behaviour of the maser emission makes them attractive targets for long-term monitoring and provides a record of accretion instabilities in the sources.

3. Masers tracing accretion bursts

Despite the fact that the sample of accretion burst sources in MYSO is very limited, methanol masers at 6.7 GHz have already shown to be the best indicator of events of this type (e.g. the 6.7 GHz maser flare reported in Fujisawa *et al.* (2015) triggered the IR observations of S255IR NIRS3). The abundance of such masers provides a large sample of target sources (e.g. Yang *et al.* 2019). Remarkably, 6.7 GHz methanol masers are known to be associated exclusively with high-mass protostars (Minier *et al.* 2003). The low frequency and usually high fluxes allow monitoring even using single-dish telescopes with a small and imperfect active surface. Methanol masers at 6.7 GHz show consistent spectra with stable or periodic fluxes, thus any sudden change can be identified in monitoring data (e.g. Szymczak *et al.* 2018a).

However, the 6.7 GHz transition is just one example of the many radiatively pumped class II methanol masers that arise during accretion bursts. Due to the increase in incident photons during accretion bursts, masers in the vicinity of a bursting source exhibit an increased flux. This gives rise to a very specific type of maser flare - class II methanol masers show extraordinary fluxes with appearance of new spectral features (e.g. Fujisawa *et al.* 2015; Sugiyama *et al.* 2019). During the latest discovered accretion event in G358.93−0.03, multiple maser transitions flared, and rare, previously undiscovered maser species and transitions were found to arise (e.g. Breen *et al.* 2019a). The energy of the accretion burst ignited previously undetected class II methanol masers, including the first ever discovered torsionally excited methanol masers (Breen *et al.* 2019a; Brogan *et al.* 2019; MacLeod *et al.* 2019). More than 30 maser lines were detected in a wide range of frequencies from 6.18 GHz to 361.2 GHz.

Simultaneous VLA observations of several different methanol masers detected in G358.93−0.03 during the burst showed that they all trace the same region around the central protostar (Bayandina *et al.* 2022a). However, the distribution of the methanol masers changed drastically from the burst epoch to post-burst epoch (Bayandina *et al.* 2022a; Burns *et al.* 2020b). A similar profound change in the location of the 6.7 GHz maser emission within the source was found in S255 after the burst (Moscadelli *et al.* 2017). And in the case of NGC6334I, a strong 6.7 GHz methanol masers was detected towards the bursting source MM1 at the burst epochs, while no masers had ever been seen towards it before (Hunter *et al.* 2018). Another common feature of flaring 6.7 GHz maser emission in all the burst sources is the presence of an extended component of methanol emission which is largely resolved with VLBI arrays but can be studied with compact arrays (Moscadelli *et al.* 2017; Hunter *et al.* 2018; Burns *et al.* 2020b).

The extreme energy of accretion bursts allows us to study fine structure of accretion disks with the help of methanol maser emission. The VLA images obtained for G358.93−0.03 during the flare hinted at the presence of spiral arm structures within the accretion disk but the low resolution of the compact array data limited the extent of the interpretation (Bayandina *et al.* 2022a). The theory was confirmed in the multi-epoch VLBI observations of the 6.7 GHz maser in G358.93−0.03 presented in Burns *et al.* (2023). In a series of VLBI observations, the thermal radiation ("heatwave") from the accretion burst was caught propagating with subluminal velocities outwards from the central accreting high-mass protostar to the outer radii of the accretion disk (Burns *et al.* 2023). Combining together the images of the heatwave propagation and fitting the resulting map, Burns *et al.* (2023) were able to infiltrate a four-arm spiral structure in the Keplerian disk around the bursting source in G358.93−0.03.

Apart from methanol masers, some other masers species in vicinity of the bursting sources have been detected. Notably, the newly discovered molecular maser species of HDO, HNCO, and $^{13}CH_3OH$ detected with the VLA were discovered to trace spiral-arm accretion flows in G358.93−0.03 (Chen *et al.* 2020a,b). The 6.7 GHz methanol maser flare in NGC6334I MM1 was accompanied by increased flux density of OH masers not only at 1665 MHz but also at 4660 and 6031 MHz (MacLeod *et al.* 2018). Accretion burst sources therefore appear to be the most promising laboratories for future discoveries of new maser species.

Accretion bursts and methanol maser flares are typically followed by an H_2O maser flare (Brogan *et al.* 2018; Hirota *et al.* 2021; Bayandina *et al.* 2022b) triggered by the light from the burst scattered by the dust in the outflow cavities (the so-called "light echo") (Caratti o Garatti *et al.* 2017). For example, the VLA images of the 22 GHz water maser emission in G358.93−0.03 showed a significant change in both the morphology and velocity gradient of the maser associated with the bursting source MM1. In addition, a bigger region seems to be affected by the accretion event as the emission of the water masers associated with other point sources in the region appeared to be suppressed at the post-burst epoch (Bayandina *et al.* 2022a).

4. Maser Monitoring Organisation

Discussion of the role of maser in the study and search of accretion bursts is inadequate without mentioning the activities of Maser Monitoring Organisation (M2O) (Burns *et al.* 2022). In order to utilise the potential of maser flares as indicators of accretion bursts, maser monitoring programs from all around the globe came together in 2017 (soon after the first accretion burst discovery) and created the M2O. M2O is a global cooperation of maser monitoring programs that searches for maser flares and manages their follow-up interferometric studies. In the few first years of the M2O, dozens of maser flares were discovered and even though not all of them were associated with accretion bursts, each flare provided valuable insights into the accretion/ejection process in MYSO (see the publication list on the project website).

The most significant achievement of the Maser Monitoring Organisation to date is the organisation and management of the follow-up observations of G358.93−0.03 after the 6.7 GHz methanol maser flare. The coordinated actions of the various scientific groups within the organisation (i.e. single-dish monitoring; IR, compact array, and VLBI follow-up; theoretical modelling) allowed for the most comprehensive and diverse observations of the accretion disk produced so far (Breen *et al.* 2019a; Brogan *et al.* 2019; MacLeod *et al.* 2019; Burns *et al.* 2020a; Chen *et al.* 2020a,b; Volvach *et al.* 2020; Stecklum *et al.* 2021; Bayandina *et al.* 2022a,b; Burns *et al.* 2023). It should also be noted that such a large variety of data was obtained thanks to extensive preparatory work, in anticipation of the discovery of such a transitive event, the M2O has obtained triggered and ToO observing time with a number of the most significant astronomical facilities. The successful study of the accretion burst in G358.93−0.03 inspires hope that the next accretion event in MYSO will get even more coverage and disclose many more secrets of the early stages of massive star formation.

5. Perspectives

The termination of the Stratospheric Observatory for Infrared Astronomy (SOFIA) mission presented a major setback for the study of accretion bursts in MYSO. The telescope provided critical confirmation of accretion bursts in the IR range (e.g. citealp-Caratti2017) which no ground-based telescope can access. The loss of the SOFIA telescope is supposed to be compensated by the activities of the James Webb Space Telescope (JWST), however, at the moment, the demand for the instrument's observing time is too great and such short-lived transient events as accretion bursts would most likely be missed. Although this is not an optimal solution, we still can use NIR spectral imaging to study NIR-bright accretion bursts (such as S255IR NIRS3 or G323.46-0.0) and sub-mm/mm data to study the NIR dark ones (such as G358.93-0.03 or NGC6334I) to confirm the burst and recover some physical characteristics.

Considering the degradation of the IR research, masers have taken on even greater importance. The activities of the M2O showed the importance of maser monitoring with single-dish telescopes. The accretion burst cases detected so far showed a wide variety of flaring masers at different frequencies. In this context, upgrading of single-dish telescopes and VLBI arrays with new high-frequency receivers can open up new perspectives for the research. For example, the INAF radio telescopes (64-m Sardinia, 32-m Medicina and 32-m Noto Radio Telescopes, Italy) are planned to be equipped with the new high-frequency receivers simultaneously operating in K, Q, and W frequency bands (central frequencies: 22, 42, and 100 GHz) by the end of 2023 (Bolli et al. in print). Note that many of the extremely rare, never-before-detected methanol masers found in G358.93−0.03 fall into the frequency range of the planned high-frequency receivers of the Italian VLBI network (e.g. 20.9, 23.12, 45.8, and 94.8 GHz in Breen *et al.* 2019a; MacLeod *et al.* 2019). The simultaneous observation of all these maser lines is critical because it can reveal the overall picture of the bursting region and unfold the place of rare masers in it. In terms of response to an accretion burst, high-frequency receivers have another advantage. H_2O masers at 22 GHz are known to flare later than methanol masers, which gives ample time to prepare for possible observations (while observations of low-frequency methanol masers have to be done immediately).

The VLBI study of the accretion burst in G358.93−0.03 provided a glimpse into the fine structure of the accretion disk around the massive star, for the first time revealing a strong evidence of the presence of spiral arms in it (Burns *et al.* 2023). It is noteworthy that the success of the "heatwave mapping" method used in Burns *et al.* (2023) depends more on the observation cadence than on spatial resolution. Given the short time scale of accretion bursts and associated maser flares, a great degree of flexibility is required of a VLBI facility in order to capture the "golden hour" of maser flares and accretion bursts. An improvement in the response times of VLBI arrays, possibly related to greater automation of the observational preparation process, could be of great importance for the study of new bursts.

Although VLBI facilities provide the highest resolution insight into flaring masers (e.g. Burns *et al.* 2020a), the VLA is indispensable for providing an overall picture of the region, observing different masers and cm-continuum simultaneously, and detecting weak (and often rare) masers not accessible to VLBI imaging (e.g. Bayandina *et al.* 2019; Chen *et al.* 2020a; Bayandina *et al.* 2022a,b). Another important moment is that the VLA requires shorter observing times and provides the data quicker than any VLBI array, which is essential in case of transient events. Thus, future observations of potential sources of accretion bursts depend heavily on the availability of the observing time with the VLA.

In summary, masers are a sensitive and highly informative probe of star formation. They follow different stages of stellar evolution and alert us to such short-lived and rare events as accretion bursts. The recent years since the discovery of the first accretion burst have shown that the key to the successful search and study of accretion bursts with maser sources is collaboration. The collaboration between experts from different fields (e.g. IR, VLBI, theory) as well as between different facilities (from single-dish monitoring stations to VLBI arrays) has already brought new, impressive results described in this article. The accumulated experience and new upcoming facilities allow us to be optimistic that there is more to come.

Acknowledgements

O.B. acknowledges financial support from the Italian Ministry of University and Research - Project Proposal CIR01_00010.

References

Bayandina, O. S., Burns, R. A., Kurtz, S. E., *et al.* 2019, *ApJ*, 884, 140
Bayandina O. S., Brogan, C. L., Burns, R. A. *et al.* 2022a, *AJ*, 163, 83
Bayandina O. S., Brogan, C. L., Burns, R. A. *et al.* 2022b, *A&A*, 664, A44
Breen S. L., Sobolev A. M., Kaczmarek J. F., *et al.* 2019a, *ApJ* (Letters), 876, L25
Brogan C. L., Hunter, T. R., Cyganowski, C. J., *et al.* 2018, *ApJ*, 866, 87
Brogan C. L., Hunter, T. R., Towner, A. P. M., *et al.* 2019, *ApJ*, 881, L39
Burns R. A., Sugiyama, K., Hirota, T., *et al.* 2020a, *Nature Astronomy*, 4, 506
Burns R. A., Orosz, G., Bayandina, O., *et al.* 2020b, *MNRAS*, 491, 4069
Burns R. A., Kobak, A., Caratti o Garatti A., *et al.* 2022, EVN Mini-Symposium 2021, 19
Burns R. A., Uno, Y., Sakai, N., *et al.* 2023, *Nature Astronomy*, 7, 557
CHIME/FRB Collaboration, *et al.* 2021, *ApJS*, 257, 59
Caratti o Garatti A., Stecklum, B., Garcia Lopez, R., *et al.* 2017, *Nature Physics*, 13, 276
Chen X., Sobolev, A. M., Ren, Z.-Y., *et al.* 2020a, *Nature Astronomy*, 4, 1170
Chen X., Sobolev, A. M., Breen, S. L., *et al.* 2020b, *ApJ* (Letters), 890, L22
Chen Z., Sun W., Chini R., *et al.* 2021, *ApJ*, 922, 90
Ellingsen S. P., Voronkov M. A., Cragg D. M., *et al.* 2007, *Astrophysical Masers and their Environments*, 213
Fujisawa K., Yonekura Y., Sugiyama K., *et al.* 2015, *ATel*, 8286, 1
Greif T. H. 2015, *Computational Astrophysics and Cosmology*, 2, 3
Hirota T., Cesaroni, R., Moscadelli, L., *et al.* 2021, *A&A*, 647, A23
Hosokawa T., Yoshida N., Omukai K., *et al.* 2012, *ApJ* (Letters), 760, L37
Hunter T. R., Brogan, C. L., MacLeod, G., *et al.* 2017, *ApJ* (Letters), 837, L29
Hunter T. R., Brogan, C. L., MacLeod, G., *et al.* 2018, *ApJ* (Letters), 854, 170
MacLeod, G. C., Gaylard, M. J. 1996, *MNRAS*, 280, 868
MacLeod G. C., Smits, D. P., Goedhart, S., *et al.* 2018, *MNRAS*, 478, 1077
MacLeod G. C., Sugiyama, K., Hunter, T. R., *et al.* 2019, *MNRAS*, 489, 3981
Meyer D. M. A., Vorobyov, E. I., Elbakyan, V. G., *et al.* 2019, *MNRAS*, 482, 5459
Minier V., Ellingsen S. P., Norris R. P., *et al.* 2003, *A&A*, 403, 1095
Moscadelli L., Sanna, A., Goddi, C., *et al.* 2017, *A&A*, 600, L8
Moscadelli L., Sanna A., Goddi C., *et al.* 2020, *A&A*, 635, A118
Olech M., Szymczak, M., Wolak, P., *et al.* 2020, *A&A*, 634, A41
Proven-Adzri E., MacLeod G. C., Heever S. P. v. d., *et al.* 2019, *MNRAS*, 487, 2407
Sobolev A. M., Moran, J. M., Gray, M. D., *et al.* 2018, *ApJ*, 856, 60
Stecklum, B., Caratti o Garatti, A., Hodapp, K., *et al.* 2018, Proceedings of the IAU Symposium 336, 37
Stecklum B., Wolf, V., Linz, H., *et al.* 2021, *A&A*, 646, A161
Sugiyama K., Saito Y., Yonekura Y., *et al.* 2019, *ATel*, 12446, 1
Szymczak M., Olech M., Sarniak R., *et al.* 2018a, *MNRAS*, 474, 219
Szymczak M., Olech, M., Wolak, P., *et al.* 2018b, *A&A*, 617, A80
Tapia M., Roth M., Persi P., 2015, *MNRAS*, 446, 4088
Volvach A. E., Volvach L. N., Larionov M. G., *et al.* 2020, *MNRAS*, 494, L59
Yang K., Chen, X., Shen, Z.-Q., *et al.* 2019, *ApJS*, 241, 18
Zinnecker H., Yorke H. W., 2007, *A&A*, 45, 481

Cosmic Masers: Proper Motion toward the Next-Generation Large Projects
Proceedings IAU Symposium No. 380, 2024
T. Hirota, H. Imai, K. Menten, & Y. Pihlström, eds.
doi:10.1017/S1743921323002594

Maser Tracers of Gas Dynamics near Young Stars New Perspectives

Alberto Sanna[1] **and Luca Moscadelli**[2]

[1]INAF, Osservatorio Astronomico di Cagliari, via della Scienza 5, 09047, Selargius, Italy.
email: alberto.sanna@inaf.it

[2]INAF, Osservatorio Astrofisico di Arcetri, Largo E. Fermi 5, 50125 Firenze, Italy

Abstract. The protostellar environment where young stars form has physical conditions suitable to excite a number of molecular maser lines that have traditionally provided an unique probe of star formation kinematics, at the highest angular resolution of radio very long baseline interferometry (VLBI) observations. In the following, we will discuss a number of recent results on our understanding of the gas dynamics traced by masers in the vicinity of young forming stars. These findings provide direct clues on how our community can substantially contribute to the field of star formation in the next decade.

Keywords. Masers, Stars: formation, ISM: kinematics and dynamics

1. Introduction

The protostellar environment where young stars form has physical conditions suitable to induce population inversions between specific pairs of energy levels in several abundant molecules, including water (H_2O), methanol (CH_3OH), hydroxyl (OH), ammonia (NH_3), silicon monoxide (SiO), and formaldehyde (H_2CO). The resulting non-thermal maser emission in the corresponding spectral transitions provides a beacon whose brightness temperature (typically $> 10^8$ K) far exceeds that of the more commonly-excited thermal emission lines. Maser maps typically show a large number of bright and compact spots that mark parcels of gas (or cloudlets) where amplified emission is beamed almost along the line-of-sight. Their enhanced brightness temperature has provided the sole target for spectroscopic Very Long Baseline Interferometry (VLBI) observations, whose milli-arcsecond beam size is inversely proportional to the expected measurement sensitivity. Consequently, since early '80 (e.g., Genzel *et al.* 1981; Gwinn *et al.* 1992), iterated VLBI observations of maser spots, which are excited in the inner circumstellar regions ($\lesssim 1000$ au), have been used to trace the displacement in time of individual cloudlets (namely, their proper motions), which, together with the doppler-shift of the maser lines, provide the full-space motion of local gas. In this context, maser lines have traditionally provided an unique probe of star formation, so far at centimeter wavelengths especially.

When talking about gas dynamics near young forming stars, one commonly is reminded of two major dynamical structures, namely, what we generally call *disk-like structures* and the *jets and winds* launched from these circumstellar regions. In the literature, two molecular masers, water and methanol, have traditionally been used to study these complementary environments, whose association with either one or the other maser species directly follows from the different excitation mechanisms. Firstly, water masers in the 22.2 GHz transition, whose pumping mechanism is predominantly collisional, were identified as tracer of shock phenomena in the vicinity of stars (e.g., Elitzur *et al.* 1989, 1992; Kaufman & Neufeld 1996; Hollenbach *et al.* 2013), either very young or evolved, and their observations has been key to reveal the dynamics of fast shocked layers of ejected gas, even in exceptional environments such as magneto-hydrodynamic disk winds (e.g., Moscadelli *et al.* 2022). Subsequently, methanol masers in the 6.7 GHz transition, whose pumping mechanism is predominantly radiative, were identified as tracer of warm and quiescent gas in the inner (massive) circumstellar regions where the requirements of an IR field and high methanol column densities are satisfied (e.g., Cragg *et al.* 1992, 2005; Sobolev *et al.* 1997).

In the following, we are going to highlight some recent results related to our understanding of water and methanol masers in star formation, and their local environments, whose developments allow us to draw a number of useful predictions for maser observations with next generation facilities, such as the next-generation VLA (ngVLA) and the Square Kilometer Array (SKA).

2. Water masers associated with proto-stellar jets and winds

2.1. *The POETS survey*

Since the last maser symposium, we have spent much efforts on a large maser project called POETS, which stands for the "Protostellar Outflows at the EarliesT Stages" survey. This project has been grounded on a survey of combined water masers and radio continuum observations at the highest angular resolution available towards a large number of young and massive stars, which we have presented in a series of papers between 2018 and 2020, with a pilot study in 2016 (Sanna *et al.* 2018, 2019b; Moscadelli *et al.* 2016, 2019, 2020).

In Fig. 1, we present an example of the type of studies we did, by mapping the water maser distribution (and motions) relative to the position and morphology of the ionised gas around each target of the sample, that had to fulfill four requirements:

- star-forming sites showing evidence of being prior to an ultra-compact H II region phase, which statistically means that a source is typically younger than ten-thousand years;
- massive regions, with luminosities of several thousands times that of our Sun;
- targets associated with bright 22.2 GHz H_2O masers, with flux densities of several tens of Jy and rich spectra of several distinct maser features;
- targets having accurate trigonometric distance measurements, belonging to the "Bar and Spiral Structure Legacy Survey" (Reid *et al.* 2014).

For our final sample of 36 targets, the dataset was made up of: 1) phase-referencing, multi-epoch, Very Long Baseline Array (VLBA) spectral line observations at 22 GHz, and 2) Very Large Array (VLA) continuum observations conducted at three frequency bands, 6, 15, and 22 GHz in the A- and B-array configurations, with sensitivities down to a few μJy beam^{-1}.

Water maser and continuum emissions allow us to trace fast shocked layers of gas and (thermal) free-free from ionized material of jets and winds, the latter generally referred to as "radio thermal jets". For each target, we aimed to quantify the morphology and dynamics of primary outflows, where primary outflows are those whose material is initially (gravitationally) bound to the driving star.

2.2. *Results: correlation among luminosities*

A first general result of POETS follows from a comparison of the (isotropic) water maser luminosity with respect to the radio continuum luminosity of the targets. In Fig. 2 (left), one can appreciate a strong linear correlation between these two quantities, which

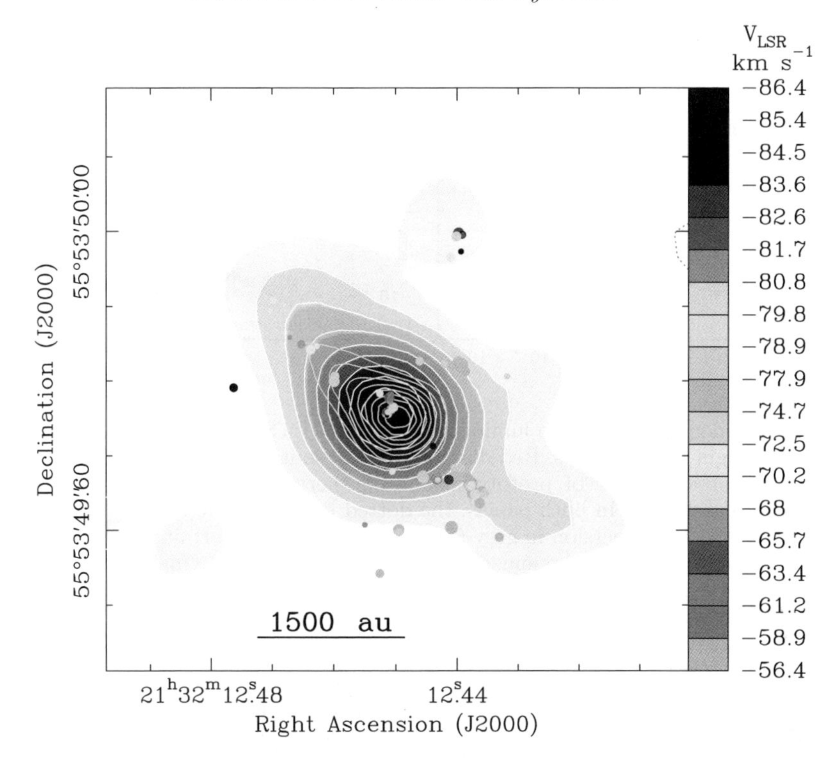

Figure 1. Example of POETS observations for the target source G097.53+03.18. This map shows a typical combination of radio continuum emission, with grey shades and cyan contours for 12 and 22 GHz emission respectively, and 22.2 GHz water maser cloudlets, marked with colored dots according to the line-of-sight velocity scale to the right. Adapted from Moscadelli *et al.* (2016).

is also associated with a one-to-one detection rate of faint ($\lesssim 100\mu$Jy) radio continuum emission towards water masers sites.

This correlation can be better understood when looking at a second correlation (Fig. 2, right), that between the same radio continuum luminosity and the bolometric luminosity of the source, where only targets whose luminosity is dominated by a single object have been considered. In this well-known diagram, the Lyman curve sets the level of radio luminosity that can be produced by Lyman photons from a young star of a given spectral type, which have energies high enough to ionize hydrogen atoms. At a first order, this curve divides the stellar population into low-mass objects to the left and high-mass objects to the right of the plot. The fact that low-mass objects cannot photo-ionize their associated radio continuum, it has long been recognized as evidence of a shock ionization mechanism at work, and this dependence yields the observed linear trend.

The fact that the same linear correlation extends also at higher star masses, for young stars associated with water masers, suggests the following conclusion – *compact radio continuum emission associated with H_2O masers is dominated by shock ionization.* From this evidence, a first corollary would state that, by means of next-generation centimeter interferometers, we expect to simultaneously detect, at milli-arcsecond resolution, masers and (thermal) continuum that are spatially coincident, allowing for an accurate registration of their relative positions.

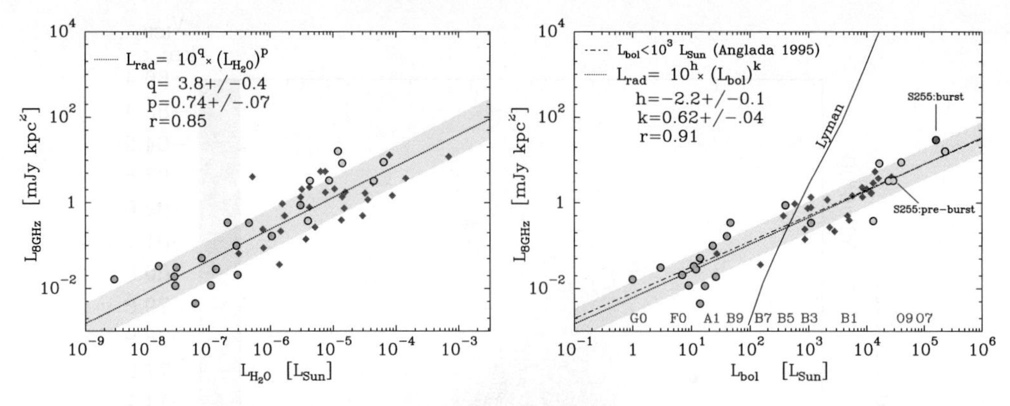

Figure 2. Dependence of the radio luminosity of the POETS sample on the H_2O maser (left) and bolometric (right) luminosities. Red diamonds mark sources from the POETS sample, cyan circles are used for a number of prototypical radio thermal jets, and green circles for low-luminosity H_2O maser sites. In both panels, the dotted bold line traces the χ^2 fit to the sample distribution with the $1\,\sigma$ dispersion in grey (fit parameters and uncertainties on top left). In the right panel, the solid line traces the ionized flux expected from the Lyman continuum of ZAMS stars earlier than B8 and spectral types, corresponding to a given luminosity, are indicated in blue near the lower axis. More details in Fig. 5 of Sanna *et al.* (2018).

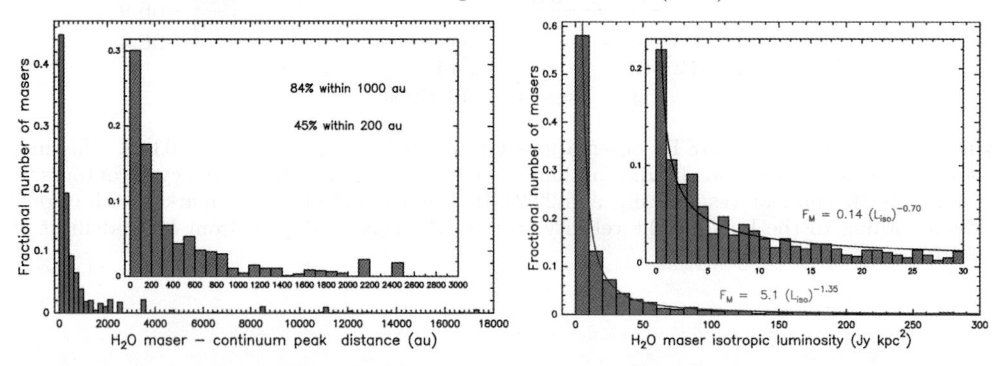

Figure 3. Normalized histograms of the number of H_2O masers found at different distances from their stars (left) and with respect to their intrinsic brightness (right). Adapted from Moscadelli *et al.* (2020).

2.3. *Results: distribution of H_2O masers from the star position*

A second general result of POETS follows from the spatial distribution of water maser cloudlets with respect to the position of the nearby radio continuum peak, generally assumed to pinpoint the young star position when its spectral index is partially optically thick. In the left panel of Fig. 3, we built up a histogram of the number of H_2O masers found at different distances from their stars, considering either spatial scales of several 1000 au (outer plot) and the inner few 1000 au (inset) from the origin. This distribution exponentially increases when the distance from the radio continuum peak decreases, allowing us to statistically state that – *about 80% of the observed water maser cloudlets are found within 1000 au of their associated star, and almost half of them within only a few 100 au.*

Alternatively, this evidence suggests that a given water maser distribution can be used as an accurate proxy of the star position, whenever a significant number of maser cloudlets are detected. Notably, this statistical result, derived for the first time by means of high-resolution and high-sensitivity observations, is at variance with past claims that water masers are sampling large distances of several 1000 au from their associated star.

2.4. *Results: a H_2O maser brightness function – how many maser spots are we missing?*

A third general result of POETS follows from the analysis of the number of water maser detections as a function of their intrinsic brightness, plotted in terms of their isotropic (normalized) luminosity. Their histogram in the right panel of Fig. 3 shows an exponential increase which sharply peaks towards low maser luminosities, similarly to the left panel.

The conclusion of this statistical analysis has a direct implication for our spectroscopic VLBI observations, meaning that – *at sub-mJy sensitivities, we expect to double the current number of H_2O maser detections, that rely on sensitivities an order of magnitude higher.* Alternatively, the combined results of Sect. 2.3 and 2.4 suggest another useful corollary for next-generation centimeter interferometers, that will open up a new era of water maser observations, allowing us to study new phenomena very close to massive young stellar objects (e.g., Moscadelli *et al.* 2022).

3. Methanol masers associated with proto-stellar disk-like structures

Since their discovery in the early '90 (Menten 1991), methanol masers at 6.7 GHz have been associated with the innermost gas surrounding young stars several times more massive than our Sun, due to the physical conditions predicted for the maser excitation (e.g., Cragg *et al.* 2005). A typical methanol maser spectrum at 6.7 GHz shows spectral features within about $10\,\mathrm{km\,s^{-1}}$ from the systemic velocity of nearby gas, suggesting that methanol masers are kinematic tracers of more quiescent gas than water masers are. This characteristic property has a counterpart in the milli-arcsecond regular structure of 6.7 GHz maser cloudlets which persists over many years (Minier *et al.* 2002; Sanna *et al.* 2010a; Moscadelli *et al.* 2011b). Proper motion measurements of 6.7 GHz maser cloudlets published since 2010 have confirmed velocity components on the plane-of-the-sky of the same magnitude of the line-of-sight velocities (Sanna *et al.* 2010a,b).

In the years, following this evidence, methanol masers have been generally referred to be associated with "disk-like structures", although this definition is vague and does not provide specific clues about the environment where masers are excited. For instance, if CH_3OH masers are expected to be quenched above volume densities of approximately $10^8\,\mathrm{cm^{-3}}$, they would not be excited in the inner few 100 au of young and massive stars where gas likely attains Keplerian equilibrium, but would only arise in the less dense outskirts where a slower rotating envelope (or toroid) is predicted. In this respect, are 6.7 GHz maser proper motions consistent with these expectations and what have we learned in the past decade?

In the following, we want to highlight some guidelines for future studies of methanol masers at 6.7 GHz that can be useful to improve our definition of "disk-like" environment.

3.1. *Physical conditions of methanol maser gas*

A first instructive example comes from a well-known star-forming region, G023.01–00.41, and the environment surrounding its most massive star ($\sim 20\,M_\odot$). In Fig. 4, we show the appearance of its disk-like structure as imaged with the Atacama Large Millimeter Array (ALMA) by means of a methanol line near 349 GHz (Sanna *et al.* 2021). For clarity, this image has been rotated by –33° with respect to the equatorial reference frame, in order to display the disk plane and outflow axis along the horizontal and vertical directions, respectively.

Thanks to powerful facilities like ALMA, one can nowadays measure temperature profiles and column (and volume) densities at angular scales comparable to maser cloudlets. For instance, we accurately derived the physical conditions per beam (of few 100 au) across the color map in Fig 4 by fitting methyl-cyanide (CH_3CN) emission lines (insets),

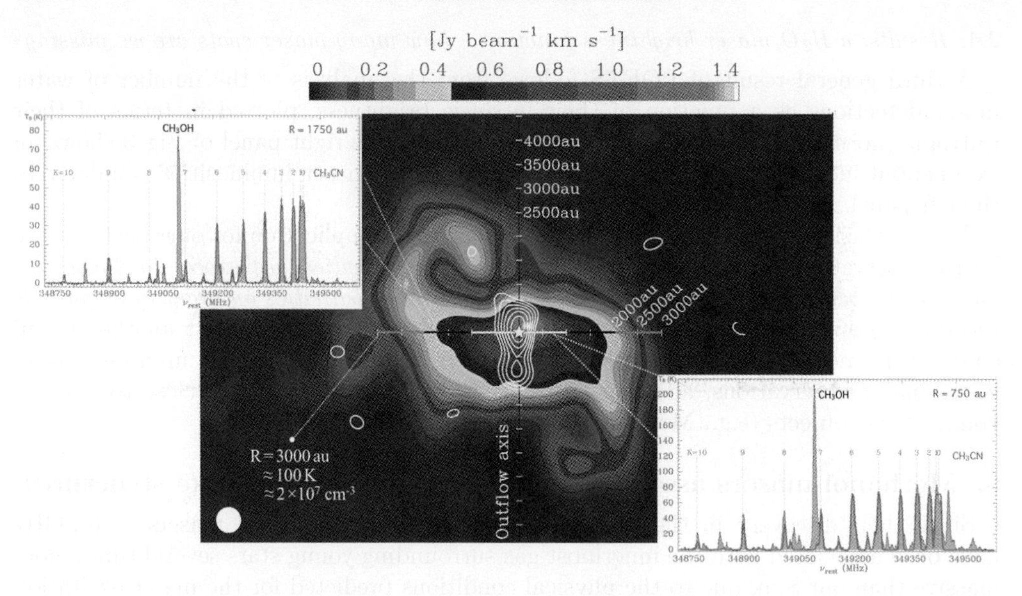

Figure 4. Appearance of the disk-like structure (color map) around the most massive object (star symbol) in the star-forming region, G023.01–00.4, as seen by ALMA through CH_3OH emission at 349.107 GHz (synthesized beam on the bottom left). The equatorial plane of the disk-like emission is drawn from the star along the horizontal axis up to radii of 3000 au, with white ticks at steps of 500 au. White contours along the vertical axis draw the radio thermal jet emission imaged with the VLA at 44 GHz, with comparable angular resolution than ALMA. As an example, insets show two ALMA spectra taken along the disk plane at different distances of 1750 au and 750 au from the central star. These spectra were used to derive the physical conditions of local gas (indicated for a radius of 3000 au), by fitting the line emission from the CH_3CN K-ladder (red profile). Adapted from Sanna *et al.* 2021.

inferring that gas is warmed up to temperatures of 90-100 K at a projected distance (R) of 3000 au from the star. At this distance, gas and dust condense to particle densities of approximately 10^7 cm^{-3}. Moving closer to the central star, the temperature rises exponentially ($R^{-0.4}$) to almost 300 K at a projected distance of 250 au, where densities reach several 10^9 cm^{-3} (Sanna *et al.* 2021).

In Fig. 5, we present the same ALMA image with overlaid the positions of individual maser cloudlets emitting at 6.7 GHz, as we measured in 2010 and registered for the secular motion (Sanna *et al.* 2010b). Their distance with respect to the young star covers a range of projected radii from 2000 au down to about 500 au, avoiding the inner densest regions and in agreement with a quenching threshold of the order of 10^8 cm^{-3}. This example shows we are in a position to draw detailed comparisons of maser positions with thermal line emission and to eventually constrain very local maser conditions. Therefore, we prompt our community for increasing the number of similar studies in the next years, which is crucial information for improving current models of maser excitation.

Another example of the synergies between interferometric observations in the millimeter and centimeter windows is related to the comparison of maser lines from the same molecular species, that are are spatially coincident but are emitted at different frequencies.

In Fig 4, we show two ALMA spectra near 349 GHz extracted at projected distances of 1750 au and 750 au from the central star. The red profile draws the best-fitted spectrum of the methyl-cyanid emission used to derive the local gas conditions, with the lower-energy (K = 1–4) lines that are optically-thick and set the maximum brightness temperature expected under local thermodynamic equilibrium. Notably, at the center of the bands a

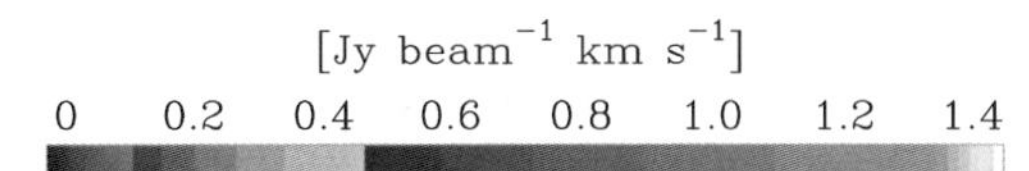

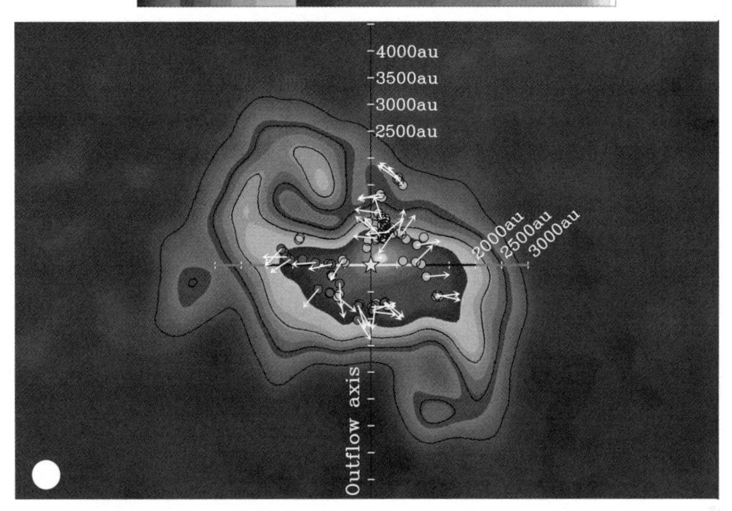

Figure 5. Same background color map as in Fig. 4 with overlaid the VLBI maser emission of methanol at 6.7 GHz (dots). Red and cyan dots indicate red- and blue-shifted line-of-sight maser velocities, respectively, and white arrows draw the local direction of the proper motion vector for individual maser cloudlets. Adapted from Sanna *et al.* 2021.

methanol line with an excitation temperature of 260 K, similar to the CH_3CN K = 4 line, stands much brighter, and its emission was indeed used for the background color map.

Since methanol and methyl-cyanid are thought to be well mixed with each other for gas densities above 10^6 cm^{-3}, these spectra can be interpreted as evidence that the methanol emission at 349.107 GHz has a strong maser contribution, in addition to a plateau of thermal emission. Moreover, this behavior is observable up to distances of about 2000 au from the central star, similar to the 6.7 GHz maser distribution. It is also worth noting that the same millimeter methanol maser was reported to be excited in another massive young star by Zinchenko *et al.* (2018). Following these findings, we further prompt our community to increase the number of millimeter masers studies, whose combination with their known centimeter counterparts can provide a unique information to improve both, maser models and our definition of "disk-like" environment.

3.2. *Expanding motions vs. infalling motions*

A final remark concerns the information coming from the analysis of full-space motions of methanol masers. From a kinematic point of view, the most characteristic pattern of a disk-like structure is expected to be rotation about an outflow axis, although a rotational component can be mixed to expanding or infalling gas motions depending on the spatial region where maser are excited. Is there any clear indication coming from maser proper motions so far?

As an example, we refer again to the star-forming region, G023.01–00.41. In Fig 5, the full-space motions of methanol masers can be decomposed in two main components, one along the line-of-sight and one on the plane of the sky. On the one hand, the former component bear the signs of (sub-Keplerian) rotation, with the redshifted and blueshifted masers to the west and east of the star position, respectively. On the other hand, the latter component traces gas pushed away from the central source, both along the disk plane and the perpendicular direction. This scenario of gas slowly expanding (and rotating) near a young star has been recurrently observed via 6.7 GHz maser proper motion measurements (e.g., Sanna *et al.* 2010b; Moscadelli *et al.* 2011a, 2013; Bartkiewicz *et al.* 2020).

Then, are 6.7 GHz masers preferentially tracing gas expansion? Or, alternatively, can 6.7 GHz masers be also associated with gas infall and, if yes, under which conditions? Interestingly, two examples exist where the full-space motions of methanol masers provide evidence of gas infall towards the central star, in the star-forming regions AFGL 5142 and Cepheus A (Goddi *et al.* 2011; Sugiyama *et al.* 2014; Sanna *et al.* 2017). Although this scenario seems less common so far, our observations might be biased towards more luminous and likely more evolved targets, where mass accretion (thus gas infall) might be reduced significantly.

Nevertheless, the chance of a direct measurement of gas infall by means of 6.7 GHz methanol masers appears very promising and would provide an extremely valuable information to the star formation community (otherwise difficult to obtain). For this reason, we prompt our community to search for methanol maser targets with evidence of infall in the next years, to increase the number of detections and eventually allow a detailed statistical analysis.

References

Bartkiewicz, A., Sanna, A., Szymczak, M., *et al.* 2020, *A&A*, 637, A15
Cragg, D. M., Johns, K. P., Godfrey, P. D., *et al.* 1992, *MNRAS*, 259, 203
Cragg, D. M., Sobolev, A. M., & Godfrey, P. D. 2005, *MNRAS*, 360, 533
Elitzur, M., Hollenbach, D. J., & McKee, C. F. 1989, *ApJ*, 346, 983
Elitzur, M., Hollenbach, D. J., & McKee, C. F. 1992, *ApJ*, 394, 221
Genzel, R., Reid, M. J., Moran, J. M., *et al.* 1981, *ApJ*, 244, 884
Goddi, C., Moscadelli, L., & Sanna, A. 2011, *A&A*, 535, L8
Gwinn, C. R., Moran, J. M., & Reid, M. J. 1992, *ApJ*, 393, 149
Hollenbach, D., Elitzur, M., & McKee, C. F. 2013, *ApJ*, 773, 70
Kaufman, M. J. & Neufeld, D. A. 1996, *ApJ*, 456, 250
Menten, K. M. 1991, *ApJL*, 380, L75
Minier, V., Booth, R. S., & Conway, J. E. 2002, *A&A*, 383, 614
Moscadelli, L., Cesaroni, R., Rioja, M. J., *et al.* 2011, *A&A*, 526, A66
Moscadelli, L., Sanna, A., & Goddi, C. 2011, *A&A*, 536, A38
Moscadelli, L., Li, J. J., Cesaroni, R., *et al.* 2013, *A&A*, 549, A122
Moscadelli, L., Sánchez-Monge, Á., Goddi, C., *et al.* 2016, *A&A*, 585, A71
Moscadelli, L., Sanna, A., Goddi, C., *et al.* 2019, *A&A*, 631, A74
Moscadelli, L., Sanna, A., Goddi, C., *et al.* 2020, *A&A*, 635, A118
Moscadelli, L., Sanna, A., Beuther, H., *et al.* 2022, *Nature Astronomy*, 6, 1068
Reid, M. J., Menten, K. M., Brunthaler, A., *et al.* 2014, *ApJ*, 783, 130
Sanna, A., Moscadelli, L., Cesaroni, R., *et al.* 2010, *A&A*, 517, A71
Sanna, A., Moscadelli, L., Cesaroni, R., *et al.* 2010, *A&A*, 517, A78
Sanna, A., Moscadelli, L., Surcis, G., *et al.* 2017, *A&A*, 603, A94
Sanna, A., Moscadelli, L., Goddi, C., *et al.* 2018, *A&A*, 619, A107
Sanna, A., Moscadelli, L., Goddi, C., *et al.* 2019, *A&A*, 623, L3
Sanna, A., Giannetti, A., Bonfand, M., *et al.* 2021, *A&A*, 655, A72
Sobolev, A. M., Cragg, D. M., & Godfrey, P. D. 1997, *A&A*, 324, 211
Sugiyama, K., Fujisawa, K., Doi, A., *et al.* 2014, *A&A*, 562, A82
Zinchenko, I., Liu, S.-Y., Su, Y.-N., *et al.* 2018, *IAU Symposium*, 332, 270

Cosmic Masers: Proper Motion toward the Next-Generation Large Projects
Proceedings IAU Symposium No. 380, 2024
T. Hirota, H. Imai, K. Menten, & Y. Pihlström, eds.
doi:10.1017/S1743921323002132

Snapshot of a magnetohydrodynamic disk wind traced with water masers

Luca Moscadelli[1], **Alberto Sanna**[2,3], **Henrik Beuther**[4], **André Oliva**[5,6] **and Rolf Kuiper**[7]

[1]INAF-Osservatorio Astrofisico di Arcetri, Largo E. Fermi 5, Firenze 50125, Italy.
email: luca.moscadelli@inaf.it

[2]INAF-Osservatorio Astronomico di Cagliari, Via della Scienza 5, Selargius (CA) 09047, Italy

[3]Max-Planck-Institut für Radioastronomie, Auf dem Hügel 69, Bonn 53121, Germany

[4]Max Planck Institute for Astronomy, Königstuhl 17, Heidelberg 69117, Germany

[5]Institut für Astronomie und Astrophysik, Auf der Morgenstelle 10, Tübingen 72076, Germany

[6]Space Research Center (CINESPA), Ciudad Universitaria Rodrigo Facio, San José 11501, Costa Rica

[7]Faculty of Physics, University of Duisburg-Essen, Lotharstraße 1, Duisburg 47057, Germany

Abstract. Disk-jet systems are common in astrophysical sources of different nature, from black holes to gaseous giant planets. The disk drives the mass accretion onto a central compact object and the jet ejects material along the disk rotation axis. Magnetohydrodynamic disk winds can provide the link between mass accretion and ejection, which is essential to ensure that the excess angular momentum is removed and accretion can proceed. However, up to now, we have been lacking direct observational proof of disk winds. This work presents a direct view of the velocity field of a disk wind around a forming massive star. Achieving a very high spatial resolution of 0.05 au, our water maser observations trace the velocities of individual streamlines emerging from the disk orbiting the forming star. We find that, at low elevation above the disk midplane, the flow co-rotates with its launch point in the disk, in agreement with magneto-centrifugal acceleration. Beyond the co-rotation point, the flow rises spiraling around the disk rotation axis along a helical magnetic field. We have performed (resistive-radiative-gravito-) magnetohydrodynamic simulations of the formation of a massive star and record the development of a magneto-centrifugally launched jet presenting many properties in agreement with our observations.

Keywords. ISM: jets and outflows, ISM: kinematics and dynamics, Stars: formation, Masers

---

1. Introduction

Accretion and ejection of matter are intimately related in different astrophysical objects covering a wide spectrum of masses, from black holes to gaseous giant planets. Magnetohydrodynamic (MHD) disk winds (Blandford and Payne 1982; Pudritz and Norman, 1983) can provide the link between mass accretion and ejection, but, so far, a direct observational proof of their existence has been lacking. The "Protostellar Outflows at the Earliest Stages" (POETS) survey (Moscadelli et al. 2016; Sanna et al. 2018) has recently imaged the disk/outflow interface on scales of 10–100 au in a statistically significant sample (37) of luminous YSOs, employing multi-frequency Jansky Very Large Array (JVLA) observations to determine the spatial structure of the ionized

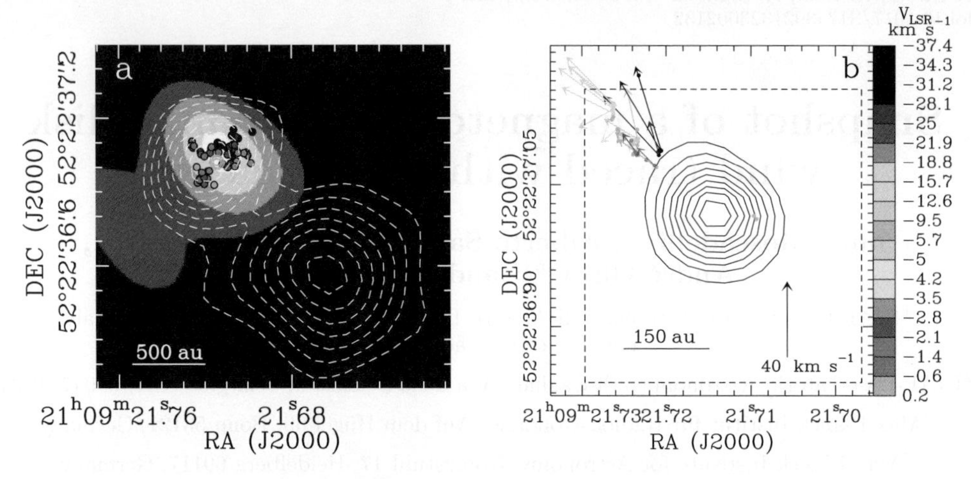

Figure 1. Previous NOEMA, JVLA and VLBA observations towards IRAS 21078+5211. (a) The gray-scale (from 10 to 35 mJy beam^{-1}) map reproduces the NOEMA 1.37 mm continuum emission (Moscadelli et al. 2021). The colored dots represent the channel emission peaks of the CH_3CN $J_K = 12_K$-11_K (K = 3–6) and HC_3N J = 24-23 lines, with colors denoting the channel $V_{\rm LSR}$: blue for [−13.6, −10] and red [−2.5, 0.5] km s^{-1}. The green (70%, 80%, and 90% of 0.50 mJy beam^{-1}) and white (from 30% to 90%, in steps of 10% of 0.096 mJy beam^{-1}) contours show the JVLA A-Array continuum at 1.3 cm and 5 cm, respectively. (b) Colored dots and arrows give absolute positions and proper motions of the 22 GHz water masers determined with multi-epoch (2010–2011) VLBA observations (Moscadelli et al. 2016), with colors denoting the maser $V_{\rm LSR}$. The black contours (from 10% to 90%, in steps of 10% of 0.50 mJy beam^{-1}) indicate the JVLA A-Array continuum at 1.3 cm. The dashed rectangle delimits the field of view plotted in Fig.2a.

emission, and multi-epoch Very Long Baseline Array (VLBA) observations to derive the three-dimensional (3D) velocity distribution of the 22 GHz water masers. One of the most interesting POETS targets is the YSO IRAS 21078+5211 ($L_{\rm bol} \sim 5\ 10^3\ L_\odot$ at a distance of 1.63±0.05 kpc, $M_{\rm YSO} = 5.6\pm2\ M_\odot$; Moscadelli et al. 2021). On scales of ~1000 au (see Fig. 1a), the location of the YSO is marked by the peaks of the Northern Extended Millimeter Array (NOEMA) 1.3 mm, and JVLA 1.3 and 5 cm continuum emissions, well aligned in position within the observational errors. The 5 cm continuum presents two slightly resolved components oriented along SW-NE, and traces the radio jet powering the collimated molecular outflow revealed with the NOEMA SO emission at larger scales. The NOEMA CH_3CN and HC_3N channel emission centroids are red- and blue-shifted to SE and NW of the YSO, respectively, and their velocity pattern is best interpreted in terms of rotation, most likely within an accretion disk. The 22 GHz water masers emerge at separations between 100 and 200 au from the YSO (see Fig. 1b), pinpointed by the 1.3 cm continuum peak. The maser spatial distribution and proper motions are collimated along the jet/outflow direction, clearly indicating that the water masers are tracing the base of the radio jet. The analysis of the 3D maser motions, specifically the local standard of rest (LSR) velocity ($V_{\rm LSR}$) gradient transversal to the jet axis and the constant ratio between the toroidal and poloidal velocities, first suggested that the jet could be launched from a MHD disk wind.

2. Results

In October 2020, we have re-observed the water maser emission in IRAS 21078+5211 with global Very Long Baseline Interferometry (VLBI) observations achieving a sensitivity of 0.7 mJy beam^{-1}, which has allowed us to detect weaker masers and sample a

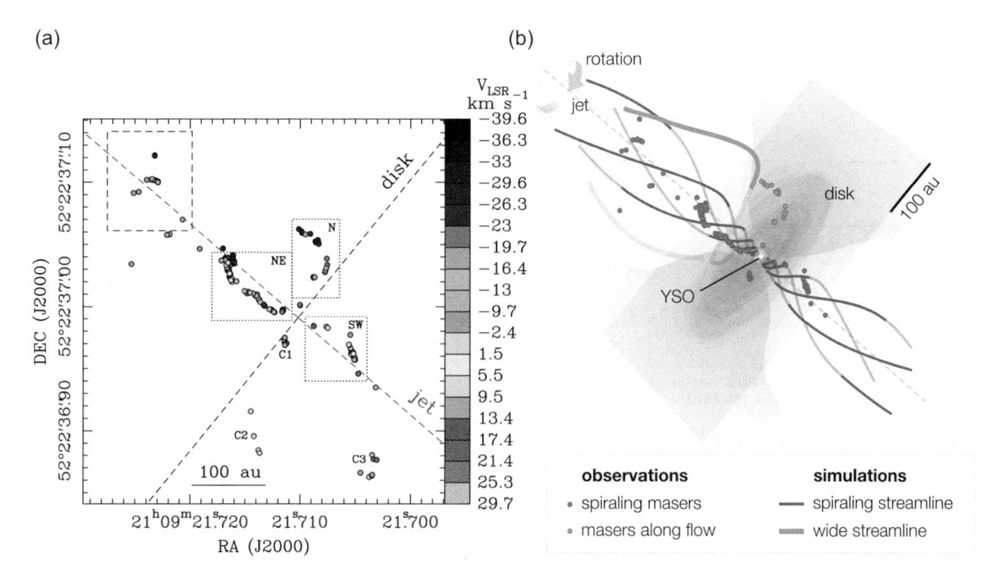

Figure 2. Global VLBI observations of the 22 GHz water masers and 3D view of the proposed kinematical interpretation (Moscadelli et al. 2022). (a) Colored dots give absolute positions of the 22 GHz water masers, with colors denoting the maser $V_{\rm LSR}$. The black dotted rectangles encompass the three regions, to the N, NE and SW, where maser emission concentrate. The blue dashed rectangle delimits the area of maser emission in the previous VLBA observations. The red and black dashed lines mark the sky-projected jet and disk axis, respectively. (b) Observed maser positions (red and orange dots) overlaid on top of streamlines (blue curves) computed from resistive-radiative-gravito-MHD simulations of a jet around a forming massive star. The streamlines close to the rotation axis show significant spiraling motion, in agreement with the kinematic signature of the masers observed in the NE and SW regions. The wide streamline from the simulation illustrates the outflowing trajectory of material from the outer disk, similar to the observed masers in the N region. For context, the protostar, the disk and the outflow cavity have been sketched in gray, based on the density structure obtained in the simulations.

region closer to the YSO (Moscadelli et al. 2022). The maser emission concentrates in three regions to NE, N, and SW, inside the three dotted rectangles of Fig.2a. The jet (the dashed red line in Fig.2a) and disk (dashed black line) axes provide a convenient coordinate system to refer the maser positions to. The interpretation of the maser kinematics is based on the analysis of the three independent observables: z, the elevation above the disk plane (or offset along the jet), R, the radial distance from the jet axis (or transversal offset), and the maser $V_{\rm LSR}$. The accuracy of the maser positions is ≈ 0.05 au, and that of the maser $V_{\rm LSR} \approx 0.5$ km s^{-1}. We can express the maser velocities as the sum of two terms, one associated with the toroidal component or rotation around the jet axis, $V_{\rm rot}$, and the other associated with the poloidal component including all the contributions owing to non-rotation, $V_{\rm off}$. Since, as indicated by previous observations (Moscadelli et al. 2021), the jet axis is close to the plane of the sky and we observe the rotation close to edge-on, we can write:

$$V_{\rm LSR} = V_{\rm off} + V_{\rm rot} = V_{\rm off} + \omega\ \Sigma\ \sin(\phi) \quad (1)$$

$$R = \Sigma\ \sin(\phi) \quad (2)$$

where ϕ is the angle between the rotation radius Σ and the line of sight, and ω is the angular velocity.

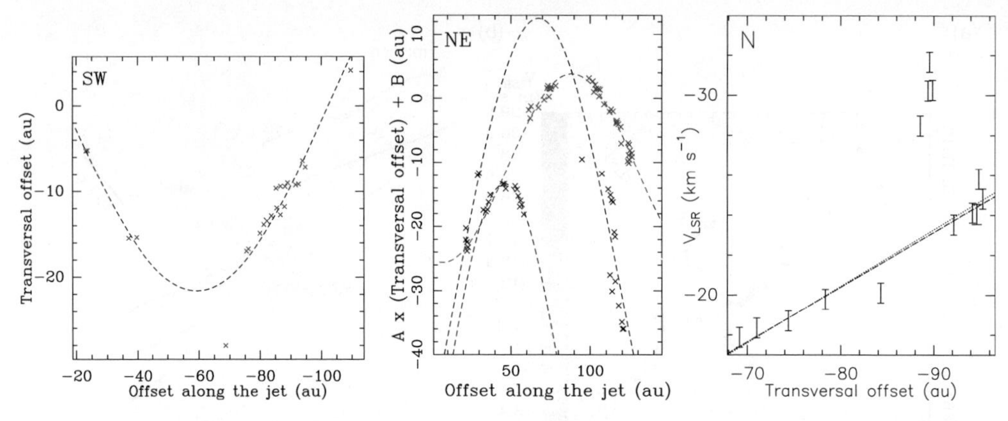

Figure 3. The SW and NE spiral motions, and the magneto-centrifugally accelerated N stream. (SW) Plot of the coordinates R versus z for the water masers in the SW region. The positional error is smaller than the cross size. The black dashed curve is the fitted sinusoid. (NE) Plot of the linear transformation of R versus z for the water masers in the NE region. Black, red, and blue colors refer to masers belonging to the NE-1, NE-2, and NE-3 streams, respectively. We plot the linear transformation of the radii to reduce the overlap and improve the visibility of each of the three streams. The dashed curves are the fitted sinusoids. (N) Plot of maser $V_{\rm LSR}$ versus R in the N region. Errorbars denote the maser $V_{\rm LSR}$ and corresponding 2σ errors, and the black dashed and dotted lines show the best linear fit and the associated uncertainty, respectively. The linear fit of $V_{\rm LSR}$ versus R has been performed considering only the masers with $V_{\rm LSR} \geq -27$ km s^{-1}.

Figs. 3 SW and 3NE show the remarkable finding that the spatial coordinates z and R of the maser emission in both the SW and NE regions satisfy the relation:

$$R = C\ \sin(f_z(z - z_0)) \tag{3}$$

where C, the amplitude of the sinusoid, f_z, the spatial frequency, and z_0, the position of zero phase, are fitted constants. While in the SW region the water masers draw a single sinusoid, in the NE region they belong to three different sinusoidal streams (labeled NE-1, NE-2 and NE-3). The comparison of Eqs. 2 and 3 leads to a straightforward interpretation of the sinusoidal relation between the coordinates by taking: 1) $\Sigma = C$, and 2) $\phi = f_z\ \|z - z_0\|$. The former equation indicates that the rotation radius is the same for all the masers, the latter shows that the motions of rotation around and streaming along the jet axis are locked together, which is the condition for a spiral motion.

The spatial distribution of the masers in the N region presents an arc-like shape (see Fig. 2a), and Fig. 3N shows that the maser $V_{\rm LSR}$ increases linearly with the rotation radius R. The relatively large separation from the jet axis and radial extent suggests that the N stream is observed close to the plane of the sky. In this case, the maser $V_{\rm LSR}$ should mainly trace rotation. Then, the good linear correlation between $V_{\rm LSR}$ and R indicates that the masers co-rotate at the same angular velocity, $\omega_{\rm N} = 0.274 \pm 0.005$ km s^{-1} au^{-1}. A simple interpretation is in terms of a magneto-centrifugally accelerated stream of gas emerging from a point of the disk. A disk in Keplerian rotation around an YSO of about 5.6 $M_\odot$ attains an angular velocity equal to $\omega_{\rm N}$ at $R \approx 40$ au. Remarkably, the linear extrapolation to lower elevation and smaller radii of the arc traced by N stream intercepts the disk axis close to 40 au (Moscadelli et al. 2022, see Fig. 6a), as expected if the gas, launched from the disk, first streams approximately along a straight line.

In conclusion, the analysis of the kinematics of the water masers in IRAS 21078+5211 strongly indicates that a MHD disk wind is a natural frame to explain both the spiral motions traced by the masers close to the jet axis in the NE and SW regions and the gas co-rotation along the N stream. Our interpretation is supported by (resistive-radiative-gravito-) MHD simulations of the formation of a massive star that lead to a magneto-centrifugally launched jet whose streamlines closely reproduce the maser patterns (see Fig. 2b). Presently, these results provide one of the best evidence for a MHD disk wind. Since water maser emission is widespread in YSOs, sensitive VLBI observations of water masers can be a valuable tool to investigate the physics of disk winds.

References

Blandford, R. D. & Payne, D. G. 1982, *M*NRAS, 199, 883–903.
Moscadelli, L., Beuther, H., Ahmadi, A., *et al.* 2021 *A*&A, 647, A114.
Moscadelli, L., Sánchez-Monge, Á., Goddi, *et al.* 2016, *A*&A, 585, A71.
Moscadelli, L., Sanna, A., Beuther, H., *et al.* 2022, *N*ature Astronomy, 6, 1068–1076.
Pudritz, R. E. & Norman, C. A. 1983, *A*pJ, 274, 677–697.
Sanna, A., Moscadelli, L., Goddi, C., *et al.* 2018, *A*&A, 619, A107.

Cosmic Masers: Proper Motion toward the Next-Generation Large Projects
Proceedings IAU Symposium No. 380, 2024
T. Hirota, H. Imai, K. Menten, & Y. Pihlström, eds.
doi:10.1017/S1743921323003216

The water and methanol masers in the face-on accretion system around the high-mass protostar G353.273+0.641

Kazuhito Motogi[1], Tomoya Hirota[2], Masahiro N. Machida[3], Kei E. I. Tanaka[4] and Yoshinori Yonekura[5]

[1]Graduate School of Sciences and Technology for Innovation, Yamaguchi University, Yamaguchi, Yamaguchi 753-8512, Japan. email: kmotogi@yamaguchi-u.ac.jp

[2]Mizusawa VLBI Observatory, National Astronomical Observatory of Japan, Oshu, Iwate 023-0861, Japan

[3]Department of Earth and Planetary Sciences, Faculty of Sciences, Kyushu University, Fukuoka, Fukuoka 819-0395, Japan

[4]Department of Earth and Planetary Sciences, Tokyo Institute of Technology, Meguro, Tokyo, 152-8551, Japan

[5]Center for Astronomy, Ibaraki University, Mito, Ibaraki 310-8512, Japan

Abstract. We report on a direct comparison of VLBI maser data and ALMA thermal-emission data for the high-mass protostar G353.273+0.641. We detected a gravitationally-unstable disk by dust and a high-velocity jet traced by a thermal CO line by ALMA long-baselines (LB). 6.7 GHz CH_3OH masers trace infalling streamlines inside the disk. The innermost maser ring indicates another compact accretion disk of 30 au. Such a nested system could be caused by angular momentum transfer by the spiral arms. 22 GHz H_2O masers trace the jet-accelerating region, which are directly connecting the CO jet and the protostar. The recurrent maser flares imply episodic jet ejections per 1–2 yr, while typical separation of CO knots indicates a variation of outflow rate per 100 yr. Our study demonstrates that VLBI maser observations are still a powerful tool to explore detailed structures nearby high-mass protostars by combining ALMA LB.

Keywords. ISM:jets and outflows, accretion disks, stars: massive, stars: protostars

1. Introduction

Astronomical masers in high-mass star-forming regions have been known as a convenient tracer of several compact circumstellar structures that were only resolved by a Very Long Baseline Interferometer (VLBI). However, it was sometimes difficult to confirm what each maser traces because a resolution for complementary thermal emission was basically insufficient. Nowadays, the Atacama Large Millimeter/submillimeter Array (ALMA) long baseline (LB) provides extremely sensitive and high-resolution thermal data that allow us to directly compare a VLBI maser distribution and thermal emissions. This paper reports on a detailed comparison of H_2O/CH_3OH masers associated with a disk-jet system detected by ALMA LB toward a very young high-mass protostar.

1.1. *Target: G353.273+0.641*

The target source, G353.273+0.641 (hereafter G353), is a relatively nearby ($\sim$ 1.7 kpc) high-mass protostar located in the NGC6357 region (Motogi *et al.* 2016). The current stellar mass (M_*) is $\sim$ 10 $M_\odot$ (Motogi *et al.* 2017). The accretion system in G353 was found to be oriented nearly face-on base on the VLBI kinematics of H_2O masers (Motogi *et al.* 2016). Motogi *et al.* (2019) have conducted the first ALMA LB observation at 150 GHz. They have resolved an infalling rotating envelope. An accretion rate $\dot{M}$ on to the stellar surface has been estimated as $\sim 3.0\times10^{-3}$ $M_\odot$ yr^{-1}. This $\dot{M}$ suggested that the accretion age of G353 ($t_{\rm acc} = M_*/\dot{M}$) is only $\sim 3\times10^3$ yr. Motogi *et al.* (2019) have also found a compact Keplerian disk of 250 au in radius. The disk is very massive ($>$ 2 $M_\odot$) with significant substructures, which indicates the gravitationally unstable nature. These facts indicate that G353 is still in the early evolutionary stage.

1.2. *Masers in G353*

The 6.7 GHz CH_3OH masers in G353 were reported by Motogi *et al.* (2016). The masers showed spiral-like distribution. Their kinematic analysis suggested that these masers trace infalling streamlines down to 15 au in radius. The 22 GHz H_2O masers were associated with a compact bipolar molecular jet almost along the line-of-sight (Motogi *et al.* 2016). The H_2O masers showed episodic maser flare activity. A typical time scale of the flare is $\sim$ 1 yr (Motogi *et al.* 2016). These maser flares were accompanied by drastic changes of a maser distribution and recurrent acceleration of molecular gas. These facts have indicated episodic shock propagations and maser excitations via jet-launching events in such a short time.

2. New ALMA follow-up observations

We have performed higher-resolution observations at 230 GHz in the ALMA Cycle 6. We have aimed to completely resolve the unstable face-on disk by dust continuum emission. We also observed ^{12}CO (J = 2–1) emission to study a thermal counterpart of the maser jet. We have obtained a synthesized beam of 30 and 140 milli-arcsecond (mas) for the dust continuum and the CO line, respectively. The 1-σ image noise levels achieved were 90 and 500 μJy beam^{-1} for continuum and line, respectively. We adopted a circular beam of 30 mas for the continuum as in Motogi *et al.* (2019) in order to highlight any non-axisymmetric structure. This beam size was slightly smoothed compared to the original beam of 28$\times$22 mas^2.

3. Direct comparison of thermal disk-jet system and masers

3.1. *Spiral disk and CH_3OH masers*

We have successfully resolved the compact disk in G353. The disk has two prominent spiral arms and further clumpy substructures (see Figure 1). Although such a spiral disk has recently been reported in a few high-mass protostars (*e.g.*, Johnston *et al.* 2020; Burns *et al.* 2023), the disk in G353 is the most compact, and hence, expected to be the youngest one. The CH_3OH masers are clearly connecting the outer spiral arms and the innermost compact emission. The maser kinematics indicates the infalling streamlines, as mentioned above (Motogi *et al.* 2017). We suggest that the infalling maser streams can be caused by the removal of angular momentum via the gravitational torque of the spiral arms. The innermost compact substructure is complementary to the semi-ring-like maser distribution. This structure may trace the innermost accretion shocks at the edge of a smaller nested disk.

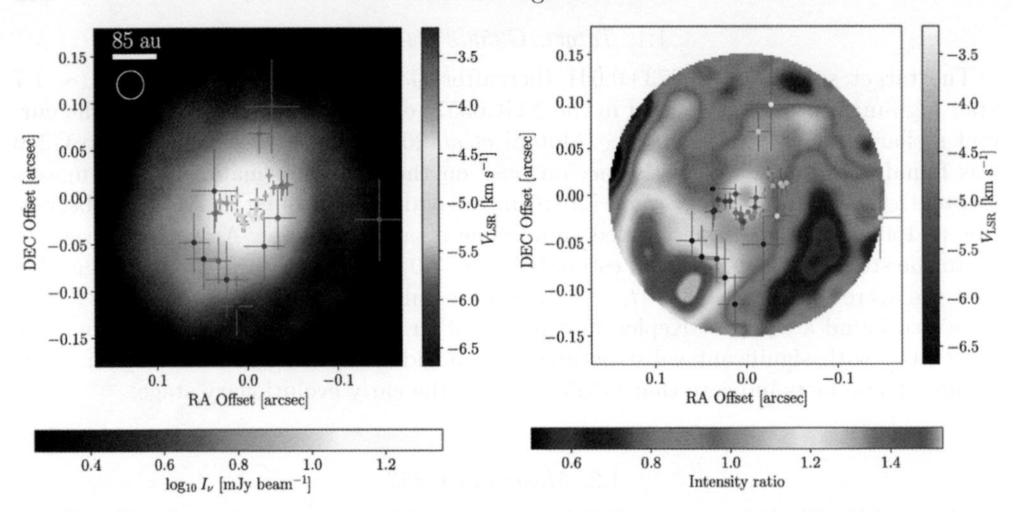

Figure 1. The left panel is the dust continuum image of the accretion disk at 230 GHz in the log scale. The color points indicates positions of 6.7 GHz CH_3OH masers reported in Motogi *et al.* (2017). The synthesized beam is presented in upper left corner. A color of each point indicate the Local Standard of Rest (LSR) velocity of each maser feature. The color in the right panel shows intensity excess map, where the original image was divided by the azimuthally-averaged value in each radius. The outside of centrifugal radius ($\sim$ 250 au) is masked. The grey points show the CH_3OH masers again.

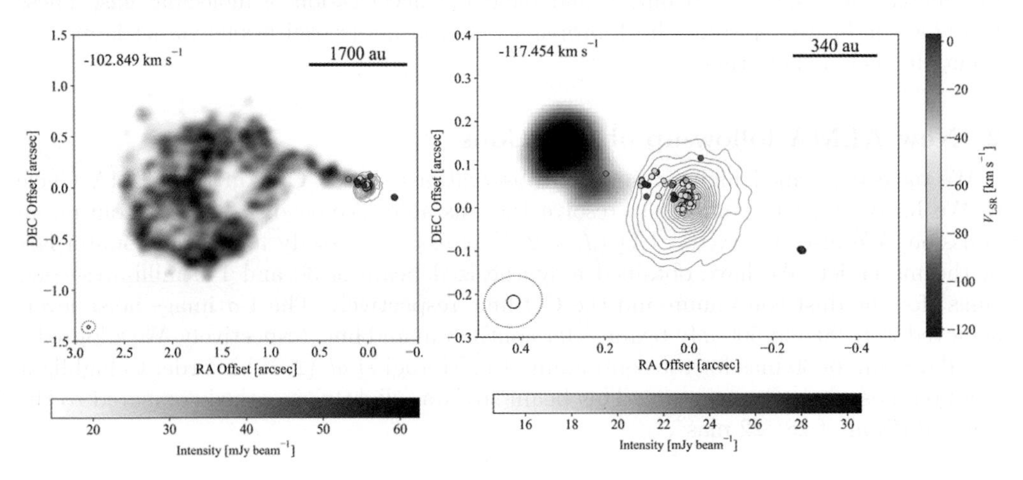

Figure 2. The grey scale in the left panel presents a channel map of the ^{12}CO jet at the LSR velocity shown in upper left corner. The right panel shows the zoom-up view of the most blue-shifted CO jet. The green contours in both panels show the dusty disk in Figure 1, which are from 10% to 99% with a step of 5% of the peak intensity. The synthesized beams for dust and CO lines are presented in lower left corners (see main text). The color points in both panels indicates the positions and LSR velocities of the H_2O masers in Motogi *et al.* (2016).

3.2. *CO jet and H_2O maser jet*

Figures 2 and 3 show channel maps of the extremely blue/red-shifted CO emission. The velocity extent of the CO jet exceeds $\pm$ 100 km s^{-1} with respect to the systemic velocity of -5 km s^{-1}. The velocity range of the blue-shifted side is up to -120 km s^{-1}. This is very similar to that of the blue-shift dominated H_2O maser emission in G353. However, the CO and H_2O masers are not spatially overlapped. The blue-shifted CO jet is distributed over an extension of the H_2O maser jet toward the East. The right panel

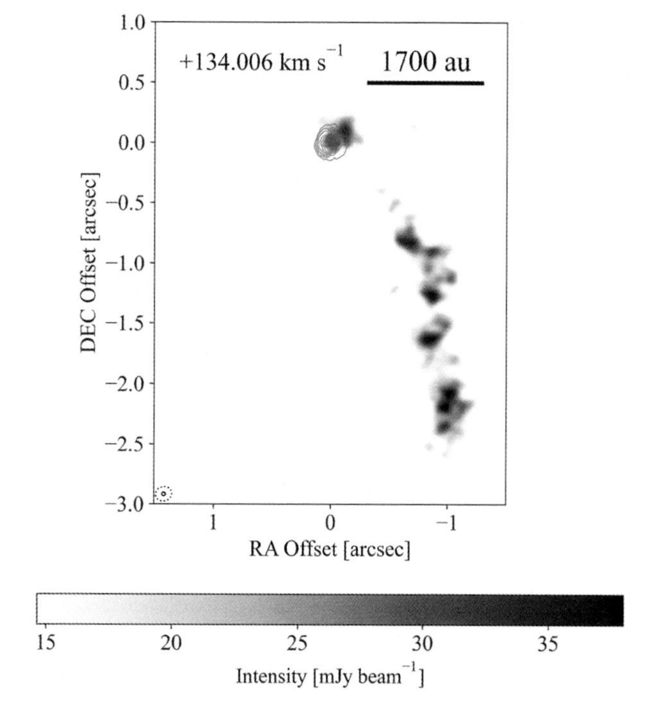

Figure 3. The grey scale shows a channel map of the red-shifted CO jet. The green contours are the same as Figure 2.

of Figure 2 shows the close-up view of the most blue-shifted CO emission. It is evident that the root of the CO jet is connected to the head of the H_2O maser jet. This geometry and the similar maximum velocity are consistent with the idea that the H_2O masers are recurrently excited in the acceleration region of the molecular jet within 100-au scale as suggested by Motogi *et al.* (2016).

Although the red-shifted CO jet is gradually bending toward the South, the overall structures are simpler compared to the blue-shifted jet, which creates a large knot at the Eastern end. A clumpy structure of the red-shifted jet suggests episodic mass ejection. A typical separation of each clump is 200–300 mas corresponding to an ejection interval of 100 yr. This ejection time is comparable to the rotation time scale of the disk at 50 au in a radius around a 10 $M_{\odot}$ protostar. But the launching point expected from the escape velocity is much smaller ($\sim$ 1 au). This fact could imply that the variation of CO jet is caused by episodic accretion (*e.g.*, Machida & Basu 2019). In this case, the infall rate could become non-steady at the scale of the CH_3OH streamlines. On the other hand, the shock propagation intervals traced by H_2O maser flares are much shorter ($\sim$ 1 yr). It may reflect some dynamical perturbations at the jet launching point ($\sim$ 1 au) or typical driving interval of a magnetohydrodynamic jet via amplification and relaxation of magnetic pressure.

Our combined ALMA-VLBI study has confirmed the association of masers with the innermost disk-jet system. Moreover, the masers reveal numerous suggestive substructures that elude even the high-resolution capabilities of the ALMA LB. We conclude by emphasizing that this combined approach significantly enhances the value of both VLBI and ALMA data for investigating high-mass star-forming regions.

References

Burns, R. A., Uno, Y., Sakai, N., *et al.* 2023, *Nature Astronomy*, 7, 557
Johnston, K. G., Hoare, M. G., Beuther, H., *et al.* 2020, *A&A*, 634, L11
Machida, M. N. & Basu, S. 2019, *ApJ*, 876, 149
Motogi, K., Sorai, K., Honma, M., *et al.* 2016, *PASJ*, 68, 69
Motogi, K., Hirota, T., Sorai, K., *et al.* 2017, *ApJ*, 849, 23
Motogi, K., Hirota, T., Machida, M. N., *et al.* 2019, *ApJ*(Letters), 877, L25

Cosmic Masers: Proper Motion toward the Next-Generation Large Projects
Proceedings IAU Symposium No. 380, 2024
T. Hirota, H. Imai, K. Menten, & Y. Pihlström, eds.
doi:10.1017/S1743921323002351

Monitoring of the polarized H_2O maser emission around the massive protostars W75N(B)-VLA 1 and W75N(B)-VLA 2

Gabriele Surcis[1], Wouter H.T. Vlemmings[2], Ciriaco Goddi[1,3] and José-María Torrelles[4,5]

[1]INAF - Osservatorio Astronomico di Cagliari, Via della Scienza 5, I-09047, Selargius, Italy.
email: gabriele.surcis@inaf.it

[2]Department of Space, Earth and Environment, Chalmers University of Technology, Onsala Space Observatory, SE-439 92 Onsala, Sweden

[3]Dipartimento di Fisica, Università degli Studi di Cagliari, SP Monserrato-Sestu km 0.7, I-09042 Monserrato, Italy

[4]Institut de Ciències de l'Espai (ICE, CSIC), Can Magrans s/n, E-08193, Cerdanyola del Vallès, Barcelona, Spain

[5]Institut d'Estudis Espacials de Catalunya (IEEC), Barcelona, Spain

Abstract. Several radio sources have been detected in the high-mass star-forming region W75N(B), among them the massive young stellar objects VLA 1 and VLA 2 are of great interest. These are thought to be in different evolutionary stages. In particular, VLA 1 is at the early stage of the photoionization and it is driving a thermal radio jet, while VLA 2 is a thermal, collimated ionized wind surrounded by a dusty disk or envelope. In both sources 22 GHz H_2O masers have been detected in the past. Those around VLA 1 show a persistent linear distribution along the thermal radio jet and those around VLA 2 have instead traced the evolution from a non-collimated to a collimated outflow over a period of ~20 years. The magnetic field inferred from the H_2O masers showed a rotation of its orientation according to the direction of the major-axis of the shell around VLA 2, while it is immutable around VLA 1.

We further monitored the polarized emission of the 22 GHz H_2O masers around both VLA 1 and VLA 2 over a period of six years with the European VLBI Network for a total of four epochs separated by two years from 2014 to 2020. We here present the results of our monitoring project by focusing on the evolution of the maser distribution and of the magnetic field around the two massive young stellar objects.

Keywords. Water masers, polarization, star formation, magnetic field

---

1. Introduction

W75N(B) is an active high-mass star-forming region at a distance of 1.30±0.07 kpc (Rygl *et al.* 2012), where several massive young stellar objects at different evolutionary stages have been mapped (e.g., Torrelles *et al.* 1997; Rodríguez-Kamenetzky *et al.* 2020). Among them, VLA 1 and VLA 2 are of great interest because they are separated by only $\sim 0''.8$ (~ 1000 au) and because they host H_2O masers, whose distribution is immutable in VLA 1 and very dynamic in VLA 2 (Surcis *et al.* 2014). The spectral index analysis performed in the radio wavelength indicates that VLA 1, whose radio continuum is persistently elongated northeast-southwest, is at the early stage of the photoionization and it is driving a thermal radio jet (Rodríguez-Kamenetzky *et al.* 2020). The radio continuum emission of VLA 2 showed a morphology variation between 1996 and 2014 from a compact roundish source (≤ 160 au; Torrelles *et al.* 1997), resembling an ultra compact H II region, to an extended source that is elongated in the northeast-southwest direction ($220\times \leq 160$ au, $PA = 65°$, Carrasco-González *et al.* 2015). The spectral index analysis indicates that VLA 2 is a thermal, collimated ionized wind surrounded by a dusty disk or envelope (Carrasco-González *et al.* 2015). VLA 2 was also suggested to be the powering source of the large-scale high-velocity CO-outflow, oriented northeast-southwest ($PA = 66°$), that was detected from W75N(B) (Shepherd *et al.* 2003). However, the main powering source of this CO-outflow still remains unknown (e.g., Qiu *et al.* 2008).

The 22 GHz H_2O masers detected around VLA 1 and VLA 2 have been widely studied and monitored both with single dishes and interferometers, revealing high-intensity variations, with extreme maser flares (up to 10^3 Jy) and important variations in the maser distribution (e.g., Lekht & Krasnov 2000; Surcis *et al.* 2014; Krasnov *et al.* 2015). Over a period of 16 years, thanks to very long baseline interferometry (VLBI) observations, it has been observed that the evolution of the H_2O masers around VLA 1 and VLA 2, despite their close separation, is completely different. While the H_2O masers around VLA 1 are always linearly distributed ($PA \approx 43°$) along the thermal radio jet, the masers around VLA 2 are instead tracing an expanding shell (Surcis *et al.* 2014) that evolved from a quasi-circular (Torrelles *et al.* 2003; Surcis *et al.* 2011) to an elliptical structure (Kim *et al.* 2013; Surcis *et al.* 2014), following the morphology change in the continuum emission observed by Carrasco-González *et al.* (2015). Therefore, in VLA 2, the H_2O masers might be tracing the passage from a non-collimated to a collimated outflow (Surcis *et al.* 2014). In addition, Surcis *et al.* (2014) also performed polarimetric studies of the 22 GHz H_2O maser emission and they found that the magnetic field around VLA 1 has not changed between 2005 and 2012 and it is always oriented along the direction of the thermal radio jet. On the other hand, the orientation of the magnetic field around VLA 2 has changed in a way that is consistent with the new direction of the major-axis of the shell-like structure that is now aligned with the thermal radio jet of VLA 1.

In order to follow the evolution of the H_2O maser distribution around VLA 2 and to monitor any variation of the magnetic field, we performed VLBI monitoring observations toward W75N(B) of the 22 GHz H_2O maser emission in full polarization mode every two years from 2014 to 2020 for a total of four epochs. The presence of the quasi-immutable H_2O maser emission around VLA 1 will be of great importance in reinforcing the findings around the more dynamic H_2O maser source VLA 2.

2. Observations

W75N(B) was observed at 22 GHz in full polarization spectral mode and phase-reference mode with several European VLBI Network† (EVN) antennas on four epochs separated by two years: 17 June 2014 (epoch 2014.46), 12 June 2016 (2016.45), 09 June 2018 (2018.44), and 25 October 2020 (2020.82). The observational and data reduction details can be found in Surcis *et al.* (2023). In this proceeding, we just mention that the data were calibrated using the Astronomical Image Processing Software package (AIPS) by following the standard calibration procedure (e.g., Surcis *et al.* 2011) and analyzed accordingly to the procedure reported in Surcis *et al.* (2011). Briefly, this consists to identify the H_2O maser features and to measure their mean linear polarization fraction (P_l), their mean linear polarization angle (χ), and their circular polarization fraction (P_V); by modeling their polarized emission with the full radiative transfer method code (FRTM code; Surcis *et al.* 2011) we estimated the intrinsic maser linewidth (ΔV_i), the

† The European VLBI Network is a joint facility of European, Chinese, South African and other radio astronomy institutes funded by their national research councils.

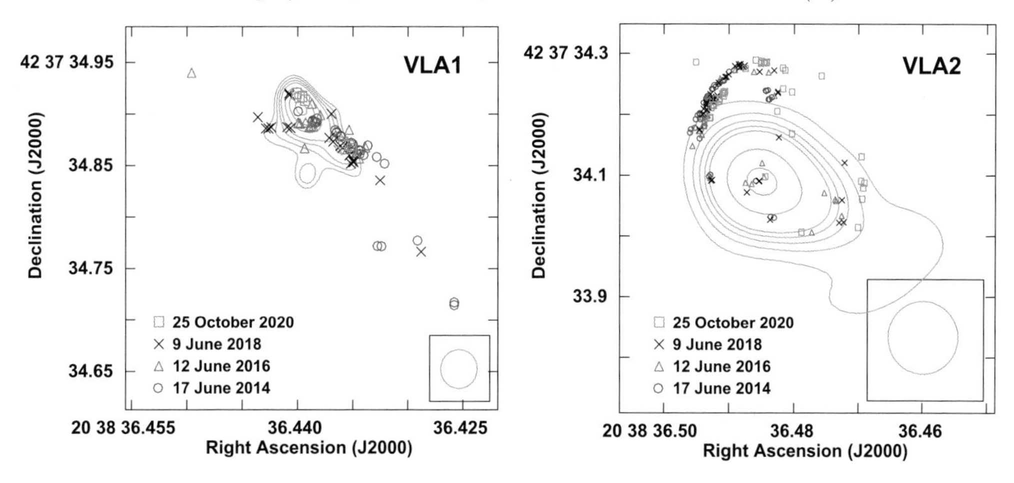

Figure 1. Comparison of the H_2O maser features detected toward VLA 1 (*left panel*) and VLA 2 (*right panel*) in the four EVN epochs and superimposed to the uniform-weighted continuum map at Q-band (Rodríguez-Kamenetzky *et al.* 2020) and to the natural-weighted continuum map at K-band (Carrasco-González *et al.* 2015), respectively, obtained both with the VLA. The light gray contours are 5, 6, 7, 8, 10, 12, 14 times $\sigma = 100$ μJy beam^{-1} in the case of VLA 1 and 5, 10, 15, 20, 25, 50, and 75 times $\sigma = 10$ μJybeam^{-1} in the case of VLA 2. The VLA beam is shown on the bottom right corner of both panels. The size of the symbols are ten times the uncertainties of the absolute positions of the maser features (see Table 2 of Surcis *et al.* 2023). For more details see Surcis *et al.* (2023).

emerging brightness temperature ($T_b\Delta\Omega$, where $\Delta\Omega$ is the maser beaming), the angle between the maser propagation direction and the magnetic field (θ), and the magnetic field strength along the line of sight ($B_{||}$).

3. Results

We detected H_2O maser features around VLA 1 and VLA 2 in all the four EVN epochs. Figure 1 shows the H_2Omaser features overplotted to the respective continuum emission at Q- and K-band. Here, the maser features in VLA 1 exhibit a linear distribution along the thermal radio jet but they do no show any hint of proper motions. On the other hand, the elliptical distribution of the maser features in VLA 2 suggests that the gas is not moving symmetrically around it as assumed in previous works (e.g., Surcis *et al.* 2014). Indeed, in the northeast we see that the maser features of epoch 2014.46 are generally slightly at northeast of those of the next epochs suggesting the presence of an obstacle (i.e., a denser medium) that might prevent the expansion of the gas, while we measure outward proper motions in the other directions around VLA 2. In particular, in the northwest the gas is moving outward with a velocity around 26–28 km s^{-1}; while in the center, the motion is toward southeast with a velocity around 38 km s^{-1}; in the east, the motion of the gas is toward south with a velocity of 12 km s^{-1}; in the south, the gas is expanding southward with a velocity of 8 km s^{-1}; and in the southwest the gas is the fastest one with a westward velocity of 78 km s^{-1}. These proper motions are visually reported in Figure 2. If we average the magnitude of all the velocities that we estimated, we obtain $|V| \approx 33$ km s^{-1}, similar to the symmetric expanding velocity of 30 km s^{-1} measured by Surcis *et al.* (2014).

Figure 2 shows the estimated magnetic field vectors and strengths. Here, we notice that a punctual comparison between the magnetic field vectors measured in the different epochs is more appropriate than a comparison of the averaged orientations as done previously (Surcis *et al.* 2014). Indeed, the magnetic field vectors estimated in similar

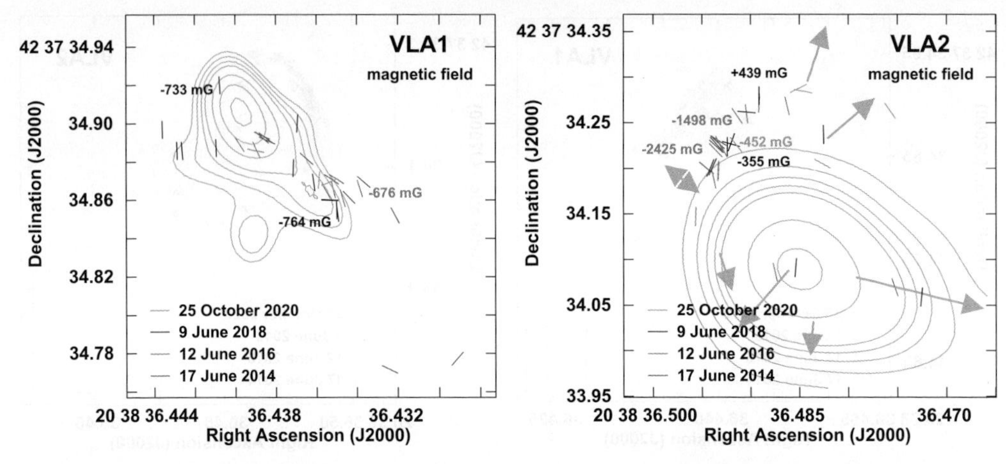

Figure 2. Magnetic field vectors and strength along the line of sight (next to the corresponding maser feature) as estimated from all the linearly and circularly polarized maser features detected around VLA 1 (*left panel*) and VLA 2 (*right panel*) during the four EVN epochs (2014.46, 2016.45, 2018.44, and 2020.82). The vectors are superimposed to the correspondig continuum map (see the caption for Figure 1). The arrows on the right panel represent the direction of the measured proper motions on the plane of the sky and their length is proportional to their magnitudes as reported in Tables 4, 5, and 6 of Surcis *et al.* (2023). The double arrow in the northeast indicates the uncertainties of the motion direction, due to the presence of an inward motion or of an outward motion of the gas.

location, but in different epochs, seem to represent a quasi-static magnetic field. We can therefore consider the magnetic field vectors estimated from the H_2O maser features in one epoch as representative of the magnetic field in those locations rather than just in that time. Consequently, we can gather the magnetic field vectors of all the epochs and consider them as measurements carried out at the same time. The magnetic field vectors in VLA 1 seem to follow the morphology of the continuum emission at Q-band, it is along the thermal jet and it bends toward south at the southwest end of the jet and toward the north at the northeastern end. In the case of VLA 2, we found that the magnetic field is generally perpendicular to the proper motions all around the source, but in the northeast it becomes parallel after encountering the supposed denser medium. This can be explained by considering that the magnetic field component perpendicular to the shock velocity is compressed and dominate the parallel component, which remains unaffected; consequently, the resulting magnetic field is along the shock front. However, if the compression of the gas in the northeast due to the passage of the shock is inefficient, then the perpendicular component of the magnetic field is no longer able to dominate over the parallel component. The presence of the denser medium in the northeast is further suggested by the strength of the magnetic field that is higher towards northeast. A more complete analysis of the results is found in Surcis *et al.* (2023).

References

Carrasco-González, C., Torrelles, J. M., Cantó, J. *et al.* 2015, *Science*, 348, 114
Kim, J.-S., Kim, S.-W., Kurayama, T. *et al.* 2013, *ApJ*, 767, 86
Krasnov, V. V., Lekht, E. E., Rudnitskii, G. M. *et al.* 2015, *Astron. Lett.*, 41, 517
Lekht, E. E. & Krasnov 2000, *Astron. Lett.*, 26, 38
Qiu, K., Zhang, Q., Megeath, S. T. *et al.* 2008, *ApJ*, 685, 1005
Rodríguez-Kamenetzky, A., Carrasco-González, C., Torrelles, J. M. *et al.* 2020, *MNRAS*, 496, 3128
Rygl, K. L. J., Brunthaler, A., Sanna, A. *et al.* 2012, *A&A*, 539, 79

Shepherd, D. S., Testi, L., & Stark, D. P. 2003, *ApJ*, 584, 882
Surcis, G., Vlemmings, W. H. T., Curiel, S. *et al.* 2011, *A&A*, 527, A48
Surcis, G., Vlemmings, W.H.T., van Langevelde, H.J. *et al.* 2014, *A&A*, 565, L8
Surcis, G., Vlemmings, W.H.T., Goddi, C. *et al.* 2023, *A&A*, 673A, 10S
Torrelles, J.M., Gómez, J.F., Rodríguez, L.F. *et al.* 1997, *ApJ*, 489, 744
Torrelles, J. M., Patel, N. A., Anglada, G. *et al.* 2003, *ApJ*, 598, L115

Cosmic Masers: Proper Motion toward the Next-Generation Large Projects
Proceedings IAU Symposium No. 380, 2024
T. Hirota, H. Imai, K. Menten, & Y. Pihlström, eds.
doi:10.1017/S1743921323002296

Simultaneous observations of exited OH and methanol maser - coincidence and magnetic field

Agnieszka Kobak

Institute of Astronomy, Nicolaus Copernicus University in Torun.
email: akobak@astro.uni.torun.pl

Abstract. We present the first results of simultaneous observations of the 6.035 GHz exited OH and 6.7 GHz methanol masers toward a sample of 10 high-mass young stellar objects (HMYSOs), observed using eMERLIN in 2020 and 2022. Searching for the coincidence and avoidance of these two maser transitions, we estimate physical conditions around central protostars. We identify Zeeman-splittings of the OH emission and determine the strength of the magnetic field. Combining it with linear polarization, we derive the magnetic field structure in these high-mass star-forming regions.

Keywords. methanol, exited OH, maser, magnetic field

1. Introduction

Astrophysical masers are one of the most important tools to study high-mass young stellar objects (HMYSOs). Due to masers, we can study, obscured by dense gas, regions around HMYSO. The most common species is methanol maser (6.7 GHz), which was widely studied by single-dish long-term observations (Szymczak *et al.* 2002); Szymczak *et al.* (2018); (Yang *et al.* 2019) as well as by interferometric observations during a few epochs (Bartkiewicz *et al.* 2009,K). Another kind of maser is exited OH (6.035 GHz, so-called ex-OH). These two transitions require specific physical conditions (Cragg *et al.* 2002). Methanol masers need a dust temperature (T_D) above 100 K and a gas temperature (T_K) below 200 K, whereby the maser diminishes when T_K exceed T_D. Masers appear in gas densities (ρ_{H_2}) between $10^{5.5}$ to $10^{7.5}$ cm^{-3}. Whereas the ex-OH maser arises in the cooler region at T_D above 20 K and T_K below 70 K, and these two parameters are independent. The gas density required for the existence of the 6.035 GHz transition is bigger, 10^6-$10^{8.5}$ cm^{-3}.

An important phenomenon during star formation is a magnetic field, which influences the star formation rate and fragmentation process (Krumholz & Federrath 2019). The strength of the magnetic field can be also a signature of the gas density according to the dependence B$\sim$n$^{0.4}$ (Garay *et al.* 1996). To estimate the strength of the magnetic field, the splitting of spectral lines (Zeeman splitting) is used. Because the OH molecule is paramagnetic, the splitting between velocity in left- and right-hand circular polarization (LHCP, RHCP) (V_Z) is significant and relatively easy to measure. To calculate the value of the magnetic field (B) we used the formula $\Delta V_Z/B = 0.056$ Baudry *et al.* (1997), with a value of 0.056 for the g-factor for the OH molecule. Using this method the magnetic field was examined for example for source W75N (Bartkiewicz *et al.* 2005) and for ON1 (Green *et al.* 2007).

In this paper, we present the result of the two sources G43.149+0.013 (hereafter G43) and G48.990−0.299 (hereafter G48). The observations were a part of the project CY10206, and were made using e-Merlin in 2020 and 2022. G43 is part of the vast star-forming region W49N. The far kinematic distance for this source is 10.96±0.35 kpc with 99% probability, for a central velocity 12 km s^{-1}(calculated using the parallax-based distance calculator V2 from Reid *et al.* (2019)). The G48 source lies at the kinematic distance of 5.44±0.47 kpc Reid *et al.* (2019).

2. Results

Results for these sources are presented in Figure 1. Both sources characterize more extended spatial emission of ex-OH than that of the methanol masers.

2.1. *G43.149+0.013*

In this source ex-OH maser emission covers an area of 280×250 mas (3070×2740 au) and a range of velocity 9.5–14.5 km s^{-1}. The methanol maser emission appears in the red-shifted part of the 6.035 GHz spectrum in the range from 13 km s^{-1}to 14.2 km s^{-1}and is spread over 150×110 mas (1650×1200 au). The spatial distribution of methanol maser is more compact than that of ex-OH, but it is more than 5 times brighter. The 6.7 GHz emission emerges in three groups, and these all coincide with the 6.035 GHz emission in both velocity and angular position, within 0.4 km s^{-1}and 25 mas. On the southeast side only extended ex-OH emission appears with a linear structure.

We identify eight Zeeman pairs at the 6.035 GHz transition. Pairs in the east have a positive magnetic field value equal to +3.2 mG and +3.7 mG, indicating that the magnetic field vectors are directed away from us. The remaining pairs show negative values from −1.3 mG to −6.1 mG, corresponding to vectors of the magnetic field directed towards us. Five Zeeman pairs have spots polarized linearly with degrees from 8% to 48%.

2.2. *G48.990−0.299*

This source shows simple methanol maser emission consisting of two spots at velocity 71.5 km s^{-1}and 71.6 km s^{-1}and a brightness of 2.8 Jy beam^{-1}. The ex-OH emission is extended over the area 290×135 mas (1580×735 au) and appears in the velocity range from 66.7 km s^{-1}to 69.6 km s^{-1}. The maximum brightness is 2.6 Jy beam^{-1}. We do not notice any coincidence between the transitions in either velocity or position.

We identify four Zeeman pairs of ex-OH masers. Two eastern, blue-shifted groups have positive values of the magnetic field +2.3 mG and +2.7 mG and the vectors are directed away from us. Two western, red-shifted groups have negative values of the magnetic field −0.7 mG and −4.8 mG, and the vectors are directed towards us. The two brightest pairs have spots with linear polarization with degrees from 14% to 31%.

3. Discussion and Summary

G43 lies southwest of the large star-formation region W49N (De Buizer *et al.* 2021). At the south of the 6.7 and 6.035 GHz emission there are peaks of 20 μm and 37 μm continuum emission (De Buizer *et al.* 2021) and a few peaks of 3.6 cm radio-continuum emission (De Pree *et al.* 1997). Closer to the maser is a region of 226 GHz (1.3 mm) continuum emission (Miyawaki *et al.* 2022). This peak lies at a distance of ~80 mas (~876 au) towards the north from the brightest methanol maser spot. Close to the source an H_{II} region is also observed. The peak is ~195 mas (~2140 au) towards the west from the brightest methanol maser spot (Hu *et al.* 2016).

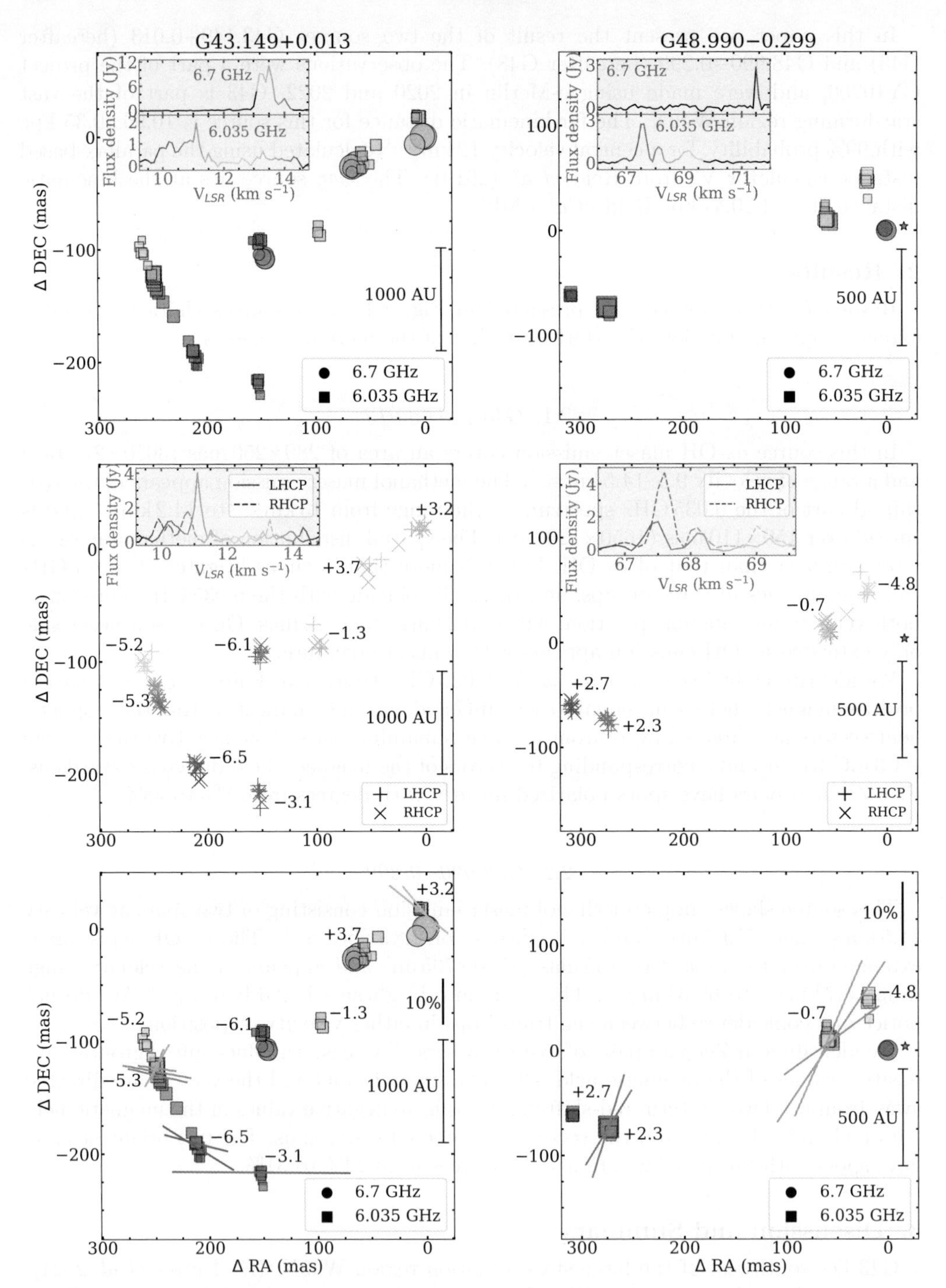

Figure 1. The 6.7 GHz methanol and 6.035 GHz ex-OH masers in G43.149+0.013 (left) and G48.990−0.299 (right). The size of the marks indicates the square root of the intensity, and the color indicates the V_{LSR} shown at the spectra. The grey star indicates the peak of H_{II} emission (Cooper *et al.* 2013). The (0,0) point corresponds to the brightest methanol maser spot. **Top:** Spectra and distributions of methanol (circle) and ex-OH (squares) maser spots. **Middle:** Spectra and distribution of left- and right-handed circular polarization (LHCP, RHCP). The numbers list the magnetic field strength in mG estimated based on the Zeeman splittings. **Bottom:** Distribution of maser spots with the values of magnetic field and percentage of linear polarization marked by colorful lines.

The distributions of two kinds of masers suggest that the whole masing region has T_K below 70 K. Regions with only 6.035 GHz emission can come from higher gas density or lower dust temperatures. Considering that: 1) the ex-OH red-shifted masers indicate a stronger magnetic field (−6.1 mG), 2) B∼$n^{0.4}$, 3) ex-OH and methanol masers coincide, we can assume that the T_D is lower than T_K and it is the reason that the blue-shifted ex-OH masers avoid methanol masers. Furthermore, that indicates the red-shifted masers lie closer to the expected protostar traced by the 3.6 cm radio-continuum emission, and T_D exceed 100 K. In the region with blue-shifted masers, the dust temperature declined faster than the kinetic temperature with the distance from the protostar and declined below the kinetic temperature.

G48 is associated with the star-forming region W51 with H_{II} emission, with a peak lying close (∼17 mas corresponding to ∼94 au) to the brightest methanol maser spot (see Figure 1) (Cooper *et al.* 2013). Like in the previous source, H_{II} emission occurs close to the methanol maser emission but for that source without ex-OH counterparts which indicates the kinetic temperature above 70 K and dust temperature above 100 K in the red-shifted part of the source. The 6.035 GHz emission lay father from H_{II} emission so, as in the case of G43, we assume that the temperatures decrease (T_K and T_D) to the values that stop the occurrence of methanol masers.

Acknowledgements

e-MERLIN is a National Facility operated by the University of Manchester at Jodrell Bank Observatory on behalf of STFC, part of UK Research and Innovation. We acknowledge support from the National Science Centre, Poland through grant 2021/43/B/ST9/02008.

References

Baudry, A., Desmurs, J.F., Wilson, T.L., Cohen, R.J., 1997, *A&A*, 325, 255

Bartkiewicz A., Szymczak M., Cohen R.J., Richards A.M.S., 2005, *MNRAS*, 361, 623–632

Bartkiewicz A., Szymczak M., van Langevelde H.J., Richards A.M.S., Pihlström Y.M., 2009, *A&A*, 502, 155

Cooper, H.D.B., Lumsden, S.L., Oudmaijer, R.D., Hoare, M.G., Clarke, A.J., Urquhart, J.S., Mottram, J.C., Moore, T.J.T., Davies, B., 2013, *MNRAS*, 430, 1125–1157

Cragg, D.M., Sobolev, A.M., Godfrey, P.D. 2002, *MNRAS*, 331, 521–536

De Buizer, J.M., Lim, W., Liu, M., Karnath, N., Radomski, J.T., 2021, *ApJ*, 923, 198

De Pree, C.G., Mehringer, D.M., Goss, W.M., 1997, *ApJ*, 482, 307

Garay, G., Ramirez, S., Rodriguez, L.F., Curiel, S., Torrelles, J.M., 1996, *ApJ*, 459, 193

Green, J.A., Richards, A.M.S., Vlemmings, W.H.T., Diamond, P., Cohen, R.J., 2007, *MNRAS*, 382, 770–778

Kobak, A., Bartkiewicz, A., Szymczak, M., Olech, M., Durjasz, M., Wolak, P., Chibueze, J.O., Hirota, T., Eislöffel, J., Stecklum, B., Sobolev, A., Bayandina, O., Orosz, G., Burns, R.A., Kim, K.T. and van den Heever, S.P, 2023, *A&A*, 671, A135

Krumholz, M.R., Federrath, C., 2019, *Frontiers in Astronomy and Space Sciences*, 6, 7

Reid, M., Menten, K.M., Brunthaler, A., 2019, *ApJ*, 885, 131

Miyawaki, R., Hayashi, M., Hasegawa, T., 2022, *PASJ*, 74, 705–737

Szymczak, M., Kus, A.J., Hrynek, G., Kepa, A., Pazderski, E., 2002, *A&A*, 392, 277–286

Szymczak, M., Olech, M., Sarniak, R., Wolak, P., Bartkiewicz, A., 2018, *MNRAS*, 474, 219–253

Hu, B., Menten, K.M., Wu, Y., Bartkiewicz, A., Rygl, K., Reid, M.J., Urquhart, J.S., Zheng, X., 2016, *ApJ*, 833, 18

Yang, K., Chen, X., Shen, Z.Q., Li, X.Q., Wang, J.Z., Jiang, D.R., Li, J., Dong, J., Wu, Y.J., Qiao, H.H., 2019, *ApJS*, 241, 39

Cosmic Masers: Proper Motion toward the Next-Generation Large Projects
Proceedings IAU Symposium No. 380, 2024
T. Hirota, H. Imai, K. Menten, & Y. Pihlström, eds.
doi:10.1017/S1743921323002703

High resolution VLBI observations of 6.7GHz periodic methanol masers

Mateusz Olech

Space Radio-Diagnostic Research Center, Faculty of Geoengineering, University of Warmia and Mazury ul. Oczapowskiego 2, 10-719 Olsztyn, Poland. email: mateusz.olech@uwm.edu.pl

Abstract. In the last 20 years a small group of 6.7GHz methanol maser sources displaying periodic variability have been identified. This variability is thought to reflect local processes linked to star formation. A number of models have been proposed e.g. colliding wind binary, protostellar pulsation, accretion on binary system. Recent studies of known sources as well as non-periodic flaring masers suggest an episodic accretion as a driving mechanism. We present the results of VLBI observation program aimed at studying known periodic methanol maser sources. High resolution maps of emission, source morphology and evolution in time will be discussed. Those results will help us fully understand the nature of maser periodicity in star forming regions.

Keywords. masers, stars: formation, ISM: clouds, radio lines: ISM

---

1. Introduction

Periodicity in class II 6.7 GHz methanol masers was first reported in Goedhart *et al.* (2004). Since then 28 sources have been identified with periods ranging from 24 to 1260 days (Tanabe *et al.* 2023). Those masers display varied types of flare profiles: sinusoidal, asymmetric or gaussian like, with large range of length of activity compared to overall period and varied relative amplitude. Some of the sources have shown correlation in flux brightness changes of IR and 6.7 GHz emission (Olech *et al.* 2022). Due to variety of types of flare profiles and source characteristics different models have been proposed to explain this phenomena. Those can be divided into models where variability is caused by modulation of seed photon flux e.g. colliding wind binary (van der Walt *et al.* 2009) or models that consider changes in pumping efficiency of masing region. The latter can be caused by different processes e.g. modulated accretion on binary protostar (Araya *et al.* 2010), stellar pulsations (Inayoshi *et al.* 2013) or rotating spiral shocks (Parfenov & Sobolev 2014) to name a few.
In order to differentiate between proposed models and better understand behavior of periodic methanol masers, high resolution interferometric observations are needed. Those proceedings present the initial results of project using European VLBI Network (EVN) observations of selected 6.7 GHz periodic sources.

2. Observations and initial results

2.1. *Observations*

Source sample selected for Very Large baseline Interferometry (VLBI) observation program consisted of 13 periodic 6.7 GHz masers that were monitored by Torun 32m antenna. Observations were conducted in phase-referencing mode with a switching cycle of 180s + 120s (maser + phase calibrator), total on source time for each target was 2h. Due to time constraints of EVN observation campaign and quiescent nature of periodic sources

Table 1. Summary of observations.

Source	Date	Period (d)	Synthesized beam (mas x mas)	1 $\sigma_{\rm rms}$ (mJy beam^{-1})
G12.89+0.489	4.06.2019	29	10.02 x 3.14	30.0
G22.357+0.066	6.03.2020	178	6.04 x 3.09	9.3
G24.148−0.090	1.11.2019	185	5.28 x 4.15	9.2
G25.411+0.105	1.11.2019	243	4.95 x 4.31	12.4
G30.400−0.296	4.06.2019	220	8.31 x 2.89	13.3
G33.641−0.228	4.06.2019	570	8.10 x 4.80	33.5
G36.70+0.090	1.11.2019	52	4.74 x 3.67	10.7
G37.550+0.200	1.11.2019	247	5.09 x 3.60	10.3
G45.473+0.134	6.03.2020	198	5.89 x 2.57	10.9
G59.63−0.192	6.03.2020	149	4.34 x 2.72	9.5
G73.06+1.80	4.06.2019	159	5.59 x 2.91	18.4
G107.298+5.639	6.03.2020	34	4.58 x 3.02	6.1
G108.76−0.99	6.03.2020	163	5.22 x 3.51	5.9

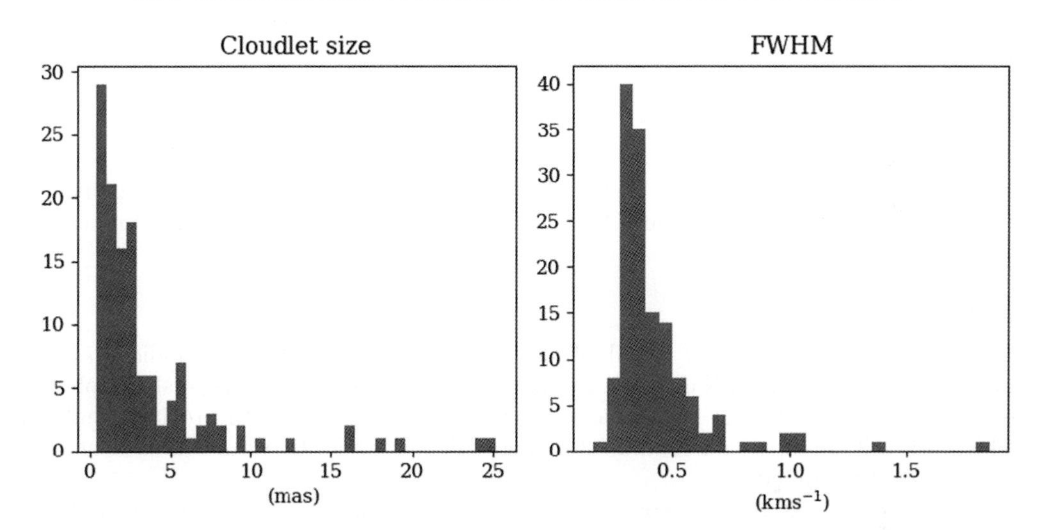

Figure 1. The statistics of measured *cloudlets* for all observed sources.

we divided the project into 3 epochs. This ensured that each source was observed during its active phase. Summary of observations is presented in Table 1.

Data was calibrated with AIPS package using with standard spectral line reduction procedure.

2.2. *Results*

For each source the *spot* emission of individual velocity channels was measured. For more physical representation of masering gas structure those *spots* were grouped into *cloudlets*: emission that's continuous in velocity and spatially within a beamwidth of adjacent channel. If there was more than 4 *spots* in group, gaussian function was fitted to the brightness distribution. This approach resulted in identification of 141 maser regions with the average *cloudlet* size of 3.61 ± 4.3 mas and 127 gaussian profiles with an average FWHM of 0.43 ± 0.22 kms^{-1}, results are shown in Figure 1. Those values are consistent with previous VLBI studies of 6.7 GHz methanol masers.

Maps of sources show varied morphology (Figure 2). Most prominent type is incomplete ring-like structure suggesting that the emission is located in well-defined region of accretion disk or at the interference between disk and envelope. Morgan *et al.*

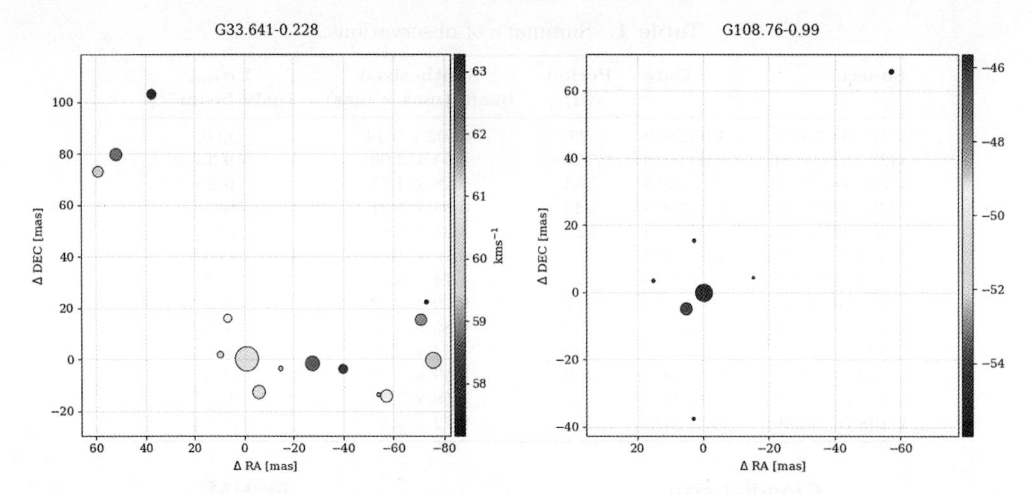

Figure 2. Example of 6.7 GHz emission maps for two periodic sources G33.641−0.228 (left) and G108.76−0.99 (right). Position is measured relatively to the brightest *spot*, color shows velocity and size of the circles scales logarithmically with *cloudlet* brightness.

(2021) argues that orientation of the disk on the sky-plane might have significant impact on observed flare profile. Other sources show core-halo or more complicated morphology.

3. Conclusions

We successfully conducted VLBI observations of 13 well known periodic 6.7 GHz methanol masers using EVN network. Initial results show mostly ring-like/disk morphologies with characteristics of individual masing regions consistent with previous studies. This work will be important to understanding of dependencies between source structure and observed periodicity.

The author thanks the Ministry of Education and Science of the Republic of Poland for support and granting funds for the Polish contribution to the International LOFAR Telescope (arrangement no. 2021/WK/02) and for maintenance of the LOFAR PL-612 Baldy (MSHE decision no. 28/530020/SPUB/SP/2022).

References

Araya, E. D., Hofner, P., Goss, W. M., *et al.* 2010, *ApJ* (Letters), 717, L133
Goedhart, S., Gaylard, M. J., *et al.* 2004, *MNRAS*, 355, 553
Inayoshi, K., Sugiyama, K., Hosokawa, T., *et al.* 2013, *ApJ* (Letters), 769, L20
Morgan, J., van der Walt., *et al.* 2021, *MNRAS*, 507, 1138
Olech, M., Durjasz, M., *et al.* 2022, *A&A*, 661, A114
Parfenov, S. Y. & Sobolev, A. M., 2014, *MNRAS*, 444, 620
Tanabe, Y., Yonekura, Y., & MacLeod, G. C., 2023, *PASJ*, 75, 351
van der Walt, D. J., Goedhar, S., & Gaylard, M. J., 2009, *MNRAS*, 398, 961

Cosmic Masers: Proper Motion toward the Next-Generation Large Projects
Proceedings IAU Symposium No. 380, 2024
T. Hirota, H. Imai, K. Menten, & Y. Pihlström, eds.
doi:10.1017/S174392132300193X

Detection of the longest periodic variability in 6.7 GHz methanol masers with iMet

Yoshihiro Tanabe and Yoshinori Yonekura

Center for Astronomy, Ibaraki University, 2-1-1 Bunkyo, Mito, Ibaraki 310-8512, Japan.
email: yoshihiro.tanabe.ap@vc.ibaraki.ac.jp

Abstract. Long-term monitoring observations of the 6.7 GHz methanol masers by Hitachi 32-m operated by Ibaraki University, which are named as "the Ibaraki 6.7 GHz Methanol Maser Monitor (iMet)", have revealed that the periods of the flux variability of 6.7 GHz methanol masers in the five high-mass star-forming regions G05.900−0.430, G06.795−0.257, G10.472+0.027, G12.209−0.102 and G13.657−0.599 are over 1000 days. These periods are approximately twice the longest known period of 6.7 GHz methanol masers of 668 days for G196.45−1.68. The facts that the flux variation patterns show symmetric sine curves and that the luminosity of the central protostar and periods of maser flux variation are consistent with the expected period-luminosity (PL) relation suggest that the mechanism of maser flux variability of G05.900−0.430, G10.472+0.027 and G12.209−0.102 can be explained by protostellar pulsation instability. From the period-luminosity relation, central stars of these three sources are expected to be very high-mass protostars with a mass of $\sim$40 $M_{\odot}$ and a mass accretion rate of $\sim 2\times10^{-2}$ $M_{\odot}\mathrm{yr}^{-1}$. On the other hand, G06.795−0.257 and G13.657−0.599 have luminosities that are an order of magnitude smaller than that expected from PL relation, and the variation patterns are intermittent, suggesting a variation mechanism of these sources originated from binary system.

Keywords. ISM: masers, stars: formation, stars: massive

---

1. Introduction

The Class II methanol maser is a well established tracer of high-mass star-forming regions (HMSFRs) and several astronomical masers have presented their characteristic periodic flux variations. Goedhart et al. (2003) first discovered a periodic flux variation of 243.3 d in the 6.7 GHz methanol masers associated with HMSFR G9.62+0.20E. By 2020, 28 periodic methanol maser sources have been reported (Tanabe *et al.* 2023 and references therein). Their periods range from 23.9 d (for G14.23−0.50 reported by Sugiyama et al. 2017) to 668 d (for G196.45−1.68 reported by Goedhart et al. 2004). Several explanations of mechanism for these maser periodicity have been proposed; colliding wind binary (CWB) system (van der Walt 2011), protostellar pulsation (Inayoshi et al. 2013), spiral shock in a circumbinary system (Parfenov & Sobolev 2014), periodic accretion in a circumbinary system (Araya et al. 2010), and eclipsing binary system (Maswanganye et al. 2015). These models can explain the maser flux variations of some sources well, however, there is no clear consensus on the general mechanism of maser periodicity. In this paper, we present the new discovery of periodicity in 6.7 GHz methanol maser source in five HMSFRs, and discuss the mechanism of their flux variability.

2. Observation

Monitoring observations of the 6.7 GHz methanol maser were made with the Hitachi 32m telescope of Ibaraki station, a branch of the Mizusawa VLBI Observatory of

Table 1. List of sources.

Source	R.A. (J2000)	Dec. (J2000)
G05.900−0.430	18 00 40.86	−24 04 20.80
G06.795−0.257	18 01 57.75	−23 12 34.90
G10.472+0.027	18 08 38.20	−19 51 50.10
G12.209−0.102	18 12 40.24	−18 24 47.50
G13.657−0.599	18 17 24.26	−17 22 12.50

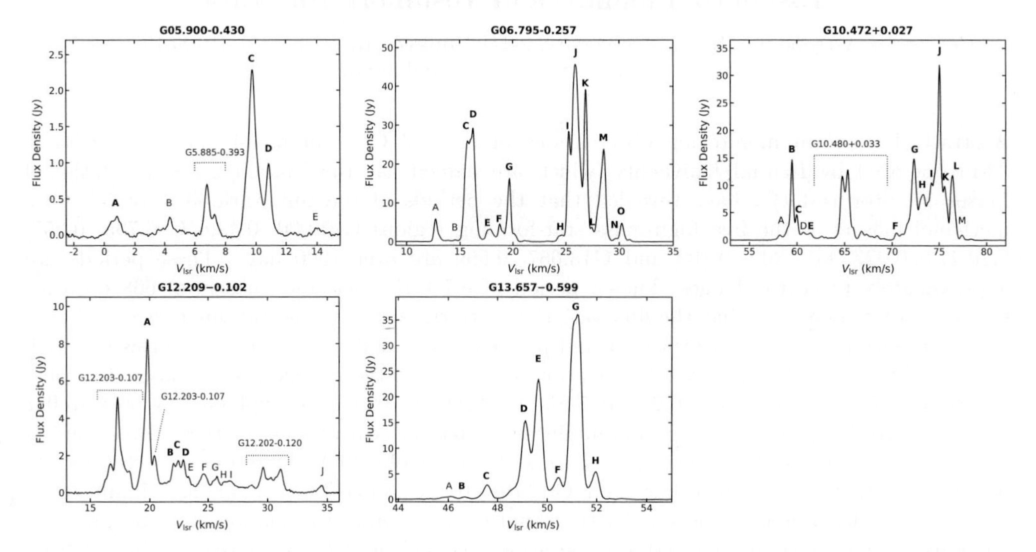

Figure 1. Averaged spectra of the 6.7 GHz methanol maser associated with five sources. Labels indicate each velocity feature and for features in bold, the periodicity is detected in this paper.

the National Astronomical Observatories Japan (NAOJ), operated jointly by Ibaraki University and NAOJ (Yonekura et al. 2016). The data presented in this paper are as a part of the Ibaraki 6.7 GHz Methanol Maser Monitor (iMet) program†.

Monitoring observations began on 2013 Jan. The cadence of observations is once per every ∼10 d from the start of the monitoring observations to 2015 Aug., and once per every ∼5 d from 2015 Sep. to the present. Observations after 2014 May were made at about the same azimuth and elevation angle to minimize intensity variations due to systematic telescope pointing errors. The half-power beam width of the telescope is $\sim 4.6'$ with the pointing accuracy better than $\sim 30''$ (Yonekura et al. 2016). The coordinates of target sources adopted in our observations are summarized in Table 1. The reference for these coordinates is Green et al. (2010). Observations are made by using a position-switching method. The OFF position is set to ΔR.A. = + 60′ from the target source. The integration time per observation is 5 minutes for both the ON and OFF positions. After averaging over 3 channels, the 1-sigma root-mean-squares noise level is approximately 0.3 Jy and the velocity resolution is 0.13 km s^{-1}.

3. Result

The averaged spectra integrated all scans of each source are presented in Figure 1. We detected several velocity features in each source and in Figure 1, features shown in bold, the periodicity is detected in this paper. The periodicity was estimated by employing the

† http://vlbi.sci.ibaraki.ac.jp/iMet/

Table 2. Detected periodicity and protostellar parameter.

Source	$v_{\rm lsr}$ (km s^{-1})	P (d)	Luminosity* $\log(L/L_\odot)$	Distance (kpc)	Distance† ref
G05.900−0.430	9.77	1265 (266)	4.8	5.9	A
G06.795−0.257	19.68	1002 (120)	4.0	3.0	A
G10.472+0.027	59.48	1652 (346)	5.7	8.6	B
G12.209−0.102	22.97	1280 (239)	5.5	13.4	A
G13.657−0.599	49.12	1258 (213)	4.4	4.5	A

*Urquhart et al. (2018). †A; Reid et al. (2016) estimated the shapes of the spiral arm by the Bayesian approach. B; Sannna et al. (2014) estimated by trigonometric parallax.

Lomb-Scargle (LS) periodogram method (Lomb 1976 and Scagle 1982). The LS method can give us false detection of a period, and thus we adopted the false alarm probability functions which is probability of judging the value corresponding to noise as a signal. In this paper, if the peak value of the power spectrum is larger than the 0.01% false alarm level, we decided the obtained period is reliable. The error of periods obtained by the LS method are estimated as the half width of half maximum (HWHM) of each peak in the periodogram.

In this paper, we present the time series and LS periodogram for one velocity feature which has the period with the largest LS power in each source. The results of periodic analysis are summarized in Table 2 and Figure 2. For all the sources, the detected periods are over 1000 days. Harmonic peaks are seen in the periodogram for G06.795−0.257 and G13.657−0.599. The flux variation pattern is classified into two types: continuous (G05.900−0.430, G10.472+0.027, and G12.209−0.102) and intermittent (G06.795−0.257 and G13.657−0.599).

4. Discussion

Inayoshi et al. (2013) proposed that the high-mass protostars with large mass accretion with rates of $\dot{M}_* \gtrsim 10^{-3}$ $M_\odot$ yr^{-1} become pulsationally unstable. This instability is caused by the κ mechanism in the He$^+$ layer. The luminosity of the protostars varies periodically and as the temperature of the surrounding material rises to and falls from a temperature suitable for maser pumping, resulting in that the maser fluxes increase and decrease. In this model, the flux variation pattern of the associated maser is expected to be continuous, thus the protostellar pulsation instability model may well describe the variation in G05.900−0.430, G10.472+0.027, and 12.209+0.102. Inayoshi et al. (2013) also derived the period-luminosity (PL) relation

$$\log\frac{L}{L_\odot} = 4.62 + 0.98 \log\frac{P}{100 \text{ days}}, \tag{1}$$

where L is luminosity of the protostar and P is period expected from the maser variation.

Figure 3 plots the estimated periods versus protostellar luminosity for each source. The solid line represents the PL relation given by equation (1). In figure 3, the central protostar luminosity and periods of maser flux variation are roughly consistent with the expected PL relation suggest that the mechanism of maser flux variability of G05.900−0.430, G10.472+0.027 and G12.209−0.102 can be explained by protostellar pulsation instability. If the pulsationally unstable model is applied for these sources, the protostellar mass and mass accretion rate for these three sources estimated from the detected period using the equations (2) and (4) in Inayoshi et al. (2013) are $M_* \sim 4 \times 10$ $M_\odot$ and $\dot{M}_* \sim 2{\times}10^{-2}$ $M_\odot$ yr^{-1}, respectively.

On the other hand, the sources which has a variable pattern with intermittent (G06.795−0.257 and G13.657−0.599) show an order of magnitude smaller than those luminosity expected from PL relation. Thus a variation mechanism of these sources may

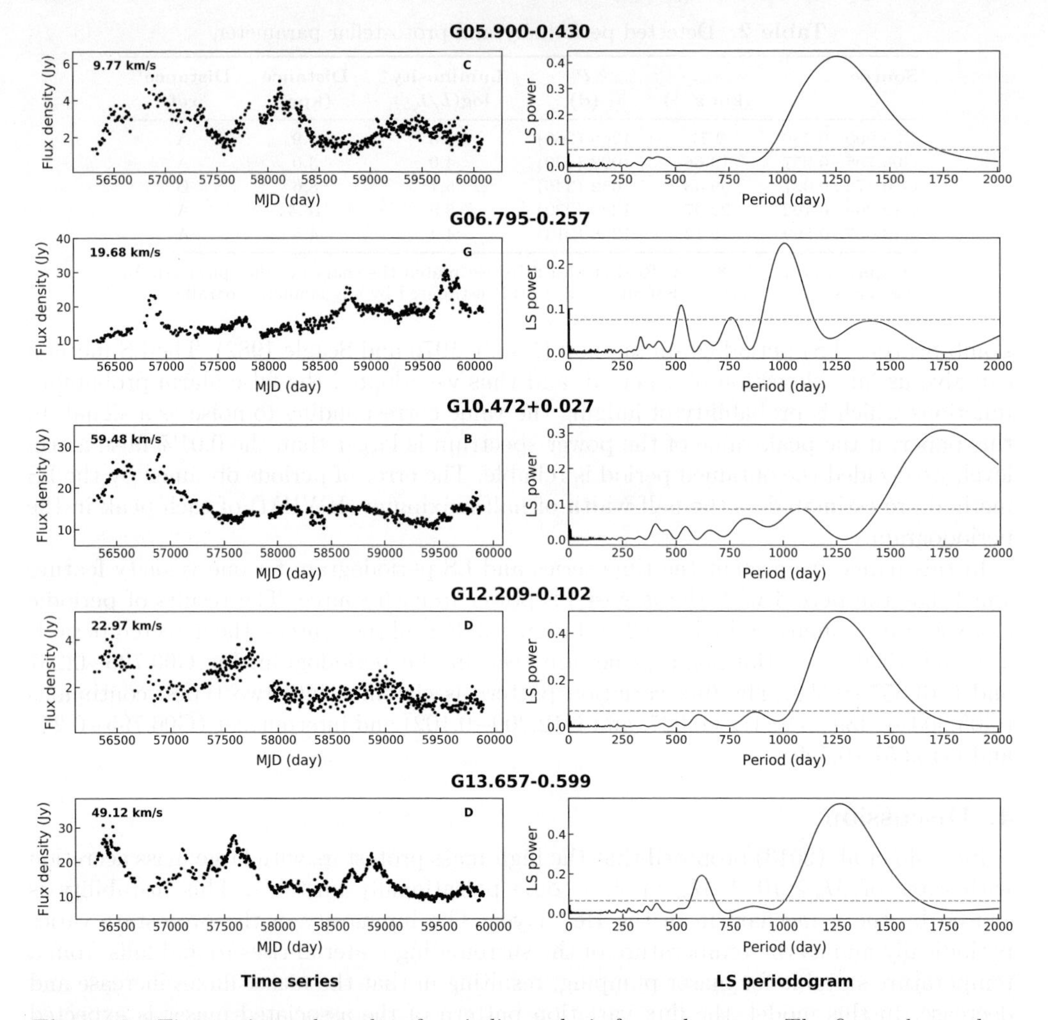

Figure 2. Time series and results of periodic analysis for each source. The first column show the time series of flux density of each velocity feature. We excluded the data points whose flux densities are less than 3 σ. The second column show the LS power spectra plots. The dotted lines in second column represent the 0.01% false alarm probability levels. In left panels for each source, labels of alphabet represent velocity features shown in Figure 1.

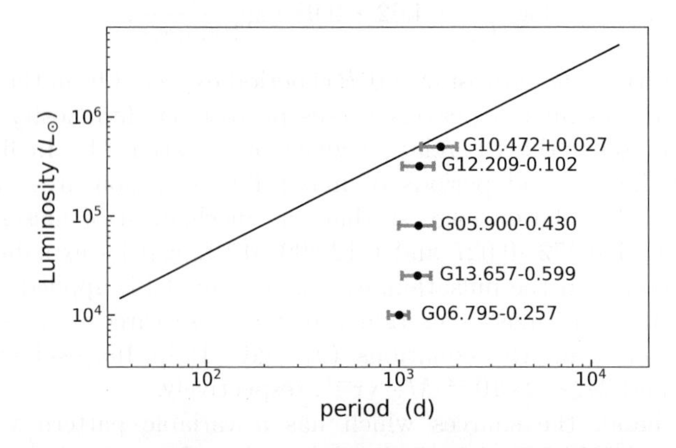

Figure 3. Protostellar luminosity versus estimated maser periods. The solid line represents the PL relation given by Inayoshi et al. (2013). According to Urquhart et al. (2018), the total uncertainty of luminosity due to measurement error and SED fitting is ~100%, which does not include uncertainty of distance of sources.

from binary system. VLBI observations with high spatial resolution will lead to a better understanding of the nature of these sources.

Acknowledgements:

This work is partially supported by the Inter-university collaborative project "Japanese VLBI Network (JVN)" of NAOJ. This study benefited from financial support from the Japan Society for the Promotion of Science (JSPS) KAKENHI program (21H01120 and 21H00032).

References

Araya, E. D., *et al.* 2010, *ApJL* 717, L133.
Goedhart, S., Gaylard, M. J., & van der Walt, D. J. 2003, *MNRAS*, 339, L33
Goedhart, S., Gaylard, M. J., & van der Walt, D. J. 2004, *MNRAS*, 355, 553
Green, J. A., *et al.* 2010, *MNRAS*, 409, 913
Inayoshi, K., *et al.* 2013, *ApJ*, 769, L20
Lomb, N. R. 1976, *Ap&SS*, 39, 447
Maswanganye, J. P., *el al.* 2015, *MNRAS*, 446, 2730
Parfenov, S. Y. & Sobolev, A. M. 2014, *MNRAS*, 444, 620
Reid, M. J., Dame, T. M., Menten, K. M., & Brunthaler, A., 2016, *ApJ*, 823, 77
Sannna, A., *et al.* 2014, *ApJ*, 781, 108
Scargle, J. D. 1982, *ApJ*, 263, 835
Sugiyama K., *et al.* 2017, *PASJ*, 69, 59
Tanabe, Y., Yonekura, Y., & MacLeod, G. C. 2023, *PASJ*, 75, 351
Urquhart, J. S., *et al.* 2018, *MNRAS*, 473, 1059
van der Walt, D. J. 2011, *AJ*, 141, 152
Yonekura, Y., *et al.* 2016, *PASJ*, 68, 74

Cosmic Masers: Proper Motion toward the Next-Generation Large Projects
Proceedings IAU Symposium No. 380, 2024
T. Hirota, H. Imai, K. Menten, & Y. Pihlström, eds.
doi:10.1017/S1743921323002636

Maser Activity of Large Molecules toward Sgr B2 North

Ci Xue[1], Anthony Remijan[2], Alexandre Faure[3] and Brett McGuire[1,2]

[1]Massachusetts Institute of Technology, Cambridge, MA 02139, USA. email: cixue@mit.edu

[2]National Radio Astronomy Observatory, Charlottesville, VA 22903, USA

[3]Université Grenoble Alpes, CNRS, IPAG, F-38000 Grenoble, France

Abstract. Single-dish observations at centimeter wavelengths have suggested that the Sgr B2 molecular cloud at the Galactic Center hosts weak maser emission from several large molecules. Here, we present the interferometric observations of the Class I methanol (CH_3OH) maser at 84 GHz, the methanimine (CH_2NH) maser at 5.29 GHz, and the methylamine (CH_2NH_2) maser at 4.36 GHz toward Sgr B2 North (N). We use a Bayesian approach to quantitatively assess the observed masing spectral profiles and the excitation conditions. By comparing the spatial origin and extent of maser emission from several molecular species, we find that the new maser transitions have a close spatial relationship with the Class I masers, which suggests a similar collisional pumping mechanism.

Keywords. ISM: molecules, Interstellar masers, Spectral line identification

1. Introduction

Common species exhibiting maser emission include, but are not limited to, OH, SiO, H_2O, H_2CO, NH_3, and CH_3OH. An empirical taxonomy based on pumping mechanisms classifies molecular masers into two categories (Cyganowski et al. 2009); Class I CH_3OH and H_2O masers are most often excited via collisions and tend to reside in shocked material associated with outflows and expanding HII regions, while Class II CH_3OH and OH masers are excited via radiation and are generally close to young stellar objects and luminous infrared sources. Recently, new maser species, such as HNCO and CH_2NH, have been discovered to exhibit compelling roles in inferring fine-scale structures and gas dynamics in the maser region (Chen et al. 2020; Gorski et al. 2021). The pumping mechanism of these new maser species, however, is less constrained, as is their relation to other common maser species. Observations of these species in maser-active regions that contain both known Class I and Class II masers are therefore critical as a first step in understanding their pumping mechanisms.

The high-mass star-forming region Sgr B2 (N) is one of the richest known sources for not only molecular discoveries but also maser activity from a diverse set of molecules (e.g. Walsh et al. 2014). In addition to the common maser species, Sgr B2 displays rare maser activity from complex molecules. The single-dish PRIMOS line-survey project has identified numerous species exhibiting maser activity at centimeter wavelengths (e.g. HNCNH, McGuire et al. 2012). The radiative transfer calculations also revealed that most emission lines of both $HCOOCH_3$ and CH_2NH detected in PRIMOS are weak masers amplifying the background radiation of Sgr B2(N) at cm wavelengths (Faure et al. 2014, 2018). However, the large $\sim 2.5' \times 2.5'$ field of view of PRIMOS at 5 GHz includes both the

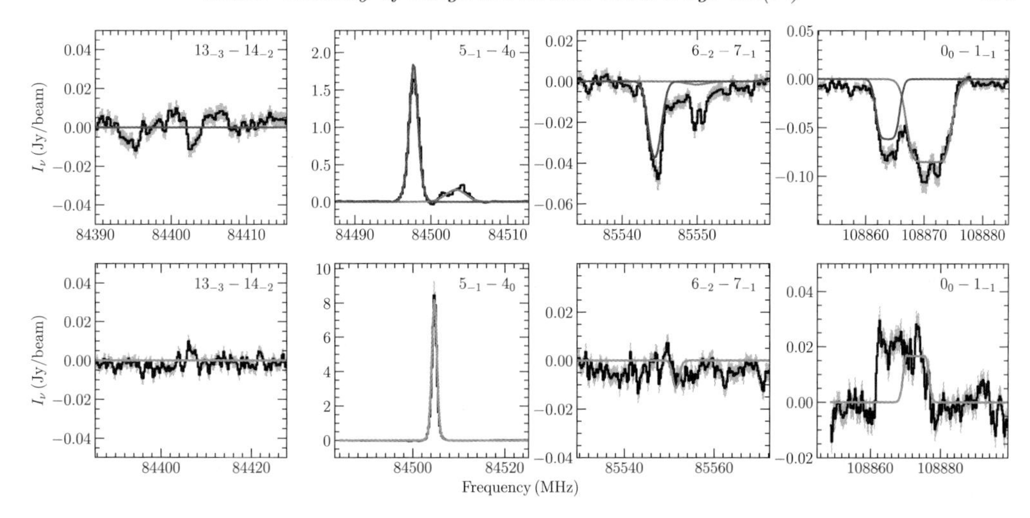

Figure 1. Observed and synthetic spectra of the E-type CH_3OH transitions. The observed spectra are shown in black with noise levels shaded in gray. The top panel shows the spectra extracted toward the CH_3OH-83 km s^{-1} spot, and the bottom panel shows the spectra extracted toward the brightest CH_3OH maser spot. The best-fit synthetic spectra are overlaid in colors, with the red trace representing the 83 km s^{-1} component, the blue trace representing the 64 km s^{-1} component, and the purple trace representing the 58 km s^{-1} components.

North and Main components of Sgr B2, complicating the interpretation of the spectral profiles. Therefore, it is crucial to have interferometric observations to disentangle in detail the various possible spatial components of maser emission.

We present a rigorous imaging study of maser emission from CH_3OH, CH_2NH, and CH_2NH_2 toward Sgr B2(N). The first is the spectra and spatial distribution per maser species toward the HII regions; the second involves the CH_3OH maser emission toward a region with weak background continuum radiation. In addition, we present a tool for modeling and fitting the unsaturated molecular maser signals with non-LTE radiative transfer models and Bayesian analysis using the Markov-Chain Monte Carlo (MCMC) approach.

2. Maser Emission associated with H II Regions

2.1. *CH_3OH $5_{-1} - 4_0$ Transition at 84 GHz*

We present the first detection of the 84 GHz Class I CH_3OH masers toward Sgr B2 with interferometric ALMA observations at a resolution of 1.5″. The brightest maser spot located ~3.6″ west of the K4 cm-wave core with a $V_{\rm lsr}$ of 58 km s^{-1} has a peak brightness temperature ($T_{\rm B,\,peak}$) of 389.2 ± 0.2 K. In addition to the brightest spot, another maser knot associated with the K6 HII region is also identified with a $T_{\rm B}$ of 75.4 ± 0.4 K at a $V_{\rm lsr}$ of 83 km s^{-1} (hereafter referred to as CH_3OH-83 km s^{-1} spot).

The spectra extracted toward the CH_3OH-83 km s^{-1} maser spot are shown in Fig. 1 (top) with two distinct velocity components being identified. A chained two-component model is constructed to iteratively generate synthetic spectra, each component being described with five parameters, $T_{\rm k}$, $n_{\rm c}$, $V_{\rm lsr}$, dV and $N_{\rm col}$. This maser spot is aligned with the K6 HII region, which provides background radiation to be amplified by the stimulated emission. Taking into account $T_{\rm cont}$, there are 11 free parameters in total to be adjusted in the MCMC analysis. All parameters converge to primarily Gaussian distributions in our model. The nominal mean for each parameter from the MCMC inference is used to construct the synthetic spectra shown in Fig. 1 (top), which fit satisfactorily with the observed spectra when noise levels are taken into account. The excitation temperature and opacity of the $5_{-1} - 4_0$ maser transition are found to be negative, with a T_{ex} and τ being −5.6 K and −2.5 for the high velocity component.

2.2. *CH_2NH $1_{1,0} - 1_{1,1}$ Transition at 5.29 GHz*

The maser activity of the CH_2NH transitions at 5.29 GHz has been revealed toward both the Galactic Center and external galaxies as point source emission (Faure et al. 2018; Gorski et al. 2021). With the VLA observations, we resolve this maser emission spatially into three velocity components with a V_{lsr} of 64, 73, and 83 km s^{-1}. The 83 km s^{-1} and 64 km s^{-1} components are both associated with the K6 HII region, with the brightest 83-km s^{-1} component displaying a compact distribution with a $T_{B,\,peak}$ of 121.8 ± 8.6 K. The VLA observations are insufficient to simultaneously constrain all the physical characteristics so the CH_2NH MCMC analysis requires a confined choice of priors on some parameters. We opted to use CH_3OH discussed in Sec. 2.1 for physical constraints as the 83 km s^{-1} components of CH_3OH and CH_2NH are spatially coherent and are both associated with K6 HII region. We therefore extracted the CH_2NH spectrum toward the same position of the CH_3OH-83 km s^{-1} maser spot. The posteriors on T_k and n_{H_2}obtained from CH_3OH are adopted as the priors for the CH_2NH model. The samplers converge at a N_{col} of $\sim 5 \times 10^{14}$ cm^{-2}. The six hyperfine components of the $1_{1,0} - 1_{1,1}$ transition have T_{ex}'s in the range of -0.75 – -0.55 K, τ's in the range of -0.03 – -0.006.

2.3. *CH_2NH_2 $2_0 - 1_1$ E Transition at 4.36 GHz*

The CH_2NH_2 transition at 4.36 GHz displays maser activity similar to the 5.29 GHz CH_2NH transition. Three velocity components were resolved for the CH_2NH_2 maser emission, with a high degree of morphological consistency with the CH_2NH maser. However, a lack of molecular collisional data for CH_2NH_2 prevents us from interpreting the spectral profiles. Future quantum mechanical calculations of the cross sections and rate coefficients of CH_2NH_2 will provide vital information for constraining its excitation conditions and pumping mechanism.

2.4. *Spatial Origin and Pumping Mechanism*

By scrutinizing the VLA and ALMA observations, we found that the K6 HII region harbors maser emission of Class I CH_3OH at 84 GHz, CH_2NH at 5.29 GHz, and CH_2NH_2 at 4.36 GHz (Fig. 2). The 83-km s^{-1} spot of the CH_2NH and CH_2NH_2 masers at K6 remarkably resembles those of the Class I CH_3OH maser. Their consistent morphology serves as strong evidence to support a similar collisional pumping mechanism for the CH_2NH and CH_2NH_2 masers with the Class I CH_3OH maser as well as a common physical condition for triggering these masers. In addition, these maser activities toward the K6 region might also imply the occurrence of energetic events in the masing regions, considering that K6 was recognized as the ionization front of the HII regions and located far from strong infrared sources (Gaume et al. 1995). By understanding the conditions that are necessary for maser activities to occur, we can also better understand how they can be used to monitor astronomical objects. For example, since these new masers are pumped by intense collisions, we can expect them to be good candidates for monitoring regions with time-dependent gas dynamics, such as episodic outflows from massive protostars and accretion disks in extra-galactic environments.

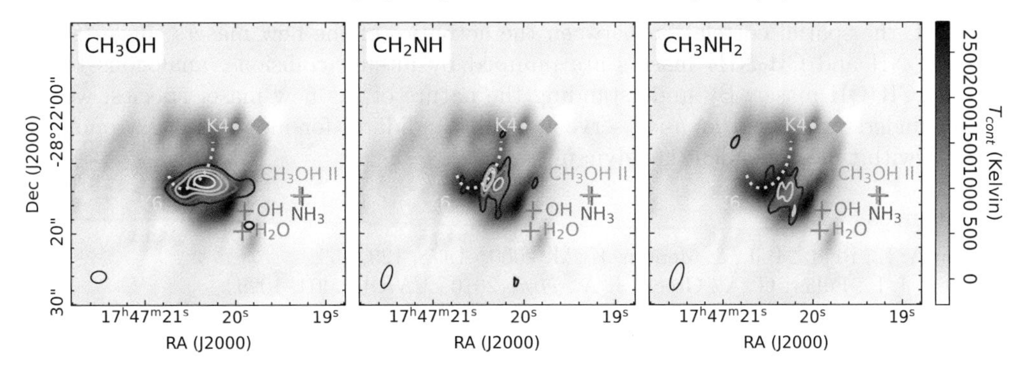

Figure 2. The 83-km s^{-1} components of the 84 GHz Class I CH_3OH maser, 5.29 GHz CH_2NH maser, and 4.36 GHz CH_2NH_2 maser are shown with contours and superimposed on the free-free continuum emission at 6.8 cm. The brightest 84 GHz Class I CH_3OH maser spot is marked with a diamond symbol while the brightest maser spots of OH, H_2O, NH_3, and Class II CH_3OH masers are marked with + symbols (Argon et al. 2000; Caswell et al. 2010; Walsh et al. 2014; Yan et al. 2022). The arc-like K6 H II region is marked with a yellow dotted line.

3. Maser Emission against Weak Background Radiation

In contrast to the CH_3OH-83 km s^{-1} maser spot, there is no significant mm-wave continuum source at the brightest CH_3OH maser spot (shown with a diamond symbol in Fig. 2). While this spot also exhibits a strong CH_3OH emission at 44 GHz (Mehringer & Menten 1997), it shows no continuum emission at cm-wave either. The closest continuum source, K4, is offset by $\sim 3.6''$ which corresponds to $\sim$0.14 pc at the distance of Sgr B2 (8.2 kpc, Reid et al. 2019). In addition, the nearest bright infrared source observed with both the 2MASS and the GLIMPSE program is located at $\sim$5.2$''$ to the west of the maser spot.

The model we consider here consists of double masers, where two masing clumps along the line of sight overlap in velocity. The weak continuum radiation is first amplified by the background masing clump, which is then further amplified by the foreground masing clump. This double-masing-clump model could account for the strong maser emission with an absence of a bright continuum source. It is analogous to the "self-amplification" model of two masing clouds that has been used to interpret water megamasers observed in circumnuclear disks in active galactic nuclei (Kartje et al. 1999). We processed the spectral analysis assuming that the only source of continuum is the cosmic microwave background. Each of the spectral features consists of two individual velocity components, which have coherent but not identical $V_{\rm lsr}$ of $\sim$58 km s^{-1} although spatially separated. As shown in Fig. 1 (bottom), the constructed profiles fit reasonably well with the observed spectra for both the maser and the optically thick emission features. The $T_{\rm ex}$ and τ of the $5_{-1} - 4_0$ maser transition are found to be -271.4 K and -0.3 for the background clump and -7.9 K and -1.4 for the foreground clump.

4. Conclusion

We report interferometric observations of maser activity from large molecules toward Sgr B2 using ALMA and VLA. Enabled by the ALMA observation, we report the first detection of the Class I CH_3OH maser at 84 GHz toward Sgr B2. Multiple 84 GHz CH_3OH maser knots are resolved with a $T_{\rm B,\,peak}$ of 389.2 ± 0.2 K at a velocity of 58 km s^{-1}. With the VLA observations, we characterize the spatial origin of the 5.29 GHz CH_2NH maser and the 4.36 GHz CH_2NH_2 maser. We found a definite association between the maser emission from these large molecules and the UC HII region. They are offset from the masing regions of the radiatively pumped maser species and the infrared sources. In contrast, the spatial correlation between the activities of the new masers suggests that the CH_2NH and CH_2NH_2 masers are pumped by intense collisions, analogous to the Class I CH_3OH maser. By understanding the nature of the new maser species, we can better understand how they may serve as good candidates for monitoring astronomical objects with time-dependent gas dynamics.

References

Argon, A. L., Reid, M. J., & Menten, K. M. 2000, *ApJS*, 129, 159
Caswell, J. L., Fuller, G. A., Green, J. A., *et al.* 2010, *MNRAS*, 404, 1029
Cyganowski, C. J., Brogan, C. L., Hunter, T. R., *et al.* 2009, *ApJ*, 702, 1615
Chen, X., Sobolev, A. M., Ren, Z.-Y., *et al.* 2020, *Nature Astronomy*, 4, 1170
Faure, A., Remijan, A. J., Szalewicz, K., *et al.* 2014, *ApJ*, 783, 72
Faure, A., Lique, F., & Remijan, A. J. 2018, *The Journal of Physical Chemistry Letters*, 9, 3199
Kartje, J. F., Königl, A., & Elitzur, M. 1999, *ApJ*, 513, 180
Gaume, R. A., Claussen, M. J., de Pree, C. G., *et al.* 1995, *ApJ*, 449, 663
Gorski, M. D., Aalto, S., Mangum, J., *et al.* 2021, *A&A*, 654, A110
Mehringer, D. M. & Menten, K. M. 1997, *ApJ*, 474, 346
McGuire, B. A., Loomis, R. A., Charness, C. M., *et al.* 2012, *ApJL*, 758, L33
Reid, M. J., Menten, K. M., Brunthaler, A., *et al.* 2019, *ApJ*, 885, 131
Walsh, A. J., Purcell, C. R., Longmore, S. N., *et al.* 2014, *MNRAS*, 442, 2240
Yan, Y. T., Henkel, C., Menten, K. M., *et al.* 2022, *A&A*, 666, L15

Cosmic Masers: Proper Motion toward the Next-Generation Large Projects
Proceedings IAU Symposium No. 380, 2024
T. Hirota, H. Imai, K. Menten, & Y. Pihlström, eds.
doi:10.1017/S1743921323002090

Feature prospects of IRAS 20126+4104 maser studies

Artis Aberfelds[1], **Anna Bartkiewicz**[2], **Jānis Šteinbergs**[1] **and Ivar Shmeld**[1]

[1]Engineering Research Institute "Ventspils International Radio Astronomy Center", Ventspils University of Applied Sciences, Inzenieru Str. 101, Ventspils, LV-3601, Latvia.
email: artis.aberfelds@venta.lv

[2]Institute of Astronomy, Faculty of Physics, Astronomy and Informatics, Nicolaus Copernicus University, Grudziadzka 5, 87-100 Torun, Poland

Abstract. IRAS 20126+4104 is an extensively studied high-mass star-forming region with many astrophysical maser lines from methanol and water molecules. The brightest and highly variable is the 6.7 GHz methanol maser transition. We present follow-up studies on this target including the monitoring with the Irbene radio telescope with high-cadence data and European VLBI Network imaging extending the VLBI monitoring to 19 years. We also plan to study the target in the future based on its variability, both in the radio domain (EVN observations are planned for June 2023) and in infra-red with the 1 meter Kagoshima University Telescope operated by Amanogawa Galaxy Astronomy Research Center (AGARC).

Keywords. masers – stars, massive – instrumentation, interferometers – stars, formation – astrometry

---

1. Introduction

The formation of high-mass stars is still an actual topic in modern astrophysics. Accretion processes which allow protostar to evolve into a massive star are still debated. There are two most promising scenarios are: a global collapse or competitive accretion (e.g. Zinnecker 2007 as a review). Astronomical masers have emerged as a powerful tool to study the high-mass star-forming regions (HMSFRs), especially the 6.7 GHz methanol maser transition which is exclusively associated with early stages of massive star formation (Menten 1991).

IRAS 20126+4104, also known as G78.122+3.633, is well studied HMSFR: a protostar is estimated to be 7 $M_{\odot}$. with a Keplerian disk seen at several hundred GHz molecular lines with Pico Veleta telescope and Plateau de Bure interferometer (Cesaroni *et al.* 1997). The 22 GHz water masers are related to a jet, while the 6.7 GHz methanol masers are located in the disk (Moscadelli *et al.* 2011). The distance based on the parallax measurement is $1.64^{+0.30}_{-0.12}$ kpc (Reid *et al.* 2019). Single-dish monitoring suggested low and high activity periods of blue-shifted components relative to -6.1 km s^{-1} line (Szymczak *et al.* 2018).

2. Further investigations

We selected this target for further monitoring using the Irbene radio telescopes. We have observed it every week since April 2017. Between August and October 2020. We have confirmed that a rapid variability of the spectral feature at the LSR velocity of

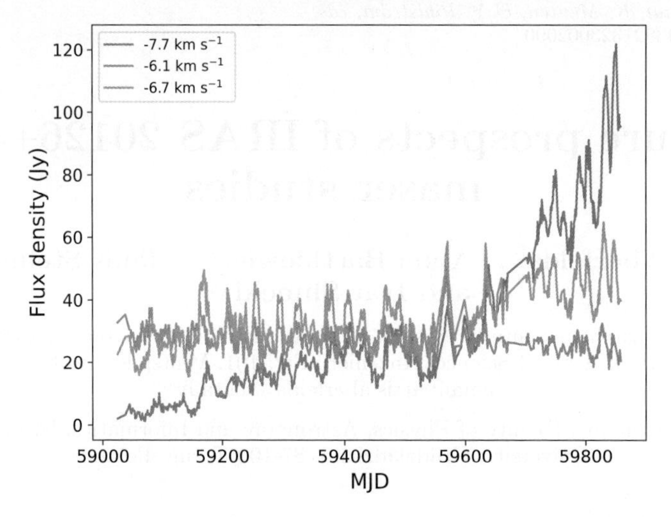

Figure 1. Time series of three IRAS 20126+4104 spectral components seen using the Irbene 16-m dish.

-7.7 km s^{-1} (Figure 1). New image of the target obtained using EVN† extended the time-baseline of the VLBI monitoring presented by Moscadelli *et al.* (2011). With the use of proper motion studies, we confirmed a view that the maser emission from the western part tracks the gas lifted from the disc near the base of outflow/jet evidenced by the 22 GHz water masers. Overall source morphology over 19 years seems to be unchanged.

The 6.7 GHz methanol maser variability is related to the variation in infrared pumping radiation (e.g. Kobak *et al.* 2023). Therefore, we have initiated a feasibility study to monitor infrared flux fluctuations employing the Amanogawa Galaxy Astronomy Research Center (AGARC) with their 1-meter infrared telescope operated by the Kagoshima University in Japan. Preliminary data from the archive of WISE All-Sky Survey suggest fluctuations in the brightness of the infra-red emission, however, too rare observations were carried out so far and no detailed conclusions can be made.

We will also continue the Irbene methanol maser monitoring program on a daily cadence for IRAS 20126+4104. From June 2020 till May 2023, we obtained ca. 700 spectra using Irbene 16-m radio telescope. Note that some of the maser components will be still in the active phase as of 2023 June. There is a need to cooperate with other long-term monitoring programs (like from Torun and Ibaraki) and evaluate the properties of the variability. It will be important to search for any periodicity - both, long- and short-term. As one can notice in Figure 1. most of lines show short-term fluctuations, which is an interesting point, of investigation to search of possible periodicity.

IRAS 20126+4104 is still an interesting source for follow-up studies. This might answer several important questions: What mechanism is responsible for the activity of the blue-shifted methanol masers?, How do individual cloudlets evolve?, Do we have a correlation of infra-red emission with the maser transition? We aim to obtain a clear image of the ongoing scenario in IRAS 20126+4104.

† The European VLBI Network is a joint facility of independent European, African, Asian, and North American radio astronomy institutes. Scientific results from data presented in this publication are derived from the following EVN project code: EA063

Acknowledgements

We acknowledge support by the European Regional Development Fund project No. 1.1.1.5/18/I/009 "Support to the Ventspils University of Applied Sciences in preparation of international cooperation projects for research and Innovation".

References

Cesaroni R., Felli, M., Testi, L., *et al.* 1997, *A&A*, 325, 725
Kobak A., Bartkiewicz A., Szymczak M., *et al.* 2023, *A&A*, 671, A135
Menten K. M. 1991, *ApJ*, 380, L75
Moscadelli L., Cesaroni R., Rioja M.J., *et al.* 2011, *A&A*, 526, A66
Reid M., Menten K. M., Brunthaler A., *et al.* 2019, *ApJ*, 885, 131
Szymczak M., Olech M., Sarniak R., Wolak P., *et al.* 2018, *MNRAS*, 474, 1
Zinnecker, H. and Yorke, H. 2007, *ARAA*, 45

Cosmic Masers: Proper Motion toward the Next-Generation Large Projects
Proceedings IAU Symposium No. 380, 2024
T. Hirota, H. Imai, K. Menten, & Y. Pihlström, eds.
doi:10.1017/S1743921323003198

The Dynamics of the Outflow Structure in W49 N

Kitiyanee Asanok[1], M.D. Gray[1], T. Hirota[2,3], K. Sugiyama[1], M. Phetra[4], B.H. Kramer[1,5], T. Liu[6,7], K.T. Kim[7] and B. Pimpanuwat[8]

[1]National Astronomical Research Institute of Thailand, The Ministry of Higher Education, Science, Research and Innovation, 260 Moo 4, Tambol Donkaew, Amuphur Maerim, Chiangmai 50180, Thailand. email: kitiyanee@narit.or.th

[2]Mizusawa VLBI Observatory, National Astronomical Observatory of Japan, 2-21-1 Osawa, Mitaka, Tokyo 181-8588, Japan

[3]Department of Astronomical Sciences, SOKENDAI (The Graduate University for Advanced Studies), 2-21-1 Osawa, Mitaka-shi, Tokyo 181-8588, Japan

[4]Graduate School, Department of Physics and Material Science, Faculty of Science, Chiang Mai University, 239 Huaykaew road, Tambol Suthep, Amphur Muang Chiangmai 50200, Thailand

[5]Max-Planck-Institut für Radioastronomie, Auf dem Hügel 69, Bonn 53121, Germany

[6]Shanghai Astronomical Observatory, Chinese Academy of Sciences, 80 Nandan Road, Shanghai 200030, China

[7]Korea Astronomy and Space Science Institute, 776 Daedeokdae-ro, Yuseong-gu, Daejeon 34055, Republic of Korea

[8]Jodrell Bank Centre for Astrophysics, School of Physics and Astronomy, The University of Manchester, Alan Turing Building, Oxford Rd M13 9PY, United Kingdom

Abstract. In this current study, we report only the preliminary result of the SiO $v=0$ $J=5\rightarrow4$ emission toward W49 N at 230 GHz, observed using the ALMA telescope on September 29, 2018. The position–velocity diagram of the SiO emission shows a structure of a bipolar outflow and has a face-on orientation with an inclination angle of 36.4±0.4 degrees with respect to the line of sight. Here we summarize the calculated physical properties of its outflow.

Keywords. masers – stars: formation, ISM: HII regions, ISM: jets and outflows, ISM: molecules, stars: individual: W49 N

---

1. The calibrated data of W49 N at 230 GHz taken from JVO portal

We have studied the star-forming region W49 N with the proper motions of 22 GHz water masers found with the KaVA telescope during February to May 2017 (Asanok *et al.* 2023). In the present work, we report the preliminary analysis results of the physical properties of the SiO outflow towards W49 N by using the calibrated data of the ALMA archive. These data were taken in Band 6 and are archived in the Japan Virtual Observatory (JVO) portal service under the project code ALMA#2016.1.00620.S.

2. The SiO $v=0$ $J=5\rightarrow4$ outflow

PV diagram: the position–velocity diagram (PV) of the SiO $v=0$ $J=5\rightarrow4$ line emission in the W49 N region shows a clear structure of a bipolar outflow with a compact size

Table 1. Physical properties of the outflow derived from the SiO $v=0$ $J=5\rightarrow 4$.

	Red lobe:	**Blue lobe:**
Velocity Interval ($\mathrm{km\,s^{-1}}$):	[+10.0 : +74.7]	[−54.7 : +8.6]
$N_{\rm SiO}$ ($10^{14}\mathrm{cm^{-2}}$):	1.4	1.8
$L_{\rm SiO}$ ($10^{-2}L_{\odot}$):	7.4	9.1
$M_{\rm out}$ ($M_{\odot}$):	41.5±5.9	51.1±7.2
$\dot{M}_{\rm out}$ ($10^{-4}M_{\odot}\mathrm{yr^{-1}}$):	39.5±4.6	7.0±1.0
P ($M_{\odot}\mathrm{km\,s^{-1}}$):	2180±308	884±125
$V_{\rm char}$ ($\mathrm{km\,s^{-1}}$):	52.6	17.3
$\dot{P}$ ($M_{\odot}\mathrm{km\,s^{-1}yr^{-1}}$):	2.0±0.2	0.4±0.1
E (10^{47}erg):	11.4±1.6	1.5±0.2
$L_{\rm mech}$ ($10^{2}L_{\odot}$):	84.0±11.1	4.8±0.6
$l_{\rm lobe}$ (10^{3}au):	12.2	9.2
$t_{\rm dyn}$ (yr):	1104±9	2532±28

of diameter (1.5×0.5) arcsec^2 in each lobe. These lobes are formed by gas ejected from the outflow core that subsequently reaches speeds sufficient to generate shocks.

Physical properties of the SiO outflow: we have adapted the equations which were taken from Nguyen-Lu'o'ong *et al.* (2013) and Liu *et al.* (2015) and computed the physical properties of the SiO $v=0$ $J=5\rightarrow 4$ line emission (see Table 1) as follows. (1) The column densities within a main beam $N_{\rm SiO}=1.6\times 10^{11}\mathrm{cm^{-2}}\times \frac{(T_{\rm ex}+0.35)\exp\{31.26/T_{\rm ex}\}}{\exp\{10.4/T_{\rm ex}\}-1}\times \frac{1}{J_\nu(T_{\rm ex})-J_\nu(T_{\rm bg})}\int T_{\rm mb}\mathrm{d}\nu$ where $J_\nu(T)=\frac{h\nu/k}{\exp(h\nu/kT)-1}$, $T_{\rm ex}$ and $T_{\rm bg}$ are the excitation temperature of the gas and background radiation; (2) the luminosity $L_{\rm SiO}\simeq 2.3\times 10^{-4}\mathrm{L_{\odot}}\times \left(\frac{\mathrm{d}}{6\,\mathrm{kpc}}\right)^2\frac{\int T_{\rm mb}\mathrm{d}\nu}{1\,\mathrm{K\,km\,s^{-1}}}$, where d is the distance of W49 N, $L_{\odot}$ the luminosity of the sun and $\int T_{\rm mb}\mathrm{d}\nu$ the velocity integrated intensity inside the main beam; (3) the gas mass in the outflow $M_{\rm out}=N_{\rm SiO}\left[\frac{\mathrm{H_2}}{\mathrm{SiO}}\right]\mu_{\rm g}m_{\rm H}d^2\Omega_{\rm A}$, where $\Omega_{\rm A}=\frac{\pi}{4\ln(2)}\theta^2_{\rm FWHM}$, $\mu_{\rm g}=1.36$ is the mean molecular mass per hydrogen atom, $m_{\rm H}$ is the hydrogen atom mass, and the ratio $\left[\frac{\mathrm{H_2}}{\mathrm{SiO}}\right]$ has a large uncertainty which depends on the SiO abundance and types of sources; (4) if we assume the outflow is powered by wind driven therefore the mass lost rate $\dot{M}_{\rm out}=\frac{\mathrm{P}}{t_{\rm dyn}V_{\rm wind}}$ where $V_{\rm wind}$ is the wind velocity; (5) the momentum $P=M_{\rm out}\times V_{\rm char}$ where the characteristic outflow velocity ($V_{\rm char}$) is defined as $V_{\rm char}=V_{\rm flow}-V_{\rm sys}$, $V_{\rm sys}$ the systemic velocity, and $V_{\rm flow}$ the intensity-weighted velocity of high-velocity emission corrected with the projection effect; (6) the momentum rate $\dot{P}=\frac{P}{t_{\rm dyn}}$; (7) the kinetic energy $E=\frac{1}{2}M_{\rm out}\times V^2_{\rm char}$; (8) the mechanical luminosity $L_{\rm mech}=E/t_{\rm dyn}$; and (9) the dynamical time scale ($t_{\rm dyn}$), respectively.

Supplementary material

To view supplementary material for this article, please visit http://dx.doi.org/10.1017/S1743921323003198

References

Asanok, K., Gray, M. D., Hirota, T., et al. 2023, *ApJ*, 943, 79
Liu, T., Kim, K.-T., Wu, Y., et al. 2015, *ApJ*, 810, 147
Nguyen-Lu'o'ng, Q., Motte, F., Carlhoff, P., et al. 2013, *ApJ*, 775, 88
Smith, N., Whitney, B. A., Conti, P. S., De Pree, C. G., Jackson, J. M. 2009, *MNRAS*, 399, 952

Cosmic Masers: Proper Motion toward the Next-Generation Large Projects
Proceedings IAU Symposium No. 380, 2024
T. Hirota, H. Imai, K. Menten, & Y. Pihlström, eds.
doi:10.1017/S1743921323002624

ALMA observations of the environments of G301.1364-00.2249A

Zh. Assembay,[1] T. Komesh,[1,2] G. Garay,[3] A. Omar,[1] J. Esimbek,[4] N. Alimgazinova,[1] M. Kyzgarina[1] and Sh. Murat[5]

[1]IETP, Al-Farabi Kazakh National University, Almaty 050040, Kazakhstan.
toktarkhan.komesh@nu.edu.kz

[2]Energetic Cosmos Laboratory, Nazarbayev University, Astana 010000, Kazakhstan

[3]Departamento de Astronomia, Universidad de Chile, Camino el Observatorio 1515, Las Condes, Santiago, Chile

[4]Xinjiang Astronomical Observatory, Chinese Academy of Sciences, Urumqi 830011, PR China

[5]General Secondary School No.137, Shymkent 160024, Kazakhstan

Abstract. *Context:* Theoretical scenarios describe the phenomenon of mass accumulation by high-mass young stellar objects (HMYSOs) through disk accretion. *Aims:* To find out whether the rotation of the core around hyper-compact (HC) HII regions is common *Methods:* The molecular core G301.1364-00.2249A was selected as the subject of investigation. Observations were carried out on CH_3CN, SO_2, H29α radio recombination line and continuum emission. Analysis involved the "Moment 0, 1, 2" and "Population-diagram" methods. *Result:* The structures of G301.1364-00.2249A exhibited clear definition. CH_3CN moment 0 images revealed multiple enhanced emission regions and an absorption area. The moment 1 image depicted a velocity gradient from southeast to northwest. *Conclusion:* The molecular gas has a rotational motion in the direction from the southeast to northwest. The rotational temperature was measured at 293 K.

Keywords. ISM: individual objects: G301.14AB—ISM: molecules—ISM: clouds—ISM: cores—stars: formation—stars: massive—ISM: kinematics and dynamics

---

1. Introduction

It is assumed that for an O-type star, 50% of the total mass is accreted after Kelvin-Helmholtz contraction (Hosokawa & Omukai 2009; Zhang, Tan, & Hosokawa 2014). These are the initial stages of ionizing radiation, the prerequisites for the emergence of the HC HII region. Therefore, in order to identify the phenomenon of mass accumulation by HMYSO, it is necessary to study the surrounding ionized region at an early stage. We observed the hot core G301.1364-00.2249A associated with HC HII region through the ALMA 6 band (256.302035 GHz – 259.599448 GHz), dust continuum and molecular line emission arising from two molecules, CH_3CN (J=14→13) ladder and SO_2 ($30_{4,26} - 30_{3,27}$ and $32_{4,28} - 32_{3,29}$).

The source G301.1364-00.2249 was catalogued as high-mass young stellar object (HMYSO) candidate by Urquhart *et al.* (2007). Due to its rising spectral index between 6.67 GHz and 8.64 GHz, Guzmán *et al.* (2012) considered this source as a jet candidate. The G301.1364-00.2249 core consists of two cores: G301.1364-00.2249A and G301.1364-00.2249B. Their peak positions are at (RA, Dec) (J2000) = ($12^h35^m35.13^s$, -63°02′31.7″)

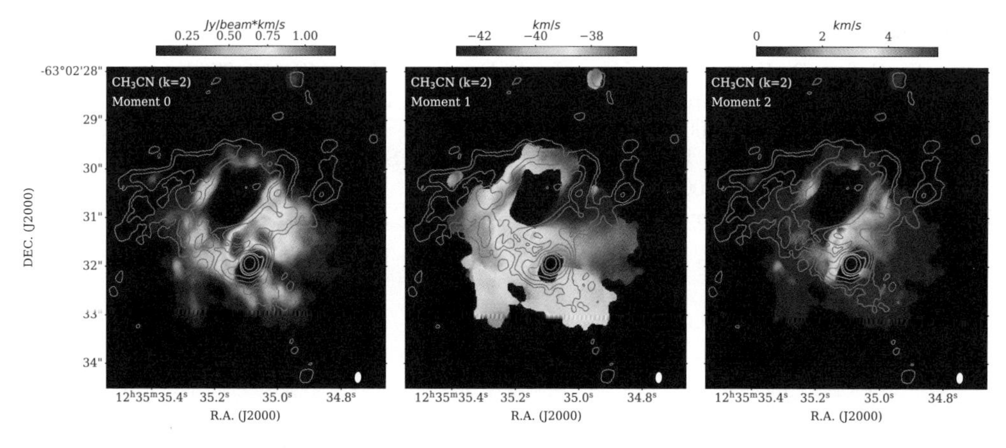

Figure 1. Moment 0, 1, 2 (colorscale) images of CH_3CN (K=2) toward G301.1364-00.2249A. Superimposed are contours of the continuum emission. Contour levels are 0.00448, 0.00896, 0.02689, 0.06274, 0.11652, 0.17926 and 0.44816 Jy beam^{-1}. The white ellipse shown at the bottom right corner indicates the beam size ($0.256'' \times 0.198''$).

and (RA, Dec) (J2000) = ($12^h35^m35.19^s$;-63°02′24.0″), respectively. Both of these cores are detected emission in the H29α radio recombination line (RRL) in our observations. In the following sections, we show the moment maps and rotational temperature analysis of G301.1364-00.2249A core.

2. Molecular gas dynamics

Moment 0 map in Figure 1 shows several enhanced emission areas (~1 Jy beam^{-1} km s^{-1}), and an absorption area which corresponds to the peak of the continuum emission. Moment 1 map shows a velocity gradient from south-east to north-west. In the image of moment 2, the velocity dispersion showing a structure at the north-east (~4 km s^{-1}) of the continuum peak. The images of moments 1, 2 determines the direction of rotation motion of the molecular gas, which is directed from south-east to north-west.

3. Rotation diagram analysis: estimation of gas temperature

Using the rotation transitions of the CH_3CN and the population-diagram method (see Araya *et al.* 2005), it is possible to determine the rotation temperature $T_{\rm rot}$=293 K and the column density $N_{\rm CH3CN}$=6,716·10^{16} cm^{-2}.

4. Future studies

In our follow-up studies, we will analyze the ionized gas, as well as the kinematics and dynamics of this source.

5. Acknowledgments

This research was funded by the Science Committee of the Ministry of Science and Higher Education of the Republic of Kazakhstan (Grant Nos. AP13067768). JE acknowledges support from the Regional Collaborative Innovation Project of Xinjiang Uyghur Autonomous Region grant 2022E01050.

References

Araya E., Hofner P., Kurtz S., Bronfman L., DeDeo S., 2005, *ApJS*, 157, 279

Guzmán A. E., Garay G., Brooks K. J., Voronkov M. A., 2012, *ApJ*, 753, 51

Hosokawa T., Omukai K., 2009, *ApJ*, 691, 823

Urquhart J. S., Busfield A. L., Hoare M. G., Lumsden S. L., Clarke A. J., Moore T. J. T., Mottram J. C., *et al.* 2007, *A&A*, 461, 11

Zhang Y., Tan J. C., Hosokawa T., 2014, *ApJ*, 788, 166

Cosmic Masers: Proper Motion toward the Next-Generation Large Projects
Proceedings IAU Symposium No. 380, 2024
T. Hirota, H. Imai, K. Menten, & Y. Pihlström, eds.
doi:10.1017/S1743921323002466

Methanol and excited OH masers in W49N as observed using EVN

Anna Bartkiewicz[1], **Marian Szymczak**[1], **Agnieszka Kobak**[1] and **Mirosława Aramowicz**[2]

[1]Institute of Astronomy, Faculty of Physics, Astronomy and Informatics, Nicolaus Copernicus University, Grudziadzka 5, 87-100 Torun, Poland. email: annan@astro.umk.pl

[2]Astronomical Institute, Department of Physics and Astronomy, University of Wrocław, Kopernika 11, 51-622 Wrocław, Poland

Abstract. We imaged the excited OH maser line at 6.035 GHz associated with the 6.7 GHz methanol masers in a selected sample of high-mass young stellar objects using the European VLBI Network. The excited OH emission was found in a survey of methanol maser sources carried out since 2018 with the Torun 32-m telescope. The overlap of radial velocities of spectral features of methanol and excited OH suggested that both lines arose in the same volume of gas, therefore, we verified this hypothesis with the interferometric data. Here, we present the first images at the milliarcsecond scale of both maser transitions and identify the Zeeman pairs at the ex-OH line estimating the strength of the magnetic field in G43.149+00.013 (W49N).

Keywords. methanol maser, exited OH maser, magnetic field, Zeeman pairs

---

1. Introduction

Significant contributions to our knowledge of the formation of high-mass young stellar objects (HMYSOs) came from VLBI observations of cosmic masers, mainly methanol and water transitions. Since a large survey of the 6.7 GHz methanol maser line using the Torun 32-m dish (Szymczak *et al.* 2018) was completed by the survey at the excited OH maser line at 6.035 GHz (ex-OH hereafter) (Szymczak *et al.* 2020), we were encouraged to start to follow up studies using European VLBI Network (EVN)† C-band receivers but now at the opposite edge to the methanol line. As a pilot project, we selected the brightest source G43.149+00.013 (known as W49N), where the radial velocities of spectral features of methanol and ex-OH overlapped suggesting that both lines arise in the same volume of gas.

2. Observations

We observed the HMYSO W49N on 2019 June 3 at the 6.035 GHz OH transition in the phase-referencing mode using J1912+0518 as a phase-calibrator. Useful data were obtained from seven EVN antennas: Effelsberg, Irbene, Jodrell Bank, Medicina, Onsala, Torun, Westerbork. The rms and synthesized beams on the final images in two circular polarizations (RHC, LHC) were 8 mJy and 15×5 mas with a position angle of $-5°$, respectively. The spectral resolution was 0.097 km s^{-1}.

† The European VLBI Network is a joint facility of independent European, African, Asian, and North American radio astronomy institutes. Scientific results from data presented in this publication are derived from the following EVN project codes: EB073.

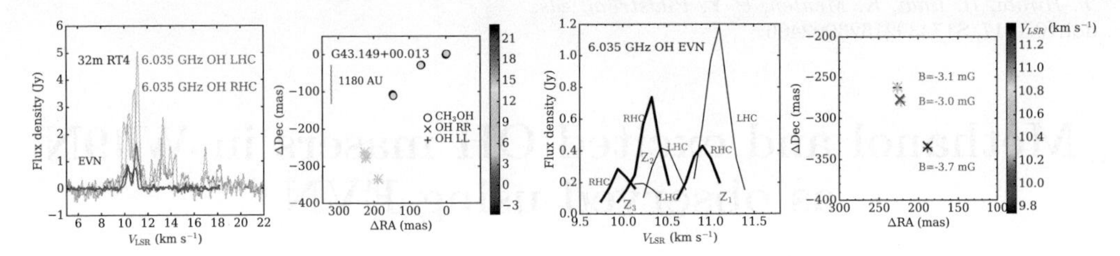

Figure 1. The ex-OH and 6.7 GHz methanol masers in G43.149+00.013 region of W49N. From left to right panels, we present 1) the 32-m Torun dish and EVN spectra, 2) the distributions of both masers, 3) the spectrum of LHC and RHC polarised emission, and 4) the distribution of the polarised masers. The colours of maser spots correspond to the LSR velocities as indicated on the vertical bars.

3. Results and Discussion

We imaged the ex-OH masers related to the southern component of W49N, the G43.149+00.013 source (Fig. 1). This source showed intermediate brightness of the 6.7 GHz methanol masers compared with other components of W49N, i.e.: G43.165+00.013, G43.171+00.004, and G43.167−00.004. According to the evolutionary sequence for masers proposed by (Breen et al. 2010), the G43.149+00.013 would be the most evolved region in W49N. However, the single-dish observations show some emission at the higher LSR velocities (from 12 to 15 km s^{-1}) that was not recovered in imaging and extensive and sensitive searching for ex-OH emission is required.

The single-dish monitoring indicates that the 6.7 GHz methanol line is non-variable (Szymczak *et al.* 2018), so we can assume the stability of methanol maser morphology. The displacement of both maser species is clear, about 200 mas, corresponding to 2200 AU for a distance of 11.1 kpc (Zhang *et al.* 2013). No co-existence was detected indicating the diversity in densities of the maser-emitting regions (Cragg *et al.* 2002). However, we note the 6.7 GHz methanol maser images were obtained in 2010 (Bartkiewicz *et al.* 2014), therefore, simultaneous observations are required (see Kobak *in this volume*).

The interesting result, for which we aimed, is the detection of three Zeeman pairs; only the interferometric data can verify the existence of Zeeman pair components. We detected three maser clouds, where LHC and RHC 6.035 GHz OH emission coincide and their velocity profiles are shifted (Fig. 1). We calculated the magnetic field strength (B) as ca. 3.5 mG oriented towards us taking the coefficient of 0.056 km s^{-1} mG^{-1} (Baudry *et al.* 1997).

4. Summary

We present the use of EVN to study gas properties in the environment of HMYSO using the 6.7 GHz methanol and 6.035 GHz ex-OH masers. The first results for W49N indicate the avoidance of both maser lines at the mas scales and the diversity in densities of the maser-emitting regions (Cragg *et al.* 2002). We confirm the existence of Zeeman pairs as it was proposed in the single-dish studies (Szymczak *et al.* 2020).

Acknowledgements

We acknowledge support from the National Science Centre, Poland through grant 2021/43/B/ST9/02008.

Supplementary material

To view supplementary material for this article, please visit http://dx.doi.org/10.1017/S1743921323002466

References

Bartkiewicz A., Szymczak M., van Langevelde H. J. 2014, *A&A*, 564, A110
Baudry, A., Desmurs, J. F., Wilson, T. L., & Cohen, R. J. 1997, *A&A*, 325, 255
Breen, S.L., Ellingsen, S.P., Caswell, J.L., Lewis, B.E. 2010, *MNRAS*, 401, 2219
Cragg, D. M., Sobolev, A. M., Godfrey, P.D. 2002, *MNRAS*, 331, 521
Szymczak, M., Olech, M., Sarniak, R., Wolak, P., Bartkiewicz, A. 2018, *MNRAS*, 474, 219
Szymczak, M., Wolak, P., Bartkiewicz, A., Aramowicz, M., Durjasz, M. 2020, *A&A*, 642, A145
Zhang, B., Reid, M.J., Menten, K.M., Zheng, X.W., Brunthaler, A., Dame, T.M., Xu, Y. 2013, *ApJ*, 775, 79

Cosmic Masers: Proper Motion toward the Next-Generation Large Projects
Proceedings IAU Symposium No. 380, 2024
T. Hirota, H. Imai, K. Menten, & Y. Pihlström, eds.
doi:10.1017/S1743921323002260

Catching unusual phenomena with extensive maser monitoring

Michał Durjasz

Institute of Astronomy, Faculty of Physics, Astronomy and Informatics, Nicolaus Copernicus University, Grudziadzka 5, 87-100 Torun, Poland. email: md@astro.umk.pl

Abstract. High brightness and low interstellar extinction allow the 6.7 GHz methanol (CH_3OH) masers to carry the information about what happens in the vicinity of the High-Mass Young Stellar Objects (HMYSOs). Monitoring this transition provides an only one opportunity to catch rare, unusual phenomena. In this paper, I describe three of them: quasi-periodic flares of the red-shifted emission in Cep A HW2, accretion burst in S255-NIRS3 and reappearance of the methanol maser flare in G24.329+0.144.

Keywords. masers, stars: massive, stars: formation, ISM: molecules, radio lines: ISM

---

1. Introduction

The 6.7 GHz CH_3OH maser is the second brightest interstellar emission line (Menten 1991). It originates in the vicinity of high-mass young stellar objects (HMYSOs) and is highly sensitive to variations of the physical conditions in their environment. Theoretical models (Sobolev *et al.* 1997; Cragg *et al.* 2005) show that the maser emission comes from the regions of number density lower than $10^9\,\mathrm{cm}^{-3}$ and dust temperatures higher than 100 K. There are theoretical and recent observational evidences that IR photons are essential in the pumping of the class II methanol masers (e.g. Olech *et al.* 2020, 2022). These attributes make the 6.7 GHz methanol maser a fairly good indicator of the protostellar activity.

2. Observations

The Torun maser research team has been using the 32-meter radio telescope of the Nicolaus Copernicus University, located in Piwnice to monitor regularly the 6.7 GHz methanol masers since June 2009. The number of observed sources varied between 70 to 250 throughout the monitoring period. The main parameters of the spectral observations are as follows: the RMS pointing error is estimated to be 10″, T_{sys} in a range from 25 to 40 K, spectral resolution: $0.09\,\mathrm{km\,s}^{-1}$, the typical ($1\sigma$) noise level was 0.25 Jy.

3. Results

Here, we describe some results from Torun's 13-year monitoring program for selected targets:

Cep A HW2. This well-known HMYSO lies at the distance of 700 pc (Moscadelli *et al.* 2017) and shows 5-year quasi-periodic variability of the most red-shifted spectral features at -0.5 and $-1.3\,\mathrm{km\,s}^{-1}$. Anti-correlation between -4.7 and $-2.6\,\mathrm{km\,s}^{-1}$ features during the periods of activity of the blue-shifted component is also visible. Such behaviour suggests a radiative connection between the masing cloudlets as proposed by Cesaroni

(1990). The measured delay between the cloudlets was 4.5 d, corresponding to a distance of ~800 AU. The anti-correlation was presented initially by Sugiyama *et al.* (2008), and new data confirm it (see also Szymczak *et al.* 2014 and Durjasz *et al.* 2022). Red-shifted flaring features originate near the edge of a dust disc, reported in Patel *et al.* (2005). Proper motions of the flaring cloudlets show sub-Keplerian orbital velocity combined with infall motion. These results confirm previous reports (Sugiyama *et al.* 2014; Sanna *et al.* 2017). We expect the next flares of -0.5 and $-1.3\,\mathrm{km\,s^{-1}}$ features to occur in the first quarter of 2025 and the increase of the $-4.7\,\mathrm{km\,s^{-1}}$ feature activity in the last quarter of 2027. The monitoring will be continued.

S255-NIRS3. The distance is estimated to be ~1.8 kpc (Burns *et al.* 2016) and the central star mass to ~20 $M_{\odot}$ (Zinchenko *et al.* 2015). The accretion burst at the 6.7 GHz transition started in 2015 and lasted for 2 years. It is one of the most extensively studied HMYSO accretion burst to date (Caratti o Garatti *et al.* 2017; Szymczak *et al.* 2018; Moscadelli *et al.* 2017). Our 13-year monitoring shows that the overall 6.7 GHz profile has not recovered to the pre-burst state from before 2015. Detection of the emission lines, typical for the young eruptive low-mass stars (Caratti o Garatti *et al.* 2017) suggest a rapid increase in the accretion rate.

G24.329+0.144. The 6.7 GHz methanol flare was detected in 2011 and it reappeared in 2019 (Wolak *et al.* 2019). The overall emission profile recovered after each burst to its previous shape. Also, the weakest features in the pre-burst spectra appear to be the strongest when a flare occurs. Interferometric observations of diverse maser lines and dust continuum towards this source were described in Kobak *et al.* (2023) and Hirota *et al.* (2022). It is important to monitor this target and confirm the ca. 8 year periodicity.

4. Conclusions

Extensive monitoring of the 6.7 GHz transition not only provides a significant amount of information about variations of the physical conditions in the vicinity of the emerging, massive stars but also extends scientific output from the high-resolution VLBI observations and allows for early detection of the accretion bursts.

Acknowledgment

We acknowledge support from the National Science Centre, Poland through grant 2021/43/B/ST9/02008.

References

Burns *et al.* 2016, *MNRAS* , 460(1), 283–290.
Caratti o Garatti *et al.* 2017, *Nature Physics*, 13(3), 276–279.
Cesaroni, R. 1990, *A&A* , 233, 513.
Cragg *et al.* 2005, *MNRAS*, 360, 533–545.
Durjasz *et al.* 2022, *A&A* , 663, A123.
Hirota *et al.* 2022, *PASJ* , 74(5), 1234–1262.
Kobak *et al.* 2023, *A&A* , 671, A135.
Menten, K. M. 1991, *ApJ*, 380, L75.
Moscadelli *et al.* 2017, *A&A*, 600, L8.
Olech *et al.* 2020, *A&A* , 634, A41.
Olech *et al.* 2022, *A&A* , 661, A114.
Patel *et al.* 2005, *Nature* , 437(7055), 109–111.
Sanna *et al.* 2017, *A&A* , 603, A94.
Sobolev *et al.* 1997, *A&A*, 324, 211–220.
Sugiyama *et al.* 2008, *PASJ* , 60, 1001.

Sugiyama *et al.* 2014, *A&A* , 562, A82.
Szymczak *et al.* 2014, *MNRAS* , 439(1), 407–415.
Szymczak, M., Olech, M., Wolak, P., Gérard, E., & Bartkiewicz, A. 2018, *A&A* , 617, A80.
Wolak *et al.* 2019, *The Astronomer's Telegram*, 13080, 1.
Zinchenko *et al.* 2015, *ApJ* , 810(1), 10.

Cosmic Masers: Proper Motion toward the Next-Generation Large Projects
Proceedings IAU Symposium No. 380, 2024
T. Hirota, H. Imai, K. Menten, & Y. Pihlström, eds.
doi:10.1017/S1743921323002089

Water maser flare and potential accretion burst in NGC 2071-IR

Andrews Dzodzomenyo[1], **James O. Chibueze**[1,2,3] and **Stefanus van den Heever**[4]

[1]Centre for Space Research, North-West University, Potchefstroom, South Africa
email: andrews.dzodzomenyo@gmail.com

[2]Department of Physics and Astronomy, University of Nigeria, Nsukka, Nigeria

[3]Department of Mathematical Sciences, University of South Africa, Cnr Christian de Wet Rd and Pioneer Avenue, Florida Park, 1709, Roodepoort, South Africa

[4]South African Radio Astronomy Observatory, South Africa

Abstract. We monitored 22 GHz water masers in NGC 2071-IR using the Hartebeesthoek 26-m telescope and identified a significant flare (up to 4722 Jy) originating from the 14.4 $\mathrm{km\,s^{-1}}$ feature associated with the protostellar core NGC 2071-IRS1. To determine if the maser flare resulted from an accretion burst, we analyzed related signatures such as simultaneous flaring of other maser species and an increase in infrared luminosity. Near-infrared (Ks-band) observations conducted on 28 December 2019 during the flare, using the Kanata/HONIR telescope, exhibited a 0.2 magnitude increase in comparison to the 2MASS magnitude obtained from observations conducted on 10 October 1999. However, our findings indicate that the flare was attributed to mechanisms other than an accretion burst.

Keywords. Masers, stars: formation, ISM: jets and outflows, ISM: individual objects (NGC 2071)

---

1. Introduction

The NGC 2071-IR is a star-forming region in the Orion B molecular cloud which is associated with low and intermediate-mass protostars and it hosts protostars (IRS 1 and IRS 3) driving high-velocity bipolar outflows (Cheng *et al.* 2022). Accretion bursts are characterized by a rapid increase in protostellar accretion rate and an increase in luminosity (Stamatellos *et al.* 2011). Accretion burst can drive H_2O maser flares (Bayandina *et al.* 2022) and long-term monitoring observations enable the detection of flares (Burns *et al.* 2022). We used the Hartebeesthoek Radio Astronomy Observatory's (HartRAO) 26-m radio telescope to monitor the 22 GHz H_2O masers in NGC 2071-IR from January 2019 to May 2022 and detected a strong flare in November 2019. During the flare, the Kanata/HONIR telescope made near-infrared observations which revealed the Ks-band magnitude increased by 0.2 mag compared to the 2MASS magnitude. Although short-term water maser flares indicate physical changes in the associated star-forming regions, our understanding of the mechanisms underlying these flares still remains incomplete.

2. An accretion burst in NGC 2071-IR?

The $V_{LSR} = 14.4\,\mathrm{km\,s^{-1}}$ feature peaked at 4722 ± 4 Jy on MJD 58837. The flare profile exhibited a gradual rise over 341 days and a rapid decay over 79 days. By comparing the features in the HartRAO spectra with the interferometric spot maps shown in Figure 1,

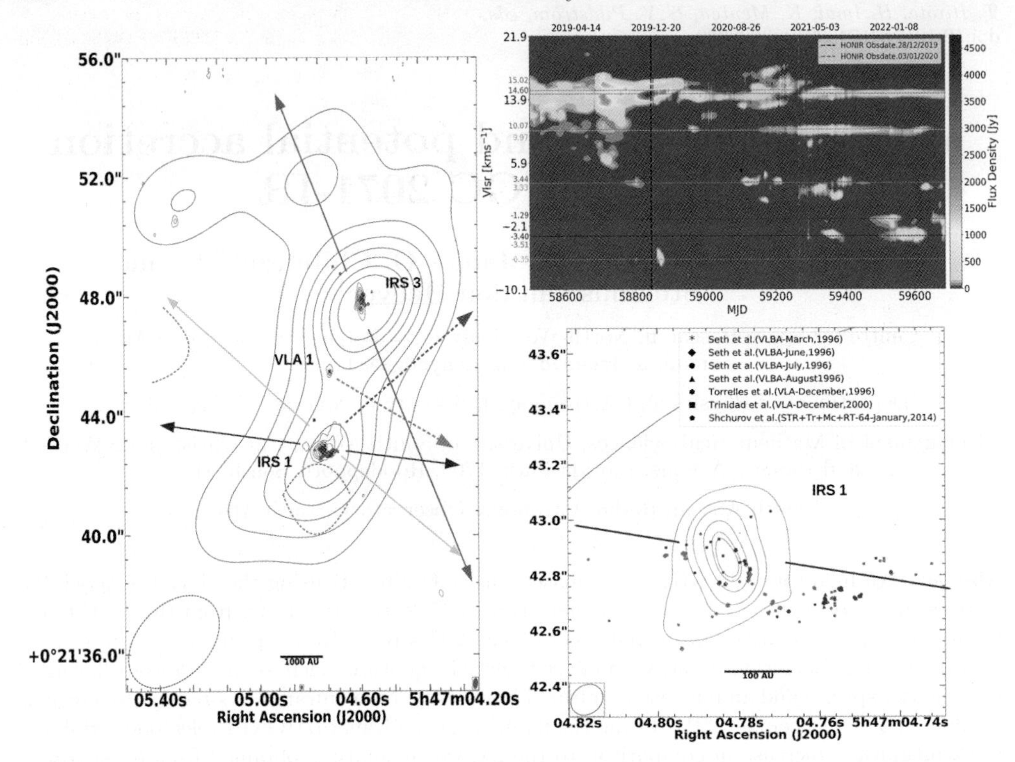

Figure 1. (*Top right*) Dynamic spectra of water masers. The vertical lines represent the Kanata/HONIR observations and the horizontal lines indicate the velocities of the main features. (*Left*) The layout of NGC 2071-IR region with ALMA 0.8 mm continuum (green contours) and the 3.6 cm VLA continuum (magenta contours) overlaid on the ALMA-ACA 0.8 mm continuum (black contours). The arrows represent the outflows in the region and their respective directions, with dashed arrows indicating potential or inferred outflows. (*Bottom right*) A zoom-in on NGC 2071-IRS1. The maser positions from the studies listed in the legend of the bottom right image are coloured blue or red to indicate blue-shifted or red-shifted masers with respect to the cloud velocity ($9.5\,\mathrm{km\,s^{-1}}$).

we determined that the flared feature is associated with IRS 1. Overlaying the water maser positions onto the IRS 1 dust emission revealed a maser distribution that resembles an outflow shell, resembling a "water spout" (Burns *et al.* 2015). To gain insights into the possible mechanism behind the flare, we studied the structure and morphology of the NGC 2071-IR region using archival millimetre and sub-millimetre observations from ALMA and the VLA.

Although significant increases in water maser flux density and luminosity characterize an accretion burst, the attained increment levels by the flare (4722 Jy) and luminosity (0.2 mag) are not significant in this context. Furthermore, the flare profile (gradual rise and sharp decay) deviates from the expected profiles observed in accretion burst sources, which typically exhibit an exponential rise followed by a gradual or steep decay (MacLeod *et al.* 2018). Moreover, the expected increase in luminosity during an accretion burst is typically by a factor of $5 \sim 10$ (Audard *et al.* 2014). Considering the multiple outflows observed in the region (Figure 1 *left*), shocks resulting from the collision of entrained ejecta in the outflow with the ambient cloud may provide a better explanation for the flare.

We acknowledge funding from DARA (Development in Africa with Radio Astronomy: Newton Fund -UKs (STFC grant ST/R001103/1) and support from HartRAO and the M2O.

References

Cheng, Y., Tobin, J., Yang, L., *et al.* 2022, *ApJ*, 933, 178.

Bayandina, O., Brogan, C., Burns, R., *et al.* 2022, *A&A*, 664, A44.

Stamatellos, D., Whitworth, A., & Hubber, D. 2011, *ApJ*, 730, 32.

Burns, R., Kobak, A., Garatti, A. o., *et al.* 2022, EVN Symposium and Users' Meeting 2021, 2021, 19.

Burns, R., Imai, H., Handa, T., *et al.* 2015, *MNRAS*, 453, 3163.

MacLeod, G., Smits, D., Goedhart, S., *et al.* 2018, *MNRAS*, 478, 1077.

Audard, M., Ábrahám, P., Dunham, M., *et al.* 2014, Protostars and Planets VI, 387.

Cosmic Masers: Proper Motion toward the Next-Generation Large Projects
Proceedings IAU Symposium No. 380, 2024
T. Hirota, H. Imai, K. Menten, & Y. Pihlström, eds.
doi:10.1017/S1743921323003423

Discovery of circular polarization of the 6.7 GHz methanol maser in G33.641-0.228

Kenta Fujisawa

The Research Institute of Time Studies, Yamaguchi University, Yamaguchi, Yamaguchi 753-8511, Japan. email: kenta@yamaguchi-u.ac.jp

Abstract. The 6.7 GHz methanol maser emitted by a high-mass star forming region G33.641-0.228 is known to exhibit fast flux density variability on timescales of less than one day. The mechanism of this variability, called burst, has not been known. We observed the circular polarization of this maser. As a result, we found that only the spectral components representing burst exhibit strong circular polarization exceeding 10%. This suggests that the two phenomena of the burst and circular polarization are related to each other.

Keywords. ISM:jets and outflows, accretion disks, stars: massive, stars: protostars

1. Introduction

High-mass star forming region G33.641-0.228 emits the 6.7 GHz methanol maser with multiple spectral peaks (Szymczak *et al.* 2000). One of them ($V_{\rm lsr}$ = 59.6 km s^{-1}) shows fast variability with a time scale of less than one day (Fujisawa *et al.* 2012, 2014), and this variability is called burst. The physical mechanism by which occurs the burst has not been known. The profile of the burst is characterized by an increase in a short period followed by a gradual decrease. Since this feature is similar to those of solar radio bursts or flares of a T Tau star (Phillips *et al.* 1996), the burst could be caused by the same mechanism as solar radio bursts or flare of T Tau stars. Since circular polarization is observed in such burst or flares, we searched circular polarization in the maser spectrum of G33.641-0.228.

2. Observation of circular polarization

Yamaguchi 32m radio telescope (Fujisawa *et al.* 2002) was used for observation. The observation system and parameters are described in Fujisawa *et al.* (2012). Data analysis was performed independently for left and right circular polarization. Observations were made intermittently from 2009 to 2016. Here we report the observation results on January 6, 2016. The observation time was 02:10 UT, the integration time was 595 seconds, and the 1 σ noise level was 0.56 Jy. Spectra obtained with left and right circularly polarization are shown in Figure 1. Filled circles are spectra of LHCP and open circles are spectra of RHCP.

Six peaks are seen in the spectrum. Although the spectra of LHCP and RHCP are almost identical, the flux density of RHCP (67.7 Jy) is clearly larger than that of LHCP (53.7 Jy) only at $V_{\rm lsr}$=59.6 km s^{-1}, and the circular polarization is 11.5 %. Other components have a polarization of 0 within the error range. The spectral peak at $V_{\rm lsr}$=59.6 km s^{-1} shows the burst. This spectral component alone exhibits burst and circular polarization, suggesting that the two phenomena are related to each other.

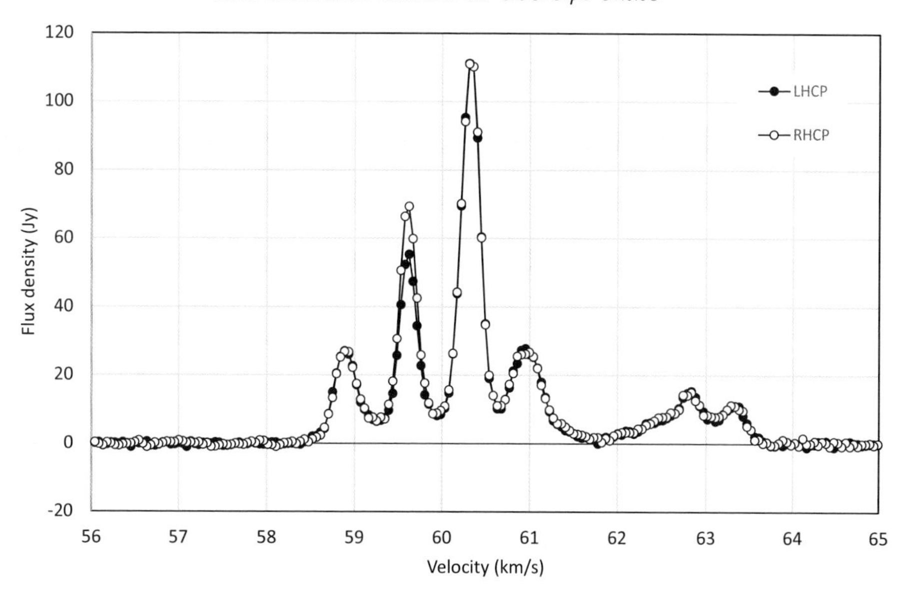

Figure 1. The maser spectra of LHCP and RHCP.

References

Fujisawa, K., Mashiyama, H., Shimoikura, T., *et al.* 2002, *8th Asian-Pacific Regional Meeting, Volume II*,3

Fujisawa, K., Sugiyama, K., Aoki, N., *et al.* 2012, *PASJ*, 64, 17

Fujisawa, K., Takase, G., Kimura, S., *et al.* 2014, *PASJ*, 66, 78

Phillips, R. B., Lonsdale, C. J., Feigelson, E. D., *et al.* 1996, *AJ*, 111, 918

Szymczak, M., Hrynek, G., & Kus, A. J. 2000, *A&AS*, 143, 269

Cosmic Masers: Proper Motion toward the Next-Generation Large Projects
Proceedings IAU Symposium No. 380, 2024
T. Hirota, H. Imai, K. Menten, & Y. Pihlström, eds.
doi:10.1017/S1743921323003332

Jet and Outflows in Massive Star Forming Region: G10.34−0.14

Jihyun Kang[1], Mikyoung Kim[2], Kee-Tae Kim[1], Hirota Tomoya[3] and KaVA SF team

[1]Korea Astronomy and Space Science Institute, Daejeon 34055, Republic of Korea.
email: jkang@kasi.re.kr

[2]Otsuma Women's University, Chiyoda-ku 102-8357, Tokyo, Japan

[3]Mizusawa VLBI Observatory, National Astronomical Observatory of Japan, Mitaka-shi, Tokyo 181-8588, Japan

Abstract. The ALMA observations of the high-mass star-forming region G10.34−0.14 reveal the existence of three massive hot cores. The most massive of these cores, core S1, exhibits both high and low-velocity jet/outflow in the CO, SiO, and CH_3OH. It is associated with water and Class I methanol masers. The core N shows a low-velocity CO outflow and is associated with an Extended Green Object, along with Class I and II methanol masers. The characteristics of the outflows and masers in these two cores suggest they are in different stage of evolution and varying physical conditions.

Keywords. stars: formation, stars: winds, outflows, stars: jets

1. Kinematics in G10.34−0.14

The massive star forming region G10.34−0.14, situated a distance of 2.9kpc away (Li *et al.* 2022), consists of 3 protostars with core masses of 4 – 10 Ms, all currently in hot molecular core stages (Baek *et al.* 2022). By analyzing the ALMA data (2015.1.01288.S: P.I. M.-K. Kim), we report the detections of jet and outflows in the high mass star forming region, G10.34−0.14, in the SiO ($5-4$, 217.104980 GHz), CO ($2-1$, 230.538 GHz), and CH_3OH ($8_{-1}-7_0$ E, 229.758811 GHz) molecular transitions.

1.1. *Core S1/S2 system*

Core S1 (RA, Dec = 18:08:59.98, −20:03:39.1) and S2 constitute a binary system within the same envelope. They are located in the southern part of the observed field. Core S1 is the brightest in the continuum emission, and the collimated jet and outflow originating from Core S1 are detected in SiO and CO emission at velocities ($|V - V_{\rm sys}| < 70$ km s^{-1}), where $V_{\rm sys} = +11$ km s^{-1} (see the left panel of Fig. 1). The high-velocity (HV; $30 < |V - V_{\rm sys}| < 70$ km s^{-1}) CO/SiO jet is well-collimated and manifests as a regularly spaced bullet-like feature in the position-velocity diagram cut along the jet axis, suggesting episodic ejections. The dynamic age of the B1 bullet as shown in Fig. 1 is approximately 500 yrs, assuming an inclination of 45°. The dynamic timescale of the outflow, estimated from the lobe extent and terminal speed of outflows, is about 6600 yrs. The mass loss rate estimated from the CO emission is 8.8×10^{-5} Ms/yr, which is 10-100 times greater than that of low mass protostellar objects, which ranges $0.1 - 5.5\times^{-6}$ Ms/yr (Li *et al.* 2020; Dutta *et al.* 2022).

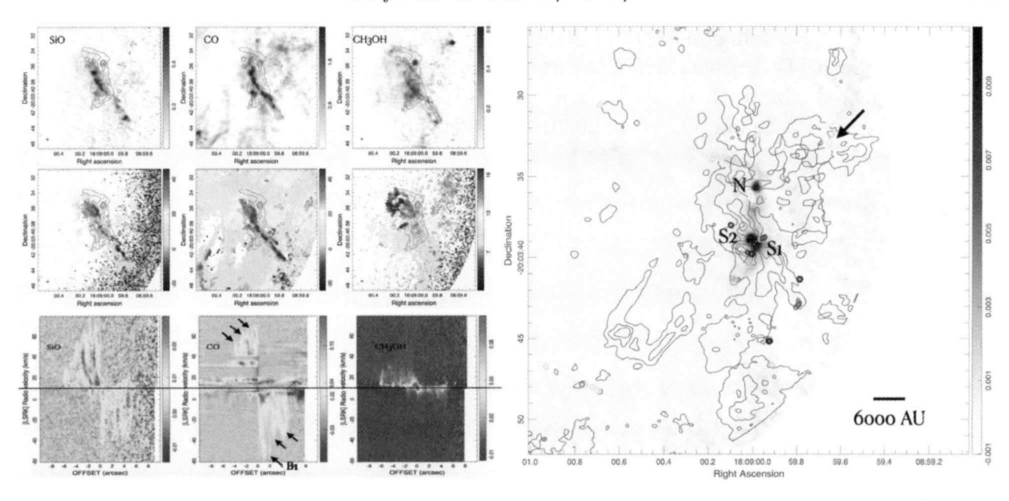

Figure 1. (Left) The top and middle images display moment 0 and 1 maps in SiO, CO, and CH_3OH from left to right. The continuum emission in green contours is overlaid on the moment maps of each molecular transitions. The HV CO jet ($30 < |V - V_{sys}| < 70$ km s^{-1}) is presented with blue and red contours in the moment 0 map of CH_3OH. The PV diagrams along the collimated CO jet for each molecular lines are shown at the bottom. Note that the jet axis shown in the moment 1 map of CO differs slightly between the blue- and red-shifted components. The systemic velocity of +11 km s^{-1} is marked by a black horizontal line. The bullet-like features are indicated by black arrows. (Right) Blue- and red-shifted LV CO outflows ($7 < |V - V_{sys}| < 15$ km s^{-1}) are overlaid on the 1.3 mm dust continuum emission. The open circles represent the positions of the 44 GHz methanol masers, adopted from Cyganowski *et al.* (2009). The symbol area corresponds to the intensity and the color indicates the velocity, ranging from +7.8 km s^{-1} (purple) to +19.7 km s^{-1} (red).

The low-velocity (LV; $|V - V_{sys}| < 20$ km s^{-1}) outflow lobes are also observed in the CH_3OH emission, and the rim of CH_3OH outflow encapsulates the HV CO jet (as presented in the moment 0 map of CH_3OH in Fig. 1), suggesting that the LV outflow could be induced by the energetic jet. The Class I 44 GHz methanol masers are also observed near Core S1 and at the tip of shocked outflow regions (right panel in Fig. 1)

1.2. *Core N*

Core N is positioned in the northern part of the field (RA, Dec = 18:08:59.98, −20:03:35.7). The LV blue-shifted CO lobe in the southeast and the red-shifted CO blob in northwest (right panel of Fig. 1) possibly originate from Core N. The red-shifted blob appears to compress the surrounding medium and generate strong methanol masers at 44, 95, and 229 GHz (Kang *et al.* 2016, this paper), as indicated by an arrow in the right panel of Fig. 1.

We note that the 6.7 GHz Class II methanol maser is situated at the center of Core N, and the Extended Green Object, a 4.5 μ m infrared excess resulting from the high-temperature CO outflow or shocked H2 gas, is located in core N and extends in the SE-NW direction (Cyganowski *et al.* 2009), which aligns with the direction of the CO outflow.

References

Baek, G., Lee, J.-E., Hirota, T., *et al.* 2022, *ApJ*, 939, 84
Cyganowski, C. J., Brogan, C. L., Hunter, T. R., & Churchwell, E. 2009, *ApJ*, 702, 1615

Dutta, S. Lee, C.-F., Johnstone, D., *et al.* 2022, *ApJ*, 925,11
Kang, J.-h., Byun, D.-Y., Kim, K.-T., *et al.* 2016, *ApJS*, 227, 17
Li, J. J, Immer, K, M. J. Reid, M. J., *et al.* 2022, *ApJS*, 262, 42
Li, S., Sanhueza, P., Zhang, Q., *et al.* 2020, *ApJ*, 903, 119

Cosmic Masers: Proper Motion toward the Next-Generation Large Projects
Proceedings IAU Symposium No. 380, 2024
T. Hirota, H. Imai, K. Menten, & Y. Pihlström, eds.
doi:10.1017/S1743921323003204

Multiple scales of view for outflow driven by a high-mass young stellar object, G25.82–W1

Jungha Kim[1], Mikyoung Kim[2], Tomoya Hirota[3,4], Minho Choi[1], Miju Kang[1], Kee-Tae Kim[1] and KaVA working group for star formation

[1]Korea Astronomy and Space Science Institute, Yuseong, Daejeon 34055, Republic of Korea.
email: junghakim@kasi.re.kr

[2]Otsuma women's university, Faculty of Home Economics, Department of Child Studies, Chiyoda, Tokyo 102-8357, Japan

[3]Mizusawa VLBI Observatory, National Astronomical Observatory of Japan, Oshu, Iwate 023-0861, Japan

[4]Department of Astronomical Science, SOKENDAI (The Graduate University for Advanced Studies), Mitaka, Tokyo 181-8588, Japan

Abstract. We are investigating the extended outflow from G25.82–W1, which is one of the members of the high-mass protocluster G25.82–0.17. The aim is to study the star-forming environment of G25.82–W1. To identify the outflow, we obtained CO $2-1$ data using the Atacama Large Millimeter/submillimeter Array.

We have identified several spatial and spectral outflows, including: 1) an extended N1–S1 CO outflow, driven by a high-mass young stellar object (HM-YSO) named G25.82–W1; 2) an elongated SE–NW outflow powered by G25.82–W2; 3) a compact and curved N2–S2 CO outflow originating from G25.82–E; and 4) a pair of knotty lobes centered on G25.82–W.

Furthermore, the innermost region of the N1–S1 CO outflow, traced by the 22 GHz H_2O maser, reveals a complex spatial and velocity structure within a 2" from its launching point.

To accurately calculate the properties of the N1–S1 CO outflow, we have utilized an accurate distance measurement of $d = 4.5$ kpc, derived from the annual parallax of the H_2O masers. The outflow rate and force are comparable to those observed in outflows from other HM-YSOs. The physical properties of the N1–S1 CO outflow follow a trend connecting the low and high-mass regimes, supporting the idea that the star-forming mode in G25.82–W1 is likely a scaled-up version of low-mass star formation.

Keywords. ISM: kinematics and dynamics - ISM: jets and outflows - stars: formation - stars: massive

1. Introduction

G25.82–0.17 is one of the high-mass star-forming regions where high-mass protocluster formation is in progress (Kim *et al.* 2020). For simplicity, hereafter, we will refer to G25.82–0.17 as G25.82. Multiple 1.3 mm continuum sources have been revealed in G25.82 through observations conducted by the Atacama Large Millimeter/submillimeter Array (ALMA), indicating the presence of young stellar objects (YSOs) at different evolutionary stages. Among the YSOs in this region, G25.82–W1 is classified as a high-mass YSO (HM-YSO) having a disk-outflow system. Our aim is to understand the star-forming environment of G25.82–W1 through the analysis of multi-scale views of the outflow from this target.

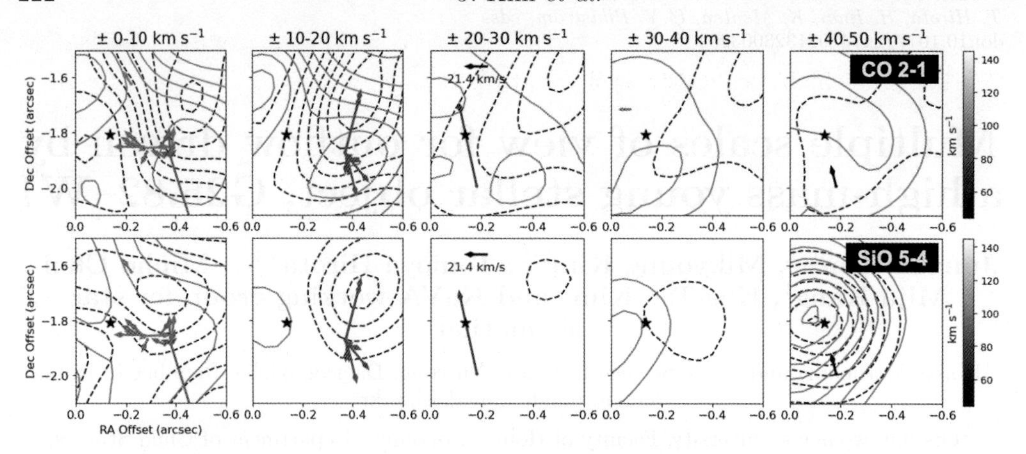

Figure 1. Proper motions of the H_2O masers are shown along with channel maps of CO 2–1 and SiO 5–4. A star indicates the position of G25.82–W1. Each contour map represents the integration of emission over 10 km s^{-1} velocity intervals. The velocity offset from the systemic velocity (93.7 km s^{-1}; Wienen *et al.* 2012) is presented in the upper left corner of each panel. Dashed contours represent the blue-shifted components, while solid contours represent the red-shifted components. In the upper panels, dashed contours are set at [20, 40, 80, 160]σ with 1σ =5.2 mJy beam$^{-1}\cdot$ km s^{-1}, while solid contours are set at [20, 40, 80, 160, 320]σ with 1σ =5.0 mJy beam$^{-1}\cdot$ km s^{-1}. Contours in the lower panels are set at [20, 35, 50, 65, 80, 95, 110, 125]σ with 1σ =3.4 mJy beam$^{-1}\cdot$ km s^{-1} for dashed lines, and 1σ =3.7 mJy beam$^{-1}\cdot$ km s^{-1} for solid lines.

2. Observation and Results

Observations: To investigate the 3D velocity structure of the outflow, H_2O maser monitoring observations were carried out toward G25.82 using two VLBI arrays: VLBI Exploration of Radio Astrometry (VERA) and KaVA (Korean VLBI Network (KVN) and VERA array).

While masers are effective for tracing velocity fields with high resolution, they can only provide sparsely distributed spatial structures. Therefore, it is crucial to study the outflow structures traced by thermal lines in order to bridge the gap between central HM-YSOs and the distribution of maser emissions in the outflows. Through complementary observations, it is also possible to derive quantitative physical properties. For these reasons, we also analyzed CO data, a well-known outflow tracer, obtained using the ALMA.

Results: An extended CO outflowing gas extending in the north-south direction is identified, driven by G25.82–W1 (referred to as the N1–S1 CO outflow). The total velocity range is from 5.9 to 178.6 km s^{-1}. The morphology of both the blue- and red-shifted lobes is characteristic. In the case of the blue-shifted lobe, it is launched toward the west from G25.82–W1 and then extends to the north, while the red-shifted lobe is initially driven toward the east and propagates to the south. The root of the N1–S1 CO outflow is likely connected to the H_2O maser outflow.

In total, 31 H_2O maser features were identified. Most of the identified H_2O maser features have blue-shifted velocities or velocities similar to the systemic velocity of G25.82 (93.7 km s^{-1}; Wienen *et al.* 2012). They are all located within approximately 2" from G25.82–W1.

As shown in Figure 1, H_2O maser features exhibit a close spatial association with the blue-shifted component of the larger-scale outflows observed in CO 2–1 and SiO 5–4 emissions. Dominant proper motions of H_2O maser features are observed in the north-south and east-west directions, where the E–W CO distribution and N1–S1 CO outflow coexist. This suggests that H_2O maser features are tracing the innermost region of the N1–S1 CO outflow.

We also fitted the measured absolute positions of the maser features to calculate the annual parallax. The combined parallax is 0.222 ± 0.021 mas, corresponding to a distance of $4.5^{+0.47}_{-0.39}$ kpc. This is comparable to the kinematic distance of 5 kpc from a previous study (Green & McClure-Griffiths 2011). The outflow properties of the N1–S1 CO outflow follow a trend that connects the low and high-mass regimes, supporting the idea that the star-forming mode in G25.82–W1 is likely a scaled-up version of low-mass star formation.

References

Kim, Jungha, Hirota, T., Kim, K., *et al.* 2020, *ApJ*, 896, 127
Green, J. A., & McClure-Griffiths, N. M., 2011, *MNRAS*, 417, 2500
Wienen, M., Wyrowski, F., Schuller, F., *et al.* 2012, *A&A*, 544, A146

Cosmic Masers: Proper Motion toward the Next-Generation Large Projects
Proceedings IAU Symposium No. 380, 2024
T. Hirota, H. Imai, K. Menten, & Y. Pihlström, eds.
doi:10.1017/S1743921323003587

A Multiwavelengh study towards Galactic HII region G10.32-0.26

Mi Kyoung Kim[1], Tomoya Hirota[2], Kee-Tae Kim[2,3] and KaVA SFR sub Working Group

[1]Otsuma Women's University, 12 Sanban-cho, Chiyoda-ku, Tokyo, 102-8357, Japan.
email: mikyoung.kim@otsuma.ac.jp

[2]National Astronomical Observatory of Japan, Oshu-shi, Iwate 023-0861, Japan

[3]The Graduated University for Advanced Studies

[4]Korea Astronomy & Space Science Institute, 776 Daedeok-daero, Yuseong-gu, Daejeon 34055, Republic of Korea

Abstract. We present the results of high-resolution continuum and molecular line observations towards the Galactic HII region, G10.32-0.26. The continuum map with ALMA reveals the five cores with masses ranging from 2.5–9.2 $M_{\odot}$. The results show that the brightest peak, Peak 1, is an HCHII region with an excitation temperature of ~12000 K, an electron density of 3.4×10^{7} cm^{3}, and a radius of 14 au. The central object is estimated to be a B0.5 star. The class II 6.7 GHz CH_3OH maser coincides with Peak 1, implying class II CH_3OH maser is associated with a later evolutionary stage of star formation. Using KaVA and KVN observation, we detect the class I 44 and 95 GHz CH_3OH masers near the weakest peak, Peak 5. We successfully imaged class I 95 GHz CH_3OH masers with VLBI for the first time. In Peak 5, the high-velocity SiO emission also exists. The continuum emission can be modelled with grey-body dust emission with $T_{\rm d}$ ~30 K, and the molecular species are poor near Peak 5, suggesting Peak 5 is in an early stage of star formation.

Keywords. Star formation, HII region, Masers

1. Introduction

Class I and II CH_3OH masers are one of the common maser species in star-forming regions. High-resolution observations of Class II CH_3OH masers show that they are associated with disk/outflow around massive YSO (De Buizer 2003; Bartkiewicz *et al.* 2009). In contrast, the detailed kinematics and physical condition of the class I CH_3OH masers are poorly understood due to a lack of high-resolution observations. We conducted VLBI and ALMA observations towards G10.32−0.26 to reveal the physical conditions of the star-forming region where the bright class I and II CH_3OH masers arise.

The 44 GHz class I CH_3OH masers were observed using KaVA (KVN and VERA array). Additionally, the 95 GHz class I CH_3OH masers were observed simultaneously using the multi-frequency receiving system of KVN antennas. The 6.7 GHz CH_3OH maser observation was conducted with JVN (Japanese VLBI Network), including VERA 4 stations, Hitachi, Yamaguchi, and Kashima.

For high angular resolution continuum and molecular line observations, we conducted ALMA observations in band 3 and band 6. We also conducted EVLA observation at 44 GHz for continuum and class I CH_3OH maser polarization.

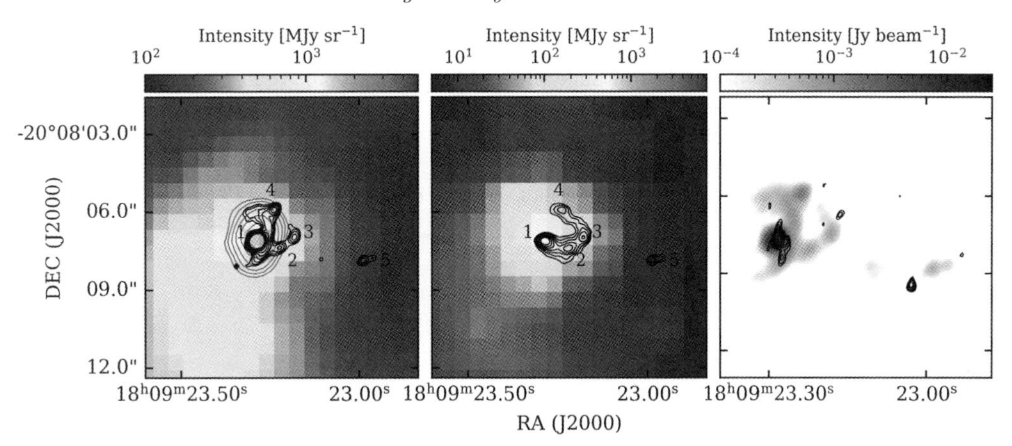

Figure 1. Left: 102 GHz ALMA continuum image (black contour) overlayed with the 5 GHz continuum emission from the CORNISH survey (grey contour) and the *Spitzer* 8 μm emission (color). Center: 223 GHz ALMA continuum image (black contour) overlaid on the *Spitzer* 4.5 μm emission (color). Right: 46 GHz continuum emission (black contour) overlaid on the ALMA band 3 continuum emission (grey).

2. Results

With ALMA band 3 and 6 observations, we identified five continuum peaks over $10'' \times 10''$ area (Figure 1). On the other hand, only Peak 1 is detected with 44 GHz EVLA observation. The masses of cores are estimated to be 2.5–9.2 $M_{\odot}$, assuming optically thin dust emission.

The compact thermal CH_3OH emission is detected at Peak 1 and 4 in the velocity range of $V_{\rm lsr}$ ~32–38 km s^{-1}, while Peak 2 and 3 show weak thermal CH_3OH emission with $V_{\rm lsr} \sim$ 32 km s^{-1}. The compact H30α emission was found in the vicinity of Peak 1. Sulfur-bearing molecules and DCN with $V_{\rm lsr}$ ~32–34 km s^{-1} are detected in the area between Peak 1 and Peak 2, and in the vicinity of Peak 4. The 6.7 GHz CH_3OH maser was detected at Peak 1 in the $V_{\rm lsr}$ ~36–39 km s^{-1} velocity range and shows the velocity gradient in the east-west direction. Assuming Keplerian rotation, we estimate the central mass to be ~14 $M_{\odot}$.

In Peak 5, SiO v=0 emission ($V_{\rm lsr}$ ~20–60 km s^{-1}) and thermal CH_3OH emission ($V_{\rm lsr}$ ~32–36 km s^{-1}) were detected. Using a fringe-rate map, we confirmed that class I 44 and 95 GHz CH_3OH masers are located at Peak 5. The positions and velocity distributions of masers are consistent with thermal CH_3OH emission.

To characterize the physical properties of dust cores, we compared the observed continuum flux with the modelled SED and determined the physical parameters by trial and error. Based on the observational results, we adopted free-free emission with a truncated power-law density distribution for the modelled SED of Peak 1 (Olnon 1975). A model of the HCHII region with an excitation temperature of 12000 K, a radius of 14 au, and electron density $n_0 \sim 3.4 \times 10^7$ cm^{-3} matches the observed continuum emissions. The excitation parameter $U \sim 7.1$ pc cm^{-2} implies that the central object is a B0.5 ZAMS star with a total luminosity $L \sim 10^4$ $L_{\odot}$ (Panagia 1973).

For Peak 2–5, we adopted an optically thin, modified grey body for SED fitting (Rathborne *et al.* 2010). The results show they are embedded in cold dust ($T_{\rm d}$ ~30–50 K). In Peak 5, we detected 44 and 95 GHz class I CH_3OH masers and elongated SiO emission, which is a tracer of shocks and collimated jets in star-forming regions. In contrast, a deficiency of molecular lines and a low dust temperature of ~30 K suggests that Peak 5 is in the very early phase of star formation.

References

Bartkiewicz, A., Szymczak, M., van Langevelde, H. J., *et al.* 2009, *A&A*, 502, 155
De Buizer, J. M. 2003, *MNRAS*, 341, 277
Olnon, F. M. 1975, *A&A*, 39, 217
Panagia, N. 1973, *AJ*, 78, 9
Rathborne, J. M., Jackson1, J. M., Chambers, E. T., *et al.* 2010, *ApJ*, 715, 310

Cosmic Masers: Proper Motion toward the Next-Generation Large Projects
Proceedings IAU Symposium No. 380, 2024
T. Hirota, H. Imai, K. Menten, & Y. Pihlström, eds.
doi:10.1017/S1743921323002363

Yamaguchi interferometer survey of protostellar outflows embedded in 70-μm dark infrared dark cloud

Keita Kitaguchi[1], Kazuhito Motogi[1,2], Kenta Fujisawa[2], Kotaro Niinuma[1,2] and Ryotaro Fujiwara[1]

[1]Graduate School of Sciences and Technology for Innovation, Yamaguchi University, Yamaguchi, Yamaguchi 753-8512, Japan. email: kmotogi@yamaguchi-u.ac.jp

[2]The Research Institute of Time Studies, Yamaguchi University, Yamaguchi, Yamaguchi 753-8511, Japan

Abstract. Recent ALMA observations detected protostellar outflows in 70-μm dark infrared dark clouds (IRDCs). These sources are candidates for the initial stages of high-mass star formation. We launched a new survey for free-free emission from outflow shocks using the Yamaguchi Interferometer (YI) at 8 GHz. We aim to catalog proto-high-mass protostar candidates that are still in the low to intermediate-mass phase. We selected starless-like clumps without any 70-μm point source from Traficante et al. (2015). We currently detected 82 sources from 167 clumps. 37 of them are fainter than 20 mJy (down to a few mJy). They tend to associate with colder and denser clumps that are suitable for star formation. This fact suggests that, at least, some of them trace star-formation activities. The highest-density clumps are, in fact, associated with several masers and molecular outflows. Furthermore, some of them have already shown a signature of ongoing cluster formation.

Keywords. ISM:jets and outflows, accretion disks, stars: massive, stars: protostars

1. Introduction

A number of evolved accretion disks in high-mass protostellar phase (*e.g.,* Motogi *et al.* 2019; Maud *et al.* 2019, etc) has been found in the last decade, indicating that high-mass stars are formed by disk mediated accretion similar to that of lower mass stars. The star-formation efficiency of a molecular core (i.e., total stellar mass over the natal core mass) is known to be 50 % (*e.g.,* Machida & Hosokawa 2013). Typical mass of high-mass starless cores is 10–30 $M_\odot$. This is too small to form O-type stars implying additional accretion from natal clamps and/or filaments (*e.g.,* Kong *et al.* 2021). How massive initial virialized core can become depends on pre-collapse condition of natal massive clump (e.g., temperature, density, and virial parameter, etc). It is essential to study a practical environment of natal clump and core just before gravitational collapse for quantitatively understanding high-mass star formation.

70-μm dark infrared dark clouds (IRDCs) are starless-like and considered to include the earliest stage of high-mass star formation. Recent ALMA observations detected protostellar outflows in such 70-μm dark IRDCs (*e.g.,* Feng *et al.* 2016; Pillai *et al.* 2019). These outflow sources are the best candidates of "Proto-high-mass protostars" (PHPs), which are still in low–intermediate mass stage but with high accretion rate. They could be extremely young seeds of high-mass protostars and still hold initial environment without significant feedback.

Table 1. Categories of surrounding environment of faint sources.

Category	Signs of star-formation	Clump numbers
1	Only the continuum detected by YI	14
2	Outflow and/or maser within the clump	3
3	Signature of cluster formation around the clump	4
Uncategorized	No other observation	16*

*One maser source associated with a foreground AGB star is included.

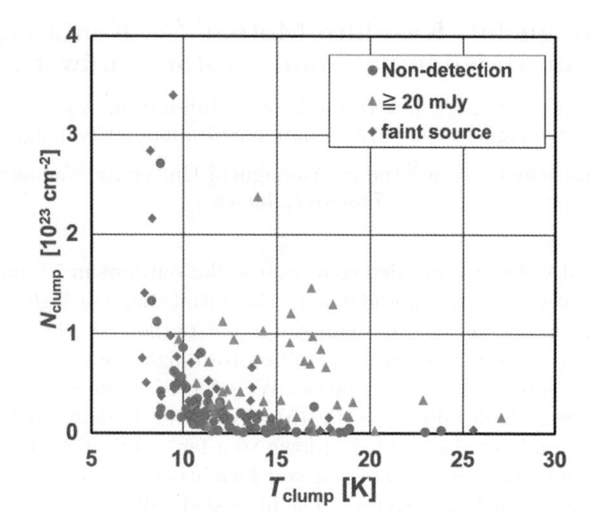

Figure 1. Dust temperatures versus H_2 surface densities for all the targeted clumps. Blue, green, and red markers show non-detection clumps, bright sources, and faint sources (< 20 mJy), respectively.

2. YI survey of protostellar outflows in 70-μm dark IRDCs

We conducted new 8 GHz continuum survey towards 70-μm dark IRDC clumps by using the Yamaguchi Interferometer (YI) (Fujisawa *et al.* 2022). We aimed to search for outflow shocks associated with PHPs. We observed 167 starless-like IRDC clumps cataloged by Traficante *et al.* (2015).

We detected 81 continuum sources and 37 of them are fainter than 20 mJy with the minimum flux density of 1.9 mJy. Figure 1 shows comparison of dust temperatures and H_2 surface densities of natal IRDC clumps. These faint sources tend to associate with relatively colder and higher surface density clumps, i.e., suitable condition for star formation. This fact suggests that some of them do trace star-formation actives such as young, deeply embedded outflow shock. We then searched for evidences of star-formation activity for these faint sources. We found that 7 clumps were associated with masers and/or molecular outflows, 14 clumps were only detected by YI, and there were no maser or outflow observation for other 16 clumps. Table 1 summarizes the environment of the faint clumps. These categories possibly reflect different evolutionary stages at 70-μm dark phase.

Supplementary material

To view supplementary material for this article, please visit http://dx.doi.org/10.1017/S1743921323002363

References

Feng, S., Beuther, H., Zhang, Q., *et al.* 2016, *ApJ*, 828, 100
Fujisawa, K., Aoki, T., Kanazawa, S., *et al.* 2022, *PASJ*, 74, 1415
Kong, S., Arce, H. G., Shirley, Y., *et al.* 2021, *ApJ*, 912, 156
Machida, M. N. & Hosokawa, T. 2013, *MNRAS*, 431, 1719
Maud, L. T., Cesaroni, R., Kumar, M. S. N., *et al.* 2019, *A&A*, 627, L6
Motogi, K., Hirota, T., Machida, M. N., *et al.* 2019, *ApJ*(Letters), 877, L25
Pillai, T., Kauffmann, J., Zhang, Q., *et al.* 2019, *A&A*, 622, A54
Traficante, A., Fuller, G. A., Peretto, N., *et al.* 2015, *MNRAS*, 451, 3089 Brown, S. F., Semel, M., *et al.* 1992, *A&A*, 265, 682

Cosmic Masers: Proper Motion toward the Next-Generation Large Projects
Proceedings IAU Symposium No. 380, 2024
T. Hirota, H. Imai, K. Menten, & Y. Pihlström, eds.
doi:10.1017/S1743921323003290

Early Star Formation Traced by Water Masers

Dmitry Ladeyschikov

Astronomical Observatory, Institute for Natural Sciences and Mathematics, Ural Federal University, 19 Mira street, Ekaterinburg, 620002, Russia. email: dmitry.ladeyschikov@urfu.ru

Abstract. In this study, the correlation between 22 GHz water masers and other maser species with far infrared/submillimeter (FIR/sub-mm) sources is investigated. Comparing luminosity to mass ratio (L/M) of FIR/sub-mm clumps linked to different maser species, 22 GHz water masers have significantly lower L/M values than 6.7 GHz methanol and 1665 MHz OH masers. This suggests 22 GHz water masers may precede them in the evolution timeline of SFRs. The close association between water masers and FIR/sub-mm sources provides insight into maser pumping conditions and evolutionary stages.

Keywords. Masers, star formation region, Hi-GAL, ATLASGAL, evolution timeline

---

1. Introduction

Water maser emission has been widely recognized as a valuable tool for investigating high-mass and low-mass star formation in the Galaxy. Extensive studies have established the presence of a collisional mechanism responsible for the interstellar 22 GHz and other H2O maser lines, as well as their association with shocks. Such shocks can be generated by various mechanisms, including protostellar jets, large-scale shocks, and disks. In contrast, class I methanol (cIM) masers, which are also shock-driven, tend to appear at a distance from radiation sources and trace the edges of outflows in star-forming clumps.

This study presents a comparative analysis between 22 GHz water maser emission and infrared/submillimeter sources from the Herschel infrared Galactic Plane Survey (Hi-GAL) and APEX Telescope Large Area Survey of the Galaxy (ATLASGAL). The majority of water maser sources associated with star formation regions are found to be connected to submillimeter and infrared sources.

2. Maser evolution in star formation region

In a study by Ellingsen (2007), a model called the "straw-man" model was proposed to depict the evolutionary sequence of masers in star-forming regions (SFRs). According to this model, methanol masers (both class I and II) are associated with the earliest stage of evolution, followed by water masers, and OH masers appear only in evolved sources with H II regions. Billington et al. (2020) examined this model using the luminosity-to-mass ratio (L/M) of ATLASGAL clumps. The L/M and dust temperature serve as indicators of the evolutionary state of star-forming clumps.

In this study the water maser archive from MaserDB database (Ladeyschikov et al. 2022) were used, consisting of 1007 masers associated with Hi-GAL sources and 960 associated with ATLASGAL sources. The sample of cIM masers at 95 GHz, class II methanol (cIIM) masers at 6 and 12 GHz and OH masers with 383, 678, 388 and 126 ATLASGAL-associated sources were also used, respectively.

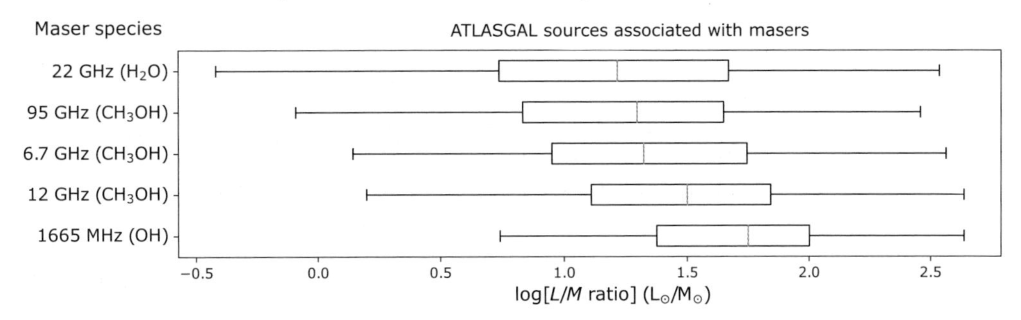

Figure 1. Box plot presenting the central 95 per cent distributions of luminosity to mass ratio for the ATLASGAL sources associated with different maser species and transitions ordered according to their mean value with lowest values at the top and largest value at the bottom.

To ensure reliable associations, the author applied certain constraints for maser associations. The beam size of the maser observations had to be smaller than 70", and the maximum distance between a maser and an ATLASGAL or Hi-GAL source was set at 30". These criteria aimed to exclude false-positive associations resulting from detections using large beam sizes, which could lead to less reliable associations.

The results of the L/M ratio analysis depicted in Figure 1, revealed that 22 GHz H2O masers appear earlier in the evolutionary sequence than cIIM masers at 6.7, 12 GHz, and 1665 MHz OH masers. The K-S test showed significant ($\sigma < 0.0013$) differences in L/M between sources associated with 22 GHz water masers and 6 GHz CH_3OH, 12 GHz CH_3OH, and 1665 MHz OH masers.

From the presented data author conclude that 22 GHz water masers arise before 6.7 GHz methanol masers in the evolutionary sequence. This conclusion is consistent with study Breen & Ellingsen (2011), but differed from several other studies (Ellingsen et al. 2007; Breen et al. 2010; Jones et al. 2020) that suggested water masers appear after the onset of 6.7 GHz methanol masers in the evolution timeline.

Water and cIM masers may reside in shock waves of the outflows from the protostars, indicating the earliest evidence of ongoing outflow activity in a star-forming region. Thus collisionally pumped water and cIM masers should exist before radiatively pumped cIIM masers. However, cIM and water masers exhibit significant differences in their variability timescales.

It is important to note that the results of this analysis could be influenced by the sample size and selection, and the detection of more masers in the future may alter the appearance of the presented figures. Additionally, limitations in sensitivity and the availability of interferometric positions for all known H2O masers might introduce some false associations, particularly in crowded regions.

This work was supported by the Ministry of Education and Science of Russia (the basic part of the State assignment, project no. FEUZ-2023-0019).

References

Billington, S. J., Urquhart, J. S., König, C., *et al.* 2020, *MNRAS*, 499, 2744
Ellingsen S. P., 2007, *MNRAS*, 377, 571
Breen, S. L., Ellingsen, S. P., 2011, *MNRAS*, 416, 178
Ellingsen, S. P., Voronkov, M. A., Cragg, D. M., *et al.* 2007, *IAUS*, 242, 213
Breen, S. L., Ellingsen, S. P., Caswell, J. L., Lewis, B. E., 2010, *MNRAS*, 401, 2219
Jones, B. M., Fuller, G. A., Breen, S. L., *et al.* 2020, *MNRAS*, 493, 2015
Ladeyschikov, D. A., Sobolev, A. M., Bayandina, O. S., Shakhvorostova, N. N., 2022, *AJ*, 163, 124

Cosmic Masers: Proper Motion toward the Next-Generation Large Projects
Proceedings IAU Symposium No. 380, 2024
T. Hirota, H. Imai, K. Menten, & Y. Pihlström, eds.
doi:10.1017/S1743921323001904

Water Maser Zeeman Splitting in the Ionized Jet IRAS 19035+0641 A

Tatiana M. Rodríguez[1], Emmanuel Momjian[2], Peter Hofner[1,2], Anuj P. Sarma[3] and Esteban D. Araya[4,1]

[1]New Mexico Tech, 801 Leroy Pl., Socorro, NM 87801, USA.
email: tatiana.rodriguez@student.nmt.edu

[2]National Radio Astronomy Observatory, P.O. Box O, 1003 Lopezville Road, Socorro, NM 87801, USA.

[3]Physics and Astrophysics Department, DePaul University, 2219 N. Kenmore Ave., Chicago, IL 60614, USA.

[4]Physics Department, Western Illinois University, 1 University Circle, Macomb, IL 61455, USA.

Abstract. A key ingredient in the earliest evolutionary phase of high-mass ($M > 8\,M_\odot$) star formation (HMSF) is the presence of a jet/outflow system. To study its role in HMSF, we have carried out high resolution ($0.1''$) VLA K-band (18−26.5 GHz) observations toward IRAS 19035+0641 A, identified as a high-mass protostellar jet candidate based on previous cm continuum data. Our observations resolve the continuum emission into an elongated structure in the NE-SW direction, confirming that the K-band continuum arises from an ionized jet. Furthermore, we detected several 22.2 GHz H_2O maser spots aligned in a direction consistent with the jet axis. Zeeman splitting was detected in the strongest maser spot. In this paper, we present our results and discuss the implications of our findings.

Keywords. Radio jets, Water masers, Massive stars, Magnetic fields

1. Introduction

The nature and role of ionized jets and magnetic fields in high-mass ($M > 8\,M_\odot$) star formation are still poorly understood, mainly due to lack of observations. The 22.2 GHz water (H_2O) maser transition arises in shocked gas regions and is therefore an excellent tool to study jet dynamics. This transition is also sensitive to Zeeman splitting, allowing us the ability to detect and study magnetic fields. We carried out a high resolution ($0.1''$) VLA 1.3 cm continuum survey toward 23 compact sources with a rising spectral index from Rosero et al. (2016), classified as jet candidates (Rodríguez et al. in prep.). Furthermore, our spectral set up allowed for simultaneous 22.2 GHz water maser observations. Here, we present and discuss our results for one of the target sources, IRAS 19035+0641 A.

2. Jet and Masers

We resolved the 1.3 cm continuum emission into an elongated structure shown in color in Figure 1 (left panel), as expected for an ionized jet. We identify a total of three continuum peaks aligned in a NE-SW direction, labeled as A1, A2, and A3 in the figure. The spectral index α (with $S_\nu \propto \nu^\alpha$) of the ionized gas is $\alpha = 0.9 \pm 0.1$. We also detected seven water maser spots well aligned with the jet, as shown in Figure 1 (right panel).

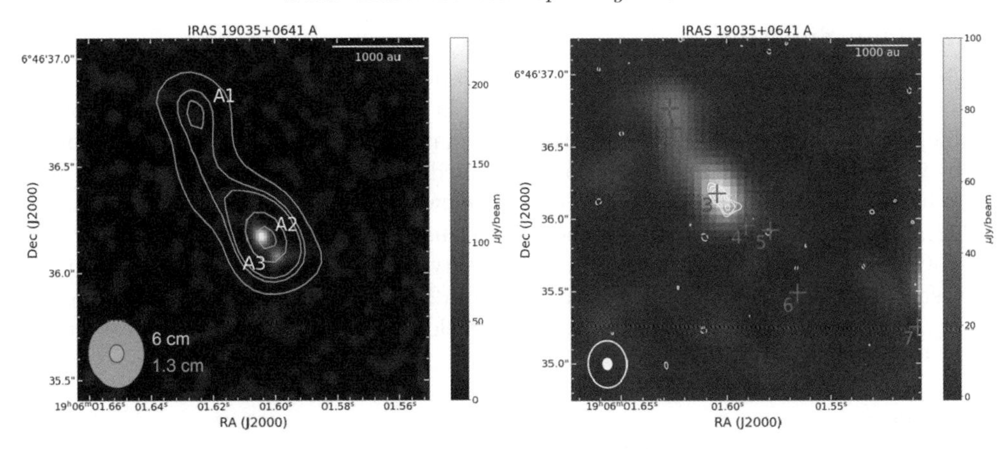

Figure 1. *Left*: 6 cm (contours, Rosero et al. 2016) and 1.3 cm continuum emission (color) toward IRAS 19035+0641 A. The three continuum peaks identified (i.e., A1, A2, A3) are labeled. *Right*: the 6 and 1.3 cm continuum emission is shown in color and contours (respectively), while the position of the water masers detected in the jet are marked with red + symbols and numbered.

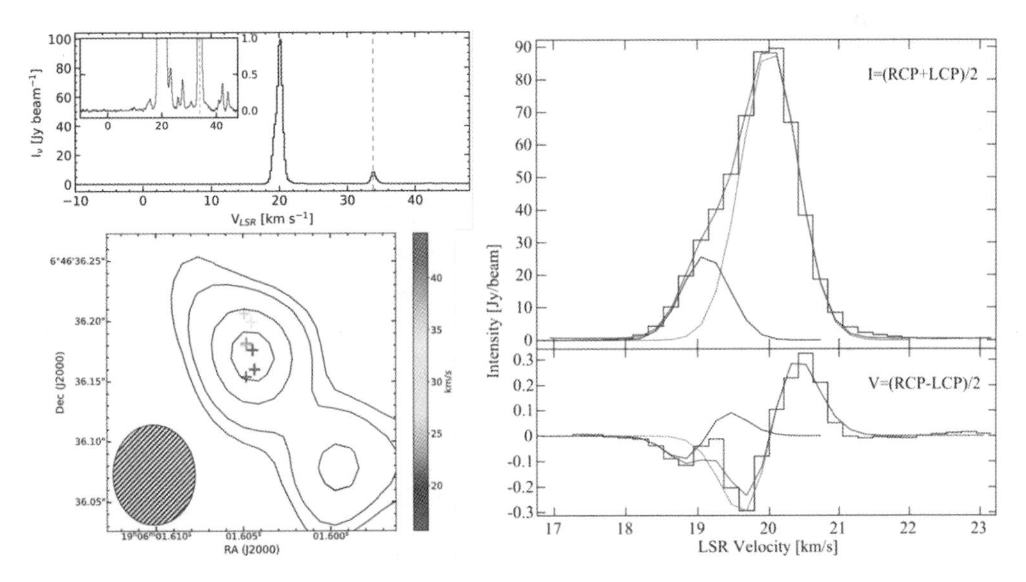

Figure 2. *Top-left*: Spectrum of maser #3, where the vertical dashed line marks the systemic velocity (33.8 km s^{-1}, Araya et al. 2005) and the inset shows a zoom-in to a 1 Jy beam^{-1} maximum. *Bottom-left*: Position obtained from 2-D Gaussian fits to each peak in the top spectrum, color coded by velocity, and overlaid on the 1.3 cm continuum contours. *Right*: Stokes I (top) and V (bottom) profile of the strongest maser #3 peak. The blue and green lines show the two Gaussian components fitted to the profile (top) and their derivatives scaled by their magnetic fields (bottom), while the red line is their sum.

Maser #3 is coincident with the continuum peak A2, is extremely bright ($\sim$100 Jy), and covers about 35 km s^{-1} in velocity with multiple peaks (Figure 2, top-left panel). We fit 2-D Gaussian to each peak in the maser #3 spectrum and found a NS velocity gradient (Figure 2, bottom-left panel). This gradient could be tracing expanding or rotating gas, although proper motion observations are needed to determine its nature.

3. Zeeman Splitting

The Stokes I profile of the brightest maser #3 peak was fit with two Gaussian components, represented with a green and a blue line in Figure 2 (right panel). The Stokes V profile, which shows the S-shape typical of Zeeman splitting, is well represented by the derivatives of the two Gaussian components scaled by their respective line-of-sight magnetic fields $B_{\rm los}$, which are 135 and 156 mG. Following Crutcher (1999) and Sarma et al. (2002), we estimate a pre-shock gas density and magnetic field values of $\approx 10^7$ cm^{-3} and ≈ 7 mG, respectively. The magnetic energy density and the kinetic energy density in the post-shock gas were calculated as in Sarma et al. (2002), and we found that the magnetic energy is higher. This indicates the magnetic field is playing an important role in this post-shock region.

References

Araya, E. D., Hofner, P., Kurtz, S., Bronfman, L., DeDeo, S. 2005, *ApJS*, 157, 279.
Crutcher, R. M. 1999, *ApJ*, 520, 706.
Rosero, V., Hofner, P., Claussen, M., Kurtz, S., Cesaroni, R. *et al.* 2016, *ApJS*, 227, 25.
Sarma, A. P., Troland, T. H., Crutcher, R. M., Roberts, D. A. 2002, *ApJ*, 580, 928.

Cosmic Masers: Proper Motion toward the Next-Generation Large Projects
Proceedings IAU Symposium No. 380, 2024
T. Hirota, H. Imai, K. Menten, & Y. Pihlström, eds.
doi:10.1017/S174392132300279X

Multi-scale observational study of G45.804−0.355 star-forming region

M. Seidu[1], J. O. Chibueze[1,2,3], G. A. Fuller[4,5], A. Avison[4,6] and N. A Frimpong[4,7]

[1]Centre for Space Research, North-West University, Potchefstroom, 2520, South Africa.
email: 28281942@nwu.ac.za

[2]Department of Mathematical Sciences, University of South Africa, Roodepoort, 1709, South Africa

[3]Department of Physics and Astronomy, University of Nigeria, Nsukka, 410001, Nigeria

[4]Jodrell Bank Centre for Astrophysics, Department of Physics and Astronomy, School of Natural Science, The University of Manchester, Manchester, M13 9PL, UK

[5]I. Physikalisches Institut, University of Cologne, Zülpicher Str. 77, 50937 Köln, Germany

[6]SKA Observatory, Jodrell Bank, Lower Withington, Macclesfield, SK11 9FT

[7]Ghana Space Science and Technology Institute, Ghana Atomic Energy Commission, LG80, Ghana

Abstract. This is a multi-wavelength study to examine the G45.804−0.355 massive star-forming region (SFR) and its environs. Using MeerKAT with angular resolution (θ) of $8''$ at 1.28 GHz, we identify for the first time, a faint radio continuum emission core in G45.804−0.355. At 1.3 mm, ALMA observations ($\theta\sim0''.7$) resolved the core into multiple dust continuum condensations including MM1 which was found to be the primary massive dust dense core in the region (mass $M_{\rm c} \sim 54.3\,M_\odot$). The dust continuum shows an arm-like extended emission within which other dense cores are situated. The velocity gradient of the MM1 core indicates that the source is associated with a rotation gas motion. The red- and blue-shifted lobes overlap at the position of MM1. The compact morphology of the 4.5 μm IR emission, the presence of spiral arms and overlapping of the red- and blue-shifted lobes suggest a face-on geometry of G45.804−0.355.

Keywords. stars: formation, (ISM:) HII regions, ISM: jets and outflows, ISM: kinematics and dynamics, ISM: individual (AGAL045.804–00.356), instrumentation: interferometers

1. Introduction

The molecular clouds in which massive young OB stars form have sizes varying from 2–15 pc for dark clouds, 0.3–3 pc for clumps and 0.03–0.2 pc for cores (Bergin & Tafalla 2007). Using molecular lines and various maser emission, we can probe the kinematics and physical properties of the environments where these massive young stellar objects (MYSOs) form. This study reports the millimetre and centimetre interferometric observations of the star-forming region, G45.804−0.355, which is a bright IR source. It is located at a parallax distance of ~ 7.3 kpc and has a clump luminosity of $\sim 1.9{\times}10^4\,L_\odot$, corresponding to a B0.5-spectral type star (Rivera-Ingraham *et al.* 2010). The IR source, G45.804−0.355 is associated with periodic 6.7 GHz methanol (CH_3OH) maser and an extended green object (EGO), giving an indication of ongoing star formation activities (Cyganowski *et al.* 2008; Olech *et al.* 2022).

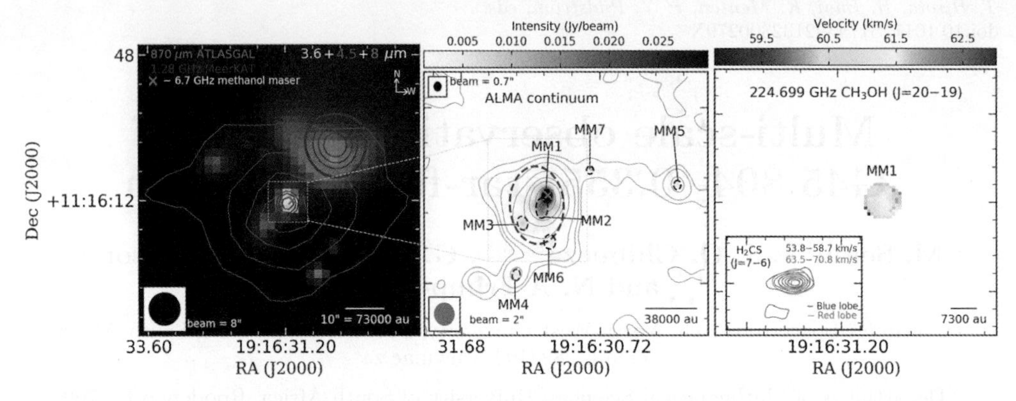

Figure 1. *Left*: IR Image of the G45.804−0.355 field overlaid with contours of ATLASGAL dust and 21 cm radio continuum emission. The black filled ellipse is the synthesized beam of MeerKAT. *Middle*: ALMA 1.3 mm continuum image, exhibiting multiple dust cores. The extended emission was achieved by convolving the dust image with a larger beam shown as a grey-filled ellipse ($2'' \times 1''.8$). The black-filled circle at the top left panel represents the ALMA synthesized beam. *Right*: First moment map and the velocity-integrated intensity map (insert) of MM1. The contours start from $3\times$ 11.9 μJy/beam for MeerKAT, $4\times$ 68 mJy/beam for ATLASGAL and $3\times$ 0.6 mJy/beam for the ALMA dust emission. Contours of the red and blue lobes start from 3σ, where σ is 14 mJy/beam km/s and 11 mJy/beam km/s for the red and blue lobes.

2. Observations

The radio continuum emission at 1.28 GHz (~21 cm) was observed in 2018 under the MeerKAT Galactic Plane Survey Legacy project (Goedhart *et al.* 2023, in prep.). The survey covered a wide field of view with each sky map covering ~1°. The continuum sensitivity was rms $\sim$ 10 μJy/beam and the angular resolution was $\theta \sim 8''$.

The data of dust continuum and molecular line emission at 1.3 mm were taken with ALMA in 2016 (project code 2015.1.01312.S, PI: Gary A. Fuller). The observed frequency range for the molecular line emission was between 224.2 GHz and 242.8 GHz. The angular resolution was $\theta \sim 0''.7$ and the velocity resolution was 1.41 km/s.

3. Results

The left panel of Figure 1 shows a faint ($S_\nu \sim 281 \pm 11$ μJy) 21 cm radio continuum emission coinciding with the 6.7 GHz methanol maser, the "fuzzy" green IR 4.5 μm emission and the sub-mm ATLASGAL dust continuum. The ALMA image shows a bright and massive central dust component (MM1). The integrated flux density and mass of MM1 are $S_\nu \sim$ 87 mJy and $M_c \sim 54.3\,M_\odot$, respectively. MM1 is surrounded by an extended emission which has an arm-like morphology and a physical size of 0.25 pc × 0.18 pc. The other sub-condensations (labelled MM2 – MM3) have masses ranging from $M_c \sim 35\,M_\odot$ to $1.5\,M_\odot$. The MM1 core is rich in thermal CH_3OH molecular lines and the moment one map reveals a rotating structure (see panel 3 of Figure 1). No separation is found between the blue and red-shifted lobe peaks.

4. Discussions and Conclusions

The high abundance of thermal methanol lines, spatially coincident radio continuum and the excess 4.5 μm, along with the association of 6.7 GHz methanol maser suggest the MM1 to be the main powering source hosting a central MYSO. Of all the identified cores, the MM1 core has the highest average-beamed column density ($\sim 3 \times 10^{22}$ cm^{-2}).

The estimated luminosity function value is $2.18\times10^3\,L_\odot$, which corresponds to an early B-type star.

The MM1 core has excess $4.5\,\mu$m IR emission, which is a good indicator of outflowing gas. However, there is no separation between the peaks of the blue- and red-shifted lobes at the position of MM1. The velocity gradient reveals a rotating envelope with spiral arms around the MM1 core. Since the morphology of the IR emission at $4.5\,\mu$m is compact and not extended like regular EGOs, we suggest the G45.804−0.355 SFR to have a face-on geometry. This is confirmed by the presence of arms and the overlapping of the red- and blue-shifted lobes.

References

Bergin, E. A., & Tafalla, M. 2007, *ARAA*, 45, 339
Cyganowski, C. J., Whitney, B. A., Holden, E., Braden, E., *et al.* 2008, *AJ*, 136, 2391
Olech, M., Durjasz, M., Szymczak, M., & Bartkiewicz, A. 2022, *A&A*, 661, A114
Rivera-Ingraham, P. A. R., Bock, J. J., Chapin, E. L., Devlin, M. J., *et al.* 2010, *ApJ*, 723, 915

Cosmic Masers: Proper Motion toward the Next-Generation Large Projects
Proceedings IAU Symposium No. 380, 2024
T. Hirota, H. Imai, K. Menten, & Y. Pihlström, eds.
doi:10.1017/S1743921323003460

Fine structure and refractive scattering of the H_2O maser in star-forming region W49N

N. N. Shakhvorostova[1], J. M. Moran[2], A. V. Alakoz[1], H. Imai[3], C. R. Gwinn[4] and A. M. Sobolev[5]

[1]Astro-Space Center of the P.N. Lebedev Institute of RAS, Russia.
email: nadya.shakh@gmail.com

[2]Center for Astrophysics | Harvard & Smithsonian, USA

[3]Center for General Education, Institute for Comprehensive Education, Kagoshima University, Japan

[4]University of California at Santa Barbara, USA

[5]Astronomical Observatory, Ural Federal University, Russia

Abstract. We used the unprecedented resolution of ~25 μas of the VLBI array formed with the RadioAstron satellite to study the structures of H_2O maser spots in the star forming region W49N. We found that anisotropic diffractive scattering of the ISM dominates the images of the maser spot, but does not completely blur them. The refractive scattering floor is about 0.001 in visibility at a baseline of 8 Gλ.

Keywords. Water masers, interstellar scattering, space VLBI

---

1. Observations of the W49N region with RadioAstron

W49N is a well-known massive star-forming region located in the Perseus arm in the Galactic direction *(l, b)* = (43.16°, 0.01°) at a distance 11.1±0.8 kpc from the Sun, 1.9 pc above the galactic plane (Zhang et al. 2013). W49N hosts the most luminous H_2O maser in the Galaxy, and the maser spots are distributed over ~ 2" across the sky. In this work we have focused on one particular cluster of masers having a diameter of about 10 mas (110 AU) in the range of velocities of −73.2 to −45.7 km/s located at the extreme western edge of the W49N complex, where the brightest masers were found. A summary of observations is given in Table 1.

2. Structure of compact maser features

The individual H_2O maser spots in the velocity range studied have an average size of 240×150 μas at a position angle (PA) of about 100 degrees. Note that observations of OH masers at 1665 MHz from a separate star-formation site (offset by 0.13 pc) associated with W49N show that they have sizes of about 40×20 mas at a PA of about 107 degrees (Deshpande et al. 2013). Comparison of these parameters, especially the close correspondence of the position angles, suggest that both species of maser spots are substantially blurred by diffractive interstellar scattering, which scales as the frequency ratio squared. The OH maser spot image parameters, scaled to 22.2 GHz by a factor of 180, are 220×110 μas. The substantially larger value of the H_2O spot image minor axis compared to relevant values for OH spot images suggests a model wherein the OH maser spots are completely dominated by scattering, whereas the H_2O masers are only partially dominated by scattering and have intrinsic sizes of roughly 100 μas. We determined that the

Table 1. Observing sessions with RadioAstron.

Obs. code	Epoch	Time, hours	Baselines Mλ	Antennas	Spectral resolution	Velocity range, km/s
ragsl5a	03.07.2015	11	25 – 5000	RA, Ef, Ys, Nt, Tr, Sh, Hh	0.12 km/s (7.8 kHz	−98, +117
ragsl1ay	22.05.2015	1	25 – 450 7800 – 8100	RA, Gb, Ef, Mc, Tr	over 16 MHz)	−190, +25

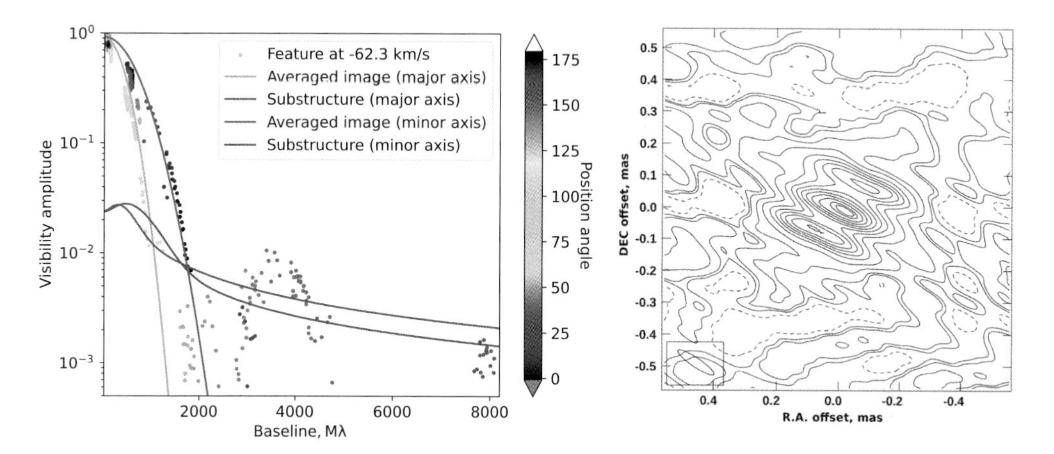

Figure 1. Left panel. *Dots*: relative visibility amplitude vs. baseline length for the brightest maser spot at $V_{LSR} = -62.3$ km/s. *Lines*: theoretical visibility curves for diffractive image and refractive noise. **Right panel**. Image of the spot at −62.3 km/s obtained with RadioAstron. The peak brightness is 2100 Jy/beam and the contours are plotted at −1, 1, 2, 5, 10, 20, 30, 40, 50, 60, 70, 80, 95, and 95% of the peak brightness.

sizes of several H_2O spots increase with velocity away from the line center, in accordance with theories of partially saturated masers. This also supports the idea that the H_2O maser structure is not completely masked by the scattering.

3. Refractive scattering of an H_2O maser emission

We determined that the relative fringe visibilities are between 0.001 and 0.01 in the projected baseline range of 1500 to 8000 Mλ (left panel of Figure 1), and this phenomenon can be attributed to refractive scattering. W49N is the first H_2O maser where refractive scattering is detected.

The magnitude of the refractive floor with respect to the magnitude of diffractive scattering is consistent with the H_2O masers being slightly resolved. Visibility curves for diffractive image and refractive noise predicted by theory (Johnson & Gwinn 2015) are also presented in the left panel of Figure 1. The refractive curves are calculated for a point source. In the case of a finite source size, the refractive floor will be lowered.

It is important that the refractive visibility curve at long space baselines changes for different spectral features, because we are viewing through different parts of the screen.

The image of the brightest maser spot at −62.3 km/s obtained with RadioAstron is presented in the right panel of Figure 1. The structure in the image shows the speckles produced by refractive scattering.

References

Zhang, B., Reid, M. J., Menten, K. M., *et al.* 2013, *ApJ*, 775, 1

Deshpande, A. A., Goss, W. M., Mendoza-Torres, J. E. 2013, *ApJ*, 775, 36

Johnson, M. D., Gwinn, C. R. 2015, *ApJ*, 805, 180

Cosmic Masers: Proper Motion toward the Next-Generation Large Projects
Proceedings IAU Symposium No. 380, 2024
T. Hirota, H. Imai, K. Menten, & Y. Pihlström, eds.
doi:10.1017/S1743921323003125

Observations of Possibly New OH Excited Rotational State Masers

Ivar Shmeld[1], **Artis Aberfelds**[1] and **Oleksey Patoka**[2]

[1]Engineering Research Institute "Ventspils International Radio Astronomy Center", Ventspils University of Applied Sciences, Latvia. email: ivarss@venta.lv

[2]Institute of Radio Astronomy of the National Academy of Sciences of Ukraine, Ukraine

Abstract. In order to search for new 6.035 GHz excited OH masers 272 star-forming regions visible from the northern hemisphere with known active methanol masers were observed with the 32 m and 16 m radio telescopes of the Ventspils International Radio Astronomy Center (VIRAC). Three possibly new excited OH maser sources at 6.035 GHz were seen.

Keywords. star forming regions, methanol and excited OH masers

---

In order to search for new 6.035 GHz excited OH masers we selected 272 sources with declination $\delta > -7.5$ deg (to avoid problems with low observation elevations) from Torun methanol maser catalog and The 6-GHz Multibeam Maser Survey. The observation campaign took place during the years 2018 – 2022 in two phases. In the first phase, during the years 2018 – 2020, 78 objects were studied, and excited OH masers were searched for at frequencies of 6.035 GHz and 6.031 GHz. The results of this phase, were published in Patoka *et al.* (2021).

During the second phase remaining 194 objects were checked, search only for 6.035 GHz masers was performed with Ventspils 32 and 16 m (RT-32 and RT-16 respectively) radio telescopes in the observation setup with a total band width of 1.5625 MHz and a channel separation of 0.381 kHz or 0.019 km s^{-1}. The system temperature was 28 – 34 K and the typical observation time was about 4 hours and 3σ noise level 0.1 – 0.2 Jy. For details we refer the reader to Patoka *et al.* (2021); here we only highlight some preliminary results of our observation campaign, particularly our detections of three, to our knowledge, new excited OH maser sources.

During the observations in the years 2018 – 2021 methanol 6.035 GHz maser sources were checked. Thirty two already known objects were detected, a tentative signal (under 3σ) was detected from eighteen sources already known as exited OH masers, ten already known sources seem possibly variable. We also confirmed two new sources from Szymczak *et al.* (2020). More detailed results will be published later elsewhere. Three objects from our positive detections – G33.641–0228, G212.06–00.74 and G43.089–0.011, whose spectra are displayed in Figure 1 to 3, may be new excited OH masers.

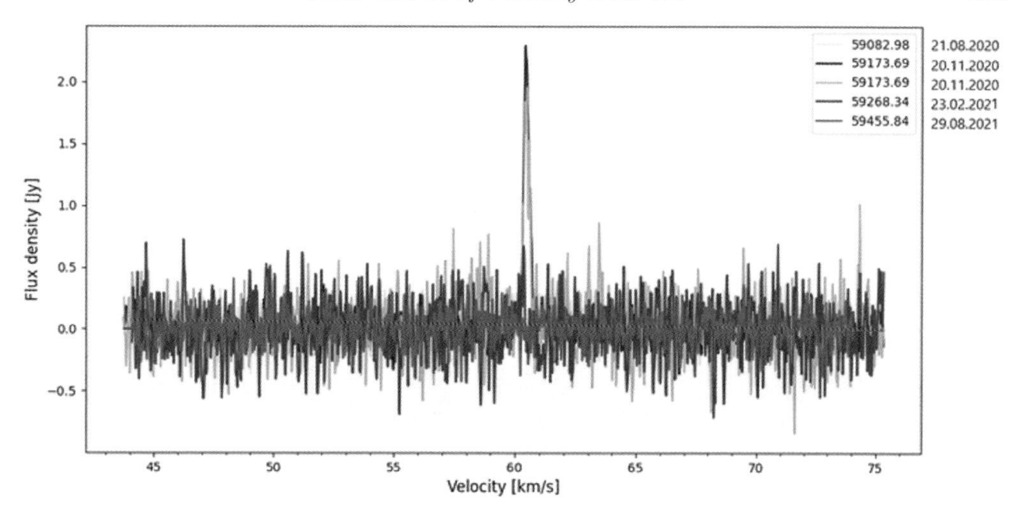

Figure 1. Spectra of 6.035 GHz excited OH maser emission toward G33.6410228. This maser was detected for the first time on 21.08.2020 with Irbene 32 m telescope (RT-32), thereafter with Irbene both telescopes RT-32 and RT-16. Five Gaussian components are displayed with fitted color lines.

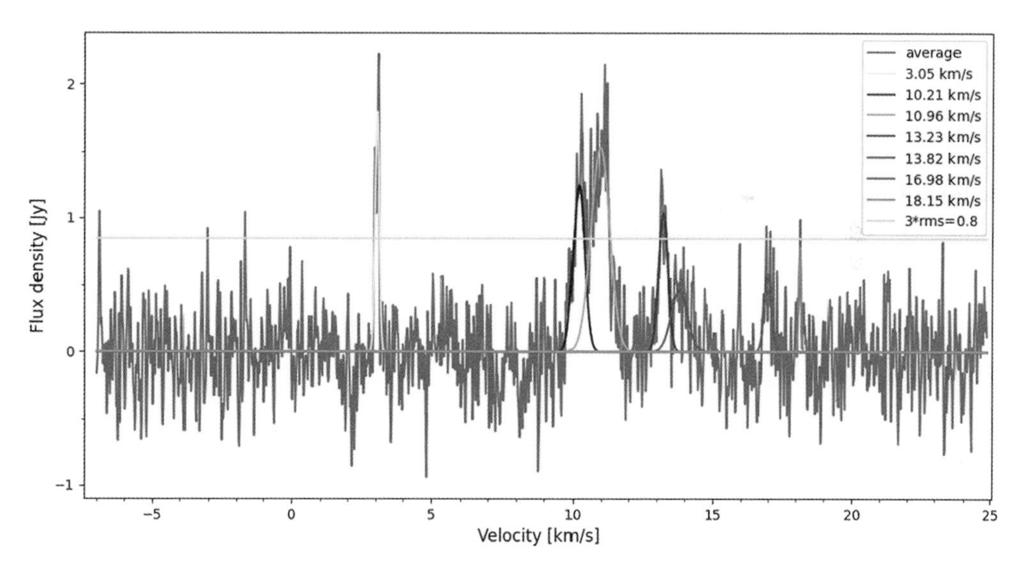

Figure 2. Same as Figure 1 but for G43.0890.011. This maser was detected for the first time on 11.07.2020 with RT32, thereafter 22.11.2020 with both telescopes. Seven Gaussian components are shown with fitted color lines.

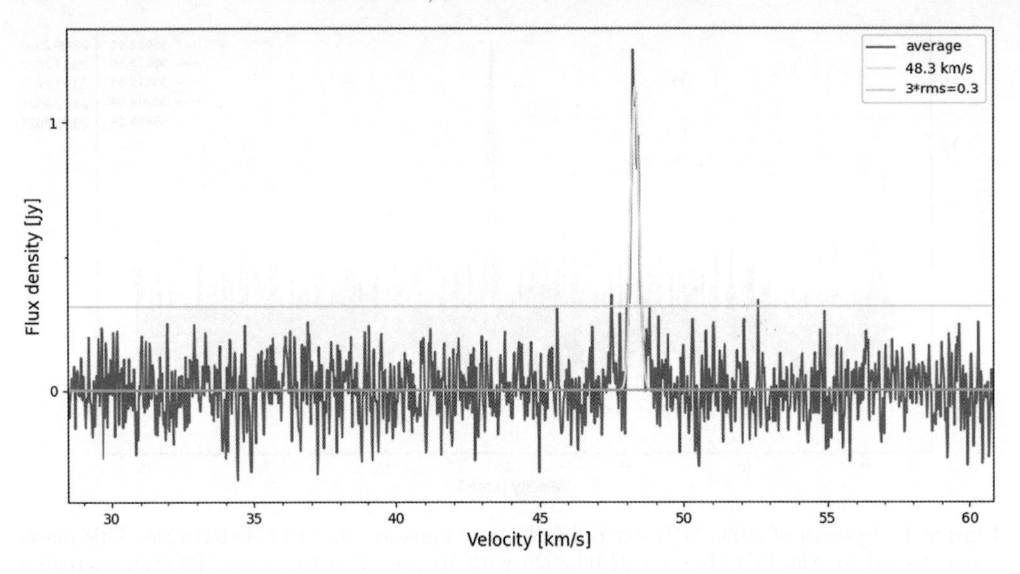

Figure 3. Same as Figure 1 but for G212.06-0074 detected with RT-32 28.03.2020.

Acknowledgements

This work was supported by the European Regional Development Fund project No. 1.1.1.5/18/I/009.

Supplementary material

To view supplementary material for this article, please visit http://dx.doi.org/10.1017/S1743921323003125

References

Patoka, O., Antyufeyev, O., Shmeld, I., *et al.* 2021, *A&A*, 653, A17

Szymczak, M.,Wolak, P., Bartkiewicz, A., *et al.* 2020, *A&A*, 642, A145

Cosmic Masers: Proper Motion toward the Next-Generation Large Projects
Proceedings IAU Symposium No. 380, 2024
T. Hirota, H. Imai, K. Menten, & Y. Pihlström, eds.
doi:10.1017/S1743921323003472

Intensity monitor of water maser emission associated with massive YSOs

Kazuyoshi Sunada[1], Tomoya Hirota[1], Mikyoung Kim[2] and Ross Burns[3]

[1]National Astronomical Observatory of Japan, Oshu-shi, Iwate 023-0861, Japan.
email: kazu.sunada@nao.ac.jp

[2]Otsuma Womens University,12 Sanban-cho,Chiyoda-ku,Tokyo 102-8357, Japan

[3]RIKEN Cluster for Pioneering Research, Wako-shi, Saitama, 351-0198, Japan

Abstract. We are carrying out the intensity monitoring of the water maser emission associated with massive young stellar objects (YSOs) using the VLBI Exploration of Radio Astrometry (VERA) antennas. We are currently monitoring 108 sources. During our long monitoring period, we could find flares on several YSOs. As an example, we show the results from *IRAS* 16293−2422 and NGC2071 IRS1 here. We also show the results of monitoring of the water maser emission from G36.115+0.552, which the methanol maser flare was reported. We observed it as often as possible, usually once a day.

Keywords. masers, starts:massive

1. Introduction

We report our activities of the monitoring observations of the H_2O maser. Our aim of this study is to observe the variation and detect the maser flare. Various follow-up observations toward the flared objects will reveal the origin of flux variations and the circumstellar environment of YSOs.

2. Observations

We are monitoring the 108 sources, which are mainly massive YSOs. We observed the maser lines of H_2O ($6_{16} - 5_{23}$, 22.23508 GHz). We use three VERA stations: Mizusawa, Ogasawara, and Ishigakishima. The beam width (FWHM) and aperture efficiency of the telescope at the observation frequency were $145''$ and 0.45, respectively. The pointing accuracy was better than 10". For the backend, we use the spectrometers which have a frequency resolution of 31.25 kHz and a bandwidth of 32 MHz. This corresponds to a velocity resolution of 0.4 km s^{-1} and a velocity coverage of $\pm$210 km s^{-1}. The on-source integration time is 6 minutes for each source. Our detection criteria is that the peak intensity exceeds the more than 4σ. The worst sensitivity of our observations is 16 Jy (4σ).

We are monitoring most of the sources since 2020. Some of the 108 sources have started the observations from 2015. Since 2020, we are observing at intervals of one month. Before 2020, we observed at intervals of two months.

When we detect maser flare from our target sources, we report it to Maser Monitoring Organisation (M2O). Both the flare sources which we detected and the flare sources which the other M2O members detected, we try to observe water maser emission and monitor it as often as possible, usually once a day.

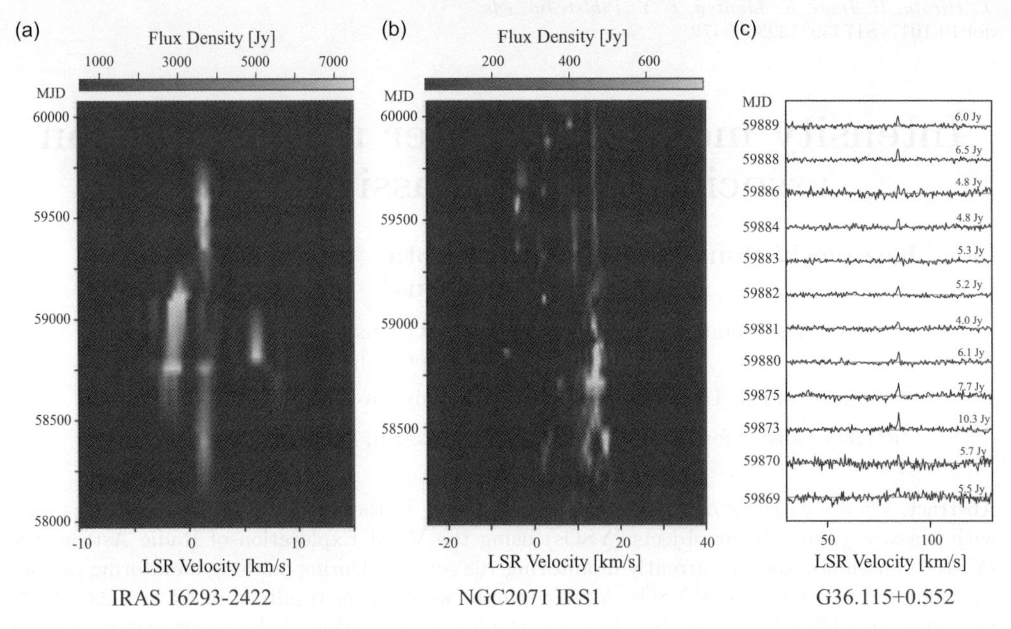

Figure 1. (a) Dynamic spectrum of H_2O masers in *IRAS* 16293−2422.(b) Same as but for NGC2071 IRS1. (c) H_2O maser spetra for G36.115+0.552. From bottom to top the spectra are shifted for each observation day.

3. Results

We present two examples, which we detected the maser flare, and one result which we carried out the follow-up observation of the methanol maser flare source.

3.1. *IRAS 16293−2422*

Figure 1(a) shows the results for *IRAS* 16293−2422 (RA(J2000) = 16h32m22.9s, Dec(J2000) = −24°28′36″). We have detected the increase in the flux of the 1.3 km s^{-1} component on MJD = 58277 (from 500 Jy to 5820 Jy). The flux density of this velocity component reached its maximum on MJD = 58762 and it gradually decreased. At the same time, the flux density of the −0.8 km s^{-1} component increased on MJD = 58587 and reached its maximum of 25700 Jy on MJD = 59088. Following this event, it is interesting to note that the flux density of the 1.3 km s^{-1} component increased again. It reached the maximum of 15990 Jy on MJD = 59517.

3.2. *NGC2071 IRS1*

Figure 1(b) shows the results for NGC2071 IRS1 (RA(J2000) = 5h47m04.8s, Dec(J2000) = +0°21′43″). The brightening of the flux density started on MJD = 58256 (109 to 226 Jy) and reached the maximum flux density of 2200 Jy on MJD = 58818. Ten months later, the flux density of the 2.6 km s^{-1} component showed the flare and reached its maximum flux density of 1260 Jy. Then, more 15 months later, the flux density of the −3.8 kms^{-1} component also flared the maximum flux density of 1000 Jy.

3.3. *G36.115+0.552*

We have also carried out the follow-up observations of flaring sources reported by other groups. The report showed that a flux increase of the 6.668 GHz Class II methanol maser of G36.115+0.552 (RA(J2000) = 18h55m16.8s, Dec(J2000) = +3°05′05″) started on MJD = 59790 (Tanabe and Yonekura 2002). We monitored the flux of the H_2O maser daily as far as possible. We show the result in Figure 1(c). The water maser peaked on MJD = 59873. This day corresponds about a week later than the date on which they observed the maximum flux of the methanol maser.

Reference

Tanabe, Y. and Yonekura, Y., 2022, *Atel, #15680*

Cosmic Masers: Proper Motion toward the Next-Generation Large Projects
Proceedings IAU Symposium No. 380, 2024
T. Hirota, H. Imai, K. Menten, & Y. Pihlström, eds.
doi:10.1017/S1743921323003174

H_2O masers and host environments of FU Orionis and EX Lupi type low-mass eruptive YSOs

Zsófia Marianna Szabó[1,2,3,4], Yan Gong[1], Wenjin Yang[1], Karl M. Menten[1], Olga S. Bayandina[5], Claudia J. Cyganowski[2], Ágnes Kóspál[3,4,6,7], Péter Ábrahám[3,4,6], Arnaud Belloche[1] and Friedrich Wyrowski[1]

[1]Max-Planck-Institut für Radioastronomie, Auf dem Hügel 69, 53121 Bonn, Germany.
email: zszabo@mpifr-bonn.mpg.de

[2]Scottish Universities Physics Alliance (SUPA), School of Physics and Astronomy, University of St Andrews, North Haugh, St Andrews, KY16 9SS, UK

[3]Konkoly Observatory, Research Centre for Astronomy and Earth Sciences, Eötvös Loránd Research Network (ELKH), Konkoly-Thege Miklós út 15-17, 1121 Budapest, Hungary

[4]CSFK, MTA Centre of Excellence, Budapest, Konkoly Thege Miklós út 15-17., H-1121, Hungary

[5]INAF - Osservatorio Astrofisico di Arcetri, Largo E. Fermi 5, 50125 Firenze, Italy

[6]ELTE Eötvös Loránd University, Institute of Physics, Pázmány Péter sétány 1/A, H-1117 Budapest, Hungary

[7]Max-Planck-Institut für Astronomie, Königstuhl 17, D-69117 Heidelberg, Germany

Abstract. The FU Orionis (FUor) and EX Lupi (EXor) type objects are rare pre-main sequence low-mass stars undergoing accretion outbursts. Maser emission is widespread and is a powerful probe of mass accretion and ejection on small scales in star forming region. However, very little is known about the overall prevalence of water masers towards FUors/Exors. We present results from our survey using the Effelsberg 100-m telescope to observe the largest sample of FUors and EXors, plus additional Gaia alerted sources (with the potential nature of being eruptive stars), a total of 51 targets, observing the 22.2 GHz H_2O maser, while simultaneously covering the NH_3 23 GHz.

Keywords. Stars: pre-main sequence, Stars: low-mass, Masers

1. Introduction

Low-mass young stellar objects in their early stellar evolution can undergo accretion-driven episodic outbursts. By studying this phenomena we are able to gather crucial information on the formation and the evolution of Sun-like stars. The member of both FUor and EXor classes experience major increase in their brightness observed in the optical and near-infrared wavelengths. FUors can brighten up to 5-6 magnitudes in the optical and stay in a high-accretion state for decades, but likely centuries (e.g., Fischer *et al.* 2022, and references therein), while EXors brighten between 1-5 magnitudes and remain in a bright state for months or years, and the outbursts are recurring (e.g., Audard *et al.* 2014; Cruz-Sáenz de Miera *et al.* 2022). Masers have been substantially used to probe low- and high-mass star formation regions (e.g., Abraham *et al.* 1981; Omodaka *et al.* 1999;

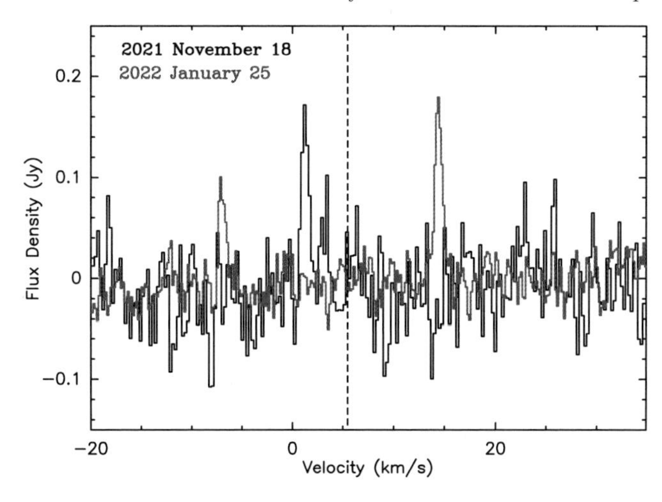

Figure 1. Detected H_2O masers towards HH 354 IRS.

Hirota *et al.* 2011; Furuya *et al.* 2003, 2001), currently still little information exists on masers in FUors/EXors.

2. Results

2.1. *H_2O masers associated with eruptive stars*

We detected H_2O masers towards 5 sources, but only 3 of the detections are likely associated with FUors/EXors, which include one EXor: V512 Per (Class I source) commonly known as SVS 13A, and two FUors: HH 354 IRS (Class 0/I) and Z CMa (Class I). The maser towards HH 354 IRS is the first to be reported. The maser component in Z CMa detected in our survey is a new component yet to be reported (see Fig. 1, Szabó *et al.* 2023b).

2.2. *Serendipitous detections towards Class 0 protostars*

Towards V512 Per (SVS 13A), multiple variable maser features were detected arising from a nearby source H_2O(B) (Class 0 object, known for H_2O masers, Haschick *et al.* 1980), which is within the beam. The source was in an active flare, contaminating the spectra of V512 Per. The peak of the emission was found to be associated with H_2O(B), but one maser feature at $>11\,\mathrm{km\,s^{-1}}$ is likely to be associated with V512 Per (see Fig. 2). Water masers were also detected towards the FUor binary RNO 1B/1C, but they are most likely arising from the molecular outflow of IRAS 00338+6312, located 4″ from the FUors (see e.g., Fiebig 1995; Fiebig *et al.* 1996).

2.3. *Discussion and conclusions*

The detection rate of our survey of FUors/Exors is only 6%, surprising in light of the close connections between H_2O maser emission and mass accretion. Possible explanations include:

1. Evolutionary effect: H_2O maser detection rate generally decreases from Class 0 to Class II sources (e.g., Furuya *et al.* 2003).
2. Low luminosities in low-mass star formation regions: low bolometric luminosities result in lower flux densities (e.g., Urquhart *et al.* 2011).
3. Rapid time variation: time variability is known and evident from our study, however masers can be in quiescence for ∼5 years (Claussen *et al.* 1996). Many of our targets

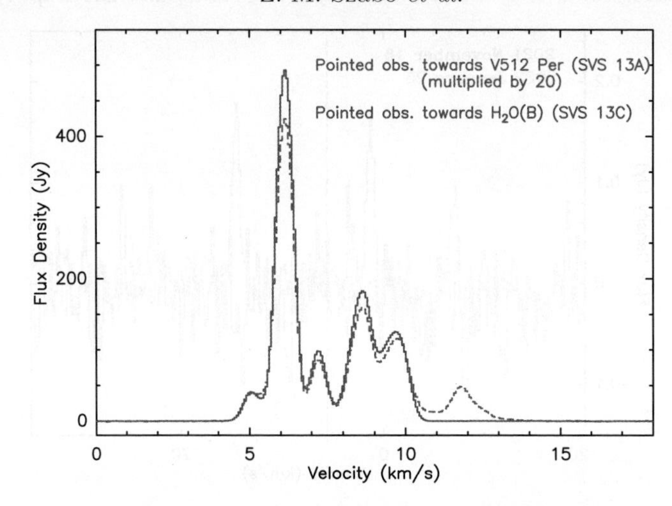

Figure 2. Pointed H_2O spectra towards V512 Per and H_2O(B) observed in 2022 February.

during the survey might have been inactive, despite showing maser emission in the past (see also Szabó *et al.* 2023b).

References

Abraham, Z., Cohen, N. L., Opher, R., *et al.* 1981, *A&A*, 100, L10

Audard, M., Ábrahám, P., *et al.* 2014, *Protostars and Planets VI*, 387

Claussen, M. J., Wilking, B. A., Benson, P. J., *et al.* 1996, *ApJS*, 106, 111

Cruz-Sáenz de Miera, F., Kóspál, Á., Ábrahám, P., *et al.* 2022, *ApJ*, 927, 125

Fiebig, D., 1995, *A&A*, 298, 207

Fiebig, D., Duschl, W. J., Menten, K. M., Tscharnuter, W. M., 1996, *A&A*, 310, 199

Fischer, W. J., Hillenbrand, L. A., *et al.* 2022, *Protostars and Planets VII*, eprint arXiv:2203.11257

Furuya, R. S., Kitamura, Y., Wootten, H., *et al.* 2001, *ApJ*, 559, L143

Furuya, R. S., Kitamura, Y., Wootten, A., *et al.* 2003, *ApJS*, 144, 71

Haschick, A. D., Moran, J. M., *et al.* 1980, *ApJ*, 237, 26

Hirota, T., Tsuboi, M., Fujisawa, K., *et al.* 2011, *ApJ*, 739, L59

Omodaka, T., Maeda, T., Miyoshi, M., *et al.* 1999, *PASJ*, 51, 333

Urquhart, J. S., Morgan, L. K., Figura, C. C., *et al.* 2011, *MNRAS*, 418, 1689

Szabó, Zs. M., Gong, Y., Yang, W., *et al.* 2023, *A&A*, 672, A158

Szabó, Zs. M., Gong, Y., Yang, W., *et al.* 2023, *A&A*, 674, A202

Cosmic Masers: Proper Motion toward the Next-Generation Large Projects
Proceedings IAU Symposium No. 380, 2024
T. Hirota, H. Imai, K. Menten, & Y. Pihlström, eds.
doi:10.1017/S174392132300340X

HMSFR G024.33+0.14: A possible new discovery in the making

S. P. van den Heever[1], M. Szymczak[2], M. Durjasz[2], A. Bartkiewicz[2], M. Olech[3] and P. Wolak[2]

[1]South African Radio Astronomy Observatory, Hartebeesthoek, 1740, South Africa.
email: sp.vandenheever@gmail.com

[2]Institute of Astronomy, Faculty of Physics, Astronomy and Informatics, Nicolaus Copernicus University, Grudziadzka 5, 87-100 Torun, Poland

[3]Space Radio-Diagnostics Research Centre, University of Warmia and Mazury, ul. Oczapowskiego 2, 10-719 Olsztyn, Poland

Abstract. Since 2017, and the formation of the maser monitoring organisation (M2O), we have observed several intriguing events. These events have included possible accretion bursts, strong jets, periodicity after a flare, a heat-wave of radiation travelling outward at a fraction of the speed of light, to name a few. In September 2019 the M2O was notified of another source showing flaring behavior, and here we present the possibility of the first discovery of long-term maser periodicity from the high-mass star formation region (HMSFR) G024.33+0.14, with a period of about 3000 days.

Keywords. masers, stars: formation, masers: flares, ISM: individual (G024.33+0.14)

1. Introduction

After the Maser Monitoring organization (M2O) was established, the international community of single-dish telescopes started working together more closely. Since then, and even some prior to the M2O, the community observed several very intriguing events, several possible accretion bursts from HMSFRs, e.g., the discovery of strong jets (Caratti o Garatti *et al.* 2017), periodicity after a flare (Proven-Adzri *et al.* 2019), and a heat-wave of radiation travelling radially outward Burns *et al.* (2020). This has also led to the discovery of roughly 30 new maser transitions, e.g., Brogan *et al.* (2019); MacLeod *et al.* (2019); Hunter *et al.* (2020), and the disentanglement of the sub-structures of the accretion disk around the protostar G358-MM1 Chen *et al.* (2020); Burns *et al.* (2023). In September 2019, the community was notified about another flaring event from the HMSFR, G024.33+0.14 (hereafter G024.33) Wolak *et al.* (2019), which has shown flaring previously in 2011. Archival data from Parkes (between 1992 and 1993) Caswell *et al.* (1995), seem to suggest evidence for long-term periodic variability with a period of $\sim$3000 days ($\sim$8.2 years).

2. HMSF region G024.33: periodic maser source with the longest period

Figure 1 shows the time series of seven spectral features of the 6.7 GHz methanol maser source G024.33 (Torun and HartRAO). With an assumption of a 3000 day recurrence (the average time difference between the flux density peaks of seven velocity components shown in Figure 1), the Parkes data from 1992 to 1993 have been extrapolated to the

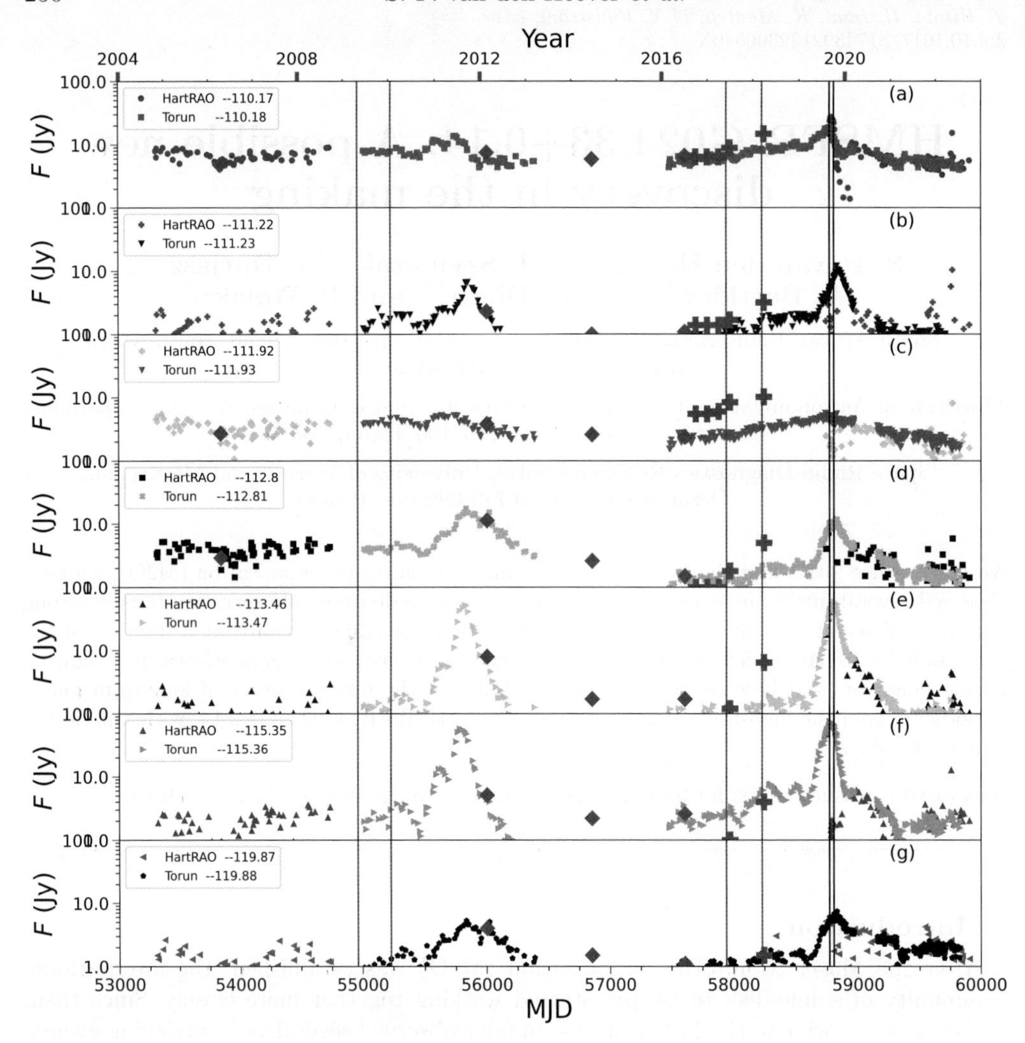

Figure 1. Time series of the flux density of the seven velocity features at; $V_{\rm LSR} = 110.16$ km s^{-1}, $V_{\rm LSR} = 111.22$ km s^{-1}, $V_{\rm LSR} = 111.92$ km s^{-1}, $V_{\rm LSR} = 112.80$ km s^{-1}, $V_{\rm LSR} = 113.46$ km s^{-1}, $V_{\rm LSR} = 115.35$ km s^{-1}, and $V_{\rm LSR} = 119.88$ km s^{-1}. The red pluses show the Parkes data extrapolated to the most recent flare, and the vertical blue and red lines demarcate the last two epochs from Parkes, extrapolated to 2009-10 and 2017-18. The vertical black lines show the peak flux density of panel (e) and (f).

most recent flare that was observed in 2019 (plotted with the red pluses). In addition, the last two epochs from Parkes (Dec 1992 and Sep 1993) are also demarcated with vertical blue and red lines for both of the flares observed 2010 and 2019, and the red diamonds demarcate various observations from 2003 onward. The similarities in the behavior of the flux density between the Parkes and the Torun data can be seen most prominently in panels (e) and (f), where increases in the flux density seem to correspond remarkably well. Although, the increase in flux density seen in these panels are not identical prior to the peak flux densities in 2010 and 2019, it can still be suggested that G024.33 is possibly the longest periodically flaring 6.7 GHz methanol maser. Therefore we predict that the next flare will happen in 2027, and the first activity prior to the flares will start in 2025, with the inferred 3000-day recurrence.

References

Caratti o Garatti, A., *et al.* 2017, *Nature Physics*, 13, 3
Brogan, C. L., *et al.* 2019, *ApjL*, 881, 2
Proven-Adzri, E., *et al.* 2019, *MNRAS*, 487, 2, 2407
Burns, R., *et al.* 2020, *Nature Astronomy*, 4, 506
MacLeod, G.C., *et al.* 2019, *MNRAS*, 489, 3, 3981
Hunter, T., *et al.* 2020, *SAN*, 29, 2
Chen, Xi., *et al.* 2020, *Nature Astronomy*, 4, 1170
Burns, R., *et al.* 2023, *Nature Astronomy*, 7, 557B
Wolak, P., *et al.* 2019, *The Astronomer's Telegram*, #13080
Caswell, J., *et al.* 1995, *PASJ*, 12, 37

Cosmic Masers: Proper Motion toward the Next-Generation Large Projects
Proceedings IAU Symposium No. 380, 2024
T. Hirota, H. Imai, K. Menten, & Y. Pihlström, eds.
doi:10.1017/S1743921323001953

Interferometric study of the class I methanol masers at 104.3 GHz

M. A. Voronkov[1], S. L. Breen[2], S. P. Ellingsen[3], A. M. Sobolev[4] and D. A. Ladeyschikov[4]

[1]CSIRO Space & Astronomy, PO Box 76, Epping, NSW 1710, Australia.
email: Maxim.Voronkov@csiro.au

[2]SKAO, Jodrell Bank, Lower Withington, Macclesfield, Cheshire SK11 9FT, UK

[3]School of Natural Sciences, University of Tasmania, Private Bag 37, Hobart, Tasmania 7001, Australia

[4]Ural Federal University, 19 Mira street, 620002 Ekaterinburg, Russia

Abstract. The Australia Telescope Compact Array (ATCA) has been used for an interferometric follow up observation at 104.3 GHz of the targets where either this or the 9.9-GHz maser was previously detected. We confirm the significant difference (by more than 1.5 orders of magnitude) from source to source of the flux density ratio for these two maser transitions. Based on the morphology and location of continuum sources, the most likely explanation of this discrepancy is the difference in the flux density of the seed radiation at the two frequencies. We also report absolute positions (with arcsec accuracy) for all detected 104.3 GHz masers.

Keywords. masers, stars: formation, HII regions

---

1. Introduction

Methanol masers tend to form series of transitions with different members of the series sharing similar observational properties such as whether they can be detected in a particular region of high-mass star formation (e.g., Cragg *et al.* 1992, and references therein). This work is about the $J_{-1}-(J\text{-}1)_{-2}$ E transition series which is responsible for the rare class I (or collisionally pumped) methanol masers at 9.9 (J=9) and 104.3 GHz (J=11). Despite the scarcity of the 104.3-GHz observations, there were reasons to suspect that significant variations from one source to another may be present in the flux density ratio of the two transitions. In particular, no 104.3-GHz maser has been detected towards 9.9-GHz maser in G331.13−0.24. On the other hand, the 9.9-GHz maser in G305.21+0.21 was only a marginal detection despite the presence of the 104.3-GHz maser in this source. In addition, the majority of 104-GHz data were obtained using assorted single dish facilities and no accurate positions have been measured. Therefore, we used the ATCA at 104.3 GHz to image all six targets known to have a maser in this series at the time of our observations (note, four additional 104.3-GHz masers were since reported, see Yang *et al.* 2023, and also a paper in this volume).

2. Results

Fig. 1 shows the environment, including the distribution of the widespread class I methanol masers at 36 and 44 GHz from Voronkov *et al.* (2014), for two selected sources, G305.21+0.21 and G331.13−0.24. As expected, the rare J_{-1}-$(J\text{-}1)_{-2}$ E masers appear in a single location at this scale. The noteworthy feature of these maps is that the

Table 1. Sources observed at 104.3 GHz in this project and the 9.9 GHz flux density from the literature for comparison. The measurement uncertainties are given in the brackets and expressed in the units of the least significant figure. For the two sources marked with (*), this result is the first measurement of the absolute position with an arcsecond accuracy for any maser in this series.

	Absolute position (J2000)		Peak	Peak flux density	
	Right Ascension	Declination	V_{LSR}	104 GHz	9.9 GHz
Source	(h:m:s)	(deg:arcmin:arcsec)	km s^{-1}	Jy	Jy
G305.21+0.21*	13:11:10.83 (7)	−62:34:38.5 (5)	−42.3 (1)	21.3 (6)	$\leq$ 0.2[1]
G331.13−0.24	Non-detection at 104.3 GHz, single channel rms noise is 0.84 Jy				~3[1,2]
G343.12−0.06	16:58:16.46 (5)	−42:52:25.5 (5)	−31.7 (1)	14.9 (7)	9.5 (3)[3]
G357.97−0.16*	17:41:20.06 (4)	−30:45:18.3 (5)	−5.0 (1)	15.8 (9)	69 (2)[4]
G012.80−0.19	18:14:10.90 (4)	−17:55:58.8 (5)	+32.9 (1)	10 (2)	4.3 (1)[1]
G019.61−0.23	18:27:37.49 (4)	−11:56:38.6 (5)	+41.3 (1)	9 (1)	3.3 (1)[1]

Source of the 9.9 GHz data: [1] Voronkov *et al.* (2010), [2] Voronkov *et al.* (this volume), [3] Voronkov *et al.* (2006), [4] Voronkov *et al.* (2011)

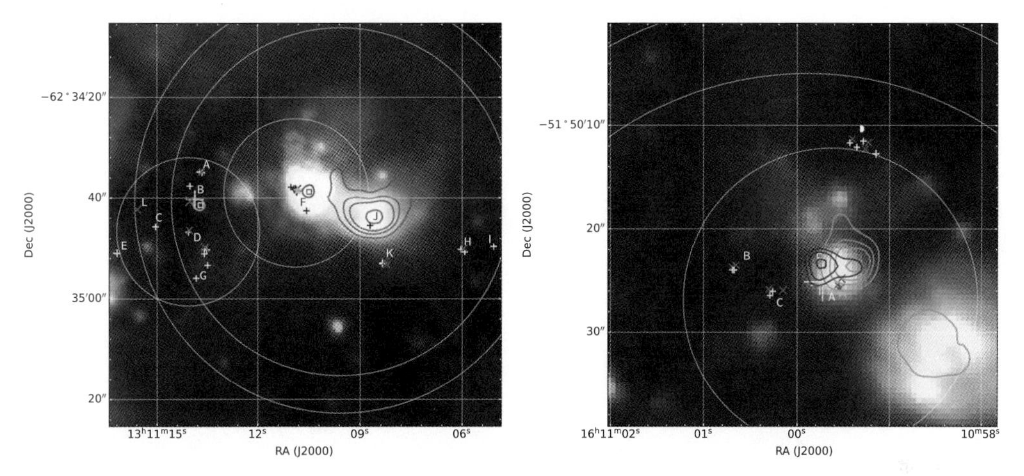

Figure 1. Two selected sources from the project: G305.21+0.21 (left) and G331.13−0.24 (right). The images are derived from that of Voronkov *et al.* (2014) and have the measured position of a J_{-1}-$(J\text{-}1)_{-2}$ E maser (either 104 or 9.9 GHz, as appropriate) shown by the star symbol along with the 3mm and 3cm continuum emission (small and large contours, respectively). As in the original images, the location of the 36 (crosses) and 44 GHz (pluses) masers are shown on top of the Spitzer 3-colour background (8.0, 4.5 and 3.6 μm images are shown as red, green and blue, respectively). The large circles represent the primary beam FWHM and pointing directions for 36, 44 and 104 GHz observations (the 9.9 GHz beam is too wide to show, with the whole field being well within the FWHM).

104.3-GHz maser in G305.21+0.21 is located near the 3-mm continuum source and far away from the 3-cm one. The opposite is observed in G331.13−0.24, where the 9.9-GHz maser is seen projected onto an HII region prominent at 3-cm, but offset from the 3-mm continuum source. Therefore, we conclude that it is the difference in the seed radiation at notably different frequencies of these two maser transitions (the emission at these two frequencies is produced by an HII region or a dust emission source which are not necessarily co-located) which is responsible for the observed relative flux densities in these two sources and may be a contributing factor for the other sources (Table 1). It is worth noting that no continuum emission at either of these frequencies has been detected at a position of a very strong 9.9-GHz maser in G357.97−0.16, quite possibly due to an insufficient sensitivity.

Supplementary material

To view supplementary material for this article, please visit http://dx.doi.org/10.1017/S1743921323001953

References

Cragg, D. M., Johns, K. P., Godfrey, P. D., Brown, R. D. 1992, *MNRAS*, 259, 203
Voronkov, M. A., Brooks, K. J., Sobolev, A. M. *et al.* 2006, *MNRAS*, 373, 411
Voronkov, M. A., Caswell, J. L., Ellingsen, S. P., Sobolev, A. M., 2010, *MNRAS*, 405, 2471
Voronkov, M. A., Walsh, A. J., Caswell, J. L., *et al.* 2011, *MNRAS*, 413, 2339
Voronkov, M. A., Caswell, J. L., Ellingsen, S. P., *et al.* 2014, *MNRAS*, 439, 2584
Yang, W., Gong, Y., Menten, K. M. *et al.* 2023, *A&A*, in press (arXiv:2305.04264)

Cosmic Masers: Proper Motion toward the Next-Generation Large Projects
Proceedings IAU Symposium No. 380, 2024
T. Hirota, H. Imai, K. Menten, & Y. Pihlström, eds.
doi:10.1017/S1743921323001941

Ultra-precise monitoring of a class I methanol maser

M. A. Voronkov[1], S. L. Breen[2], S. P. Ellingsen[3], J. A. Green[4], A. M. Sobolev[5], S. Yu. Parfenov[5] and D. J. van der Walt[6]

[1]CSIRO Space & Astronomy, PO Box 76, Epping, NSW 1710, Australia.
email: Maxim.Voronkov@csiro.au

[2]SKAO, Jodrell Bank, Lower Withington, Macclesfield, Cheshire SK11 9FT, UK

[3]School of Natural Sciences, University of Tasmania, Private Bag 37, Hobart, Tasmania 7001, Australia

[4]SKAO, SKA-LOW Science Operations Centre, ARRC Building, 26 Dick Perry Avenue, Technology Park, Kensington WA 6151 Australia

[5]Ural Federal University, 19 Mira street, 620002 Ekaterinburg, Russia

[6]Centre for Space Research, North-West University, Potchefstroom 2520, South Africa

Abstract. We report the results of a 7-year monitoring program using the Australia Telescope Compact Array (ATCA) for the 9.9 GHz class I methanol maser in G331.13-0.24 where a periodic class II methanol maser is present. The great deal of the project was to control systematics at an unprecedented level. Although no periodic flux variation was found, the maser shows a very stable decline of 166 ± 7 μJy/day. The radial velocity of the maser is stable down to 1 m/s level. We also report a marginal periodic signal in radial velocity (comparable to the level of systematics) of about 20 ± 7 cm/s with the period of 475 ± 22 days, close to that of the 6.7-GHz maser in the source. No hyperfine split was detected which suggests preferential excitation of a single hyperfine transition.

Keywords. masers, stars: formation, HII regions

1. Introduction

Periodic variability of radiatively pumped (or class II) methanol masers has recently become a hot topic as limited number of astrophysical phenomena can give rise to periodic flux variations (e.g., Goedhart *et al.* 2014; Parfenov & Sobolev 2014, and references therein). The collisionally pumped (or class I) methanol masers are often present in the same region, although usually believed not to originate in the same volume of gas (e.g., Voronkov *et al.* 2014). We carried out a long-term monitoring campaign of G331.13−0.24, observing it simultaneously at 6.7 and 9.9-GHz (convenient transitions with different pumping for ATCA monitoring observations) approximately every 20 days in order to probe the effects of the seed radiation on maser variability (both masers amplify the emission of an HII region, see the map of Voronkov *et al.* in this volume, and would vary in sync if the flux density of the HII region varies). The 6.7-GHz maser in the selected source has a period of about 500 days (Goedhart *et al.* 2014). No results of long term monitoring of the 9.9-GHz maser (or any other class I maser in any source) have previously been reported.

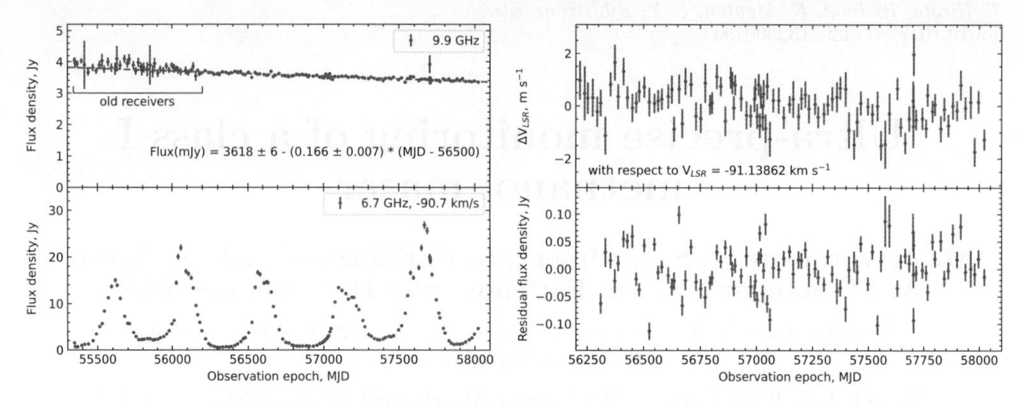

Figure 1. Left panel: flux density time series for the 9.9-GHz maser (top) and the most periodic feature of the 6.7-GHz maser (bottom). Note, the two masers are not co-located, but seen projected onto the same HII region (see Voronkov *et al.* in this volume for a map of the region). Right panel: time series of the radial velocity (top) and de-trended flux density (bottom). There could be a periodic wobble with the period close to that of the 6.7-GHz maser in the radial velocity. The weak variations in flux density are believed to be instrumental and may have a component with one year period.

2. Results

No periodic variability was found for the 9.9-GHz maser at a similar period to the 6.7-GHz maser (there could be a weak annual variation believed to be instrumental). Instead, the maser showed a steady decline in flux density (see Fig. 1). In addition, the 9.9-GHz spectrum turned out to be rather simple with just a single feature well described by a Gaussian. Combined with high 488 Hz spectral resolution (<15 m/s at 9.9 GHz) and a good signal-to-noise ratio this enabled very accurate radial velocity measurement (down to m/s accuracy, see the right-hand panel of Fig. 1). On the other hand, we can confidently exclude hyperfine split for the 9.9-GHz transition ($9_{-1} - 8_{-2}$ E) at kHz level. The split expected following Belov *et al.* (2016) would be comparable to the FWHM of the maser line and, therefore, easily detectable. This suggests that preferential pumping of a single hyperfine transition may take place.

The predictable behaviour and simplicity of the maser along with a large number of epochs allowed us to reveal many low-level systematics in the data. This investigation resulted in a number of software fixes (bugs in MIRIAD's Doppler regridding code), workarounds (external barycentric correction model) and improvements to the data reduction procedure (new methods for the absolute flux scale calibration and flux estimation which are less sensitive to weather effects, gain-elevation correction). The resulting radial velocity time series (top right in Fig. 1) has a marginal periodic wobble (20 ± 7 cm/s). It is comparable to the level of systematics but has the period (475 ± 22 days) close to that of the 6.7-GHz maser. Otherwise, the radial velocity is remarkably stable at the level of m/s, and the lack of observed acceleration over the 7-year period implies the minimum distance between the maser and the YSO ranging from 120 to 4300 au for a 10 $M_\odot$ YSO depending on the orientation (the strongest limit is when the maser and the central mass are aligned along the line of sight), provided there is no other mechanism at play counteracting gravity in this system. The residual (after the linear trend is removed) flux density variations (bottom right in Fig. 1) are likely to be dominated by systematics of unknown nature (resulting in a weak spurious annual periodicity).

Supplementary material

To view supplementary material for this article, please visit http://dx.doi.org/10.1017/S1743921323001941

References

Belov, S. P., Golubiatnikov, G. Yu., Lapinov, A. V. *et al.* 2016, *J. Chem. Phys.*, 145, 024307

Goedhart, S., Maswanganye, J. P., Gaylard, M. J., van der Walt, D. J. 2014, *MNRAS*, 437, 1808

Parfenov, S., Yu., Sobolev, A. M. 2014, *MNRAS*, 444, 620

Voronkov, M. A., Caswell, J. L., Ellingsen, S. P., *et al.* 2014, *MNRAS*, 439, 2584

Cosmic Masers: Proper Motion toward the Next-Generation Large Projects
Proceedings IAU Symposium No. 380, 2024
T. Hirota, H. Imai, K. Menten, & Y. Pihlström, eds.
doi:10.1017/S1743921323003162

Spatio-kinematics of water masers in the HMSFR NGC6334I before and during an accretion burst

Jakobus M. Vorster[1,2], **James O. Chibueze**[2,3,4], **Tomoya Hirota**[5,6] **and Gordon C. MacLeod**[7,8]

[1]University of Helsinki, Finland,

[2]North-West University, South Africa, email: jakobus.vorster@helsinki.fi

[3]Department of Mathematical Sciences, University of South Africa, Roodepoort, South Africa,

[4]University of Nigeria, Nigeria,

[5]National Astronomical Observatory of Japan, Japan,

[6]SOKENDAI (The Graduate University for Advanced Studies), Japan,

[7]The Open University of Tanzania, Tanzania,

[8]SARAO Hartebeesthoek Astronomical Observatory, South Africa.

Abstract. In 2015, the high-mass star-forming region NGC6334I-MM1 underwent an accretion burst. Using VERA, we monitored 22 GHz water masers before and during the accretion burst to observe the changes in the maser spatial and velocity distributions in the region. The masers in CM2-W2 and MM1-W1 displayed variability that could be attributed to the accretion burst. The bright masers in CM2-W2 were found to better trace the shock structure as the epochs progressed. The mean 3D speeds derived from the proper motions were 50 km/s and 54 km/s for the pre-burst and burst epochs respectively. High-velocity proper motions were found at the southern edges of the N-S ($\sim$80 km s^{-1}) and NW-SE ($\sim$150 km s^{-1}) bipolar outflows. The precise mechanism of the flaring of the water masers due to the accretion burst has yet to be investigated.

Keywords. masers, stars: formation, ISM: jets and outflows

---

1. Introduction

The high-mass star-forming region NGC6334I recently underwent a period of high-mass accretion (Hunter *et al.* 2017). This accretion burst was accompanied by the flaring of multiple maser species, such as 6.7 GHz methanol and 22 GHz water masers (MacLeod *et al.* 2018). High-resolution VLBI observations of the evolution of water masers during the rapid onset of the accretion burst give information on the time dependence of the water masers, which is influenced by the burst's evolution.

2. Observations and Data Reduction

We conducted seven observations of 22 GHz water masers in NGC6334I with the VLBI Exploration for Radio Astrometry (VERA) array. The observations were done before and during the burst on 2014.72, 2014.90, 2015.08, 2015.28, 2015.88, 2016.11 and 2016.19. The spectral resolution was 0.4 km s^{-1} and the beam size was 1.3×3.3 milliarcseconds. The data were calibrated and imaged with standard imaging techniques

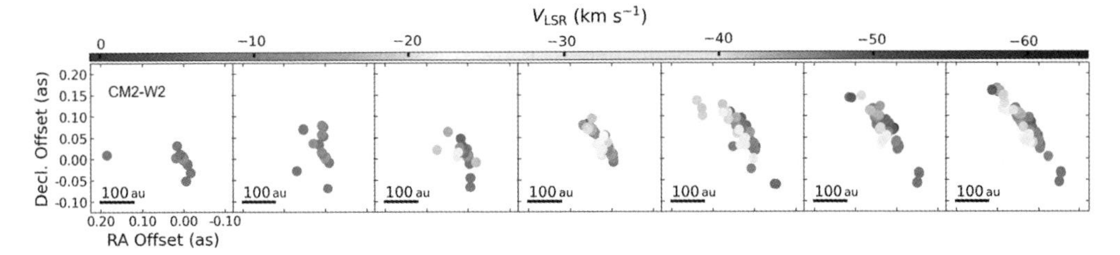

Figure 1. Top: Water maser positions and radial velocities over all seven observed epochs in CM2-W2.

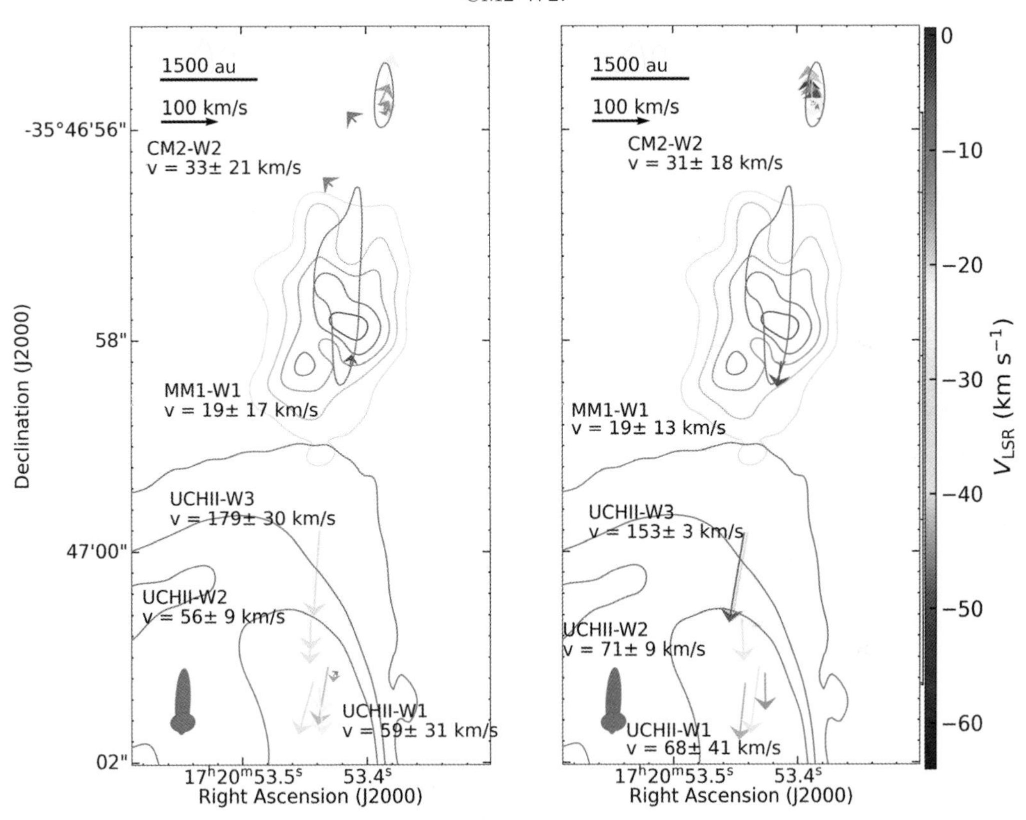

Figure 2. Proper motions of water masers in NGC6334I before (2014.7 – 2015.3) and during (2015.9 – 2016.2) the accretion burst. The mean velocities of each region is shown in black text. The brown contours represent the ALMA 1.3 mm continuum and the grey contours the JVLA 5 cm continuum, as in Brogan *et al.* (2018).

with the Astronomical Image Processing System (AIPS). The data were self-calibrated on the brightest persistent maser spot. Proper motions were calculated by identifying persistent groups of maser spots showing a Gaussian spectral distribution, which were termed maser features. The position of the feature in each epoch was calculated with the flux-weighted mean position of the spots in the feature.

3. Results and Discussion

Figure 1 shows the positions of maser features in CM2-W2, the northern bow shock, 2800 au from the accretion bursting source MM1B (Chibueze *et al.* 2021). The maser features trace a bow shock structure more clearly as the burst progresses. The linear size of the maser feature at (0,0) of Figure 1 also increased with the accretion burst. The brightening of the feature was also seen in its spectral distribution. Figure 2 shows the proper motions. of maser appearing to trace the bipolar outflow in NGC6334I. New high-velocity features were excited in the southern regions.

The 22 GHz water masers are understood to be collisionally pumped in turbulent and shocked regions (Hollenbach *et al.* 2013). The changes in the large (~ 100 au) and small scale (~ 1 au) spatial distribution of the masers indicate that the variability in the masers can be attributed to the accretion burst rather than hydrodynamic variability which is intrinsic to water masers. Investigating the precise mechanism of how the burst affected the water masers requires three observations to be explained: The better tracing of the bow shock, the increase of the linear size of single features and the dramatic flare observed in single dish monitoring observations.

Supplementary material

To view supplementary material for this article, please visit http://dx.doi.org/10.1017/S1743921323003472

References

Brogan, C.L., Hunter, T.R., Cyganowski, C.J., *et al.* 2018, *ApJ*, 866, 87
Hunter, T.R., Brogan, C.L., MacLeod, G.C., *et al.* 2017, *ApJ*, 837, L29
MacLeod, G.C., Smits, D.P., Goedhart, S., *et al.* 2018, *MNRAS*, 478, 1077
Hollenbach, D., Elitzur, M. and McKee, C.F. 2013, *ApJ*, 773, 70
Chibueze, J.O., MacLeod, G.C, Vorster, J.M., *et al.* 2021, *ApJ*, 908, 175

Cosmic Masers: Proper Motion toward the Next-Generation Large Projects
Proceedings IAU Symposium No. 380, 2024
T. Hirota, H. Imai, K. Menten, & Y. Pihlström, eds.
doi:10.1017/S174392132300220X

Multi-wavelength maser observations of the Extended Green Object G19.01–0.03

Gwenllian M. Williams[1], Claudia J. Cyganowski[2], Crystal L. Brogan[3], Todd R. Hunter[3], John D. Ilee[1], Pooneh Nazari[4] and Rowan J. Smith[5]

[1]School of Physics & Astronomy, The University of Leeds, Leeds, LS2 9JT, UK
email: G.M.Williams@leeds.ac.uk

[2]Scottish Universities Physics Alliance (SUPA), School of Physics and Astronomy, University of St Andrews, North Haugh, St Andrews, KY16 9SS, UK

[3]National Radio Astronomy Observatory (NRAO), 520 Edgemont Rd, Charlottesville, VA 22903, USA

[4]Leiden Observatory, Leiden University, PO Box 9513, 2300 RA Leiden, NL

[5]Jodrell Bank Centre for Astrophysics, Department of Physics and Astronomy, University of Manchester, Oxford Road, Manchester, M13 9PL, UK

Abstract. We report the detections of NH_3(3,3) and 25 GHz and 278.3 GHz Class I CH_3OH maser emission associated with the outflow of the Extended Green Object G19.01–0.03 in sub-arcsecond resolution Atacama Large Millimeter/submillimeter Array (ALMA) and Karl G. Jansky Very Large Array (VLA) observations. For masers associated with the outer outflow lobes ($> 12.5''$ from the central massive young stellar object; MYSO), the spatial distribution of the NH_3(3,3) masers is statistically indistinguishable from that of previously known 44 GHz Class I CH_3OH masers, strengthening the connection of NH_3(3,3) masers to outflow shocks. In sub-arcsecond resolution VLA observations, we resolve the 6.7 GHz Class II CH_3OH maser emission towards the MYSO into a partial, inclined ring, with a velocity gradient consistent with the rotationally supported circumstellar disc traced by thermal gas emission.

Keywords. masers, stars: individual: G19.01–0.03, stars: formation, stars: massive, stars: protostars, techniques: interferometric

1. Introduction

Masers are well-established tracers of massive young stellar objects (MYSOs) and their outflows, with high-resolution Karl G. Jansky Very Large Array (VLA) and Atacama Large Millimeter/(sub)millimeter Array (ALMA) observations ongoing over the past decade continuing to lead to the identification of new maser lines in these environments. In this proceeding, we summarise our recent sub-arcsecond resolution ALMA and VLA observations of maser emission towards the Extended Green Object (EGO) G19.01-0.03, an MYSO that drives a well-collimated bipolar outflow (Williams *et al.* 2022, 2023).

2. Outflow-tracing maser emission, and a 6.7 GHz Class II methanol maser ring

The $CH_3OH(9_{-1,9} - 8_{0,8})$ line at 278.3 GHz is known to exhibit maser behaviour (e.g. Yanagida *et al.* 2014). Our ALMA peak intensity map of this line (Figure 1a) shows it generally traces the morphology of the bipolar outflow driven by the central MYSO

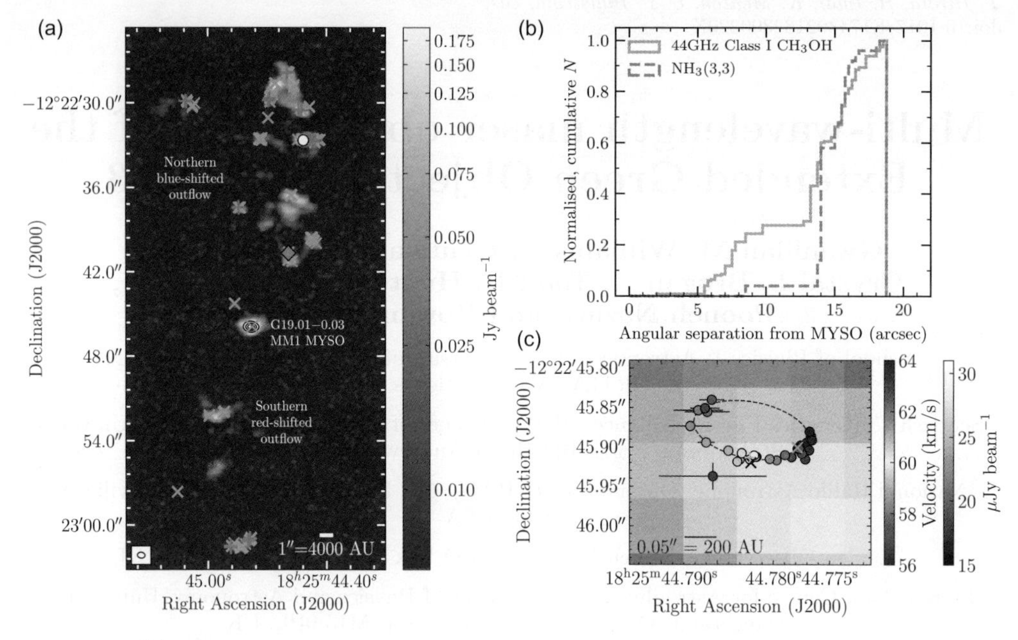

Figure 1. **(a)** ALMA peak intensity map of the $CH_3OH(9_{-1,9} - 8_{0,8})$ line at 278.3 GHz towards G19.01−0.03. The white circle marks the position of peak emission. ALMA 1.05 mm continuum contours are shown in black towards the MYSO at 64, 256 and 600σ. The orange cross marks the intensity-weighted 6.7 GHz Class II CH_3OH maser position (as in Fig.1c). Blue pluses denote new $NH_3(3,3)$ maser spots, and pink crosses denote 44 GHz Class I CH_3OH masers (Cyganowski *et al.* 2009). The green diamond marks the 25 GHz Class I CH_3OH maser. **(b)** Histogram of the angular separation of the $NH_3(3,3)$ and 44 GHz Class I CH_3OH maser spots (blue and pink respectively; as in Fig.1a) from the MYSO's ALMA 1.05 mm continuum peak (Williams *et al.* 2023). **(c)** Fitted 6.7 GHz Class II CH_3OH maser positions, coloured by velocity, overplotted on the VLA 5.01 cm continuum (adapted from Williams *et al.* 2022). The black cross and yellow and cyan pluses show the 1.05 mm, 5.01 cm and 1.21 cm continuum peak positions respectively.

(previously observed in ^{12}CO, SiO and HCO^+ by Cyganowski *et al.* 2011). We classify the brightest 278.3 GHz CH_3OH emission as a candidate maser, due to (i) its narrow spectral profile, (ii) its brightness temperature being an order of magnitude higher than that of thermal CH_3OH lines in our spectral tuning, and (iii) its coincidence, both spatially and kinematically, with known 44 GHz Class I CH_3OH masers (Cyganowski *et al.* 2009) and newly identified $NH_3(3,3)$ masers.

$NH_3(3,3)$ is known to exhibit maser behaviour thought to arise due to outflow-induced shocks (e.g. Brogan *et al.* 2011). With our $\sim 0.5''$ resolution VLA observations, we detect 50 $NH_3(3,3)$ maser emission spots to the $> 5\sigma$ level in the outflow (Figure 1a). These spots are coincident spatially and kinematically with the outflow-tracing 278.3 GHz CH_3OH emission. At large angular separation from the MYSO ($> 12.5''$), their spatial distribution is statistically indistinguishable from that of known 44 GHz Class I CH_3OH masers (Figure 1b).

The 25 GHz CH_3OH ($5_{2,3} - 5_{1,4}$) line can exhibit thermal and/or Class I maser emission towards EGOs (e.g. Towner *et al.* 2017). We tentatively detect (i) thermal emission towards the MYSO ($5.5\sigma \sim 36.8$ K), and (ii) candidate maser emission towards the outflow ($7.8\sigma \sim 52.3$ K) that is positionally and kinematically coincident with 44 GHz and 278.3 GHz Class I CH_3OH maser emission (Figure 1a).

With VLA 2.8″ resolution observations, Cyganowski *et al.* (2009) detected 6.7 GHz Class II CH_3OH maser emission towards the central MYSO. With our new ∼0.6″ resolution VLA observations, the maser spot distribution exhibits a partial ellipse, consistent with an inclined ring (Figure 1c). The strongest maser emission is blue-shifted with respect to the systemic velocity of the MYSO (∼60 km s^{-1}; Cyganowski *et al.* 2011), and the maser velocities follow that of the Keplerian circumstellar disc traced by thermal gas emission (Williams *et al.* 2022).

References

Brogan, C. L. *et al.* 2011, *ApJ*, 739, L16
Cyganowski, C. J. *et al.* 2009, *ApJ*, 702, 1615
Cyganowski, C. J. *et al.* 2011, *ApJ*, 729, 124
Towner, A. P. M. *et al.* 2017, *ApJS*, 230, 22
Yanagida, T. *et al.* 2014, *ApJ*, 794, L10
Williams, G. M. *et al.* 2022, *MNRAS*, 509, 748
Williams, G. M. *et al.* 2023, *MNRAS*, 525, 6146

Cosmic Masers: Proper Motion toward the Next-Generation Large Projects
Proceedings IAU Symposium No. 380, 2024
T. Hirota, H. Imai, K. Menten, & Y. Pihlström, eds.
doi:10.1017/S1743921323002946

Torun methanol maser monitoring program

P. Wolak[1], M. Szymczak[1], A. Bartkiewicz[1], M. Durjasz[1], A. Kobak[1] and M. Olech[2]

[1]Institute of Astronomy, Faculty of Physics, Astronomy and Informatics, Nicolaus Copernicus University, Grudziadzka 5, 87-100Torun, Poland. email: wolak@astro.umk.pl

[2]Space Radio-Diagnostic Research Center, Faculty of Geoengineering, University of Warmia and Mazury, Oczapowskiego 2, PL-10-719 Olsztyn, Poland

Abstract. Since 2009, the Torun 32 m radio telescope has been used to monitor a sample of ∼140 sources of the 6.7 GHz methanol maser emission. In 2022, the sample was extended to about 250 targets. Approximately three-quarters show variability greater than 10% on timescales of a few weeks to several years. The most significant results are detecting a few flare events and discovering about a dozen periodic variables with periods ranging from a month to a few years. Here, we present the preliminary analysis of the properties of periodic masers.

Keywords. masers, radio lines: ISM, stars: formation

1. Introduction

It is well known that the emission of class II 6.7 GHz methanol masers occurs in high-mass star-forming regions and is a powerful tool for studying processes during the early stage of star formation. One way to better understand star formation is to monitor maser line variability. A significant achievement in this field was the detection of maser outbursts caused by a rapid increase in the rate of accretion and the discovery of periodic sources. Up to now periodic variability of 6.7 GHz methanol masers were reported in 28 sources, excluding the recent detection of ∼1800 days quasi-periodic variations in Cep A (Durjasz *et al.* 2022). The periods range from 24 to 1260 days. Several mechanisms that explain the variety in flare profiles and periods have been hypothesized. Some models, such as the colliding wind binary model, consider seed photon flux changes as the driving mechanism (van der Walt 2011). Others consider the change in the pumping efficiency to be the main mechanism responsible for flares. The pump rate can vary due to periodic modulation of the accretion rate in binary systems (Araya *et al.* 2010) and spiral shocks (Parfenov & Sobolev 2014). Inayoshi *et al.* (2013) proposed the κ mechanism to explain high-mass star pulsation in the phase of high accretion. There is a claim that the orientation of the disc-outflow system in the plane of the sky might have a significant impact on observed flare profiles (Morgan *et al.* 2021).

2. Data and results

The Torun 32-m radio telescope has been monitoring a sample of ∼140 objects since 2009 (Szymczak *et al.* 2018) and 12 periodic sources were detected so far. In the last 5 years, we have discovered 7 objects; G24.148−0.009, G30.400−0.296, G33.641−0.228, G45.804−0.35 G49.043−1.079, G59.633−0.192 and G108.76−0.99 (Olech *et al.* 2019, 2022). We updated Table A3 from Olech *et al.* (2019) and performed a statistical analysis of parameters such as the period, timescale of variability (FWHM) and the ratio of rise to decay time ($R_{\rm rd}$) for 28 periodic maser sources. Histograms summarizing the results

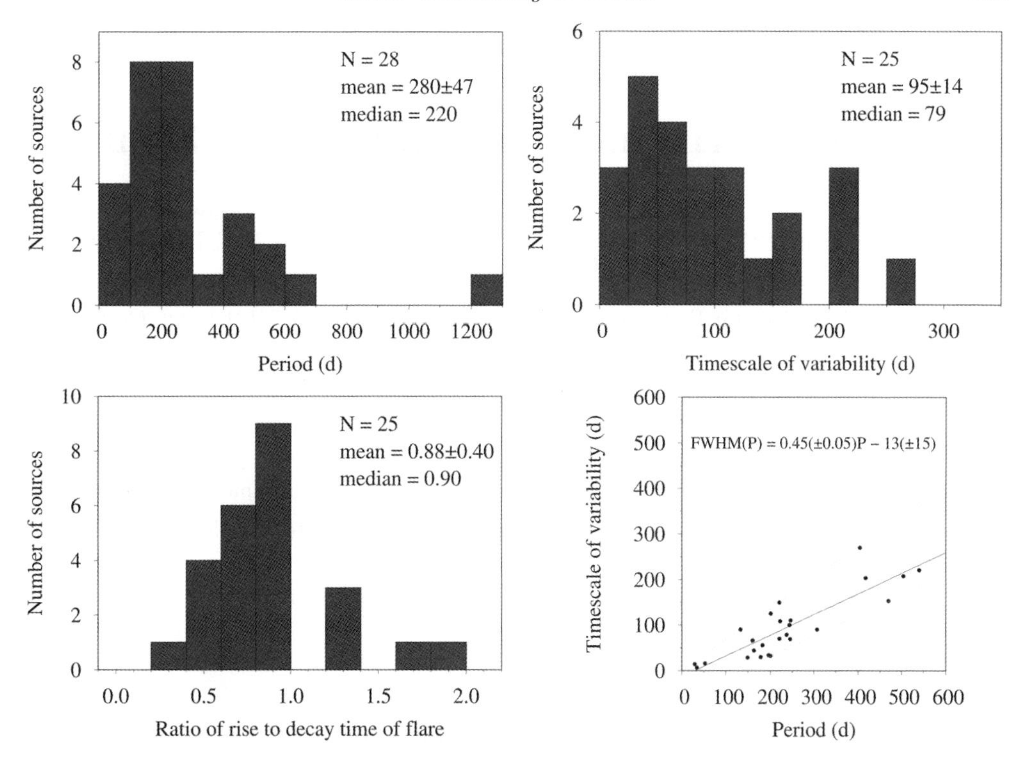

Figure 1. Distributions of selected variability parameters of known periodic sources from Table A3 in Olech *et al.* 2019 updated with new objects reported by Olech *et al.* (2022) and Tanabe *et al.* (2023).

and the average and median values are given in Fig. 1. The sources with a period in the range of 133–245 days are 57% (16/28) of the sample. For 64% (16/25) of the sources the FWHM ranges from 6.4 to 100 days. The profile of flare is asymmetric; for about 80% (20/25) of the sources $R_{\rm rd}$ is from 0.35 to 1.0. The ratio of FWHM to period can be a measure of the duty cycle. It is the smallest (0.16) in G75.76+0.34 and the largest (0.68) in G358.460−0.391. This suggests that the mechanisms that trigger periodic outbursts can be quite diverse.

This work is supported by NCN grant 2021/43/B/ST9/02008.

References

Araya, E. D., Hofner, P., Goss, W. M., *et al.* 2010, *ApJ*, 717, L133
Durjasz, M., Szymczak, M., Olech M., Bartkiewicz A. 2022, *A&A*, 663, A123
Inayoshi, K., Sugiyama, K., Hosokawa, T., Motogi, K., & Tanaka, K. E. I. 2013, *ApJ*, 769, L20
Morgan, J., van der Walt, D. J., Chibueze, J. O., & Zhang, Q. 2021, *MNRAS*, 507, 1138
Olech, M., Szymczak, M., Wolak, P., Sarniak, R., & Bartkiewicz, A. 2019, *MNRAS*, 486, 1236
Olech, M., Durjasz, M., Szymczak, M., Bartkiewicz. 2022, *A&A*, 661, A114
Parfenov, S. Y. & Sobolev, A. M. 2014, *MNRAS*, 444, 620
Szymczak, M., Olech, M., Sarniak, R., Wolak, P., Bartkiewicz, A. 2018, *MNRAS*, 474, 219
Tanabe, Y., Yonekura, Y., MacLeod, G C. 2023, *PASJ*, 75, 2
van der Walt, D. J. 2011, *AJ*, 141, 152

Cosmic Masers: Proper Motion toward the Next-Generation Large Projects
Proceedings IAU Symposium No. 380, 2024
T. Hirota, H. Imai, K. Menten, & Y. Pihlström, eds.
doi:10.1017/S1743921323002569

ATLASGAL: methanol masers at 3 mm

W. Yang[1], Y. Gong[1], K. M. Menten[1], F. Wyrowski[1], J. S. Urquhart[2], C. Henkel[1,3,4], T. Csengeri[5], S. P. Ellingsen[6], A. R. Bemis[7] and J. Jang[1]

[1]Max-Planck-Institut für Radioastronomie, Auf dem Hügel 69, D-53121 Bonn, Germany
email: wjyang@mpifr-bonn.mpg.de

[2]Centre for Astrophysics and Planetary Science, University of Kent, Canterbury CT2 7NH, UK

[3]Astronomy Department, Faculty of Science, King Abdulaziz University, PO Box 80203, Jeddah, 21589, Saudi Arabia

[4]Xinjiang Astronomical Observatory, Chinese Academy of Sciences, Urumqi 830011, PR China

[5]Laboratoire d'astrophysique de Bordeaux, Univ. Bordeaux, CNRS, B18N, allée Geoffroy Saint-Hilaire, 33615 Pessac, France

[6]School of Natural Sciences, University of Tasmania, Private Bag 37, Hobart, Tasmania 7001, Australia

[7]Leiden Observatory, Leiden University, P.O. Box 9513, 2300 RA Leiden, The Netherlands

Abstract. We analyzed the 3 mm wavelength spectral line survey of 408 clumps from the APEX telescope large area survey of the Galaxy, focusing on the methanol maser transitions. The main goals of this study are (1) to search for new methanol masers, (2) to statistically study the relationship between class I masers and shock tracers, (3) to study the properties between methanol masers and their host clumps, also as a function of their evolutionary stages and, (4) to better constrain the physical conditions using multiple co-spatial line pairs.

Keywords. Masers, star: formation, ISM: molecules, radio lines: ISM

1. Introduction

Methanol (CH_3OH) masers are common phenomena in star-formation regions, providing numerous maser transitions with inverted populations highlighting conditions with enhanced activity (e.g., Menten 1991). These CH_3OH masers are divided into two categories (Batrla *et al.* 1987; Menten 1991). Class I CH_3OH masers are thought to be tracers of shocked regions and are produced by collisional pumping (e.g., Leurini *et al.* 2016). Class II CH_3OH masers are found in close proximity to infrared sources, OH masers and ultracompact HII regions, and are believed to be radiative pumped (e.g., Cragg *et al.* 2005).

The APEX telescope large area survey of the Galaxy (ATLASGAL) is an unbiased 870 μm sub-millimetre continuum survey of the inner Galaxy (Schuller *et al.* 2009), providing a large inventory of dense molecular clumps. A number of 408 ATLASGAL sources were selected to cover a large range of evolutionary stages, and were observed using the IRAM-30 m with a frequency coverage of 83.8 – 115.7 GHz (Csengeri *et al.* 2016). This well characterized sample allows us to search for new methanol masers and perform a statistical analysis of the properties of detected masers and their associated star-forming clumps at different evolutionary stages.

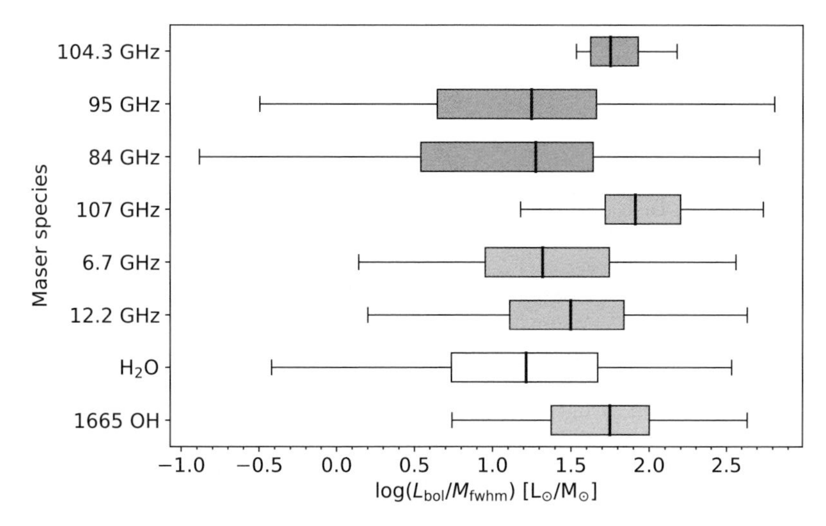

Figure 1. The box plot shows the distributions of bolometric luminosity-to-mass ratios for clumps associated with different masers. Except for 3-mm CH_3OH masers, the plotting data for H_2O, OH, and class II methanol masers at 6.7 and 12.2 GHz are taken from Fig. 7(A) in Ladeyschikov *et al.* (2022).

2. Detection

• A total of 54, 100 and 4 sources were detected to show class I maser emission at 84 ($J_k = 5_{-1} - 4_0\,E$), 95 ($8_0 - 7_1\,A^+$) and 104.3 GHz ($11_{-1} - 10_{-2}\,E$), respectively. Among them, fifty 84 GHz masers, twenty nine 95 GHz masers, and four 104.3 GHz masers are new discoveries. Our work increases the rare class I maser at 104.3 GHz from five to nine (Yang *et al.* 2023).

• Eleven sources were detected to harbor class II methanol maser emission at 107 GHz ($3_1 - 4_0\,A^+$), and the known number of detections has increased from 25 to 33. No sources show maser emission at 85.5, 86.6, 86.9, 104.1 and 108 GHz (Yang *et al.* in prep).

• We also detected 19 sources with CH_3OH absorption features at 107 GHz, which could be anti-inversion, since analyses show that their continuum background is dominated by the Cosmic Microwave Background.

3. Results

• We found that (1) more and stronger class I CH_3OH masers were detected towards sources showing SiO line wings than towards sources without SiO wings; (2) the total integrated intensity of class I masers is positively correlated with SiO integrated intensity and FWZP of SiO (2–1) emission. These facts strongly suggest that the properties of class I masers are regulated by shock properties also traced by SiO.

• The properties of class I CH_3OH masers show positive correlations with the following properties of associated ATLASGAL clumps (Urquhart *et al.* 2022): bolometric luminosity, clump mass and peak H_2 column density. There is no statistically significant correlation between the luminosity of class I CH_3OH and the luminosity-to-mass ratio, dust temperature, or mean H_2 volume density.

• Through studying the evolutionary stages of different masers during star formation (see Fig. 1), we found that CH_3OH masers at 104.3 (107) GHz appear to trace a short and evolved stage compared to the other class I (II) CH_3OH masers.

• Our statistical equilibrium calculations show that physical conditions can be better constrained in regions with multiple class I CH_3OH masers (more details in Yang *et al.* 2023).

Supplementary material

To view supplementary material for this article, please visit http://dx.doi.org/10.1017/S1743921323002569

References

Batrla, W., Matthews, H. E., Menten, K. M., *et al.* 1987, *Nature*, 326, 49
Cragg, D. M., Sobolev, A. M., & Godfrey, P. D. 2005, *MNRAS*, 360, 533
Csengeri, T., Leurini, S., Wyrowski, F., *et al.* 2016, *A&A*, 586, A149
Ladeyschikov, D. A., Gong, Y., Sobolev, A. M., *et al.* 2022, *ApJS*, 261, 14
Leurini, S., Menten, K. M., & Walmsley, C. M. 2016, *A&A*, 592, A31
Menten, K. M. 1991, *Atoms, Ions and Molecules: New Results in Spectral Line Astrophysics*, 16, 119
Schuller, F., Menten, K. M., Contreras, Y., *et al.* 2009, *A&A*, 504, 415
Urquhart, J. S., Wells, M. R. A., Pillai, T., *et al.* 2022, *MNRAS*, 510, 3389
Yang, W., Gong, Y., Menten, K. M., *et al.* 2023, *A&A*, in press, arXiv:2305.04264

Cosmic Masers: Proper Motion toward the Next-Generation Large Projects
Proceedings IAU Symposium No. 380, 2024
T. Hirota, H. Imai, K. Menten, & Y. Pihlström, eds.
doi:10.1017/S174392132300251X

High-cadence 6.7 GHz methanol maser monitoring observations by Hitachi 32-m radio telescope

Yoshinori Yonekura[1], Yoshihiro Tanabe[1] and Ren Moriizumi[2]

[1]Center for Astronomy, Ibaraki University, Mito, Ibaraki 310-8512, Japan.
email: yoshinori.yonekura.sci@vc.ibaraki.ac.jp

[2]Graduate School of Science and Engineering, Ibaraki University, Mito, Ibaraki 310-8512, Japan

Abstract. We started high-cadence monitoring observations of 6.7 GHz methanol masers from Dec. 2012 using Hitachi 32-m radio telescope (Yonekura *et al.* 2016). Observations have been conducted basically every day. On average, 13 hours of observations have been made per day, amounting to 4000–5000 hours per year. The cadence varies by sources: one observation in 1–50 days. In addition to already known 29 sources (Tanabe et al. 2023 and references therein), we have newly identified ∼20 sources with periodic flux variability. We have also detected 5 sources with sudden flux rises in 2019–2022, including G358.93−0.03 which was confirmed to be associated with the accretion burst.

Keywords. masers, stars:formation, stars:massive, stars:flare

---

1. Introduction

Almost all 6.7 GHz methanol masers are known to be associated with high-mass protostellar objects. After the discovery of this maser in 1991, more than 1000 objects associated with this maser are detected. Because this maser is pumped by the radiation from the central star, the flux density of this maser was thought to be somewhat stable. In 2003, the periodic flux variability for this maser sources G9.62+0.20E is reported. Until now, 29 maser sources with periodic flux variability have been found. Five mechanisms for the flux variability are suggested. In three of which, the temperature of the dust grains can be changed: (i) rotation of spiral shock wave in the gap of the circumbinary accretion disk, (ii) periodic accretion of material from the circumbinary disc, and (iii) pulsation of the central star. In the remaining two, the flux of seed photons can be periodically changed: (iv) a colliding wind binary, and (v) an eclipsing binary. Aside from the periodic flux variability, non-periodic sudden flux rises with the factor of 10–1000 were detected for 4 maser sources (S255IR-IRS3, NGC 6334 I-MM1, G358.93−0.03, G24.33+0.14). These sudden flux rises were confirmed to be the results of the accretion burst from the observational results of the brightening of the central sources in NIR/FIR/(Sub-)mm.

We started high-cadence monitoring observations of 6.7 GHz methanol masers from Dec. 2012 using Hitachi 32-m radio telescope named "Ibaraki 6.7 GHz class II methanol maser database (iMet)†", in order to detect periodic flux variations and sudden flux rises. In this paper, we briefly summarize the observations and results obtained by iMet.

† All the data are available at iMet (Ibaraki 6.7 GHz class II methanol maser database) web at http://vlbi.sci.ibaraki.ac.jp/iMet/

Table 1. Input catalogs for the "master catalog".

Name of the Input Catalog	Reference	Region
Parkes methanol multibeam survey	Caswell *et al.* 2010	$345 < L < 6$
	Green *et al.* 2010	$6 < L < 20$
	Green *et al.* 2012	$186 < L < 330$
Arecibo Methanol Maser Galactic Plane Survey	Pandian *et al.* 2007	$35 < L < 54$
Compilation catalog at 2009	Xu *et al.* 2009	

2. Observations

2.1. *Target selection*

At first, we have compiled the "master catalog" from the papers listed in Table 1. Then, we have selected 442 sources with Decl. ≥ -30 deg as targets for the monitoring observations at Ibaraki.

2.2. *Observations*

Observations are made using the Hitachi 32-m radio telescope of Ibaraki station, a branch of the Mizusawa VLBI Observatory of NAOJ. Integration time is 5 min. Bandwidth is 8 MHz, centered on 6668 MHz (the rest frequency of the methanol maser is 6,668.5192 MHz). This corresponds to the velocity coverage of $\sim$360 km s^{-1}. Note that the velocity coverage is not centered at $V_{\rm lsr} = 0$ km s^{-1} because the doppler corrections are not performed during the observation. The bandwidth of 8 MHz is divided into 8192 ch, resulting the velocity resolution of $\sim$0.044 km/s. The system noise temperature including atmosphere is $\sim$25 K (zenith) – $\sim$40 K (EL = 15 deg). The typical 1-sigma rms noise level is $\sim$0.3 Jy. The half-power beam width is $\sim$4.6 arcmin with the pointing accuracy better than $\sim$30 arcsec. No real-time pointing corrections are applied. In order to minimize the reduction of the measured flux density due to the pointing error, observations of each sources are executed at the same azimuth and elevation angle. Observations are made by using a position-switching method with the OFF position of ΔR.A. = 60 arcmin or -60 arcmin.

The observation have started from Dec. 2012. Observations are divided into 3 periods: (1) 2012/Dec./30 – 2015/Aug./24: 442 sources are divided into 9 groups (1,2,...,9) and each group is observed once per $\sim$9 days (2) 2015/Sep./18 – 2017/Mar./07: 154 sources showing variability are selected from the results of period (1) and divided into 4 groups (A,B,C,D). Each group is observed once per $\sim$4 days. The rest 288 source are not observed. (3) 2017/Jun./14 – now: 442 sources are observed with the hybrid of (1) and (2), i.e., using the sequence of [1ABCD2ABCD3ABCD...8ABCD9ABCD].

3. Results

We have newly identified $\sim$20 sources with periodic flux variability in addition to already known 29 sources. We have also detected 5 sources with sudden flux rises in 2019–2022, including G358.93−0.03 which was confirmed to be associated with the accretion burst.

4. Acknowledgement

This work is partially supported by the Inter-university collaborative project "Japanese VLBI Network (JVN)" of NAOJ. This study benefited from financial support from the Japan Society for the Promotion of Science (JSPS) KAKENHI program (21H01120 and 21H00032).

References

Caswell, J.L., *et al.* 2010, *MNRAS*, 404, 1029
Green, J.A., *et al.* 2010, *MNRAS*, 409, 913
Green, J.A., *et al.* 2012, *MNRAS*, 420, 3108
Pandian, J.D., *et al.* 2007, *ApJ*, 656, 255
Tanabe, Y., Yonekura, Y., & MacLeod, G.C. 2023, *PASJ*, 75, 351
Xu, Y., *et al.* 2009, *AA*, 507, 1117
Yonekura, Y., *et al.* 2016, *PASJ*, 68, 74

Group photo of the Maser Monitoring Organisation (M2O). Taken by Ka-Yiu Shum.

Chapter 5
Pulsation and Outflows in Evolved Stars

Cosmic Masers: Proper Motion toward the Next-Generation Large Projects
Proceedings IAU Symposium No. 380, 2024
T. Hirota, H. Imai, K. Menten, & Y. Pihlström, eds.
doi:10.1017/S1743921323002788

Mass Loss in Evolved Stars

Lynn D. Matthews

Massachusetts Institute of Technology Haystack Observatory, 99 Millstone Road, Westford, MA 01886 USA. email: lmatthew@mit.edu

Abstract. Intense mass loss through cool, low-velocity winds is a defining characteristic of low-to-intermediate mass stars during the asymptotic giant branch (AGB) evolutionary stage. Such winds return up ∼80% of the initial stellar mass to the interstellar medium and play a major role in enriching it with dust and heavy elements. A challenge to understanding the physics underlying AGB mass loss is its dependence on an interplay between complex and highly dynamic processes, including pulsations, convective flows, shocks, magnetic fields, and opacity changes resulting from dust and molecule formation. I highlight some examples of recent advances in our understanding of late-stage stellar mass loss that are emerging from radio and (sub)millimeter observations, with a particular focus on those that resolve the surfaces and extended atmospheres of evolved stars in space, time, and frequency.

Keywords. stars: AGB – stars: mass loss – stars: winds, outflows – masers

---

1. Introduction

Asymptotic giant branch (AGB) represent the final thermonuclear burning stage in the life of low-to-intermediate mass stars, including stars like the Sun. The AGB marks the second ascent of the red giant branch for these stars, following the depletion of their core hydrogen supply and the completion of core helium burning. The internal changes to the structure of the star during the AGB cause the effective temperature to cool to ∼2000-3000 K, while to maintain hydrostatic equilibrium, the star expands to several hundred times its previous size—reaching a diameter of several astronomical units (AU) across ($\sim 6 \times 10^{13}$ cm). At the same the resulting stellar luminosity increases to ∼5000–10,000 $L_{\odot}$. AGB stars become unstable to pulsations and typically undergo radial pulsations with periods of order 1 year, accompanied by significant changes in the visible light output of the star (as high as $\Delta m_V \sim$8 mag). A general overview of the properties of AGB stars can be found in Habing & Olofsson (2003).

A consequence of the low effective temperatures of AGB stars is that molecules and dust are able to form and survive in their extended atmospheres. Importantly, the dust that forms helps to drive copious rates of mass loss ($\dot{M} \sim 10^{-8}$ to $10^{-4} M_{\odot}$ yr^{-1}) through cool, dense, low-velocity winds ($V_{\rm outflow} \sim$ 10–20 km s^{-1}). These winds are thus over a million times stronger than the current solar wind.

The dramatic mass loss that occurs during the AGB evolutionary phase has implications for wide range of problems in astrophysics. The details of AGB mass loss (including its duration, as well as the fraction of the initial stellar mass that is shed) dramatically impact stellar evolutionary tracks, affecting the maximum luminosity a given star will reach and the type of stellar remnant that it will ultimately leave behind (e.g., Rosenfield *et al.* 2014; Kaliari *et al.* 2014). The mass lost by AGB stars accounts for $\gtrsim$50% of the dust and heavy element enrichment in the Galaxy, thus providing a primary source of raw material for future generations of star and planets (Tielens *et al.* 2005; Karakas

2010). And for extragalactic astronomy and cosmology, accurate prescriptions for AGB mass loss are crucial for stellar population synthesis calculations (e.g., Salaris *et al.* 2014; Villaume *et al.* 2015), for understanding dust production and composition in external galaxies (e.g., Narayanan *et al.* 2021), for interpreting the integrated starlight of distant galaxies (e.g., McGaugh & Schombert 2014), and for devising prescriptions of gas recycling and chemical evolution in galaxy models (e.g., Leitner & Kravtsov 2011; Gan *et al.* 2019).

This article does not attempt a comprehensive review of AGB mass loss (see instead, Höfner & Olofsson 2018; Decin 2021). Its main focus is to highlight some of the unique insights that can be gained from observations at cm and (sub)mm wavelengths that resolve AGB stars in space, time, and frequency.

2. Challenges to Understanding AGB Winds and Mass Loss

In contrast to luminous hot stars where the winds are driven by atomic line opacity (e.g., Lamers & Cassinelli 1999), AGB winds are thought to be primarily dust-driven, with radiation pressure on dust grains transferring momentum to the gas through absorption and/or scatting, resulting in material being dragged outward to power a quasi-steady wind. This basic theoretical framework for AGB winds was established roughly half a century ago (e.g., Wickramasinghe *et al.* 1966; Kwok 1975). However, despite decades of effort, we still lack a complete and fully predictive theory of AGB mass loss (see Höfner & Olofsson 2018).

To first order, dust driving appears to work relatively well for subsets of AGB stars with carbon-rich atmospheres (C/O>1), as the carbonaceous grains that are present tend to have high opacity to stellar radiation, enabling efficient momentum transfer and wind driving. However, more generally, this model has limitations. For example, growing empirical evidence suggests that real AGB winds may often deviate significantly from the idealized picture of steady, spherical symmetric outflows (e.g., Nhung *et al.* 2015; Le Bertre *et al.* 2016; Decin *et al.* 2020). Furthermore, the majority of AGB stars have oxygen-rich chemistries (C/O<1), and the silicate-rich grains that form in their extended atmospheres generally have insufficient infrared opacity to drive the winds with the efficiency needed to account for the observed mass-loss rates. Höfner *et al.* (2016) showed that the effects of photon scattering may help to alleviate this problem. Nonetheless, a persistent conundrum is that grains require sufficiently cool temperatures ($\sim$1000–1500 K) and low densities to form and survive, but such conditions are typically not reached interior to $r \sim 2-3R_{\star}$ (i.e., $r \sim$6–7 AU) around a typical AGB star. Thus some additional process is required to transport material from the stellar "surface" into the wind launch region.

It is now widely believed that pulsation and/or convection play key roles in facilitating AGB mass loss (e.g., Willson & Bowen 1985; Höfner 2016; McDonald *et al.* 2018). In broad terms, the interplay between pulsation and convection produces shock waves in the extended atmosphere, pushing gas outward; dust formation subsequently occurs in the wake of the shock; and finally, radiation pressure on the resulting grains drags material outward to power the wind (see Figure 2 of Höfner & Olofsson 2018). However, the underlying physics is highly complex, and many details are poorly understood and poorly constrained observationally.

3. Insights from Studies of Large-scale Circumstellar Ejecta

For decades, a primary means of studying AGB mass loss has been through observations of the spatially extended circumstellar envelopes (CSEs) of chemically enriched gas and dust that are a ubiquitous feature of these stars. These CSEs may be observed using a wide variety of tracers, including molecular line emission, such as CO (Knapp *et al.* 1998; De Beck *et al.* 2010) or other thermal lines (Patel *et al.* 2011; Claussen *et al.* 2011); far-infrared emission from dust (Young *et al.* 1993; Cox *et al.* 2012), and in some cases, scattered optical light (Mauron & Huggins 2006); far-ultraviolet continuum (Martin *et al.* 2007; Sahai & Stenger 2023), or H I 21-cm line emission from atomic hydrogen (Gérard & Le Bertre 2006; Matthews *et al.* 2013). Historically, AGB CSEs were typically envisioned and modeled as spherically symmetric shells, but many of the aforementioned studies show clearly that CSE morphologies can be extraordinarily diverse. Depending on the age of the central star, its mass-loss rate, and the particular observational tracer, the observed extent of the CSE can range from tens of thousands of AU to a parsec or more, and properties of the CSE can be dramatically shaped by the presence of (sub)stellar companions (Maercker *et al.* 2012; Aydi & Mohamed 2022) or the star's motion through the surrounding interstellar medium (e.g., Cox *et al.* 2012; Martin *et al.* 2007; Villaver *et al.* 2012; Matthews *et al.* 2013).

Global studies of CSEs supply a wide array of fundamental information on the mass-loss properties of evolved stars, including measurements of the mass-loss rate and outflow speed. In addition, they can provide clues on the nature of the central star (age, temperature, initial mass), the timescale of the mass-loss history, and the mass-loss geometry (spherical, bipolar, etc.). Despite the long history of studies of AGB CSEs, observations using the latest generation of radio telescopes continue to yield new insights and surprises. One recent example is the ATOMIUM project†, an Atacama Large Millimeter/submillimeter Array (ALMA) Large Project that targeted a sample of AGB stars and red supergiants in the 214–270 GHz range with the goal of obtaining a better understanding of the chemical and physical processes that govern red giant winds (Decin *et al.* 2020; Gottlieb *et al.* 2022). Results to date show that asphericity appears to be the norm among AGB ejecta and that there is a correlation between the morphology of AGB ejecta and the current mean mass-loss rate. This program has also added to growing evidence that long-period companions ($P >$1 yr) commonly play a role in shaping CSEs, and that a common mechanism controls the wind morphology of both AGB stars and planetary nebulae (Decin *et al.* 2020).

Another Large ALMA Project aimed at studying AGB ejecta is DEATHSTAR‡, which has used the ALMA Compact Array to obtained spatially resolved CO measurements and line profiles for a sample of ~70 chemically diverse AGB stars. Results to date show that large-scale asymmetries and complex velocity profiles are common. Future radiative transfer modeling is underway to determine accurate mass-loss rates and temperature distributions of the gas for the sample (Ramstedt *et al.* 2020; Andriantsaralaza *et al.* 2021). Meanwhile, the NESS§ program has been conducting a volume-limited survey of ~850 evolved stars in CO and in the sub-mm continuum using the APEX and JCMT telescopes, with the goal of measuring outflow parameters, gas-to-dust ratios, and other information critical for characterizing the mass-loss histories of a large sample of stars (Sciclula *et al.* 2022). It is worth emphasizing that single-dish projects like NESS remain a valuable complement to interferometric surveys such as ATOMIUM and DEATHSTAR, owing to their ability to target larger samples of stars and to characterize spatially extended and diffuse molecular emission in CSEs which can be resolved out in interferometric measurements.

4. Advances in Atmospheric Modeling of AGB Stars

The complex physics of AGB star atmospheres makes modeling them both challenging and computationally expensive. Approximations of local thermodynamic equilibrium

† `https://fys.kuleuven.be/ster/research-projects/aerosol/atomium`
‡ `https://www.astro.uu.se/deathstar/index.html`
§ `https://evolvedstars.space`

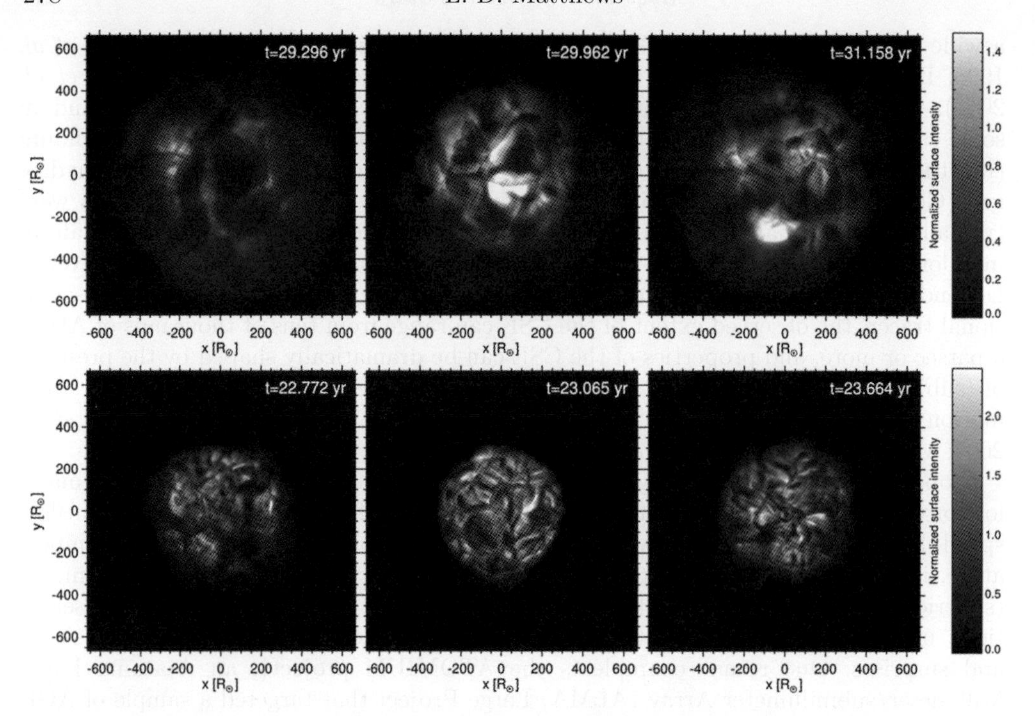

Figure 1. Time sequences of bolometric surface intensity for AGB stars from the 3D CO5BOLD hydrodynamic models of Freytag & Höfner (2023). A 1.0 $M_\odot$ model is shown in the top row and a 1.5 $M_\odot$ model in the bottom row. Snapshots are spaced by 8 and 14 months (top) and 3.5 and 7 months (bottom), respectively, from the starting frame. The size of each box is ~5.6 AU across.

(LTE) break down in the dynamic, time-varying conditions of AGB atmospheres, and a wide range of physics needs to be included (pulsation, convection, dust formation, etc.) to produce meaningful results. Furthermore, because of the enormous spatial extents of AGB star atmospheres and outflows, the relevant spatial scales required in the model can span many orders of magnitude, ranging from the scales of shock regions and sub-surface convective cells ($\ll R_\star$) to scales of $> 1000 R_\star$ ($> 10^{16}$ cm) as required to fully trace the evolution of temperature, density, and composition of the wind. Until recently, these challenges have often meant relying on 1D models with simplified physics (e.g., Ireland *et al.* 2011; Liljegren *et al.* 2018). While instructive for some applications, these models have important limitations. For example, since the effects of convection are inherently 3D, the result is a blurring of the distinction between pulsation and convective processes in 1D models (see Freytag & Höfner 2023). Fortunately, computational advances have begun enabling sophisticated new 3D radiation-hydrodynamical models that are able to overcome such limitations by incorporating a wide range of relevant physics, including radiative transfer, frequency-dependent opacities, pulsation, convection, shocks, dust formation, grain growth and evaporation, and wind acceleration (e.g., Freytag *et al.* 2017; Freytag & Höfner 2023).

Figure 1 shows two time sequences of images of bolometric surface intensity from hydrodynamic simulations of AGB stars from Freytag & Höfner (2023). The frames separated by a few months. Both models have a similar luminosity, but the $1 M_\odot$ model (top) exhibits a lower surface gravity, a more extended atmosphere, and more efficient dust formation, while the $1.5 M_\odot$ model (bottom) displays a smaller radius, a better defined surface, and smaller, more granular surface features. While we may not yet have a fully predictive theory of stellar mass loss, models such as these are now giving us incredibly detailed predictions that can be confronted with observations.

5. Zooming into the Action

Studies of the large-scale CSEs (Section 3) remain an invaluable tool for characterizing AGB mass loss. However, directly confronting the types of highly detailed models described above, and solving many of the outstanding puzzles related to the launch and geometry of AGB winds and their relationship to stellar pulsations, shocks, convection, and other dynamic phenomena, demands additional types of observations. In particular, there is a need for observations that:

(1) *Spatially* resolve the stellar atmosphere on relevant physical scales (i.e., $r \lesssim 10$ AU, or even $r \ll R_\star$) to probe the photosphere, surrounding molecular layers, and the dust formation zone.
(2) *Temporally* resolve processes on relevant dynamical timescales (which for AGB stars can range from days to years to decades).
(3) *Spectrophotometrically* distinguish different layers of the atmosphere to trace changes in physical conditions and chemistry.
(4) Directly measure gas motions.

Fortunately, observations at radio (cm through sub-mm) wavelengths using the latest generation of radio telescopes provide a variety of means to achieve these objectives. The remainder of this review highlights some examples of recent progress in our understanding of AGB mass-loss through radio observations of both molecular masers and thermal continuum emission that resolve the dynamic behavior of AGB atmospheres in space and/or time. I close by briefly highlighting the prospects of future observational facilities for additional progress in these areas.

6. Molecular Masers as a Tool for Understanding AGB Mass Loss

The first detection of molecular maser emission associated with the extended atmosphere of an evolved star was made by Wilson & Barrett (1968), who detected masing OH lines at 1612, 1665, and 1667 MHz from the red supergiant NML Cyg using the Green Bank 140-ft telescope. This was followed a few years later by the detection of H_2O (at 22.2 GHz) and SiO (at 43.1 GHz) masers, respectively, in other O-rich red giants (Sullivan 1973; Thaddeus *et al.* 1974). Today, molecular masers from various transitions and isotopologues of OH, H_2O, and SiO have been detected in hundreds of O-rich Galactic AGB stars and red supergiants (e.g., Engels & Lewis 1996; Pardo *et al.* 1998; Kim *et al.* 2010; Rizzo *et al.* 2021), providing a unique resource for probing the gas dynamics and physical conditions of their atmospheres. While C-rich AGB stars generally do not give rise to masers from O-bearing species, they may show maser activity from other molecules, including HCN or SiS (Henkel 1983; Omont *et al.* 1989; see also Section 6.3 below).

More general overviews of the topic of stellar masers can be found in e.g., Humphreys & Gray (2004); Kemball (2007); Colomer (2008); Gray 2012; and Richards (2012). Here I highlight just a few examples of recent results that illustrate the role of maser studies for addressing the objectives outlined in Section 5. I focus primarily on SiO masers, which tend to arise inside the wind launch region of AGB stars and close to the dust formation zone. In contrast, H_2O and OH masers tend to arise at successively larger radii ($\gtrsim 10^{14}$ cm and $\gtrsim 10^{15}$ cm, respectively), beyond the wind launch zone (e.g., Dickinson 1978). Different transitions and isotopologues of SiO are further segregated according to the specific combinations of temperature and density that are necessary to produce masing in each respective line. In addition to their favorable location in the atmosphere, the compact sizes and high brightness temperatures (often $> 10^6$ K) of SiO masing regions provide the advantage of enabling observations of the masers with extraordinarily high angular resolution (<1 mas) using very long baseline interferometry (VLBI) techniques. For example, the longest baseline of the Very Long Baseline Array (VLBA) of ∼8600 km gives an angular resolution of ∼0.5 mas at 43 GHz, corresponding to a spatial resolution of ∼0.1 AU ($\sim 0.05 R_\star$) for a star at 200 pc.

6.1. *Spatial Distributions*

Thanks to VLBI studies of SiO masers in a number of AGB stars that have been undertaken since the 1990s, it is now well established that SiO masers in AGB stars are typically found to lie (in projection) in ring-like or partial ring-like structures, with a mean radius of roughly twice that of the stellar photosphere (e.g., Diamond *et al.* 1994; Cotton *et al.* 2006; Imai *et al.* 2010). Evidence for spatial segregation is observed between different SiO transitions and isotopologues, allowing them to be used as probes of changes in physical conditions and gas kinematics over scales $\ll R_\star$ (e.g., Desmurs *et al.* 2000; Wittkowski *et al.* 2007). This information also provides important constraints on maser pumping models (e.g., Humphreys *et al.* 2002; Gray *et al.* 2009). Unfortunately, a persistent challenge has been that owing to bandwidth limitations of previous generations of instruments, it was generally not possible to observe different transitions strictly simultaneously. This, coupled with the typical use of self-calibration procedures (which can erase absolute astrometric information), meant that there has historically been uncertainty regarding the astrometric alignment between different transitions, as well as their locations relative to the central star. While approximate methods can be used to align the different measurements (e.g., Desmurs *et al.* 2000), in many cases, lingering uncertainties can be as high as several mas and potentially result in ambiguities in the interpretation (see, e.g., Soria-Ruiz *et al.* 2004).

One example of important progress in overcoming this challenge has been achieved using the Korean VLBI Network (KVN) Multi-Band System (Han *et al.* 2008) together with the so-called frequency phase transfer (FPT) technique (Dodson *et al.* 2014). This approach has enabled simultaneous observations of up to five SiO and H_2O transitions in several evolved stars (e.g., Dodson *et al.* 2014; Yoon *et al.* 2018; Kim *et al.* 2018). Currently, maximum KVN baselines are ∼450 km. However, future extension of this technique to longer baselines would be highly desirable to achieve even finer resolution of individual maser-emitting clumps and to enable improved astrometric precision for following their proper motions over time.

6.2. *Magnetic Field Measurements*

Another type of investigation that is possible through observations of stellar masers, including SiO masers, is the study of polarization and magnetic fields in circumstellar environments (see the overview by Vlemmings *et al.* 2012). For example, in a VLBA study of SiO masers in the OH/IR star OH 44.8-2.3, Amiri *et al.* (2012) measured linear polarization of up to 100% in individual maser clumps, enabling them to map out the magnetic field vectors surrounding the star. For the brightest maser clump they also found evidence of circular polarization, enabling estimation of the magnetic field strength (1.5±0.3 G). Intriguingly, both the distribution of the SiO maser clumps and the orientation of the magnetic field vectors surrounding this star point to a preferred outflow direction for the stellar wind, hinting that a dipole magnetic field may play a role in shaping and defining the outflow. These findings in support of a non-spherically symmetric outflow complement other recent findings pointing to similar trends based on

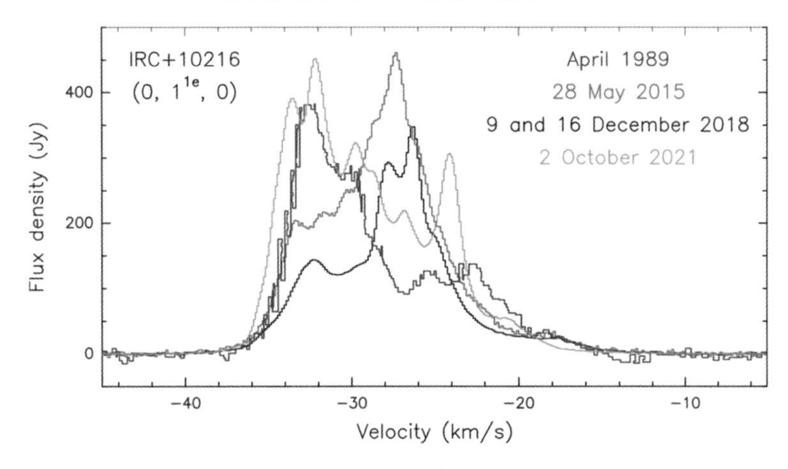

Figure 2. Observed variability of the HCN $(0,1^{1e},0)$ $J=2-1$ maser at 177.2 GHz in the carbon star IRC+10216. The different colored lines show the results from different observing dates. From Jeste *et al.* (2022).

the study of molecular line emission in AGB stars on larger scales (e.g., Decin *et al.* 2020; Hoai *et al.* 2022b; Winters *et al.* 2022).

6.3. *Masers in Carbon-rich AGB Stars*

As noted above, SiO masers are generally absent in AGB stars with carbon-rich chemistries. However, masing action in C-type stars has been observed in a few other species, including HCN (e.g., Omont *et al.* 1989; Izumiura *et al.* 1987; Bieging 2001; Menten *et al.* 2018). HCN is the most common molecule in the atmospheres of C stars after H_2 and CO, although its masing properties have been relatively little studied to date compared with SiO masers in O-rich stars.

If we wish to study the inner regions of the CSEs of C-type stars using HCN masers (in a manner analogous to what is possible using SiO masers in O-type AGB stars) it is helpful to target higher J, vibrationally-excited states of HCN, where the opacity is lower. Studies of these transitions have been limited until now, owing to a dearth of observational facilities equipped with receivers covering the necessary frequency range ($\nu >$176 GHz). However, recently this has begun to change. For example, Jeste *et al.* (2022) surveyed a sample of 13 C-type stars using the APEX telescope and bands centered at 180 GHz, 230 GHz, and 345 GHz, respectively, providing access to 26 different HCN transitions. Masing was observed in several different transitions, including the HCN $(0,1^{1e},0)$ $J=2-1$ (ν_0=177.2 GHz) line, which was detected in 11 targets, suggesting that it is a common feature of carbon stars. Furthermore, the observed velocity extents of theses masers indicate that the lines are originating in the acceleration zone where dust is forming, implying they have the potential to serve as an important new diagnostic tool for the study of wind launching in C-type AGB stars. For stars with multi-epoch observations, clear changes in the line profile were seen with time (e.g., Figure 2), including in some cases, over the course of only a few days.

6.4. *The Time Domain*

6.4.1. Global Measurements

Adding a temporal dimension of maser studies significantly expands what we can learn about the time-varying atmospheres of AGB stars compared with single-epoch observations. As noted in Section 1, radial pulsations are a defining characteristic of

AGB stars, and these commonly have periods of order 1 year. These are accompanied by changes in the visible light output of the star by factors of up to a thousandfold (e.g., Reid & Goldston 2002 and references therein). In the case of SiO masers, it has long been recognized that variations in the SiO line profiles are correlated with the pulsation cycle. For example, for Mira-type variables, the study of Pardo *et al.* (2004) found a correlation between the integrated intensity of the SiO v=1, $J = 1 - 0$ line at 43.1 GHz and both the infrared and optical light curves. The masers tend to vary in phase with the infrared, but lag the optical light curve phase by $\sim$0.05 to 0.2. The authors cited this as evidence that the SiO masers must be radiatively pumped. However, secular variations of the SiO masers not obviously linked with the pulsation cycle were also seen.

While there have been numerous time-domain studies of SiO and other stellar masers over the past few decades, our ability to fully exploit and interpret the results for understanding AGB star atmospheres and mass-loss (as well as the underlying maser physics) has been hampered by several limitations of these studies. Among these are: (i) limited instrumental bandwidths (which have precluded simultaneously monitoring multiple maser transitions); (ii) limited spectral resolution (which obscures the complex velocity structure of the line and may make it impossible to discern subtle changes with time); (iii) limited signal-to-noise ratios (preventing the detection of weaker lines); (iv) sample selection biases [e.g., exclusion of semi-irregular and irregular (non-Mira-type) variables]; (v) and monitoring programs which are either short-lived (a few years or less) and/or sparsely sampled (observing cadences of $\geq$1 month), thus producing observations which are unable to sample all relevant dynamical timescales for the stars. Fortunately, recent progress has been made in nearly all of these areas.

One example of the power of simultaneously monitoring multiple lines with a wide frequency band is the study by Rizzo *et al.* (2021), which surveyed 67 O-rich AGB stars and red supergiants between λ7 mm and λ1 mm, targeting SiO rotational transitions between $J = 1 - 0$ and $J = 5 - 4$, vibrational numbers v=0 to 6, and 3 different isotopologues (^{28}SiO, ^{29}SiO, and ^{30}SiO). This study resulted in the detection of several new SiO lines in many of the targets, thus revealing the fascinating complexity of their multi-wavelength SiO spectra. Among these was first detection of an SiO v=6 line. Additionally, dramatic variations in the line profiles of some targets were seen on timescales as short as $\sim$2 weeks.

Evidence for SiO maser variability over even shorter timescales was reported by Gómez-Garrido *et al.* (2020). These authors performed daily monitoring of a sample of 6 stars and found evidence of rapid ($\sim$1 day) intensity variations of $\sim$10–25% in multiple SiO lines in two semi-regular variables (RX Boo and RT Vir). Similar variations were not seen in the Mira-type variables in the sample. The authors postulated that the semi-regular variables may have intrinsically smaller maser-emitting clumps and more chaotic shock behaviors in their atmospheres. However, high-cadence VLBI monitoring observations of semi-regular variables will be needed to test these ideas and improve our understanding of these phenomena.

6.4.2. Spatially and Temporally Resolved Imaging Spectroscopy

As described above, multi-epoch maser measurements provide important insights into the time-varying behavior of AGB star atmospheres. When this is combined with spatially resolved measurements (particularly with VLBI resolution), our ability to interpret the results in a physically meaningful way is significantly enhanced (e.g., Richards *et al.* 1999; Gray & Humphreys 2000; Phillips *et al.* 2001; Wittkowski *et al.* 2007).

Undoubtedly one the most spectacular examples of the power of spatially resolved maser monitoring observations for the study of evolved stars is the 78-epoch study of the

SiO v=1, $J = 1 - 0$ masers in the Mira variable TX Cam undertaken by Gonidakis *et al.* (2013). Using the VLBA, these authors observed the star on a ~2–4 week cadence over the course of nearly 5 years, resulting in a dramatic "movie" of the star's evolving atmosphere. This study confirmed that the proper motions of SiO maser clumps can be used to trace gas motions close to the stellar photosphere, revealing both the expansion and infall of gas. The width and boundary of the SiO maser "ring" in TX Cam (actually a shell seen in projection) was found to vary with stellar pulsation phase, and evidence for the creation of shocks with velocity of ~7 km s^{-1} was observed during each pulsation cycle. These shocks in turn affected the intensity and variability of the masers. Importantly, the TX Cam observations showed no evidence of strong shocks (>10 km s^{-1}), in agreement with past analysis of radio continuum light curves of other AGB stars (Reid & Menten 2007; see also Section 7). This supports a model where stronger shocks are damped by the time they reach $r \sim 2R_{\star}$. Additionally, the distribution and velocity structure of the masers are strongly suggestive of a bipolar outflow (Figure 3), adding to evidence that such geometries are in fact commonplace for AGB mass loss (see also Sections 3 and 6.2). Although the TX Cam movie is now a decade old, it is worth highlighting again here to emphasize the incredible scientific richness of data sets of this kind for understanding the physics of AGB star atmospheres and to underscore the importance of undertaking similar observations in the future for additional AGB stars spanning a range of properties.

7. Studies of Radio Photospheres

Despite the many advantages of masers for probing the atmospheric properties and mass-loss physics of evolved stars, such studies suffer from certain limitations. For example, maser emission is not observed in all AGB stars, and for some AGB stars, the maser emission may become at times too weak to detect. In addition, the interpretation of changes in the maser emission over time can be challenging in cases where the spatial distribution of the maser clumps is not spatially resolved, or where only a single observing epoch is available. In such instances, it can be difficult to distinguish changes resulting from varying physical conditions (e.g., changes in temperature or density) from changes caused by motions of the maser-emitting gas. Fortunately, recent advances in other observational techniques can help to provide complementary information, including observations of thermal continuum emission from the atmospheric region known as the *radio photosphere*.

The existence of so-called radio photospheres in AGB stars was first established by Reid & Menten (1997). These authors examined a sample of nearby AGB stars and found that the flux densities at cm through far-infrared wavelengths were systematically higher than predicted from a simple blackbody model based on the known stellar effective temperatures. This led Reid & Menten to postulate that the stars must have an optically thick layer (i.e., a radio photosphere) lying at $r \sim 2R_{\star}$. They developed a model for the radio photosphere in which the opacity arises primarily from interactions of free electrons with neutral H and H_2. For a typical O-rich AGB star, the τ=1 surface of the radio photosphere lies at r ~2–3 AU and the spectral index of the emission is slightly shallower than a blackbody ($\alpha \approx 1.86$).

Figure 4 shows schematically where the radio photosphere lies in relation to the other atmospheric layers in a typical O-rich AGB star. Crucially, the radio photosphere resides in the zone between the classical photosphere and the wind launch region. A consequence is that the properties of the radio photosphere will be impacted by the shocks, pulsation, convection, and other key physical processes that are believed to be responsible for helping to transport material into the wind launch region at $r \sim 10R_{\star}$ (Reid & Menten 1997; Gray *et al.* 2009; see also Section 2).

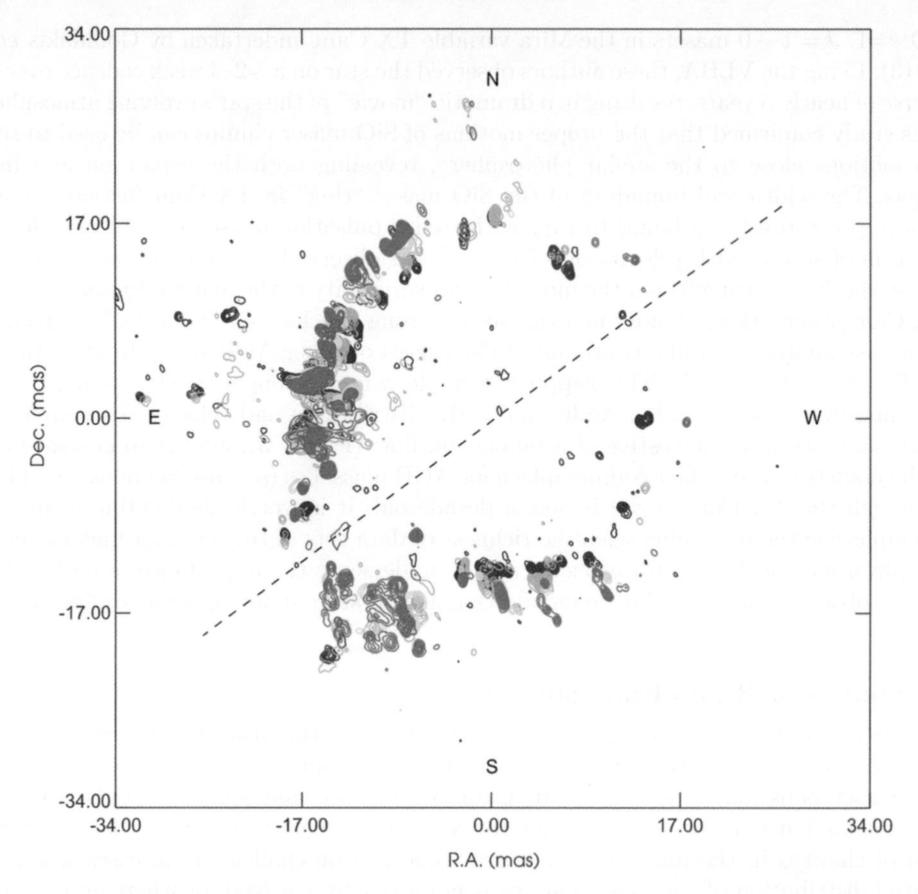

Figure 3. Contour maps of the velocity-integrated SiO v=1, $J=1-0$ maser emission in TX Cam, as observed during four different epochs over the span of ~7 months. Data from each epoch are indicated by a different color. Based on the proper motions of the maser spots between epochs, it is apparent that the expansion velocity is higher along the SE-NW axis (dashed line) compared with the NE-SW axis, indicating a bipolar geometry. From Gonidakis *et al.* (2013).

The emission from radio photospheres is thermal, and its brightness temperature is too low to be studied at ultra-high angular resolution using VLBI techniques. However, the radio photospheres of nearby AGB stars ($d \lesssim 200$ pc) can be resolved with the longest baseline configurations of the Very Large Array (VLA) and ALMA. Using λ7 mm observations obtained with the legacy VLA, Reid & Menten (2007) and Menten *et al.* (2012) produced the first spatially resolved images of the radio photospheres of 4 nearby AGB stars (the O-rich stars Mira, W Hya, R Leo, and the carbon star IRC+10216, respectively). The three O-rich stars also exhibit SiO masers, and Reid & Menten (2007) used simultaneous observations of the λ7 mm continuum and the SiO maser emission to establish unambiguously for the first time that the SiO masers are distributed in a shell exactly centered on the stellar photosphere.

Another key finding to emerge from the above studies was that some of the radio photospheres showed clear evidence for deviation from spherical symmetry. However, with only a single measurement epoch, it was impossible to discern whether these shapes were static or time-varying. Taking advantage of the order-of-magnitude boost in continuum sensitivity of the upgraded Karl G. Jansky VLA, Matthews *et al.* (2015, 2018) reobserved the stars studied by Reid & Menten (2007) and Menten *et al.* (2012). The resulting observations confirmed that asymmetric shapes are a common feature of radio photospheres.

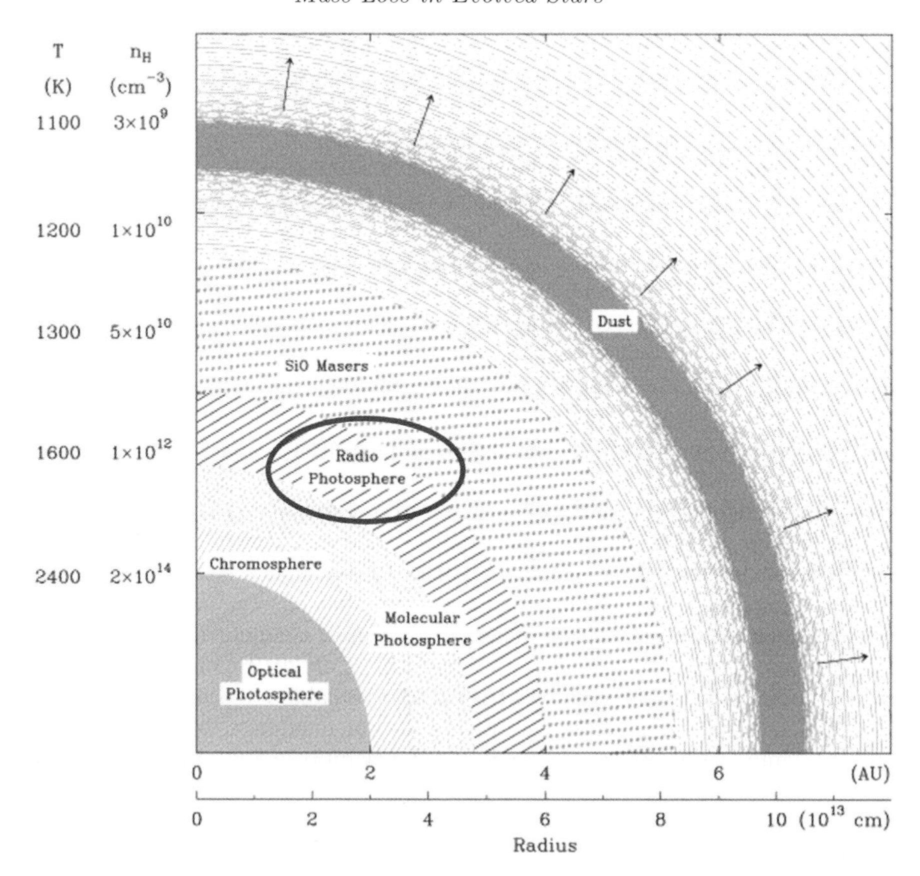

Figure 4. Schematic cross-section of the atmospheric layers in a typical O-rich AGB star. A radio photosphere lies at $\sim 2R_\star$, just interior to the dust formation zone and wind launch region. The properties of the radio photosphere are therefore susceptible to the underlying physical processes that help to launch the wind, including pulsation, convective flows, and shocks. The radio photosphere is also adjacent to the region that gives rise to SiO maser emission in many AGB stars. Adapted from Menten & Reid (1997).

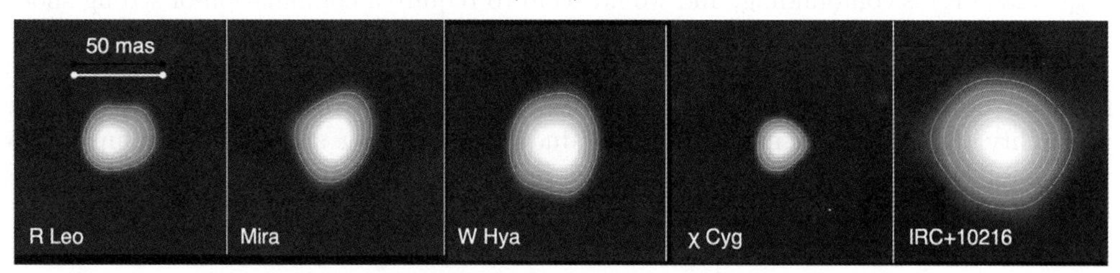

Figure 5. Spatially resolved images of radio photospheres of nearby AGB stars at λ7 mm, obtained using the Jansky VLA. The images were produced using RML imaging techniques, which enabled a modest level of super-resolution (see text). Adapted from Matthews *et al.* (2018) and Matthews *et al.* in prep.

Furthermore, secular shape changes were discernible in observations taken several years apart. This latter finding suggests that the observed non-spherical shapes most likely result from a combination of pulsation and/or convective effects rather than rotation or the tidal effects of a companion.

As part of their analysis, Matthews *et al.* (2018) showcased how the interpretation of marginally spatially resolved observations of radio photospheres can be further enhanced through the application of a class of radio imaging techniques known as regularized maximum likelihood (RML) methods. These imaging algorithms have been exploited recently to meet the challenges of VLBI imaging at mm wavelengths using sparse arrays, where traditional CLEAN deconvolution tends to perform poorly (see overview by Fish *et al.* 2016). However, many of the same challenges apply to stellar imaging, and applying RML methods to their λ7 mm VLA data, Matthews *et al.* found that it was possible to achieve robust, super-resolved images with resolution as fine as $\sim 0.6\times$ the diffraction limit. This enabled clearly discerning evidence of brightness asymmetries and non-uniformities in the radio photospheres observed with VLA resolution which were not visible in CLEAN images (Figure 5). The observed photospheric features appear qualitatively consistent with the giant convective cells originally predicted to occur in red giant atmospheres by Schwarzschild (1975) and that are seen in the bolometric intensity images produced by recent 3D hydrodynamic simulations (e.g., Figure 1). The formation and dissipation of these cells is suspected of playing an important role in AGB mass loss (e.g., Höfner & Olofsson 2018).

The wavelength dependence of the opacity in radio photospheres (Reid & Menten 1997) implies that shorter wavelengths probe successively deeper layers of the atmosphere. This means that the different wavelength coverages of the VLA and ALMA are highly complementary for the study of radio photospheres, and that observations of a given star at multiple wavelengths can be used to measure the run of temperature with depth in its atmosphere (e.g., Matthews *et al.* 2015; Vlemmings *et al.* 2019; O'Gorman *et al.* 2020). The higher frequencies available at ALMA are also valuable for providing an additional boost in angular resolution. While the VLA's 35 km maximum baselines and highest frequency (λ7 mm) receivers provide a FWHM resolution of $\theta \sim$40 mas (sufficient to marginally resolve nearby AGB stars within $d \lesssim$200 pc), the combination of ALMA's longest baseline configuration (16 km maximum baselines) and Band 7 (λ0.89 mm) receiver can achieve $\theta \sim$12–20 mas, sufficient to supply several resolution elements across a nearby AGB star.

Using one such high-resolution ALMA data set at λ0.89 mm, Vlemmings *et al.* (2017) reported evidence for a "hot spot" on the surface of the AGB star W Hya which they interpreted as evidence for a pocket of chromospheric gas with a brightness temperature $T_B >$53,000 K. The presence of such hot plasma associated with such a cool star ($T_{\rm eff} \approx$2300 K) is confounding, and would seem to require a combination of strong shock heating and long post-shock cooling times, at odds with current pulsation and convection models. On the other hand, a re-analysis of the same data by Hoai *et al.* (2022a) seem to show no evidence for the presence of this hot spot on W Hya, suggesting the possibility that its origin may have been due to an imaging artifact. Follow-up observations of W Hya and other similar stars are clearly of interest to investigate these findings.

8. Prospects for the Study of AGB Mass Loss with Next Generation Radio Arrays

Section 7 described several examples of recent results that illustrate what is possible to achieve from spatially resolved imaging of the thermal continuum of evolved stars at cm and (sub)mm wavelengths using current state-of-the-art observational facilities. While these results are both groundbreaking and scientifically valuable, we can anticipate an enormous leap in such capabilities in the coming decade thanks to planned next-generation radio facilities, including the Next Generation Very Large Array (ngVLA; Murphy *et al.* 2018) and the Square Kilometer Array (SKA; e.g., Schilizzi 2004; Braun *et al.* 2019).

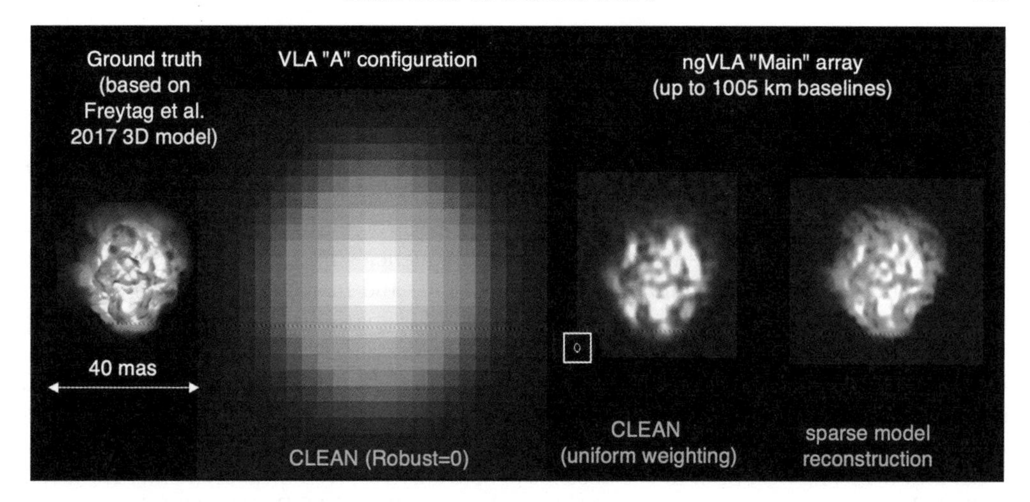

Figure 6. Simulated observations of a radio photosphere at λ7 mm (46 GHz) for an AGB star at d=200 pc. A bolometric surface intensity model from Freytag *et al.* (2007; left) was used as a proxy for the expected appearance of the thermal radio emission. The second image shows a simulated observation of the model with the current VLA 'A' configuration (35 km maximum baselines). The right two panels show 1-hour simulated observations with the ngVLA Main Array (1000 km maximum baselines), imaged using two different methods: traditional CLEAN, and a regularized maximum likelihood (sparse modeling) method. For additional information, see Akiyama & Matthews (2019).

The ngVLA will be built in the United States and Mexico, and its "Main Array" is expected to have ~218 dishes of 18 m diameter spread over an area several hundred km across. With its combination of frequency coverage (1.2–116 GHz), thermal sensitivity (~0.2–0.7 μJy beam^{-1} hr^{-1}), and angular resolution (~1 mas at 100 GHz), the ngVLA will be a game-changer for stellar imaging (e.g., Figure 6) and for the study of evolved stars and their CSEs over all relevant spatial scales, ranging from $\ll R_\star$ to $\gtrsim 10^6 R_\star$ (Matthews & Claussen 2018; Carilli *et al.* 2018; Akiyama & Matthews 2019).

At the highest angular resolutions of the ngVLA Main Array, some examples of science related to AGB stars and their mass loss that will be enabled include:

• The ability resolve radio surfaces to beyond $d \gtrsim 1$ kpc (thus expanding samples of resolved AGB stars by ×300).

• Resolution of radio surfaces over two decades in frequency for nearby stars ($d \lesssim 200$ pc).

• Simultaneous, astrometrically registered studies of photospheric continuum and multiple maser lines.

• Ability to undertake detailed comparison with (contemporaneous) optical/infrared images from facilities such as CHARA and the VLT (see Paladini *et al.* 2018; Ridgway *et al.* 2019).

One of the most exciting prospects of the ngVLA for stellar science will be the ability for the first time, to make "movies" of the evolving radio photospheres of nearby stars over the course of their pulsation cycles and to quantitatively characterize changes in stellar properties over time (Akiyama & Matthews 2019). Currently, images of radio photospheres made with the VLA and ALMA have insufficient angular resolution and imaging fidelity to discern subtle changes in parameters such as stellar radius and brightness temperature with time, or to chronicle the evolution of surface features that are predicted to occur over timescales of weeks or months (see Figure 1; see also, e.g., Figure 3 of Freytag *et al.* 2017). However, as shown by Akiyama & Matthews (2019), this will change dramatically with the ngVLA. Indeed, time-lapse movies of the thermal emission should provide exquisite levels of detail comparable to what can now be seen in time-lapse movies of SiO masers with VLBI resolution (cf. Section 6.4.3). Furthermore, it is worth noting that simultaneous studies of both thermal and maser emission should provide unprecedented levels of detail for helping to reveal further insights into the mass-loss process of these dynamic and fascinating stars.

The SKA mid-frequency array will be built in the Karoo desert of South Africa, and its initial design (SKA1-Mid) is expected to cover frequencies from 350 MHz to 15.4 GHz. Because of its shorter maximum baselines ($\leq$150 km) and more limited frequency coverage, SKA1-Mid will not be able to rival the ngVLA for spatially resolved stellar imaging, though it will be able to moderately resolve radio photospheres at $d \lesssim 200$ pc. In addition, it will be a powerful tool for obtaining sensitive radio light curves of hundreds of evolved stars. As shown by Reid & Menten (1997; see also Reid & Goldston 2002), the measurement of radio light curves supplies valuable information on the amplitudes of shocks in AGB star atmospheres, even in the absence of spatially resolved measurements. However, radio light curves are currently available for only a handful of AGB stars. Obtaining useful light curves requires a combination of good sensitivity (typical flux densities are $\lesssim$1 mJy in cm bands), accurate calibration (better than $\sim$10%), and both frequent and long-term temporal sampling (every 1–2 weeks over timescales of many months). For these reasons, such measurements are technically and logistically challenging with current arrays. However, SKA should be able to produce the most accurate radio light curves for AGB stars to date, with quasi-simultaneous coverage across a wide range of frequencies (see also Marvel 2004).

9. Summary

The mass loss that occurs during the late stages of stellar evolution has implications for a wide range of problems in astrophysics. However, we do not yet have a complete and fully predictive theory. Sophisticated new 3D hydrodynamic models for AGB star atmospheres are now available—including pulsation, convection, and other vital physics—that make highly detailed predictions that can be confronted with observations. However, rigorously testing such models requires access to observations that spatially resolve the stellar atmosphere and wind launch region on scales ($r \lesssim 10$ AU), and temporally resolve relevant dynamical timescales (which can span days to months to years). In this review, I have highlighted examples of recent cm and (sub)mm wavelength observations, including observations of molecular masers and thermal continuum emission, that are making progress in these areas and helping to advance our understanding of late-stage stellar mass loss. Even greater advances are anticipated in the next decade when a new generation of radio telescopes, including the ngVLA and SKA, comes online.

Acknowledgment

This work was supported by award AST-2107681 from the National Science Foundation.

References

Akiyama, K. & Matthews, L. D. 2019, ngVLA Memo 66, https://library.nrao.edu/public/memos/ngvla/NGVLA_66.pdf

Amiri, N., Vlemmings, W. H. T., Kemball, A. J., & van Langevelde, H. J. 2012, *A&A*, 538, A136

Andriantsaralaza, A., Ramstedt, S., Vlemmings, W. H. T., *et al.* 2021, *A&A*, 653, 53

Aydi, E. & Mohamed, S. 2022, *MNRAS*, 513, 4405

Bieging, J. H. 2001, *ApJ*, 549, 125

Braun, R., Bonaldi, A., Bourke, T., Keane, E., & Wagg, J. 2019, eprint arXiv:1912.12699
Carilli, C. L., Butler, B., Golap, K., Carilli, M. T., & White, S. M. 2018, in: E. Murphy (ed.), Science with a Next Generation Very Large Array, Astronomical Society of the Pacific Conference Series, 517, (San Francisco:ASP), p. 369
Claussen, M. J., Sjouwerman, L. O., Rupen, M. J., *et al. ApJ* (Letters), 739, L5
Colomer, F. 2008, *PoS*, id.34, `http://pos.sissa.it/cgi-bin/reader/conf.cgi?confid=72`
Cotton, W. D., Vlemmings, W. H. T., Mennesson, B., *et al.* 2006, *A&A*, 456, 339
Cox, N. J. L., Kerschbaum, F., van Marle, A.-J., *et al.* 2012, *A&A*, 537, 35
De Beck, E., Decin, L, de Koter, A. *et al.* 2010, *A&A*, 523, 18
Decin, L. 2021, *ARA&A*, 59, 337
Decin, L., Montargès, M., Richards, A. M. S., *et al.* 2020, *Science*, 369, 1497
Desmurs, J. F., Bujarrabal, V., Colomer, F., & Alcolea, J. 2000, *A&A*, 360, 189
Diamond, P. J., Kemball, A. J., Junor, W., *et al.* 1994, *ApJ*, 430, 61
Dickinson, D. F. 1978, *Sci. Amer.*, 238, 90
Dodson, R., Rioja, M. J., Jung, T.-H., *et al.* 2014, *AJ*, 148, 97
Engels, D. & Lewis, B. M. 1996, *A&AS*, 116, 117
Fish, V., Akiyama, K., Bouman, K., *et al.* 2016, *Galaxies*, 4, 54
Freytag, B. & Höfner, S. 2023, *A&A*, 669, A155
Freytag, B., Liljegren, S., & Höfner, S. 2017, *A&A*, 600, A137
Gan, Z., Choi, E., Ostriker, J. O., Ciotti, L., & Pellegrini, S. 2019, *ApJ*, 875, 109
Gérard, E. & Le Bertre, T. 2006, *AJ*, 132, 2566
Gómez-Garrido, M., Bujarrabal, V., Alcolea, J., *et al.* 2020, *A&A*, 642, 213
Gonidakis, I., Diamond, P. J., & Kemball, A. J. 2013, *MNRAS*, 433, 3133
Gottlieb, K., Decin, L., Richards, A. M. S., *et al.* 2022, *A&A*, 660, 94
Gray, M. 2012, Maser Sources in Astrophysics, (Cambridge: Cambridge University Press)
Gray, M. D. & Humphreys, E. M. L. 2000, *New Astron.*, 5, 155
Gray, M. D., Wittkowski, M., Scholz, M., Humphreys, E. M. L., Ohnaka, K., & Boboltz, D. 2009, *MNRAS*, 394, 51
Habing, H. J. & Olofsson, H. 2003, Asymptotic Giant Branch Stars, (Berlin: Springer)
Han, S.-T., Lee, J.-W., Kang, J., *et al.* 2008, *IJIMW*, 29, 69
Henkel, C., Matthews, H. E., & Morris, M. 1983, *ApJ*, 267, 184
Hoai, D. T., Nhung, P. T., Darriulat, P., *et al.* 2022a, *Vietnam Journal of Science, Technology, and Engineering*, accepted (arXiv:2203.09144)
Hoai, D. T., Nhung, P. T., Darriulat, P., *et al.* 2022b, *MNRAS*, 510, 2363
Höfner, S. 2016, in: 19th Cambridge Workshop on Cool Stars, Stellar Systems, and the Sun (CS19), 10.5281/zenodo.154673
Höfner, S., Bladh, S., Aringer, B., & Ahuja, R. 2016, *A&A*, 594, 108
Höfner, S. & Olofsson, H. 2018, *A&ARev*, 26, 1
Humphreys, E. M. L. & Gray, M. D. 2004, in: R. Bachiller, F. Colomer, J. F. Demurs, & P. de Vincente (eds.), Proceedings of the 7th European VLBI Network Symposium, p. 177
Humphreys, E. M. L., Gray, M. D., Yates, J. A., *et al.* 2002, *A&A*, 386, 256
Imai, H., Nakashima, J.-I., Deguchi, S., *et al.* 2010, *PASJ*, 62, 431
Ireland, M. J., Scholz, M., & Wood, P. R. 2011, *MNRAS*, 418, 114
Izumiura, H., Ukita, N., Kawabe, R., *et al.* 1987, *ApJ*, 323, 81
Jeste, M., Gong, Y., Wong, K. T., *et al.* 2022, *A&A*, 666, A69
Kaliari, J. S., Marigo, P., & Tremblay, P.-E. 2014, *ApJ*, 782, 17
Karakas, A. I. 2010, *MNRAS*, 403, 1413
Kemball, A. J. 2007, in: J. M. Chapman & W. A. Baan, Astrophysical Masers and their Environments, IAU Symposium 242, p. 236
Kim, J., Cho, S.-H., Oh, C. S., & Byun, D.-Y. 2010, *ApJS*, 188, 209
Kim, D.-J., Cho, S.-H., Yun, Y., *et al.* 2018, *ApJ*, 866, 19
Knapp, G. R., Young, K., Lee, E., & Jorissen, A. 1998, *ApJS*, 117, 209
Kwok, S. 1975, *ApJ*, 198, 583
Lamers, H. J. G. L. M. & Cassinelli, J. P. 1999, Introduction to Stellar Winds, (Cambridge: Cambridge University Press)

Le Bertre, T., Hoai, D. T., Nhung, P. T., & Winters, J. M. 2016, in: C. Reylé *et al.* (eds.), SF2A-2016: Proceedings of the Annual meeting of the French Society of Astronomy and Astrophysics, 433

Leitner, S. N. & Kravtsov, A. V. 2011, *ApJ*, 734, 48

Liljegren, S., Höfner, S., Freytag, B., & Bladh, S. 2018, *A&A*, 619, A47

Maercker, M., Mohamed, S., Vlemmings, W. H. T., *et al.* 2012, *Nature*, 490, 232

Martin, D. C., Seibert, M., Neill, J. D., *et al.* 2007, *Nature*, 448, 780

Marvel, K. 2004, *New Astron. Revs*, 48, 1349

Matthews, L. D. & Claussen, M. J. 2018, in: E. Murphy (ed.), Science with a Next Generation Very Large Array, Astronomical Society of the Pacific Conference Series, 517, (San Francisco: ASP), p. 281

Matthews, L. D., Le Bertre, T., Gérard, E., & Johnson, M. C. 2013, *AJ*, 145, 97

Matthews, L. D., Reid, M. J., & Menten, K. M. 2015, *ApJ*, 808, 36

Matthews, L. D., Reid, M. J., Menten, K. M., & Akiyama, K. 2018, *AJ*, 156, 15

McDonald, I., De Beck, E., Zijlstra, A. A., & Lagadec, E. 2018, *MNRAS*, 481, 4984

McGaugh, S. S. & Schombert, J. M. 2014, *AJ*, 148, 77

Menten, K. M., Reid, M. J., Kamiński, T., & Claussen, M. J. 2012, *A&A*, 543, A73

Menten, K. M., Wyrowski, F., Keller, D., & Kamiński, T. 2018, *A&A*, 613, 49

Murphy, E. J. 2018, in: A. Tarchi, M. J. Reid, & P. Castangia (eds.), Astrophysical Masers: Unlocking the Mysteries of the Universe, IAU Symposium 336, p. 426

Narayanan, D., Turk, M., Robitaille, T., *et al.* 2021, *ApJS*, 252,12

Nhung, P. T., Hoai, D. T., Winters, J. M., *et al.* 2015, *A&A*, 583, 64

Omont, A., Guilloteau, S., & Lucas, R. 1989, in: S. Torres-Peimbert (ed.), Planetary Nebulae, IAU Symposium 131, p. 453

O'Gorman, E., Harper, G. M., Ohnaka, K., *et al.* 2020, *A&A*, 638, A65

Paladini, C., Baron, F., Jorissen, A., *et al.* 2018, *Nature*, 553, 310

Pardo, J. R., Alcolea, J., Bujarrabal, V., *et al.* 2004, *A&A*, 424, 145

Pardo, J. R., Cerincharo, J., Gonzales-Alfonso, E., & Bujarrabal, V. 1998, *A&A*, 329, 219

Patel, N. A., Young, K. H., Gottlieb, C. A., *et al.* 2011, *ApJS*, 193, 17

Phillips, R. B., Sivakoff, G. R., Lonsdale, C. J., & Doeleman, S. S. 2001, *AJ*, 122, 2679

Ramstedt, S., Vlemmings, W. H. T., Doan, L., *et al.* 2020, *A&A*, 640, 133

Reid, M. J. & Goldston, J. E. 2002, *ApJ*, 568, 931

Reid, M. J. & Menten, K. M. 1997, *ApJ*, 476, 327

Reid, M. J. & Menten, K. M. 2007, *ApJ*, 671, 2068

Richards, A. M. S. 2012, in: R. S. Booth, W. H. T. Vlemmings, & E. M. L. Humphreys (eds.), Cosmic Masers - from OH to H_0, IAU Symposium 287, p. 199

Richards, A. M. S., Cohen, R. J., Bains, I., & Yates, J. A. 1999, in: T. Le Bertre, A. Lebre, & C. Waelkens (eds.), Asymptotic Giant Branch Stars, IAU Symposium 191, p. 315

Ridgway, S., Akeson, R., Baines, E., *et al.* 2019, *BAAS*, 51, 332

Rizzo, J.R., Cerincharo, J., & García-Miró, C. 2021, *ApJS*, 253, 44

Rosenfield, P., Margio, P., Girard, L., *et al.* 2014, *ApJ*, 790, 22

Sahai, R. & Stenger, B. 2023, *AJ*, 165, 229

Salaris, M., Weiss, A., Cassarà, L. P., Piovan, L., & Chiosi, C. 2014, *A&A*, 565, 9

Schilizzi, R. T. 2004, in: J. M. Oschmann, Jr. (ed.), Ground-based Telescopes, *Proceedings of the SPIE*, 5489, 62

Schwarzschild, M. 1975, *ApJ*, 195, 137

Sciclula, P., Kemper, F., McDonald, I., *et al.* 2022, *MNRAS*, 512, 1091

Soria-Ruiz, R., Alcolea, J., Colomer, F., *et al.* 2004, *A&A*, 426, 131

Sullivan, W. T. III, 1973, *ApJS*, 25, 393

Thaddeus, P., Mather, J., Davis, J. H., & Blair, G. N. 1974, *ApJ*, 192, 33

Tielens, A. G. G. M, Waters, L. B. F. M., & Bernatowicz, T. J. 2005, in: A. N. Knot, E. R. D. Scott, & B. Reipurth (eds.), Chondrites and the Protoplanetary Disk, Astronomical Society of the Pacific Conference Series, 341, (San Francisco: ASP), p. 605

Villaume, A., Conroy, C., & Johnson B. D. 2015, *ApJ*, 806, 82

Villaver, E., Manchado, A., & García-Segura, G. 2012, *ApJ*, 748, 94
Vlemmings, W. H. T. 2012, in: R. S. Booth, W. H. T. Vlemmings, & E. M. L. Humphreys (eds.), Cosmic Masers - from OH to H_0, IAU Symposium 287, p. 31
Vlemmings, W. H. T., Khouri, T., O'Gorman, E., *et al.* 2017, *Nature Ast.*, 1, 848
Vlemmings, W. H. T., Khouri, T., & Olofsson, H. 2019, *A&A*, 626, 81
Wickramasinghe, N. C., Donn, B. D., & Stecher, T. P. 1966, *ApJ*, 146, 590
Willson, L. A. & Bowen, G. W. 1985, in: Relations Between Chromospheric-Coronal Heating and Mass Loss in Stars, p. 127
Wilson, W. J. & Barrett, A. H. 1968, *Science*, 161, 778
Winters, J. M., Hoai, D. T., Wong, K. T., *et al.* 2022, *A&A*, 658, 135
Wittkowski, M., Boboltz, D. A., Ohnaka, K., Driebe, T., & Scholz, M. 2007, *A&A*, 470, 191
Yoon, D.-W., Cho, S.-H., Yun, Y., *et al.* 2018, *Nature Comm.*, 9, 2534
Young, K., Phillips, T. G., & Knapp, G. R. 1993, *ApJ*, 409, 725

Cosmic Masers: Proper Motion toward the Next-Generation Large Projects
Proceedings IAU Symposium No. 380, 2024
T. Hirota, H. Imai, K. Menten, & Y. Pihlström, eds.
doi:10.1017/S1743921323002958

Masers in evolved stars; the Bulge Asymmetries and Dynamical Evolution (BAaDE) Survey

Loránt O. Sjouwerman[1], Ylva M. Pihlström[2], Megan O. Lewis[3], Rajorshi Bhattacharya[2], Mark J Claussen[1] and BAaDE Collaboration

[1]National Radio Astronomy Observatory, Socorro New Mexico, USA 87801.
email: lsjouwer@nrao.edu

[2]University of New Mexico, Albuquerque New Mexico, USA 87131

[3]Nicolaus Copernicus Astronomical Center, Polish Academy of Sciences, Warsaw, Poland 00-716

Abstract. The Bulge Asymmetries and Dynamic Evolution (BAaDE) project attempts to improve our knowledge about the structure and dynamics of the inner Milky Way galaxy by sampling tens of thousands of infrared color-selected evolved stars with circumstellar envelopes (CSEs). The SiO masers in these CSEs yield the object's line-of-sight velocity instantly and accurately, and together provide a sample of point-mass particles that are complementary to high-mass star formation masers typically found in the Galactic Disk as well as to optical samples that cannot reach into the Galactic Plane and Bulge due to extremely high visual extinction. This presentation introduces the BAaDE survey and highlights current results.

Keywords. Milky Way galaxy, Galaxy structure, Asymptotic Giant Branch stars, masers, circumstellar material

---

1. Introduction

Currently the general picture of the structure and dynamics of the Milky Way galaxy is that it is a spiral galaxy consisting of a disk characterized by a nearly flat rotation curve and a stellar bar embedded in a bulge, rotating like a solid feature. Whereas this simple symmetric model seems sufficient to explain the major building blocks, only the fine details hidden in observations may reveal the origins and perturbations that have shaped the Milky Way as we see it today. Obtaining a more complete view of the stellar populations and stellar kinematics allow to better disentangle the Milky Way's history, which then can be applied to understand galaxy formation elsewhere.

Surveys to determine the shape and composition of the Milky Way have been performed since the development of optical instrumentation some centuries ago. It was only in the early 1900s that a biased view due to the existence of interstellar extinction was revealed, explaining for example the "zone of avoidance". A one-degree wide region with low extinction toward the Galactic Bulge, the so-called *Baade's window*†, has since been recognized as one of several "windows" for its ability to provide the least extinction-affected observational views of Bulge stars in the optical, albeit located several degrees

† Wilhelm Heinrich Walter ("Walter") Baade, German astronomer, 1893-1960

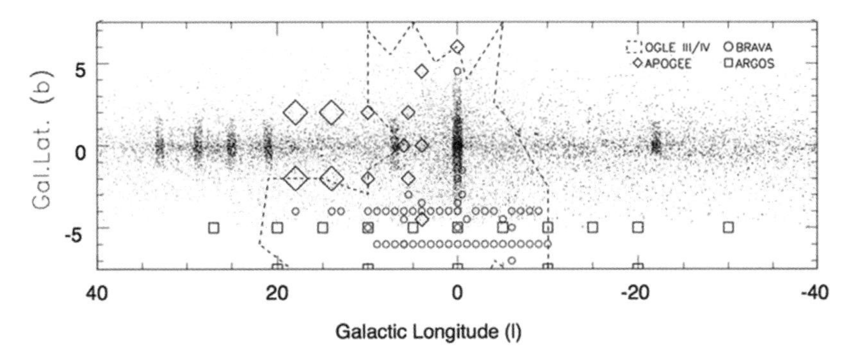

Figure 1. The BAaDE sample compared to sky coverage footprints of ground-based optical and NIR surveys toward the Bulge. BAaDE is sampling the inner Galaxy where other surveys are lacking coverage. Note that MSX observed a few regions with higher sensitivity resulting in a higher density of (fainter) targets in some directions, including the Galactic Center.

away ($l = 1.02°, b = -3.92°$) from the most inner part of the Galaxy. Observations in the near-infrared (NIR) and mid-infrared (MIR) wavebands, where extinction is reduced by an order of magnitude, only recently became feasible with the development of high angular resolution detectors that are able to (mostly) circumvent confusion due to the high stellar density in the Plane and Bulge areas. Nevertheless, these surveys typically either lack the footprint (Figure 1) or kinematic observations to obtain a more complete and homogeneous coverage of the inner Galaxy.

A major step forward towards identifying objects suitable for studies of the inner Bulge used the newly derived predictive properties of MIR color-color diagrams. In particular, specific colors derived from the Infrared Astronomical Satellite (IRAS) identify relatively thick circumstellar envelopes (CSEs) conducive to sustain 1612 MHz ground-state hydroxyl (OH) masers (van der Veen & Habing (1988)). The maser radiation, in the radio wavebands, is not affected by interstellar extinction and is luminous enough to be detected at great distances, covering the entire Milky Way. Once the spectrum is observed to show a typical double-peaked profile, the stellar line-of-sight velocity is known. Several surveys combined have revealed about 4000 stellar OH masers, with about two thirds in the inner Galaxy. The resulting kinematics have been studied and used in modeling of the dynamics of the Galaxy. Even though the OH masers are found in the optically obscured regions in the Plane and Bulge, indeed also in the very center of the Galaxy, their collective number is insufficient to perform a detailed analysis of Galactic structure and asymmetries. Adding the known sample of another couple of thousand stellar silicon-monoxide (SiO) masers, mostly found by Japanese research groups, only expands the data partially as many OH and SiO masers share the same stellar host.

To significantly improve the situation, a new bold approach must be employed. Large-scale optical and NIR ground-based surveys like 2MASS, APOGEE, BRAVA and OGLE, whether or not primarily with the goal of obtaining data for the purpose of studying Galactic dynamics, have yielded incredible amounts of useful data. However, these surveys still carry the aforementioned biases with incomplete coverage and therefore do not reach all the way into the Plane and Bulge. On the other hand, recent sensitive and high angular resolution space-based multi-band NIR and MIR surveys (MSX, GLIMPSE, WISE, etc.) have enabled selecting large samples, less affected by interstellar extinction and closer to the zero-latitude Plane and Galactic Center, based on color and color-magnitude properties of specific types of objects.

1.1. *The Bulge Asymmetries and Dynamic Evolution project*

To reach all the way into the mid-Plane, the above mentioned NIR and MIR surveys have enabled us to select objects with specific properties probing deep into the Galactic Plane and Bulge. For example, the MIR-bright low-to-intermediate mass (0.8-8 $M_{\odot}$) long-period variable Miras occupy a certain region in the MIR color-color space. These stars on the Asymptotic Giant Branch (AGB) have substantial yet relatively thin CSEs, the ones that are likely to harbor SiO masers. Specifically, we drew our target sample from the Midcourse Space Experiment (MSX) point source catalog as it covers the entire Plane and provides colors that promise a high SiO maser detection rate (Sjouwerman *et al.* (2009)). The total sample consists of about 28 000 objects of which two-thirds, north of Declination $-35°$, can be observed with the Karl G. Jansky Very Large array (VLA) and the remaining southern sources with the Atacama Large Millimeter/submillimeter Array (ALMA). These radio interferometers are sufficiently sensitive to obtain a meaningful observation in a very short time interval, such that observing thousands of sources does not impose a major impact on telescope resources. Requiring a minimal spectral resolution to recognize maser lines and requesting a detection limit realistically needed for follow-up Very Long Baseline Interferometry (VLBI) campaigns, less than a minute observing time per target source is needed.

Unfortunately, the VLA and ALMA are not identical instruments. The VLA receivers are capable of observing up to a frequency of 50 GHz, whereas ALMA at that time did not have observing capabilities below a frequency of 80 GHz. Fortunately, on the other hand, the SiO molecule has rotational transitions that can be observed with the VLA around 43 GHz (J=1-0) as well as with ALMA around 86 GHz (J=2-1). Furthermore, it appears that observations for individual objects in the J=1-0 transitions are interchangeable for observations in the J=2-1 transitions (and vice-versa) when used for dynamical modeling purposes (Stroh *et al.* (2018)). This ensures that an SiO maser survey, combining the VLA and ALMA results to analyze the northern and southern parts of the Milky Way, should at most only suffer from minimal observational bias in the data taken between the two different instruments. That is, if there is bias, it would be in the uncorrected (for line-of-sight dependent interstellar extinction) infrared color selection and limiting magnitude of the MSX point source catalog.

By selecting MIR sources and observing them in the radio regime, the traditional optically selected small-area Baade's window may symbolically be re-defined as *BAaDE's* window; a new view of the Milky Way that is unrestricted in extent and shares the same low-to-no interstellar extinction requirement to study stars in the Galactic Plane and Bulge.

The Bulge Asymmetries and Dynamic Evolution (BAaDE) project aims to significantly improve models of the structure and dynamics of the inner Galaxy. The goal is to probe into regions not reachable with optical and NIR sampling of the Galactic Bulge and Plane by performing an SiO maser survey in evolved AGB stars. These CSE masers reveal the stellar line-of-sight kinematics and can be used as point-mass particles in dynamical modeling representing the older stellar populations. The survey will be complementary to many other surveys, either because of very limited overlap (e.g., sampling different types of objects) or by providing additional information (e.g., providing velocities). The project also includes novel studies to obtain relatively accurate stellar distances in order to derive general stellar properties like bolometric and maser luminosities. The BAaDE survey, by itself or in combination with approximate distances, will yield a wealth of data allowing many different studies and statistical analysis on AGB stars, CSEs and SiO maser modeling and occurrence. For the latter see the contribution by Lewis *et al.* in these proceedings. In the remainder, we will showcase our data and highlight some of the early results published elsewhere.

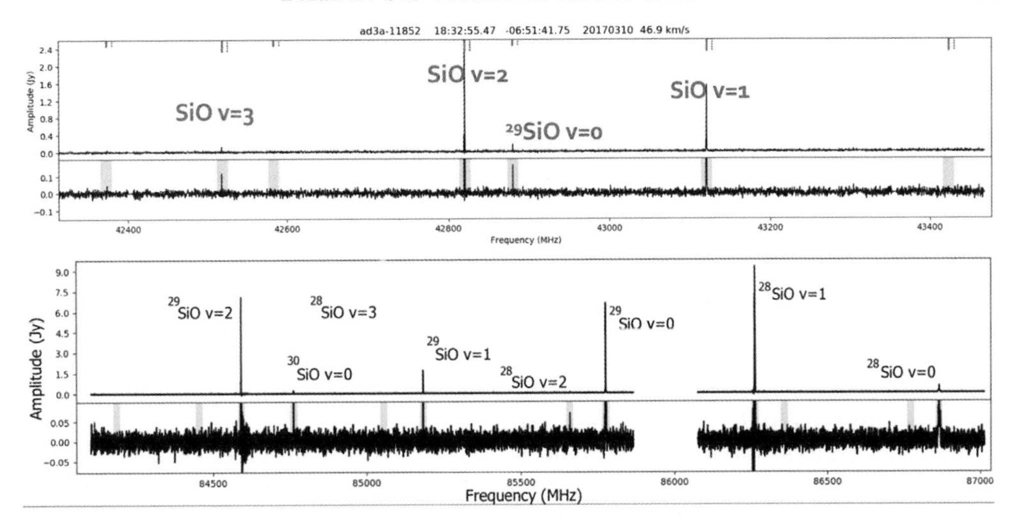

Figure 2. Typical spectra for the BAaDE survey detections. Up to seven transitions can be detected using the VLA (top) for sources north of Declination -35^{d} and up to eleven using ALMA (bottom) for the more southern sources.

2. Observations

The VLA observations started in 2013 and finished in 2017 (using Semesters 13A through 17A) whereas the ALMA observations took place from 2015 through 2022 (during Cycles 2-8) and included about 19 000 versus 9 000 targeted observations at 43 GHz versus 86 GHz. It should be noted that ALMA was still adding new antennas to the array during our campaign, resulting in reduced time on source in the later years. The spectral resolutions used were about 2 and 1 km/s, respectively, and both survey observations achieved an RMS channel noise of about 15 mJy/beam. All maser detections were assumed to be point-like using the baseline lengths involved, which drastically simplified our data calibration procedures; specifically we used self-calibration on the bright masers (Sjouwerman et al. in preparation).

For a survey this large, multiple observing proposals in different proposal cycles or semesters are the norm. The effects of proposing for cumulative observations are — at least in the case of the BAaDE SiO maser survey — that 1) publications based on preliminary analysis on part of the data appear before the final data set is in hand, that 2) different observing priorities between the allocations resulted in some strategy and data quality trade-offs and that 3) acquired knowledge based on the preliminary data may alter the original observing strategy and target list, potentially affecting homogeneity. That is, whereas the VLA survey was completed mostly homogeneously covering up to seven SiO (isotopologue) transitions for ∼19000 objects, the VLA and early ALMA data allowed to recognize contaminating sources (carbon-rich evolved stars, young stellar objects) using infrared colors and removing them from our ALMA target list. This opened up to reassign correlator resources originally intended to identify these sources and reallocate that bandwidth to cover additional potential SiO (isotopologue) transitions. The 6000 ALMA targets observed after 2018 therefore cover eleven transitions compared to ∼1400 observations covering only the four above ∼85.5 GHz. It should also be noted that the BAaDE science 'data' consists of spectra (Figure 2); it has never been the intention to create images nor image cubes other than for checking purposes, which is somewhat different from the regularly adopted ALMA "image science" philosophy.

3. Results

As at this time all observations have just been completed, the final data reduction and combination of the data sets will be reported on elsewhere. From the early data, however, several results have been obtained and are summarized here.

Pihlström *et al.* (2018) have investigated the positional agreement of the input catalog, the MSX PSC 2.3, with phase-referencing VLA A-array configuration maser positions. In addition, the maser positions were matched with possible WISE, 2MASS, and Gaia counterparts. The Gaia positions, due to optical extinction and after careful matching procedure only reliably available for a small subset of the sources, typically agree with the radio maser position to within $0.01''$ (10 mas). The more readily available 2MASS counterparts, generally used for the data reduction, match well within $0.2''$. The WISE and MSX positions are less accurate and typically deviate around a quarter and one arcsecond from the radio position, respectively.

Additionally, as so far is the case in all the BAaDE reobservations of known detections (i.e., in the below), the re-detection rate is around 80%. This is shockingly similar to the instantaneous detection rate in genuine Mira-type AGB stars in the BAaDE survey, which suggests the speculation that SiO masers are present in all these sources with a "maser-on" time of 80% of the stellar pulsation cycle.

Stroh *et al.* (2018) resolved the potential detection rate bias in the SiO 43 GHz (J=1-0) versus the 86 GHz (J=2-1) transitions in (near) simultaneous observations of bright — likely nearby — SiO masers. At least 75% of the 43 GHz detections were also detected at 86 GHz, with a fraction of the non-detections attributed to bad weather conditions. This supports the suggestion of Sjouwerman *et al.* (2004) that 43 GHz and 86 GHz SiO masers typically co-exist (simultaneously) in the same source, with the same line-of-sight velocity of the maser. They find that, in terms of detectability for the BAaDE survey, the 86 GHz masers are on average a factor 1.36 fainter than the 43 GHz masers. Interestingly the ratio is influenced by the presence of the 43 GHz $v = 3$ transition: it reaches unity when the line is present and a factor two when not. Their work likely also demonstrates the validity the radiative pumping model that includes a water-overlap transition (Desmurs *et al.* (2014)). Finally, given the limited average difference and large spread in J=1-0 to J=2-1 maser line ratios, detecting the 86 GHz maser on the remote side of the Galaxy should not be much harder than detecting the 43 GHz maser, justifying BAaDE's pragmatic application of observational constraints on the southern and northern hemisphere to probe the entire Milky Way galaxy.

Trapp *et al.* (2018) performed an early kinematic analysis using the pilot VLA and ALMA data. They showed that there is a clean division into two separate kinematic populations: a "cold", low dispersion (σ ~50 km/s) foreground (and background!) Galactic Disk component consisting of the younger and more massive AGB stars and a "hot", high dispersion (σ ~100 km/s) component reminiscent of older and lower mass AGB stars that would be typically found in the Galactic Bulge or Bar (Figure 3). Interestingly the higher dispersion stars, apart from deviating more from circular motions that can be expected from the low dispersion disk stars, also show signs of cylindrical rotation at the higher latitudes, as predicted by bar models (Shen *et al.* (2010)). That is, based on the pilot BAaDE SiO maser data, the inner Galaxy shows evidence of a (tri-axial) bar and is not completely described by an elliptical shape and kinematics of an ensemble of fully independent orbits.

Stroh *et al.* (2019) have published the early 86 GHz ALMA data, about 1400 sources observed for four SiO transitions and the CS-line. The latter was included to distinguish non-detections in oxygen-rich sources from carbon-rich sources. In particular, when imaged the carbon-rich (evolved) AGB stars would clearly show a point-like source whereas the more extended structures could be identified with young stellar objects

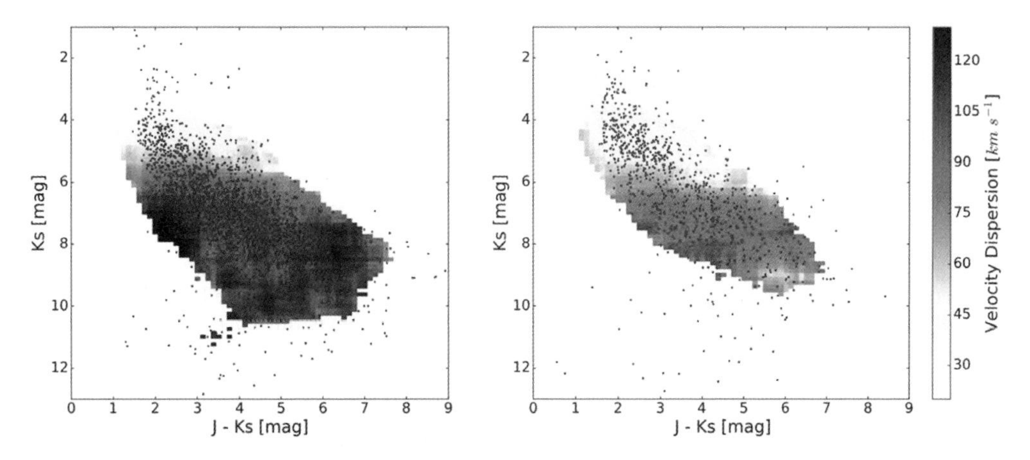

Figure 3. The BAaDE pilot data reveals the separation of the Disk and Bulge populations, where the start of the distinction using the far-side ALMA apparent K_s-magnitude (~6.5 mag) is fainter than the near-side VLA apparent K_s-magnitude (~5.5 mag). See Trapp *et al.* (2018).

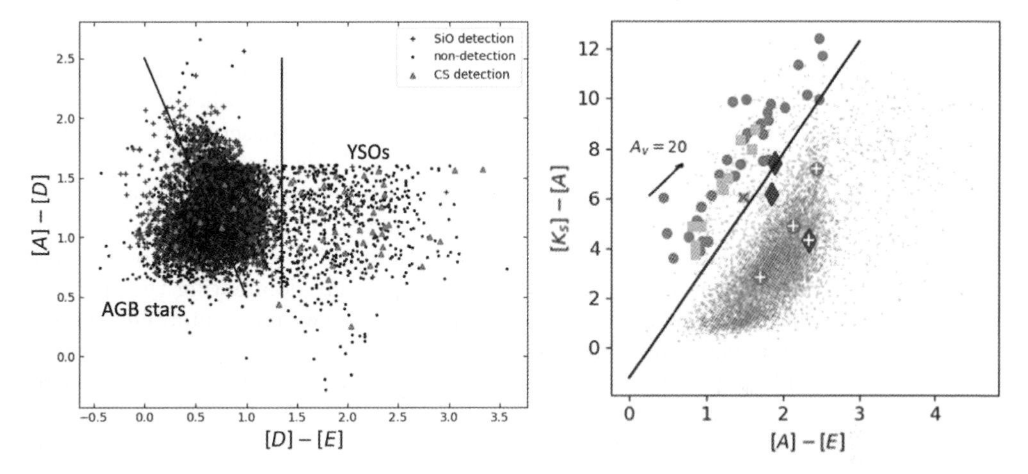

Figure 4. Left: MSX colors alone can distinguish between evolved and young stellar objects. Right: To further distinguish between carbon-rich and oxygen-rich evolved stars, the combined 2MASS/MSX $[K_s]-[A]$ color is required. See Lewis *et al.* (2020a) and Lewis *et al.* (2020b).

(YSOs) in their embedding molecular clouds, or alternatively planetary nebulae (PNe) or compact HII regions. From this, and the associated line-of-sight velocity information of the detections, they were able to decontaminate the sample and refine the genuine oxygen-rich AGB stars that were the objects of interest using MIR colors, specifically MSX [D]−[E]. This was crucial to understand part of the non-detections, also in the 43 GHz VLA data (see Lewis (2021)), and allowed to remove unlikely SiO maser detections in the remaining ALMA observations; the VLA observations had completed by that time and were refined to oxygen-rich AGB stars afterwards.

Lewis *et al.* (2020a) have expanded the MSX-only color criteria of Stroh *et al.* (2019) with 2MASS K_s data to distinguish sources that are questionable genuine oxygen-rich AGB stars (Figure 4). Where the MSX [D]−[E] color (or [D]−[18], where [18] is taken from AKARI) typically separates the evolved from the young, pre-main sequence stars, the combination of MSX and 2MASS photometry is necessary to distinguish among the evolved stars: the carbon-rich versus the oxygen-rich. This ensures that the BAaDE survey data can be properly assigned to the correct type of object and that any analysis

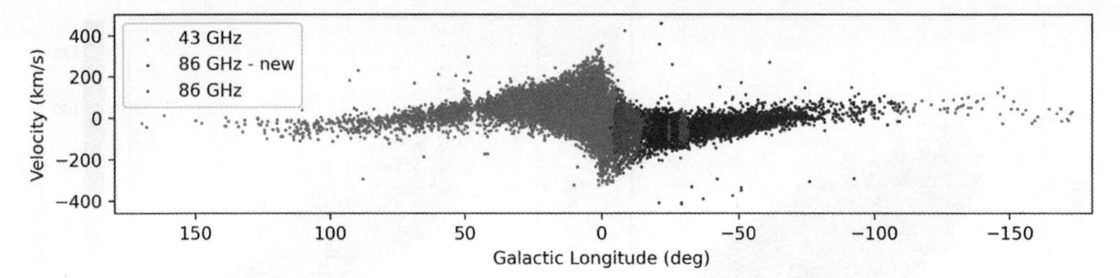

Figure 5. The BAaDE survey has recently been completed by including the line-of-sight velocities of the southern ALMA part ($-15° < l < -100°$) providing a full coverage of SiO maser velocities in AGB stars over Galactic longitude.

is not hampered by contamination. The availability of a large sample observed over the entire Galactic Plane also yielded insight in the relative distribution: carbon-rich AGB stars seem to be spread relatively uniformly over the inner Galaxy when compared to the much more peaked distribution of oxygen-rich stars, centered on the Galactic Center.

Lewis *et al.* (2020b) confirmed the separation in Lewis *et al.* (2020a) and resolved a final ambiguity: that of bright single-line detections near 42.9 GHz. This frequency is near a carbon transition (of HC_3N) as well as the (oxygen) ($v = 0$) transition of the ^{29}SiO isotopologue. Detecting the latter transition without detecting any other SiO maser, as well as detecting the HC_3N line as a narrow (few km/s), perhaps maser feature were both considered unlikely. Nevertheless, with reobserving these single-line objects Lewis *et al.* (2020b) have shown that these detections should be attributed to the ^{29}SiO isotopologue, and that it is indeed possible that an isotopologue line can be (much) brighter, or even the only line detected in oxygen-rich AGB stars. Furthermore, this behavior can be reversed, with the typical, traditional 42.8 and 43.1 GHz ^{28}SiO masers dominating the spectrum within a few years. This should fuel the development of new maser pumping models, which would need to explain this observed behavior.

Dike *et al.* (2021) have paid specific attention to the 43 GHz J=1-0 ground-vibrational state emission. This transition, when detected, is both seen as a wide feature as well as a narrow feature in the individual spectra. The wider features are attributed to thermal emission and used to measure SiO outflow velocities which typically range within 15-30 km/s and are similar to the OH and CO outflow velocities. The narrow profiles are attributed to maser emission in this transition and, for composite thermal/maser profiles, removed when fitting the outflow velocities. It should be noted that the thermal lines typically are seen in the more-nearby sources, suggestive of a sensitivity limited result.

4. Summary

The preliminary analysis of the BAaDE survey data has already yielded a wealth of new information, both on the individual sources as well as the distribution and kinematics of the different population components in the Milky Way galaxy. Whereas there are many more topics to address, we think that we understand the potential limitations and possible confusion in the data and are well positioned to study the Bulge Asymmetries and Dynamical Evolution of the Galaxy. This is even more true now that we have completed the results of the final ALMA observations and have a complete coverage of the Galactic Bulge (Figure 5).

5. Next Challenges

One of the next moves is to include the BAaDE survey results in realistic Galactic dynamics models. These models will be helped by additional information such as source distances and proper motion measurements. Whereas the latter is probably logistically not feasible for the entire sample, we are investigating an approach that may be successful for a reasonable number of sources. Either way, obtaining proper motions will require an investment of several years.

Measuring distances to a large number of objects may also be unrealistic. That is, the Gaia mission has measured the (geometric) parallax distances to only a fraction of (optically visible) BAaDE targets. However, we have our reservations about blindly assigning these distances to our sources, partly because we challenge the optical counterparts, partly because the stellar photosphere is the size of the anticipated parallax. Geometric parallaxes measured using VLBI techniques on the maser emission in principle is possible, but again unrealistic for a large sample. We are therefore exploring methods using existing infrared photometry (from e.g., MSX, 2MASS, etc.) and existing pulsation cycle period measurements (from e.g., OGLE, VVV, ZTF, etc.); see also the contribution of Bhattacharya *et al.* elsewhere in these proceedings. We are in the early experimental phase of these methods and expect to be reporting on them at a future occasion.

References

Desmurs, J.-F., Bujarrabal, V., Lindqvist, M., *et al.* 2014, *A&A*, 565, 127
Dike, V., Morris, M. R., Rich, R. M., *et al.* 2021, *AJ*, 161, 111
Lewis, M. O. 2021, *PhD Thesis*, University of New Mexico, USA
Lewis, M. O., Pihlström, Y. M., Sjouwerman, L. O., *et al.* 2020, *ApJ*, 892, 52
Lewis, M. O., Pihlström, Y. M., Sjouwerman, L. O., *et al.* 2020, *ApJ*, 901, 98
Pihlström, Y. M., Sjouwerman, L. O., Claussen, M. J, *et al.* 2018, *ApJ*, 868, 72
Shen, J., Rich, R. M., Kormendy, J., *et al.* 2010, *ApJL*, 720, 72
Sjouwerman, L. O., Messineo, M. & Habing, H. J. 2004, *PASJ*, 56, 45
Sjouwerman, L. O., Capen, S. M. & Claussen, M. J 2009, *ApJ*, 705, 1554
Stroh, M. C., Pihlström, Y. M., Sjouwerman, L. O., *et al.* 2018, *ApJ*, 862, 153
Stroh, M. C., Pihlström, Y. M., Sjouwerman, L. O., *et al.* 2019, *ApJS*, 244, 25
Trapp, A. C., Rich, R. M., Morris, M. R., *et al.* 2018, *ApJ*, 861, 75
van der Veen, W. E. C. J., & Habing, H. J. 1988, *A&A*, 194, 125

Cosmic Masers: Proper Motion toward the Next-Generation Large Projects
Proceedings IAU Symposium No. 380, 2024
T. Hirota, H. Imai, K. Menten, & Y. Pihlström, eds.
doi:10.1017/S174392132300296X

Properties of pulsating OH/IR stars revealed from astrometric VLBI observation

Akiharu Nakagawa[1], Tomoharu Kurayama[2], Hiroshi Sudou[3] and Gabor Orosz[4]

[1]Graduate School of Science and Engineering, Kagoshima University, 1-21-35 Korimoto, Kagoshima, Kagoshima 890-0065, Japan. email: nakagawa@sci.kagoshima-u.ac.jp

[2]Teikyo University of Science, 2-2-1 Senju-Sakuragi, Adachi-ku, Tokyo 120-0045, Japan

[3]Department of General Engineering, National Institute of Technology, Sendai College, Natori Campus, 48 Nodayama, Medeshima-Shiote, Natori-shi, Miyagi 981-1239, Japan

[4]Joint Institute for VLBI ERIC (JIVE), Oude Hoogeveensedijk 4, 7991 PD, Dwingeloo, Netherlands

Abstract. We aim to reveal properties of evolution stages in AGB phase; Mira, OH/IR stars, and non-variable OH/IR stars. We presented results of our VLBI observations of four stars; NSV17351, OH39.7+1.5, IRC−30363, and AW Tau. We used the VERA VLBI array to observe 22 GHz H_2O masers. Parallaxes of the four sources were obtained to be 0.247±0.035 mas (4.05±0.59 kpc), 0.54±0.03 mas (1.85±0.10 kpc), 0.562±0.201 mas (1.78±0.73 kpc), and 0.449±0.032 mas (2.23±0.16 kpc). Determination of pulsation period of NSV17351 was done for the first time. We revealed the position and kinematics of NSV17351 in our Galaxy and found that NSV17351 is located in an interarm region. A new period-magnitude relationship was indicated in the infrared region. Various other properties based on the distance measurements are also discussed. We have to emphasize that the VLBI astrometry is effective and the only way for parallax measurements of dust obscured OH/IR stars.

Keywords. VLBI, Astrometry, Masers, AGB stars, OH/IR stars

---

1. Introduction

1.1. *Evolution of AGB stars*

In the last stage of stellar evolution with initial masses of 0.8-10 $M_\odot$, they go through a phase called Asymptotic Giant Branch (AGB) stars (Karakas & Lattanzio 2014). That is to say, almost all stars will spend a period of their life as the AGB stars. The AGB stars return various elements into interstellar space by stellar winds. So, they are important objects that contribute to the chemical composition of the universe and Galaxy. They often show thick circumstellar dust shells and long pulsation periods. A closer look at the evolution on the AGB phase reveals more detailed stages of their growth. On the early AGB phase, stars have relatively thin dust layers, so we can observe them in optical and infrared bands. Mira type variables are thought to be at this early phase. However, as they progress, they become faint and unobservable in optical band due to the absorption by circumstellar dust shells. Instead, they become brighter in the infrared band due to re-radiation from the outer dust layer. At this stage, many objects represent OH masers in their outermost layer, therefore they are referred to as OH/IR stars. The OH/IR stars, a sub-class of AGB stars, are thought to be at a late stage of the AGB phase before they evolve to planetary nebulae (te Lintel Hekkert *et al.* 1991).

1.2. *AGB stars with various initial mass and detailed evolution in AGB phase*

Stellar properties of the AGB stars such as ages, mass loss rates, and pulsation periods depend on their initial masses. There are sub-classes in the OH/IR stars. Massive OH/IR stars with higher luminosity and longer period are classified as a super AGB phase (Karambelkar *et al.* 2019). The OH/IR stars with intermediate mass ($>5\,M_\odot$) and large mass loss rates ($>10^{-4}\,M_\odot$/yr) are classified as extreme-OH/IR stars (Justtanont *et al.* 2015). Thus, there are considered to be several stages in the AGB phase depending on its evolution. Mira type variables are at early stage of the AGB phase. They have relatively thin dust shells and frequently show SiO and H_2O masers. After the Mira phase, the H_2O molecules are transported to outer side of the circumstellar envelope and then photodissociated to produce OH maser. Due to an excess of infrared emission from the circumstellar dust shell and the presence of OH maser emission, they will be recognized as the OH/IR stars. Before the post-AGB phase, stellar pulsation gradually diminishes, and they experience a phase called non-variable OH/IR stars (Kamizuka *et al.* 2020). To understand the sequential evolution from early to late AGB phases, studies of the AGB stars with various properties are necessary.

1.3. *Pulsation period and mass of AGB stars*

The AGB stars with longer pulsation periods are considered to be more massive than ones with shorter periods. According to Feast (2009), AGB stars with pulsation period of 1000 days are considered to have masses of 3-4 $M_\odot$. Mean density and periods of pulsating stars couple each other (Cox 1980). And masses and periods also couple each other (Takeuti *et al.* 2013). For studies of AGB stars with various masses, we need observations of stars with wide period range. In our previous observations with the VERA from 2003 to 2017, we have conducted many VLBI observations toward dozens of AGB stars. Most of the sources are classified as Mira variables and semiregular (SR) variables, so their pulsation periods are shorter than 400 days. From the studies, we have revealed structures and kinematics of circumstellar matters (e.g. (Nyu *et al.* 2011; Nakagawa *et al.* 2016, 2018; Matsuno *et al.* 2020)). However, there are very few VLBI studies which focusing on OH/IR stars. A wide coverage of their pulsation period is necessary to collect AGB stars on various evolutional phases and masses. Therefore, after 2017, we have started VLBI observations toward OH/IR stars with longer period range. Scope of our long-term VLBI study is to reveal astrometric and physical properties of OH/IR stars (parallax based distance, proper motion, internal maser motion, luminosity, mass loss rate and so on). We think comparisons of these properties between OH/IR stars and Mira variables can bring clues to understand evolutional relation. On this report, we mainly present current status and results from astrometric VLBI observations. Preliminary results of parallaxes and proper motions are presented.

1.4. *Period-Magnitude relation of AGB stars with wide period range*

Using a distance of a source determined from its parallax, we can estimate absolute magnitude of the source. In Nakagawa *et al.* (2018), we reported a relation between their pulsation periods and K-band absolute magnitude (Mk) of the Galactic long period variables based on our VLBI observations with VERA. As we have mentioned in the previous section, the period coverage of our previous observations is shorter than $\sim$400 days. With our new target sources, we can explore the relation in the longer period range and can extend the relation.

1.5. *Advantage of VLBI on parallax measurements of AGB and OH/IR stars*

Chiavassa *et al.* (2018) used three-dimensional radiative hydrodynamics simulation of convection and explored the impact of the convection-related surface structure in AGB stars on the photometric variability. They extracted parallax errors in Gaia DR2 for SR variables in the solar neighborhood and compared it to the synthetic predictions of photocenter displacements. As a result, they reported that position of the photocenter displays temporal excursion between 0.077 – 0.198 au (5 to 11% of the corresponding stellar radius), depending on the simulation considered. At the distance of 1 kpc, this excursion corresponds to 0.077 – 0.198 mas. Since distances of the VLBI target AGB stars in our study are hundred to a few kpc, size of this excursion can be estimated to be 0.1 – 1 mas. Time variation of the surface degrades accuracy of parallax measurements based on optical image. Gaia measurements can be suffered from this effect.

Mira variables which considered to be at an early stage of the AGB phase are bright in both optical and infrared. Recently, the Gaia Data Release 3 (Gaia DR3; Gaia Collaboration *et al.* (2022))† provided a huge amount of astrometric measurements. Parallax of large number of Mira variables are available in the Gaia database. However, if the central star is surrounded by a thick dust layer, the source become faint and they cannot be observable with Gaia. For example, the OH127.8+0.0 is known as an OH/IR star with high mass loss rate and long pulsation period of 1994 days. The OH127.8+0.0 is very luminous in infrared, but we cannot find the source in Gaia DR3 because it is faint in optical band due to circumstellar extinction by the dust layer. In the astrometric observations of such stars, the VLBI can be a powerful and promising tool.

2. Observation

2.1. *Single dish monitoring of H_2O and SiO maser*

We have been observing H_2O and SiO maser emissions using the 20 m aperture telescope at VERA Iriki station in order to obtain its spectra and variability. Since the pulsation periods of large number of dusty OH/IR stars are not found in the literature or databases, we have to determine the pulsation period from single dish monitoring by ourselves. Integration time of our single-dish observations are 10 to 40 minutes to reduce noise levels (antenna temperature in K) in each observation to less than 0.05 K. Time intervals between the single-dish observations are not necessarily constant, but are approximately one month interval. The conversion factor from antenna temperature to the flux density is 19.6 Jy K^{-1}. A 32 MHz bandwidth data with 1024 spectral channels gives a frequency resolution of 31.25 kHz, which corresponds to a velocity resolution of 0.42 km s^{-1} at 22 GHz and 0.21 km s^{-1} at 43 GHz. We adopted a signal-to-noise ratio (S/N) of 3 to 5 as a detection criterion in our single-dish observations.

2.2. *VLBI observations and data reduction*

We use the VLBI Exploration of Radio Astrometry (VERA) to observed 22 GHz H_2O and 43 GHz SiO maser emissions of Mira variables and OH/IR stars. The VERA consists of four 20 m aperture radio telescopes at Mizusawa, Iriki, Ogasawara, and Ishigaki-jima. The longest baseline is 2270 km between Mizusawa and Ishigaki-jima stations. Each antenna of VERA is equipped with a dual-beam system (Kawaguchi *et al.* 2000) with which we can simultaneously observe a target maser source and an extragalactic continuum source within a separation angle between 0.3° and 2.2°. The extragalactic sources are used as position references. Using the dual-beam system, we can calibrate short-term tropospheric fluctuations with the phase-referencing technique (Honma *et al.*

† Gaia Data Release 3; https://www.cosmos.esa.int/web/gaia/data-release-3

Table 1. Gaia DR3 parallaxes of our VLBI target sources.

Source name (2017–2022)†	Π_{DR3} [mas]	**Err.** [%]	**Period** [days]	**Source name (2023–)**†	Π_{DR3} [mas]	**Err.** [%]	**Period** [days]
NSV 17351	0.088±0.147	166	1122	V697 Her	1.029±0.129	13	497
OH 127.8−0.0	na	na	1994	NSV 23099	0.209±0.102	49	431
NSV 25875	na	na	1535	OH 358.667−0.044	0.207±0.142	69	300
RAFGL 5201	−0.131±0.253	−194	600	OH 358.23+0.11	−0.061±0.190	−313	704
OH 83.4−0.9	0.836±0.556	66	1428	OH 0.66−0.07	na	na	na
OH 141.7+3.5	na	na	1750	IRAS 18039−1903	na	na	na
CU Cep	0.231±0.057	25	700	OH 9.097−0.392	0.261±0.232	89	634
RAFGL 2445	−1.548±0.369	−24	na	RAFGL 1686	1.053±0.359	34	500
OH 39.7+1.5	na	na	1260	IRAS 18176−1848	2.404±0.618	26	na
OH 26.5+0.6	na	na	1589	OH 44.8−2.3	0.918±0.631	69	na
OH 42.3−0.1	na	na	1650				
IRC−30363	0.241±0.130	54	720				
IRC+10322	0.553±0.183	33	570				
IRC+10451	0.818±0.196	24	730				
OH 26.2−0.6	na	na	1330				
OH 51.8−0.1	na	na	1270				
OH 358.16+0.49	na	na	1507				
V1018 Sco	na	na	na				

†: Duration of our VLBI observations using VERA.

2008). Then, relative position of the target maser spots can be determined with respect to the position reference source with an accuracy of less than 0.1 mas. By tracking the skyplane motions of the maser spots, we derive annual parallax of the target.

The signals of left-handed circular polarization from the target and position reference source were acquired with a total data recording rate of 1 giga-bit per second (Gsps). It gives a total bandwidth of 256 MHz. The data were recorded onto the hard disks of the "OCTADISK" system (Oyama *et al.* 2016). This entire bandwidth is divided into 16 IF channels. Each IF channel then has a width of 16 MHz. Then one IF channel (16 MHz) was assigned for the maser source and the remaining 15 IF channels (16 MHz $\times$ 16 = 240 MHz) were assigned to the position reference sources. Correlation processing was done with the Mizusawa software correlator at Mizusawa VLBI observatory, NAOJ. In the final output from the correlator, the 16 MHz bandwidth data of H_2O or SiO masers was divided into 512 channels with a frequency resolution of 31.25 kHz. This corresponds to a velocity resolution of 0.42 km s^{-1} at 22 GHz and 0.21 km s^{-1} at 43 GHz.

2.3. *Target sources*

From 2017, we started observations of OH/IR stars in addition to Mira variables. In Table 1, we show target sources of our study. For the sources in the left part of the table, astrometric VLBI observations were made between 2017 and 2022. Although successful observations of all the sources were difficult, data acquisition for some stars has been completed and data reduction is currently underway. In 2023, we proposed VLBI observations of the new sources presented in the right part of the table. Parallaxes from the Gaia DR3 (Π_{DR3}) and its relative errors are also presented. For the sources for which no parallax values were found in the Gaia DR3, we indicated "na" in the columns of Π_{DR3} and Err. In some sources, parallaxes show negative values. In addition, several other sources have errors larger than 100 %. As a result, the table shows that it is very difficult for Gaia to determine the accurate parallax of OH/IR stars. Therefore, the VLBI method is still important in parallax measurements of these dust-obscured OH/IR stars. We can also find various pulsation periods in the table. For the sources at low galactic latitudes and indicating longer period, we can expect that they are young and more massive than typical Mira variables.

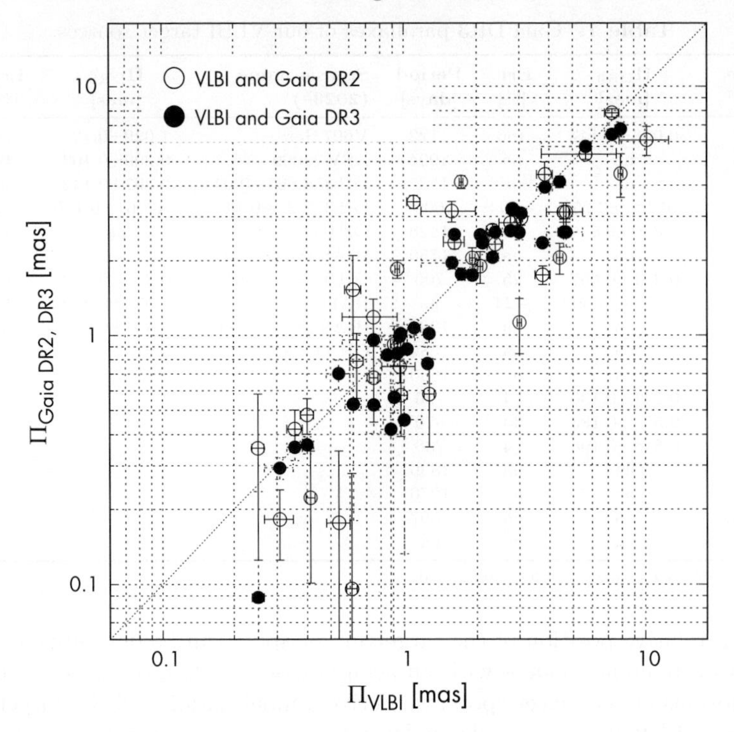

Figure 1. Annual parallaxes determined from VLBI (horizontal axis) and Gaia (vertical axis) in logarithmic scale. Open circles correspond to the comparison between VLBI and Gaia DR2, filled circles correspond to the comparison between VLBI and Gaia DR3.

3. Results and discussion

3.1. *Comparison between VLBI and Gaia*

In Figure 1, we present parallaxes of AGB stars determined from VLBI ($\Pi_{\rm VLBI}$) and Gaia DR2/DR3 ($\Pi_{\rm Gaia\,DR2}$, $\Pi_{\rm Gaia\,DR3}$). Horizontal and vertical axes represent parallax values in logarithmic scales. Open (filled) circles represent comparison between VLBI and Gaia DR2 (Gaia DR3). A dotted line shows a relation of $\Pi_{\rm VLBI} = \Pi_{\rm Gaia}$. The dispersion of the filled circles is clearly smaller that of open circles, which indicates that in many sources the two parallax measurements are closer to the same value in VLBI and Gaia DR3. However, we can find that there is still a large difference for a filled circle at the VLBI parallax of 0.247 mas (a filled circle at the bottom). We know that this data point corresponds to an OH/IR star NSV17351. Although we are curious about other OH/IR stars as well, they could not be presented on this figure because Gaia parallaxes of many dust-obscured OH/IR stars are not available.

3.2. *Results of individual sources*

3.2.1. NSV17351 (OH/IR star)

We determined pulsation period of NSV17351 from our single dish monitoring of H_2O maser at 22GHz. From our least-squares analysis assuming a simple sinusoidal function, the pulsation period of NSV17351 was solved to be 1122±24 days (Nakagawa *et al.* 2023). The model fitting is presented with a solid curve in the left panel of Figure 2. As we cannot find any information on the period in the literature or online databases, this

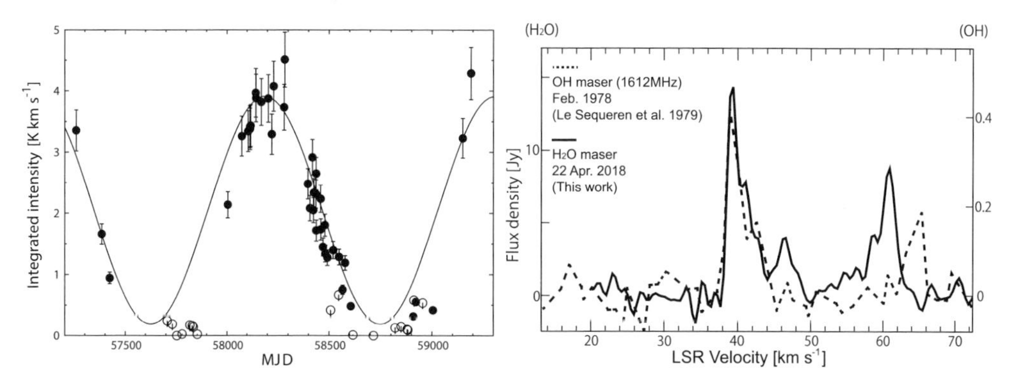

Figure 2. (Left): Time variation of the integrated H_2O maser intensities of NSV17351. Filled circles represent results of successful detection. In the case of non-detection, we put open circles with downward arrows as representatives of detection upper limits. Solid line is the model indicating a pulsation period of 1122±24 days. (Right): Superpositions of H_2O maser (solid line) and OH maser (dotted line) of NSV17351 obtained in 2018 and 1978, respectively. Cut off velocity of the blue-shifted side seems to be exactly same in both spectra.

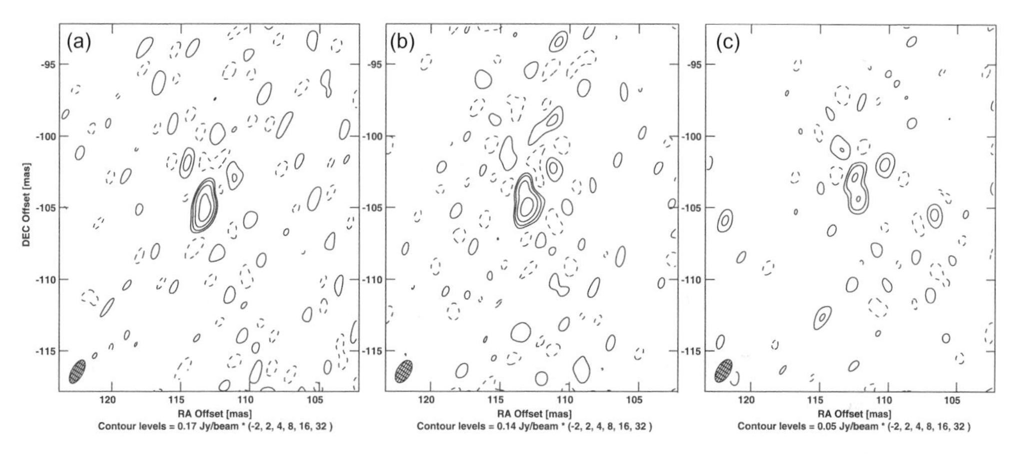

Figure 3. VLBI images of maser spots of NSV17351 at $V_{\rm LSR}$ of 39.15 km s^{-1} detected on (a) 16 April 2018, (b) 1 November 2018 and (c) 12 March 2019 (Nakagawa *et al.* 2023). The synthesized beams are presented at bottom left of each map.

is the first time we have measured the periodicity. We think NSV17351 is a candidate of extreme OH/IR stars because of its long periodicity.

In the right panel of Figure 2, we superposed H_2O maser spectrum on 22 April 2018 (solid line) and 1612 MHz OH maser spectra in February 1978 (dotted line). We can find that the cut off velocity in the blue-shifted component shows exactly same velocity (38 km s^{-1} to 40 km s^{-1}). Since it is thought that OH molecules are supplied by photodissociation of H_2O molecules carried to the outer part of the circumstellar envelope, we comprehend that the H_2O molecules were carried to the outermost region and the H_2O gas has accelerated to the terminal velocity.

Using VERA observation of H_2O maser at 22GHz from 2018 to 2019, we derived a parallax of 0.247±0.035 mas, which corresponds to a distance of 4.05±0.59 kpc. In Figure 3, examples of maser spot images at the same velocity channel are presented. As the shape of the spot gradually changes with time, we carefully examined the maser structure, its time variation and continuity, then we concluded that the southern components in the

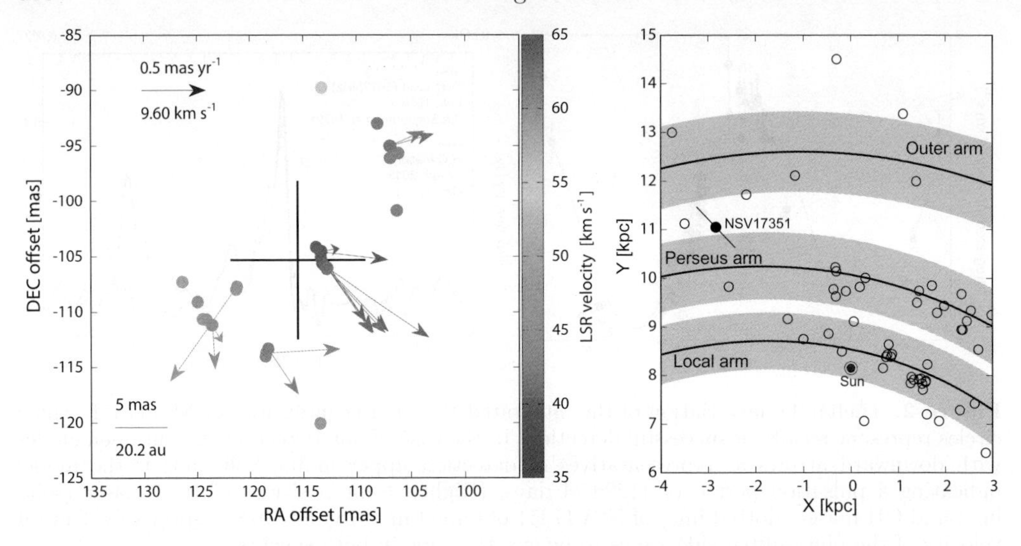

Figure 4. (Left): Skyplane distribution and expanding motions of H_2O maser spots of NSV17351 Nakagawa *et al.* (2023). Filled circles indicate maser spots and arrows indicate their internal motions. A cross symbol indicates an estimated position of the central star. (Right): Location of NSV17351 on the face-on view of the Milky Way. The Galactic center is at (0, 0) kpc and the Sun is indicated with the symbol ($\odot$) at (0, 8.15) kpc. The filled circle with an error bar indicates the position of NSV17351. Open circles indicate maser sources which have Galactocentric distances of > 7 kpc. Solid lines and grey regions indicate centers of three spiral arms and their widths reproduced from a study by Reid *et al.* (2019).

maps (b) and (c) are identical to the peak in map (a). Stellar system proper motion of $(\mu_\alpha \cos\delta, \mu_\delta) = (-1.19 \pm 0.11, 1.30 \pm 0.19)$ mas yr^{-1} was also obtained. Circumstellar distribution and motions of H_2O masers in 80×120 au area is presented in the left panel of Figure 4. We derived an outward expansion velocity of the H_2O masers to be 15.7 ± 3.3 km s^{-1}. Estimated stellar positions are indicated by cross symbols, the lengths of which represent the position errors. Bluest maser spots overlap estimated position of central star.

In the right panel of Figure 4, we can find that NSV17351 is located slightly outside the Perseus arm. The location of NSV17351 can be understood by considering the age. Feast (2009) reported that Mira variables showing a period of 1000 d have initial masses of 3 to 4 $M_\odot$. By assuming this mass, we can estimate age of NSV17351 to be 1.6×10^8 to 3.5×10^8 years. The age of NSV17351 is two orders of magnitude larger than the typical age of high mass star forming regions associated with spiral arms, and we can say that we are observing a state that NSV17351 leaves the arm where it was born, but not yet sufficiently dynamically relaxed.

3.2.2. OH39.7+1.5, IRC−30363 (OH/IR star) and AW Tau (Mira)

We observed H_2O masers in OH39.7+1.5, IRC−30363 (OH/IR star), and AW Tau (Mira) at 22GHz. For OH39.7+1.5, using two maser spots at radial velocities of 34.6 and 8.6 km s^{-1}, a parallax of 0.54 ± 0.03 mas was obtained. This corresponds to a distance of 1.85 ± 0.10 kpc. Averaged proper motion of the two maser spots is $(\mu x, \mu y) = (-0.22\pm0.13, -1.53\pm0.13)$ mas yr^{-1}. For IRC−30363, using a maser spots at radial velocities of 9.72 km s^{-1}, a parallax of 0.562 ± 0.201 mas was obtained. This corresponds to a distance of 1.78 ± 0.73 kpc. For AW Tau, using a maser spots at radial velocities of -9.54 km s^{-1}, a parallax of 0.449 ± 0.032 mas was obtained. This corresponds to a distance of 2.23 ± 0.16

kpc. Pulsation periods of OH39.7+1.5, IRC−30363 (OH/IR star) and AW Tau are 1260, 720, and 672 days, respectively. All sources shows longer pulsation periods than typical Mira variables.

3.3. *Absolute magnitude in near-IR and mid-IR*

We estimated absolute K-band magnitudes (Mk) of three AGB stars, AW Tau, IRC−30363, and OH39.7+1.5 based on parallax measurements in the previous section. When we compare the three Mk values with a period-Mk diagram of the Galactic long period variables, we found that Mk of OH39.7+1.5 is far below the expected value. If we assume thick dust layer or high mass loss rate for the OH/IR star OH39.7+1.5, we can possibly explain this darkening due to circumstellar absorption.

We also estimated absolute magnitudes of the three sources in mid-infrared using WISE W3 band. The central wavelength of the W3 band is 12 μm (Wright *et al.* 2010). The absolute magnitudes M_{W3} of AW Tau, IRC−30363, and OH39.7+1.5 were obtained to be -10.71 ± 0.16, -12.35 ± 0.81, and -12.94 ± 0.12, respectively. When considered in conjunction with our preliminary results obtained so far, absolute magnitudes M_{W3} of various AGB stars seems to represent a unified relation along pulsation period in mid-infrared. To confirm the indication of the period-M_{W3} relation, we have to continue astrometric VLBI observations of OH/IR stars with wide period range.

4. Summary

We are observing Miras and OH/IR stars with astrometric VLBI. Acquired distances and stellar parameters helps us to understand evolutional relation of sub-classes in AGB phase. The VERA was used to conduct all the VLBI observations of H_2O (22 GHz) and SiO (43 GHz) masers. Phase referencing technique was used to measure parallaxes. Results of NSV17351, AW Tau, IRC−30363, and OH39.7+1.5 were presented here. Absolute magnitudes in near-infrared and mid-infrared bands were derived based on the obtained parallaxes. Indication of a new period-magnitude relation in mid-infrared (WISE W3 band) band can be found. For further understanding, we need more detailed measurements of the circumstellar masers of AGB stars with various types, pulsation periods, and masses.

References

Chiavassa, A., Freytag, B., & Schultheis, M. 2018, *A&A*, 617, L1
Cox, J. P. 1980, *Theory of Stellar Pulsation. (PSA-2)*, Volume 2. John P. Cox. ISBN: 9781400885855. Princeton University Press, 1980
Feast, M. W. 2009, *AGB Stars and Related Phenomena*, 48
Gaia Collaboration, Vallenari, A., Brown, A. G. A., *et al.* 2022, arXiv:2208.00211
Honma, M., Kijima, M., Suda, H., *et al.* 2008, *PASJ*, 60, 935
Justtanont, K., Barlow, M. J., Blommaert, J., *et al.* 2015, *A&A*, 578, A115
Kamizuka, T., Nakada, Y., Yanagisawa, K., *et al.* 2020, *ApJ*, 897, 42
Karakas, A. I. & Lattanzio, J. C. 2014, *PASA*, 31, e030
Karambelkar, V. R., Adams, S. M., Whitelock, P. A., *et al.* 2019, *ApJ*, 877, 110
Kawaguchi, N., Sasao, T., & Manabe, S. 2000, *Proc.SPIE*, 4015, 544
te Lintel Hekkert, P., Caswell, J. L., Habing, H. J., *et al.* 1991, *A&AS*, 90, 327
Matsuno, M., Nakagawa, A., Morita, A., *et al.* 2020, *PASJ*, 72, 56
Nakagawa, A., Kurayama, T., Matsui, M., *et al.* 2016, *PASJ*, 68, 78
Nakagawa, A., Kurayama, T., Orosz, G., *et al.* 2018, *Astrophysical Masers: Unlocking the Mysteries of the Universe*, 336, 365
Nakagawa, A., Morita, A., Sakai, N., *et al.* 2023, *PASJ*, 75, 529

Nyu, D., Nakagawa, A., Matsui, M., *et al.* 2011, *PASJ*, 63, 63
Oyama, T., Kono, Y., Suzuki, S., *et al.* 2016, *PASJ*, 68, 105
Reid, M. J., Menten, K. M., Brunthaler, A., *et al.* 2019, *ApJ*, 885, 131
Takeuti, M., Nakagawa, A., Kurayama, T., *et al.* 2013, *PASJ*, 65, 60
Wright, E. L., Eisenhardt, P. R. M., Mainzer, A. K., *et al.* 2010, *AJ*, 140, 1868

Cosmic Masers: Proper Motion toward the Next-Generation Large Projects
Proceedings IAU Symposium No. 380, 2024
T. Hirota, H. Imai, K. Menten, & Y. Pihlström, eds.
doi:10.1017/S1743921323002582

(Sub)mm Observations of Evolved Stars

Elizabeth Humphreys[1,2], Suzanna Randall[3], Yoshiharu Asaki[2,4] and Per Bergman[5]

[1]European Southern Observatory (ESO) Vitacura, Alonso de Cordova 3107, Vitacura, Santiago, Chile. email: elizabeth.humphreys@alma.cl

[2]Joint ALMA Observatory (JAO), Alonso de Cordova 3107, Vitacura, Santiago, Chile

[3]ESO Headquarters, Karl-Schwarzschild-Strasse 2, Garching, Germany

[4]National Astronomical Observatory of Japan, Osawa 2-21-1, Mitaka, Tokyo 181-8588, Japan

[5]Department of Space, Earth and Environment, Chalmers University of Technology, Onsala Space Observatory, 43992 Onsala, Sweden

Abstract. Evolved stars on the asymptotic giant branch and red supergiants have multiple processes that can be studied in the (sub)mm, including stellar surfaces, circumstellar thermal gas and dust, and masers. Telescopes such as APEX and ALMA have opened the possibility to perform studies that are revealing new information on these, as well as on the role of binaries in shaping stellar winds and the evolution to planetary nebulae. Here, we discuss some recent results for (sub)mm observations towards evolved stars focusing particularly on masers. This includes SiO and water masers, as well as ALMA high angular resolution observations of HCN masers towards a carbon-rich star.

Keywords. Masers, stars: AGB and post-AGB, submillimeter

1. Introduction

Since the last maser IAU in 2017, there have been multiple exciting results for (sub)mm masers. In addition to those reported elsewhere in these proceedings, a CO maser was detected using ALMA towards oxygen-rich AGB star W Hya (Vlemmings *et al.* 2021). For water masers, results have included those for terahertz masers (Neufeld *et al.* 2021) and 658 GHz masers (Baudry *et al.* 2018). For HCN masers, recent results have included those of Menten *et al.* (2018), Wong *et al.* (2019), Fonfría *et al.* (2019) and Jeste *et al.* (2022). SiS masers have been imaged using ALMA towards IRC+10216 (Fonfría *et al.* 2018). Using the ALMA Large Programme ATOMIUM dataset, multiple interesting results have arisen for SiO and water masers (Homan *et al.* 2020, Homan *et al.* 2021, Etoka *et al.* 2022, Richards *et al.* 2022). In the remainder of this article we focus on three (sub)mm observations of evolved stars since the last maser IAU related to SiO masers/thermal CO, HCN masers and water masers respectively.

2. ALMA CO and SiO Observations towards GX Mon

Using ALMA, Randall *et al.* (2020) observed Mira variable GX Mon with the aim of targeting CO (2-1) emission and SiO maser emission. The CO map (Figure 1) reveals a complex circumstellar spiral arc structure. This structure is consistent with hydrodynamical models for an AGB star with a binary companion in an eccentric orbit (orbital separation 7 – 61 AU). While several other species (including SiO, SiS, SO_2, and CS) are detected in the data, only the SO (5–4) map shows a similar, but much weaker,

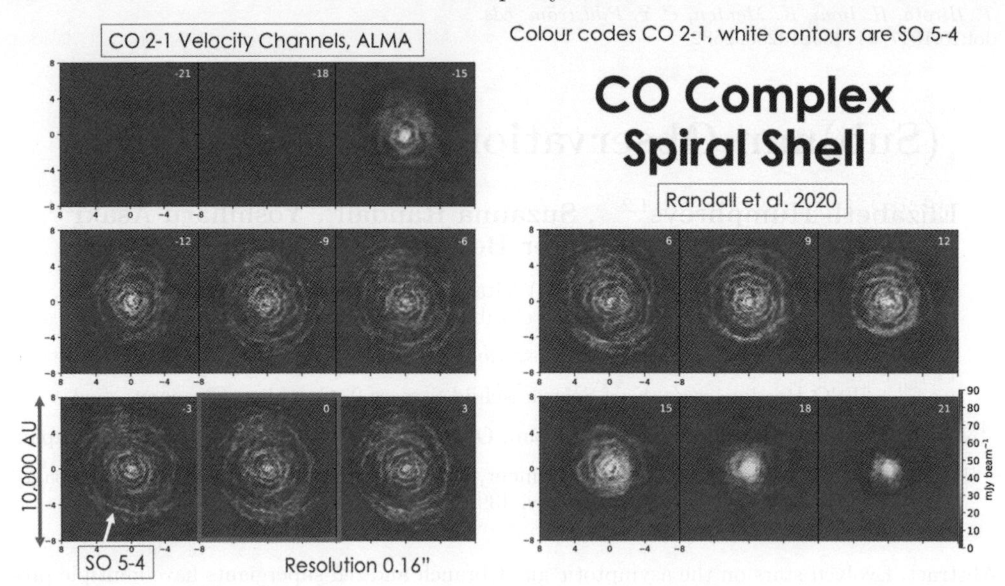

Figure 1. ALMA CO (2-1) observations of GX Mon with 0.16 arcsecond resolution after Randall *et al.* (2020). The different panels display velocity channels with that of the systemic velocity (adjusted to zero) outlined by the red box. Colour codes the CO 2-1 emission and white contours are SO 5-4 emission (seen more easily in Figure 2). The SO emission appears to be enhanced in an arc-like structure at about 5 arcseconds south of the central star. At the position of the SO arc, there is also an arc-like filament in the ACS/HST image in the F606W and F814W bands.

distribution as that imaged for the CO (Figure 2). Component fitting to the v=1 J=5-4 SiO maser line indicated maser emission distributed in a ring-type morphology near ($<$ 5 R_*) to the central star. There was no evidence for maser emission tracing material shaped by the binary interaction unlike e.g., Humphreys (2018), Homan *et al.* (2020).

3. ALMA High-angular Resolution HCN Maser Observations

Schilke, Mehringer and Menten (2000) and Schilke & Menten (2003) discovered HCN maser lines at 804.751 and 890.761 GHz towards carbon-rich stars. These lines can be very strong, for example towards IRC+10216 in the CSO observations of Schilke & Menten (2003) the peak line flux density was about 800 Jy at 805 GHz and about 8000 Jy at 891 GHz. Energy levels for the 891 GHz maser are >4000 K above ground state, therefore these masers should probe the inner circumstellar envelope. First imaging of these lines by Wong *et al.* (2019) using ALMA with a resolution of 0.1 arcseconds confirmed this. In those observations the extents of the HCN maser regions were found to be ~10 - 30 AU for V Hya and IRC+10216. New ALMA test data that were taken towards carbon-rich star R Lep, of the 891 GHz HCN maser emission towards that target (Asaki et al. submitted), are soon to be publicly released. The data have an angular resolution of 5 milli-arcseconds, making the image the highest angular resolution image ever made using ALMA to date (ALMA Configuration 10, Band 10).

4. APEX Observations of Water Masers at 437, 439, 471 & 474 GHz

Deguchi (1977) was the first paper, or among the first papers, to make specific predictions about which (sub)mm water lines should be observable. Among the lines predicted is the 474 GHz which was then discovered by Menten *et al.* (2008) and is one of the lines studied by Bergman & Humphreys (2020) towards a sample of 11 evolved stars using

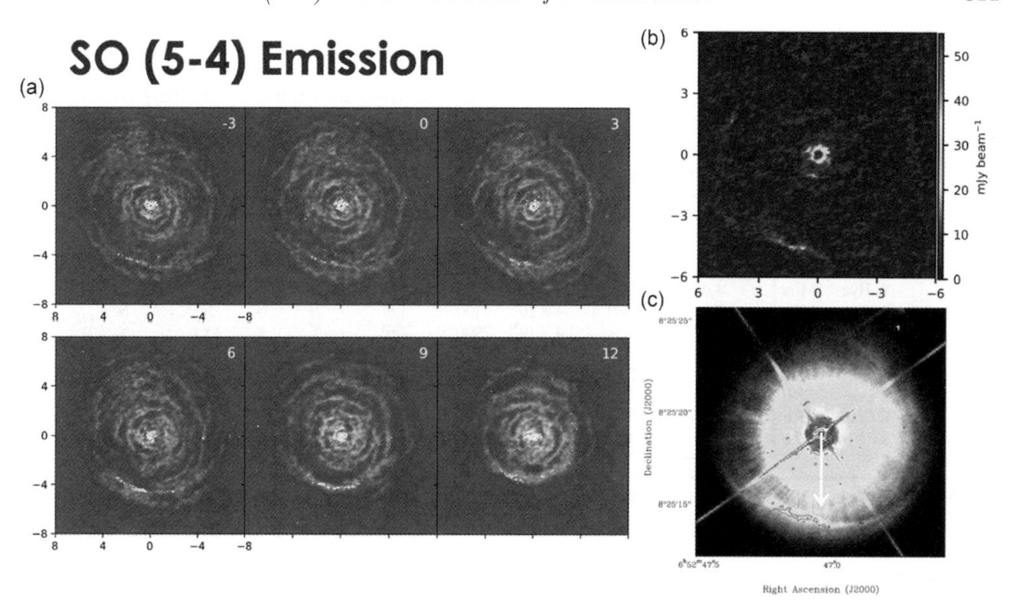

Figure 2. SO (5-4) observations of GX Mon after Randall *et al.* (2020). *a)* Selected velocity channels of CO 2-1 emission (colourscale) and SO 5-4 emission (white contours) from Figure 1; *b)* SO (5-4) image for velocity range -11.4 to -4.9 km s^{-1}, chosen to emphasise the morphology of the weak ring and arc-like emission; *c)* Composite ACS/HST image of GX Mon in the F606W and F814W bands, overlaid with the SO (54) emission from the above panel. The white arrow indicates the direction of the stellar proper motion (from Gaia DR2 via Simbad).

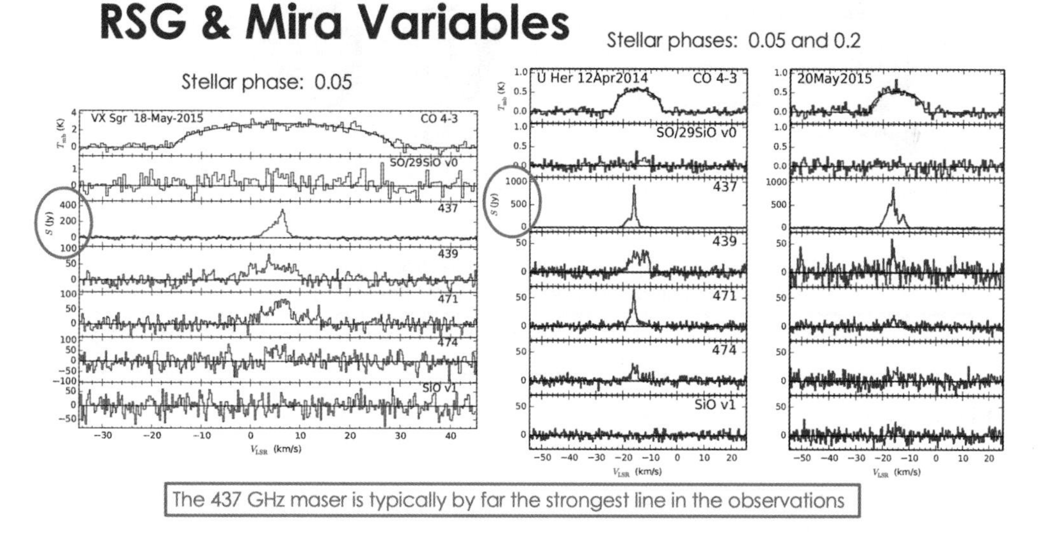

Figure 3. APEX observations towards Red Supergiant VX Sgr and Mira Variable U Her (2 epochs) after Bergman & Humphreys (2020). The observations of CO, SO, and ^{29}SiO are reported in main beam temperature using a velocity resolution of 0.5 km s^{-1}. Observations of the water lines and SiO v = 1 J = 11 - 10 are reported in flux density and are shown with a spectral resolution of 0.25 km s^{-1}. Of the water masers studied here towards 11 objects - at 437, 439, 471 & 474 GHz - the 437 GHz maser is typically by far the strongest in the observations which is not reproduced in the modelling.

Target List

Derived mass-loss rates are on the whole fairly similar

Star	Type	Magnitude [a] variation range (ΔV)	Period (days)	L_* ($L_\odot$)	T_* (K)	Distance [b] (pc)	Water masers detected in this work 437 GHz	439 GHz	471 GHz	474 GHz
VX Sgr	RSG	7.5	732	1.0×10^5	3500	1600	Y	Y	Y	Y
U Her	Mira	7.0	404	4.4×10^3	2700	266	Y	Y	Y	Y
RR Aql	Mira	6.7	395	7.3×10^3	2500	633	Y	N	N	N
W Hya	Mira	4.0	390	4.5×10^3	3100	104	Y	Y	Y	Y
R Hya	Mira	7.4	380	7.4×10^3	2100	124	N	N	N	N
RS Vir	Mira	7.6	354	4.4×10^3	2900	610	Y	Y	(Y)	Y
R Leo	Mira	6.9	310	2.5×10^3	2000	95	N	N	N	N
R Aql	Mira	6.5	270	4.4×10^3	2700	422	Y	Y	Y	N
R Dor	SRb	1.5	172	6.5×10^3	2400	59	N	N	(Y)	N
R Crt	SRb	...	160	2.1×10^4	2800	261	N	N	N	N
RT Vir	SRb	1.6	158	4.0×10^3	2800	226	N	N	N	N

Tentative finding that the presence of strong masers in these lines correlates with the amplitude of magnitude variation i.e., strong shocks could be needed to create the pumping conditions

Figure 4. Target list from Bergman & Humphreys (2020) indicating which sources have maser detections. There is a tentative finding that strong masers in these lines correlates with the amplitude of the stellar magnitude variation.

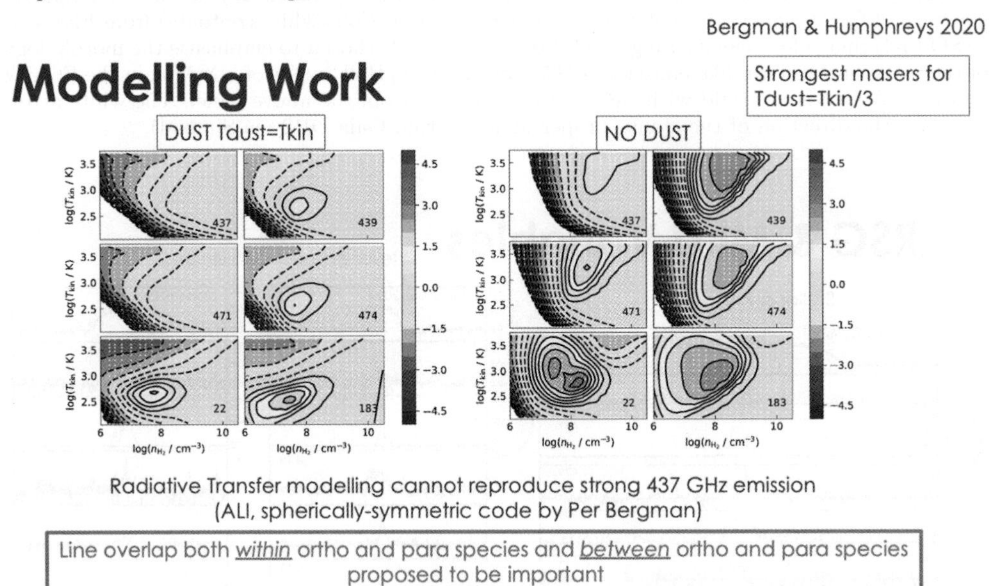

Figure 5. Example modelling results from Bergman & Humphreys (2020). We note that a case with no dust is not realistic for evolved star environments.

APEX (Figure 3). These lines originate from the so-called water "transposed backbone". From the sample of 11 evolved stars, 7 display one or more of the masers at 437, 439, 471, and 474 GHz. The fact that the maser lines are detected near the stellar velocity indicates that they are likely to originate from the inner circumstellar envelopes of the targets. Bergman & Humphreys (2020) tentatively link the presence of masers to the degree of variability of the target star, that is, masers are more likely to be present in Mira variables than in semi-regular variables (Figure 4). Typically, the 437 GHz line is the strongest maser line observed among those studied but that could not be reproduced in radiative transfer modelling (Figure 5). See also Yates, Field & Gray (1997) and Gray *et al.* (2016) for radiative transfer modelling results including these lines. Bergman & Humphreys (2020) propose that line overlap *between* ortho and para water may need to be incorporated in models in order to reproduce observations. Some potential overlaps of interest are identified in that paper.

5. Conclusions

Since the last maser IAU in 2017, work on masers is increasingly being performed in the (sub)mm wavelength range. However, there is much we still do not know about these masers in evolved stars and more imaging studies to determine locations in circumstellar envelopes are needed. With ALMA Cycle 10, Band 1 and flexible tuning mmVLBI are offered which may bring new opportunities for evolved star maser study.

References

Bergman, P. & Humphreys, E. M. L. 2020, *A&A*, 638, A19
Baudry, A., Humphreys, E. M. L., Herpin, F., *et al.* 2018, *A&A*, 609, A25
Decin, L., Montargs, M., Richards, A. M. S., *et al.* 2020, *Science*, 369, 1497
Deguchi, S. 1977, *PASJ*, 29, 669
Etoka, S., Baudry, A., Richards, A. M. S., *et al.* 2022, IAU Symposium, 366, 199
Fonfría, J. P., Fernández-López, M., Pardo, J. R., *et al.* 2018, *ApJ*, 860, 162
Fonfría, J. P., Fernández-López, M., Pardo, J. R., *et al.* 2019, IAU Symposium, 343, 398
Gray, M. D., Baudry, A., Richards, A. M. S., *et al.* 2016, *MNRAS*, 456, 374
Jeste, M., Gong, Y., Wong, K. T., *et al.* 2022, *A&A*, 666, A69
Homan, W., Montargs, M., Pimpanuwat, B., *et al.* 2020, *A&A*, 644, A61
Homan, W., Pimpanuwat, B., Herpin, F., *et al.* 2021, *A&A*, 651, A82
Humphreys, E. 2018, in Imaging of Stellar Surfaces, 13
Menten, K. M., Lundgren, A., Belloche, A., Thorwirth, S., & Reid, M. J. 2008, *A&A*, 477, 185
Menten, K. M., Wyrowski, F., Keller, D., & Kaminski, T. 2018, *A&A*, 613, A49
Neufeld, D. A., Menten, K. M., Durán, C., *et al.* 2021, *ApJ*, 907, 42
Randall, S. K., Trejo, A., Humphreys, E. M. L., *et al.* 2020, *A&A*, 636, A123
Richards, A. M. S., Assaf, K. A., Baudry, A., *et al.* 2022, IAU Symposium, 366, 204
Schilke, P., Mehringer, D. M., & Menten, K. M. 2000, *ApJ*, 528, L37
Schilke, P. & Menten, K. M. 2003, *ApJ*, 583, 446
Vlemmings, W. H. T., Khouri, T., & Tafoya, D. 2021, *A&A*, 654, A18
Wong, K. T. 2019, in ALMA2019: Science Results and Cross-Facility Synergies, 55
Yates, J. A., Field, D., & Gray, M. D. 1997, *MNRAS*, 285, 303

Cosmic Masers: Proper Motion toward the Next-Generation Large Projects
Proceedings IAU Symposium No. 380, 2024
T. Hirota, H. Imai, K. Menten, & Y. Pihlström, eds.
doi:10.1017/S1743921323002375

SiO maser line ratios in the BAaDE Survey

Megan O. Lewis[1], **Ylva M. Pihlström**[2] **and Loránt O. Sjouwerman**[3]

[1]Nicolaus Copernicus Astronomical Center, Polish Academy of Sciences, Warsaw, Poland 00-716. email: mlewis@camk.edu.pl

[2]University of New Mexico, Albuquerque, NM, USA 87131

[3]National Radio Astronomy Observatory, Socorro, NM, USA 87801

Abstract. Multi-transition SiO maser emission has been detected in over 10 thousand evolved stars across the plane of the Milky Way by the Bulge Asymmetries and Dynamical Evolution (BAaDE) survey. In addition to the large source catalog of the survey, the frequency coverage is also unprecedented: the J=1–0 (43 GHz) data cover seven separate transitions of SiO, and the J=2–1 (86 GHz) data cover ten SiO transitions. In contrast, most other SiO maser data only probe the SiO v=1 and v=2 at 43 GHz and/or the v=1 at 86 GHz. Our extended range allows for the derivation of SiO line ratios for a huge population of evolved stars, including those derived from rare transitions associated with ^{29}SiO and ^{30}SiO isotopologues. We examine how these ratios are affected by the specific combinations of transitions that are detected in a single source. Furthermore, we present a class of 'isotopologue dominated' sources where the ^{29}SiO transitions are the brightest in the 43 GHz spectrum. Finally, using Optical Gravitational Lensing Experiment (OGLE) light curves of our maser stars, changes in line ratios as a function of stellar phase are discussed.

Keywords. SiO masers, asymptotic giant branch stars, stellar variability, circumstellar envelopes

1. Introduction

Prior to observations conducted for the Bulge Asymmetries and Dynamical Evolution (BAaDE) survey about 4,000 stars hosting SiO masers had been identified in the literature, whereas the BAaDE SiO maser survey has so far identified over 9,800 Galactic SiO masers with the VLA, and over 5,000 with ALMA (see Sjouwerman's contribution to these proceedings). Therefore, the BAaDE sample is the largest and most uniform database of SiO masers, making it the most comprehensive set of observations with which to compare SiO maser modeling. Here we present line ratios and relationships between line properties and stellar phase for ∼9,800 BAaDE masers from the VLA portion of the survey. These results can be used to constrain models to study pumping mechanisms, and to tie observed line ratios to properties in the circumstellar envelope (CSE). With an unprecedented number of detected masers sources, many of which lie deep in the Galactic bulge, a major goal of the BAaDE survey is to establish these masers as sign-posts signaling CSE conditions in stars throughout the Galaxy. Radio observations are unique in their ability to probe these sources in the Galactic plane and bulge. For results on the ALMA portion of the survey see Stroh *et al.* (2019); for results of the thermal SiO v=0 transition see Dike *et al.* (2021).

The VLA portion of the BAaDE survey utilizes a spectral setup covering seven separate SiO transitions, six of which are maser transitions. The seventh, the SiO v=0, can be

Table 1. BAaDE detections for individual transitions.

J=1−0 transition	Frequency (GHz)	Number of detections	Percent (entire sample)	Percent (O-rich AGBs)
SiO v=0	43.424	118	0.6%	1.5%
SiO v=1	43.122	9297	51.2%	62.1%
^{29}SiO v=0	42.880	1461	8.0%	15.1%
SiO v=2	42.820	9177	50.5%	61.1%
^{29}SiO v=1	42.584	19	0.1%	0.2%
SiO v=3	42.519	2034	11.2%	17.1%
^{30}SiO v=0	42.373	223	1.2%	2.6%

either a maser or a thermal transition. Table 1 shows the detection rates of each of these transitions in both the full sample and in a slightly smaller sample which has been filtered for C-rich asymptotic giant branch (AGB) stars and other potential contaminating sources.

2. Ratio results

We calculate average line ratios for all pairs of transitions detected in the survey as well as for pairs of lines when a third line is present in the same spectrum; the presence of a third line often shifts the average line ratios, sometimes drastically. These shifts may indicate a change in CSE conditions or the presence of a line overlap, leading to deviations in the level populations associated with the masers. Results regarding a few specific ratios follow.

2.1. *Primary lines (^{28}SiO v=1 and v=2)*

The ^{28}SiO v=1 and v=2 lines are nearly always detected as a pair. Out of 9,727 detected sources 8,660 sources show both lines (89%), 583 are detected in v=1 without v=2, and 462 in v=2 without v=1 in our sample. These primary lines are almost always the brightest in a given source, and as such are nearly always present when any additional lines are detected. One notable exception is when the ^{29}SiO v=0 emission is the brightest (or only) emission in a source (see Lewis *et al.* 2020).

When both lines are detected in a typical O-rich AGB source they have similar flux densities, with the v=2 line measuring typically 93% of the v=1 flux density. The v=1 line being slightly brighter than the v=2 is consistent with other populations of Mira variables (Alcolea, Bujarrabal, & Gomez-Gonzalez 1990; Nyman *et al.* 1993).

2.2. *SiO v=3*

The ^{28}SiO v=3 line is also a maser line but is considerably fainter than the v=1 or 2 lines and is detected in 2,034 sources (11%). As compared to the v=1 line, the v=3 line is about 6 times dimmer. Sources that show the SiO v=1, 2, and 3 lines generally have $log(\frac{F(^{28}SiOv=2)}{F(^{28}SiOv=1)})$ values that are higher than sources that only show the SiO v=1 and the v=2. The fit to the distribution of all sources with v=1 and v=2 lines yields an average value of $log(\frac{F(^{28}SiOv=2)}{F(^{28}SiOv=1)}) = -0.034$ and width of 0.22, while the fit to the distribution of sources also containing the v=3 line gives an average of $log(\frac{F(^{28}SiOv=2)}{F(^{28}SiOv=1)}) = 0.010$ and a width of 0.17.

As it has been shown that the SiO v=1 transition is relatively brighter than the SiO v=2 in Mira variables, and the SiO v=2 is brighter in OH/IR stars, these line ratio trends may indicate that the SiO v=3 may be more common in thicker shells (like those of OH/IR stars).The idea that the SiO v=3 line is formed in denser environments is also

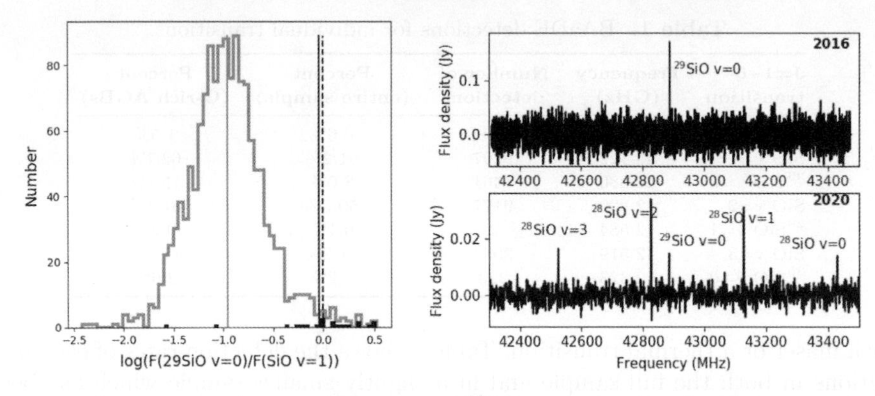

Figure 1. Left: Line ratios between SiO v=1 and ^{29}SiO v=0, showing that this ratio is typically around 10:1. Black histogram bins mark where the ^{29}SiO v=0 line is also detected, demonstrating that the line ratios typical of these detections have much higher ^{29}SiO v=0 flux than average. Right: Two spectra of the same source taken four years apart demonstrating that 43 GHz isotopologue dominated spectra can drastically change ratios to more typical behavior.

supported by the modeling of Desmurs *et al.* (2014), and by observations of its increased detection rate at stellar maximum as will be discussed in Section 3.

2.3. ^{29}*SiO v=0*

The ^{29}SiO v=0 line is a maser line and is detected in 1,475 sources (8%). This line has likely been detected in fewer than 100 sources outside of the BAaDE survey; it is reported as detected in 30 sources by the online maser resource maserdb (Sobolev *et al.* 2019). As compared to the v=1 line it is about 10 times dimmer, but shows a wide spread of ratios and is occasionally brighter than the primary lines.

We find an average of $log(\frac{F(^{28}SiOv=2)}{F(^{28}SiOv=1)}) = -0.067$ and a width of 0.19 when the ^{29}SiO v=0 is present. Therefore, unlike the SiO v=3 line, the ^{29}SiO v=0 line is preferentially found in sources where the $log(\frac{F(^{28}SiOv=2)}{F(^{28}SiOv=1)})$ value is slightly lower than average.

Although ^{29}SiO v=0 and SiO v=3 masers show preferences for sources with different $log(\frac{F(^{28}SiOv=2}{F(^{28}SiOv=1)})$ ratios (^{29}SiO v=0 preferentially occurs when this value is low, while SiO v=3 preferentially occurs when it is high), this tendency is marginal. There is considerable overlap in the primary line ratios which harbor each of these secondary transitions. There are 538 instances of ^{29}SiO v=0, and SiO v=1, 2, 3 all being detected in a single source. In fact, this is the most commonly detected combination of four separate transitions in one source.

2.4. ^{29}*SiO v=1*

The ^{29}SiO v=1 transition has only been reported in four sources prior to the BAaDE survey (Cho & Ukita 1995; Rizzo *et al.* 2021), and is additionally the SiO transition with the lowest detection-rate in the VLA portion of the survey. We find 19 detections (0.1%) of this transition, none of which have been reported before.

This transition is mostly detected in sources where the ^{29}SiO v=0 and/or ^{30}SiO v=0 line is relatively bright compared to the primary lines, and is the only secondary line that does not show the trend of being detected more often in sources with other bright lines. It is on average 55% as bright as the ^{29}SiO v=0, and 35% as bright as the SiO v=1. This line is detected in 1.2% of sources that also show the ^{29}SiO v=0 line and 2.7% of sources that also show the ^{30}SiO v=0, which is a considerable increase over the general 0.1%

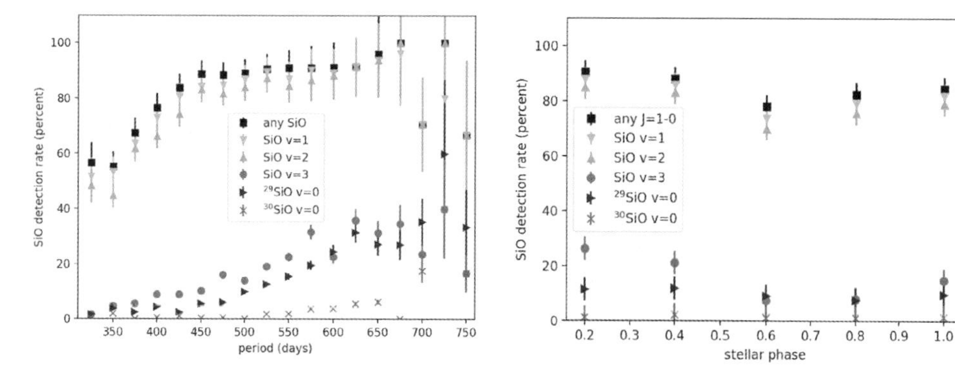

Figure 2. Left: Detection rates of the five most common SiO transitions covered by the BAaDE survey as a function of stellar period. Right: Detection rates of the five most common SiO transitions covered by the BAaDE survey as a function of stellar phase.

detection rate of this line. Sources with the ^{29}SiO v=1 line are also detected in the ^{29}SiO v=0 and ^{30}SiO v=0 lines in 94.7% and 31.6% of sources respectively. This 31.6% may be artificially low and affected by sensitivity as the ^{30}SiO v=0 is roughly half as bright as the ^{29}SiO v=0. In short, a detection of this line marks sources with high relative flux densities and especially high detection rates of the ^{29}SiO v=0 and ^{30}SiO v=0 lines.

2.4.1. Isotopologue dominated spectra

^{29}SiO v=1 detections occur in a small sample of sources where the isotopologue transitions are bright compared to the primary lines (Fig. 1), creating a stark contrast in the average ratios of $log(\frac{F(^{29}SiOv=0)}{F(^{28}SiOv=1)})$ and $log(\frac{F(^{30}SiOv=0)}{F(^{28}SiOv=1)})$ when ^{29}SiO v=1 is present as compared to when it is not. These observational results show that certain sources, perhaps only at certain stellar phases, display isotopologue-dominated spectra where line ratios are very atypical between the ^{28}SiO and other species. These sources were shown to have variable line ratios in (Lewis *et al.* 2020; see Fig. 1). The cause of 43 GHz isotopologue dominated spectra is unknown but is likely related to the maser pumping. Both high $log(\frac{F(^{29}SiOv=0)}{F(^{28}SiOv=1)})$ values and detections of the ^{29}SiO v=1 transition can be used to identify these sources.

3. Line ratios as a function of period and pulsation phase

By cross-matching BAaDE sources with Mira variable sources from OGLE we can also explore these line ratios as functions of period and phase. OGLE has observed 2785 of the BAaDE sources, 2342 of which were detected during the BAaDE survey. The detection rates of all seven SiO transitions are strong functions of stellar period as seen in Fig. 2. The detection rates for the primary lines rise as a function of period up to about 450 days before they flatten out, while the detection rates of the secondary lines continue to rise through 600 days and beyond. The physical explanation for this dependence is the subject of ongoing work.

As OGLE is a long-term monitoring survey with time-coverage overlapping the single-epoch BAaDE observations, we can often reliably check the I-band stellar phase of these cross-matched sources at the time of our maser observations. All detection rates also show a dependence on stellar phase, with maser detection rate being highest near infrared maximum and lowest near minimum (see Fig. 2). The SiO v=3 line shows the strongest dependence on phase, which fits the picture that this line forms in high densities.

Acknowledgments

The research leading to these results has received funding from the European Research Council (ERC) under the European Union's Horizon 2020 research and innovation program (grant agreement No. 951549), and the Polish National Science Centre grant MAESTRO 2017/26/A/ST9/00446.

References

Alcolea, J., Bujarrabal, V., & Gomez-Gonzalez, J. 1990, *A&A*, 231, 431

Cho, S.-H. & Ukita, N. 1995, *PASJ*, 47, 1

Desmurs, J. F., Bujarrabal, V., Lindqvist, M., Alcolea, J., Soria-Ruiz, R., & Bergman, P. 2014, *A&A*, 565, A127

Dike, V., Morris, M. R., Rich, R. M., Lewis, M. O., Quiroga-Nuñez, L. H., Stroh, M. C., Trapp, A. C., & Claussen, M. J. 2021, *AJ*, 161, 111

Lewis, M. O., Pihlström, Y. M., Sjouwerman, L. O., & Quiroga-Nuñez, L. H. 2020, *ApJ*, 901, 98

Nyman, L. A., Hall, P. J., & Le Bertre, T. 1993, *A&A*, 280, 551

Rizzo, J. R., Cernicharo, J., & Garcia-Miro, C. 2021, *ApJS*, 253, 44

Sobolev, A. M., Ladeyschikov, D. A., & Nakashima, J 2019, *Research in Astronomy and Astrophysics*, 19, 34

Stroh, M. C., Pihlström, Y. M., Sjouwerman, L. O., Lewis, M. O., Claussen, M. J., Morris, M. R., & Rich, R. M. 2019, *ApJS*, 244, 25

Cosmic Masers: Proper Motion toward the Next-Generation Large Projects
Proceedings IAU Symposium No. 380, 2024
T. Hirota, H. Imai, K. Menten, & Y. Pihlström, eds.
doi:10.1017/S1743921323002144

Patterns in water maser emission of evolved stars on the timescale of decades

Jan Brand[1], Dieter Engels[2] and Anders Winnberg[3]

[1]INAF-Istituto di Radioastronomia, Bologna, Italy; email: brand@ira.inaf.it

[2]Hamburger Sternwarte, Hamburg, Germany;

[3]Onsala Rymdobservatorium, Onsala, Sweden

Abstract. We present our past and current long-term monitoring program of water masers in the circumstellar envelopes of evolved stars, augmented by occasional interferometric observations. Using as example the Mira-variable U Her, we identify three types of variability: periodic (following the optical variation), long-term (years-decades) and short-term irregular (weeks-months). We show there are regions in the maser shell where excitation conditions are favourable, which remain stable for many years. Lifetimes of maser clouds in the wind-acceleration zone are of the order of up to a few years. Much longer lifetimes are found for the peculiar case of a maser cloud outside that zone (as in RT Vir), or in some cases where the motion of spectral features can be followed for the entire 2 decade monitoring period (as in red supergiant VX Sgr).

Keywords. AGB and post-AGB stars; water masers; monitoring

---

1. Introduction

In the circumstellar envelopes (CSEs) of O-rich evolved stars, physical conditions are often favourable for the excitation of masers of SiO, H_2O and OH, in order of increasing distance from the star. Water masers occur in that part of the CSE where the stellar wind accelerates. As the wind is not homogeneous, masing conditions change with time and with location in the CSE, and the maser emission line profiles are highly variable. Single spectra are not representative; for the study of the evolution of the stellar wind monitoring is therefore essential.

We have been carrying out monitoring campaigns of water masers in CSEs from 1987 to the present day, with some interruptions. The observations are made with the 32-m antenna in Medicina (1987-2011, 2015 and $>$ 2018) and the Effelsberg 100-m dish (1990-2002). Our sample (Brand *et al.* 2018) consists of three types of late-type stars: semi-regular variables (SRVs), Mira-variables and red supergiants (RSGs). By making use of occasional interferometric observations (our own, or taken from the literature) we can link peaks in the maser spectral profiles with their spatial counterparts, thus breaking the spatial degeneracy inherent to the line profiles.

Monitoring data of the SRVs have been discussed in Winnberg *et al.* (2008) and Brand *et al.* (2020), and papers on the Miras and RSGs are in preparation. Here we anticipate some results from those analyses.

2. Variabilities of brightness and timescales

Figure 1 illustrates the different types of variability we encounter in the monitoring data of Mira variables, using U Her as example. The left panel clearly shows a general decline in the integrated flux density of the maser emission over at least two decades. At

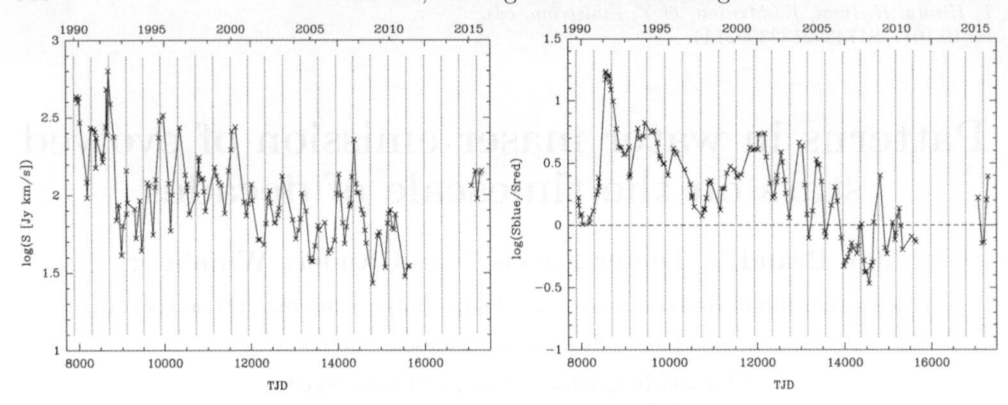

Figure 1. U Her. *left:* Total water maser flux vs. time (TJD=truncated Julian Day=JD-2440000.5). Periodic variability (vertical dotted lines indicate optical maxima), together with overall long-term decline and superimposed short-time irregularities. *right:* Ratio of total water maser flux on blue and red side (relative to the systemic velocity) of the maser spectra vs. time (from Winnberg et al. 2023, in prep.).

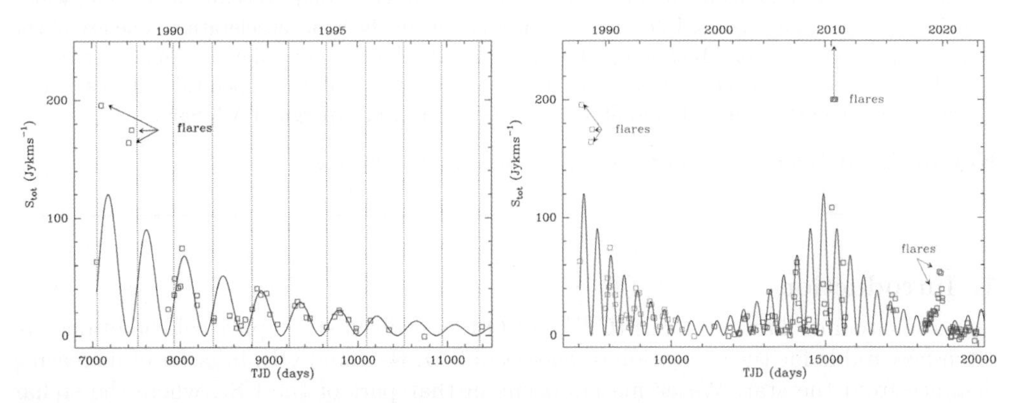

Figure 2. R Cas. *left:* 1987-1999. Dotted lines mark the nominal peaks in optical light. The fitted curve is an exponentially damped sinusoid with a period of 434 days (=P_{opt}). *right:* 1987–2022.

the same time it is evident that the maser emission responds, with a delay of ~2 months, to the optical variations of the star, as the decreasing flux is modulated with the optical period of 405 days. Superimposed on this are short-time irregular variations. The right panel shows how the emission on the blue side of the stellar velocity gradually becomes less dominant with time, in a somewhat erratic manner; periodic variation is also still recognisable. This change in the blue/red flux ratio indicates differences in excitation conditions in the front- and back caps of the maser shell. This is a consequence of the stellar wind being inhomogeneous, and leads to significant changes in the maser line profile with time. Except for the periodic component, the brightness variability properties of the SRVs (Brand *et al.* 2020) are as for the Mira variables.

Our VLA-observations 1990–1992 show that the brightest maser spots are found in the western part of a ring-like distribution in all 4 epochs (Winnberg *et al.* 2011). Maser spot maps (from the literature) taken in the 6.5 years around our observations show the same prevalence. This again points to an inhomogeneous wind, and to there being regions with favourable excitation conditions that are stable over long times. In this case, the timescale is of the order of a decade. (Winnberg et al. 2023, in prep.).

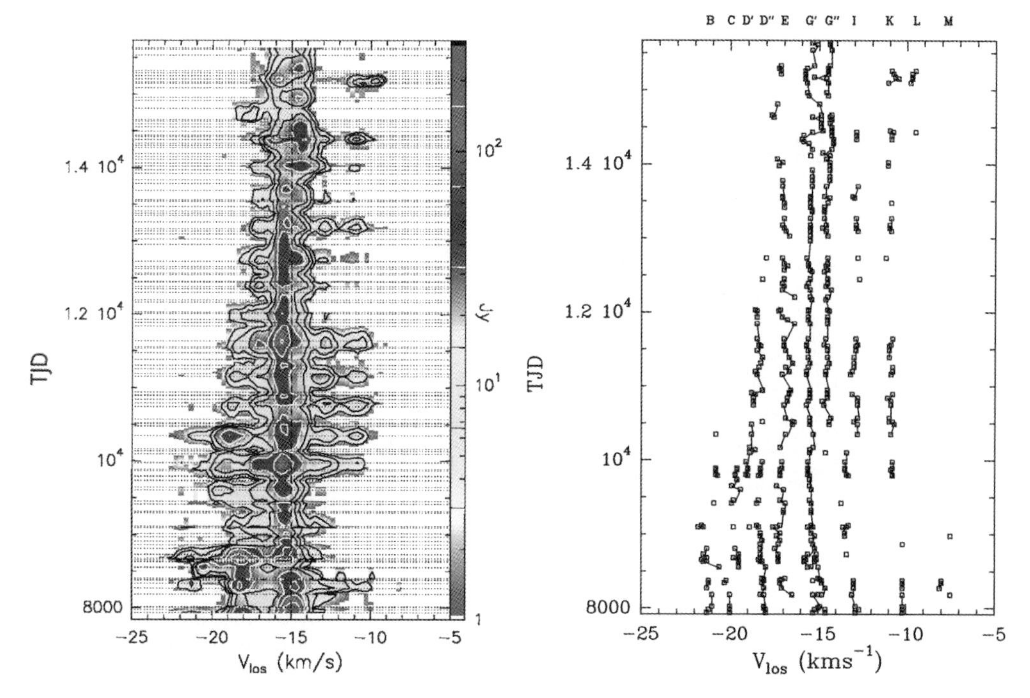

Figure 3. U Her 1990-2011. *left:* Flux density-velocity-time (FVt)-diagram. The stellar velocity is indicated with the vertical dashed line; dotted lines: days with an observation. *right:* velocities of the peaks of the Gaussian components fitted to the maser spectra, versus time (from Winnberg et al. 2023, in prep.).

3. R Cas: A unique brightness variability pattern

A unique long-term behaviour of brightness variation is shown by R Cas, which is one of the stars we continue to monitor since 1987. Figure 2 (left) shows the first decade of data (as reported by Brand *et al.* 2002), with the fitted damped sinusoid with a period equal to the optical one. Figure 2 (right) shows all data 1987-2022 and this shows a repeating pattern. We interpret this as the result of at least two maser clouds moving through a stationary region in the CSE with favourable conditions for maser excitation (cf. Sect. 2). Emission gradually increases in intensity after entry, reaching a maximum and then slowly decreases as the clouds move out of the region. The entire passage takes about 2 decades; the next maximum is expected in 2029.

4. Maser clouds: lifetimes

A remarkable result of the long-term monitoring program are the almost constant velocities observed for the major spectral features in SRVs and Mira variables, as shown for example in the FVt-diagram (Fig. 3, left) of U Her. For clarification, the plot on the right shows the velocities of the peaks in the emission profiles, from Gaussian fits. We see how some components are present for some time and then disappear, other lines are visible intermittently, while one or two components are visible more or less the whole time. All show almost no change in velocity. Knowing that the water masers originate in an accelerated wind allows one to put an upper limit to the lifetime of a maser spot. Assuming a spherically expanding wind, with parameters determined from our maser spot maps, we can calculate the maximum time between two observations before the velocity changes by more than one or two spectral resolution elements ($\lesssim$0.6 km s^{-1}). For U Her this is found to be of the order of 2-4 years. This is thus the maximum lifetime

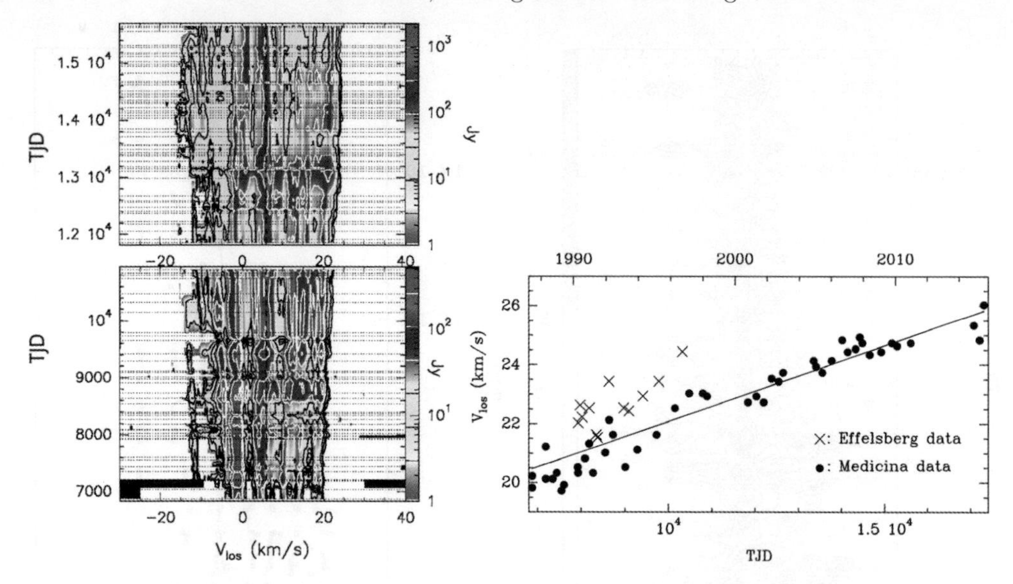

Figure 4. VX Sgr. *left:* FVt-diagram 1987-2011. Note the increase in the velocity of the component defining the red edge of the emission profiles. Dashed and dotted lines as in Fig. 3. Black areas indicate no data. *right:* Fit to sampled points along the red edge (1987-2015). Drawn line is an rms-weighted fit to the Medicina data, with a slope 0.188 ± 0.008 km s^{-1} yr^{-1} (from Engels *et al.* 2021).

of a maser spot for the velocity to remain constant. On the other hand, in RT Vir we found a maser component with a lifetime of at least 7.5 yr; in this case the maser cloud had likely already moved outside the acceleration zone (Brand *et al.* 2020).

5. Maser clouds: velocity drifts

In more extended H_2O maser shells with longer lifetimes of the emission regions, velocity drifts are indeed detected. This has been shown for example in the case of IK Tau (Brand *et al.* 2018). Other cases were found among the RSGs. The FVt-diagram of VX Sgr (Fig. 4, left) shows that the red edge of the maser emission increases in velocity with time. Figure 4 (right) shows the line-of-sight velocity of the spectral feature defining the red edge (cloud VXSgr-H_2O-1987/14), sampled at a few dates. The fitted line indicates an acceleration of (0.188 ± 0.008) km s^{-1} yr^{-1}. Using published interferometric maps (Murakawa *et al.* 2003) and a spherically expanding CSE-wind model, we find that during our 28 years of monitoring the maser cloud has moved radially outward from 140 to 230 au (Engels *et al.* 2021). The case demonstrates that single-dish monitoring with occasional interferometry can indeed trace the movement of maser-emitting clouds in regions with favourable conditions that are blown forward by the stellar wind.

6. Conclusions

Our monitoring programs show that to trace the structure variations of the stellar wind through water masers, in addition to making occasional interferometric maps, single-dish monitoring with medium-sized radio telescopes is an efficient and practical tool, as these are the only facilities to do time-domain astronomy on timescales of decades. Results from our on-going Medicina Long Project are posted on our dedicated website: https://www.ira.inaf.it/~brand/Medicina-monitoring.html.

References

Brand, J., Baldacci, L., & Engels, D. 2002, in: V. Migenes & M.J. Reid (eds.), *Cosmic Masers: From Proto-Stars to Black Holes*, Proc. IAU Symposium No. 206 (San Francisco: ASP), p. 310

Brand, J., Engels, D. & Winnberg, A. 2018, in: A. Tarchi, M.J. Reid & P. Castangia (eds.), *Astrophysical Masers: Unlocking the Mysteries of the Universe*, Proc. IAU Symposium No. 336 (CUP), p. 393

Brand, J., Engels, D., & Winnberg, A. 2020, *A&A* 644, A45

Engels, D., Brand, J., & Winnberg, A. 2021, Poster at: *The Origin of Outflows in Evolved Stars*, IAU Symposium No. 366. Leuven, Belgium. Zenodo. https://doi.org/10.5281/zenodo.5702619

Murakawa, K., Yates, J. A., Richards, A.M.S., & Cohen, R. J. 2003, *MNRAS* 344, 1

Winnberg, A., Engels, D., Brand, J., Baldacci, L., & Walmsley, C.M. 2008, *A&A* 482, 831

Winnberg, A., Brand, J., & Engels, D. 2011, in: F. Kerschbaum, T. Lebzelter & R.F. Wing (eds.), *Why Galaxies Care about AGB Stars II: Shining Examples and Common Inhabitants*, *ASP-CS* 445, p. 375

Cosmic Masers: Proper Motion toward the Next-Generation Large Projects
Proceedings IAU Symposium No. 380, 2024
T. Hirota, H. Imai, K. Menten, & Y. Pihlström, eds.
doi:10.1017/S174392132300217X

Results of KVN Key Science Program for evolved stars

Youngjoo Yun[1], Se-Hyung Cho[1], Dong-Hwan Yoon[1], Haneul Yang[1], Richard Dodson[2], María J. Rioja[2,3] and Hiroshi Imai[4]

[1]Korea Astronomy and Space Science Institute, 776 Daedeok-daero, Yuseong-gu, Daejeon 34055, Republic of Korea. email: yjyun@kasi.re.kr

[2]International Centre for Radio Astronomy Research, The University of Western Australia, 35 Stirling Highway, Western Australia, Australia

[3]Observatorio Astronómico Nacional (IGN), Alfonso XII, 3 y 5, E-28014 Madrid, Spain

[4]Center for General Education, Institute for Comprehensive Education, Kagoshima University 1-21-30 Korimoto, Kagoshima 890-0065, Japan

Abstract. We present the results of KVN Key Science Program (KSP) for evolved stars, which was launched in 2014. The first phase of KSP ended in June 2020 and the second phase started in October 2020. The goal of KSP is to study the physical characteristics of the evolved stars by observing the spatial distribution and temporal variability of the stellar masers at four frequency-bands (K, Q, W and D bands). The 22 GHz H_2O maser is usually observed from the outer part of circumstellar envelopes compared to the 43, 86, 129 GHz SiO masers, thus the kinematic links between these regions can be studied by the multi-frequency simultaneous observations of KSP along the stellar pulsation cycles. This eventually enable us to study the enormous mass-loss rate of evolved stars, and the accumulated results from KSP are expected to shed light on the study of the late stage of the stellar evolution.

Keywords. masers, radiative transfer, techniques: interferometric, stars: AGB and post-AGB, stars: circumstellar matter, stars: mass loss

---

1. Introduction

The SiO and H_2O masers are commonly observed from many oxygen-rich evolved stars and enable us to see the innermost part of circumstellar envelopes (CSEs) which consist of thick dust and gas layers ejected from the central stars. The masing conditions of the 22 GHz H_2O maser are usually fulfilled at the outer part of the CSEs compared to those of the 43, 86, and 129 GHz SiO masers (Richards *et al.* 2020). KVN can simultaneously observe the stellar masers at four frequency-bands, so the spatial distributions of individual masers can be directly compared each other in the KVN VLBI images. Therefore, the physical conditions as a function of the distance from the central stars can be estimated by KVN observations at once. Temporal variability of the intensities and the internal motions of the stellar masers can trace the movements of the gas and dust in the CSEs, which are closely related to the stellar pulsations (Höfner & Olofsson 2018). In order to study the mass-loss processes initiated in the inner part of the CSEs of evolved stars, the KVN Key Science Program (KSP) was launched in 2014. KSP monitoring observations toward the evolved stars have provided the dynamical characteristics of the stellar masers which enable us to trace the mass-loss processes and to study the origin of the morphological changes during the asymptotic giant branch (AGB) to the planetary nebula (PN). Many KSP data have shown the unprecedented results of the stellar masers, which have been already published. Here we introduce some preliminary results, which are prepared for more comprehensive studies.

2. Source selection and observations

In the first phase of KSP (2014 - 2020), 16 sources were selected, from which the strong stellar masers were detected in preceding KVN single dish surveys. Seven sources have been observed in the second phase of KSP since 2020; 3 sources are newly added and the rest are overlapped with those of the first phase KSP. Most of the sources are oxygen-rich AGB stars and red supergiants except χ Cyg which is a S-type star.

The H_2O $6_{16}-5_{23}$ (22.2 GHz) maser, the SiO v=1, 2, J=1→0 (43.1 GHz, 42.8 GHz) masers, and the SiO v=1, J=2→1, J=3→2 (86.2 GHz, 129.3 GHz) masers have been simultaneously observed toward the sources every month except the maintenance season (July and August). The receiving bandwidth is 256 MHz which consists of 16 base band channels (BBCs) yielding 1 Gbps recording rate. The velocity channel resolutions are 0.42, 0.22, 0.11, and 0.07 km s^{-1} for K, Q, W, and D bands respectively, which are achieved by the Distributed FX (DiFX) software correlator (Deller *et al.* 2007).

Basic calibration of the observed data has been done using the Astronomical Image Processing System (AIPS), and further calibration such as the phase referencing process has been applied to the multi-band data using the source frequency phase referencing (SFPR) method (Dodson *et al.* 2014). The relative spatial distributions between the masers observed at four frequency-bands can be precisely determined by the SFPR method. Large amounts of the data accumulated from the long-term monitoring observations have been effectively reduced through the pipeline processes.

3. Preliminary results of several sources

Results of the first phase KSP have been published and also combined with those of the second phase for further comprehensive studies. The combined results cover the stellar pulsation period several times for most of the target sources so they enable us to trace the dynamical evolution of the CSEs affected by the pulsation-driven shock propagation. The spatial distributions of the masers are significantly different from source to source and also show the distinct temporal variability along the pulsation period of individual stars.

Figure 1 shows the contour maps of the masers observed at four frequencies toward VY CMa which is a supergiant showing locally-developed mass ejection (Decin *et al.* 2016). The spoke-like features (Zhang *et al.* 2012) of the SiO masers show the linearly-distributed density enhancement implying the mass loss along the radial direction. Therefore, the position of the central star can be determined by finding the intersection of the extended directions of individual spoke-like features. The north-east region is crowed with the SiO maser features in overall elliptical distribution, which implies the asymmetric motions in the inner part of the CSE. The asymmetric distributions of the SiO masers around VY CMa have been consistently observed during the KSP observations. The south-east part of the 22 GHz H_2O maser appears to be associated with the interface between the stellar wind and the large dense clump (Richards *et al.* 2014).

Figure 2 shows the typical ring-like structures of the SiO masers around IK Tau, the oxygen-rich Mira variable whose pulsation period is about 470 days (Hale *et al.* 1997). The left panel of Fig. 2 shows the results observed in January 2016 and the right panel shows the results observed in February 2020. The corresponding stellar phases are about 0.3 and 0.5 respectively. The circular distributions in January 2016 changed to the elliptical distributions in February 2020. The morphological changes of the SiO maser distributions from circle to ellipse or vice versa occur during the KSP monitoring observations. This might be related to the intrinsic anisotropy of the stellar pulsation

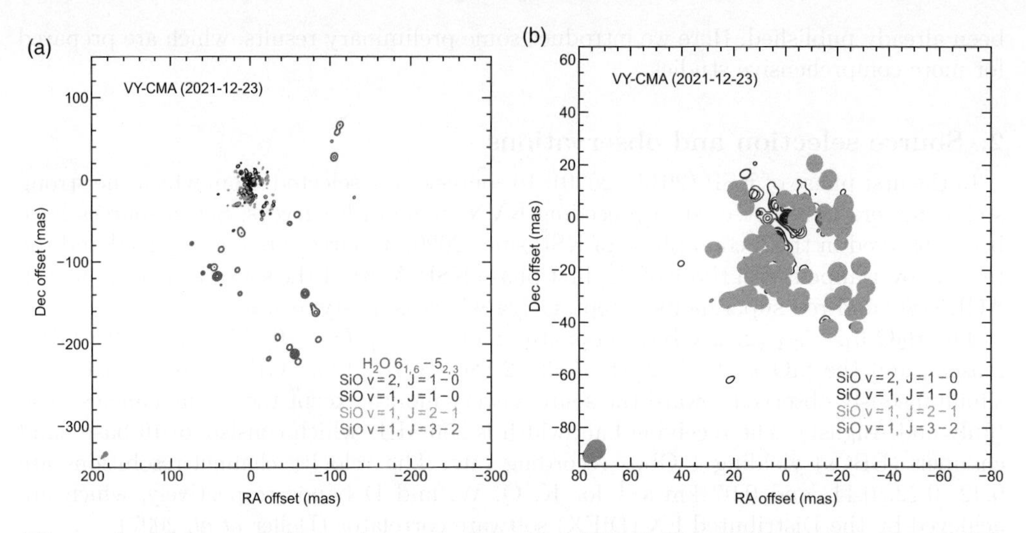

Figure 1. (a) Integrated intensity maps of the 22.2 GHz H_2O and 42.8/43.1/86.2/129.3 GHz SiO masers of VY CMa.(b) Enlarged plot of the SiO maser region.

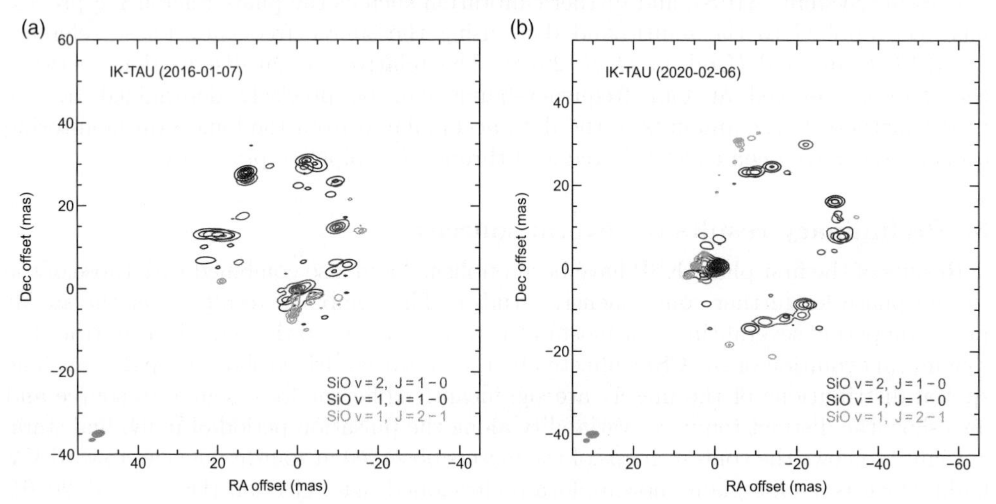

Figure 2. (a) Contour maps of 42.8/43.1/86.2 GHz SiO masers of IK Tau observed in January 2016.(b) Contour maps of 42.8/43.1/86.2 GHz SiO masers of IK Tau observed in February 2020.

and the high clumpiness of the CSE materials. Furthermore, the direction of the major axis of elliptical shape changed along the stellar pulsation, e.g., NE-SW direction in April 1996 (Boboltz & Diamond 2005) and NW-SE direction in February 2020 of KSP results. The relative spatial distributions between the different SiO masers around IK Tau show a somewhat consistent aspect, i.e., the v=2, J=1→0 SiO maser occurs in innermost region and the v=1, J=2→1 is detected in outermost region as shown in Fig. 2.

Figure 3 shows the velocity spot map of the v=1, J=2→1 SiO maser around W Hya, the oxygen-rich AGB star. Its distance is about 100 pc (van Leeuwen 2007), and the mass-loss rate is in the range between 10^{-7} and 1.5×10^{-7} $M_{\odot}\ yr^{-1}$ (Maercker *et al.* 2008; Khouri *et al.* 2014). Previous studies (Ohnaka *et al.* 2016; Khouri *et al.* 2020) found the strong visible and near-IR emissions from the clumpy dust envelopes in the north part. The v=1, J=2→1, J=3→2 SiO masers of W Hya show the bulging features in north and north-west parts of the ring-like structure in KSP observations. The north

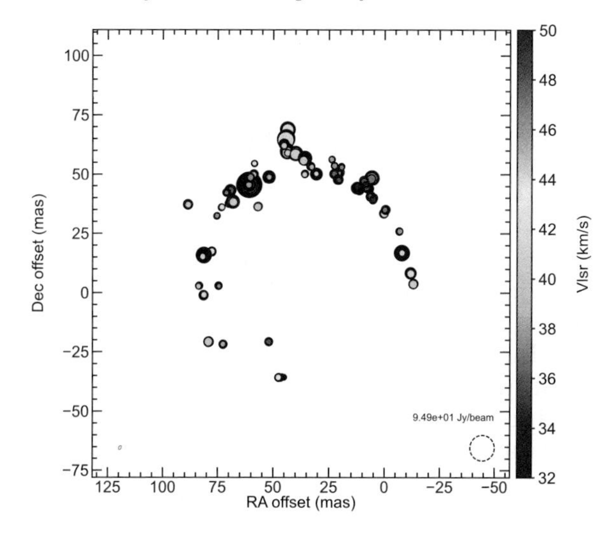

Figure 3. Velocity spot map of the v=1, J=2→1 (86.2 GHz) SiO maser emitted from the CSE of W Hya observed in December 2016. The stellar velocity is about 42 km s^{-1}. The maser intensities are represented by the circle areas varying with a logarithmic scale relative to the maximum flux density of 94.9 Jy $beam^{-1}$ indicated by a dotted circle at bottom-right corner.

clumpy structure lies at the stellar velocity of 42 km s^{-1} while the north-west bulge appears to be blue-shifted with respect to the stellar velocity. KSP observations have monitored the locally-enhanced mass ejection initiated in the inner part of the CSE of W Hya, which help us to study the mass-loss mechanism of AGB stars.

4. Summary

A large amount of data for the masers of evolved stars has been accumulated from the successful operation of KSP. The SFPR processes enable KSP to yield the astrometrically-registered maser images with a high accuracy up to sub-milliarcseconds (Dodson *et al.* 2014). The relative spatial distributions between the stellar masers have provided the crucial information to study the complicated physical environments of the CSEs of evolved stars. Furthermore, the results of long-term monitoring observations toward the stellar masers show the dynamical evolution of clumpy structures and the kinetic links between the inner part and the outer part of CSEs, thus they lead us to trace the mass-loss processes at the late stage of stellar evolution.

References

Boboltz, D. A., & Diamond, P. J. 2005, *ApJ*, 625, 978
Decin, L., Richards, A. M. S., Millar, T. J., *et al.* 2016, *A&A*, 592, A76
Deller, A. T., Tingay, S. J., Bailes, M., & West, C. 2007, *PASP*, 119, 318
Dodson, R., Rioja, M. J., Jung, T.-H., *et al.* 2014, *AJ*, 148, 97
Hale, D. D. S., Bester, M., Danchi, W. C., *et al.* 1997, *ApJ*, 490, 407
Höfner, S., & Olofsson, H. 2018, *A&AR*, 26, 1
Khouri, T., de Koter, A., Decin, L., *et al.* 2014, *A&A*, 561, A5
Khouri, T., Vlemmings, W. H. T., Paladini, C., et al. 2020, *A&A*, 635, A200
Maercker, M., Schöier, F. L., Olofsson, H., Bergman, P., & Ramstedt, S. 2008, *A&A*, 479, 779
Ohnaka, K., Weigelt, G., & Hofmann, K. H. 2016, *A&A*, 589, A9
Richards, A. M. S., Impellizzeri, C. M. V., Humphreys, E. M., *et al.* 2014, *A&A*, 572, L9
Richards, A. M. S., Sobolev, A., Baudry, A., *et al.* 2020, *AdSpR*, 65, 780
van Leeuwen, F. 2007, *A&A*, 474, 653
Zhang, B., Reid, M. J., Menten, K. M., & Zheng, X. W. 2012, *ApJ*, 744, 23

Cosmic Masers: Proper Motion toward the Next-Generation Large Projects
Proceedings IAU Symposium No. 380, 2024
T. Hirota, H. Imai, K. Menten, & Y. Pihlström, eds.
doi:10.1017/S1743921323002570

The Astrometric Animation of Water Masers toward the Mira Variable BX Cam

Shuangjing Xu[1,2], Hiroshi Imai[3], Youngjoo Yun[1], Bo Zhang[2], María J. Rioja[4,5,6], Richard Dodson[4], Se-Hyung Cho[1,7], Jaeheon Kim[1], Lang Cui[8], Andrey M. Sobolev[9], James O. Chibueze[10,11], Dong-Jin Kim[12,13], Kei Amada[3], Jun-ichi Nakashima[14], Gabor Orosz[15], Miyako Oyadomari[3], Sejin Oh[1], Yoshinori Yonekura[16], Yan Sun[2,17], Xiaofeng Mai[2,17], Jingdong Zhang[2,17], Shiming Wen[2] and Taehyun Jung[1]

[1]Korea Astronomy and Space Science Institute, 776 Daedeok-daero, Daejeon 34055, Republic of Korea. email: sjxuvlbi@gmail.com

[2]Shanghai Astronomical Observatory, Chinese Academy of Sciences, Shanghai 200030, People's Republic of China

[3]Kagoshima University, 1-21-35 Korimoto, Kagoshima 890-0065, Japan

[4]The University of Western Australia, 35 Stirling Hwy, Crawley, Western Australia, 6009, Australia

[5]CSIRO Astronomy and Space Science, PO Box 1130, Bentley WA 6102, Australia

[6]Observatorio Astronómico Nacional (IGN), Alfonso XII, 3 y 5, 28014 Madrid, Spain

[7]Department of Physics and Astronomy, Seoul National University, Seoul 08826, Republic of Korea

[8]Xinjiang Astronomical Observatory, Chinese Academy of Sciences, Urumqi 830011, People's Republic of China

[9]Ural Federal University, 19 Mira Street, 620002 Ekaterinburg, Russia

[10]Centre for Space Research, North-West University, Potchefstroom 2520, South Africa

[11]University of Nigeria, Carver Building, 1 University Road, Nsukka, Nigeria

[12]Massachusetts Institute of Technology Haystack Observatory, Westford, MA 01886, USA

[13]Max-Planck-Institut für Radioastronomie, Auf dem Hügel 69, D-53121 Bonn, Germany

[14]Sun Yat-sen University, 2 Daxue Road, Tangjia, Zhuhai, People's Republic of China

[15]Joint Institute for VLBI ERIC, Oude Hoogeveensedijk 4, 7991PD Dwingeloo, Netherlands

[16]Center for Astronomy, Ibaraki University, 2-1-1 Bunkyo, Mito, Ibaraki 310-8512, Japan

[17]University of Chinese Academy of Sciences, Shijingshan, Beijing, 100049, People's Republic of China

Abstract. We report VLBI monitoring observations of the 22 GHz H_2O masers toward the Mira variable BX Cam. Data from 37 epochs spanning ∼3 stellar pulsation periods were obtained between May 2018 and June 2021 with a time interval of 3–4 weeks. In particular, the VERA dual-beam system was used to measure the kinematics and parallaxes of the H_2O maser features. The obtained parallax, 1.79±0.08 mas, is consistent with *Gaia* EDR3 and previous VLBI measurements. The position of the central star was estimated relied on *Gaia* EDR3 data and the center position of the 43 GHz SiO maser ring imaged with KVN. Analysis of the 3D maser kinematics revealed an expanding circumstellar envelope with a velocity of 13 ± 4 km s^{-1} and significant spatial and velocity asymmetries. The H_2O maser animation achieved by our dense monitoring program manifests the propagation of shock waves in the circumstellar envelope of BX Cam.

Keywords. masers—stars: individual (BX Cam)—stars:evolved, astrometry, kinematics

---

1. Introduction

The long-period variables (LPVs), such as Mira variables, are stars of low to intermediate masses that have reached the late stage of the asymptotic giant branch (AGB) phase. They are characterized by long-period (>100 d) variations in radius, brightness, and temperature, and exhibit an intense mass-loss phenomenon. Numerous details of the physics of late-stage stellar mass loss remain poorly understood, ranging from the wind launching mechanism(s) to the geometry and timescales of mass loss. Fundamentally, material on the surface of an evolved star gets colder and forms dust as it moves into interstellar space. The newly formed dust is accelerated by the stellar radiative pressure and forms an expanding envelope. However, such processes are complicated because the related phenomena occur in a dynamically and physically unstable region, including radial pulsations, shocks, magnetic fields, opacity changes due to dust and molecule formation, and large-scale convective segments in the stellar interior. One of the major challenges in understanding AGB winds is that mass loss often exhibits significant asymmetry on very different spatial (1–10^4 AU) and temporal scales (a few months to a few 10^4 years) (see the review of Höfner & Olofsson 2018 and their references therein).

Circumstellar SiO and H_2O masers are located, respectively, at the innermost part and the acceleration zone of the circumstellar envelope (CSE) around a typical oxygen-rich star (Richards 2012). The maser structures are composed of clusters of compact maser features (isolated gas clumps with sizes of ~1 AU), enabling us to measure the three-dimensional velocity field of the CSE using Very Long Baseline Interferometry (VLBI) with ultra-high spatial resolution (Imai et al. 2003). VLBI maser observations offer one of the most powerful tools to measure the dynamics and physical conditions of the stellar atmosphere, as demonstrated by a 78-epoch "movie" of the SiO masers in a Mira variable covering three pulsation cycles (~5 yr) (Gonidakis et al. 2013). Such a maser movie needs a reference frame for registering individual maps. Since the animated SiO masers formed an almost perfect ring, the maser maps could be registered with the fitted center of the ring. However, this is very challenging for most of the CSE masers with broken structures, e.g., H_2O masers.

We are tracing the asymmetric maser motions with the astrometric approaches with VLBI and *Gaia* (Gaia Collaboration et al. 2021), as demonstrated by the first "movie" of the H_2O masers in the Mira variable star BX Camelopardalis (BX Cam) (Xu et al. 2022). A maser animation derived based on astrometric steps enables us to visualize a radial expansion of the CSE masers and identify deviations of the maser motions from constant-velocity motions, such as radial accelerations, rotations, etc.

2. Observations and Methods

The monitoring observations of H_2O and SiO masers around BX Cam are carried out with the East Asian VLBI Network (EAVN) (Akiyama et al. 2022), as a part of the EAVN Synthesis of Stellar Maser Animations (ESTEMA) project. ESTEMA aims at intensive VLBI monitoring observations of CSE masers associated with LPVs of different pulsation periods (300—1000 d) over a few stellar pulsation cycles. The time cadence for a single observational epoch is 3–4 weeks, which is approximately equal to a time resolution of 1/20 stellar pulsation cycles of BX Cam (~22 d). The reported H_2O data of 37 epochs in total were obtained from 2018 May to 2021 June, spanning approximately three stellar pulsation periods ($P = \sim 440$ d).

EAVN enables us to simultaneously monitor four to eight lines of H_2O and SiO masers at four frequency (22/43/86/129 GHz) bands with the sub-array KVN (Korean VLBI Network). The absolute positions, proper motions, and annual parallaxes are determined using the dual-beam astrometry of the sub-array VERA (VLBI Exploration of Radio Astrometry) for masers (Nagayama et al. 2020) and using Gaia for the central star. We can only measure the absolute positions of the bright masers of BX Cam with the VERA dual-beam bona-fide astrometry, then propagate this to all of the maser spots in the EAVN image by registering the bright masers in the pair of images. The stellar position with respect to the H_2O maser distribution is also astrometrically registered by a ring of SiO masers using the source-frequency phase-referencing (SFPR) technique (Dodson et al. 2014) from the KVN data. After subtracting the position offset and amotion of the central star, the asymmetric maser maps can be registered on the reference frame accurately.

3. Results

3.1. *Periodic variation of the H_2O masers*

ESTEMA observations allow us to study the variability and periodicity of maser intensity on timescales of a few weeks to a few years, investigating the possible correlation with stellar pulsation. The American Association of Variable Star Observers (AAVSO) reported a period of 486 d† for BX cam, however, we (Xu et al. 2022) and Matsuno et al. (2020) determined a period of 440 d by reanalyzing the AAVSO database. The peaks of the H_2O maser intensity exhibit an average delayed time-lag $\sim 25 \pm 9$ days or phase-lag $\sim 0.06 \pm 0.02$ related to the peaks of optical intensity. The averaged time interval of H_2O maser peak intensity is $\sim 447 \pm 11$ days, which fits the period 440d better than 486d (AAVSO).

We investigated the spatial and velocity variation of maser features at different epochs. The shapes in the maser structures reveal the complex variation of the interior structure, such as rotation, expansion of the shape, flashing sub-structure, and changes in direction. The relative proper motions are not necessarily consistent between different cycles for some maser features. Almost all of the maser features exhibit radial velocity drifts that are not consistent between different cycles, indicating gradual behaviors related to the pulsation phases. We find that the light curves of the individual maser features or groups have different time lags relative to the optical light curve.

3.2. *Astrometry*

Few maser features can be detected with VERA Bona-Fide astrometry alone. Using the brightest maser feature (MF1 in Figure 1), we found different deviations from a common constant proper motion and conservatively obtained a parallax value of 1.79 ± 0.08 mas, which is consistent with 1.76 ± 0.10 mas from *Gaia* EDR3 (Gaia Collaboration et al. 2021) and 1.73 ± 0.03 mas from the previous VERA result (Matsuno et al. 2020) within a 1-sigma level. The absolute positions of the bright and compact maser spots with VERA astrometric data were obtained and used to register the VERA and EAVN maps with an accuracy of ~ 0.1 mas. Then, we estimated the proper motion results for all maser features with the EAVN data.

Gaia EDR3 provided the stellar position in optical astrometry; however, there may exist some position uncertainty for this AGB star (Xu et al. 2019). We registered the maps of the H_2O and SiO masers with the map taken with the KVN data using the SFPR

† `https://www.aavso.org/vsx/index.php?view=detail.top&oid=4634`

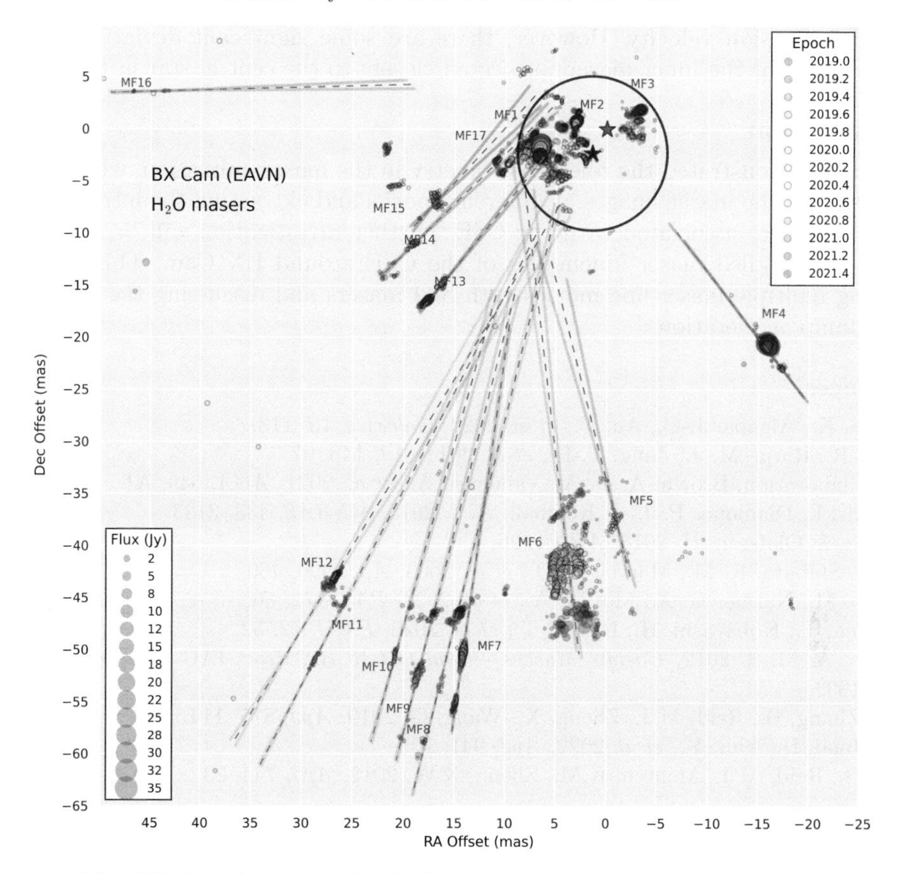

Figure 1. The BX Cam flow traced by H_2O maser features (MF#). The positional reference is *Gaia* EDR3 (red star). The black circle represents the fitted ring of SiO masers. The maser spots are shown in different colors to indicate different epochs, which have been astrometrically fixed. An auxiliary dashed line illustrates a possible trajectory of a maser feature from 2012 to 2022, assuming a constant velocity vector.

calibration technique (Dodson et al. 2014). We obtained the position of the central star, indicated as the center of the solid circle of the SiO maser ring. The differences between the estimated position and the *Gaia* EDR3 result are 1.4 mas ($\sim$3 sigma) in the east direction and 2.3 mas ($\sim$5 sigma) in the north direction on January 10, 2020.

3.3. *The expansion of the BX Cam flow traced by H_2O masers*

The constant velocity proper motions and spoke-like maser features in BX Cam 'point back' towards the central star (see Figure 1). However, the point-back directions of the maser features in different groups may converge into different areas, which may also have offsets from the location of the central star and the ring of SiO masers. Using the VERA data from 2012 to 2014, Matsuno et al. (2020) obtained a three-dimensional velocity of 14.8$\pm$1.4 km s^{-1} with three collimated flows. However, our result shows a wider range of 9 km s^{-1} (inner radii) to 19 km s^{-1} (outer radii) in different directions. These differences may be caused by the different pulsation phases originating from the characteristics of BX Cam, which trace the different expanding shells. The expansion velocity of H_2O masers is estimated to be 13.0 $\pm$ 3.7 km s^{-1}. The outer masers roughly exhibit radial expansions and are located at similar distances from the central star, with roughly a constant expansion velocity. However, there are some significant deviations from the spherical flow for the inner masers projected closely to the central star in the sky.

4. Conclusion

We have demonstrated the role of astrometry in the maser animation, which is crucial for registering the maser images at different epochs and extracting the intrinsic motions of maser features associated with the CSE. Further investigation will be conducted to explore the detailed maser kinematics of the CSE around BX Cam. This will involve analyzing multiple maser-line movies with SiO masers and discussing the possibility of maser clump acceleration.

References

Akiyama, K., Algaba, J.-C., An, T., *et al.* 2022, *Galaxies*, 10, 113.
Dodson, R., Rioja, M. J., Jung, T.-H., *et al.* 2014, *AJ*, 148, 97.
Gaia Collaboration, Brown, A. G. A., Vallenari, A., *et al.* 2021, *A&A*, 649, A1.
Gonidakis, I., Diamond, P. J., & Kemball, A. J. 2013, *MNRAS*, 433, 3133.
Höfner, S. & Olofsson, H. 2018, *ARAA*, 26, 1.
Imai, H., Shibata, K. M., Marvel, K. B., *et al.* 2003, *ApJ*, 590, 460.
Matsuno, M., Nakagawa, A., Morita, A., *et al.* 2020, *PASJ*, 72, 56.
Nagayama, T., Kobayashi, H., Hirota, T., *et al.* 2020, *PASJ*, 72, 52.
Richards, A. M. S. 2012, *Cosmic Masers - from OH to H0*, Proc. IAU Symposium No. 287, p. 199.
Xu, S., Zhang, B., Reid, M.J., Zheng, X., Wang, G. 2019, *ApJ*, 875, 114.
Xu, S., Imai, H., Yun, Y., *et al.* 2022, *ApJ*, 941, 105.
Zhang, B., Reid, M.J., Menten, K.M., Zheng, X.W. 2012, *ApJ*, 744, 23.

Cosmic Masers: Proper Motion toward the Next-Generation Large Projects
Proceedings IAU Symposium No. 380, 2024
T. Hirota, H. Imai, K. Menten, & Y. Pihlström, eds.
doi:10.1017/S1743921323001916

Water Fountain Sources Monitored in FLASHING

Hiroshi Imai[1,2], Kei Amada[3], José F. Gómez[4], Lucero Uscanga[5], Daniel Tafoya[6], Keisuke Nakashima[7], Ka-Yiu Shum[7], Yuhki Hamae[7], Ross A. Burns[8,9], Yosuke Shibata[7], Rina Kasai[7], Miki Takashima[7] and Gabor Orosz[10]

[1]Amanogawa Galaxy Astronomy Research Center, Graduate School of Science and Engineering, Kagoshima University, 1-21-35 Korimoto, Kagoshima 890-0065, Japan.
email: hiroimai@km.kagoshima-u.ac.jp

[2]Center for General Education, Institute for Comprehensive Education, Kagoshima University, 1-21-30 Korimoto, Kagoshima 890-0065, Japan; e-mail: hiroimai@km.kagoshima-u.ac.jp

[3]Graduate School of Science and Engineering, Kagoshima University, Kagoshima 890-0065, Japan

[4]Instituto de Astrofísica de Andalucía, CSIC, Granada E-18008, Spain

[5]Departamento de Astronomía, Universidad de Guanajuato, Guanajuato 36000, Mexico

[6]Department of Space, Earth and Environment, Chalmers University of Technology, Onsala 439 92, Sweden

[7]Faculty of Science, Kagoshima University, Kagoshima 890-0065, Japan

[8]National Astronomical Observatory of Japan, Mitaka, Tokyo 181-8588, Japan

[9]RIKEN, Wako, Saitama 351-0198, Japan

[10]Joint Institute for VLBI ERIC, Dwingeloo 7900, Netherlands

Abstract. We have investigated the spectral evolutions of H_2O and SiO masers associated with 12 "water fountain" sources in our FLASHING (Finest Legacy Acquisitions of SiO-/H_2O-maser Ignitions by Nobeyama Generation) project. Our monitoring observations have been conducted using the Nobeyama 45 m telescope every 2 weeks–2 months since 2018 December except during summer seasons. We have found new extremely high velocity H_2O maser components, breaking the records of jet speeds in this type of sources. Systematic line-of-sight velocity drifts of the H_2O maser spectral peaks have also been found, indicating acceleration of the entrained material hosting the masers around the jet. Moreover, by comparing with previous spectral data, we can find decadal growths/decays of H_2O maser emission. Possible periodic variations of the maser spectra are further being inspected in order to explore the periodicity of the central stellar system (a pulsating star or a binary). Thus we expect to see the real-time evolution/devolutions of the water fountains over decades.

Keywords. masers, stars: AGB and post-AGB, stars: mass loss, stars: winds, outflows

1. Introduction

The FLASHING project is dedicated to intensive single-dish monitoring observations of 22 GHz H_2O and 43 GHz SiO masers associated with "water fountain" sources (WFs), which are classified as sources with H_2O maser emission tracing high-velocity, collimated outflows or jets driven by stars in the transition of the AGB to post-AGB phase (Desmurs 2012; Gómez *et al.* 2017). The origin of the collimated outflows from dying stars is a clue

Table 1. Observation specification and setup.

Frequency range (GHz)	Receiver Name	Spectral window width (km s^{-1})	Velocity resolution (km s^{-1})	Beam size (FWHM) (arcsec)	Aperture efficiently (%)	Typical system noise temperature (K)
20 – 24	H22	±820†	0.41	73–74	61‡	100 to 70 §
42.5 – 44.5	H40	±420	0.41 – 0.43	39	55‡	150
40 – 46	Z45	±420	0.41 – 0.43	38	49¶	100

† With three spectrum windows. ‡ Aperture efficiency in the 2018 season. § Improved from 100 K to 70 K since the 2022–2023 season.
¶ Aperture efficiency in the 2022–2023 season in the H22+Z45 mode.

to explore the mechanism of their extreme stellar mass loss at a rate of up to $10^{-3}\ M_{\odot}\mathrm{yr}^{-1}$ for a period shorter than <200 yr (Khouri *et al.* 2021). The common envelope evolution is one of the most plausible scenarios of such mass loss (Khouri *et al.* 2021). Because of such possible short-lived events and periodic behaviors, it is expected to see some evolution and/or systematic variation in the WF masers over decades.

We have conducted single-dish monitoring observations of these maser sources in FLASHING with the new capability of simultaneous 22 GHz- and 43 GHz-band observations using the Nobeyama 45 m telescope (Okada *et al.* 2020). Some of them and the masers visible only in the southern hemisphere have also been observed with the Australia Telescope Compact Array (ATCA) (Uscanga *et al.* 2023). If new maser features are detected, follow-up interferometric observations are crucial for their localization (e.g., Imai *et al.* 2020; Uscanga *et al.* 2023).

2. FLASHING observations

Table 1 gives the typical specifications of the observed frequency bands in FLASHING. Spectra of 22 GHz H_2O and 43 GHz SiO (v =1, 2, 3 $J=1\rightarrow 0$) maser lines, as well as some thermal lines in the 22 GHz- and 43 GHz-bands, were simultaneously obtained in 16 spectral windows in total using the SAM45 digital spectrometer.

The FLASHING observations have been conducted since 2018 December, except during summer seasons. At first, a combination of the H22 (IEEE right- and left-hand circular polarization, or RHCP and LHCP) and the H40 (LHCP only) receivers was used. In the season of 2018 December–2019 May, the observations were conducted in the "Back-up Program", in which the opportunities of the observations were available in only when weather conditions were invalid for observations at higher frequency bands, due to too strong winds or too high humidity. However, some of the WFs could be monitored approximately every two weeks. In the seasons in 2019 December–2022 March, the observation sessions were allocated in a regular cadence (3–4 weeks). Since 2022 September, a combination of the H22 and the Z45 (two orthogonal linear polarization) receivers has been adopted. The integration time has been set so as to obtain typically a root-mean-square (rms) noise level of 0.1 Jy in the H_2O maser spectra in the velocity resolution of $\sim$1 km s^{-1}. As a result, the rms noise levels were yielded to be $\sim$0.2 Jy in the SiO maser spectra with the H40 receiver and $\sim$0.1 Jy with the Z45 receiver.

3. Early results of FLASHING

Through the FLASHING monitoring, we have tracked the evolution of H_2O maser spectra. Such evolution has been reported in IRAS 18286−0959 (Imai *et al.* 2020) and IRAS 18043−2116 (Imai *et al.* 2020; Uscanga *et al.* 2023), in which new spectral peaks of H_2O masers have broken the records of the top speed of the WF jets (up to $\sim$280 km s^{-1}). They may indicate the growth of the jet as clearly demonstrated in IRAS 18286−0959, in which the length of the distribution of H_2O maser features has doubled in a decade

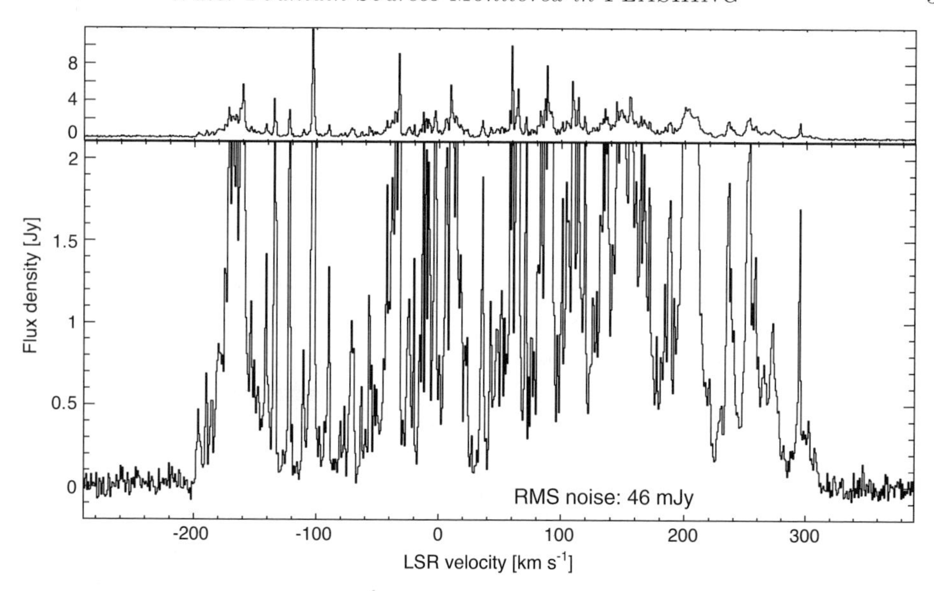

Figure 1. Spectrum of H_2O masers in IRAS 18286−0959 taken on 2023 March 30. The velocity components at the local-standard-of-rest (LSR) velocities higher than 200 km s^{-1} and lower than −100 km s^{-1} have appeared in only the recent FLASHING observations.

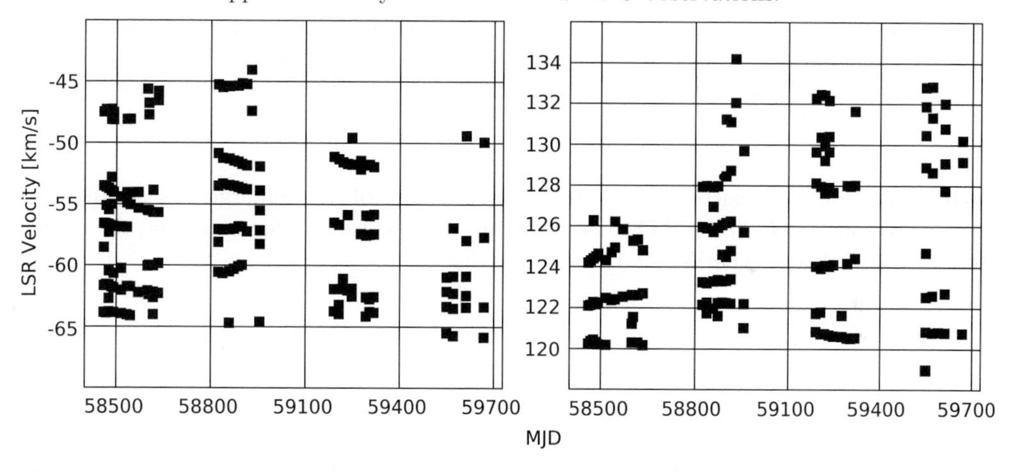

Figure 2. Spectral peak LSR velocities of the H_2O masers in the two major, high-velocity components in W 43A during 2018 December–2022 April. The drifts of the spectral peak velocities with respect to the systemic LSR velocity (∼35 km s^{-1}) indicate further accelerations of the maser clumps.

(Imai *et al.* 2020) and the total velocity width of the spectrum also has increased (Figure 1). On the other hand, due to too short visible duration of the spectral peaks, it is challenging to see the possible rapid changes in their velocities expected from the jet deceleration (Orosz *et al.* 2019).

If groups of the spectral peaks are changing their velocities in some systematic trends, they may indicate a specific dynamics of the outflow. Such an example has been found in W 43A (Figure 2), suggesting that the entrained material hosting the masers may be accelerated by the faster jet as suggested (Tafoya *et al.* 2020).

Note that we are also monitoring the devolution of the WFs, namely fading of H_2O and SiO masers: SiO masers in W 43A and H_2O masers in IRAS 17291−2147, OH 12.8−0.9, IRAS 18596+0315, IRAS 19134+2131, and K3−35 (H_2O masers in a planetary nebula).

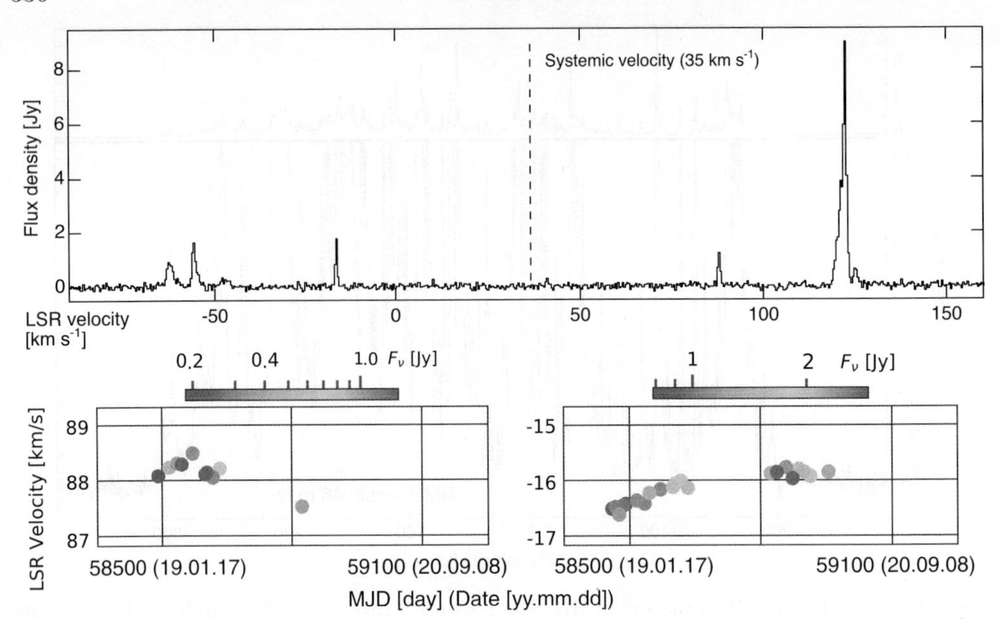

Figure 3. Top: Spectrum of H_2O masers in W 43A on 2019 April 30. One can see two or three pairs of double spectral peaks whose center velocities are well consistent with the systemic velocity of W 43A ($\sim$35 km s^{-1}, Tafoya *et al.* 2020). Bottom: Spectral peak LSR velocities of the H_2O masers in the two low-velocity components in W 43A during 2018 December–2020 May. The two components are located symmetrically with respect to the systemic LSR velocity ($\sim$35 km s^{-1}). A possible coherent flux density variation is found between the two components.

Such devolution is consistent with the predicted rate of WF emergences about every decade (Khouri *et al.* 2021). SiO masers are also unique targets for exploring the central stellar system driving the WFs (Amada *et al.* 2022).

4. Future perspectives

Periodicity of the maser spectra will be found if the central stellar system of the WF is composed of a long period variable such as an OH/IR star or a binary system (Tafoya *et al.* 2020; Khouri *et al.* 2021). The former case will be found in low velocity maser components, which may be associated with a relic of a spherical circumstellar envelope, as seen in W 43A (Figure 3) and IRAS 18286−0959 (Imai *et al.* 2013). They may indicate the WF hosted by the star that is still in the asymptotic giant branch (AGB) phase. The latter case will be found in some symmetry in the maser spectra and the spatial patterns. They have been confirmed in the VLBI maps of H_2O masers in W 43A (Tafoya *et al.* 2020), IRAS 18286−0959 (Imai *et al.* 2020), and IRAS 18113−2503 (Orosz *et al.* 2019). Localization of the dynamical center of the symmetric components will elucidate the origin of the periodicity and test the scenario of the common envelope evolution mentioned above. For such a test, imaging of thermal emission with the Atacama Large Millimeter-submillimeter Array (ALMA) will be crucial (Tafoya *et al.* 2020).

In both cases mentioned above, the complexity of the maser spectra affected by chaotic variation and the artificial periodicity due to time gaps of the monitoring program should be solved in future monitoring observations with higher cadence, say every several days. Even for a small number of the WFs ($\sim$15, Desmurs 2012; Gómez *et al.* 2017), it is challenging to sustain such observations with only one large telescope, such as the Nobeyama 45 m telescope. Therefore, it is indispensable nowadays to develop a collaboration of the large telescopes regularly monitoring the WFs, as conducted by the Maser Monitoring Organization (M2O).

References

Amada, K., Imai, H., Hamae, Y., *et al.* 2022, *AJ*, 163, 85
Desmurs, J.-F. 2012, *Cosmic Masers - from OH to* H_0, Proc. IAU Symposium 287, 217
Gómez, J. F., Suárez, O., Rizzo, J. R., *et al.* 2017, *MNRAS*, 468, 2081
Imai, H., Uno, Y., Maeyama, D., et al. 2020, *PASJ*, 72, 58
Imai, H., Nakashima, J., Yung, B. H. K., *et al.* 2013, *ApJ*, 771, 47
Khouri, T., Vlemmings, W. H. T., Tafoya, D., *et al.* 2021, *New Astron.*, 6, 275
Okada, N., Hashimoto, I., Kimura, K., *et al.* 2020, *PASJ*, 72, 7
Orosz, G., Gómez, J. F., Imai, H., *et al.* 2019, *MNRAS*, 482, L40
Tafoya, D., Imai, H., Gómez, J. F., *et al.* 2020, *ApJ*(Letters), 890, L14
Uscanga, L., Imai, H., Gómez, J. F., *et al.* 2023, *ApJ*, 948, 17

Cosmic Masers: Proper Motion toward the Next-Generation Large Projects
Proceedings IAU Symposium No. 380, 2024
T. Hirota, H. Imai, K. Menten, & Y. Pihlström, eds.
doi:10.1017/S1743921323002156

Evolution of the outflow traced by water masers in the evolved star IRAS 18043−2116

Lucero Uscanga[1], Hiroshi Imai[2], José F. Gómez[3], Daniel Tafoya[4], Gabor Orosz[5], Tiege P. McCarthy[6], Yuhki Hamae[7] and Kei Amada[8]

[1]Departamento de Astronomía, Universidad de Guanajuato, A.P. 144, 36000 Guanajuato, Gto., Mexico. email: luscag@gmail.com

[2]Center for General Education, Institute for Comprehensive Education, Kagoshima University 1-21-30 Korimoto, Kagoshima 890-0065, Japan

[3]Instituto de Astrofísica de Andalucía, CSIC, Glorieta de la Astronomía s/n, E-18008 Granada, Spain

[4]Department of Space, Earth and Environment, Chalmers University of Technology, Onsala Space Observatory, 439 92 Onsala, Sweden

[5] Joint Institute for VLBI ERIC, Oude Hoogeveensedijk 4, 7991 PD Dwingeloo, The Netherlands

[6]School of Natural Sciences, University of Tasmania, Private Bag 37, Hobart, Tasmania 7001, Australia

[7]Department of Physics and Astronomy, Faculty of Science, Kagoshima University 1-21-35 Korimoto, Kagoshima 890-0065, Japan

[8]Department of Physics and Astronomy, Graduate School of Science and Engineering, Kagoshima University 1-21-35 Korimoto, Kagoshima 890-0065, Japan

Abstract. We present the spectral and spatial evolution of H_2O masers associated with IRAS 18043−2116, a well-known water fountain hosting a high-velocity collimated jet, which has been found in the observations with the 45 m telescope of Nobeyama Radio Observatory and the Australia Telescope Compact Array. We found new highest velocity components of the H_2O masers, with which the resulting velocity spread of $\simeq 540$ km s^{-1} breaks the speed record of fast jets/outflows in this type of sources.

Keywords. masers — stars: AGB and post-AGB — stars: individuals (IRAS 18043−2116)

1. Introduction

Water fountains (WFs) are evolved stars, mostly in the post-asymptotic giant branch (post-AGB) phase that show H_2O maser emission tracing high-velocity collimated jets when they are observed at high-angular resolution (Imai 2007; Desmurs 2012). The velocity spread in their H_2O maser spectra is typically > 50 km s^{-1}, and can be as large as $\simeq$ 500 km s^{-1} (Gómez *et al.* 2011). Their H_2O maser emission thus traces significantly faster motions than the typical expansion velocities of circumstellar envelopes (CSEs) during the AGB phase (10–30 km s^{-1}; Sevenster *et al.* 1997). The short dynamical ages of the maser jets (5−100 yr; Imai 2007; Tafoya *et al.* 2020) may indicate that WFs represent one of the first manifestations of collimated mass-loss in evolved stars.

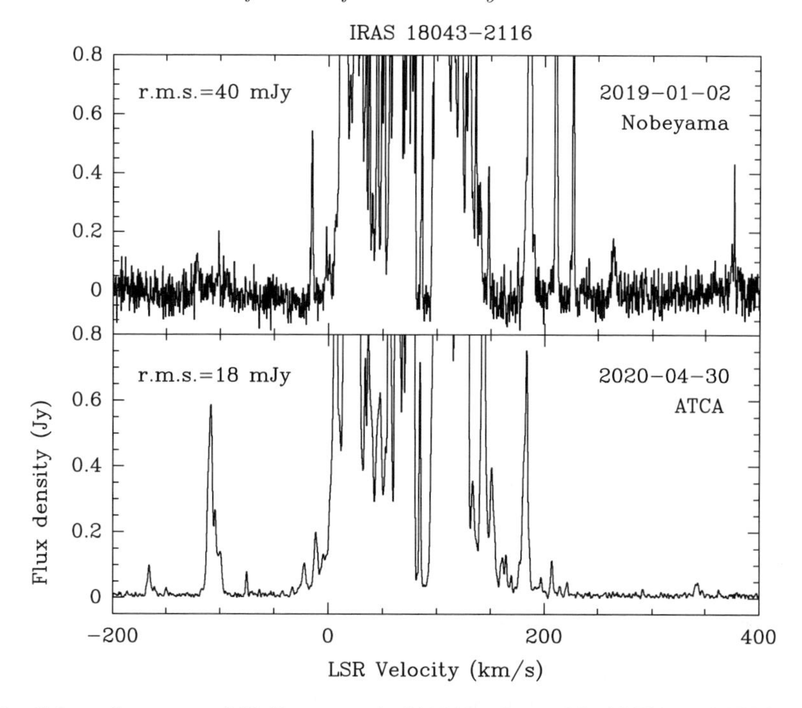

Figure 1. Selected spectra of H_2O masers in I18043 taken with ATCA and NRO telescopes. The root-mean-square (RMS) noise levels are indicated in each spectrum. Note the new highest velocity components at $V_{\rm LSR} \simeq 376$ km s^{-1} (top) and $V_{\rm LSR} \simeq -165$ km s^{-1} (bottom).

Previously, the H_2O maser spectra of IRAS 18043−2116 (hereafter I18043) presented a velocity spread of nearly 400 km s^{-1} (Walsh *et al.* 2009; Pérez-Sánchez *et al.* 2017). Here we present new detections of the highest velocity components of the masers in this WF yielded with the 45 m telescope of Nobeyama Radio Observatory (NRO) and the Australia Telescope Compact Array (ATCA), breaking the record of the top speed of the jet in I18043.

2. Observations

We observed I18043 within the monitoring program called FLASHING (Finest Legacy Acquisitions of SiO- and H_2O-maser Ignitions by the Nobeyama Generation). FLASHING is described in detail within this volume by Imai *et al.* (2023). Here we analyzed FLASHING observations of H_2O masers conducted during 2019 January – 2020 April (eight epochs), as well as a follow-up with the ATCA from 2020 April to 2021 March (four epochs). Observations with NRO had a total velocity coverage of ∼1600 km s^{-1} with a velocity resolution of 0.41 km s^{-1} while for ATCA observations the total velocity coverage was ∼3453 km s^{-1} with a similar velocity resolution of 0.42 km s^{-1}. Single-dish and interferometric observations of H_2O masers in WF sources are able to trace the evolution of the collimated jet or even measure directly the growth of the WF outflow. We complemented these studies with ALMA observations taken on 2019 January to estimate the systemic velocity of I18043 and compare the spatio-kinematical distribution of the masers with that of the molecular material.

3. Results

Figure 1 shows two spectra of H_2O masers of I18043 taken with the NRO 45 m telescope and ATCA on 2019 January 2 and 2020 April 30, respectively. Here the new high

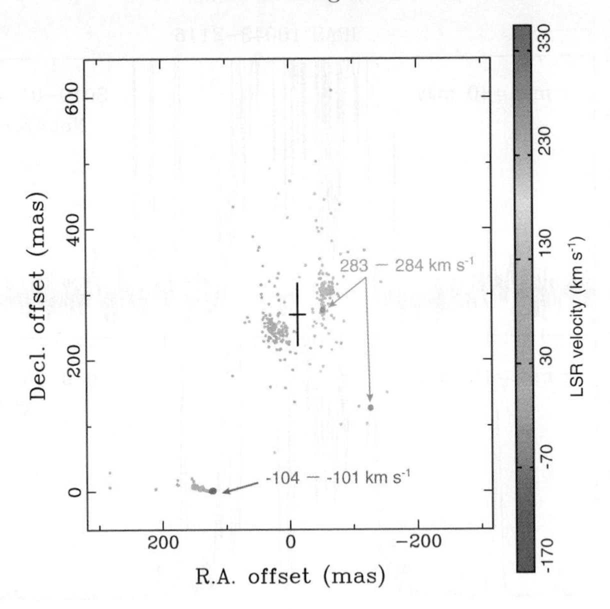

Figure 2. Map of H_2O masers in I18043 taken with ATCA on 2021 March 30. The maser distribution has a larger positional uncertainty in declination offset (up to 200 mas). The new blue-shifted components ($V_{LSR} \sim -102$ km s^{-1}) and the red-shifted components ($V_{LSR} \sim 284$ km s^{-1}) are highlighted in bigger filled circles. The position and the size of a black cross indicate the position and statistical error (3σ) of the continuum source. The synthesized beam is 1.42 arcsec×0.41 arcsec at a position angle of $-1.27°$ (Uscanga *et al.* 2023).

velocity components are clearly seen, red-shifted with respect to the systemic velocity of $V_{LSR} \simeq 87.0$ km s^{-1} km s^{-1} (Deacon, Chapman, & Green 2004) at $V_{LSR} \simeq 376$ km s^{-1} and blue-shifted at $V_{LSR} \simeq -165$ km s^{-1}. The new velocity spread ($\simeq 540$ km s^{-1}) is the largest ever detected in the spectra of H_2O masers of WFs, surpassing the case of IRAS 18113−2503 (velocity spread $\simeq$500 km s^{-1}; Gómez *et al.* 2011).

The spatial distribution of the H_2O masers observed with ATCA on 2021 March 30 is presented in Figure 2. Two main clusters of maser components can be distinguished. A continuum source toward the center of the maser distribution was also detected at 22 GHz with a flux density of $S_\nu = 0.97 \pm 0.09$ mJy. This continuum source exhibits an increase in the flux density, i.e., $S_\nu = 0.75 \pm 0.05$ mJy on 2020 April 30. Note that the new high velocity components ($V_{LSR} \sim -102$ km s^{-1}, $V_{LSR} \sim 284$ km s^{-1}) are farther away from the continuum source in right ascension (see Figure 2). The same behaviour was seen in other new high velocity components ($V_{LSR} \sim -165$ km s^{-1}, $V_{LSR} \sim 344$ km s^{-1}, Uscanga *et al.* 2023).

Figure 3 (left) shows the spectrum of the CO(J=2→1) emission observed with ALMA. The line consists of a strong central component and high-velocity wings. From the velocity of line peak of the central component, we derived a systemic velocity of $V_{LSR,sys} = 87 \pm 1$ km s^{-1} for I18043. This value is compatible with the systemic velocity obtained by Deacon, Chapman, & Green (2004) using OH maser observations. The locus of the CO(J=2→1) emission peak in each bin with a velocity width of 6 km s^{-1} is shown in Figure 3 (right). The observed dispersion of the locus in the R.A. offset indicates a wide-angle of the outflow. We note that there exists a clear velocity gradient in the R. A. offset of the CO emission from −20 to 20 km s^{-1} with respect to the systemic velocity. The wanders of the peak positions in the blue- and red-shifted CO emission components, for $V_{offset} = 10 - 70$ km s^{-1} are roughly consistent with the ones of H_2O masers. This

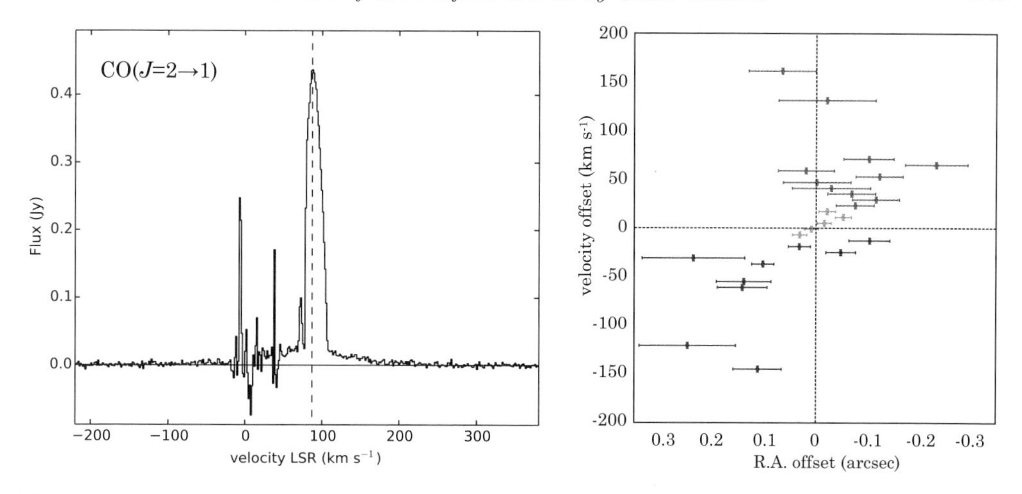

Figure 3. *Left*: ALMA spectrum of the CO(J=2→1) emission in I18043. The vertical dashed line indicates the systemic velocity of the source, $V_{\rm LSR,sys} = 87$ km s^{-1}. *Right*: Right Ascension offsets from the continuum peak position as a function of the velocity offset from the systemic velocity. The blue, green and red marks represent emission in the velocity ranges: $-170 < V_{\rm offset} < -20$ km s^{-1}, $-10 < V_{\rm offset} <$ 10 km s^{-1} and $10 < V_{\rm offset} <$ 170 km s^{-1}, respectively (Uscanga *et al.* 2023).

suggests that the H_2O masers are associated with the CO outflow itself, or possibly the shock regions in the outflow.

4. Discussion

Considering previous observations toward the WF I18043 (Walsh *et al.* 2009), we estimate that the spatial separation of the two main clusters of H_2O masers has doubled (45 mas vs 88 mas) in a period of about $\simeq$12.5 yr. Interestingly, the new highest velocity components possibly located at the outer outflow lobes are farther away from the center, indicating a rapid growth of the outflow triggered by an increase in the maximum outflow velocity.

The spatio-kinematical characteristics of the H_2O masers in this WF are different from those found recently in IRAS 18286−0959, where new high velocity components (>200 km s^{-1}) have also been discovered within FLASHING program (Imai *et al.* 2020). These components are located closer to the central stellar system than other high velocity components (50200 km s^{-1}). Overall the components present a point-symmetric distribution that could be related to a precessing jet. Among the known WFs, there is a variety of high-velocity collimated outflows traced by the H_2O masers. Monitoring observations combining single-dish and interferometric studies, including proper-motion measurements are crucial to discover the enigmas of these outflows in this particular evolutionary stage of the low-mass evolved stars.

Regarding the CO emission, the CO velocity wanders indicate a large opening angle of the outflow, within which the locations of different velocity components are scattered due to an inhomogeneity of the outflow velocity. We estimated a full opening angle of the outflow of ∼60° (Uscanga *et al.* 2023). Using VLBA data, Orosz (2017) estimated a dynamical age of the outflow associted with I18043 to be $t_{\rm jet} \lesssim 30$ yr.

CO emission has been also mapped in other WFs, such as W 43A and IRAS 15103−5754. In the first one, the CO emission traces a highly collimated jet (Tafoya *et al.* 2020); while in the second one, the CO emission is confined within a biconical structure suggesting a wide-angle outflow quite similar to the case of I18043, but with a different spatio-kinematics (Gómez *et al.* 2018). It seems that there is a wide variety of outflows exhibiting CO emission in WFs that requires further investigation.

Related to the continuum emission, from previous data of Pérez-Sánchez *et al.* (2017), we deduce that the continuum emission is consistent with a fully-ionized collimated outflow. To determine whether it is a shock- or radiative-ionized region, further observations are needed.

Acknowledgements

L.U. acknowledges support from the University of Guanajuato grant ID CIIC 090/2023.

References

Deacon, R. M., Chapman, J. M., Green, A. J. 2004, *ApJS*, 155, 595
Desmurs, J.-F. 2012, IAU Symp. 287, Cosmic Masers - from OH to H0, 287, 217
Gómez, J. F., Niccolini, G., Suárez, O., *et al.* 2018, *MNRAS*, 480, 4991
Gómez, J. F., Rizzo, J. R., Suárez, O., *et al.* 2011, *ApJL*, 739, L14
Imai, H. 2007, in IAU Symp. 242, Astrophysical Masers and their Environments, 242, 279
Imai, H., Amada, K., Gómez, J. G., *et al.* 2023, IAU Symp. 380, Cosmic Masers: Proper Motion toward the Next-Generation Large Projects, in press
Imai, H., Uno, Y., Maeyama, D., *et al.* 2020, *PASJ*, 72, 58
Orosz, G. 2017, Ph.D. Thesis, Kagoshima University, Japan
Pérez-Sánchez, A. F., Tafoya, D., García López, R., *et al.* 2017, *A&A*, 601, A68
Sevenster, M. N., Chapman, J. M., Habing, H. J., *et al.* 1997, *A&AS*, 122, 79
Tafoya, D., Imai, H., Gómez, J. F., *et al.* 2020, *ApJL*, 890, L14
Uscanga, L., Imai, H., Gómez, J. F., *et al.* 2023, *ApJ*, 948, 17
Walsh, A. J., Breen, S. L., Bains, I., *et al.* 2009, *MNRAS*, 394, L70

Cosmic Masers: Proper Motion toward the Next-Generation Large Projects
Proceedings IAU Symposium No. 380, 2024
T. Hirota, H. Imai, K. Menten, & Y. Pihlström, eds.
doi:10.1017/S1743921323001965

Nascent planetary nebulae: new identifications and extraordinary evolution

Roldán A. Cala[1], José F. Gómez[1] & Luis F. Miranda[1]

[1]Instituto de Astrofísica de Andalucía, CSIC, Glorieta de la Astronomía s/n, E-18008 Granada, Spain. email: rcala@iaa.es

Abstract. Planetary nebulae (PNe) harbouring masers of H_2O (H_2OPNe) and/or OH (OHPNe) are thought to be nascent PNe. They are extremely scarce, and so far only eight members are know to date. Here we explain our current effort to identify new H_2OPNe and/or OHPNe. We report IRAS 07027−7934 as a new bona fine OHPN. Its 1612 MHz OH spectrum seems to be changing from double- to single-peaked since the redshifted emission has vanished almost completely, and the 1667 MHz OH maser emission has disappeared. For the OHPN Vy 2-2, we found that its central star is unexpectedly carbon (C)-rich, has a low-mass progenitor, and could be a post-common envelope binary system. Moreover, we confirm Vy 2-2 as a nascent PN. We speculate that low-mass C-rich central stars in post-common envelope systems could be a common end of H_2OPNe and OHPNe.

Keywords. masers, planetary nebulae: general, stars: AGB and post-AGB, stars: carbon, stars: Wolf-Rayet

---

1. Introduction

The Sun and stars with initial masses ($M_{\rm i}$) ≤ 8 $M_\odot$ are expected to form a planetary nebula (PN) before their end as white dwarfs. Complex shapes of PNe may result from the interaction between binaries or multiples stellar systems (de Marco *et al.* 2022). To understand the entire evolution of a star or stellar system in the PN phase, both the circumstellar envelopes and central stars of PNe require to be characterized.

PNe harbouring masers of H_2O (H_2OPNe) and OH (OHPNe) are considered to be extremely young or nascent PNe (Zijlstra *et al.* 1989; Miranda *et al.* 2001; Gómez *et al.* 2018). This is because, after the AGB phase, H_2O and OH masers seem to disappear in timescales of 10^2 and 10^3 yr, respectively (Lewis 1989; Gómez *et al.* 1990), while, simultaneously, the PN phase starts in timescales of 10^3-10^4 yr, depending on the initial mass of the star (Miller Bertolami 2016). Therefore, even though both intermediate- (4–8 $M_\odot$) and low-mass ($\leq$1.5 $M_\odot$) stars can develop an oxygen (O)-rich envelope, necessary for H_2O and OH maser emission being present, it has been proposed that progenitors of maser-emitting PNe should be of intermediate mass, because they evolve faster during the post-AGB phase. Currently, there are only eight bona fide H_2OPNe and/or OHPNe confirmed to date (e.g. Uscanga *et al.* 2012, 2014; Qiao *et al.* 2016), and the onset of the PN phase remains poorly understood.

This contibution aims to present our current effort to increase the known members of H_2OPNe and OHPNe, as well as their characterization as a group.

2. New identifications

New H_2OPNe and OHPNe candidates can be identified by cross-matching the interferometric positions of maser and radio continuum emitters, considering that radio continuum

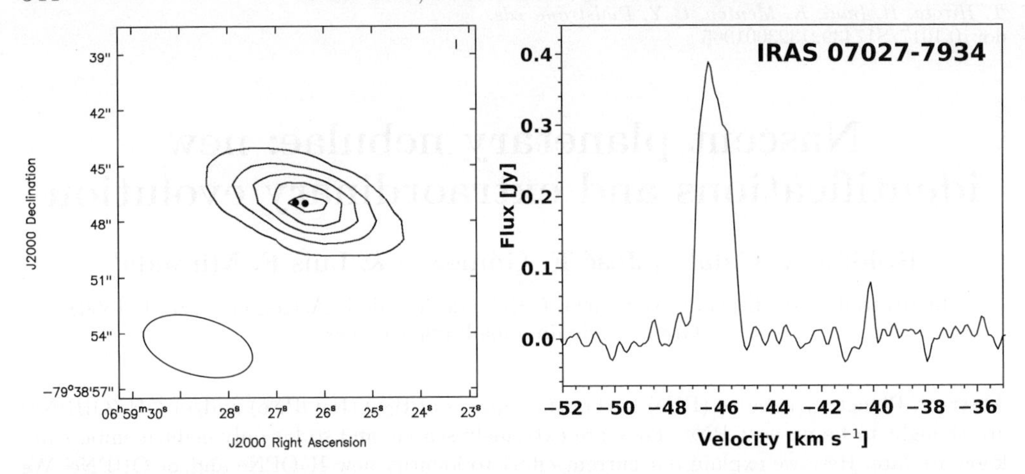

Figure 1. Contour map (left) of the radio continuum emission at 2.1 GHz of IRAS 07027−7934. The contours are 5, 15, 25, 35 and 45×1-σ (where $\sigma = 0.13\,\mathrm{mJy\,beam^{-1}}$ is the rms of the map). The beam is shown in the bottom left corner. Black dots represent the positions of the 1612 MHz OH maser features observed in the spectrum (right). The relative positional accuracy between maser and continuum emission is ∼0.5 arcsec.

emission is a common characteristic of PNe (e.g. de Gregorio-Monsalvo *et al.* 2004; Uscanga *et al.* 2012). Hence, we are carrying out a comprehensive search for new H_2OPNe and OHPNe based on radio continuum and spectral line data obtained from simultaneous interferometric observations.

We have started with the Southern Parkes Large-Area Survey in Hydroxyl (SPLASH; Dawson *et al.* 2022), which was initially carried out with single-dish observations. We have processed all the continuum data associated with the interferometric follow-up of SPLASH (see Qiao *et al.* 2020), which was carried out with Australia Telecope Compact Array. We found three previously identified bona fide OHPNe, and four new OHPNe candidates (Cala *et al.* 2022). The new OHPNe candidates present characteristics that are similar to those observed in bona fide OHPNe (see Cala *et al.* 2022). We aim to confirm the new OHPNe candidates as bona fide PNe by means of optical/infrared spectroscopy.

For this contribution we report that the single dish 1612 MHz OH maser emission detected towards the PN IRAS 07027−7934 (Zijlstra *et al.* 1991) has been interferometrically confirmed, as it presents spatial coincidence with the radio continuum emission at 2.1 GHz (Fig. 1, left). The OH spectrum at 1612 MHz displays two different spectral components (Fig. 1, right). The redishifted component (at $-40\,\mathrm{km\,s^{-1}}$) has significantly weakened with respect to the Zijlstra *et al.* (1991) observations. A possible explanation is that the (background) redshifted emission is being progressively absorbed by an expanding, optically thick photoionized nebula. Furthermore, the only component observed in the single dish 1667 OH MHz maser spectrum has also disappeared in our observations. We speculate that this variability of the OH spectra could be a usual evolutionary feature of these objects as they enter the PN phase.

3. Extraordinary evolution

To characterize maser-emitting PNe as a group, we have started studying Vy 2-2, the first OHPN identified (Seaquist & Davis 1983). Vy 2-2 consists of a very bright, compact shell of radius ∼0.24 arcsec and a faint bipolar formation of ∼2 arcsec in size (Christianto & Seaquist 1998). All available information indicates that the ejected matter from the central star in the AGB/post-AGB phase is O-rich (C/O < 1) because OH masers, silicate dust, and water ice coexist in an environment with undetectable silicon

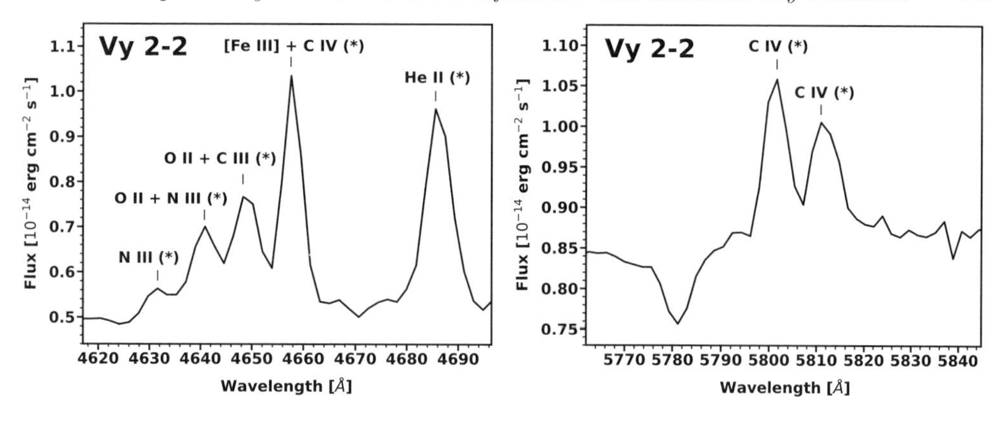

Figure 2. Optical spectra of V 2-2 in two spectral ranges. Weak stellar (*) and nebular emission lines are indicated.

carbides and polycyclic aromatic hydrocarbons (e.g. Molster *et al.* 2002; Rinehart *et al.* 2002). Furthermore, the photodissociation region (PDR) of Vy 2-2 has been detected and presents an abundance ratio C/O $<$ 1 (Liu *et al.* 2003).

We have analyzed optical spectra of Vy 2-2 obtained with CAFOS at the 2.2m telescope of the Calar Alto Observatory to study the properties of its photoionized gas. The photoionized shell has a very low value of C/O $\sim$ 0.15, similar to that of $\sim$0.26 (Wesson *et al.* 2005) and consistent with the C/O $<$ 1 in the PDR (Liu *et al.* 2003). Moreover, we obtained a N/O abundance ratio of $\sim$0.39, indicating that *hot bottom burning* has not occurred in the progenitor of the central star. This implies a $M_{\rm i}\leq 1.5$ M$_\odot$ that is contrary to the intermediate-mass stars proposed as progenitors of maser-emitting PNe.

The most intringuing result is that the optical spectra reveal weak stellar emission lines that indicate that the central star of Vy 2-2 is unexpectedly C-rich (C/O $>$ 1; Fig. 2). In particular, its C-rich surface chemistry is inferred from the C IV$\lambda\lambda$5801,5812 recombination lines (Fig. 2, right). The idea of a PN with such low values of C/O and N/O being photoionized by a C-rich central star is extraordinary in the evolution of a low-mass star. Furthermore, we confirm that Vy 2-2 is a nascent PN and could be a post-common envelope binary system (Cala et al. 2023, in preparation).

We investigate if O-rich envelopes photoionized by C-rich central stars could be a common characteristic among H_2OPNe and OHPNe. In IRAS 17347$-$3139 the H_2O masers trace a circumstellar ring (de Gregorio-Monsalvo *et al.* 2004), while the mid-infrared spectra display emission from both O-rich and C-rich dust (e.g. Jiménez-Esteban *et al.* 2006). Spectra of its (obscured) central star would be crucial to check its possible C-rich nature. In IRAS 18061$-$2505, the H_2O masers likely trace a circumstellar ring (Gómez *et al.* 2008), mid-infrared spectra show dual dust chemistry, and its central star has been classified as [WC8] (Górny *et al.* 2004). The object has been suggested to be a post-common envelope PN that could have underwent a late or very late thermal pulse (see Miranda *et al.* 2021). Finally, the newly identified OHPN IRAS 07027$-$7934 (Figure 1) shows dual dust chemistry (Cohen *et al.* 2002), and the central star has been classified as [WC11] (Menzies & Wolstencroft 1990). The formation mechanism and AGB/post-AGB progenitors of [WC] central stars of PNe are still not fully understood. However, taking into account the characteristics of Vy 2-2 and the PNe mentioned above, low-mass C-rich central stars in post-common envelope systems could be a common end of H_2OPNe and OHPNe.

4. Conclusions

We have presented our efforts to identify new maser-emitting PNe and to characterize them as a group. We have identified four new OHPN candidates that have very similar properties to those of bona fide OHPNe, and confirmed IRAS 07027−7934 as a new bona fide OHPN. We found variability in the OH spectra of IRAS 07027−7934, which could be related to the expansion of a nascent photoionized region consistent with the onset of the PN phase. In the OHPN Vy 2-2, the low values of nebular C/O and N/O abundance ratios, and the low-mass C-rich central star suggest that it is a post-common envelope binary system. Furthermore, at least two more maser-emitting PNe, IRAS 18061−2505 and IRAS 07027−7934, host C-rich central stars. Hence, we speculate that H_2OPNe and OHPNe could be a group of PNe with low-mass C-rich central stars in post-common envelope systems.

This work is partially supported by grant P20-00880 of Junta de Andalucía, grants PID2020- 114461GB-I00, CEX2021-001131-S of MCIN/AEI /10.13039/501100011033, and PRE2018-085518 funded by MCIN/AEI and by ESF Investing in your future.

References

Cala, R. A., Gómez, J. F., Miranda, L. F., *et al.* 2022, *MNRAS*, 516, 2235
Christianto, E. R., Davis, L. E., 1983, *ApJ*, 274, 659
Cohen, M., Barlow, M. J., Liu, X. W., Jones, A. F., 2002, *MNRAS*, 332, 879
Dawson, J. R., Jones P. A., Purcell, C., *et al.* 2022, *MNRAS*, 512, 3345
Gómez, J. F., Suárez, O., Gómez, Y., Miranda, L. F. *et al.* 2008, *AJ*, 135, 2074
Gómez, J. F., Niccolini, G., Suárez, O., Miranda L. F., *et al.* 2018, *MNRAS*, 480, 4991
Gómez, Y., Moran, J. M., Rodríguez, L. F., 1990, *Rev. Mexicana AyA*, 20, 55
Górny, S. K., Stasińska, G., Escudero, A. V., Costa, R. D. D., 2004, *A&A*, 427, 231
de Gregorio-Monsalvo I., Gómez, Y., Anglada, G., *et al.* 2004, *ApJ*, 601, 921
Jiménez-Esteban, F. M., Perea-Calderón, J. V., Suárez, O., Bobrowski, M., García-Lario, P. 2006, in: M. J. Barlow & R. H. Méndez, eds, *Planetary Nebulae in our Galaxy and Beyond*, Proc. IAU Symposium No. 234, 437
Lewis, B. M., 1989, *ApJ*, 338, 234
Liu, X. W., Barlow, M. J., Cohen, M., *et al.* 2001, *MNRAS*, 323, 343
de Marco, O., Akashi, M., Akras, S., *et al.* 2022, *Nature*, 6, 1421
Menzies, J. W., Wolstencroft, R. D., 1990, *MNRAS*, 247, 177
Miller Bertolami, M. M., 2016, *A&A*, 588, A25
Miranda, L. F., Gómez, Y., Anglada, G., Torrelles, J. M., 2001, *Nature*, 414, 284
Miranda, L. F., Suárez, O., Olguín, L., *et al.* 2021, *arXiv:2105.05186*
Molster, F. J., Waters, L. B. F. M., Tielens, A. G. G. M., 2002, *A&A*, 382, 222
Qiao, H. H., Walsh, A. J., Gómez, J. F., *et al.* 2016, *ApJ*, 817, 37
Qiao, H. H., Breen, S. L., Gómez, J. F., *et al.* 2020, *ApJS*, 247, 5
Rinehart, S. A., Houck, J. R, Smith, J. D., Wilson, J. C., 2002, *MNRAS*, 336, 66
Seaquist, E. R., Davis, L. E., 1983, *ApJ*, 274, 659
Uscanga, L., Gómez, J. F., Miranda, L. F., *et al.* 2014, *A&A*, 444, 217
Uscanga, L., Gómez, J. F., Suárez, O., Miranda, L. F., 2012, *A&A*, 547, A40
Wesson, R., Liu, X. W., Barlow, M. J., 2005, *MNRAS*, 362, 424
Zijlstra, A. A., te Lintel Hekkert, P., Pottasch, S., *et al.* 1989, *A&A*, 217, 157
Zijlstra, A. A., Gaylard, M. J., te Lintel Hekkert, P., *et al.* 1991, *A&A*, 243, L9

Cosmic Masers: Proper Motion toward the Next-Generation Large Projects
Proceedings IAU Symposium No. 380, 2024
T. Hirota, H. Imai, K. Menten, & Y. Pihlström, eds.
doi:10.1017/S1743921323002223

Signposts of transitional phases on the Asymptotic Giant Branch

S. Etoka

Jodrell Bank Centre of Astrophysics, Manchester, UK. Sandra.Etoka@googlemail.com

Abstract. When low- and intermediate-mass stars pass through the Asymptotic Giant Branch (AGB) they experience dramatic changes in their circumstellar shell (CSE) influenced by their mass loss, the possible presence of a (closeby) companion and the magnetic field. Masers, well spread in this environment, provide a powerful tool to reveal the CSE changes occurring when the stars undergo a transitional phase on the AGB. These can be indirect, via for instance the modification of the pumping conditions or a direct consequence of e.g. a companion and/or of the magnetic field. Evidences of such changes have been observed towards Miras, materialized by strong - both in intensity and degree of polarisation - (OH) flaring events and towards stars believed to be transitioning from the Mira to the OH/IR phase, showing an unusual high degree of polarisation. How OH maser emission can be used as a signpost of transitional phases along the AGB is explored.

Keywords. masers, stars: AGB and post-AGB, star: evolution, polarization, circumstellar matter

---

1. Introduction

When low and intermediate stars (i.e., less than 8 $M_\odot$) leave the Main Sequence, they head to the Asymptotic Giant Branch (AGB). This is associated with a dramatic increase of mass-loss rates, reaching typical values ranging from 10^{-6} to $10^{-4}\,M_\odot\ yr^{-1}$. As a consequence, the star surrounds itself with a circumstellar shell (CSE) of gas and dust of varied opacity. The AGB is populated with Miras, Semi-Regulars (SRs) and OH/IR stars, but their actual relation in terms of their original mass and evolution is still a matter of debate. In particular, it has been shown that the oxygen-rich (variable) Miras and OH/IR stars lay in a continuous sequence in the IRAS colour-colour diagram built from their 12, 25 and 60μm emission while the "non-variable" OH/IR stars (in essence the OH/IR stars leaving the AGB), having the highest [12]-[25] colour, deviate (i.e., depart) from this sequence (van der Veen & Habing 1988). Though originally interpreted as an evolutionary consequence, this "sequence" is most likely a mix of both evolution and masses to reconcile with the associated mass-ranges of Miras and OH/IR stars (Etoka & Le Squeren 2004) As it is based on FIR measurements, the IRAS colour-colour diagram and the "sequence" itself are a powerful tool to characterize the CSE (i.e., "filtering out the star") with amongst other properties, both the expansion velocity and the mass-loss rate increasing along the "sequence" (cf. Fig. 1).

OH maser is well spread in these objects and long-term monitoring of this emission provides dynamical information of the CSE (Etoka & Le Squeren 2000). OH is a paramagnetic molecule and is consequently particularly well suited for the retrieval of both the magnetic field strength and structure (Etoka & Diamond 2010).

Here we concentrate on the evolution of Miras and OH/IR stars. The transitional phase between (variable) OH/IR stars and the (non-variable) post-AGB stage is discussed in

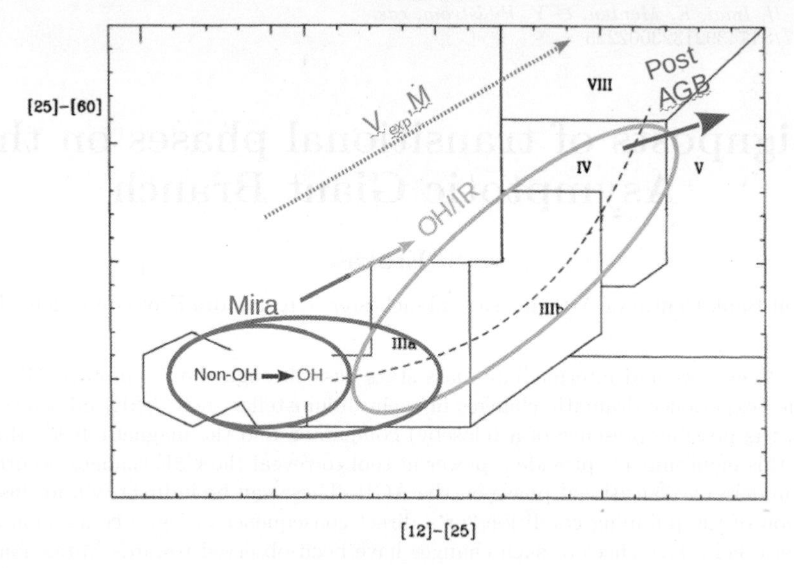

Figure 1. Schematic view of the Mira, OH/IR and post-AGB location on the IRAS colour-colour diagram.

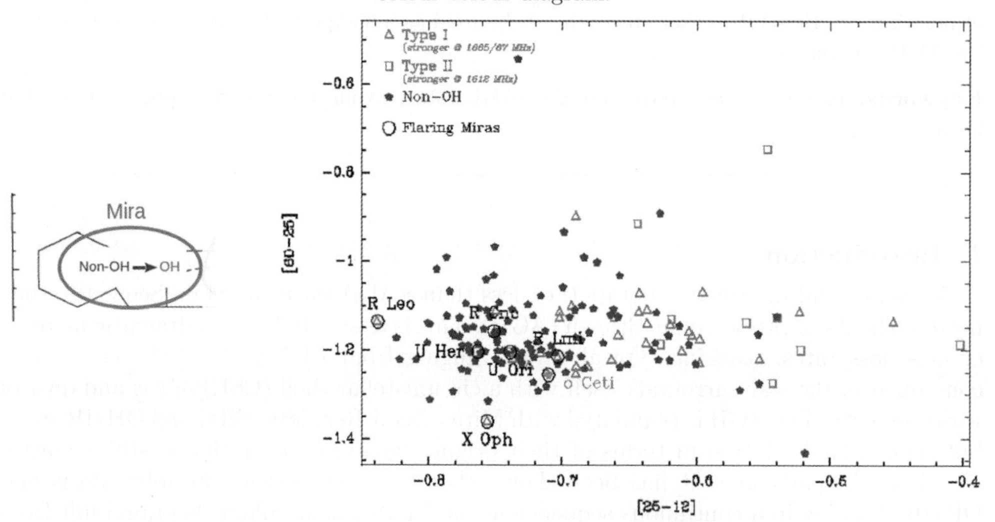

Figure 2. Non-OH → OH Mira transitional phase: Location of the flaring Miras in the IRAS-colour-colour diagram.

"The loss of OH maser emission in the early stage of Post-AGB evolution" (Etoka et al. these proceedings). The study presented here is based on OH-maser survey and long-term monitoring programmes performed at the Nançay radio telescope (NRT).

2. Non-OH→OH: Flaring Miras

The first OH flaring event in a Mira was recorded towards U Ori (Pataki & Kolena 1974). Successive detections were made toward X Oph (Etoka & Le Squeren 1996), U Her, R Leo, R LMi, R Cnc, (Etoka & Le Squeren 1997) and recurrently towards *o* Ceti (Dickinson *et al.* (1975), Gérard & Bourgois (1993), Etoka *et al.* (2017)). Figure 2 presents their location in the IRAS Colour-Colour diagram.

The flaring Miras are all found to be scattered into the non-OH maser area characterised by very thin CSEs as a clear signature of the non-OH → OH transitional phase

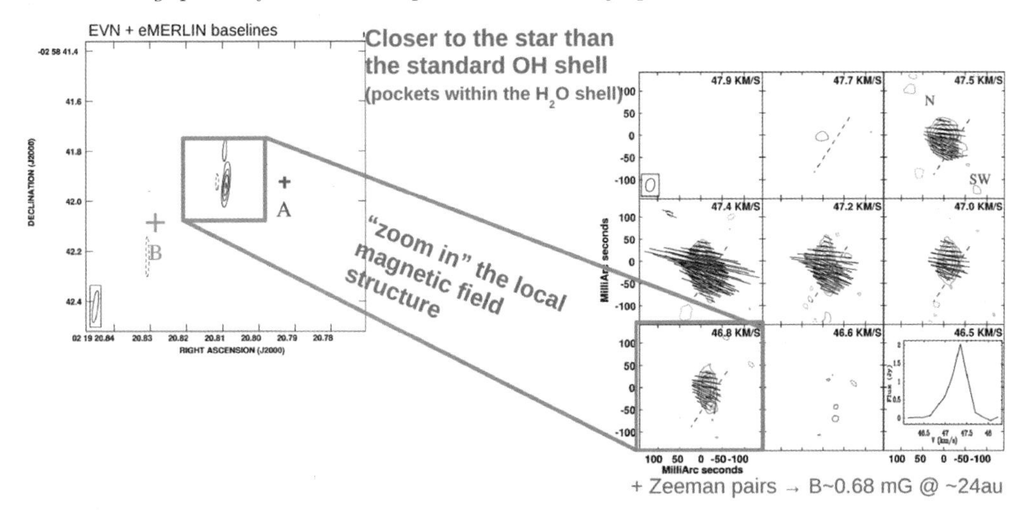

Figure 3. *o* Ceti flaring event: location of the flare with respect to the 2 stars of the binary system (i.e., Mira A & B) and polarimetric characteristics of the emission.

in which the CSE fills itself with enough dust to produce {just} enough IR radiation for the necessary radiative pumping of this emission. Added to the location of these Miras in the IRAS colour-colour diagram, all the flaring events display common features: the flaring components are spectrally compact (in terms of velocity range), peaking close to the stellar velocity and exhibiting systematically a very high degree of polarisation. Because of these characteristics, VLBI is mandatory for their mapping. The latest of such event observed towards *o* Ceti is summarized in Figure 3. It confirmed for the first time that indeed such emission is very localised and compact. The associated pocket(s) of OH emission, located much closer to the star than the standard OH masers, at H_2O region's radii, provide(s) an insight into the local magnetic fields structure of these regions. Zeeman pairs revealed line-of-sight local magnetic field strengths of $\sim$0.70 mG at $\sim$25 AU from the central star (Etoka *et al.* 2017). The long-term monitoring of this event revealed changes in the polarimetric properties of the emission as the flaring event progressed (attested by a slow change in velocity) including a dramatic flip in the strength of the dominating circular polarisation. The latter is most probably related to the complexity of magnetic field lines' direction/orientation probed by the flaring maser progressing away from its original location.

3. Mira→OH/IR

In a survey towards sources distributed along the IRAS colour-colour diagram so as to investigate the spectral evolution of the mainline transitions with the CSE thickness, 2 objects located in the lower part of the diagram with very atypical polarimetric signatures were discovered (Etoka & Le Squeren 2004).

Figure 4 presents their location in the colour-colour diagram, along with their mainlines spectra. Both objects show a wide range of very highly polarised features in the 2 mainlines with hints of potential Zeeman pairs. Because these features are spread over a wide velocity range, it is clear that a good fraction of the CSE is consequently affected. The location of these objects in the colour-colour diagram corresponds to intermediate expansion velocities and mass-loss rates and their CSE is anticipated to be between $\leq$1 and a few arcsec (hence needing "intermediate-baseline" interferometric mapping).

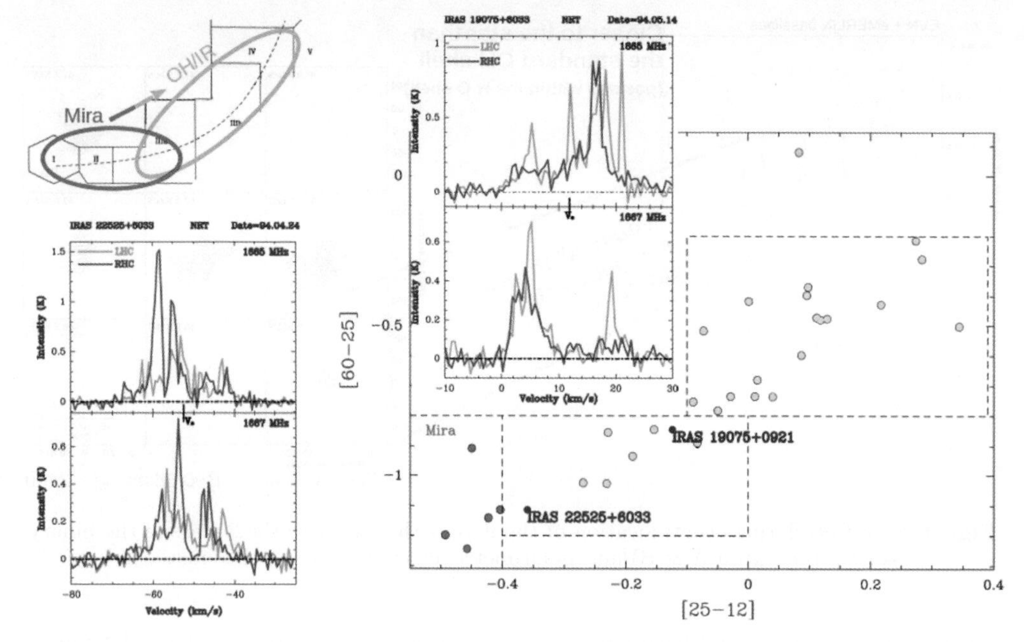

Figure 4. Mira→OH/IR: Location in the colour-colour diagram and {spectral} characteristic features.

4. Conclusion

In order to characterise the evolutionary path of a star on the AGB, it is important to understand the transitional phases that some of the stars will experience before turning into a planetary nebula. OH-maser emission has proved to be a very sensitive signpost of such objects with (a mix of) variability, spectral and polarimetric signatures, the latter allowing insight into the local magnetic field structure. Surveys and long-term monitoring are crucial for their identification and follow up. Concerted monitoring programmes with dedicated instruments located at different latitudes like e.g. the NRT and the new NARIT Thai National Radio Telescope would allow a greater number of sources and location in the Galaxy to be investigated. These need to be supplemented with interferometric observations, with VLBI for the compact events such as those associated with the Non-OH→OH Mira transitional phase and intermediate-baseline mapping (e.g. eMERLIN; SKA in the future) for the transitional phases Mira→OH/IR→post-AGB as related to more extended objects.

References

Dickinson, D.F., Kollberg, E., Yngvesson, S., 1975, *ApJ*, 199, 131

Etoka, S., Gérard, E., Richards, A.M.S., Engels, D., Brand, J., Le Bertre, T., 2017, *MNRAS*, 468, 1703

Etoka, S., Diamond, P.J., 2010, *MNRAS*, 406, 2218

Etoka, S., Le Squeren, A.M., 2004, *A&A*, 420, 217

Etoka, S., Le Squeren, A.M., 2000, *A&AS*, 146, 179

Etoka, S., Le Squeren, A.M., 1997, *A&A*, 321, 877

Etoka, S., Le Squeren, A.M., 1996, *A&A*, 315, 134

Gérard, E., Bourgois, G., 1993 *LNP*, 412, 365

Pataki, L. & Kolena, J., 1974, *BAAS*, 6, 340

van der Veen, W.E.C.J. & Habing H.J., 1988, *A&A*, 194, 125

Cosmic Masers: Proper Motion toward the Next-Generation Large Projects
Proceedings IAU Symposium No. 380, 2024
T. Hirota, H. Imai, K. Menten, & Y. Pihlström, eds.
doi:10.1017/S1743921323002442

ALMA explores the inner wind of evolved O-rich stars with two widespread vibrationally excited transitions of water

Alain Baudry[1], Ka Tat Wong[2], Sandra Etoka[3], Anita M.S. Richards[3], Malcolm D. Gray[4], Fabrice Herpin[1], Taïssa Danilovich[5], Sofia Wallström[6], Leen Decin[6], Carl A. Gottlieb[7] and the ATOMIUM consortium

[1]Laboratoire d'Astrophysique de Bordeaux, Univ. de Bordeaux, CNRS, B18N, allée Geoffroy Saint-Hilaire, 33615 Pessac, France. alain.baudry@u-bordeaux.fr

[2]Department of Physics and Astronomy, Uppsala Univ., Box 516, 75120 Uppsala, Sweden

[3]Jodrell Bank Centre for Astrophysics, The University of Manchester, M13 9PL, Manchester, UK

[4]NARIT, 260 Moo 4, Chiangmai 50180, Thailand

[5]School of Physics and Astronomy, Monash Univ., Clayton 3800 Victoria, Australia

[6]Institute of Astronomy, KU Leuven, 3001 Leuven, Belgium

[7]Harvard-Smithsonian Center for Astrophysics, Cambridge MA 02138, USA

Abstract. ALMA observations with angular resolution in the range ~20–200 mas demonstrate that emission at 268.149 and 262.898 GHz in the (0,2,0) and (0,1,0) vibrationally excited states of water are widespread in the inner envelope of O-rich AGB stars and red supergiants. These transitions are either quasi-thermally excited, in which case they can be used to estimate the molecular column density, or show signs of maser emission with a brightness temperature of $\sim 10^3$–10^7 K in a few stars. The highest spatial resolution observations probe the inner few stellar radii environment, up to ~10–12 $R_\star$ in general, while the mid resolution data probe more thermally excited gas at larger extents. In several stars, high velocity components are observed at 268.149 GHz which may be caused by the kinematic perturbations induced by a companion. Radiative transfer models of water are revisited to specify the physical conditions leading to 268.149 and 262.898 GHz maser excitation.

Keywords. stars: AGB and post-AGB, supergiants, radiation mechanisms: thermal, radiation mechanisms: non-thermal, masers, techniques: interferometric

---

1. Introduction

Water (H_2O) in the inner gas layers above the photosphere of O-rich evolved stars is an essential oxidizing agent in the formation of gas-phase clusters and, ultimately, dust grains (e.g. Gobrecht et al. 2016). Ten rotational transitions, in the ground or high-lying vibrational states of water with excitation energies ~4000–9000 K, were recently identified by Baudry *et al.* (2023) in the ALMA Large Program (LP) called ATOMIUM: 'ALMA tracing the origins of molecules in dust-forming O-rich M-tytpe stars'. ATOMIUM is the first ALMA LP in the field of stellar evolution. It included the observations of a sample

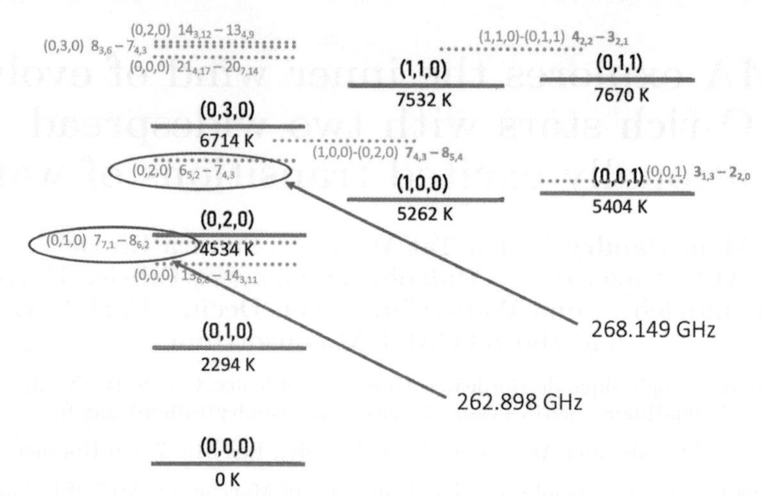

Figure 1. Energy level diagram indicating in red the band origins of the 8 lowest vibrational states of H_2O, with quantum numbers (v_1,v_2,v_3) and associated energy. The 10 transitions observed in the ATOMIUM LP are indicated by green or orange dotted lines (for pure rotational transitions or ro-vibrational transitions). The two most widespread lines of water in this work are observed at 268.149 GHz and 262.898 GHz.

of 14 Asymptotic Giant Branch (AGB) stars and three Red Supergiants (RSGs) aiming at understanding the chemistry of the dust precursors as well as the links between chemistry and the wind dynamics of these stars; see first results in e.g., Decin *et al.* (2020) and Gottlieb *et al.* (2022).

We focus this presentation on the two most widespread transitions of water that were discovered at 268.149 and 262.898 GHz during the ATOMIUM survey together with eight other transitions of H_2O (Figure 1) and several high-lying transitions of OH (Baudry *et al.* 2023).

2. H_2O results at 268.149 and 262.898 GHz

17 stars were observed as part of the ATOMIUM project (Gottlieb *et al.* 2022). The results presented here are taken from main array observations in two configurations separated by ∼8–10 months, giving ∼20–200 mas resolution, probing the wind acceleration region. Sensitivity was ∼1 mJy/beam at the highest angular resolution and our velocity resolution was ∼1.2 km/s. Complementary data at a spectral resolution of ∼0.1 km/s were acquired with the ALMA Compact Array (ACA) for a few stars 26 months later, providing additional variability information.

Nine out of the ten H_2O lines detected with ALMA are new discoveries in space. The transition at 268.149 GHz, reported earlier in VY CMa and IK Tau (Tenenbaum *et al.* 2010; Velilla-Prieto *et al.* 2017), is now detected in 15 stars of our sample. It is also detected in W Hya (Ohnaka et al. this symposium). The two strongest and most widespread transitions of water in our sample lie in the (0,2,0) and (0,1,0) vibrational states at 268.149 and 262.898 GHz (15 and 12 sources, respectively). Sources with strong 268.149 GHz emission tend to show relatively strong emission at 262.898 GHz, although there is no clear correlation of the 268.149 GHz line parameters with those of other transitions. However, eight sources in our sample, as well as VY CMa, IK Tau and R Hya, are detected at 658.007 GHz, another most widespread transition in the (0,1,0) vibrational

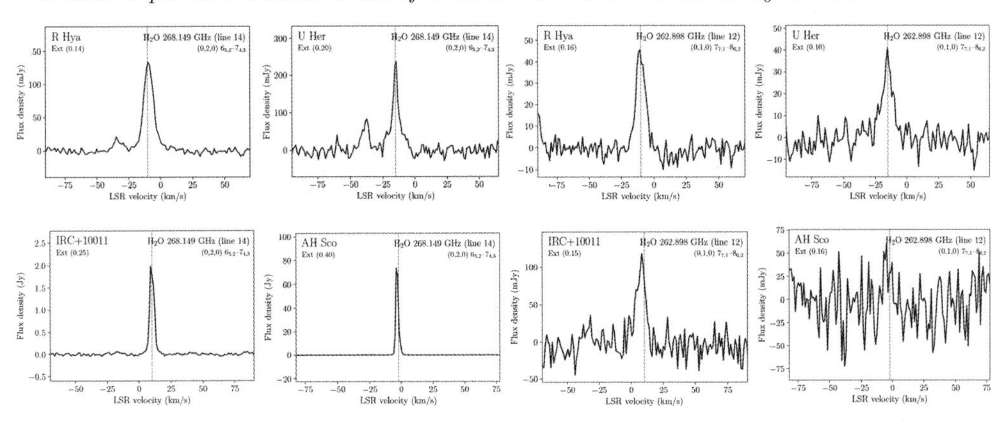

Figure 2. Spectra of the 268.149 and 262.898 GHz water lines in R Hya, U Her, IRC+10011 and AH Sco. The flux density scale at 268.149 GHz in IRC+10011 and AH Sco is expressed in Jy. The blue vertical line indicates the systemic local-standard-of-rest (LSR) velocity.

state (e.g., Baudry *et al.* 2023). Figures 2 and 3 present some spectra and channel maps, respectively, of our 268.149 and 262.898 GHz data.

3. Line excitation in the gas acceleration zone and maser models

We found that, in some stars, the 268.149 and 262.898 GHz transitions are in quasi-Local Thermal Equilibrium (LTE). This is demonstrated by constructing population diagrams from the velocity-integrated line flux densities for a range of excitation energies including at least six H_2O transitions. The 268.149 and 262.898 GHz transitions do not depart substantially from LTE in several sources and, overall, we derive H_2O column densities of $\sim$0.6–6$\times10^{20}$ cm^{-2}. However, in some sources, such as U Her, IRC+10011 and AH Sco, the lower limit of the brightness temperature, T_b, is $\sim10^4$–10^7 K at 268.149 GHz, thereby indicating the likelihood of maser action, which is supported by the narrow line profile and high flux density in IRC+10011 and AH Sco (Figure 2), time variability observed in some sources and the different angular extents mapped at two different epochs in U Her. In contrast with the 268.149 GHz observations, the 262.898 GHz transition does not exhibit strong maser action. The highest value of T_b is $\sim$1100 K, although this could just be a lower limit. However, in IRC+10011, we observed time variability (time span of $\sim$0.5 yr) in spectral features across the line profile.

The angular size of the H_2O sources was measured for all observed transitions and stars from our channel maps or integrated intensity maps. Sizes of 10–50 mas are obtained from high resolution images and the emission typically extends $\sim$ 3–12 $R_\star$ from the star, occasionally further (e.g. AH Sco). Water is observed in the wind acceleration zone defined according to the wind velocity profile analyses of multiple lines in the ATOMIUM survey (Gottlieb *et al.* 2022). In this zone, the blue and red wings of the 268.149 GHz line profile often exceeds the velocity range spanned by the CO (2–1) line. This is also observed for the SiO wings that are ($\sim$1.6$\times$) larger than the CO velocities and can be explained by turbulence due to the irregular nature of the inner wind and shocks propagating above the photosphere, or by complex kinematics (e.g. binarity). At small spatial scales, velocity gradients were also observed in R Hya and U Her against the photosphere, underlining complex gas motions.

Our maser models (Gray *et al.* 2016) were re-examined (Baudry *et al.* 2023) to show that the 268.149 GHz negative opacity increases with increasing dust temperature, T_d, as expected near the photosphere, while 658.007 GHz line inversion requires lower values of T_d and relatively high values of the kinetic temperature (T_K, see the left and middle

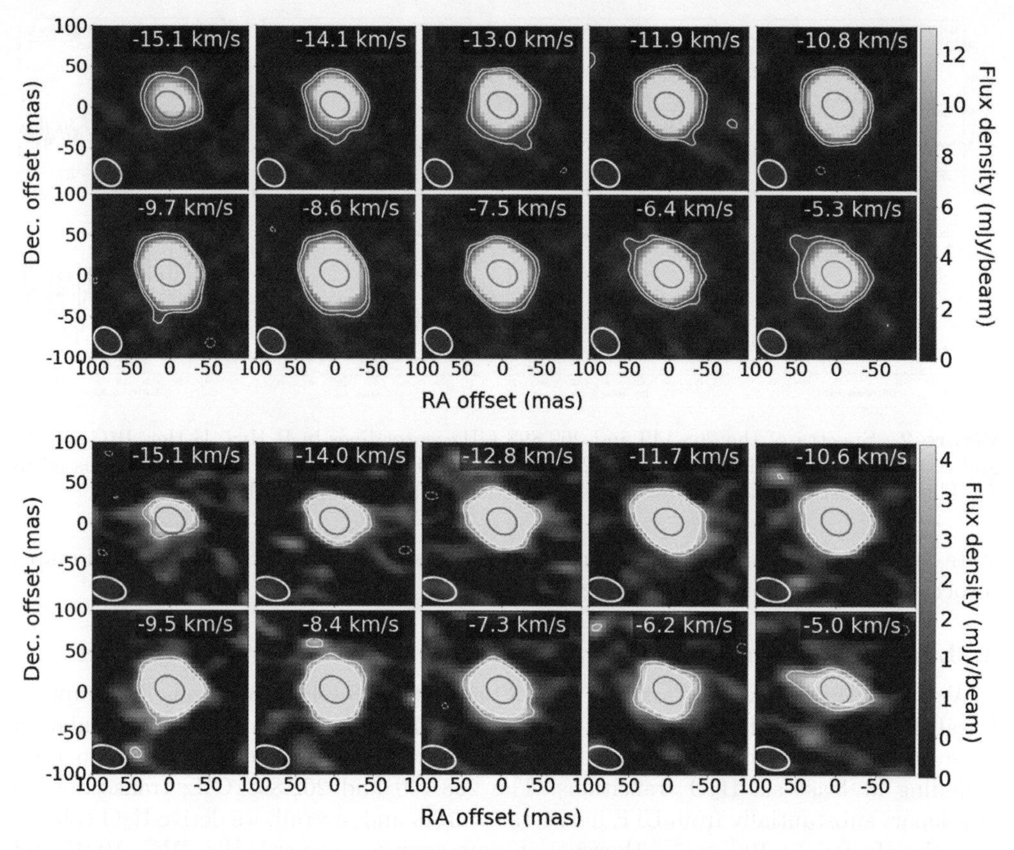

Figure 3. Channel maps of the 268.149 and 262.898 GHz lines in R Hya (upper and lower panels). The red contour near the center delineates the continuum emission at half peak intensity. The line and continuum beams are superimposed in the lower left corner of each panel and shown in white and dark-red, respectively.

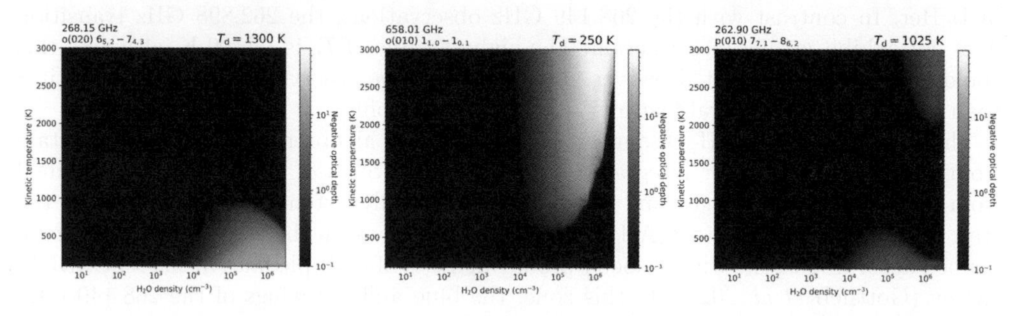

Figure 4. Negative opacity of 268.149 GHz line, in the (0,2,0) state, and 658.007 and 262.898 GHz lines in the (0,1,0) state. The first and third panels are for similar values of the dust temperature, T_d.

panels in Figure 4). The right panel of Figure 4 shows that the 262.898 GHz para-transition in the (0,1,0) state is less strongly inverted than the 268.149 GHz transition for similar physical conditions, as observed; note that the 262.898 GHz transition can also be inverted in the high T_K regime. For all lines, inversion requires H_2 number densities around a few times 10^9–10^{10}cm^{-3}.

4. Conclusion

The most common H_2O transitions in our sample of O-rich evolved stars are observed in high-lying levels at 268.149 (6040 K) and 262.898 GHz (4480 K) in the (0,2,0) and (0,1,0) vibrational states, respectively. Maser emission in the 268.149 GHz line was observed with the ALMA main array and the ACA in some stars, whereas thermal emission dominates at 262.898 GHz in most sources. Combined with the OH lines discovered during our survey and described in Baudry *et al.* (2023), we derived an OH/H_2O abundance ratio of around 10^{-2}. Our H_2O maser models were re-examined to specify the physical conditions leading to line inversion near the photosphere.

References

Baudry, A., Wong, K.T., Etoka, S., *et al.* 2023, *A&A*, 674, A125 (arXiv 2305.03171)
Decin, L., M. Montargès, M., Richards, A.M.S., *et al.* 2020, *Science*, 369, 1497
Gobrecht, D., Cherchneff, I., Sarangi, A., *et al.* 2016, *A&A*, 585, A6
Gottlieb, C.A., Decin, L., Richards, A.M.S., *et al.* 2022, *A&A*, 660, A94
Gray, M.D., Baudry, A., Richards, A.M.S., *et al.* 2016, *MNRAS*, 456, 374
Tenenbaum, E. D., Dodd, J. L., Milam, S. N., *et al.* 2010, *ApJ*, 720, L102
Velilla-Prieto, L., Sánchez Contreras, C., Cernicharo, J., *et al.* 2017, *A&A*, 597, A25

Cosmic Masers: Proper Motion toward the Next-Generation Large Projects
Proceedings IAU Symposium No. 380, 2024
T. Hirota, H. Imai, K. Menten, & Y. Pihlström, eds.
doi:10.1017/S1743921323002715

High resolution ALMA imaging of H_2O, SiO, and SO_2 masers in the atmosphere of the AGB star W Hya

Keiichi Ohnaka[1] and **Ka Tat Wong**[2]

[1]Universidad Andrés Bello, Santiago, Chile. email: k1.ohnaka@gmail.com

[2]Uppsala University, Uppsala, Sweden

Abstract. The mass-loss mechanism in asymptotic giant branch (AGB) stars is not yet fully understood. We present 20-milliarcsecond resolution ALMA imaging of the well-studied AGB star W Hya in multiple molecular lines at 250–269 GHz, including masers from SiO, H_2O, and SO_2. The images show complex plumes, arcs, and clumps over the stellar disk and in the atmosphere extending to several stellar radii. We detected prominent emission components over the stellar disk—instead of pure absorption as expected—in some $Si^{17}O$, ^{30}SiO, H_2O, and SO_2 lines. The surface emission seen in the $Si^{17}O$ and vibrationally excited H_2O lines is particularly strong, indicating maser actions. The masers seen over the stellar disk indicate radial amplification.

Keywords. Asymptotic Giant Branch, mass loss, masers, stellar imaging

1. Introduction

Despite its importance in stellar evolution and the chemical enrichment of galaxies, the mass-loss mechanism in asymptotic giant branch (AGB) stars is not yet well understood. It is often considered that the radiation pressure on dust grains due to absorption or scattering of stellar photons drives mass outflows (Höfner & Olofsson 2018). However, dust grains that form in oxygen-rich AGB stars are too transparent to absorb sufficient photons, and thus cannot drive the mass loss (Woitke 2006; Höfner 2007). As an alternative, Höfner (2008) proposes that the scattering—instead of absorption—of stellar photons by large (0.3–1.0 μm) transparent grains may drive mass outflows.

To shed light on the long-standing problem of mass loss from AGB stars, it is indispensable to probe the region within a few $R_\star$, where dust forms and the wind accelerates. For this goal, we focus on the well-studied oxygen-rich AGB star W Hya. Its light curve clearly shows a period of 389 days, and the radial velocities measured from the infrared spectral lines show pulsation (Hinkle *et al.* 1997; Lebzelter *et al.* 2005). Thanks to its proximity (98pc, Vlemmings *et al.* 2003), it has been studied with various observational techniques from the visible to the infrared to the radio (e.g., Zhao-Geisler *et al.* 2011; Khouri *et al.* 2015, and references therein). Therefore, W Hya is one of the best targets to study the mass-loss mechanism in great detail.

The recent polarimetric imaging of W Hya at five wavelengths from 645 to 820 nm with spatial resolutions of 23–30 mas using VLT/SPHERE-ZIMPOL (Ohnaka *et al.* 2016, 2017) reveals clumpy dust clouds very close to the star at an angular distance of $\sim$50 mas, corresponding to just $\sim$2 $R_\star$. Their 2-D radiative transfer modeling suggests the predominance of large (0.4–0.5 μm), transparent grains of Al_2O_3, or Mg_2SiO_4, or $MgSO_3$

in the clumpy clouds, lending support to the aforementioned scenario of the scattering-driven mass loss. While the SPHERE-ZIMPOL data do not allow us to distinguish the specific grain species, Takigawa *et al.* (2017) found excellent agreement in the spatial distribution of gas-phase AlO emission at 344 GHz as observed with ALMA and that of the clumpy dust clouds. These authors conclude that the clumpy clouds within $\sim$3 $R_\star$ predominantly consist of Al_2O_3 grains.

2. ALMA observations

We observed W Hya with ALMA at 250–269 GHz in Band 6 using the C43-10 configuration with a maximum baseline of 16 km. The observational set-up provided spatial resolutions of 16–20 mas and velocity resolutions of 1.1–1.2 km s^{-1}. Our ALMA observations took place on 2019 June 8, when W Hya was at minimum light with a variability phase of 0.53. The millimeter continuum images can be fitted with a nearly circular disk with an angular diameter of $\sim$60 mas, which is about 1.5 times larger than the stellar angular diameter of $\sim$40 mas measured in the near infrared (Woodruff *et al.* 2009). In contrast to the previous 338 GHz observation by Vlemmings *et al.* (2017), no signature of a hot spot is seen in our ALMA continuum images.

3. Detection of emission over the stellar disk

We detected a total of 40 molecular lines from the following species: ^{29}SiO, ^{30}SiO, Si^{17}O, H_2O, SO_2, $^{34}SO_2$, SO, AlO, AlOH, TiO, and HCN. The large angular diameter of W Hya and ALMA's high angular resolution allow us to spatially resolve the emission extending to $\sim$100 mas ($\sim$5 $R_\star$) as well as the absorption over the stellar disk. The emission is irregularly shaped with a plume extending in the NNW, a tail extending in the SSE, and an extended atmosphere elongated in the ENE–WSW direction. However, we found that approximately a third of the identified lines show emission—instead of pure absorption as expected—over the stellar disk. They are the lines of ^{30}SiO ($v=2$), Si^{17}O ($v=0$), H_2O ($v_2=2$), SO_2 (many lines), and AlO, although the emission of this last line is weak. It should be noted that not all lines of a given molecular species show the emission over the stellar disk. For example, while the ^{30}SiO ($v=2$) and Si^{17}O ($v=0$) lines show emission, the ^{29}SiO ($v=3$) and ^{30}SiO ($v=1$) lines show clear absorption over the stellar disk as expected. Likewise, not all detected SO_2 lines show emission over the surface, although some of them have similar energy levels. Therefore, the emission cannot be explained by the presence of hot gas in front of the star, because all lines of the same molecular species would show emission in that case.

The vibrationally excited H_2O line ($v_2=2$, $6_{5,2}$–$7_{4,3}$) at 268 GHz shows particularly strong emission over the stellar disk in spite of its high upper level energy of 6039 K. Its peak intensity is twice as strong as the continuum intensity. This strongly indicates maser actions. This line has also been reported in multiple AGB stars and red supergiants, suggesting that excitation of this line is widespread among oxygen-rich evolved stars (Baudry *et al.* 2023, and references therein). Some of the detected sources exhibit evidence of strong maser actions.

The maser emission is confined within a radius of $\sim$50 mas (= 2.4 $R_\star$). The peak of the H_2O maser emission observed over the stellar disk are found at redshifted velocities between 3 and 5 km s^{-1} with respect to the systemic velocity. However, the line profiles extracted over the stellar disk are broad, ranging from about -10 km s^{-1} to 14 km s^{-1}. This suggests the presence of a wide range of velocity components between $\sim$1.5 $R_\star$ (millimeter continuum radius) and $\sim$2.4 $R_\star$. The dynamical models for AGB stellar winds (e.g., Höfner *et al.* 2022) show that the velocity of pulsation-driven shocks can reach the magnitude of $\sim$10–15 km s^{-1}. This is in broad agreement with the observed velocity width of the H_2O masers.

Based on the radiative transfer models of Gray *et al.* (2016), Baudry *et al.* (2023, and this symposium) show that, in the presence of warm dust of $\sim$1300 K, maser emission can occur in the 268 GHz H_2O transition if the H_2O density is higher than $\sim 10^4$ cm^{-3} and the kinetic temperature is lower than $\sim$900 K. A lower kinetic temperature (< 500 K) is required for a lower dust temperature (900 K). This suggests that H_2O maser emission may trace pockets of dense, cool regions in the inner wind of oxygen-rich stars, where dust formation can take place.

The emission over the surface in other molecular lines can also be interpreted as maser actions. This naturally explains that some lines of a given molecular species show the emission over the stellar disk, while other lines of the same species show pure absorption as expected.

K.O. acknowledges the support of the Agencia Nacional de Investigación Científica y Desarrollo (ANID) through the FONDECYT Regular grant 1210652. K.T.W. acknowledges support from the European Research Council (ERC) under the European Union's Horizon 2020 research and innovation programme (Grant agreement no. 883867, project EXWINGS).

References

Baudry, A., Wong, K. T., Etoka, S., *et al.* 2023, *A&A*, 674, A125
Gray, M. D., Baudry, A., Richards, A. M. S., *et al.* 2016, *MNRAS*, 456, 374
Hinkle, K., Lebzelter, T., & Scharlach, W. W. G. 1997, *AJ*, 114, 2686
Höfner, S. 2007, *ASP-CS*, 378, 145
Höfner, S. 2008, *A&A*, 491, L1
Höfner, S., & Olofsson, H. 2018, *A&AR*, 26, 1
Höfner, S., Bladh, S., Aringer, B., & Eriksson, K. 2022, *A&A*, 657, A109
Khouri, T., Waters, L. B. F. M., de Koter, A., *et al.* 2015, *A&A*, 577, A114
Lebzelter, T., Hinkle, K. H., Wood, P. R., Joyce, R. R., & Fekel, F. C. 2005, *A&A*, 431, 623
Ohnaka, K., Weigelt, G., & Hofmann, K.-H. 2016, *A&A*, 589, A91
Ohnaka, K., Weigelt, G., & Hofmann, K.-H. 2017, *A&A*, 597, A20
Takigawa, A., Kamizuka, T., Tachibana, S., & Yamamura, I. 2017, *Science Advances*, 3, 2149
Vlemmings, W. H. T., van Langevelde, H. J., Diamond, P. J., Habing, H. J., & Schilizzi, R. T. 2003, *A&A*, 407, 213
Vlemmings, W. H., T., Khouri, T., O'Gorman, E., *et al.* 2017, *Nature Astronomy*, 1, 848
Woitke, P. 2006, *A&A*, 460, L9
Woodruff, H. C., Ireland, M. J., Tuthill, P. G., *et al.* 2009, *ApJ*, 691, 1328
Zhao-Geisler, R., Quirrenbach, A., Köhler, R., Lopez, B., & Leinert, C. 2011, *A&A*, 530, A120

Cosmic Masers: Proper Motion toward the Next-Generation Large Projects
Proceedings IAU Symposium No. 380, 2024
T. Hirota, H. Imai, K. Menten, & Y. Pihlström, eds.
doi:10.1017/S1743921323002739

Discovery of SiO masers in the "Water Fountain" source, IRAS 16552–3050

Kei Amada[1], Hiroshi Imai[1,2,3], Yuhki Hamae[4], Keisuke Nakashima[1], Ka-Yiu Shum[1], Daniel Tafoya[5], Lucero Uscanga[6], José F. Gómez[7], Gabor Orosz[8], and Ross A. Burns[9]

[1]Graduate School of Science and Engineering, Kagoshima University, Kagoshima 890-0065, Japan. email: k0501862@kadai.jp

[2]Amanogawa Galaxy Astronomy Research Center, Graduate School of Science and Engineering, Kagoshima University, 1-21-35 Korimoto, Kagoshima 890-0065, Japan

[3]Center for General Education, Institute for Comprehensive Education, Kagoshima University, 1-21-30 Korimoto, Kagoshima 890-0065, Japan

[4]Faculty of Science, Kagoshima University, 1-21-35 Korimoto, Kagoshima 890-0065, Japan

[5]Department of Space, Earth and Environment, Chalmers University of Technology, Onsala Space Observatory, 439 92 Onsala, Sweden

[6]Departamento de Astronomía, Universidad de Guanajuato, A.P. 144, 36000 Guanajuato, Gto., Mexico

[7]Instituto de Astrofísica de Andalucía, CSIC, Glorieta de la Astronomía s/n, E-18008 Granada, Spain

[8]Joint Institute for VLBI ERIC, Dwingeloo 7900, Netherlands

[9]National Astronomical Observatory of Japan, Mitaka, Tokyo 181-8588, Japan

[10]RIKEN, Wako, Saitama 351-0198, Japan

Abstract. We report new detections of SiO $v=1$ and $v=2$ $J=1\rightarrow0$ masers in the "water fountain" source IRAS 16552-3050, which was observed with the Nobeyama 45 m telescope from March 2021 to April 2023. Water fountains are evolved stars whose H_2O maser spectra trace high-velocity outflows of >100 km s^{-1}. This is the second known case of SiO masers in a water fountain, after their prototypical source, W 43A. The line-of-sight velocity of the SiO masers are blue-shifted by ~25 km s^{-1} from the systemic velocity. This velocity offset imply that the SiO masers are associated with nozzle structure formed by a jet penetrating the circumstellar envelope, and that new gas blobs of the jet erupted recently. Thus, the SiO masers imply this star to be in a new evolutionary stage.

Keywords. masers, stars: AGB and post-AGB, stars: mass loss, stars: winds, outfows

---

1. Introduction

"Water fountain" (WF) sources are dying stars, in the asymptotic giant branch (AGB) or post-AGB phases, that show bipolar jets traced by H_2O maser emission. To date, 15 WFs have been confirmed (e.g. Gómez *et al.* 2017). Interestingly, the WFs exhibit a wide variety in the spatio-kinematics of their H_2O masers as summarized in Imai *et al.* (2020). Recent molecular line observations of the WFs have suggested that the central stellar system is probably a binary forming a common envelope (e.g., Khouri *et al.* 2021).

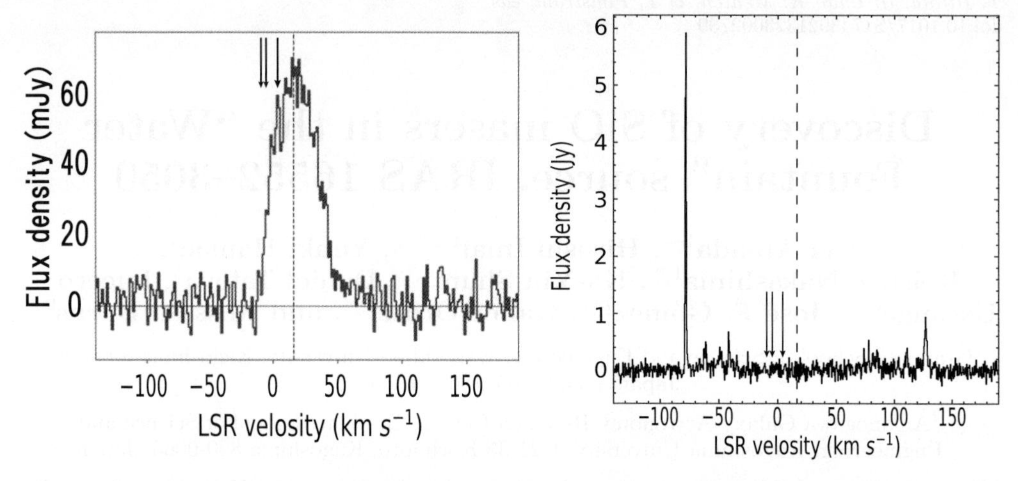

Figure 1. *Left*: Spectrum of CO ($J=1\rightarrow 0$) emission observed with ALMA. A vertical dashed line indicates the systemic velocity, V_{sys} = 16 km/s. Vertical arrows indicate the velocities of the SiO masers around $V_{LSR}\sim -9$, -5, 4. The SiO masers are blue-shifted by ~25 km/s from the V_{sys}. *Right*: Spectrum of the H_2O maser observed with Nobeyama 45m radio telescope. The vertical dashed line and vertical arrows are the same as those in left panel.

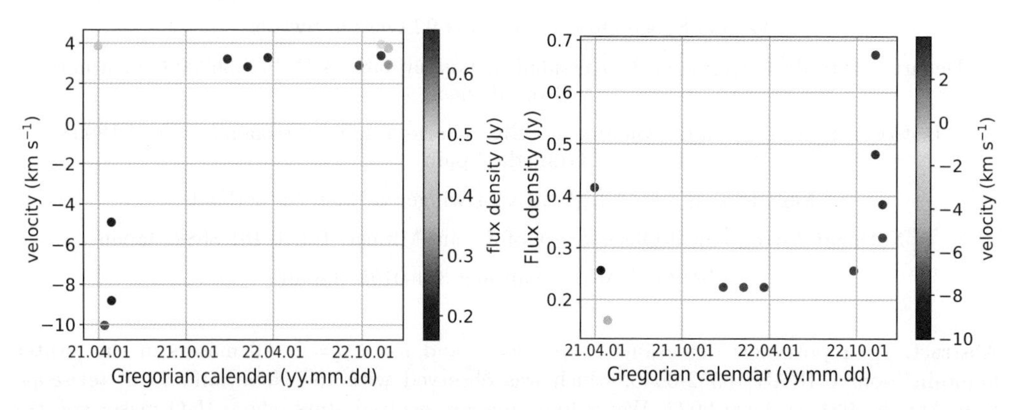

Figure 2. Time variation in the spectral peaks of SiO v = 2 masers toward IRAS 16552-3050. The left and right panels show the variation in LSR velocity and flux density for the individual spectral peaks, respective. Periodicity is not confirmed for either LSR velocity or flux density. (Our poster showed that new redshift components was also detected, but it was a fake due to a mistake of analysis.)

However, the evolutionary status of WFs in the context of stellar evolution is still mostly unknown due to the limited information on the vicinities of their central systems. To date, the presence of SiO masers has been confirmed in only one WF, W 43A (Nakashima & Deguchi 2003). Imai *et al.* (2005) mapped the W 43A SiO $v=1$ $J=1\rightarrow 0$ masers, exhibiting a biconical outflow with a wide full opening angle (~40°). However, the W 43A SiO masers have vanished.

2. SiO masers in IRAS 16552–3050

We detected the SiO $v=1$ and $v=2$ masers in IRAS 16552-3050 (hereafter abbreviated as I16552) on 2021 March. The first SiO maser observation toward IRAS I16552 had been conducted on 2011 using KVN, but the SiO maser had not been detected (Yoon *et al.* 2014). Although our intensive monitoring observations have been conducted since 2018

December using the Nobeyama 45m radio telescope, SiO maser has never been detected. Therefore, the SiO masers of I16552 were detected for the first time.

The SiO masers are blueshifted by $\sim$13–25 km s^{-1} from the systemic velocity, which is determined from the spectrum of the CO $J = 2 \rightarrow 1$ emission obtained with ALMA. This velocity offset is comparable to those in W 43A ($\sim$20 km s^{-1}, Imai *et al.* 2005). In W 43A, the SiO masers exhibited biconical expansion at a velocity less than that of H_2O masers. This supports a model in which the SiO masers were associated with the jet's nozzles formed in entrained material dragged by the fast jet (Tafoya *et al.* 2020). In fact, the velocities of the SiO masers corresponded to the spectral edge of the CO emission. The SiO masers in I16552 may resemble in the case of W 43A, but it should be confirmed in future observations by finding the red-shifted counterpart SiO masers and by spatially locating these masers.

References

Gmez, J. F., Surez, O., Rizzo, J. R., *et al.* 2017, *MNRAS*, 468, 2081

Imai, H., Uno, Y., Maeyama, D., *et al.* 2020, *PASJ*, 72, 58

Tafoya, D., Imai, H., Gmez, J. F., *et al.* 2020, *ApJ* (Letters) 890, L14

Khouri, T., Vlemmings, W. H. T., Tafoya, D., *et al.* 2021, *New Astron.*, 6, 275

Nakashima, J., & Deguchi, S. 2003, *PASJ*, 55, 229

Imai, H., Nakashima, J., Diamond, P. J., Miyazaki, A., & Deguchi, S. 2005, *ApJ* (Letters) 622, L125

Yoon, D.-H., Cho, S.-H., Kim, J., Yun, Y. J., & Park Y.-S. 2014, *AJ*, 211, 15

Cosmic Masers: Proper Motion toward the Next-Generation Large Projects
Proceedings IAU Symposium No. 380, 2024
T. Hirota, H. Imai, K. Menten, & Y. Pihlström, eds.
doi:10.1017/S1743921323002168

Interferometric Observations of the Water Fountain Candidates OH 16.3−3.0 and IRAS 19356+0754

P. Chacón[1], L. Uscanga[1], H. Imai[2,3], B. H. K. Yung[4], J. F. Gómez[5], J. R. Rizzo[6], O. Suárez[7], L. F. Miranda[5], G. Anglada[5] and J. M. Torrelles[8]

[1]DA, University of Guanajuato, A.P. 144, 36000 Guanajuato, Gto., Mexico.
email: p.chaconmartinez@ugto.mx

[2]AGARC, GSSE, Kagoshima University, 1-21-35, Korimoto, Kagoshima 890-0065 Japan

[3]CGE, ICE, Kagoshima University, Japan

[4]NCAC, ul. Rabiańska 8, 87-100 Toruń Poland

[5]IAA-CSIC, Glorieta de la Astronomía s/n, 18008 Granada, Spain

[6]CAB, INTA-CSIC, Ctra de Torrejón a Ajalvir, km 4, 28850 Torrejón de Ardoz, Madrid, Spain

[7]Observatoire de la Côte dAzur, UCA, Nice, France

[8]ICE, CSIC, Carrer de Can Magrans, s/n 08193 Barcelona, Spain

Abstract. Water Fountains (WFs), located between the AGB and PN phases of stellar evolution, may provide significant clues on the shaping process of PNe. We present new VLA observations of the WF candidates OH 16.3-3.0 and IRAS 19356+0754. We detect H_2O and OH maser and radio continuum emission towards OH 16.3-3.0. We suggest that the OH maser emission of OH 16.3-3.0 is associated with an aspherical circumstellar envelope due to its spatio-kinematics and peculiar spectral profile. We could not confirm the candidates as bona fide WFs because of the narrow velocity spread (OH 16.3-3.0) or non-detection (IRAS 19356+0754) of H_2O maser emission. Further monitoring could help to discern their nature.

Keywords. masers, stars: AGB and post-AGB

1. Introduction

WFs are evolved stars with initial masses $< 8M_\odot$, most of which are in the post-AGB phase (Imai 2007; Desmurs 2012; Gómez *et al.* 2017). Their H_2O masers trace high-velocity collimated outflows with expansion velocities larger than those in the AGB phase, which are typically 10−30 km s^{-1} (Sevenster *et al.* 1997). This collimated mass-loss could be one of the first indications of a morphological change in the circumstellar envelopes of evolved stars (Imai 2007). OH 16.3−3.0 (OH16.3) (IRAS 18286−1610) and IRAS 19356+0754 (I19356) were considered as WF candidates because they showed larger velocities in their H_2O maser spectra than those of OH, based on single-dish observations (Yung *et al.* 2013, 2014). We selected these candidates because of their relatively low H_2O maser velocity coverage compared to the 15 WFs so far known. Here we present interferometric observations to confirm their WF nature.

2. Observations and Method

Our observations were carried out with the VLA in its A configuration in 2019. We observed simultaneously H_2O maser and radio continuum emission at 22 GHz.

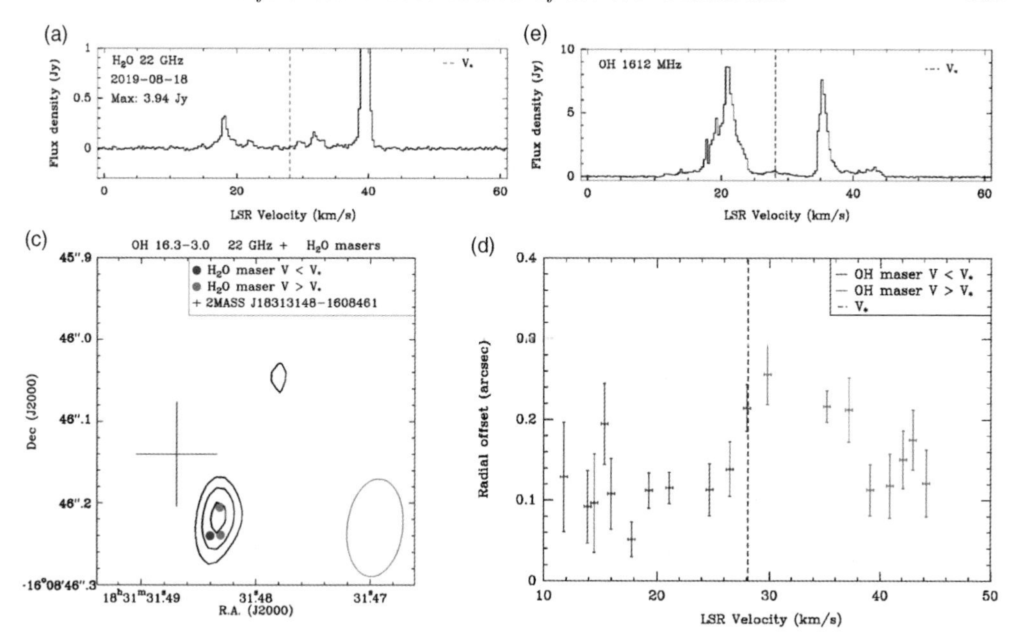

Figure 1. VLA observations toward the WF candidate OH16.3. (a) H_2O maser spectrum, zoomed in to better show the faint emission components. (b) OH maser spectrum. It shows a double-peaked profile with prominent wings. (c) Contour map of the radio continuum emission at 22 GHz. The contours are 3,4,5 times 1.3×10^{-5} Jy beam^{-1}. The positions of the H_2O maser components are indicated by blue and red dots according to their velocities. The relative positional uncertainty between the H_2O masers is < 3.5 mas. The size of the cross of the 2MASS position indicates its 1σ position error. The magenta ellipse shows the synthesized beam. (d) *V*-*R* diagram of OH16.3. Since the stellar position does not coincide with the center of the spatial distribution of the OH masers, we considered the intensity-weighted emission centroid of the OH maser positions as the zero radial offset. The dashed lines indicate the assumed stellar velocity (V_*).

Moreover, OH maser observations at 1612 MHz were performed. Positions of the masers and continuum emission were obtained by 2-D Gaussian fitting. To explore the spatio-kinematics of the OH masers, we constructed a velocity-radius (*V*-*R*) diagram, where *V* is the LSR velocity, and *R* is the angular (radial) offset from the stellar position. To interpret this diagram, we considered a geometric model consisting of an AGB shell and a biconical, symmetric outflow (Fig. 2a in Zijlstra *et al.* 2001). This model is based on the interacting-winds scenario. The relationship between *R* and *V* is given by $R = R_{\rm shell}[1 - (V - V_*)^2/V_{\rm exp}^2]^{0.5}$ (for details, see Zijlstra *et al.* 2001).

3. Results and Conclusions

In OH16.3, we detect H_2O maser emission with three distinct velocity components (with flux densities from 0.14 to 3.94 Jy, Fig. 1a), and unresolved radio continuum emission with a flux density of 0.07 mJy at 22 GHz (Fig. 1c). We also detect OH maser emission (Fig. 1b). This OH emission is expected to be associated with an aspherical circumstellar envelope because of: its peculiar OH maser spectrum profile (with prominent wings), the spatial distribution of its blue- and red-shifted masers overlapped in the plane of the sky (not shown here), and the deviations of an arc-shaped structure for a spherical AGB shell in its *V*-*R* diagram (predicted *V*-*R* relations for the geometric model can be found in Zijlstra *et al.* 2001) (Fig. 1d). OH16.3 may be a peculiar AGB or post-AGB star with a non-spherical wind. We note that the velocity spread of the H_2O

maser emission is smaller than that in OH, different from what was found on single-dish observations. The spatial distribution of the H_2O maser components (Fig. 1c) does not show conclusive evidence for a collimated jet. It is still possible that the source is a WF, but with the outflow lying close to the plane of the sky, or a different kind of WF, e.g., with a less massive progenitor or with a relatively younger age (Uscanga *et al.* 2019). We did not detect H_2O maser emission in I19356 with a 3σ detection limit of 17.7 mJy, despite a previous single-dish detection with a flux density of ~ 0.2 Jy and rms of 10 mJy (Yung *et al.* 2014). Further monitoring could help to discern the nature of these sources.

Supplementary material

To view supplementary material for this article, please visit http://dx.doi.org/10.1017/S1743921323002168

References

Desmurs, J.-F. 2012, *Proc. IAU Symp.*, 287, 217
Gómez, J. F., Suárez, O., Rizzo, J. R., *et al.* 2017, *MNRAS*, 468, 2081
Imai, H. 2007, *Proc. IAU Symp.*, 242, 279
Sevenster, M. N., Chapman, J. M., Habing, H. J., *et al.* 1997, *Astron. Astrophys. Suppl. Ser.*, 122, 79
Uscanga, L., Gómez, J. F., Yung, B. H. K., *et al.* 2019, *Proc. IAU Symp.*, 343, 527
Yung, B. H. K., Nakashima, J.-I., Imai, H., *et al.* 2013, *ApJ*, 769, 20
Yung, B. H. K., Nakashima, J.-I. & Henkel, C. 2014, *ApJ*, 794, 81
Zijlstra, A. A., Chapman, J. M., te Lintel Hekkert, P., *et al.* 2001, *MNRAS*, 322, 280

Cosmic Masers: Proper Motion toward the Next-Generation Large Projects
Proceedings IAU Symposium No. 380, 2024
T. Hirota, H. Imai, K. Menten, & Y. Pihlström, eds.
doi:10.1017/S1743921323002041

Preliminary results on SiO maser emission from the AGB binary system: R Aqr.

J.-F. Desmurs[1], J. Alcolea[1], V. Bujarrabal[1], M. Santander Garcia[1], M. Gomez-Garrido[1] and J. Mikolajewska[2]

[1]Observatorio Astronómico Nacional (OAN/IGN), Spain. email: jf.desmurs@oan.es

[2]Nicolaus Copernicus Astronomical Center of Polish Academy of Sciences

Abstract. Binary/multiple stellar systems are as abundant as single stars. They are very interesting cosmic laboratories as they are related to many astrophysical phenomena. In the case of AGB (and post-AGB) stars, binaries are the most likely explanation for the shaping of non spherical PNe, a long standing problem in stellar evolution. While many binaries are known in the PNe phase, there are few examples in the previous AGB phase. Hydrodynamical models of the binary interaction are available, but well known parameters for orbits are non-existent. We observed the AGB binary system R Aqr at 7 mm using the Global VLBI array in wide-band continuum and SiO masers. The strong SiO masers were used to self-calibrate the observations and pinpoint the location of the AGB component. We used the continuum to try to directly detect the WD companion or its close environments (non-thermal emission from accretion disks/jets). We present our preliminary results for R Aqr on the relative positions of the J=1−0 SiO masers (v=1) with respect to the white dwarf position. This opens a new way to determine orbits in binaries at the AGB phase and better study their role in late stellar evolution and PNe shaping.

Keywords. AGB, SiO masers

---

1. Introduction

R Aqr is a symbiotic system consisting in a primary AGB star (a Mira-type variable star, a pulsating red giant star, R Aqr-A) and a WD secondary (R Aqr-B), as well as an ionized nebulae around the system, located at 200-300 pc. Our ALMA observations (Bujarrabal 2018) have revealed the strong shaping of the CSE by the companion, with the detection of a two-arm spiral in molecular gas. Based on the velocity variations of the primary we derived the parameters of the orbit (Gromadzki 2009). However, it was not till very recently that the two components were finally resolved by means of VLT (Schmid 2017) and ALMA observations (Bujarrabal 2018). Using these observations and additional spectroscopic data, we have recomputed the orbital parameters of the system, which now agree with the spiral structure seen in molecular gas. We recently published (Bujarrabal 2021) a comparison of our CO observations with hydrodynamical simulations of the system, but this is a delicate subject, as some of the orbital parameters still show large errors. This is mostly due to the lack of spatially resolved data, just four epochs, and the intrinsic uncertainties in spectroscopic data: R Aqr-A is a pulsating star, and the pulsation velocity is larger than the orbital velocity; this effect can be minimized by averaging data over full pulsation periods, but some errors remain. The best way to improve the situation is by adding more data points to the relative orbit, but this is difficult as the maximum separation in the sky of the two stars is just 50 mas, and the system is close to periastron (when a stronger interaction is expected) and conjunction;

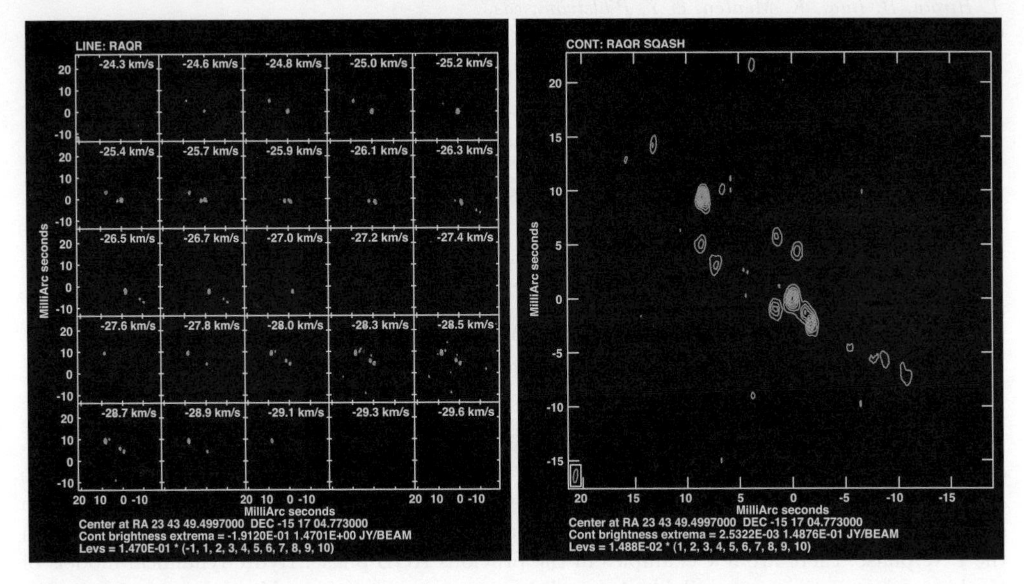

Figure 1. Global VLBI SiO v=1 J=1−0 maser emission maps at 43 GHz toward R Aqr: Left panel, channel maps of SiO emission between −24 and −30 km s^{-1}. Right panel: SiO integrated flux map.

in the next years the angular separation between the two components will be of 10-15 mas. Measuring such small separations is not an easy task even for the VLT or ALMA, while it will be very easy for a Global-VLBI experiment at 43 GHz.

2. Observations

We perform Global-VLBI observations of the R Aqr binary systems, both in spectral lines SiO v=1 J=1−0 maser at 43122.08 MHz assuming a source velocity with respect to the local standard of rest of −25 km s^{-1}, and broad band continuum in Q-band (7 mm). We recorded at maximum bit rate for maximum continuum sensitivity, in dual circular polarization, 8 sub-bands 32 MHz wide. The SiO maser pinpoints the location of the AGB component, while the continuum marks the position of the hot component in the system (the WD).

3. Preliminary results

On Figure 1, we present the synthesis image maps of the SiO v=1 J=1−0 maser emission toward R Aqr. The maps were generated following the standard line imaging and calibration techniques with the AIPS data reduction package. The synthesized resolution beam is about 1.0x0.4 mas in natural weighting. On the left panel, we show the channel maps of the SiO maser distribution. We search for emission across more than 25 km s^{-1} (from −15 km s^{-1} to −40 km s^{-1}) but we only find detections in the velocity range −24 km s^{-1} to −30 km s^{-1} (according to the predicted velocity curve). The maser peak flux ranges between 30 to 150 mJy/beam which means a brightness temperature of the order of ∼10 to 100 mega-K. On the right panel, we show the total integrated flux SiO emission map. We can notice that the maser spots do not show the typical ring distribution observed in AGB stars, also observed in this source by Cheulhong (2014). In contrast, it shows an elongated structure in the North-East to South-West direction; a similar structure was also found in the first epoch observed by Boboltz (1997). The continuum map is not shown as the data reduction is still under progress.

References

Boboltz, D.A, Diamond, P.J and A. J. Kembal, A.J, 1997, *ApJ* 487, L147
Bujarrabal, V., Alcolea, J., Mikolajewska, J., *et al.* 2018, *A&A*, 616, L3.
Bujarrabal, V., Agúndez, M., Gómez-Garrido, M. *et al.* 2021, *A&A*, 651, A4
Cheulhong Min, Naoko Matsumoto, Mi Kyoung Kim, *et al.* 2014, *PASP*, 66, Issue 2, 38
Gromadzki, M., Mikolajewska, J., 2009, *A&A*, 495, 931
Schmid, H.M., Bazzon, A., Milli, J., *et al.* 2017, *A&A*, 602, A53.

Cosmic Masers: Proper Motion toward the Next-Generation Large Projects
Proceedings IAU Symposium No. 380, 2024
T. Hirota, H. Imai, K. Menten, & Y. Pihlström, eds.
doi:10.1017/S1743921323002533

A database of circumstellar OH masers update

Dieter Engels[1] and **Belen López-Martí**[2]

[1]Hamburger Sternwarte, Universität Hamburg, Gojenbergsweg 112, 21029 Hamburg, Germany.
email: dengels@hs.uni-hamburg.de

[2]Universidad San Pablo-CEU, Campus de Montepríncipe, 28668 Boadilla del Monte, Madrid, Spain

Abstract. The database of circumstellar OH masers by Engels & Bunzel (2015) was updated to include new 1612, 1665, and 1667 MHz OH maser observations published between 2015 and 2022. A cross-correlation of the database was made with infrared catalogues (AllWISE and 2MASS) and with GAIA DR3. This led frequently to significantly improved coordinates and identified contaminations with non-stellar sources. About 40% of all OH maser-detected stars were not detected by GAIA. These are mostly representatives of the population of highly obscured stars at the end of AGB or at the beginning of post-AGB evolution.

Keywords. masers, stars: AGB and post-AGB, circumstellar matter

---

1. New maser database release v2.5 including literature until 2022

The circumstellar envelopes of AGB stars frequently host masers emitted by the OH molecule with transitions at 1612, 1665, and 1667 MHz. Many of the optically very obscured stars at the end of AGB evolution (OH/IR stars) were discovered with surveys for these masers. The number of OH emitting stars discovered increased steadily from $\sim$130 (Engels 1979), to 439 (te Lintel Hekkert *et al.* 1989) (only 1612 MHz emitters) to >2300 (Engels & Bunzel 2015). The OH maser database was compiled with the aim to allow a quick access to the literature, where OH maser observations are reported. Observations published in refereed papers are now covered until 2022. Most new objects in the updated maser database were discovered by the large recent THOR (Beuther *et al.* 2019) and SPLASH OH maser surveys (Qiao *et al.* 2020) conducted along the Galactic plane. The database version 2.5 comprises $\sim$17000 observations and >2900 different stars detected in at least one transition (+24% relative to the 2015 release). The vast majority are Asymptotic Giant Branch (AGB) stars. About 130 stars (<5% of all stars with OH masers) have interferometric follow-up observations and/or were included in monitoring programs since the beginnings in the 1970s. The database is organized in three tables. The main table contains individual observations (maser velocities and flux densities), in addition to the coordinates. The two auxiliary tables monitoring projects and interferometric follow-up observations. A detailed description of the database is given in Engels & Bunzel (2015).

Access to the database is possible over the Web: www.hs.uni-hamburg.de/maserdb

2. Cross-correlations with the AllWISE, 2MASS and GAIA catalogues

We looked for counterparts of the OH/IR stars in the GAIA DR3 release to overcome the poor coordinate precision of many members of the maser database (López Martí

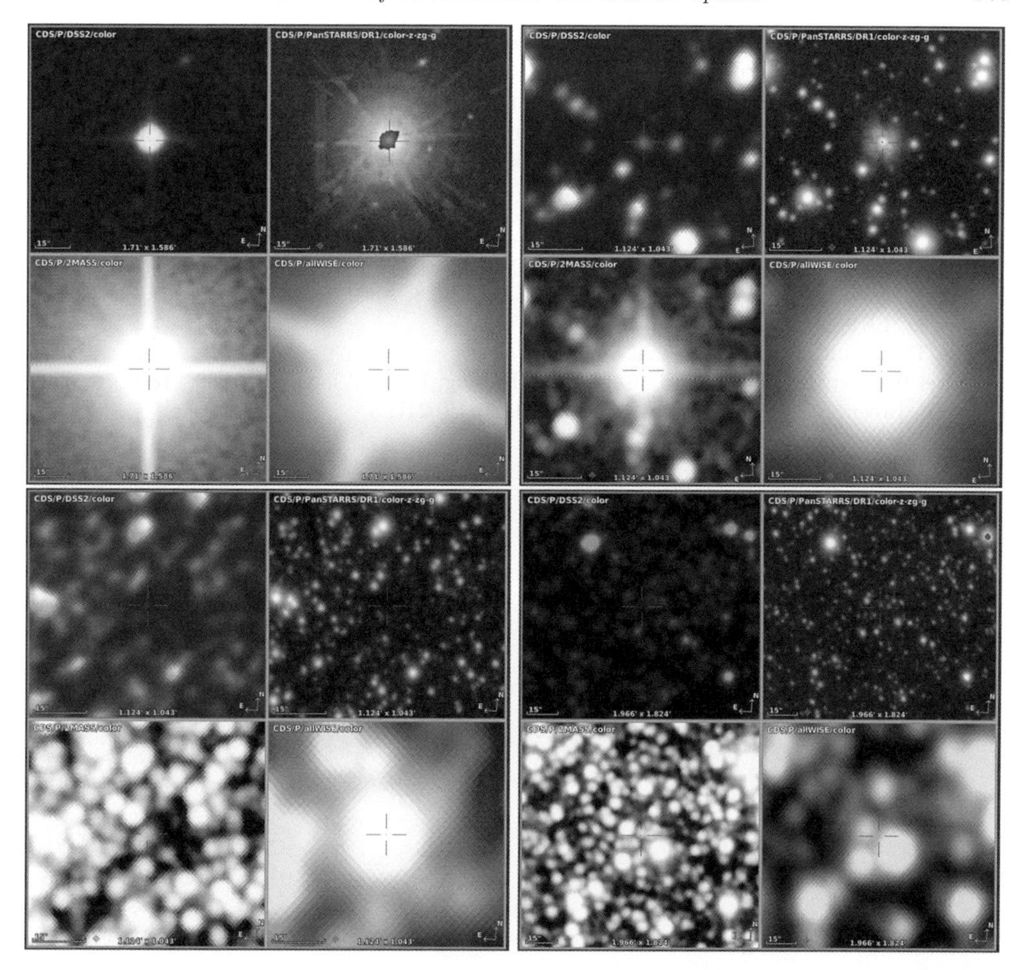

Figure 1. Displays from Aladin with counterparts on the Digital Sky Survey, on Pan-STARSS, on 2MASS (K-band; 2.2μm) and on WISE (W4 band, 22μm). The objects are a bright Mira variable T Vir (upper left), IRAS 19382+3400, an OH/IR star with optical counterpart (upper right), OH000.689+02.140, a strongly obscured OH/IR star (lower left), and OH356.524+02.526, an object without unambiguous IR identification.

et al. in preparation). Because a direct cross-match between the original coordinates and GAIA would have led to a great number of spurious detections, the process was done in several steps: First we did a cross-match of the database coordinates with the AllWISE Catalog, then we used the refined WISE coordinates to run a second cross-match with the 2MASS Point Source Catalog, and finally the 2MASS coordinates were used in a cross-match with GAIA DR3. We also used a successively decreasing cross-match radius to account for the improvement in the resolution from one survey to the next.

The cross match was visually supervised using the tool Aladin (Figure 1). Outside the Galactic plane ($|\,b\,| > 2^\circ$) unique counterparts could be found usually within $\sim 1'$ (upper panels of Figure 1) of the maser input coordinates. In the plane and especially close to the Galactic Center, the interstellar extinction and the crowding of sources made the cross-match more difficult. Often a distinguished red WISE counterpart gave a unique identification (lower left panel of Figure 1), but at times no counterpart could be found. In these cases either the input coordinates seem to be erroneous or the maser was mis-classified and is of non-stellar origin (lower right panel of Figure 1).

We could identify 93% of the OH maser sources listed in the database. 52% are present in the GAIA DR3 release, while 41% are too obscured and fainter than $G \approx 21$ mag at optical wavelengths. Significant improvements in coordinates of the IR/GAIA identified OH/IR stars could be achieved, with coordinate corrections reaching >10 arcsec for 9% of the stars. The inclusion of the improved coordinates for these stars (~ 2700 stars) is planned for the next release (Engels & López Martí, in preparation).

References

Beuther, H., Walsh, A., Wang, Y., *et al.* 2019, *A&A*, 628, A90
Engels, D. 1979, *A&AS*, 36, 337
Engels, D. & Bunzel, F. 2015, *A&A*, 582, A68
te Lintel Hekkert, P., Versteege-Hensel, H. A., Habing, H. J., Wiertz, M. 1989, *A&S*, 78, 399
Qiao, H., Breen, S. L., Gómez, J.F., *et al.* 2020, *ApJS*, 247, 5

Cosmic Masers: Proper Motion toward the Next-Generation Large Projects
Proceedings IAU Symposium No. 380, 2024
T. Hirota, H. Imai, K. Menten, & Y. Pihlström, eds.
doi:10.1017/S1743921323002004

The loss of OH maser emission in the early stage of Post-AGB evolution

S. Etoka[1], D. Engels[2], T. Ullrich[2], J.B. González[3] and B. López-Martí[4]

[1]Jodrell Bank Centre of Astrophysics, Manchester, UK, Sandra.Etoka@googlemail.com

[2]Hamburger Sternwarte, Universität Hamburg, Germany

[3]MAX IV Laboratory, Lund University, Sweden

[4]Universidad San Pablo - CEU, Madrid, Spain

Abstract. Based on the results of an on-going monitoring program of 1612 MHz OH masers in OH/IR stars, we determined a lifetime encompassing late AGB and early post-AGB evolution of at least 4500 years. Fading of the OH masers observed with the Nançay Radio Telescope is detected in several post-AGB OH/IR stars on timescales of decades, while AllWISE/NEOWISE light curves taken almost in parallel show diverse behaviours.

Keywords. masers, infrared: stars, stars: AGB and post-AGB, star: evolution,

---

1. Introduction

The transition of ~2-8 $M_\odot$ stars from the Asymptotic Giant Branch (AGB) to the post-AGB stage is accompanied by strong mass loss leading to the formation of optically-thick circumstellar shells. Variable OH/IR stars with high mass-loss rates ($\geq 10^{-6}\,M_\odot\,yr^{-1}$) become obscured at visual light and, for the most extreme cases, also in the near-infrared (NIR: 1–5μm). Their 1612 MHz OH maser light curves on the AGB show large amplitude variations with periods up to 2500 days. Other OH/IR stars with non-periodic small amplitudes variations are characteristic of post-AGB evolution. After that most of their mass is lost during the AGB phase, the mass-loss rates drop on short timescales of a few thousand years from late AGB values of 10^{-5}–10^{-4} to post-AGB values of 10^{-7}–$10^{-8}\,M_\odot\,yr^{-1}$ (Miller-Bertolami 2016). Because of the short time scales and the obscuration of the stars, the details of the transition process from the AGB to the post-AGB phase are not well constrained. There is however growing evidence accumulating that following variations of the IR emission (Kamizuka *et al.* 2020), and of the OH masers (Wolak *et al.* 2014) on timescales of decades allows to study aspects of this transition process in real time. From an ongoing 1612-MHz OH-maser monitoring program we present here post-AGB OH/IR stars with fading OH maser emission.

2. Post-AGB OH/IR stars with fading OH masers

The sample monitored comprises 114 stars located along the Galactic plane at $10 < l < 150°$, and $|b| < 4°$. The majority of the OH/IR stars are periodically pulsating AGB stars (N=78, 68%), while 31 stars (27%) show only small-amplitude irregular variations (Engels *et al.* 2019)†. For the latter we extracted infrared photometry from the ALLWISE Multiepoch Photometry Catalog (Wright *et al.* 2010) and (Cutri *et al.* 2013)

† Updates available on the project web page www.hs.uni-hamburg.de/nrt-monitoring

Table 1. Post-AGB OH/IR stars with fading OH masers. T_{end} is the predicted year at which the peak flux density should drop below 100 mJy. ΔW2 is the systematic IR brightness change between 2009 and 2021 derived from NEOWISE data.

Name	T_{end} [yr]	**ΔW2 [mag/yr]**	**Name**	T_{end} **[yr]**	**ΔW2 [mag/yr]**
OH 12.8−0.9	2031	−0.06	OH 31.0+0.0	2030	+0.03
OH 17.7−2.0	2035	+0.04	OH 37.1−0.8	2043	−0.02
OH 18.5+1.4	∼2100	< 0.01			

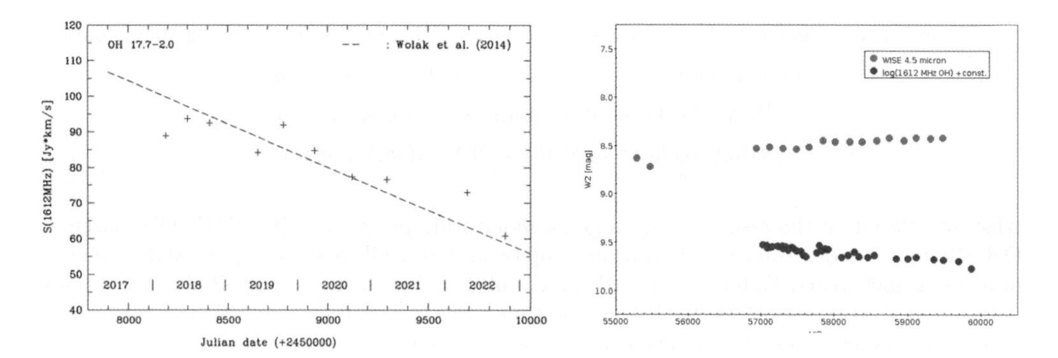

Figure 1. *Left:* 2018-2022 light curve of 1612-MHz OH-maser emission of OH 17.7−2.0. Superimposed is the predicted light curve by Wolak *et al.* (2014). *Right:* 2015–2022 1612-MHz OH maser light curve of OH 37.1 − 0.8 in logarithmic units (blue) and 2009–2021 WISE 4.5μm light curve (red). The OH maser data is shifted by an arbitrary constant.

and the 2022 NEOWISE-R Single Exposure (L1b) source table (Mainzer *et al.* 2014) to obtain light curves over the years 2009–2021. We averaged the photometry in 10 day intervals and made linear fits to the data restricting on the NEOWISE data.

Following Engels & Jiménez-Esteban (2007), we made a new estimate of the OH maser lifetime based on the re-detection of all masers in 2022. We obtained statistically a lower limit for the lifetime of > 4300 years (1σ) and a post-AGB minimum lifetime of > 1700 years (Ullrich *et al.* in preparation). Among the post-AGB OH/IR stars we identified 5 stars (namely, OH 12.8−0.9, OH 17.7−2.0, OH 18.5+1.4, OH 31.0+0.0, and OH 37.1−0.8) with continuous decline of the integrated 1612-MHz OH-maser emission. The predicted drop to peak flux densities <100 mJy is expected to occur within the next 80 years (Table 1). The most convincing case is OH 17.7−2.0. The fading was discovered by Wolak *et al.* (2014) who predicted its disappearance by 2030. Our recent observations 2018–2022 confirm the masers ongoing decline (Fig. 1, left panel).

We searched in the AllWISE/NEOWISE photometry for increasing infrared brightness evolving in parallel, following the findings of Kamizuka *et al.* (2020) at 2.2μm. We found mixed results for the systematic brightness variations in the WISE W1 and W2 bands. For two objects brightening and for two objects dimming was seen (Table 1). Also, five out of six post-AGB OH/IR stars of the sample of Kamizuka *et al.* (2020) showing IR brightening at 2.2μm, show no long-term variations of the OH maser brightness in our monitoring program. An exception is OH 37.1 − 0.8 for which the light curves are shown in Fig. 1 (right panel).

Our results do not show a clear correlation between the fading OH and long-term infrared brightness variations. However, the long-term variations show that stellar evolution in the AGB to post-AGB transition phase can be observed in real time.

References

Cutri, R. M., Wright, E. L., Conrow, T., *et al.* 2013, Expl. Suppl. to the AllWISE Data Release Products

Engels, D., Etoka, S., & Gérard, E., 2019, *IAUS*, 343, 389

Engels, D. & Jiménez-Esteban, 2007, *A&A*, 475, 941

Kamizuka, T., Nakada, Y., Yanagisawa, K., *et al.* 2020, *ApJ*, 897, 20

Mainzer, A., Bauer, J., Cutri, R. M., *et al.* 2014, *ApJ*, 792, 30

Miller-Bertolami, M. M. 2016, *A&A*, 588, 25

Wright, E.L., Eisenhardt, P. R. M., Mainzer, A. K., *et al.* 2010, *AJ*, 140, 1868

Wolak, P., Szymczak, M., Bartkiewicz, A., & Gérard, E., 2014, 12th EVN-Symp. 2014, 116

Cosmic Masers: Proper Motion toward the Next-Generation Large Projects
Proceedings IAU Symposium No. 380, 2024
T. Hirota, H. Imai, K. Menten, & Y. Pihlström, eds.
doi:10.1017/S1743921323002247

A sensitive search for SiO maser emission in planetary nebulae

José F. Gómez[1], **Roldán A. Cala**[1], **Luis F. Miranda**[1], **Hiroshi Imai**[2,3], **Mayra Osorio**[1] and **Guillem Anglada**[1]

[1]Instituto de Astrofísica de Andalucía, CSIC, E-18008 Granada, Spain. email: jfg@iaa.es

[2]Amanogawa Galaxy Astronomy Research Center, Graduate School of Science and Engineering, Kagoshima University, Kagoshima 890-0065, Japan

[3]Center for General Education, Institute for Comprehensive Education, Kagoshima University, Kagoshima 890-0065, Japan

Abstract. Eight planetary nebulae (PNe) are known to emit OH and/or H_2O masers, but there is no report of an SiO maser in this type of objects. We present a search for SiO masers in 16 confirmed and candidate PNe, carried out with the Australia Telescope Compact Array. We found no evidence of association between SiO masers and PNe in our data. Previous detections of thermal SiO emission in PNe show that these molecules can be present in gas phase in this type of objects, but it is not yet clear whether they can be found where the physical conditions are appropriate for maser pumping. We suggest that the best candidates for a new search would be PNe showing high-velocity outflows.

Keywords. masers, planetary nebulae: general, stars: AGB and post-AGB, stars: mass loss

1. Introduction

Maser emission of SiO, H_2O, and OH is widespread in Asymptotic Giant Branch (AGB) stars with Oxygen-rich circumstellar envelopes. However, it is expected that it quickly disappear after the end of strong mass-loss (up to 10^{-4} $M_\odot$ yr^{-1}, Blöcker 1995) of the AGB phase, with timescales $\simeq 10 - 10^4$ years, depending on the molecular species (Gómez *et al.* 1990). Planetary nebulae (PNe) bearing maser emission are, therefore very scarce, and are considered to be nascent PNe. So far, only eight PNe are confirmed to emit OH and/or H_2O masers (see Cala *et al.* 2022, and references therein). No SiO-maser-emitting PN has ever been found. We present here a sensitive search for SiO masers in PNe in two carefully selected samples.

2. Source selection and observations

We carried out interferometric observations of SiO masers at 43 GHz with the Australia Telescope Compact Array. Four SiO transitions (v=0,1,2,3 of J=1-0), as well as radio continuum emission (4 GHz total bandwidth) were covered simultaneously. We selected two source samples, based on different criteria, in order to maximize the probability of detection.

For the first sample, we selected all PNe with known masers of OH and H_2O, as well as some candidate PNe with masers (sources for which the PN nature is still not confirmed). The sample comprises 11 sources. The angular resolution of the observations was $\simeq 7''$.

For the second sample, we cross-matched detections of SiO masers (obtained with single-dish radio telescopes) with catalogs of known PNe and of radio continuum sources.

PNe show prominent radio continuum emission, due to free-free processes in the photoionized gas, so this emission may pinpoint a PN. We selected those cases in which a radio continuum source or a catalogued PN is within the beam of the reported single-dish SiO detection. Our sample comprised five sources. The angular resolution of these observations was $\simeq 0.5''$.

3. Results

In the first sample (confirmed and candidate PNe with masers of other species), no SiO detection was obtained, with 3σ upper limits $\simeq 7$ mJy. This is a much more stringent constraint than in previously reported SiO observations toward five of our sources ($\simeq 1$ Jy, Nyman *et al.* 1998). All sources in the sample, however, show radio continuum emission at 43-45 GHz. In some of these sources, ours is the first report of continuum emission at this frequency band.

In the second sample (single-dish SiO detections near radio sources and PNe), we detected SiO emission in four out of our five targets. In none of them we could confirm that the maser emission is associated with a radio continuum source. These four SiO emitters are most likely to be AGB stars. We obtained accurate positions for the SiO masers in these sources. The only case in which we still cannot completely rule out the association of SiO with radio continuum emission is IRAS 17390-3014. We did not detect any radio continuum emission at 7mm in that source, but there is an extended continuum source (NVSS 174218-301526), detected at 1.4 GHz (Condon *et al.* 1998), with a deconvolved size of $80'' \times 34''$, whose association is unclear and that seems to engulf the position of the evolved star. Our observations, with an angular resolution of $0.5''$ do not seem to have enough sensitivity to detect such an extended source. Further observations are needed to confirm the existence of this continuum source, its nature, and its putative association with the SiO masers.

4. Discussion

Our non-detection of SiO masers in the sampled PNe could be due to several reasons. One is that the SiO abundance in gas phase may be too low to provide the necessary column density for a detection. Even if SiO is present in gas phase, it may be located in regions where the physical conditions are not appropriate for population inversion.

The SiO abundance in gas phase tends to drop in circumstellar envelopes at increasing distances from the central star, because the molecule accretes into dust grains, but it may be released again in the presence of shocks. In the particular case of PNe, however, SiO molecules could be photodissociated by the hot, central star. There are some reports on thermal SiO emission using single-dish telescopes (e.g., Edwards & Ziurys 2014), and our ALMA observations of the young PN IRAS 15103-5754 also show thermal SiO emission located at < 400 AU from the central star (Gómez et al. in preparation). While this indicates that SiO molecules can be found in gas phase in PNe, there is still no data showing their presence at the locations where SiO masers are typically found in AGB stars (a few stellar radii). Of course, SiO masers may be pumped by different mechanisms from those of in the AGB, and thus be found farther away from the central star. We note, for instance, that SiO masers can be excited at the base of high-velocity outflows in the case of post-AGB stars (Imai *et al.* 2005), at distances $\simeq 10$ AU. Assuming the same may happen in PNe, promising candidates for a new SiO maser search could be objects showing fast outflows, which may be even more evolved than PNe harboring OH and H_2O masers.

This work is partially supported by grant P20-00880 of Junta de Andalucía, grants PID2020-114461GB-I00, CEX2021-001131-S of MCIN/AEI /10.13039/501100011033, and PRE2018-085518 funded by MCIN/AEI and by ESF Investing in your future.

References

Blöcker, T. 1995, *A&A*, 297, 727
Cala, R. A., Gómez, J. F., Miranda, L. F., *et al.* 2022, *MNRAS*, 516, 2235.
Condon, J. J., Cotton, W. D., Greisen, E. W., *et al.* 1998, *AJ*, 115, 1693
Edwards, J. L. & Ziurys, L. M. 2014, *ApJL*, 794, L27
Gómez, Y., Moran, J. M., & Rodríguez, L. F. 1990, *Rev. Mexicana AyA*, 20, 55
Imai, H., Nakashima, J.-i., Diamond, P. J., *et al.* 2005, *ApJL*, 622, L125
Nyman, L.-A., Hall, P. J., & Olofsson, H. 1998, *A&AS*, 127, 185.

Cosmic Masers: Proper Motion toward the Next-Generation Large Projects
Proceedings IAU Symposium No. 380, 2024
T. Hirota, H. Imai, K. Menten, & Y. Pihlström, eds.
doi:10.1017/S1743921323002065

A Profile-based Approach to Finding New Water Fountain Candidates using Databases of Circumstellar Maser Sources

J. Nakashima[1,2], H. Fan[1], D. Engels[3], Y. Zhang[1,2], J.-J. Qiu[1,2], H.-X. Feng[1], J.-Y. Xie[1], H. Imai[4,5] and C.-H. Hsia[6]

[1]School of Physics and Astronomy, Sun Yat-sen University, Tang Jia Wan, Zhuhai, 519082, P. R. China. email: nakashima.junichi@gmail.com

[2]CSST Science Center for the Guangdong-Hong Kong-Macau Greater Bay Area, Sun Yat-Sen University, 2 Duxue Road, Zhuhai 519082, Guangdong Province, PR China

[3]Hamburger Sternwarte, Universität Hamburg, Gojenbergsweg 112, D-21029 Hamburg, Germany

[4]Center for General Education, Institute for Comprehensive Education, Kagoshima University 1-21-30 Korimoto, Kagoshima 890-0065, Japan

[5]Amanogawa Galaxy Astronomy Research Center (AGARC), Faculty of Science, Kagoshima University 1-21-30 Korimoto, Kagoshima 890-0065, Japan

[6]The Laboratory for Space Research, Faculty of Science, The University of Hong Kong, Cyberport 4, Hong Kong SAR, China

Abstract. Water fountains (WFs) are thought to represent an early stage in the morphological evolution of circumstellar envelopes surrounding low- and intermediate-mass evolved stars. These objects are considered to transition from spherical to asymmetric shapes. Despite their potential importance in this transformation process of evolved stars, there are only a few known examples. To identify new WF candidates, we used databases of circumstellar OH (1612 MHz) and H_2O (22.235 GHz) maser sources, and compared the velocity ranges of the two maser lines. Finally, 41 sources were found to have a velocity range for the H_2O maser line that exceeded that of the OH maser line. Excluding known planetary nebulae and after reviewing the maser spectra in the original literature, we found for 11 sources the exceedance as significant, qualifying them as new WF candidates.

Keywords. Evolved stars, masers, post-AGB stars, AGB stars

---

1. Purpose of the research

The main purpose of this study is to search for new candidates for WFs (for WFs, see, e.g., Imai 2009) by cross-checking H_2O (22.235 GHz) and OH (1612 MHz) maser sources from databases of circumstellar maser sources. We then investigated the basic infrared properties of the selected sources having a larger H_2O maser velocity range ΔV_{H_2O} compared to ΔV_{OH} of the OH maser (Yung *et al.* 2013).

2. Methodology

First, we selected circumstellar maser sources in which both the H_2O maser at 22.235 GHz and the OH maser at 1612 GHz are detected. We then used WISE and other

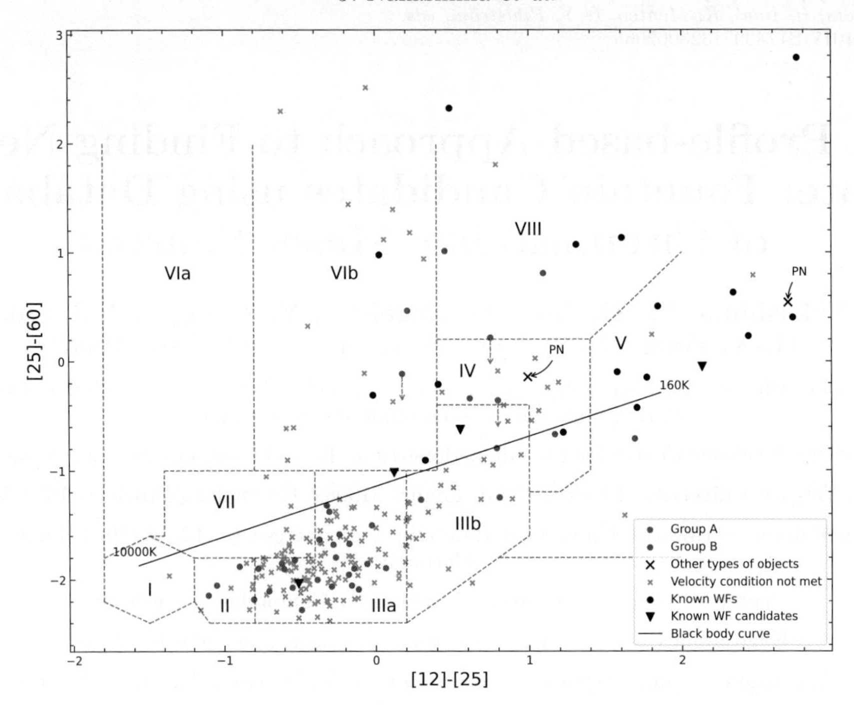

Figure 1. IRAS two-color diagram of 41 sources selected by the velocity condition $\Delta V_{\rm OH} \leq \Delta V_{\rm H_2O}$. This plot uses color indices defined as follows $[12]-[25]=2.5\log\left(\frac{F_{25\mu\rm m}}{F_{12\mu\rm m}}\right)$ and $[25]-[60]=2.5\log\left(\frac{F_{60\mu\rm m}}{F_{25\mu\rm m}}\right)$. For 11 sources the selection by the criterion is reliable (Group A), while for the rest verification is needed (Group B).

infrared archival images to remove possible YSOs. Finally, if the criterion $\Delta V_{\rm OH} \leq \Delta V_{\rm H_2O}$ is satisfied, then such maser sources were selected as candidates for WF. As a source of information on OH maser data, we used the database of Engels & Bunzel (2015) accessible from CDS/VizieR†. This database contains 8,474 OH 1612 MHz observations toward 2,195 sources. In the present analysis we used only the 1612 MHz line data. For the 22.235 GHz H_2O maser line, we used an unpublished database compiled by the same authors. The version 0.1 of the database used for our analysis contains records of 6,085 observations towards 3,642 sources.

3. Summary of initial results

From the cross-match between the OH and H_2O maser databases we extracted a sample of 229 sources with detections in both transitions. Comparison of their maser velocity ranges yielded 41 sources, which met the velocity criterion. From a detailed examination of the maser line profiles available in the literature, we concluded that the deviation of the H_2O maser velocity range was significant for 11 of the 41 initially selected sources (Group A in Fig. 1). The main results are as follows:

(1) We examined the IRAS colors of the samples and found that two of the 11 sources with a confirmed H_2O maser velocity deviation (IRAS 19069+0916 and IRAS 19319+2214) are in the color region for post-AGB stars. The H_2O maser profile of these sources are similar to that of known WFs.

(2) Of the 11 selected sources, six sources were located in the color region of the AGB stars. For two of the six sources (IRAS 19422+3506 and IRAS 22516+0838), the

† https://cdsarc.cds.unistra.fr/viz-bin/cat/J/A+A/582/A68

H_2O maser properties are different from those of typical AGB stars, and confirmation observations with radio interferometry are desired.

(3) We also confirmed the possibility that sources exhibiting the velocity deviation of the H_2O maser line could include astrophysically interesting sources other than WFs. Such objects could include, for example, peculiar planetary nebulae with maser emission and stellar merger remnants.

References

Engels, D., & Bunzel, F. 2015, *A&A*, 582, A68
Imai, H. 2009, in AGB Stars and Related Phenomena, ed. T. Ueta, N. Matsunaga, & Y. Ita, 62
Yung, B. H. K., Nakashima, J., Imai, H., *et al.* 2013, *ApJ*, 769, 20

Cosmic Masers: Proper Motion toward the Next-Generation Large Projects
Proceedings IAU Symposium No. 380, 2024
T. Hirota, H. Imai, K. Menten, & Y. Pihlström, eds.
doi:10.1017/S1743921323002740

HINOTORI and Maser Observations

Keisuke Nakashima[1], Ka-Yiu Shum[1], Hiroshi Imai[2,3] and HINOTORI Collaboration

[1]Faculty of Science, Kagoshima University, 1-21-35 Korimoto, Kagoshima 890-0065, Japan, email: k8024878@kadai.jp

[2]Amanogawa Galaxy Astronomy Research Center, Graduate School of Science and Engineering,

Kagoshima University, 1-21-35 Korimoto, Kagoshima 890-0065, Japan,

[3]Center for General Education, Institute for Comprehensive Education, Kagoshima University,

1-21-30 Korimoto, Kagoshima 890-0065, Japan

Abstract. HINOTORI (Hybrid Integration Project in Nobeyama, Triple-band Oriented) has constructed a higher sensitivity 22/43 GHz and a 22/43/86 GHz simultaneous observation systems in the Nobeyama 45-m telescope by introducing new frequency separation filters in the telescope's quasi-optics. The performance of the observation systems, such as the beam squint, the aperture efficiency, the system noise temperature when inserting the filters, and the phase stability of the signal path have been evaluated. It is indicated that the established systems have sufficient performance for single-dish and VLBI observations. The single-dish observation demonstrations using the triple-band system were successfully conducted in acquiring scientific data including multiple maser lines.

Keywords. masers, instrumentation: interferometers

---

1. Introduction

HINOTORI is a project to realize simultaneous single-dish and VLBI observations in three frequency (22, 43, and 86 GHz) bands with the Nobeyama 45 m telescope (Okada *et al.* 2020). Four receivers named H22 (22 GHz, RHCP+LHCP), H40 (43 GHz, LHCP), Z45 (43 GHz, two orthogonal linear polarizations), and TZ (86 GHz, two orthogonal linear polarizations) are used in this project for radio signals flow from the quasi-optics of this telescope (Imai *et al.* 2022). The H22+H40 simultaneous observation system has already been established and used for scientific single-dish and VLBI observations. This time, we have established new simultaneous observing systems, H22+Z45 and H22+H40+TZ, by installing another frequency-separating filters. Here, we summarize the instrumental performance of these simultaneous observation systems and the observation demonstrations in the H22+H40+TZ mode.

2. System evaluation

The beam squint measurements with the new filters have been conducted using the data of H_2O, SiO $v=1$ and 2 ($J=1\rightarrow 0$) maser lines for the H22+Z45 mode and SiO $v=1$ ($J=2\rightarrow 1$ and $1\rightarrow 0$) maser lines for the H22+H40+TZ mode. The pointing offsets of H22 and Z45 with the filters due to the bean squint were distributed around $(dAz, dEl) \sim (+3'', +7'')$, respectively (Imai *et al.* 2022). The pointing offsets of TZ in the H22+H40+TZ mode was distributed around $(dAz, dEl) \sim (+3'', +3'')$. These distributions are smaller than 1/6 of each beam size, resulting in a negligible effect on

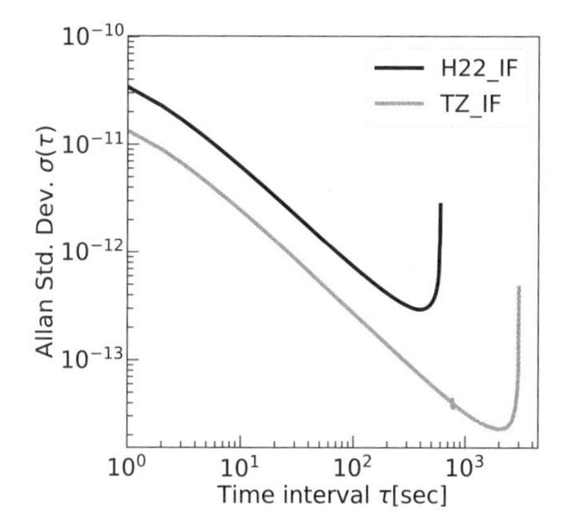

Figure 1. Phase stability of IF signals at 6 GHz from TZ and 5.435 GHz from H22 (used in VLBI). The IF signals from TZ have a phase variation less than 0.1 rad per second. The phase stability is comparable to that of H22, which is currently used for simultaneous 22/43 GHz VLBI observations, and stable for longer than the timescale of the atmospheric variability, so that the effect of phase fluctuations due to the observing system is less than thermal fluctuations, at least on that timescale.

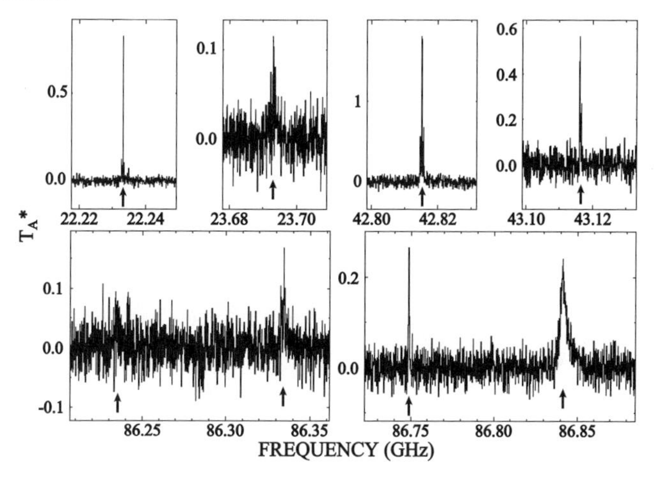

Figure 2. Spectra of IRAS 19312+1950 obtained from single-dish observations on January 17, 2023, using the H22+H40+TZ simultaneous observation system. Molecules detected and indicated by arrows are (from top left to bottom right sub-panel): H_2O, NH_3, SiO $v=2$ and 1 (J=1→0), SiO $v=1$ (J=2→1), $H^{13}CN$ (J=1→0), $H^{13}CO^+$ (J=1→0) and SiO $v=0$ (J=2→1) and also identified by Qiu *et al.* (2023).

observations of point-like sources such as maser sources. The beam sizes with the new filters are unchanged ($\sim 75''$, $\sim 41''$ and $\sim 19''$ for H22, Z45 and TZ, respectively). The aperture efficiency was $\eta_A \sim 52\%$ for Z45 with the new filter, 5 points lower than that of H40 without the filter ($\eta_A \sim 57\%$) for the former two-band mode, and $\eta_A \sim 32\%$ for TZ with filters, 11 points lower than that of TZ without the filters ($\eta_A \sim 43\%$). The decrease in η_A in the former case is consistent with that in the H22+H40 simultaneous observation system already in use for scientific observations today within 1%. The system noise temperature was $T_{\rm sys}$ ~140 K for Z45 with the filter, 20 K lower than that of H40 without the filter ($T_{\rm sys}$ ~120 K) for the former mode, and $T_{\rm sys}$ ~500 K for TZ with filters, 100 K lower than that of TZ without filters ($T_{\rm sys}$ ~400 K) for the latter triple-band mode.

The phase stability of the intermediate frequency (IF) signal path from TZ has also been measured because we have installed a new remotely controllable signal generator as the first local oscillator of TZ. It was evaluated with Allan standard deviations $\sigma(\tau)$, which were measured by observing artificial RF signals, converting them to IF signals with the new signal generator, and comparing the converted signals with those directly from another high-quality signal generator. The measured stability looked sufficient for VLBI observations (Figure 1).

3. Observation demonstrations of the H22+H40+TZ mode

Single-dish observation demonstrations in the H22+H40+TZ simultaneous observation mode toward IRAS 19312+1950 have confirmed that a variety of maser lines can be simultaneously detected (Figure 2). Thus, one can see multiple maser sources within a common antenna beam, which should be the targets of future simultaneous single-dish and VLBI observations for study on high-mass star formation and the progress and fades of copious mass loss from dying stars.

References

Imai, H., Ogawa, H., Niimura, K., *et al.* 2022, Proceedings of the 2022 Asian-Pacific Microwave Conference (APMC), 587

Okada, N., Matsumoto, T., Kondo, H., *et al.* 2020, Proceedings of SPIE 11453, 1145349

Qiu, J.-J., Zhang, Y., Nakashima, J.-i., *et al.* 2023, *A&A*, 669, A121

Cosmic Masers: Proper Motion toward the Next-Generation Large Projects
Proceedings IAU Symposium No. 380, 2024
T. Hirota, H. Imai, K. Menten, & Y. Pihlström, eds.
doi:10.1017/S1743921323002284

Fully 3D modelling of masers towards AGB stars - latest development and early results

B. Pimpanuwat[1], **M. D. Gray**[2], **S. Etoka**[1], **W. Homan**[3] and **A. M. S. Richards**[1]

[1]Jodrell Bank Centre for Astrophysics, Department of Physics and Astronomy, University of Manchester, M13 9PL, UK. email: bpimpanuwat@gmail.com

[2]National Astronomical Research Institute of Thailand, 260 Moo 4, T. Donkaew, A. Maerim, Chiangmai 50180, Thailand

[3]Institut d'Astronomie et d'Astrophysique, Université Libre de Bruxelles (ULB), CP 226, 1050 Brussels, Belgium

Abstract. We present new results from a 3D modelling code for maser flares which provides the user with control over the physical conditions; maser cloud geometry and orientation; and fast runtime via parallelisation. The statistics of simulated observables suggest that achievable amplification may be dependent on viewpoints of the source and that a randomly placed observer is likely to detect an unremarkable blue- or red-shifted maser unless the line-of-sight direction is optimal for maser amplification. A preliminary model of masers towards π^1 Gru based on SPH simulations also shows promising consistency with ALMA observations of high-J SiO transitions from the source.

Keywords. Masers, Molecular processes, Radiative transfer, Stars: AGB and post-AGB

1. Introduction

A 3D code for maser flares at VLBI scale (Gray *et al.* 2018) has been in development for constructing maser models with more realistic physical conditions e.g. 3D geometric consideration, multi-cloud systems, cloud shape and orientation, velocity profiles and multiple sources of radiation. Comparing the models to observations will shed some light on the observational conditions for maser emission in AGB stars.

2. Latest development on the 3D maser code

Since the previous reports (Gray *et al.* 2019, 2020b, 2020a), we have added new ray contributions from the central star to the calculation of population inversion and simulation of spectra and VLBI images. User control over the size and intensity (in terms of saturation intensity) of the star was introduced and the dynamical models for Miras called CODEX (Ireland *et al.* 2011) were employed for setting up 3D velocity fields. Better intrinsic depth scaling and parallelisation through OpenMP were also implemented for increased optimisation.

3. Statistics of observations of circumstellar masers

For a maser in a CSE, modelled by a group of shock-flattened maser clouds distributed in a spheroidal shell, it is important to understand what characteristics of the maser are likely to be detected in a given observation. 2000 observer's positions, defined by polar

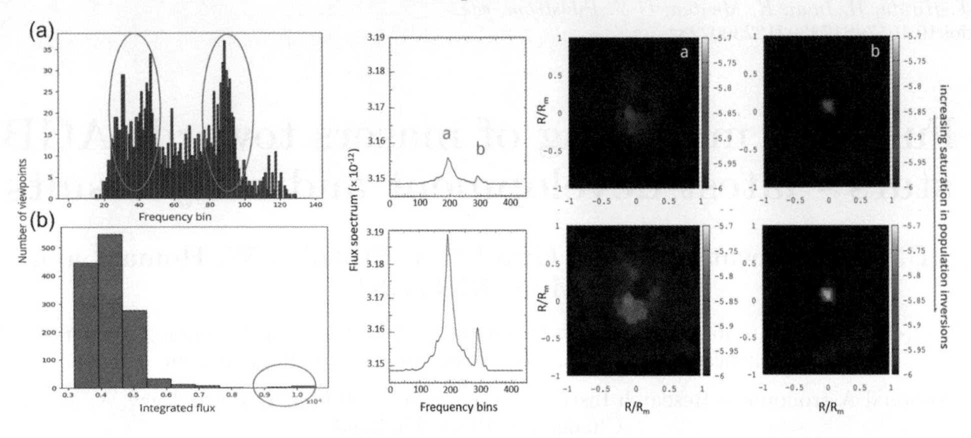

Figure 1. First simulated spectra and images of the π^1 Gru model (in model units). The red- and blue-shifted features are marked a and b, respectively. No effect from the star yet in these plots.

and azimuthal angles (with respect to the model axes), were randomly selected and flux densities were evaluated at the same distance of 1,000 model sizes from the star.

Fig. 1 shows histograms illustrating how the peak flux density per model is distributed in frequency bins (representing line-of-sight velocity) and the distribution of integrated flux. There is evidence to suggest that an observer is likely to detect the maser as being red- or blue-shifted with respect to the stellar velocity and its flux is unspectacular unless the observer's position is in the viewpoint that favours optimal maser amplification (i.e. the largest maser gain length). This is due to the geometry-dependent nature and the beaming effect of maser emission. A more extensive discussion on the maser statistics together with the code advancements in Section 2 is in progress (Pimpanuwat *et al. in prep.*).

4. Model of π^1 Gru

π^1 Gru was chosen as our first detailed attempt at a maser model for AGB stars based on the maser component maps (Homan *et al.* 2020) produced from ALMA Band 6 observations obtained as part of the ATOMIUM Large Programme (Decin *et al.* 2020; Gottlieb *et al.* 2022) and smoothed particle hydrodynamical (SPH) modelling work (e.g. Maes *et al.* 2021). The source is an S-type AGB star displaying SRb-type variability. It has a 197-day period, an effective temperature of 2300 K, a mass of 1.5 $M_{\odot}$, an angular diameter of 21 mas and a radial velocity (V_{LSR}) of -11.7 km/s (e.g. Gottlieb *et al.* 2022 and references therein).

The simulated spectra (with flux density in model units) and interferometric images of the π^1 Gru model (Fig. 1) can produce the characteristics of a typical maser flare observation. As the maser becomes more saturated, especially near and around the companion, a rise in maser intensity can be seen as per the maser component maps of high-J SiO masers reported in Homan *et al.* 2020; Fig. 11–12. Note, however, that only radiation from the background was considered and a model which includes maser pumping and seed radiation from the star is currently under investigation (Pimpanuwat *et al. in prep.*).

Supplementary material

To view supplementary material for this article, please visit http://dx.doi.org/10.1017/S1743921323002284

References

Decin, L., Montargès, M., Richards, A. M. S., *et al.* 2020, *Science*, 369, 1497.
Gottlieb, C. A., Decin, L., Richards, A. M. S., *et al.* 2022, *A&A*, 660, A94.
Gray, M. D., Mason, L., & Etoka, S. 2018, *MNRAS*, 477, 2628.
Gray, M. D., Baggott, J., Westlake, J., *et al.* 2019, *MNRAS*, 486, 4216.
Gray, M. D., Etoka, S., Travis, A., *et al.* 2020, *MNRAS*, 493, 2472.
Gray, M. D., Etoka, S., & Pimpanuwat, B. 2020, *MNRAS*, 498, L11.
Homan, W., Montargès, M., Pimpanuwat, B., *et al.* 2020, *A&A*, 644, A61.
Ireland, M. J., Scholz, M., & Wood, P. R. 2011, *MNRAS*, 418, 114.
Maes, S., Homan, W., Malfait, J., *et al.* 2021, *A&A*, 653, A25.

Cosmic Masers: Proper Motion toward the Next-Generation Large Projects
Proceedings IAU Symposium No. 380, 2024
T. Hirota, H. Imai, K. Menten, & Y. Pihlström, eds.
doi:10.1017/S1743921323002235

Investigating the inner circumstellar envelopes of oxygen-rich evolved stars with ALMA observations of high-J SiO masers

B. Pimpanuwat[1], A. M. S. Richards[1], M. D. Gray[2], S. Etoka[1] and L. Decin[3]

[1]Jodrell Bank Centre for Astrophysics, Department of Physics and Astronomy, University of Manchester, M13 9PL, UK. email: bpimpanuwat@gmail.com

[2]National Astronomical Research Institute of Thailand, 260 Moo 4, T. Donkaew, A. Maerim, Chiangmai 50180, Thailand.

[3]Institute of Astronomy, KU Leuven, Celestijnenlaan 200D, 3001 Leuven, Belgium

Abstract. We highlight a few results from ALMA Band 6 observations of high-rotational transition number (J) SiO masers towards oxygen-rich AGB and red supergiant stars carried out as part of the ATOMIUM Large Programme in 2018–2020. A search for a relationship between mass-loss rates and flux-weighted mean angular distances of maser components was inconclusive, as linear regression models for the ^{28}SiO v=1 J=5–4 and J=6–5 transitions were inconsistent. Supplementary APEX observations towards the ATOMIUM AGB stars also suggest variability at different stellar pulsation phases.

Keywords. Masers, Stars: AGB and post-AGB, Stars: circumstellar matter, Stars: mass-loss

---

1. Observations

The observations were taken with ALMA Band 6 (213.83–269.71 GHz) between Autumn 2018 and Spring 2020 in three array configurations sensitive to scales ranging from 0.4–10 arcsecs, as part of the ATOMIUM Large Programme (Decin *et al.* 2020). The data were processed in CASA using the ALMA calibration pipeline and dedicated scripts for self-calibration and imaging (Gottlieb *et al.* 2022).

We fitted two-dimensional Gaussian components to each patch of emission above $10\sigma_{rms}$ in each of the channel maps covering SiO v=1,2 J=5–4 and J=6–5 lines, using the SAD task in the AIPS package. Only the Gaussians within either 0.01 arcsec, or, if larger, the position errors (0.5×synthesised beam/signal-to-noise ratio) of their counterparts in adjacent channels were selected to ensure strongest signal pickup.

2. Mass-loss rates *vs* mean angular separations from the star

The determination of mean angular separations from the star (R) of maser components was adapted from the method described in Assaf *et al.* (2011). For each candidate position, we determined the angular distances of all SiO maser components. We then calculated the radius that encloses 90% of the total emission. A subset of sources with >10 fitted components (for that particular transition) was analysed to minimise bias due to the sporadicity of high-J SiO maser components. The uncertainty in R was taken to be the standard deviation of the mean.

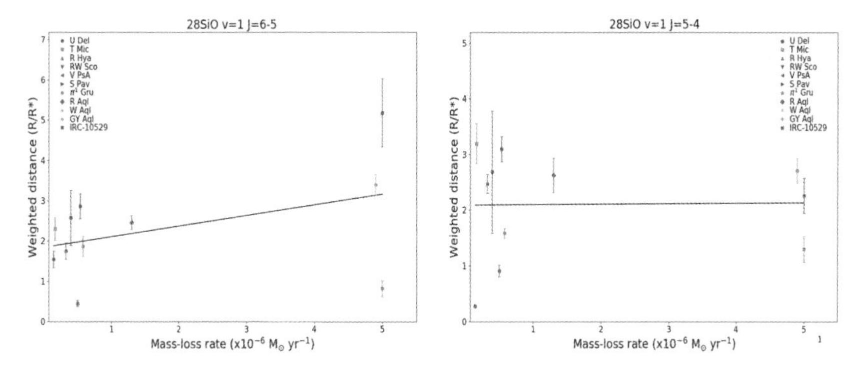

Figure 1. Plots of flux-weighted mean angular separation of fitted maser components against mass-loss rate for ^{28}SiO v=1 J=6–5 (left) and ^{28}SiO v=1 J=5–4 (right). The solid lines represent the best-fit linear regression models. Only the sources with considerable maser action are included in the analysis.

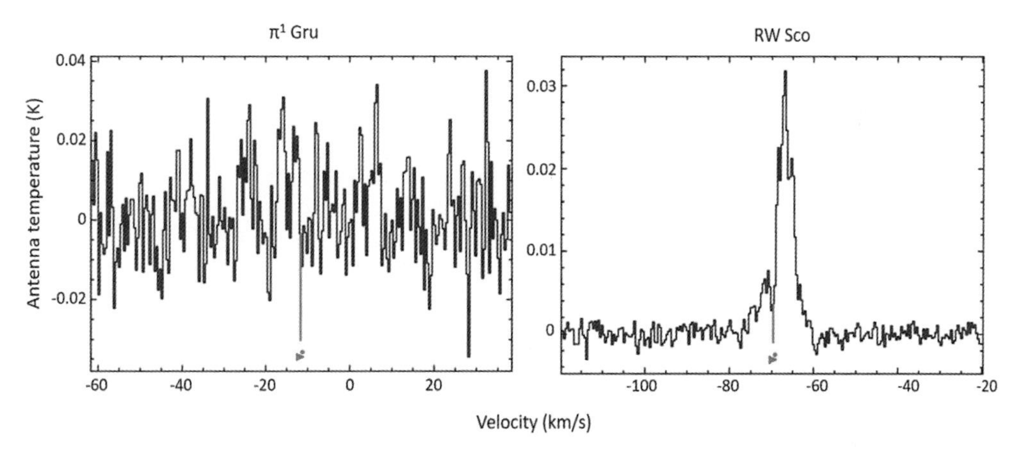

Figure 2. Preliminary spectra of ^{28}SiO v=1 J=5–4 emissions towards π^1 Gru (left) and RW Sco (right) in the first epoch of APEX observations; these show zero maser action in π^1 Gru and strong masers in RW Sco. The vertical lines labelled v_0 mark the stellar velocity. Note that the y-axis is still in units of K (courtesy of S. Etoka).

The plots of flux-weighted R against mass-loss rate for the two most frequently detected high-J SiO masers in the ATOMIUM sample are shown in Fig. 1. It is unclear whether a correlation exists between both observables as the coefficients of the linear regression model are 0.43 and 0.02 for ^{28}SiO v=1 J=6–5 and ^{28}SiO v=1 J=5–4 lines. This discrepancy could be due to intrinsic properties of the maser transitions or competitive gain (Doel *et al.* 1995 and references therein). However, the current approach lacks any information in the line-of-sight direction which skews the results towards small radii as many bright maser components lie close to or in front of the star. Further analysis is required to verify these results.

3. Variability

Two epochs of supplementary single-dish observations of the SiO v=1 J=5–4 and J=6–5 lines towards a selected subset of ATOMIUM targets were taken using APEX between June–December 2022. Fig. 2 shows two preliminary spectra of the ^{28}SiO v=1

J=5–4 line towards π^1 Gru (phase = 0.5) and RW Sco (phase = 0.0). When compared to the ATOMIUM observations (Homan *et al.* 2020; Pimpanuwat *et al. in prep.*), both sources show evidence of maser variability, especially π^1 Gru where the previously detected maser has subsided.

Supplementary material

To view supplementary material for this article, please visit http://dx.doi.org/10.1017/S1743921323002235

References

Assaf, K. A., Diamond, P. J., Richards, A. M. S., Gray, M. D. 2011, *MNRAS*, 415, 1083.
Decin, L., Montargès, M., Richards, A. M. S., *et al.* 2020, *Science*, 369, 1497.
Doel, R. C., Gray, M. D., Humphreys, E. M. L., *et al.* 1995, *A&A*, 302, 797
Gottlieb, C. A., Decin, L., Richards, A. M. S., *et al.* 2022, *A&A*, 660, A94.
Homan, W., Montargès, M., Pimpanuwat, B., *et al.* 2020, *A&A*, 644, A61.

Cosmic Masers: Proper Motion toward the Next-Generation Large Projects
Proceedings IAU Symposium No. 380, 2024
T. Hirota, H. Imai, K. Menten, & Y. Pihlström, eds.
doi:10.1017/S1743921323003095

Water masers high resolution measurements of the diverse conditions in evolved star winds

A. M. S. Richards[1], Y. Asaki[2], A. Baudry[3], J. Brand[4], L. Decin[5], S. Etoka[1], M. D. Gray[6], F. Herpin[3], R. Humphreys[7], B. Pimpanuwat[1], A. P. Singh[8], J. A. Yates[9] and L. M. Ziurys[8]

[1]JBCA, Dept. Physics & Astronomy, University of Manchester, UK.
email: a.m.s.richards@manchester.ac.uk

[2]JAO, Chile, NAOJ, Chile, Department of Astronomical Science, SOKENDAI, Japan

[3]Université de Bordeaux, Laboratoire d'Astrophysique de Bordeaux, France

[4]INAF, Bologna, Italy

[5]Institute of Astronomy, KU Leuven, Belgium

[6]NARIT, Chiang Mai, Thailand

[7]Minnesota Institute for Astrophysics, University of Minnesota, USA

[8]Dept. of Chemistry, The University of Arizona, USA

[9]Dept. Physics & Astronomy, UCL, UK

Abstract. We compare detailed observations of multiple H_2O maser transitions around the red supergiant star VY CMa with models to constrain the physical conditions in the complex outflows. The temperature profile is consistent with a variable mass loss rate but the masers are mostly concentrated in dense clumps. High-excitation lines trace localised outflows near the star.

Keywords. masers, stars: massive, mass loss

1. Water masers around evolved stars

Over 100 H_2O maser transitions are predicted, about half of which lie in ALMA bands. Fig. 1 shows the predictions of Gray *et al.* (2016) for combinations of gas temperature (T_k) and number density (n) optimising the maser amplification for lines imaged around VY CMa. Maser components in spectral channels can be fitted with accuracy ≈ (beam size)/(signal to noise ratio). Coincidence/avoidance of different masers constrains physical conditions in clumps, diffuse gas and directed outflows in the winds at an order of magnitude higher resolution than is possible with thermal lines. Only H_2O masers trace O-rich stellar winds from a few stellar radii $R_\star$, along with SiO masers, to hundreds $R_\star$, interleaving OH mainline masers.

2. Mass loss from VY CMa

VY CMa is a massive red supergiant, stellar radius ($R_\star$) 5.7 mas (Wittkowski *et al.* 2012), whose mass loss rate has varied between 5×10^{-5}–10^{-3} $M_\odot$ over centuries (Decin *et al.* 2006). Much of the wind is concentrated in dusty clumps, plumes and discrete ejecta seen by HST in scattered light e.g. Humphreys *et al.* (2021), and by ALMA e.g. Fig 2. OH mainline masers, favoured at $n < 10^{14}$ m^{-3}, T_k <500 K, interleave the outer 22-GHz H_2O maser clumps which need hotter, denser conditions, whilst OH

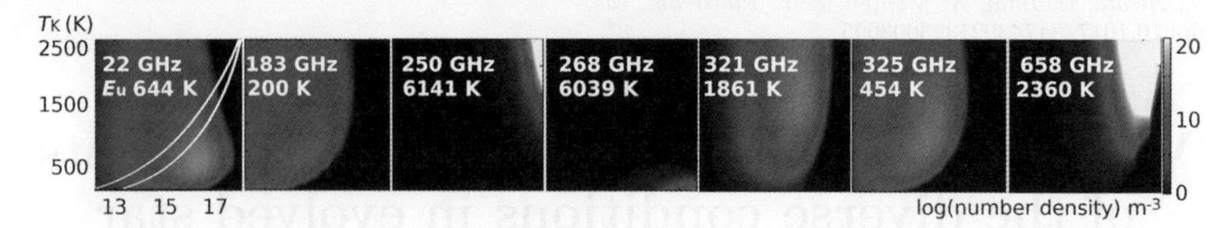

Figure 1. Colour scale: H_2O maser (negative) optical depth models (Gray *et al.* 2016). Labels: transition frequency and upper energy level. n assumes a fractional H_2O abundance $4{\times}10^{-5}$. The radiation temperature is set to 50 K except for 268 GHz, where it is 1250 K (also see Baudry, these proceedings).

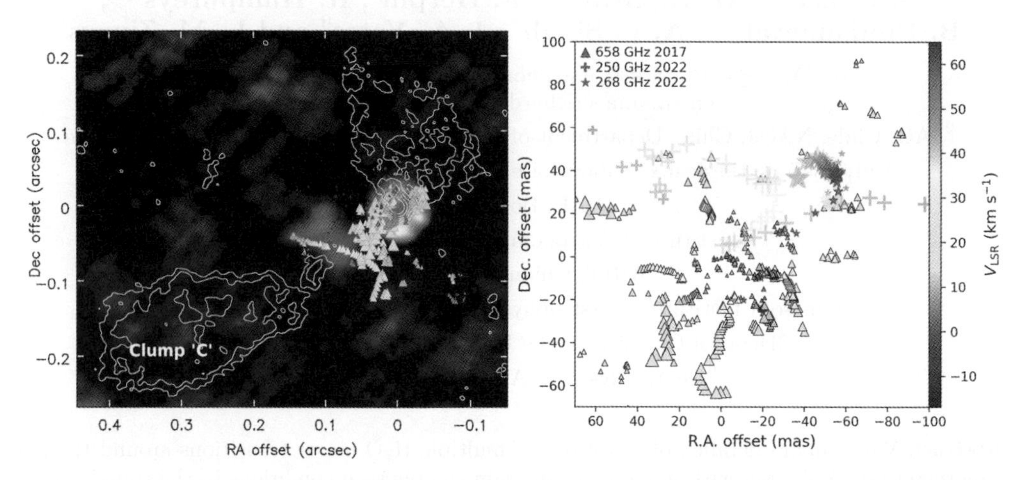

Figure 2. ALMA results: VY CMa is at [0,0]. Left: Contours: 650-GHz continuum emission. Background colour scale: integrated 658-GHz maser emission at 15-mas resolution (Asaki *et al.* 2020) observed in 2017. Components: lower-resolution 658 GHz observations (Richards *et al.* 2014) observed in 2013. Right: components fitted to the 15-mas 658 GHz masers and to lower-resolution 250 and 268 GHz masers.

and 183-GHz H_2O masers are associated with the SW clump whose dust is seen only in scattered light (Richards *et al.* 2018). Analysis of the overlaps and avoidance of 321, 325 and 658-GHz masers observed contemporaneously (Richards *et al.* 2014, 2018) broadly support the temperature – wind distance model of Decin *et al.* (2006) but suggest that much of the masing gas is concentrated in clumps with higher than average number densities. Fig 2, left shows an arc of 658-GHz masers, observed at ∼100-mas resolution in 2013, curving around clump 'C' (O Gorman *et al.* 2015). The masers appear partly resolved-out in the high-resolution 2017 data, supporting a shock origin which can lead to a large maser beaming angle even for strong amplification (Richards *et al.* 2010).

Fig 2, right, shows 'spokes' of high resolution 658-GHz masers appearing to emanate from the star in all directions, suggesting clumps with strong velocity gradients. The 250-GHz and 268-GHz emission comes from the highest-energy level H_2O masers yet imaged (see Baudry *et al.* 2023 for other stars) and is concentrated to the N and NE. The observational spectral and angular resolution was too low to confirm the maser nature of this 250-GHz emission but the 268-GHz masers have a brightness temperature $>5{\times}10^5$ K and are entirely confined to a bow-shock-like arc suggesting a recent ejection (see also Singh et al. in prep).

In future we will use multi-transition observations of a more symmetric CSE to infer wind conditions in 3D on $R_\star$ scales to inform models of chemistry and dust formation/evolution.

References

Asaki, Y., Maud, L. T., Fomalont, E. B., *et al.* 2020, *ApJS* 247, 23
Baudry, A., Wong, K. T., Etoka, S. *et al.* 2023, *A&A*, accepted
Decin, L., Hony, S., de Koter, A., *et al.* 2006, *A&A*, 456, 549
Humphreys, R. M., Davidson, K., Richards, A. M. S., *et al.* 2021, *AJ*, 161, 98
Gray, M. D., Baudry, A., Richards, A. M. S., *et al.* 2016, *MNRAS* 456, 374
O'Gorman, E., Vlemmings, W., Richards, A. M. S. *et al.* 2015, *A&A*, 573, L1
Richards, A. M. S., Elitzur, E., Yates, J. A., 2011, *A&A*, 525, A56
Richards, A. M. S., Impellizzeri, C. M. V., Humphreys, E. M., *et al.* 2014, *A&A*, 572, L9
Richards, A. M. S., Gray, M. D., Baudry, A., *et al.* 2018, Eds Tarchi, Reid & Castangia, IAUS336, 347
Wittkowski, M., Hauschildt, P. H., Arroyo-Torres, B., *et al.* 2018, *A&A*, 540, L12

Cosmic Masers: Proper Motion toward the Next-Generation Large Projects
Proceedings IAU Symposium No. 380, 2024
T. Hirota, H. Imai, K. Menten, & Y. Pihlström, eds.
doi:10.1017/S1743921323003137

Annual parallax measurement of extreme OH/IR candidate star OH39.7+1.5

Ryosuke Watanabe

Graduate School of Science and Engineering, Kagoshima University. email: k7299015@kadai.jp

Abstract. OH/IR stars are low- to intermediate-mass stars (about 0.5-8 solar masses) in the AGB phase, and are considered to be in the process of evolving from Mira variables to planetary nebulae. We aim to understand the evolutionary stages of these AGB stars by approaching them from astrometric VLBI observations with VERA. In 2017, we have started VLBI observations of several OH/IR stars including OH39.7+1.5 presented in this poster. We observed the H_2O maser of OH39.7+1.5 and obtained an annual parallax of 0.55 ± 0.03 mas (distance D = 1.81 $\pm$ 0.12 kpc). Using this annual parallax, we revealed distribution (about 35 au square), internal motions, and expansion velocities (average about 15.4 ± 5.1 km s^{-1}) of the maser spots. We compared expansion velocity of H_2O maser with that of OH maser and found to be consistent across the error range. This suggests that the radial velocity of H_2O gas around OH39.7+1.5 has reached a terminal velocity. OH39.7+1.5 has not data on annual parallax or proper motion in Gaia DR3, and this is the first time that annual parallax has been measured.

Keywords. masers, astrometry, parallaxes, star:AGB and post-AGB

---

1. Introduction

Stars with initial mass of low- to intermediate-mass experience an asymptotic giant branch (AGB) phase at the end of their evolution. OH/IR stars are considered to be in the AGB phase, the stage of evolution from Mira variables to planetary nebulae (Herwig 2005). This phase is the late AGB phase. The late AGB phase is a very short time in their lives, although the details of the stellar evolution are unclear. We intend to systematically study the evolutionary process and the circumstellar structure by estimating the distance from the annual parallaxes obtained from astrometric VLBI using the VERA and investigate the distribution and motion of the circumstellar maser.

2. Observations

2.1. *VLBI Observations*

We observed H_2O maser at 22GHz of OH39.7+1.5 using the VLBI Exploration of Radio Astrometry (VERA) at 22 epochs from October 3, 2019 to May 22, 2022 with an interval of about one month. The VERA is a VLBI array which consists of four 20 m aperture radio telescopes located at Mizusawa, Iriki, Ogasawara, and Ishigaki-jima. Each antenna of VERA has a dual beam system, which is excellent for measuring the annual parallax of dusty AGB stars like OH/IR stars. The VERA cancel out the phase fluctuation caused by atmosphere by simultaneously observing a target and position reference source (J1856+0610) with dual-beam system. The bandwidth of OH39.7+1.5 is 1 IF (16 MHz), divided into 512 channels, with a frequency resolution of 31.25 kHz. This corresponds to a velocity resolution of 0.42 km s^{-1} at 22 GHz. The bandwidth of J1856+0610 is 15 IF (240 MHz), each 16 MHz IF divided into 32 channels.

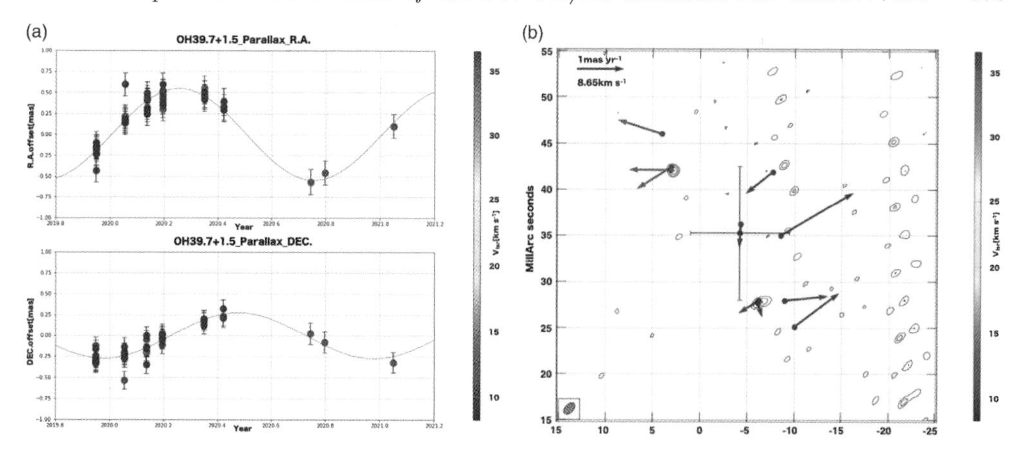

Figure 1. (a) Annual parallax of OH39.7+1.5 and (b) H_2O maser distribution and internal motion.

2.2. *Target source*

OH39.7+1.5 (also named as RAFGL 2290,V1366 Aql, and IRAS 18560+0638) is an OH/IR star with a spectral type of M II. It has emissions of SiO, H_2O, and OH masers. From previous studies, the pulsation period is known to be 1360 days from monitored observations of the OH maser, and OH 39.7+1.5 is a candidate for an Extreme-OH/IR star. OH39.7+1.5 has no data on annual parallax or proper motion in Gaia DR3.

3. Results and Discussion

3.1. *Parallax of OH39.7+1.5*

We determine an annual parallax of OH39.7+1.5 using positions of maser spots detected in the phase-referencing analysis (Figure 1(a)). From 9 out of 16 VLBI observations, we found 10 H_2O maser spots in 6 velocity channels. As a detection threshold, we adopted a signal to noise ratio of 7. We derived a parallax of 0.55 ± 0.03 mas. This corresponds to a distance of 1.81 ± 0.12 kpc. OH39.7+1.5 has no data on annual parallax or proper motion in Gaia DR3, and this is the first time that annual parallax has been measured.

3.2. *H_2O maser distribution and internal motion*

In Figure 1(b), we presented distribution of maser features integrated over 91 velocity channels (velocity width 38.2 km s^{-1}) and internal motion of H_2O maser spots. Maser spots detected three or more times were used to estimate internal motion. The map covers an area of 40 mas × 40 mas (about 72 au × 72 au). The maser spots are distributed from northeast to southwest over an area of about 35 au. The average expansion velocity is 15.4 ± 5.1 km s^{-1}, which is indicative of the expansion velocity of the H_2O maser of a typical AGB star (Höfner & Olofsson 2018). In a previous study, the expansion velocity of the OH maser of OH 39.7+1.5 was obtained to be around 16.5 km s^{-1} (Baud & Habing 1983). Since the H_2O maser is accelerated to the same degree as the OH maser, it is considered to be approaching the terminal velocity. The maser distributions of the two OH/IR stars obtained by our group show the most blue-shifted component consistent with the estimated position of the central star. The most blue-shifted component of OH39.7+1.5 is detected about 8 mas southwest of the central star. This distribution is different from that of previous studies.

References

Baud, B. & Habing, H. J. 1983, *A&A*, 127, 73
Herwig, F. 2005, *ARAA*, 43, 435
Höfner, S. & Olofsson, H. 2018, *A&AR*, 26, 1

Photographs taken during the social events. Top: Afternoon tea during the excursion, taken by Ka-Yiu Shum. Bottom: A local vinegar factory visited during the excursion, taken by Tomoya Hirota.

Photographs taken during the social events. Top: Observatory of the Sakura-jima volcano, taken by Ka-Yiu Shum. Bottom: Dancing during the banquet, taken by Ryosuke Watanabe.

Chapter 6
Theory of Masers and Maser Sources

Cosmic Masers: Proper Motion toward the Next-Generation Large Projects
Proceedings IAU Symposium No. 380, 2024
T. Hirota, H. Imai, K. Menten, & Y. Pihlström, eds.
doi:10.1017/S1743921323002764

Variability, flaring and coherence – the complementarity of the maser and superradiance regimes

Martin Houde[1], Fereshteh Rajabi[2], Gordon C. MacLeod[3,4], Sharmila Goedhart[5,6], Yoshihiro Tanabe[7], Stefanus P. van den Heever[4], Christopher M. Wyenberg[8], and Yoshinori Yonekura[7]

[1]The University of Western Ontario, London, Ontario N6A 3K7, Canada.
email: mhoude2@uwo.ca

[2]McMaster University, Hamilton, Ontario L8S 4M1, Canada

[3]The Open University of Tanzania, P.O. Box 23409, Dar-Es-Salaam, Tanzania

[4]Hartebeesthoek Radio Astronomy Observatory, P.O. Box 443, Krugersdorp, 1741, South Africa

[5]South African Radio Astronomy Observatory, 2 Fir Street, Black River Park, Observatory 7925, South Africa

[6]Center for Space Research, North-West University, Potchefstroom Campus, Private Bag X6001, Potchefstroom 2520, South Africa

[7]Center for Astronomy, Ibaraki University, 2-1-1 Bunkyo, Mito, Ibaraki 310-8512, Japan

[8]Institute for Quantum Computing and Department of Physics and Astronomy, The University of Waterloo, 200 University Ave. West, Waterloo, Ontario N2L 3G1, Canada

Abstract. We discuss the role that coherence phenomena can have on the intensity variability of spectral lines associated with maser radiation. We do so by introducing the fundamental cooperative radiation phenomenon of (Dicke's) superradiance and discuss its complementary nature to the maser action, as well as its role in the flaring behaviour of some maser sources. We will consider examples of observational diagnostics that can help discriminate between the two, and identify superradiance as the source of the latter. More precisely, we show how superradiance readily accounts for the different time-scales observed in the multi-wavelength monitoring of the periodic flaring in G9.62+0.20E.

Keywords. Radiative processes: non-thermal – masers – ISM: molecules – ISM: G9.62+0.20E

1. Introduction

The monitoring of astronomical objects harboring sources of maser radiation reveals ubiquitous variability in the measured intensities over a range of time-scales and light curve patterns. Most intriguing are strong flaring behaviours, which can consist of isolated or recurring events. Periodic flaring sources are of particular interest because of, among other things, their potential for shedding light on the nature of the engines at the root of the observed periodicities. Accordingly, a growing number of such objects have been discovered and closely monitored with several spectral transitions in the recent past.

Recurring and periodic flaring sources also often display behaviours that can seem difficult to explain theoretically. Examples can involve one spectral spectral transition (observed at different systemic velocities) or the comparison between several lines. Figure 1 shows the case of a recurring flare observed by Fujisawa et al. (2012) in the methanol

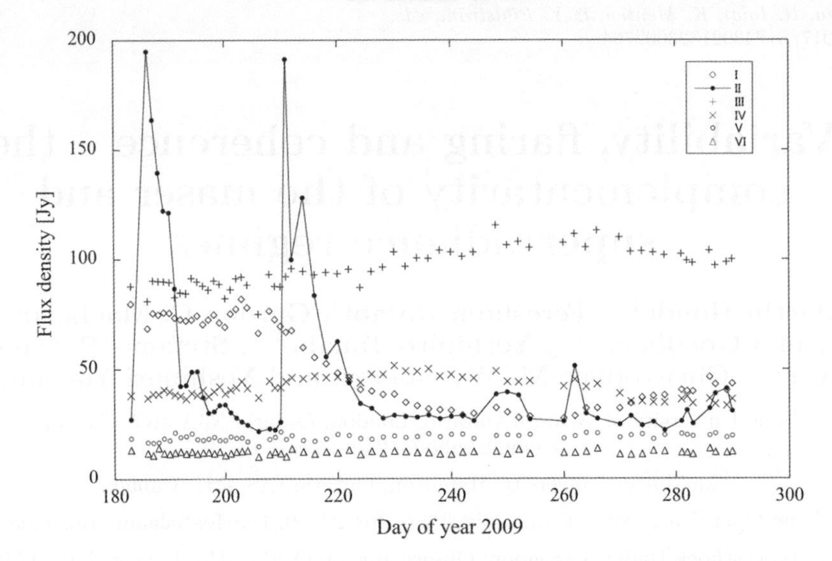

Figure 1. Flaring in the methanol 6.7 GHz transition for the G33.64–0.21 high mass star-forming region. Six different masing spectral features are detected in this line, but only one (Feature II) shows two strong flares exhibiting exceedingly fast rise times. Taken from Fujisawa et al. (2012).

6.7 GHz transition for the G33.64–0.21 high mass star-forming region. Although six different masing spectral features are detected in this line, only one (Feature II) shows two strong flares with an amplification of a factor of approximately eight relative to the quiescent flux level and exceedingly fast rise times. The challenge in making sense of these observations, besides the nature of the flares themselves and their source, resides in explaining why features from the same spectral transitions but at different velocities could exhibit such vastly different responses to a common excitation.

Figure 2 shows observations obtained by Szymczak et al. (2016) in methanol 6.7 GHz and water 22 GHz during a monitoring campaign of the G107.298+5.639 star-forming region (see also Olech et al. 2020). This source exhibits flaring in multiple spectral transitions at a common period of 34.4 d. More interestingly, and as shown in the figure, the methanol 6.7 GHz and water 22 GHz flares are seen to alternate with the flux density of one transition peaking when the other reaches a minimum. Also to be noted are the different time-scales of the flares, where the methanol 6.7 GHz flares are consistently of shorter duration than the water 22 GHz features. While the alternation of the flares between the two species can perhaps be explained by the effect of an infrared (methanol-) pumping source on the dust temperature and its quenching effect on the water 22 GHz population inversion (Szymczak et al. 2016), their different time-scales are more problematic.

There are also more fundamental questions pertaining to the physics underlying the flaring phenomenon. That is, when subjected to an excitation signal (e.g., the infrared pump source or a seed electric field signal for a maser-hosting region) a physical system will often exhibit a transient response before settling into a steady-state regime. It has long been established in the quantum optics community that for an ensemble of radiators (i.e., atoms or molecules) exhibiting a population inversion the two regimes are associated with different radiation processes. That is, stimulated emission rules the quasi steady-state regime while Dicke's superradiance is at play during the transient phase (Dicke 1954; Feld and MacGillivray 1980). One may wonder whether this dichotomy also applies to astrophysical sources and could be essential for explaining some of the characteristics of flares emanating from maser-hosting sources.

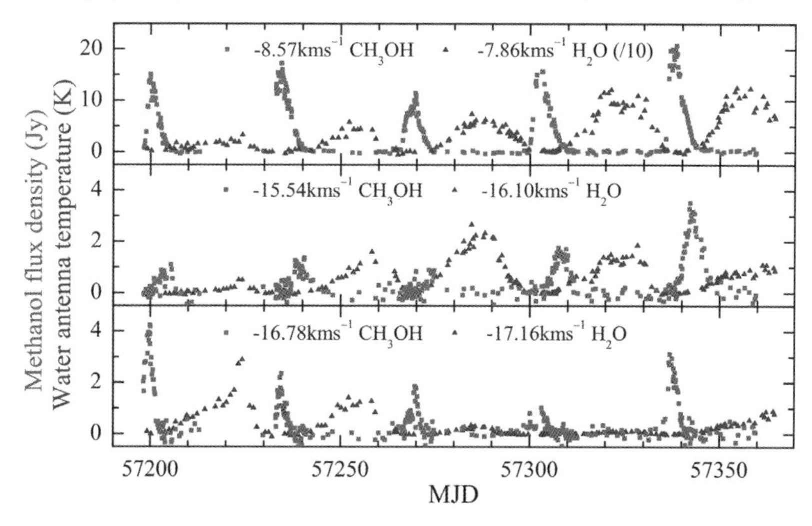

Figure 2. Flaring in the methanol 6.7 GHz and water 22 GHz transitions for the G107.298+5.639 star-forming region. The methanol 6.7 GHz and water 22 GHz flares are seen to alternate with the flux density of one transition peaking when the other reaches a minimum. Also to be noted are the different time-scales of the flares, where the methanol 6.7 GHz flares are consistently of shorter duration than the water 22 GHz features. Taken from Szymczak et al. (2016).

In this review, we will use observations from multi-transition monitoring of the G9.62+0.20E high mass star-forming region to probe the physics underlying the periodic flaring observed in this source and determine whether different processes characterize the bursting and quiescent phases. The paper is structured as follows. We first introduce Dicke's superradiance and its study at the laboratory level in Sec. 2, while we discuss the association of the transient and quasi steady-state regimes to the superradiance and maser phenomena, respectively, as well as their manifestation as two independent limits of the Maxwell-Bloch equations in Sec. 3. Finally, in Sec. 4 we model existing multi-transition data obtained for G9.62+0.20E and show that superradiance can naturally account for the diversity of time-scales observed in this source.

2. Dicke's superradiance

Given that masers were first discovered in the laboratory (Gordon et al. 1954, 1955), it may be appropriate to inquire whether scientists in the quantum optics community have studied or have performed experiments on "flaring" with such systems? The answer to this question is a resounding "yes." This is due to the introduction of the superradiance phenomenon by Dicke (1954), which, it is interesting to note, actually predates that of the maser. In his foundational paper, Dicke pointed out that molecules in a gas (atoms could also be used but we focus on molecules as we will be dealing with corresponding transitions in this review) may not always radiate independently, whereas their interaction with a common electromagnetic field will bring a quantum mechanical entanglement between them. Dicke thus treated the ensemble of molecules as a "single quantum mechanical system" and proceeded in calculating spontaneous emission rate for the gas when focusing on a single spectral transition.

Under conditions where velocity coherence and population inversion prevail, he showed that the commonly assumed independent spontaneous emission process is replaced by a transient phenomenon he termed "superradiance." With superradiance radiation from the ensemble of molecules happens on a photon cascade. For example, when the $N \gg 1$

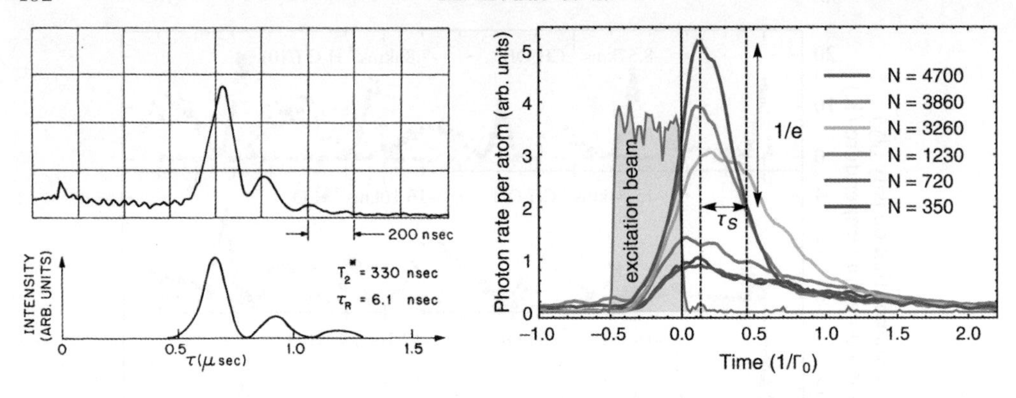

Figure 3. *Left:* First experimental realisation of superradiance by Skribanowitz et al. (1973) in a rotational transition of a vibrationally excited HF gas (top) and theoretical fit (bottom). A near complete population inversion was achieved with an intense pump pulse at $\lambda \sim 2.5\,\mu$m (the small peak at $\tau = 0$ in the top graph), after which followed a superradiance burst at 84 μm. *Right:* Recent laboratory results of superradiance experiments with different numbers of laser-cooled ^{87}Rb atoms by Ferioli et al. (2021), whose responses were probed and compared for different types of pumping conditions (the pump signal is shown in grey in the figure).

molecules are all in the excited state the emission of photons takes place at transition rates ranging from $N\Gamma_0$ (at the beginning and the end of the cascade) to $\sim N^2\Gamma_0/4$ (in the middle of the cascade), with Γ_0 the single-molecule transition rate (i.e., the spontaneous emission rate of the transition). This results in powerful bursts of radiation intensity lasting on a time-scale on the order of $T_\mathrm{R} \propto 1/(N\Gamma_0)$†, while the coherent nature of the radiation and the process can be attested through the scaling of the peak intensity with N^2. Evidently, coherence in the radiation cannot occur in all directions whenever the molecules are separated by a finite distance. In this case, Dicke showed that superradiance happens only in well-defined radiation modes with successive photons exhibiting increasing correlation in their propagation direction, and therefore results in highly collimated beams of radiation. In the process, Dicke thus introduced and described the "photon bunching" phenomenon before it became established within the physics community (Hanbury Brown and Twist 1956; Dicke 1964).

While, as previously mentioned, the theoretical introduction of superradiance preceded that of the maser in the literature, its experimental verification comparatively lagged significantly as we had to wait for almost 20 years before it could be realized in the laboratory. The left panel of Figure 3 shows the first experimental verification of superradiance obtained by Skribanowitz et al. (1973) in a rotational transition of a vibrationally excited HF gas. A near complete population inversion was achieved with an intense pump pulse at $\lambda \sim 2.5\,\mu$m (the small peak at $\tau = 0$ in the top graph), after which followed a superradiance burst at 84 μm. The time delay between the pump pulse signal and the intensity burst, as well as its "ringing," are telling characteristics of superradiance for these experimental conditions. Superradiance has since become an area of sustained and intense research in the quantum optics community. It has been tested and used to probe fundamental physics questions for a wide range of systems under a vast array of conditions (see Chap. 2 of Benedict et al. 1996 for an early summary). An example of a recent experiment by Ferioli et al. (2021) is shown in the right panel of Figure 3, where the superradiance responses from assemblies containing different numbers of laser-cooled ^{87}Rb atoms were probed and compared for different types of pumping conditions.

It is interesting to note that despite the intense amount of research accomplished in quantum optics laboratories over several decades, superradiance has remained largely

† We will soon give a more precise definition of the superradiance time-scale.

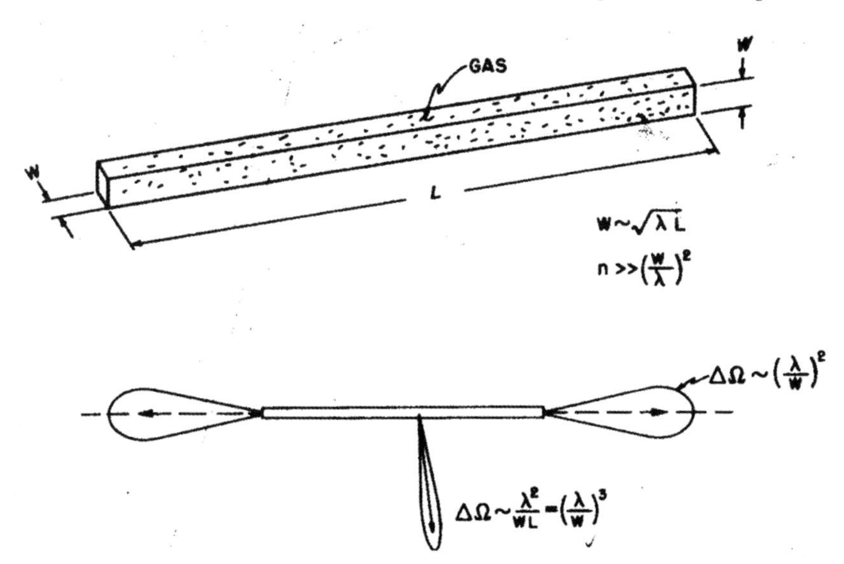

Figure 4. *Top:* A cylindrical gas configuration of length L containing a single molecular species at velocity coherence. The dimensions of the cylinder are consistent with a Fresnel number of unity (i.e., the width $w \sim \sqrt{\lambda L}$). *Bottom:* A schematic representation of the cylinder with its diffraction patterns of solid angle $\Delta\Omega$. We will investigate the radiation emitted within one lobe at the end of the cylinder, say, on the right, with $\Delta\Omega \sim (\lambda/w)^2$. Taken from Dicke (1964).

unnoticed by the astronomical community until the recent work of Rajabi and Houde (2016a,b, 2017, 2020) (see also Rajabi et al. 2019, 2023).†

3. The complementarity of the maser and superradiance regimes

The temporal evolution of a gas, whose components (e.g., molecules) are modeled as two-level systems, interacting with an ambient electromagnetic field is described by the so-called Maxwell-Bloch equations (MBE). The application of the MBE is not limited to systems hosting a population inversion but is general and applies to a wide range of conditions where numerous fundamental physical phenomena can be theoretically studied and modeled. For the problem at hand, we will consider a gas whose (single) molecular constituents exhibit velocity coherence and occupy a linear volume configuration, e.g., a circular or rectangular cylinder as shown in Figure 4, and focus on a single transition between two specific energy levels.

While the MBE can in principle be numerically solved in three dimensions, it is simpler (both analytically and computationally) and adequate to limit the analysis to a one-dimensional problem (Gross and Haroche 1982). With this approximation the only remaining spatial variable, z, is that defining the symmetry axis of the cylinder while the temporal evolution is tracked as a function of the retarded time $\tau = t - z/c$. The three-dimensional reality of the problem is recovered "by hand" after the computations are effected by imposing a Fresnel number of unity, where the width of the cylinder is set to $w = \sqrt{\lambda L}$ with L the length of the system. This condition limits the radiation intensity of the one-dimensional system to that contained to a solid angle $\Delta\Omega \sim (\lambda/w)^2$ at one end-fire of the cylinder (see the bottom schematic in Figure 4) and restricts phase coherence to a volume of physically allowable dimensions. We will refer to such system as a "sample."

† Dicke's superradiance, as discussed here, should not be confused with "black hole superradiance," which is connected to Hawking radiation; see Brito et al. 2015 for a review of black hole superradiance and its historical connection to Dicke's superradiance.

Under the slowly varying envelope (SVEA) and rotating wave approximations, the one-dimensional MBE in the rest frame of the gas take the form

$$\frac{\partial n'}{\partial \tau} = \frac{i}{\hbar}\left(P^+E^+ - E^-P^-\right) - \frac{n'}{T_1} + \Lambda_n \tag{1}$$

$$\frac{\partial P^+}{\partial \tau} = \frac{2id^2}{\hbar}E^-n' - \frac{P^+}{T_2} + \Lambda_P \tag{2}$$

$$\frac{\partial E^+}{\partial z} = \frac{i\omega_0}{2\epsilon_0 c}P^-, \tag{3}$$

where n' is (half of) the population density difference between the upper and lower energy levels, while P^+ and E^+ are the amplitudes of the molecular polarization and the electric field, respectively defined by

$$\mathbf{P}^{\pm}(z,\tau) = P^{\pm}(z,\tau)\, e^{\pm i\omega_0\tau}\boldsymbol{\epsilon}_\mathrm{d} \tag{4}$$

$$\mathbf{E}^{\pm}(z,\tau) = E^{\pm}(z,\tau)\, e^{\mp i\omega_0\tau}\boldsymbol{\epsilon}_\mathrm{d}. \tag{5}$$

The unit polarization vector $\boldsymbol{\epsilon}_\mathrm{d} = \mathbf{d}/d$ is associated with the molecular transition of dipole moment $d = |\mathbf{d}|$ at frequency ω_0. The superscript "+" in equations (1)-(5) is for the polarization corresponding to a transition from the lower to the upper level and the positive frequency component of the electric field (Gross and Haroche 1982; Rajabi and Houde 2020; Rajabi et al. 2023). Any non-coherent phenomena (e.g., collisions) that cause relaxation of the population density and de-phasing of the polarization are respectively accounted for through the phenomenological time-scales T_1 and T_2. The evolution of the system is initiated through internal fluctuations in n' and P^+ modeled with an initial non-zero Bloch angle

$$\theta_0 = \frac{2}{\sqrt{N}}, \tag{6}$$

with N the number of molecules in the sample (see Rajabi et al. 2019 for more details). The polarization pump Λ_P in equation (2) consists entirely of those fluctuations (i.e., for $P^{\pm}$). The inversion pump Λ_n in equation (1) is for an effective pump signal that directly drives a population inversion.

In this section we will endeavour to establish the complementarity between the maser action and superradiance, and demonstrate how they arise from the MBE. To do so, we will follow the discussion presented in Sec. 3.1 of Rajabi and Houde (2020) for a system that is "instantaneously inverted" at $\tau = 0$. That is, we set the inversion pump to

$$\Lambda_n(z,\tau) = \begin{cases} 0, & \tau < 0 \\ \Lambda_0, & \tau \geq 0 \end{cases} \tag{7}$$

with Λ_0 a constant level and fix an initial population inversion level $n'_0 \equiv n'(z, \tau = 0) = \Lambda_0 T_1$ at all positions z. The inversion pump signal would then compensates for any non-coherent relaxation in the weak intensity limit (i.e., when $\partial n'/\partial\tau \approx 0$), while setting a non-zero initial population inversion level is equivalent to having an instantaneous inversion at $\tau = 0$.† With this set-up, we let the system evolve and monitor $n \equiv 2n'$, P^+ and E^+ at all positions z and retarded times τ.

We show in Figure 5 the evolution at two positions of such a sample for a methanol gas at the 6.7 GHz spectral transition. For this example we set $L = 2\times10^{15}$ cm and $n_0 \equiv 2n'_0 = 3.3\times10^{-12}\ \mathrm{cm}^{-3}$, which corresponds to a population inversion of approximately $0.1\ \mathrm{cm}^{-3}$ for a $1\ \mathrm{km\,s}^{-1}$ velocity distribution. These two parameters, along with the Einstein spontaneous emission coefficient for that line ($\Gamma_{6.7\,\mathrm{GHz}} = 1.56\times10^{-9}\ \mathrm{s}^{-1}$), set

† The nature of Λ_n will change in the next section when modeling data for G9.62+0.20E.

the superradiance characteristic time-scale for this sample through

$$T_{\mathrm{R}} = \frac{8\pi}{3\lambda^2 n_0 L}\tau_{\mathrm{sp}}, \tag{8}$$

where $\tau_{\mathrm{sp}} = \Gamma^{-1}_{6.7\,\mathrm{GHz}}$ leads to $T_{\mathrm{R},\,6.7\,\mathrm{GHz}} = 4 \times 10^4$ s (Rajabi and Houde 2020). Finally, the time-scales for non-coherent relaxation and de-phasing were fixed to $T_1 = 1.64 \times 10^7$ s and $T_2 = 1.55 \times 10^6$ s, respectively. Before we study the response of the sample we note that the initial column density of the population inversion $n_0 L$ appears in the denominator of T_{R}, implying that an increased length for the system will lead to a faster response. This characteristic will be essential for understand the sample's behaviour and will lead to the notion of a critical threshold for the appearance of superradiance. We also note that in the figure the intensity $I = c\epsilon_0 \left|E^+\right|^2/2$ is normalized to NI_{nc} with the non-coherent intensity (i.e., that expected from the sample when the molecules are radiating independently) given by

$$I_{\mathrm{nc}} = \frac{2}{3}\frac{\hbar\omega_0}{AT_{\mathrm{R}}}, \tag{9}$$

where $A = \lambda L$ is the cross-section of the sample's end-fire (for a Fresnel number of unity; see Rajabi and Houde 2016b).

In Figure 5 we see that the response of the system has an entirely different character depending on the position where it is probed. At $z = 0.4\,L$ the intensity (top right) exhibits a smooth transition from $I \approx 0$ at $\tau = 0$ to its steady-state value, which is very weak at $I \sim 10^{-10} NI_{\mathrm{nc}}$. This low intensity is due to the absence of coherence in the gas, as can also be assessed from the low level of polarization $P^+ \sim 10^{-15}\,\mathrm{D\,cm^{-3}} \ll n_0 d$ (top left; $d \simeq 0.7$ D for this transition). For reasons that will soon become clear we will term this smooth transient regime as "non-superradiance" (or "non-SR" in the figure). The sample's response is completely different at its end-fire, i.e., at $z = L$, where a strong oscillatory transient regime is seen in the intensity, peaking at $I \simeq 1.6 \times 10^{-3} NI_{\mathrm{nc}}$ before reaching a steady-state value of $I \simeq 0.5 \times 10^{-3} NI_{\mathrm{nc}}$. The strong transient is also seen in n and P^+ with the latter reaching a peak value of $P^+ \sim n_0 d$. This is indicative of the presence of coherence in the gas during the transient regime, which we qualify as "SR transient."

Our usage of the "non-SR" and "SR transient" terms is based on the fact that it can be analytically shown that there exist two distinct limits or regimes discernible within the framework of the MBE for an initially inverted system such as the one considered here (Feld and MacGillivray 1980; Rajabi and Houde 2020). In the limit when the population density and the polarization are changing on a time-scale much shorter than T_1 and T_2, i.e.,

$$\frac{\partial n'}{\partial \tau} \gg \frac{n'}{T_1} \quad \text{and} \quad \frac{\partial P^+}{\partial \tau} \gg \frac{P^+}{T_2}, \tag{10}$$

the sample enters a rapid transient regime that is characterized by the presence of coherence in the gas and strong intensities scaling with N^2 (see below). This is the domain of Dicke's superradiance. The bottom graphs in Figure 5 form indeed a good example of an "SR transient." The slower, non-coherent transient regime observed in the top graphs in Figure 5 are thus "non-SR transient."

On the other hand, at the opposite limit when the population density and the polarization are varying on a time-scale much longer than T_1 and T_2, i.e.,

$$\frac{\partial n'}{\partial \tau} \ll \frac{n'}{T_1} \quad \text{and} \quad \frac{\partial P^+}{\partial \tau} \ll \frac{P^+}{T_2}, \tag{11}$$

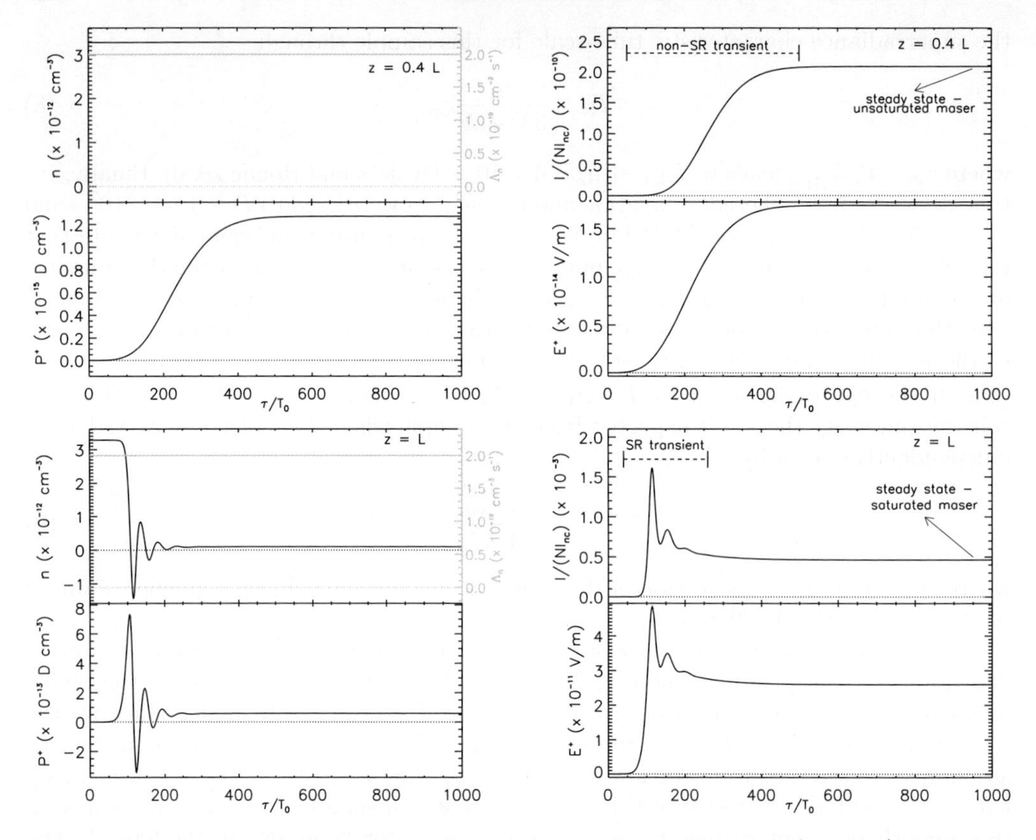

Figure 5. *Top:* Temporal evolution of the population inversion $n \equiv 2n'$ and P^+ (*left*), and the radiation intensity I and E^+ at $z = 0.4\,L$ (*right*). The intensity exhibits a smooth, non-superradiance (non-SR), transient response as the critical threshold has not been reached (see equation 12). The inversion pump signal is shown in light/cyan using the vertical scale on the right and $T_0 = 1 \times 10^5$ s. *Bottom:* Same as above but at the end-fire, i.e., at $z = L$, of the sample. Here, a strong oscillatory transient regime is seen in the intensity since the inverted column density exceeds the critical threshold for superradiance. This is accompanied by an increase in coherence as assessed by the peak polarization level $P^+_{\mathrm{max}} \sim n_0 d$ ($d \simeq 0.7$ D), which supports a picture where the molecules in the gas behave cooperatively as a single macroscopic electric dipole. Adapted from Rajabi and Houde (2020).

it can be shown that the MBE simplify to the maser equation (see Sec. 2.1 of Rajabi and Houde 2020 for a derivation). This is the so-called "steady-state limit" where the maser action (and stimulated emission) is at play. This steady-state regime is identified Figure 5 in the intensity graphs on the right. More precisely, at $z = 0.4\,L$ (top) the low intensity suggests the existence of a (steady-state) unsaturated maser while at $z = L$ (bottom) we are in the presence of a saturated maser.

The fact that the slow limit characterized by equation (11) is associated with the maser action can be readily verified by plotting the (normalized) steady-state intensity $I_{\mathrm{steady}}/\left(NI_{\mathrm{nc}}\right) \equiv I\left(\tau = 1000T_0\right)/\left(NI_{\mathrm{nc}}\right)$ for all positions z, as shown in Figure 6. We can clearly verify this association by the presence of the exponential growth unsaturated maser and linear gain saturated maser regimes, which also justify our earlier attributions for $z = 0.4\,L$ and $z = L$.

We can also verify the association between Dicke's superradiance and the fast transient limit by tracking the peak intensity, the minimum population density and the peak polarization as a function of the position z along the sample. This is shown in Figure 7,

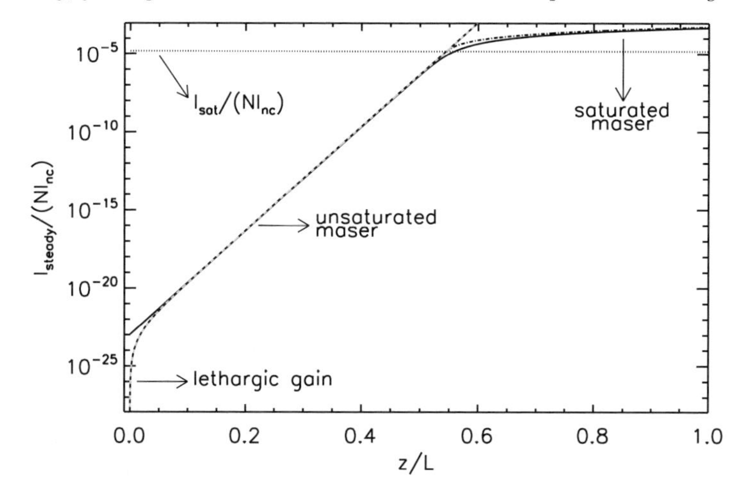

Figure 6. Steady-state (maser) intensity at the sample's end-fire ($z = L$). The unsaturated (exponential growth) and saturated (linear growth) regimes are indicated; note the logarithmic scale for the intensity. These two domains meet at the critical length $z_{\text{crit}}/L \approx 0.54$, where the intensity neighbours the saturation intensity I_{sat} (shown with the horizontal dotted line). Taken from Rajabi and Houde (2020); see this paper for a more in-depth discussion.

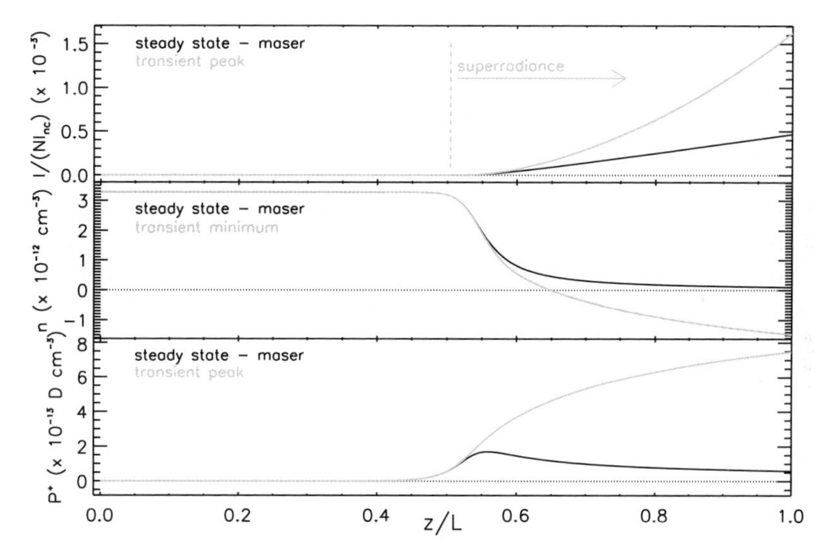

Figure 7. The superradiance/transient (light/cyan curves) and maser/steady-state (dark/black curves) intensities (top), population densities (middle) and polarization amplitudes (bottom) at all positions z/L along the sample. The superradiance regime begins at $z/L \approx 0.5$ where the critical inverted column density threshold is reached; this domain also corresponds to that of the saturated maser regime. The superradiance peak intensity scales as z^2 while the saturated maser's is linear with position ($\propto z$). The superradiance population density minimum level reaches more negative values with increasing z while the maser inversion asymptotically tends to zero. The level of coherence (i.e., the polarization amplitude) increases with distance for superradiance. Adapted from Rajabi and Houde (2020).

where these quantities are respectively plotted from top to bottom in light/cyan along with the corresponding steady-state (i.e., maser) values in black. In the top graph for the intensities we can clearly see the linear gain for the saturated maser intensity $I_{\text{steady}} \propto z \propto N$ and the quadratic behaviour for the transient superradiance regime $I_{\text{SR}} \propto z^2 \propto N^2$.

This functionality, as well as those for the population density and the polarization, are the hallmark of Dicke's superradiance.

One more important piece of information can be gathered from Figures 6 and 7. That is, it can be seen through their comparison that the transition between the unsaturated and saturated maser regimes (where $I_{\mathrm{steady}} = I_{\mathrm{sat}}$ in Figure 6, with I_{sat} the saturation intensity) and the appearance of superradiance (from the top graph of Figure 7) happen at the same position $z \equiv z_{\mathrm{crit}}$. It can be shown that this position is associated with the reaching of a critical threshold for the column density

$$n_0 z_{\mathrm{crit}} = \frac{4\pi}{3\lambda^2} \frac{\tau_{\mathrm{sp}}}{T_2} \ln\left(\frac{T_2}{T_1 \theta_0^2}\right), \tag{12}$$

which sufficiently reduces the superradiance characteristic time-scale T_{R} to reach the fast limit of equation (10) (Rajabi and Houde 2020). In other words, the appearances of superradiance in the transient response and a saturated maser in the steady-state regime are closely linked and will take place whenever $L \geq z_{\mathrm{crit}}$. Systems of shorter lengths will not host the superradiance phenomenon and will be restricted to the unsaturated maser regime in the steady state. We can also combine equations (8) and (12) to reformulate this critical threshold as

$$T_{\mathrm{R,crit}} = \frac{2T_2}{\ln\left[T_2/\left(T_1\theta_0^2\right)\right]}. \tag{13}$$

For our system we find $T_{\mathrm{R}} < T_{\mathrm{R,crit}} \approx 0.05\, T_2 \simeq 7 \times 10^4$ s, which satisfies the requirement for superradiance.

4. Modeling of G9.62+0.20E

One further characteristic of the superradiance transient response not discussed in Sec. 3 is that the shape of the intensity curve as a function of time is largely independent of that of the excitation (i.e., the pump signal or the seed electric field). An example is shown in Figure 8 where a strong 6.7 GHz methanol flare in S255IR-NIRS3 is modeled using two different inversion pump signals (Szymczak et al. 2018; Rajabi et al. 2019; Rajabi and Houde 2020). Other parameters for the model (i.e., L, T_1 and T_2) are similar to those used in Sec. 3. We thus see that the shape of the fast transient superradiance response is minimally affected by the detailed nature of the pump signal. This behaviour will be advantageous in this section as we endeavour to model existing data from the G9.62+0.20E star-forming region. That is, while we do not know the nature of the excitation responsible for the periodic flaring observed in this source, the fact that the the shape of the measured light curves are the results of the natural (or characteristic) transient responses of the corresponding systems will allow us to perform meaningful comparisons between spectral transitions from different molecular species.

G9.62+0.20E is a high mass star-forming region located 5.2 kpc away (Sanna et al. 2009). Methanol 6.7 GHz monitoring by Goedhart et al. (2003) (and subsequent studies) has revealed this source to be flaring with a main period of approximately 243 d (a second period of 52 d was recently discovered by MacLeod et al. 2022). This characteristic has since led to a significant number of studies, observational and theoretical, which resulted in increased monitoring with other molecular species/transitions and models aimed at explaining the source of the periodicity as well as the physical mechanisms behind the flaring activity (see Rajabi et al. 2023 for a recent review). The availability of multi-transition monitoring data for G9.62+0.20E is particularly interesting to us, as it will allow us to probe the transient (i.e., flaring) regimes for the different molecular species and transitions, as well as apply and test the model based on the MBE discussed in Sec. 3. Here, we will present some of the results from Rajabi et al. (2023).

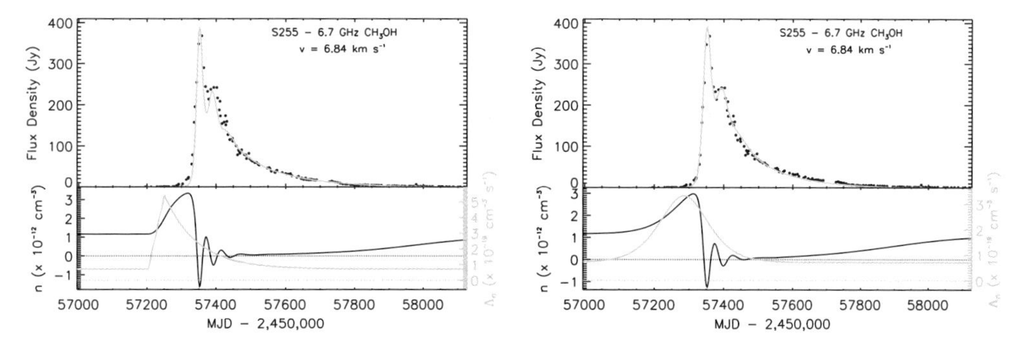

Figure 8. Superradiance model for the S255IR-NIRS3 6.7 GHz methanol flare at $v_{\rm lsr} = 6.84$ km s^{-1} (Szymczak et al. 2018; Rajabi et al. 2019; Rajabi and Houde 2020). The top panels show the superradiance fit (light/cyan) to the data (dots), while the bottom panels are for the pump (light/cyan, vertical axis on the right) and the population density (black, vertical axis on the left). *Left:* The pump pulse is modeled with a slow rising exponential function to approximately mimic the IR observations presented in Uchiyama et al. (2020). *Right:* A smoother and symmetric function is used for the pump signal. The shape of fast transient superradiance response is thus seen to be largely independent of that of the pump signal. Taken from Rajabi et al. (2023).

The data sets we used were published elsewhere in the literature prior to the work of Rajabi et al. (2023). That is, the OH 1665 MHz and 1667 MHz data were obtained with KAT-7 (Foley et al. 2016) and initially published in Goedhart et al. (2019), as were the methanol 12.2 GHz observations from the HartRAO 26-m telescope. The methanol 6.7 GHz observations are those from MacLeod et al. (2022), which resulted from the corresponding monitoring of G9.62+0.20E performed with the Hitachi 32 m telescope of the Ibaraki station from the NAOJ Mizusawa VLBI Observatory (Yonekura et al. 2016). The reader is referred to the cited references for more details concerning these data and to Rajabi et al. (2023) for a more comprehensive discussion of our analysis.

One interesting feature resulting from the multi-transition monitoring data of G9.62+0.20E uncovered by Goedhart et al. (2019) is the existence of different duration for the flares from different molecular species and transitions. An example is shown in Figure 9 where one flare from OH 1665 and 1667 MHz and methanol 6.7 and 12.2 GHz each is shown. We can clearly see from the position of the peaks that the flare duration decreases from top to bottom in the figure (i.e., from OH 1665 MHz $\rightarrow$ OH 1667 MHz $\rightarrow$ methanol 6.7 GHz $\rightarrow$ methanol 12.2 GHz). At first sight, this behaviour may seem difficult to explain as the sources for these features are likely to be subjected to a common excitation (e.g., these flares are all recurrent at the same period of 243.3 d). However, this is the kind of results we should expect within the framework of the MBE, in view of the existence of the superradiance transient regime.

More precisely, using equation (8) we can estimate the relative superradiance timescales for the different spectral lines pertaining to Figure 9 to be

$$\frac{T_{\rm R,1665\,MHz}}{T_{\rm R,6.7\,GHz}} \simeq 1.4\frac{(nL)_{\rm 6.7\,GHz}}{(nL)_{\rm 1665\,MHz}} \tag{14}$$

$$\frac{T_{\rm R,1667\,MHz}}{T_{\rm R,6.7\,GHz}} \simeq 1.3\frac{(nL)_{\rm 6.7\,GHz}}{(nL)_{\rm 1667\,MHz}} \tag{15}$$

$$\frac{T_{\rm R,12.2\,GHz}}{T_{\rm R,6.7\,GHz}} \simeq 0.6\frac{(nL)_{\rm 6.7\,GHz}}{(nL)_{\rm 12.2\,GHz}}, \tag{16}$$

which, interestingly, scale in the same manner as the observations for equal inversion column densities. Of course, there is no guarantee that all sources will share the same

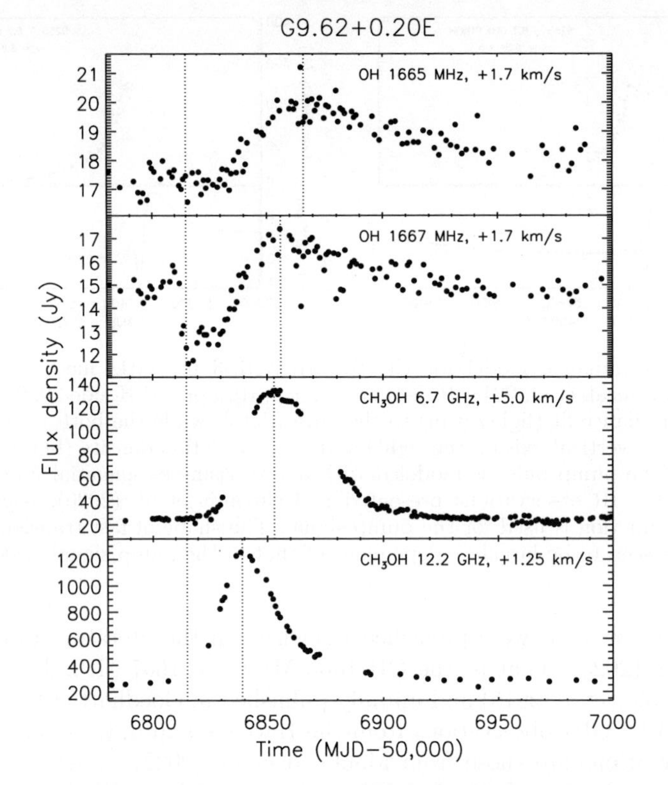

Figure 9. Comparison of profiles between OH and methanol flares in G9.62+0.20E. The vertical line at MJD 56815 denotes the start of the methanol 12.2 GHz flare, as defined by Goedhart et al. (2019). The other vertical lines indicate the occurrence of peak intensities for each transition. The methanol 6.7 GHz light curve was delayed by 15 days to align its start with that of the others. Taken from Rajabi et al. (2023).

value for the inversion column density but, still, this pattern can certainly serve as a motivation for verifying if our MBE/superradiance approach can help elucidate these observational results.

We therefore proceeded to numerically model several flaring features using the MBE in the manner explained in Rajabi et al. (2023). Results are shown in Figure 10 for features in OH 1665 MHz ($v_{\rm lsr} = +1.7\,\rm km\,s^{-1}$), methanol 12.2 GHz ($v_{\rm lsr} = +1.7\,\rm km\,s^{-1}$) and methanol 6.7 GHz ($v_{\rm lsr} = +5.0\,\rm km\,s^{-1}$ and $+8.0\,\rm km\,s^{-1}$). To do so we fixed the length of the systems to $L = 50$ au† and set the excitation from a population inversion pump to

$$\Lambda_n(z,\tau) = \Lambda_0 + \sum_{m=0}^{\infty} \frac{\Lambda_{1,m}}{\cosh^2\left[\left(\tau - \tau_0 - m\tau_1\right)/T_{\rm p}\right]}. \tag{17}$$

The constant pump rate Λ_0, the amplitude $\Lambda_{1,m}$ of pump pulse m at $\tau = \tau_0 + m\tau_1$ and the delay τ_0 to line our fit up to the corresponding data are adjusted for all models. Unchanged are the period $\tau_1 = 243.3$ d and the width of the pump pulse $T_{\rm p} = 4$ d. According to our previous discussion, the choice of the pump pulse's profile is arbitrary. While the relation and de-phasing time-scales are set independently for all transitions, little variations were observed between the different fits. More precisely, we have $T_1 = 210$ d

† The chosen value of $L = 50$ au is to some extent arbitrary since the inversion column density is the relevant parameter for setting the superradiance time-scale (see equation 8). That is, any variation in L could be compensated by applying the opposite change in the population inversion density n.

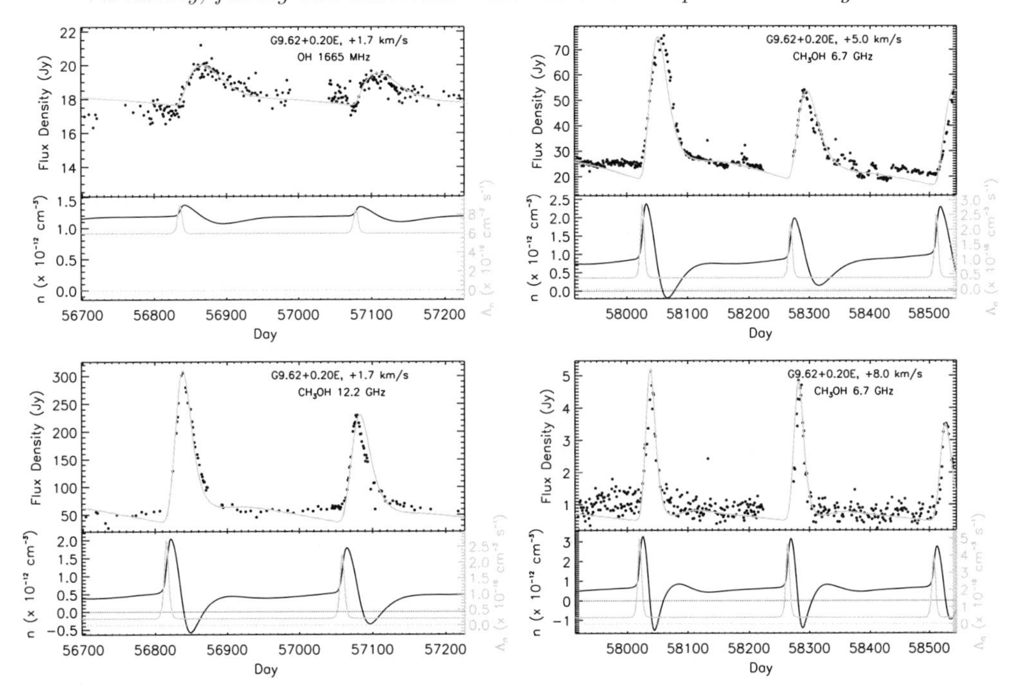

Figure 10. Model fits using the MBE for flaring features in OH 1665 MHz ($v_{\rm lsr} = +1.7\,{\rm km\,s^{-1}}$; top left), methanol 12.2 GHz ($v_{\rm lsr} = +1.7\,{\rm km\,s^{-1}}$; bottom left) and methanol 6.7 GHz ($v_{\rm lsr} = +5.0\,{\rm km\,s^{-1}}$ and $+8.0\,{\rm km\,s^{-1}}$; top and bottom right, respectively). Each graph shows the model fit (light/cyan; top) to the data (black dots), the population density (black curve; bottom, left vertical scale) and the inversion pump signal Λ_n (cyan curve; bottom, right vertical scale). The inversion pump is defined by equation (17) with a common pump pulse duration $T_{\rm p} = 4$ d at a period of 243.3 d for all features. The length of the systems is set at $L = 50$ au, while $T_1 = 210$ d and $T_2 = 13.5$ d for OH 1665 MHz, and $T_1 = 220$ d and $T_2 = 12.6$ d for methanol 6.7 GHz (both velocities) and 12.2 GHz. Adapted from Rajabi et al. (2023).

and $T_2 = 13.5$ d for OH 1665 MHz, and $T_1 = 220$ d and $T_2 = 12.6$ d for methanol 6.7 GHz (both velocities) and 12.2 GHz.

Despite the small changes in parameters the resulting fits in Figure 10 are found to recover the observed time-scales well by simply modulating the amplitude of the pump components Λ_0 and $\Lambda_{1,m}$. This is because in doing so we are in effect changing the inversion column density nL, which in turn directly affects the characteristic time-scale of superradiance $T_{\rm R}$ (see equation 8). This also explains, for example, why the methanol 6.7 GHz can exhibit significantly varying time-scales at different velocities, as seen for the corresponding $v_{\rm lsr} = +5.0\,{\rm km\,s^{-1}}$ and $+8.0\,{\rm km\,s^{-1}}$ features in Figure 10. More models to observations are presented in Rajabi et al. (2023). Most notable are the cases of unusually shaped flares in methanol 6.7 GHz and OH 1667 MHz; we refer the reader to the corresponding section in that paper.

Although the flaring periodicity in G9.62+0.20E necessitates an excitation that is different from the case of instantaneous inversion discussed in Sec. 3, the same type of behaviour is observed. Most notable is the clear disconnect between the duration of the inversion pump pulse (i.e., 4 d) and that of the bursts (e.g., from ~ 40 d for methanol 6.7 GHz at $v_{\rm lsr} = +8.0\,{\rm km\,s^{-1}}$ to on the order of 100 d for OH 1665 MHz). This is a clear manifestation of a superradiance transient response in the gas hosting the flaring events. As seen in Figure 10, while the interaction of the fast and brief pump pulses drives a rapid increase in the population inversion level and subsequently stimulates a

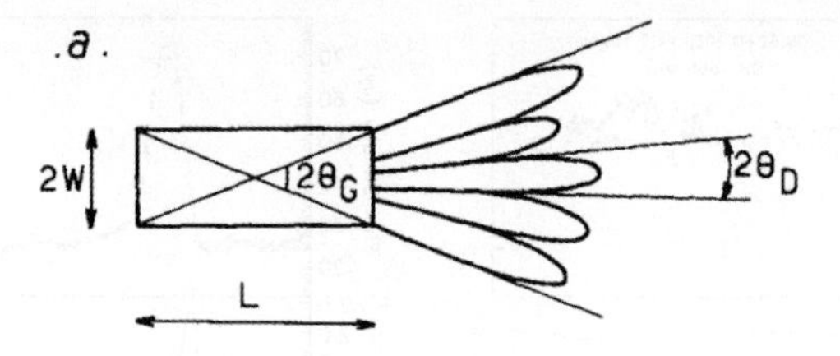

Figure 11. Schematic for a region of length L and a cross-section of width $2W$ exceeding that of a single sample, as defined in Sec. 3. This region has a Fresnel number greater than unity and supports a geometrical angle $\theta_{\mathrm{G}} = W/L$ at its end-fire. It will therefore contain several independent samples of diffraction angle $\theta_{\mathrm{D}} = \lambda/W$ during a superradiance event. Although radiating coherently individually, these samples combine in a non-coherent manner to make up the total intensity measured at the output. For an astronomical maser-hosting source, this intensity is likely to appear non-coherent to an observer. Taken from Gross and Haroche (1982).

transient superradiance response in the system, the duration of the superradiance burst bears no relation to that of the excitation. We also note that adopting a fixed value for the duration of the pump pulses is a reasonable assumption, as it is likely that a single source is at the root of the 243 d periodicity of the flares. Time-scales intrinsic to the pump (i.e., the period and pulse duration) should be the same for all transition/species used for the monitoring. On the other hand, the coupling of the inversion pump signal is bound to vary with position within G9.62+0.20E and wavelength. The dependence of the superradiance response on the strength of the excitation and the parameters characterizing the molecular transition/species used to probe the regions under study renders it an attractive and powerful mechanism to make sense of the diverse flaring time-scales and profiles observed.

Finally, we end with a word concerning the potential effects that the coherent nature of superradiance might be suspected to have on the characteristics of observations such as those presented in this section. That is, we have already stated that superradiance bursts scale in intensity with the square of the number of molecules involved in the process and inversely with that number in duration (Dicke 1954; Feld and MacGillivray 1980; Gross and Haroche 1982; Rajabi and Houde 2020). However, one should not necessarily expect coherence to be readily detected in astronomical sources hosting flares such as those analyzed here. The reason for this is presented in Figure 11 and rests on the relative size of a coherent superradiance sample and an astronomical maser-hosting region. Considering as an example the OH 1665 MHz sample resulting from the fit presented in Figure 10, we find that its radius is limited to $\sim 10^{-5}$ au to ensure a Fresnel number of unity. Evidently, this size is several orders of magnitude smaller than the spot size of a maser-hosting region. For example, a region of 10 au undergoing a flaring event would necessarily break up in a very large number of independent and uncorrelated samples, in the manner shown in Figure 11. Although individually radiating coherently, these samples combine in a non-coherent manner to make up the total intensity measured at the output. For an astronomical maser-hosting source, this intensity is thus likely to appear non-coherent to an observer. Our superradiance OH 1665 MHz sample of Figure 9 outputs a peak flux density of $\approx 5 \times 10^{-36}$ erg s^{-1} cm^{-2} at 5.2 kpc (or $\sim 10^{-5}$ Jy in the bandwidth of $\sim 10^{-7}$ Hz approximately associated to the duration of a flare). Only a small fraction of a typical maser-hosting is therefore needed to account for the detected radiation intensities.

Acknowledgements

M.H.'s research is funded through the Natural Sciences and Engineering Research Council of Canada Discovery Grant RGPIN-2016-04460. The 6.7 GHz methanol maser data obtained by the Hitachi 32-m telescope is a part of the Ibaraki 6.7 GHz Methanol

Maser Monitor (iMet) program. The iMet program is partially supported by the Inter-university collaborative project 'Japanese VLBI Network (JVN)' of NAOJ and JSPS KAKENHI grant no. JP24340034, JP21H01120, and JP21H00032 (YY).

References

Benedict, M. G. *et al.* 1996, *Super-radiance: Multiatomic Coherent Emission*
Brito, R., Cardoso, V., & Pani, P. In *Superradiance, New Frontiers in Black Hole Physics* 2015, volume 971 of *Lecture Notes in Physics, Berlin Springer Verlag*, 199
Dicke, R. H. 1954, *Phys. Rev.*, 93, 99
Dicke, R. H. 1964, *Quantum electron.*, 1, 35
Feld, M. S. & MacGillivray, J. C. In *Coherent Nonlinear Optics* 1980, 7
Ferioli, G., Glicenstein, A., Robicheaux, F., Sutherland, R. T., Browaeys, A., & Ferrier-Barbut, I. 2021, *Phys. Rev. Lett.*, 127, 243602
Foley, A. R., Alberts, T., Armstrong, R. P., *et al.* 2016, *MNRAS*, 460, 1664
Fujisawa, K., Sugiyama, K., Aoki, N., *et al.* 2012, *PASJ*, 64, 17
Goedhart, S., Gaylard, M. J., & van der Walt, D. J. 2003, *MNRAS*, 339, L33
Goedhart, S., van Rooyen, R., van der Walt, D. J., Maswanganye, J. P., Sanna, A., MacLeod, G. C., & van den Heever, S. P. 2019, *MNRAS*, 485, 4676
Gordon, J. P., Zeiger, H. J., & Townes, C. H. 1954, *Phys. Rev.*, 95, 282
Gordon, J. P., Zeiger, H. J., & Townes, C. H. 1955, *Phys. Rev.*, 99, 1264
Gross, M. & Haroche, S. 1982, *Phys, Rep.*, 93, 301
Hanbury Brown, R. & Twist, R. Q. 1956, *Nature*, 178, 1447
MacLeod, G. C., Yonekura, Y., Tanabe, Y., *et al.* 2022, *MNRAS*, 516, L96
Olech, M., Szymczak, M., Wolak, P., Gérard, E., & Bartkiewicz, A. 2020, *A&A*, 634, A41
Rajabi, F. & Houde, M. 2016a *ApJ*, 826, 216
Rajabi, F. & Houde, M. 2016b *ApJ*, 828, 57
Rajabi, F. & Houde, M. 2017, *Sci. Adv.*, 3, e1601858
Rajabi, F. & Houde, M. 2020, *MNRAS*, 494, 5194
Rajabi, F., Houde, M., Bartkiewicz, A., Olech, M., Szymczak, M., & Wolak, P. 2019, *MNRAS*, 484, 1590
Rajabi, F., Houde, M., MacLeod, G. C., *et al.* 2023, *arXiv e-prints,*, arXiv:2303.08793
Sanna, A., Reid, M. J., Moscadelli, L., *et al.* 2009, *ApJ*, 706, 464
Skribanowitz, N., Herman, I. P., MacGillivray, J. C., & Feld, M. S. 1973, *Phys. Rev. Lett.*, 30, 309
Szymczak, M., Olech, M., Wolak, P., Bartkiewicz, A., & Gawroński, M. 2016, *MNRAS*, 459, L56
Szymczak, M., Olech, M., Wolak, P., Gérard, E., & Bartkiewicz, A. 2018, *A&A*, 617, A80
Uchiyama, M., Yamashita, T., Sugiyama, K., *et al.* 2020, *PASJ*, 72, 4
Yonekura, Y., Saito, Y., Sugiyama, K., *et al.* 2016, *PASJ*, 68, 74

Cosmic Masers: Proper Motion toward the Next-Generation Large Projects
Proceedings IAU Symposium No. 380, 2024
T. Hirota, H. Imai, K. Menten, & Y. Pihlström, eds.
doi:10.1017/S1743921323002259

Recombination lines and maser effects

Zulema Abraham[1]

[1]Instituto de Astronomia, Geofísica e Ciências Atmosféricas, Universidade de São Paulo
Rua do Matão 1226, CEP 05508-090, São Paulo, Brazil. email: zulema.abraham@iag.usp.br

Abstract. Maser effects occur in recombination lines when the plasma departs from local thermodynamic equilibrium (LTE). Its consequence is not as dramatic as that found in molecular masers, and therefore it is more difficult to recognize. Besides, it occurs in compact high density regions, and its lines fall at millimeter and submillimeter wavelengths, only recently available with good angular resolution. This review will focus on the theoretical aspects of maser recombination lines and on the recent detection of these masers in different astronomical objects.

Keywords. masers, radio lines, stars: mass loss, HII regions, jets and outflows

1. Introduction

Radio recombination lines are the result of transitions between atomic energy levels of high quantum number n. The electrons populate these levels after recombination, and then decay into lower levels, producing the lines.

The first recombination line (H90α) towards HII regions was detected at cm wavelengths in 1964 from Pushchino (Dravskikh & Dravskikh 1964) and in 1965 (109α) from NRAO (Hoglund and Mezger 1965). During the first years, the velocity measured from the radio recombination lines was used to study the spiral structure of our Galaxy.

The intensity of the recombination lines, together with that of the bremsstrahlung continuum radiation was used to determine the physical conditions of the clouds (density and temperature) assuming local thermodynamic equilibrium (LTE) conditions of the gas. It was soon seen that the temperatures obtained from radio recombination lines in large HII regions were smaller than those obtained from optical lines (eg. Mezger and Hoglund 1967). The reason could be inhomogeneity in the temperature distribution or the assumption of LTE not being valid, implying maser effects (Goldberg 1966).

The discovery of the first maser in recombination lines occurred in 1989 in the massive star MCW349A (Martin-Pintado *et al.* 1989) and later in the massive and evolved star η Carinae (Cox *et al.* 1995).

At the time of our last maser meeting in Cagliari, other maser candidates were reported: the source IRS2 in MonR2 (Jiménez-Serra *et al.* 2020), and later the Planetary Nebula Mz3, in data obtained with Hershel (Aleman *et al.* 2018).

The question is why it took so long for these discoveries, and the answer is that it was necessary to look at the right lines and the right objects.

2. Recombination Line Maser Theory

The discovery of masers in recombination lines led to the study of the physical conditions necessary for maser amplification. The first comprehensive study was presented by Strelnitski *et al.* (1996). He considered the radiative transfer across a plane parallel slab of width s.

The equation of radiative transfer can be written as:

$$\frac{dI_\nu}{d\tau_\nu} = -I_\nu + \frac{j_\nu}{\kappa_\nu}, \tag{2.1}$$

where I_ν is the specific intensity at frequency ν (energy of the radiation per unit surface, time, frequency and solid angle), j_ν is the emission coefficient, κ_ν the absorption coefficient per unit length, and τ_ν the optical depth, given by:

$$\tau_\nu = \int_0^s \kappa_\nu \, ds. \tag{2.2}$$

The solution of equation 2.1 is:

$$I_\nu = I_\nu(0)e^{-\tau_\nu} + S_\nu \left[1 - e^{-\tau_\nu}\right], \tag{2.3}$$

with:

$$S_\nu = \frac{j_\nu}{\kappa_\nu}. \tag{2.4}$$

S_ν is called the source function and in thermodynamic equilibrium is equal to the Planck function $B_\nu(T)$. In general, the source function contains all the physics: scattering, stimulated emission, non-uniformity, etc.

Let us consider no background radiation, and transition between two levels $n_1 \rightarrow n_2$ along a slab of depth L. The source function can be written as:

$$S_\nu = \eta_\nu B_\nu(T), \tag{2.5}$$

with:

$$\eta_\nu = \frac{\kappa_c^* + \kappa_L^* b_2}{\kappa_c^* + \kappa_L^* b_1 \beta_{12}}, \tag{2.6}$$

and

$$\beta_{12} = \frac{1 - \frac{b_2}{b_1}\exp(-h\nu_0/kT_e)}{1 - \exp(-h\nu_0/kT_e)}, \tag{2.7}$$

b_n represents the population of the level of quantum number n relative to its LTE value, β_{12} is the ratio between the correction factor for stimulated emission and its LTE value and ν_0 is the rest frequency of the transition.

The level population of the H atoms, relative to their LTE values was calculated with different degrees of accuracy. Initially, only the quantum number n was considered (Brockelhurs 1970, Walmsley 1990). Storey & Hummer (1995) included later the angular momentum l, and finally the contribution of the continuum radiation was introduced (Prozesky A. & Smits 2018, Zhu *et al.* 2019).

Depending on the ratio b_2/b_1, β_{12} can be negative. The same is valid for η_ν, which also depends on value of the absorption coefficients, which can be written as:

$$\kappa_c^* = 6.64 \times 10^{-8} \frac{N_e N_1}{\nu^2} \left(\frac{10^4}{T_e}\right)^{1.5} g_{\text{ff}} \;\; \text{pc}^{-1}, \tag{2.8}$$

$$\kappa_L^* = 1.06 \times 10^{12} \frac{N_e N_i(v)}{\nu} \left(\frac{10^4}{T_e}\right)^{2.5} mK(m) \exp\left(\frac{\xi_n}{kT_e}\right)(\nu\phi_\nu), \tag{2.9}$$

N_e and N_i are the electron number density of electrons and ions, respectively, T_e the electron temperature, g_{ff} the Gaunt factor for continuum bremsstrahlung, $mK(m)$ is an approximation to the oscillator strength given by Menzel (1968), ξ_n the energy of level n

and $\phi(v)$ the line profile, which takes into account the Doppler, turbulence and pressure broadening.

The total optical depth of the slab to radiation of frequency ν is:

$$\tau_{tot} = \tau_c + \tau_L = (\kappa_c^* + \kappa_L^* b_1 \beta_{12}) L \tag{2.10}$$

The intensity at the line center, relative to the continuum is expressed by:

$$r(\nu_0) = \frac{I(\nu_0) - I_c(\nu_0)}{I_c(\nu_0)} = \eta \frac{1 - \exp(-\tau_{tot})}{1 - \exp(-\tau_c)} - 1, \tag{2.11}$$

or

$$r(\nu_0) = \eta \frac{1 - \exp(-\tau_c - \tau_L)}{1 - \exp(-\tau_c)} - 1 \tag{2.12}$$

Depending on the sign of η and the relative values of τ_c and τ_L we can have different values for $r(\nu_0)$. In particular, when η and τ are large and negative,

$$r(\nu_0) \approx \frac{|\eta|}{1 - \exp(-\tau_c)} \exp(|\tau_{tot}|). \tag{2.13}$$

Note that η depends only on the physical conditions of the emitting region, while the optical depth depends on the velocity distribution of the material, implying that the length L can be different for the continuum and for the line, which requires frequency coherence.

Since τ_{tot} is negative, and it is formed by the sum of τ_c and τ_L, there will be a large difference between the intensity of the recombination lines when the region becomes optically thin, as observed in MCW349A, between the lines H41α and H30α.

3. Maser Recombination Lines in η Carinae

η Carinae (hereafter η Car) is the object that presents the strongest recombination line masers detected until now (Cox *et al.* 1995, Abraham *et al.* 2014, 2022). The star, which possibly is in the LBV (Luminous Blue Variable) phase of evolution, underwent a strong episode of mass loss (> 30 $M_\odot$) in 1840; the ejected mass expanded, cooled down and formed the well-known Homunculus Nebula. η Car is part of a binary system in an extremely eccentric orbit, with period of 5.52 years, as revealed by the strictly periodic light curves, observed from radio frequencies to X-ray energies (Damineli 1996, Corcoran *et al.* 2017). The companion star, as well as the photosphere of η Car are not directly visible, because they are embedded in the massive slow wind of η Car, but the companion (probably a Wolf-Rayet star) is believed to have a fast wind of its own; the wind-wind collisions produce shocks that are the origin of the X-ray emission.

Interferometric observations of the binary system at cm wavelengths obtained with the Australian Compact Array (ATCA) revealed extended continuum emission from the surroundings of the binary system, coincident with the Homunculus Nebula, and a disk like structure at the core, with an extension and intensity that varied with the phase of the binary orbit (Duncan & White 2003). This structure was interpreted as gas ionized by the hot companion star, which provided a number of ionizing photons that varied with the orbital phase.

Although cm wave recombination lines were also observed together with the continuum emission, only mm wave lines, observed with the SEST radiotelescope with arcsec resolution, presented strong maser effects, first attributed to the dense wind of η Car (Cox *et al.* 1995). These maser sources presented a single line, contrary to what is observed in the other maser sources, which were modeled as a combination of a Keplerian disk and wind ejected from the disk (Báez-Rubio *et al.* 2013, Jiménez-Serra *et al.* 2020).

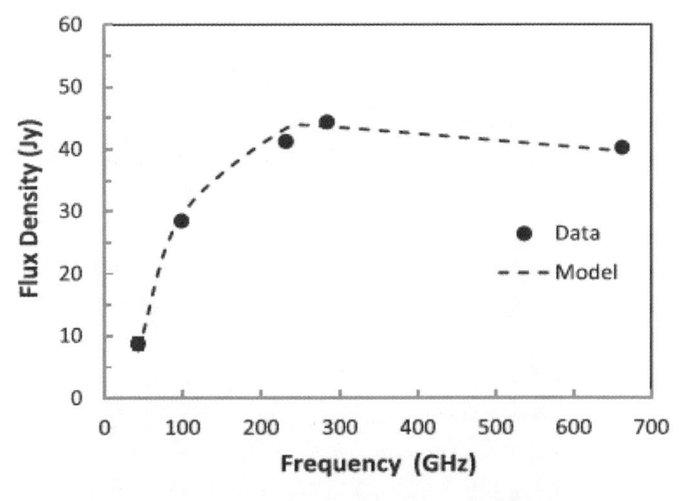

Figure 1. Continuum pectrum of η Car obtained with ALMA on 2014 (Abraham *et al.* 2014).

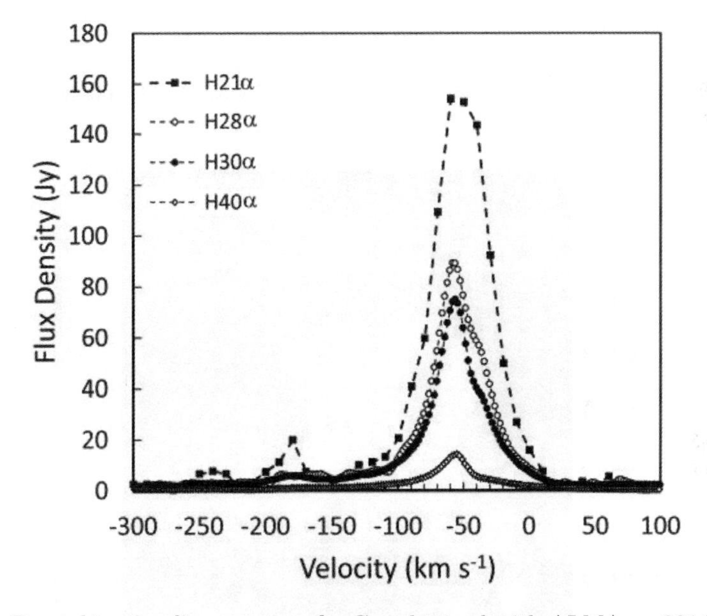

Figure 2. Recombination line spectra of η Car obtained with ALMA on 2014 (Abraham *et al.* 2014).

η Car was observed with ALMA on 2012 (cycle 0) in the recombination lines H42α, H40α, H30α, H28α, and H21α (Abraham *et al.* 2014), but it was not resolved, even at the higher frequencies, where the beam size was 0.35×0.42 arcs (at the distance of η Car 50 mas correspond to 110 au).

The continuum spectrum is compatible with that of an HII region, with a turnover frequency close to 300 GHz, as can be seen in Figure 1.

In Figure 2 we show the superimposed spectra of the H40α, H30α, H28α, and H21α lines.

Based on the cm wave observations, which were modeled as originated in an expanding source, the mm wave continuum and lines were modeled as arising from an expanding shell of radius $R = 6.6 \times 10^{15}$ cm, width $\Delta R = 0.1R$, electron density $N_e = 1.3$ km s^{-1} and temperature $T_e = 1.7 \times 10^4$ K, expanding with a bulk velocity of 53 km s^{-1}, with

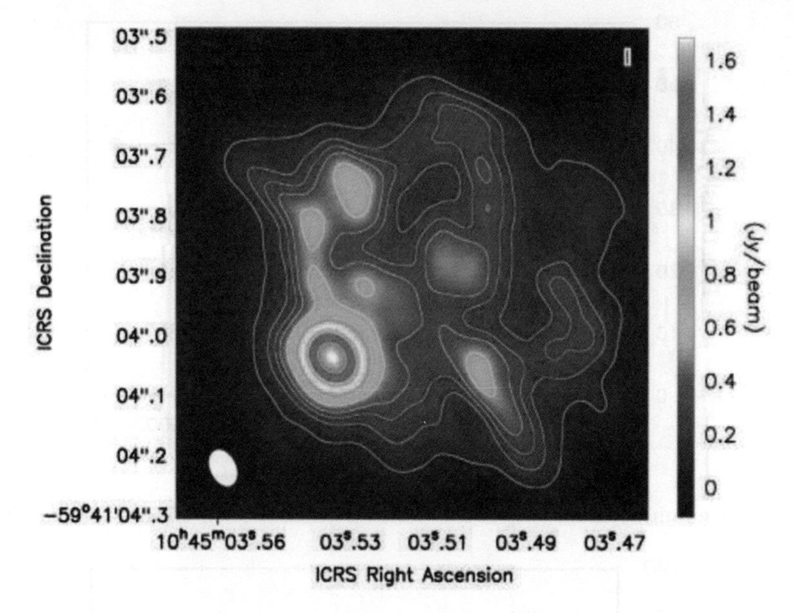

Figure 3. 230 GHz continuum image of η Car obtained with ALMA on 2017 (Abraham *et al.* 2020).

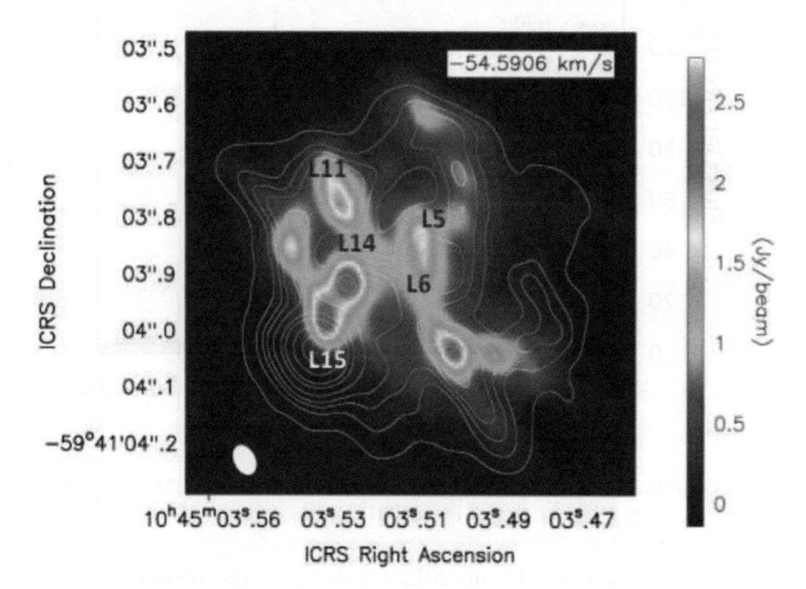

Figure 4. Iso-velocity image of the recombination line H30α, centered at the velocity of -54.6 km s^{-1} and resolution of 1.3 km s^{-1}, of η Car obtained with ALMA on 2017 (Abraham *et al.* 2020).

the velocity ranging from 60 km s^{-1} to 20 km s^{-1} in the inner and outer borders of the shell, respectively.

However, η Car was observed again with ALMA in the 230 GHz continuum and in the H30α recombination line on 2017, with resolution 65×43 mas, and the expanding shell model was not confirmed.

In fact, the continuum emission showed a compact source, centered on η Car and extended emission in the NW direction, as can be seen in Figure 3.

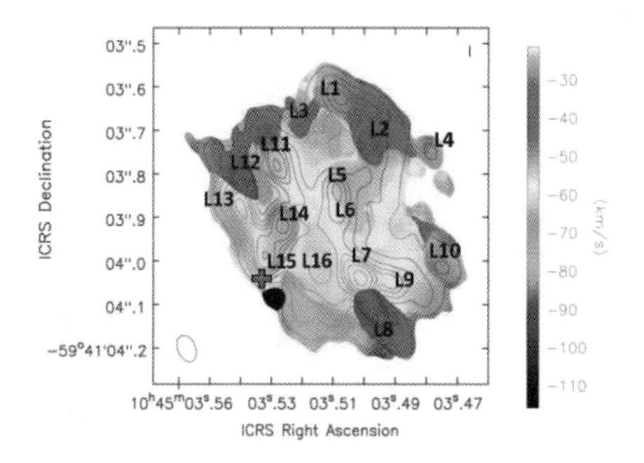

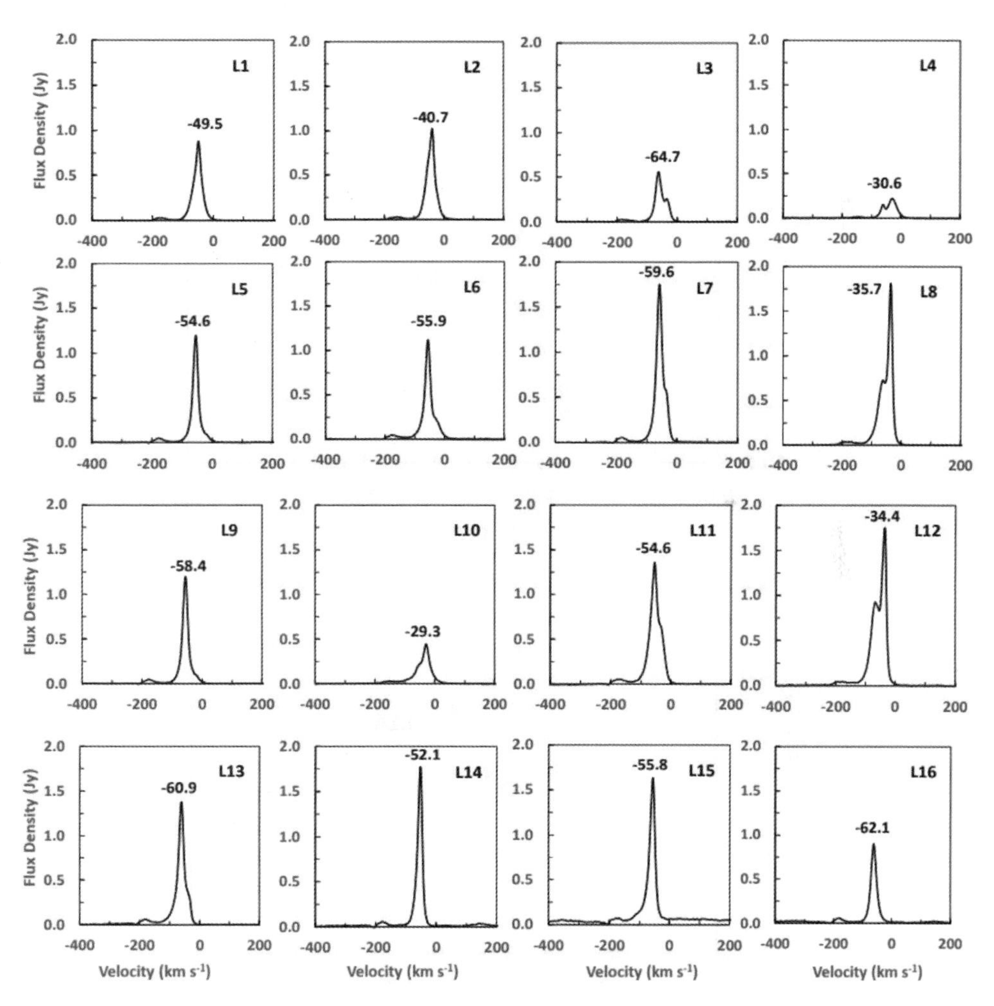

Figure 5. Spectra of the 16 compact component identified in the H30α images of η Car detected with ALMA in 2017. (Abraham *et al.* 2020).

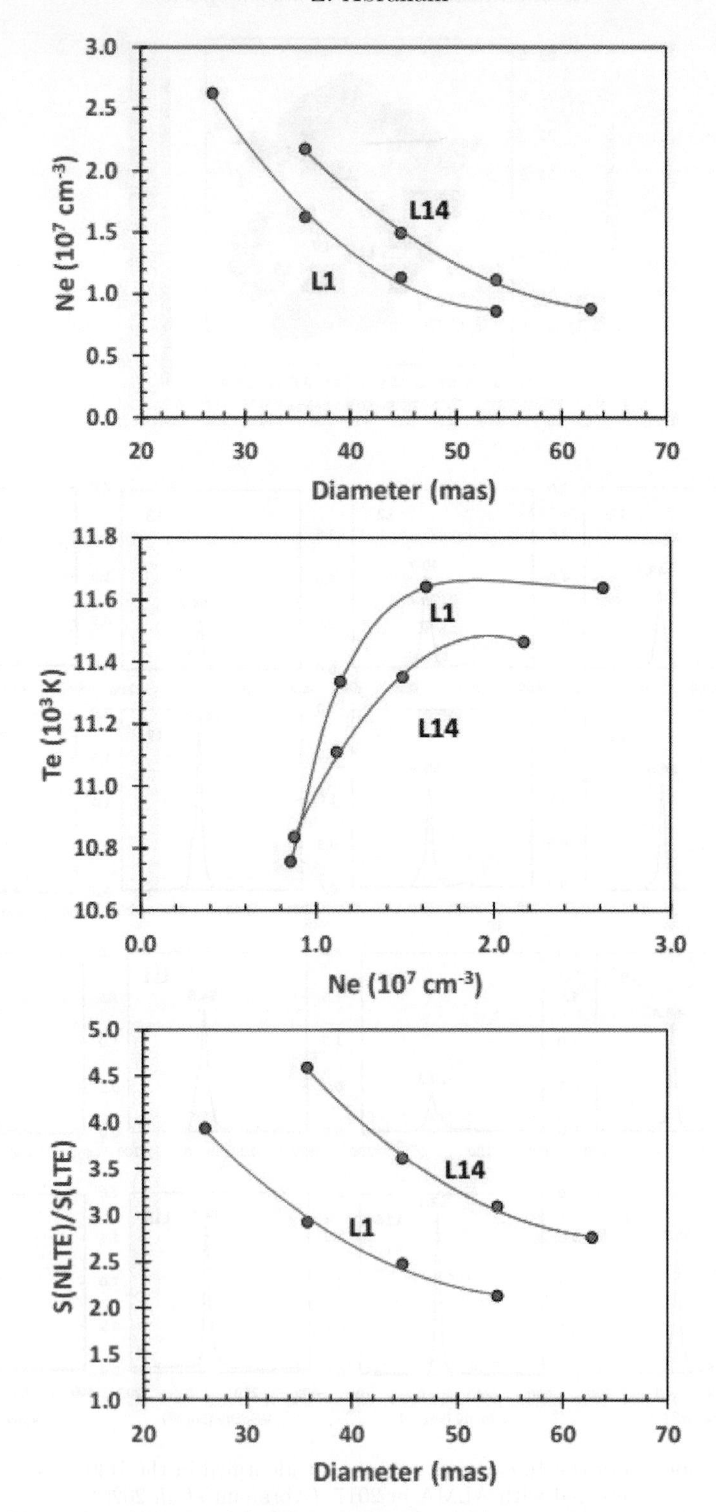

Figure 6. Physical conditions of the clumps that fit the spectra of the H30α recombination line of clumps 1 and 14, as identified in Figure 5.

The iso-velocity intensity maps of the different velocity channels, with width of 1.3 km s^{-1}, revealed several compact (not resolved) components with different velocities on top of the 230 GHz continuum emission. One of these maps, centered at the velocity of maximum emission is shown in Figure 4.

In Figure 5 we show the profiles of the different clumps, identified in the image of the first intensity momentum superimposed on the raster image of the line velocity. We can see that the positions and velocities of the clumps are not compatible with the model of an expansion shell.

Since the maser sources are not resolved, the physical conditions of two of them, labeled as 1 and 14, were investigated for different values of the radius (smaller than 50 mas or 110 ua), assumed to be spherical and homogeneous clumps. The results are shown in Figure 6.

References

Abraham, Z., Beaklini, P. P. B., Cox, P., *et al.* 2020, *MNRAS*, 499, 2493

Abraham, Z., Falceta-Gonçalves D., Beaklini P. P. B., 2014, *ApJ*, 791, 95

Abraham Z., Beaklini P. P. B., Cox, P., Falceta-Gonçalves D., Nyman L-A, 2020, *MNRAS*, 499, 2493

Aleman, I. *et al.* 2018, *MNRAS*, 477, 4499

Bez-Rubio, A., Martn-Pintado J., Thum C., Planesas P., 2013, *A&A*, 553,45

Brockelhurst, M. 1979, *MNRAS*, 148, 417

Corcoran, M. F., Liburd, J., Morris, D., *et al.* 2017, *ApJ*, 838, 45

Cox P. *et al.* 1995, *A&A*, 295, L39

Damineli, A. 1996, *ApJ*, 460, L49

Dravskikh, Z. V. & Dravskikh, A. F., 1964, *Astron. Tsirk.*, 282, 2

Duncan, R. A., White, S. M., 2003, *MNRAS*, 338, 425

Goldberg, L., 1966, *ApJ*, 144, 1225

Hoglund, B. and Mezger, P. G., 1965, *Science*, 150, 339.

Jimnez-Serra, I., Bez-Rubio, A., Martn-Pintado, J., Zhang, Q., Rivilla, V. M., 2020, *ApJ*, 897, 33

Martin-Pintado, J., Bachiller, R., Thurn, C., Walmsley, C. M., 1989, *A&A*, 215, L13

Menzel, D. H., 1968, *Nature*, 218, 756.

Mezger, P. G. & Hoglund, B., 1967, *ApJ*, 147, 579

Prozesky, A. & Smits, D. P., 2018, *MNRAS*, 478, 2766

Storey, P. J., Hummer, D. G., 1995, *MNRAS*, 272, 41

Strelnitski, V. S., Ponomarev, V. O., Smith, H. A. 1996, *ApJ*, 470, 1118

Walmsley, C. M. 1990, *A&AS*, 82, 201

Zhu, F.-Y., Zhu, Q.-F., Wang, J.-Z., Zhang, J.-S., 2019, *ApJ*, 881, 14

Cosmic Masers: Proper Motion toward the Next-Generation Large Projects
Proceedings IAU Symposium No. 380, 2024
T. Hirota, H. Imai, K. Menten, & Y. Pihlström, eds.
doi:10.1017/S1743921323002399

Flaring Masers and Pumping

M. D. Gray[1], S. Etoka[2], B. Pimpanuwat[2], A. M. S. Richards[2] and F. J. Cowie[2]

[1]National Astronomical Research Institute of Thailand, 260 Moo 4, T. Donkaew, A. Maerim, Chiangmai 50180, Thailand. email: malcolm@narit.or.th

[2]Jodrell Bank Centre for Astrophysics, Department of Physics and Astronomy, University of Manchester, M13 9PL, UK.

Abstract. We briefly consider the history of maser variability, and of flaring variability specifically. We consider six proposed flare generation mechanisms, and model them computationally with codes that include saturation and 3-D structure (the last mechanism is modelled in 1-D). Fits to observational light curves have been made for some sources, and we suggest that a small number of observational parameters can diagnose the flare mechanism in many cases. The strongest flares arise from mechanisms that can increase the number density of inverted molecules in addition to by geometrical effects, and in events where unsaturated quiescent masers become saturated during the flare.

Keywords. masers, molecular processes, radiative transfer, radio lines: ISM

1. Introduction

A significant advance in the study of variability in astrophysical masers towards star-forming regions was the long-term, high-cadence, monitoring of 6.7-GHz methanol masers over more than 4 yr by Goedhart *et al.* (2004). The time resolution of 1–2 weeks for most sources was enough to show that the majority vary, and do so with a variety of patterns, including monotonic rise or fall, and aperiodic, quasi-periodic and periodic oscillations. Periods of the sources in the dataset in Goedhart *et al.* (2004) covered 132–520 d. Since then, there have been extensions to both longer (longest at time of writing 1260 d, Tanabe *et al.* 2023) and shorter (for example, 23.9 d, Sugiyama *et al.* 2017) periods, and other variability phenomena have been studied. These include variability correlations with other methanol maser transitions, such as 12.2 GHz (Goedhart *et al.* 2009), with transitions in other species, for example OH and H_2O (MacLeod *et al.* 2018), and with infra-red (IR) continuum radiation, likely involved in maser pumping (Moscadelli *et al.* 2017; Kobak *et al.* 2023). Anti-correlated behaviour between flares in 22-GHz H_2O and 6.7-GHz methanol masers, for example, Olech *et al.* (2020), suggests these transitions are pumped by different mechanisms. Although there is no accepted definition of flaring variability, it can perhaps be considered as a more than usually powerful change in the flux density of one or more spectral features. Periodicity, if found, links maser flares to astrophysical properties of the system, such as orbital parameters (Parfenov and Sobolev 2014; van der Walt *et al.* 2016) and stellar pulsation (Inayoshi *et al.* 2013). Useful analysis is very reliant on good flare monitoring data, such as that provided by the M2O organization† (Burns *et al.* 2022), and multi-band data that can identify IR or radio continuum correlations with the maser emission.

† www.masermonitoring.com/

The analysis in the present work relies on approximating the maser system as a series of time-independent snapshots. Therefore, events on timescales $< 1 - 5\,\mathrm{d}$, dependent on the mechanism, are beyond the scope of this model. In particular, we cannot address variability that requires fully time-dependent models of the sort required to model superradiance (Rajabi *et al.* 2019), pump fluctuations (Clegg and Cordes 1991), or other very short timescales (hours, minutes or less). Notwithstanding these limitations, we consider the variations in flare properties from the following mechanisms: (1) rotation of aspherical clouds, (2) variation in the energy density of pumping radiation, (3) variation in the intensity of the amplified background radiation, (4) shockwave impact on an initially spherical cloud, (5) cloud overlap in the line of sight, and (6) catastrophic release of saturation under the assumption of complete velocity redistribution (CVR: see Section 1.6).

1.1. *Rotating Clouds: Rotors*

Prolate and oblate rotors were considered by Gray *et al.* (2019) in a model that includes full saturation of the maser. The distortion algorithm applied ensured that the prolate and oblate objects preserved the volume of a 'parent' sphere. Average light-curves from many viewpoints show variability indices F/F_0 of typically a few to few tens, where F is the line-centre flux density at flare maximum and F_0 is the quiescent flux density in the same channel. Stability considerations make long-term periodicity and short flares, with a timescale shorter than the sound-crossing time of the cloud, very unlikely. In common with many mechanisms, saturation reduces the variability index, but if saturation is not achieved at the peak of the flare, the flaring object may not be bright enough to be observable. The duty cycle, the fraction of time that the light curve is above half its peak value is modest, typically in the range $0.1 - 0.3$.

1.2. *Variation in the Pumping Radiation*

In this mechanism, the pumping radiation follows a light curve that we shall call the 'driving function'. The light curve of the maser is then a distortion of the driving function, highly modified by amplification and saturation (see Figure 1). In most cases, the maser light curves emphasize the brighter parts of the driving function and suppress the weaker parts, so that the duty cycle of the maser light curve is the lower: almost always < 0.3 and typically ~ 0.1. The driving functions used range from a basic sinusoid to representations of stellar pulsation and periodic orbital effects based on accretion and colliding binary winds. Details are given in the caption of Figure 1. The highest variability indices from this mechanism can exceed 10^7, but to achieve such a value typically requires a very low optical depth in the quiescent cloud, and the maximum flux density in the flare is only modest. Some compromise is therefore required between the variability index and the maximum flux density achieved, noting that it is this latter parameter that controls the observability of the flare (Gray *et al.* 2020b). The ability of this mechanism to produce powerful flares arises as a natural consequence of generating more inverted molecules, rather than depending simply on geometry, though aspherical clouds were also considered in Gray *et al.* (2020b).

1.3. *Variation in the Background Radiation*

If the level of background radiation is sufficient to generate a saturated maser, then further increase in the background intensity does little to the output; only a significant fall in the background, that lifts saturation, produces a large effect on the maser (Gray *et al.* 2020b). The light curve from this mechanism is therefore characteristically different from what is typical of those from Section 1.2. A maser light curve driven by background

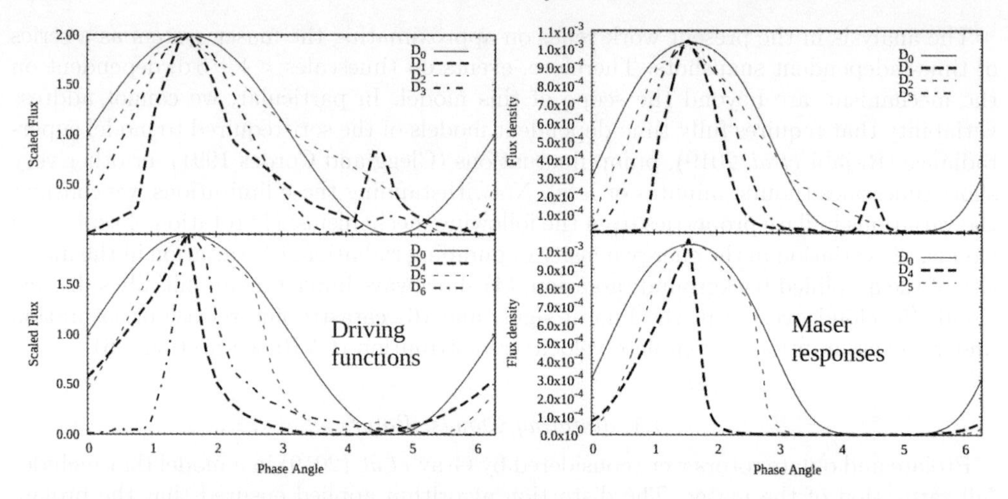

Figure 1. Sample periodic driving functions (left panel) and maser responses (right panel). The driving functions are, D_0: sinusoid, D_1: IR continuum digitized from Stecklum *et al.* (2018), D_2: theoretical Cepheid light curve from Bhardwaj *et al.* (2017), D_3–D_5: light curves for stars only, shock only, and combined effect by van der Walt *et al.* (2016) from the spiral shock model of Parfenov and Sobolev (2014), D_6: background radiation light curve from the colliding wind eccentric binary model by van der Walt *et al.* (2009). The responses were generated from models with a maser depth range of 5.0–10.0 between quiescent and flare peaks, and for a prolate cloud viewed along its long axis.

variability (driving function D_6) typically has a long duty cycle (> 0.5), increasing with saturation, as demonstrated in Figure 2 (left panel), though it shares with the variable pump a natural periodicity if the driving function is also periodic. The typical shape of the light curve with this mechanism, and strong saturation has been referred to as an 'anti-flare'. Candidate objects from Goedhart *et al.* (2004) include G338.92−0.06 and G351.78−0.54.

1.4. *Shock Compression*

The key to this mechanism is a compression of the gas and the establishment of lines of sight parallel to the shock front, and within the shocked material, that exhibit lengths comparable to the diameter of the original cloud, but contain significantly more material. A very rapid initial rise in the maser output, over the first 20–50 d, is followed by a slower rise as the shock progresses (Figure 2 upper right). Relaxation has been approximated as an exponential decay, based on the sound crossing time in the shocked material. Unlike the variable radiation mechanisms discussed above, shock-generated maser flares are therefore highly directional (Gray *et al.* in preparation) as shown in Figure 2, lower right graph. Chemical effects may also increase the actual number of inverted molecules within the cloud in the case of H_2O molecules (Kaufman and Neufeld 1996), and populations can also be increased by shock-induced release from grain surfaces, for example in Class 1 methanol masers (Sobolev 1993). The great structural change imposed by the shock makes periodicity impossible for a single cloud and the pattern of decay of the flare very difficult to predict, but a periodically generated shock wave could cause a quasi-periodic flare if it struck a number of similar clouds on each new passage. In the ideal isothermal hydrodynamic case, this mechanism provides the largest variability indices, of order 10^8, but values of 10^5 can still be achieved by more realistic MHD shocks.

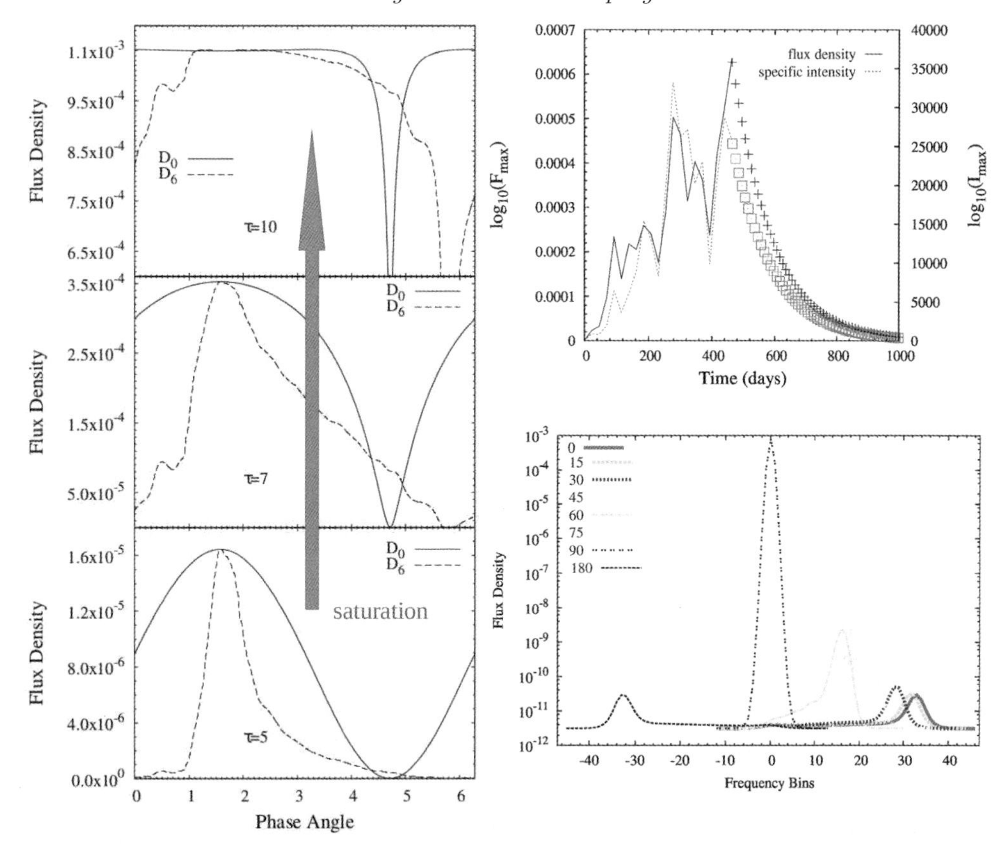

Figure 2. Left panel: periodic light curves for the variable background mechanism driven by the driving functions D_0 and D_6 (see Figure 1); the most saturated model is at the top. Upper right panel: the light curve generated by a hydrodynamic isothermal shock of speed 7.5 km s^{-1} striking an initially spherical cloud of radius 1 au containing 10.9 inverted ortho-H_2O molecules cm^{-3}. Curves are shown for the flux density in the brightest spectral channel and the specific intensity in the brightest pixel (right-hand y-axis). Lower right panel: the effect of varying the viewing angle on the maser spectrum in the fully shocked cloud. Angles vary from 0°, facing the approaching shock, through 90°, parallel to the shock front to 180° (behind the shock). All the graphs in this figure use a specific intensity scaled to the saturation intensity. With a background level of 10^{-6} and an observer at a distance of 1000 domain units, the flux density then has a background level of $\pi \times 10^{-12}$.

1.5. *Line of Sight Overlap*

The idea of a continuous Keplerian disc as the source of varying masers in a star forming region goes at least as far back as the study of S255 by Cesaroni (1990). For discrete clouds, considerable preliminary work on discs has been carried out for a massive star-formation scenario (Cowie *et al.* in preparation). The duration of an overlap then depends on the viewing angle (ϕ, see Figure 3, top left). The effect of varying ϕ, and consequently the velocity-coherent gain path, is explored in Figure 3 (top right). Behaviour also strongly depends on the orbital radii. Two clouds with a small separation in orbital radius generate a flare of very long duration (thousands of days), but rise and decline on order of a year is typical of orbital radii of a few hundred au, with 100 au orbital spacing, orbiting a central object of $>10\,M_\odot$. Light curves show considerable sub-structure if one, or both, clouds are aspherical and rotating with a rotation period similar to the duration of the overlap (Figure 3, bottom left). The variability index for overlaps appears limited

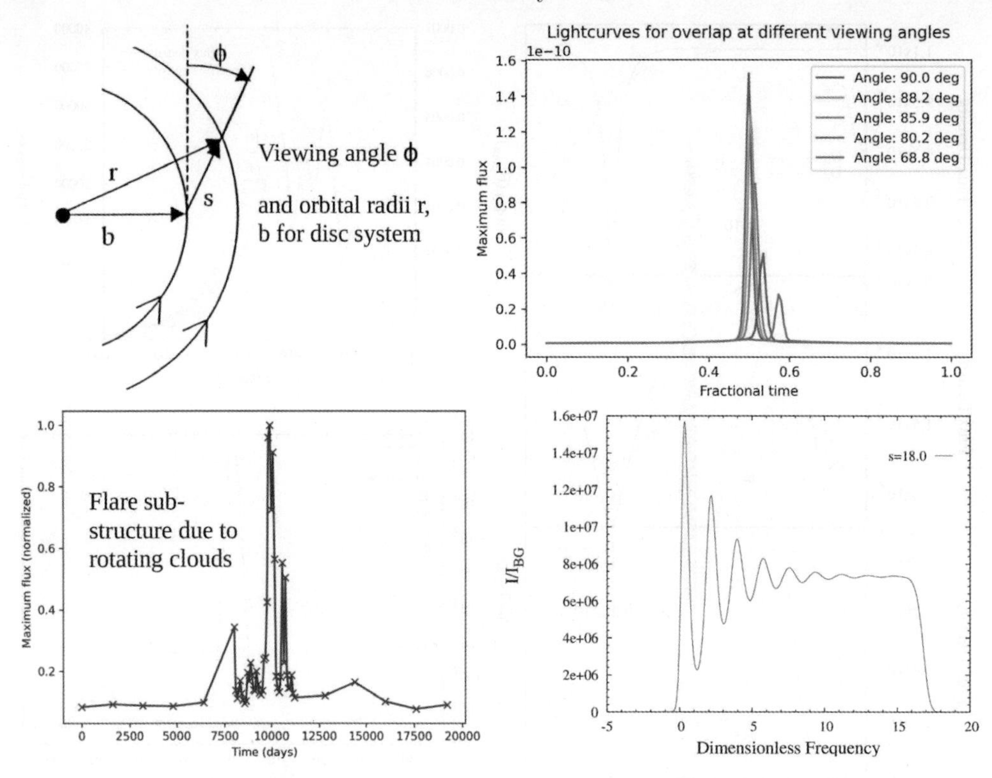

Figure 3. Top left: Angle and radius definitions for overlap in a disc system (Watson and Wyld 2000). Top right: Effect of viewing angle on overlap-based flare light curves. Bottom left: substructure in flare light curve due to rotating clouds: $r = 53$ au, $b = 50$ au, $\phi = 90°$. Flux density scale as in Figure 2, except that the observer is at 10000 domain units. Bottom right: Spectrum of a CVR catastrophe flare: intensity scaled to the background for a 1-D model is plotted against dimensionless frequency, scaled to the Doppler width, at a dimensionless distance, s, or velocity shift in Doppler units, of 18.0.

to $\lesssim$500, and is highest for a cloud pair with orbital radii of 1850 and 1300 au, and where the individual clouds are unsaturated, but the pair shows significant saturation. Much faster flares can of course be generated by random hyperbolic encounters. Periodicity is unlikely, but possible in a discrete-cloud disc system.

1.6. *Saturation Catastrophe*

This is a more subtle effect that requires a maser system with CVR, where transfer of population between the velocity subgroups of a quantum-mechanical energy level, that happens to be the upper or lower level of a maser transition, is maintained at a rate comparable to the maser pump rate. Such rapid redistribution operates through a network of optically thick IR transitions (Goldreich and Kwan 1974). Under CVR, the molecular response must saturate as a whole, and remains approximately Gaussian, even under very strong saturation. With a velocity gradient in the line of sight, a very strong maser line can maintain saturation even as the molecular response in the maser transition begins to shift away from the centre of the line. At a certain shift, the saturation is catastrophically released, and a new maser feature appears at the new, shifted line centre of the molecular response. A system can display several 'echoes' of this initial phenomenon, but the amplitude generally decreases for the second and subsequent maser peaks (see Figure 3, bottom right). The velocity spacing between peaks is not quite periodic. For the optimum parameter choices, the variability index on the initial maser peak can be as high as several million.

2. Fits to Data

At least for the mechanisms based on IR and background variability, just three observational parameters, namely the duty cycle, variability index and flux density at peak may be plotted as sets of contours on a diagram with axes showing two computational parameters, the maser optical depth under quiescent conditions, and the change in the maser optical depth (or the background intensity) during the flare (Gray *et al.* 2020b). In this way a fit can be found from the crossing places of the contours, leading to a fitted computational light curve. The flux density achieved is used to check for consistency between modelled and observed sizes of the maser objects, or 'clouds', noting that, owing to beaming, observed maser sizes can be considerably smaller than the objects from which the emission arises. This fitting technique has been applied to a spectral component of the flare in G107.298+5.65, and to two components of the flare in S255-NIRS3 with satisfactory results (Gray *et al.* 2020a). Source choice for the fitting procedure was strongly based on the availability of contemporaneous, high-cadence IR or continuum background data, which is vital for establishing a driving function.

It is likely that the maximum flux density and variability index can also be used as key observational parameters in the analysis of other mechanisms. However, in the absence of periodicity, the duty cycle requires replacement with a more general temporal parameter, such as the rise time of the flare or the asymmetry, between rise and decay, of the light curve.

3. The Landscape of Maser Flares

The variability index achievable by a particular flare mechanism varies considerably, from tens up to tens, or even hundreds, of millions. A strong controlling factor is the relative importance of supplying fresh inverted molecules to the system, and simple geometry. Models that rely on increasing the overall number density of inverted molecules over a significant proportion of a cloud tend to generate more powerful flares than those that are primarily geometric, and operate by increasing the column density of inverted molecules in selected directions. For example, variability of a radiative pump increases the inverted number density, as does the shock mechanism (which also selects for high column-density directions). Both of these can be very powerful; by contrast, rotors are purely geometric, and provide comparatively unspectacular flares. Line of sight overlaps also increase only the column density of inverted molecules at selected directions and velocities. The CVR mechanism appears, perhaps, not to fit this pattern, but it effectively provides a continuingly increasing number of inverted molecules from a velocity range that would normally be inaccessible, producing secondary gain in the maser (Field *et al.* 1994) until the catastrophe point is reached in velocity separation between the centres of the radiation line and molecular response. The variable background mechanism is an outlier in terms of the shape of the flare light curves it produces. It does change the number density of inverted molecules, but does so solely through saturation.

Given the variety of mechanisms considered in Section 1.1 to Section 1.6 above, we have attempted to use the different characteristic light curve shapes, timescales and variability indices to construct a diagnostic chart for deducing the underlying mechanism of each maser flare from a small number of observational parameters. Initially, we consider a plot of variability index against timescale for the mechanisms where periodicity is unlikely, in Figure 4, noting that the IR pump and background mechanisms that have been omitted for simplicity can appear anywhere on the plot to the right of the prohibited band, and

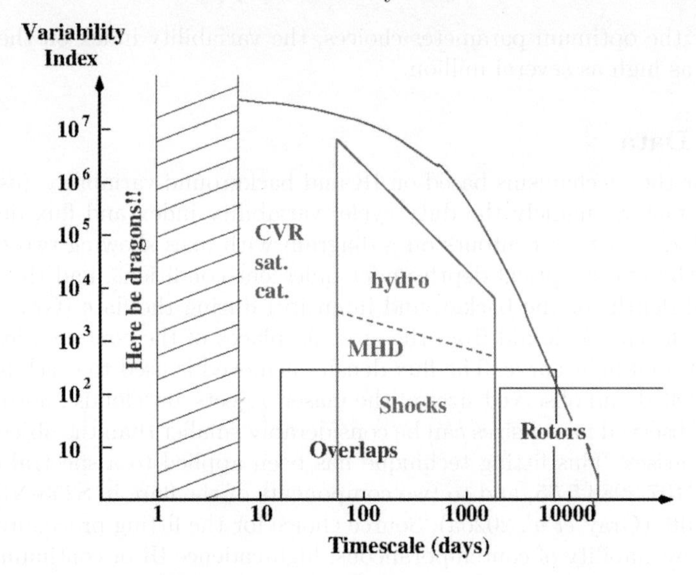

Figure 4. Zones in the variability index / timescale plane accessible to the shock, CVR, overlap and rotor mechanisms. The shaded region marks an approximate boundary between timescales accessible to the current model, and shorter times that would require a fully time-dependent model of the maser with simultaneous coupling to the gas dynamics.

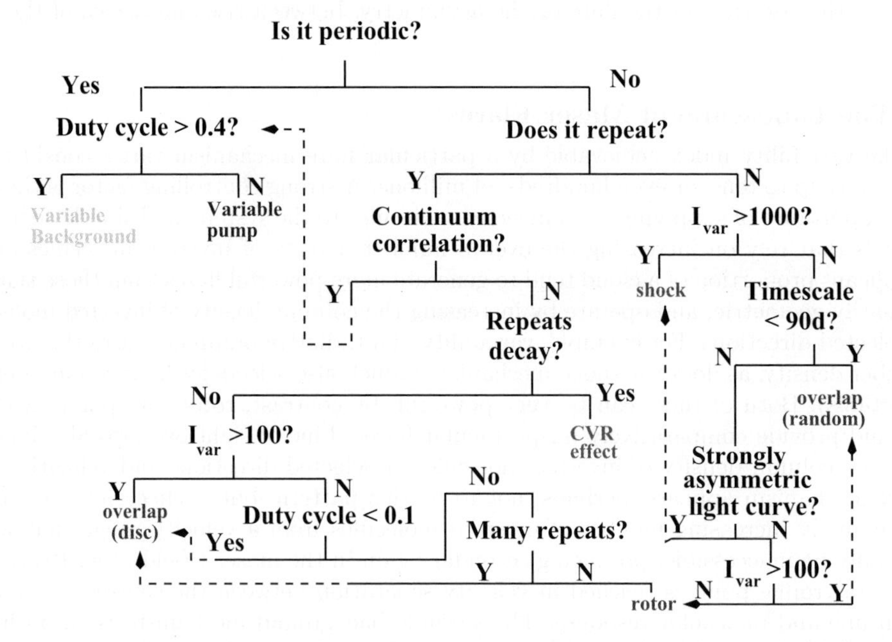

Figure 5. Our first attempt at a chart designed to identify the cause of a maser flare from observational properties of the light curve. The most fundamental choice is between periodic and non-periodic flares.

with suitable variability indices. Timescales for these two mechanisms are controlled by the driving functions.

With the loci of each mechanism in Figure 4 in mind, and an initial division into periodic and non-periodic sources, we have devised the diagnostic chart in Figure 5 that utilizes variability index (I_{var}) values, timescales and certain special features of light curves, such as the duty cycle for periodic sources, in order to derive flare mechanisms from observational parameters.

4. Conclusions

Details of the shape of the rise and decay in the light curve of a maser flare provide useful diagnostics of the flare mechanism. In many cases, a small number of parameters, the variability index, maximum observed flux density, and a temporal parameter, are sufficient to identify the mechanism responsible. The most powerful flares are generated by mechanisms that change the number density of the maser molecule in addition to geometrically increasing the column density along certain lines of sight, and optimum conditions include a maser that is unsaturated in the quiescent state, but becomes saturated during the flare.

Accurate analysis depends crucially on the availabiliy of high-cadence, multi-transition monitoring, as provided by the M2O organization, and also on the availability of contemporaneous continuum data in the radio and IR wavebands; the *NEOWISE* satellite mission (Nugent *et al.* 2015) deserves particular mention in this respect.

References

Bhardwaj, A., Kanbur, S. M., Marconi, M., Rejkuba, M., *et al.* 2017, *MNRAS*, 466, 2805
Burns, R. A., Kobak, A., Garatti o Caratti, A., *et al.* & Maser Monitoring Organization (M2O) 2022, European VLBI Network Mini-Symposium and Users' Meeting 2021, 12–14 July, 2021, 19
Cesaroni, R. 1990, *A&A*, 233, 513
Clegg, A. W. & Cordes, J. M. 1991, *ApJ*, 374, 150
Field, D., Gray, M. D., & de St. Paer, P. 1994, *A&A*, 282, 213
Goedhart, S., Gaylard, M. J., & van der Walt, D. J. 2004, *MNRAS*, 355, 553
Goedhart, S., Langa, M. C., Gaylard, M. J., & van der Walt, D. J. 2009, *MNRAS*, 398, 995
Goldreich, P. & Kwan, J. 1974, *ApJ*, 190, 27
Gray, M. D., Baggott, J., Westlake, J., & Etoka, S. 2019, *MNRAS*, 486, 4216
Gray, M. D., Etoka, S., & Pimpanuwat, B. 2020a, *MNRAS*, 498, L11
Gray, M. D., Etoka, S., Travis, A., & Pimpanuwat, B. 2020b, *MNRAS*, 493, 2472
Inayoshi, K., Sugiyama, K., Hosokawa, T., Motogi, K., & Tanaka, K. E. I. 2013, *ApJ*, 769, L20
Kaufman, M. J. & Neufeld, D. A. 1996, *ApJ*, 456, 250
Kobak, A., Bartkiewicz, A., Szymczak, M., *et al.* 2023, *A&A*, 671, A135
MacLeod, G. C., Smits, D. P., Goedhart, S., Hunter, T. R., *et al.* 2018, *MNRAS*, 478, 1077
Moscadelli, L., Sanna, A., Goddi, C., Walmsley, M. C., *et al.* 2017, *A&A*, 600, L8
Nugent, C. R., Mainzer, A., Masiero, J., Bauer, J., *et al.* 2015, *ApJ*, 814, 117
Olech, M., Szymczak, M., Wolak, P., Gérard, E., & Bartkiewicz, A 2020, *A&A*, 634, A41
Parfenov, S. Y. & Sobolev, A. M. 2014, *MNRAS*, 444, 620
Rajabi, F., Houde, M., Bartkiewicz, A., Olech, M., Szymczak, M., & Wolak, P. 2019, *MNRAS*, 484, 1590
Sobolev, A. M. 1993, Astrophysical Masers: Proceedings of the Conference, Arlington, VA, March 9–11, 1992, 219
Stecklum, B., Caratti o Garatti, A., Hodapp, K., Linz, H., Moscadelli, L., & Sanna, A. 2018, Astrophysical Masers: Unlocking the Mysteries of the Universe, volume 336 of IAU Symposium, 37
Sugiyama, K., Nagase, K., Yonekura, Y., Momose, M., *et al.* 2017, *PASJ*, 69, 59
Tanabe, Y., Yonekura, Y., & MacLeod, G. C. 2023, *PASJ*, 75, 351
van der Walt, D. J., Goedhart, S., & Gaylard, M. J. 2009, *MNRAS*, 398, 961
van der Walt, D. J., Maswanganye, J. P., Etoka, S., Goedhart, S., & van den Heever, S. P. 2016, *A&A*, 588, A47
Watson, W. D. & Wyld, H. W. 2000, *ApJ*, 530, 207

Cosmic Masers: Proper Motion toward the Next-Generation Large Projects
Proceedings IAU Symposium No. 380, 2024
T. Hirota, H. Imai, K. Menten, & Y. Pihlström, eds.
doi:10.1017/S1743921323002429

A comprehensive model of maser polarization

Boy Lankhaar

Department of Space, Earth and Environment, Chalmers University of Technology, Onsala Space Observatory, 439 92 Onsala, Sweden. email: lankhaar@chalmers.se

Abstract. Maser polarization observations have been successfully used to characterize magnetic fields towards a variety of astrophysical objects. Circular polarization yields the magnetic field strength of the maser source, and linear polarization yields information on the magnetic field morphology. Linear polarization can be produced when the maser saturates or through its anisotropic pumping. We present a comprehensive model of the polarization of masers. In contrast to regular excitation modeling, we relax the assumption of isotropically populated level populations, and model both the total population and level alignments. Through this approach, we obtain quantitative estimates on the anisotropic pumping of a variety of maser sources. In this way, the maser polarization may be related to the gas density, temperature, geometry and the magnetic field. Using the results of our modeling, we discuss, and give predictions, of the polarization of SiO, methanol, and water (mega)masers.

Keywords. masers, polarization, magnetic fields, stars: formation, stars: evolved

1. Introduction to maser polarization

Masers are prone to produce polarized radiation. While circular polarization is produced through the Zeeman effect, linear polarization is produced through the interaction of the maser molecules with the directional radiation field that the maser gives rise to. While maser polarization is a powerful tool to trace magnetic field structures in regions and at densities where no other tracers are available, modeling efforts have not been able to fully explain the features of the emergence of this polarization in a comprehensive model.

It is usual to represent the maser system as a two-level system. Each level is split up into its magnetic sublevels, that may be shifted in energy from each other by the Zeeman effect. A level of angular momentum j is subsequently split up into $2j+1$ magnetic sublevels, that are shifted in energy $E_{jm} = E_j + \mu_N B g_j m$, where μ_N is the nuclear magneton, B is the magnetic field, and g_j is the level specific g-factor. Note that we are modeling a diamagnetic molecule in this example; for a paramagnetic molecule, the Zeeman effect would scale with the Bohr magneton μ_B. The shift in energy of the magnetic sublevels of the maser transitions due to the Zeeman effect subsequently shifts the frequency associated with the magnetic subtransitions of the maser transition $\nu(m,m') = \nu_0 + \frac{\mu_N}{h} B[g_j m - g_{j'} m']$, where ν_0 is the transition frequency at $B=0$, and j' is the lower level, with associated magnetic quantum number m'. Selection rules allow only for the $\Delta m = m - m' = \pm 1, 0$ transitions that are the $\sigma^{\pm}$ and π^0 transitions. When the magnetic field is along the line-of-sight, the $\sigma^{\pm}$ are the only allowed transitions and are associated with an adverse circular polarization. Then, the circular polarization, $V \propto \sigma^+ - \sigma^-$, or following (Lankhaar & Teague 2023), $V \simeq \frac{\mu_N}{h} \bar{g} B_{\mathrm{los}} \frac{dI}{d\nu} = \frac{z B_{\mathrm{los}}}{2} \frac{dI}{d\nu}$, where I is the Stokes I total intensity of the maser line, and z is the Zeeman coefficient that is dependent on the transition g-factor, $\bar{g}$, that may be computed from the upper and lower level g-fators (see, Vlemmings *et al.* 2020 for the procedure). The relation between the circular polarization, the z-factor and the line-of-sight component of the magnetic field has been useful to derive magnetic field strengths in star-forming regions (e.g. Lankhaar *et al.* 2018), towards evolved stars (e.g. Vlemmings *et al.* 2006) as well as in OH megamaser galaxies (e.g. Robishaw *et al.* 2008).

When the magnetic field is directed with an angle with respect to line-of-sight, both the $\sigma^{\pm}$ and the π^0 are allowed transitions. The sum of the $\sigma^{\pm}$ transitions add up to match the line-strength of the π^0 transitions. Since $\sigma^{\pm}$ and π^0 transitions are associated with linear polarization, respectively, perpendicular and parallel to the plane-of-the-sky component of the magnetic field, one expects for small Zeeman splittings (with respect to the line-width), that the total linear polarization averages to zero. This is indeed the case if the magnetic sublevels of each maser state are equally populated. However, when magnetic sublevels are populated unequally, a net linear polarization is produced.

An imbalance in the population of magnetic sublevels can be produced for masers in either of two ways. The first case is the saturated maser, whose rate of stimulated emission exceeds the rate of decay of the maser levels. In this case, the different propensities of the $\sigma^{\pm}$ and π^0 transition groups come to expression in the sublevel populations. From both analytical and numerical modeling, it turns out that when the magnetic field projection angle, θ is smaller than the magic angle, or the Van Vleck angle, $\theta_m \approx 54.7^o$, then a net polarization parallel to the magnetic field is produced, while for other angles, polarization perpendicular to the magnetic field is produced. We call the production of maser polarization through this mechanism: maser saturation polarization. The second case is a result of the differential pumping of the magnetic sublevels. In that case, polarization can be produced for any saturation degree. This polarization mechanism is called 'anisotropic pumping'. The relation between the magnetic field and the polarization vector is a function of the anisotropy of the pumping, However, first-principle calculations that quantify anisotropic pumping have not yet been developed, and rather, this mechanism is invoked from phenomenological arguments.

The most salient example of an anisotropically pumped maser are SiO masers excited in the extended atmosphere of evolved stars. The maser transitions occur primarily in the vibrationally excited states, which are excited by the IR radiation from the central star. Since this exciting radiation comes primarily from one direction, i.e. is anisotropic, excitation to the magnetic sublevels is unequal. This naturally leads to anisotropically pumped masers. Similarly, when masers are excited in an anisotropic geometry, such as a shock, the radiation field that is generated by the maser molecule is also anisotropic. Under the right conditions, the anisotropy of the radiation field can significantly manifest in the maser states, thus leading to an anisotropically pumped maser. Collisional interactions often quench the anisotropy on the account that they occur isotropically. Collisions (between neutral species) in a gas with a scalar temperature do not have a preferred direction (Lankhaar & Vlemmings 2020b).

2. A comprehensive model of maser polarization

To move past a phenomenological representation of the anisotropic pumping of masers, we construct a comprehensive model of maser polarization. In it, we divide the maser polarization modeling in two steps. First, we perform an excitation modeling, where we relax the common assumption of equally populated magnetic sublevels. Second, we perform the maser polarization radiative transfer of a two-level maser system (Lankhaar & Vlemmings 2019), where we use the relevant pumping- and decay-operators from our earlier excitation modeling.

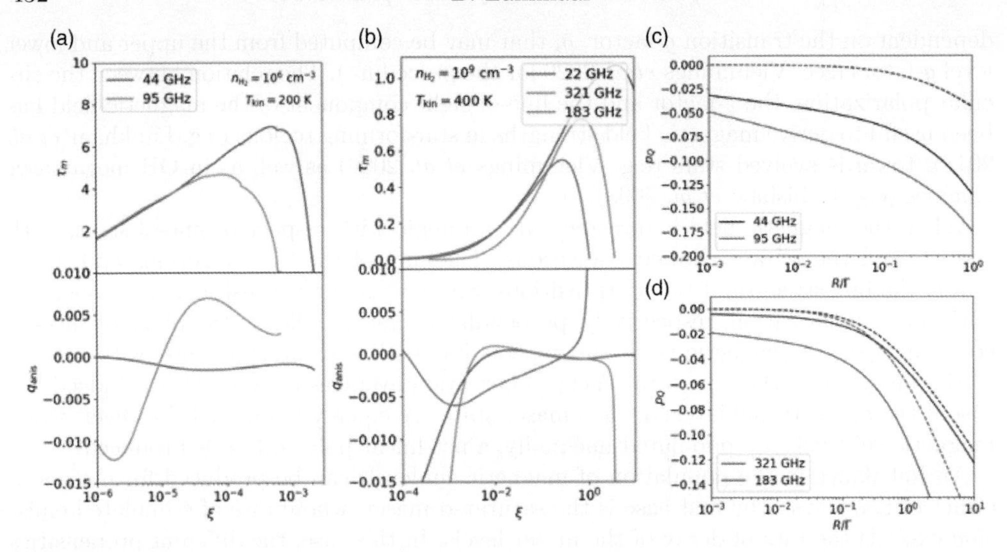

Figure 1. (Anisotropic) pumping parameters of methanol (a) and water (b) masers from polarization resolved excitation modeling. Polarization fractions of methanol (c) and water (d) maser species from polarized radiative transfer simulations using the input of polarization resolved excitation modeling.

To perform the excitation modeling we used PORTAL (Lankhaar & Vlemmings 2020b) in an anisotropic LVG (large velocity gradient) geometry, while fully modeling the magnetic sublevels of all states involved in the excitation, and the two polarization modes of the radiation field. We use an irreducible tensor formalism to afford us a reduction in dimensionality at minimal costs to the accuracy (for a discussion, see Lankhaar & Vlemmings 2020b).

The results of the excitation modeling yield (unsaturated) propagation coefficients that are characterized by isotropic and anisotropic parts. For unsaturated masers, the production of linear polarization may be easily represented by, $p_Q^{\rm anis} \simeq \tanh\left[-\tau_{\rm maser} q_{\rm anis} \sin^2\theta\right] \approx -\tau_{\rm maser} q_{\rm anis} \sin^2\theta$, where the last approximation is valid for polarization degrees $< 10\%$, $\tau_{\rm maser}$ is the maser optical depth and $q_{\rm anis}$ is the maser anisotropy factor. When the maser saturates, this approximation has to be abandoned in favor of rigorous maser polarization modeling, as can be performed with CHAMP (Lankhaar & Vlemmings 2019), that uses the input that can be extracted from the excitation modeling.

3. Simulations

We modeled the excitation of water masers, class I methanol masers, and SiO masers. We modeled all of these species in a plane-parallel like LVG geometry with an aspect ratio of 10, where the velocity gradient follows $\lambda(\mu) = \lambda[\mu^2 + (1-\mu^2)/10]$, where λ is the general velocity gradient and μ is the projection onto the symmetry axis. This is representative for the shock and expanding envelope environments that these masers are excited in. To model the SiO masers, we included the radiation field from a 2000 K stellar object, where we put the maser at 3 stellar radii away from the star.

Methanol masers: 44 GHz polarization vs 95 GHz polarization. There are many class I methanol maser transitions, but in the interest of space, we only compare our results with maser polarization observations of the 44 GHz and 95 GHz methanol masers. Kang *et al.* 2016 quantified the linear polarization of 44 GHz and 95 GHz methanol masers towards high mass star forming regions, and find that for co-spatial masers, the

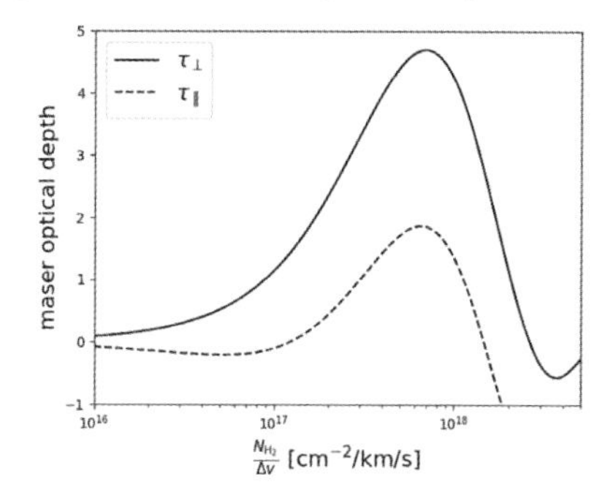

Figure 2. SiO maser optical depth of the radiation component perpendicular and parallel to the magnetic field.

95 GHz transition is more polarized, by a factor up to 3, compared to 44 GHz transition. Since most class I methanol masers are believed to be unsaturated, their polarization is likely due to anisotropic pumping as a result of the shock geometry they are excited in. In Figure 1a, we plot the anisotropic pumping factor, $q_{\rm anis}$, for typical excitation conditions of $T_{\rm kin} = 200$ K and $n_{\rm H_2} = 10^6$ cm^{-3}, for a range of specific column densities (we use the metric ξ for specific column densities, for more discussion, see Leurini *et al.* 2016). We find that the 95 GHz transition has significant anisotropic pumping, while the anisotropic pumping is rather weak for 44 the GHz transition. When we use the results of our excitation analysis as the input for a polarized maser radiative transfer simulation, we find that—for the same degree of saturation, R/Γ—that the 95 GHz maser is about twice as strongly polarized as the 44 GHz maser.

Water masers. Water masers are often found to be significantly polarized. Of the water maser transitions, the 22 GHz maser is generally the strongest, and accordingly most searched for its polarization. Some millimeter and sub-millimeter water transitions have recently been uncovered as masers, of which the 183 GHz and 321 GHz masers have been used as extragalactic circumnuclear accretion disk tracers (e.g. Pesce *et al.* 2023). We compare the polarization of these masers according to our modeling. Figure 1b, shows the anisotropic pumping factor, $q_{\rm anis}$, in our model of the maser polarization for typical excitation conditions of $T_{\rm kin} = 400$ K and $n_{\rm H_2} = 10^9$ cm^{-3}, for a range of specific column densities. We find that while anisotropic pumping is present for both the 321 GHz and 22 GHz masers, it is significantly enhanced for the 183 GHz maser. When we use the results of our excitation analysis as the input for a polarized maser radiative transfer simulation, we find that—for the same degree of saturation, R/Γ—that the 183 GHz maser is polarized stronger than the 22 GHz and 321 GHz masers. This is the result of not only the anisotropic pumping, but also of the low angular momentum of the transition, that enhances the polarization due to saturation effects. We find that the 22 GHz water maser shows similar polarization signature when anisotropic pumping is excluded from the maser radiative transfer.

SiO masers. With polarization degrees up to 100 %, SiO masers are the prototypical example of anisotropically pumped masers. We modeled the excitation of SiO masers for typical excitation conditions of $T_{\rm kin} = 1200$ K and $n_{\rm H_2} = 10^{12}$ cm^{-3}, for a range of specific column densities. Figure 2 shows a comparison between the predicted optical depth of the radiation field mode perpendicular and parallel to the projected magnetic field direction for the $J = 1 \rightarrow 0$ transition. We find a significant difference between the optical depth between both radiation modes. In fact, for a significant range of specific column densities, we find that the perpendicular mode is masing, while the parallel radiation mode is in absorption. Such a configuration predicts 100 % polarization for all degrees of saturation. Additionally, it shows the importance of modeling SiO masers comprehensively, including also the magnetic sublevels in the maser excitation, as large differences in population between them within an energy level can occur.

References

Lankhaar, B., Vlemmings, W. *et al.* 2018, *Nature Astronomy*, 2, 145-150
Lankhaar, B. & Vlemmings, W. 2019, *A&A*, 628, A14
Lankhaar, B. & Vlemmings, W. 2020, *A&A*, 636, A14
Lankhaar, B. & Vlemmings, W. 2020, *A&A*, 638, L7
Vlemmings, W., Lankhaar, B., *et al.* 2019, *A&A*, 624, L7
Lankhaar, B. & Teague, R. 2023, *accepted to A&A*
Vlemmings, W., Diamond, P. & Imai, H. 2006, *Nature*, 440, 58
Robishaw, T., Quataert, E. & Heiles, C. 2008, *ApJ*, 680, 981
Pesce, D., Braatz, J., *et al.* 2023, *ApJ*, 948, 134
Kang, J., Byun, DY., *et al.* 2016, *ApJS*, 227(2), 17
Leurini, S., Menten, K. & Walmsley, C. 2016, *A&A*, 592, A31

Cosmic Masers: Proper Motion toward the Next-Generation Large Projects
Proceedings IAU Symposium No. 380, 2024
T. Hirota, H. Imai, K. Menten, & Y. Pihlström, eds.
doi:10.1017/S1743921323003551

Maser polarization simulation in an evolving star: effect of magnetic field on SiO maser in the circumstellar envelope

M. Phetra[1,2], M. D. Gray[2], K. Asanok[2], B. H. Kramer[2,3], K. Sugiyama[2], S. Etoka[4] and W. Nuntiyakul[5]

[1]Graduate School, Chiang Mai University, Chiang Mai 50200. email: montree_ph@cmu.ac.th

[2]National Astronomical Research Institute of Thailand, Chiang Mai 50180, Thailand

[3]Max Planck Institute for Radio Astronomy, Auf dem Hügel 69, Bonn 53121, Germany

[4]Jodrell Bank Centre for Astrophysics, School of Physics and Astronomy, University of Manchester, M13 9PL, UK

[5]Department of Physics and Materials Science, Faculty of Science, Chiang Mai University, Chiang Mai 50200, Thailand

Abstract. Maser polarization changes during a pulsation in the CSE of an AGB star are related in a complicated way to the magnetic field structure. 43 GHz SiO maser transitions are useful for polarization study because of their relatively simple Zeeman splitting structure and their location. This work uses 3D maser simulation to investigate the effect of the magnetic field on maser polarization with different directions. The results show that linear polarization depends on the magnetic direction while circular polarization is less significant. The EVPA changes through $\pi/2$ at an angle of around 50 degrees, approximately the Van Vleck angle. The EVPA rotation result from 3D maser simulation is consistent with results from 1D simulations, and may explain the 90 degree change of the EVPA within a single cloud in the observational cases of TX Cam and R Cas.

Keywords. AGB star, maser polarization, maser simulation

---

1. Introduction and Method

The pulsation of CSE in an AGB star is a puzzling process, with its properties varying periodically. VLBI observations show that maser intensity and polarization in this region significantly change during pulsation. A model of the pumping process is required to help us understand the physical mechanisms behind the phenomenon of maser variability. Goldreich *et al.* (1973) showed that the magnetic field may be one of the physical conditions controlling polarization because of the Zeeman splitting effect, especially in a molecule with weak Landé splitting factors in its quantum states. The SiO molecule is a good maser polarization candidate because it is located closest to the AGB star with a strong magnetic field present (Vlemmings 2019). Observations of SiO masers at transition mode v=1, J=1-0 (43 GHz) may show a nonuniform polarization and 90-degree flip (Assaf *et al.* 2013; Tobin *et al.* 2019). Here, we consider the effect of magnetic field on maser clouds by using 3D maser polarization simulation which is modified from Gray *et al.* (2018).

The maser simulation starts with a density matrix (DM) as a function of time, position, and Doppler velocity. The DM is used to determine the population of energy levels

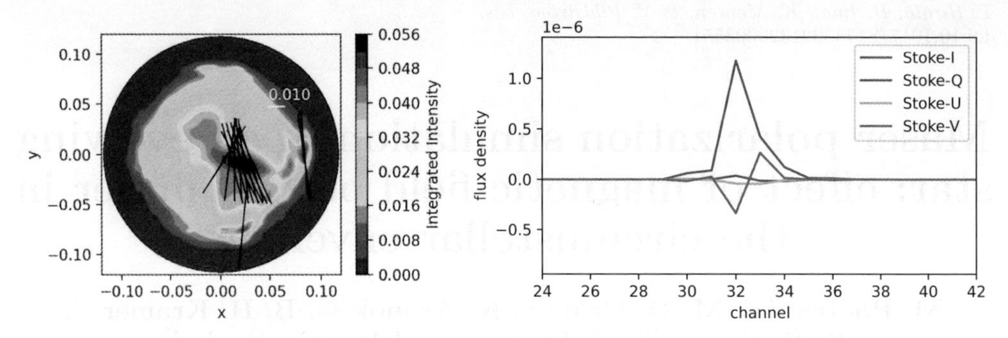

Figure 1. Stokes-I map with EVPA and Spectra of the depth multiplier = 300 for parallel MFD.

which are controlled by a Hamiltonian operator containing a magnetic perturbation, an interaction with the electric field, and an additional decay rate. Then, we use a spherical distribution of background rays, each with 2 electric field components to evolve the DM into an increasingly saturated state by using the radiative transfer equation. Finally, we amplify rays from a plane background that pass through the known DM in the observer's direction for the formal solution as image intensities and spectra.

In this work, we set the domain as a multi-node spread in one cloud with a cylindrical shape to calculate the DM. The magnetic field is applied to the domain in different directions from parallel to perpendicular with the z-axis in steps of 5 degrees. Then we have 19 solutions with different magnetic field directions (MFDs). Then, the depth multiplier is used to optically thicken the model (Phetra *et al.* 2023) until the inversions are close to 0. Finally, formal solutions from the plane background are converted to observer's Stokes parameters.

2. Results

The important simulation parameters are set by the Zeeman splitting as $g\Omega = 0.7405\,\mathrm{s}^{-1}\,(\mathrm{mG})^{-1}$, decay rate as $\Gamma = 5\,\mathrm{s}^{-1}$, the ratio of the background intensity to the saturation intensity as 10^{-6} (Western & Watson 1984), and the magnetic field strength $B = 10^4\,\mathrm{mG}$ (Herpin *et al.* 2006; Diamond *et al.* 1997) (the splitting width become $7405\,\mathrm{s}^{-1}$ and use this value for a channel width). We solve for the DM in a range of depth from 3 to 300 (in steps of 3) using the spherical radiation background with a radius of 5 domain units. For formal solutions, the plane radiation background is behind the domain by 5 units, and radiates towards the distant observer who is 1000 units in front of the domain. The example result for the observer which comes from the tube domain with parallel field is shown in Figure 1 as Stokes-I map with their EVPA and spectra.

In the Stokes-I map, color contours refer to the integrated intensity (saturation units) from all frequency channels, and the EVPA (EVPA $= 0.5\arctan(U/Q)$) of each ray is shown in a black solid line with the length represented by Stokes-I shown at the upper-right of the image, the highest integrated intensity is about 0.056. For spectra, we plot all Stokes parameters at different colors shown in the upper-right of the image. The spectra of Stokes-Q and U look small while Stokes-V clearly shows the s-shape.

The rays from the spherical background in the 3D simulation each have an electric field with a uniformly distributed random phase and a normally distributed random amplitude. This means that random features can propagate through a series of solutions at increasing values of the depth multiplier. Therefore, we make an exponential fit to the integrated flux as a function of the depth multiplier to match the general trend. The fitting parameters are used to calculate the polarization properties consisting of fractional linear polarization ($m_l = \sqrt{Q^2 + U^2/I}$), fractional circular polarization ($m_c =$

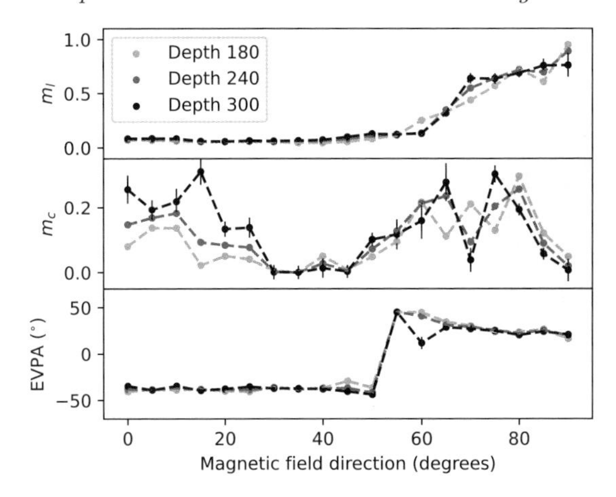

Figure 2. m_l, m_c, and EVPA with their residual values as a function of MFD for a depth multiplier 180, 240, and 300.

$|V|/I$), and EVPA. Polarization properties of each domain, at a set of depths, are plotted as a function of the MFD in Figure 2. The black vertical line at each simulation MFD represents the standard error of the computed value with respect to the fitted exponential. Uncertainty on m_l, m_c, and EVPA were obtained by standard error propagation.

The value of m_l increases rapidly from 0.13 to 0.64 over the MFD range of 60 to 70 degrees while m_c fluctuates in the range of 0.00 to 0.30. We found that the EVPA is suddenly changed at the MFD of 50 to 55 degrees from -44° to 45°.

3. Discussion and Conclusions

We use the 3D maser polarization simulation to explain the effect of MFD in CSE. The preliminary results show that the polarization properties depend on the direction of the magnetic field, as expected from 1D models. When the magnetic field is parallel with z-axis, Stokes-V shows a negative value at the low frequency, and a positive value at high frequency (Nedoluha & Watson 1994). This result agrees with 1D results in Tobin *et al.* (2023).

For the comparison of polarization properties with MFD, we found that m_l is low at low angles and then increases after a MFD of about 60 degrees while the 1D model in Tobin *et al.* (2023) shows that m_l is larger at low angles. We found a nonzero in circular polarization which the fractional is about 0.3, similar to the 1D model with a magnetic field strength of 10 G. However, Watson & Wyld (2001) and Tobin *et al.* (2023) show the dependence of unitless of circular polarization on MFD, θ, as $v/(pB\partial i/\partial v) = \cos\theta$, we need to check this relation base on our 3D simulation results. We found that the fractional polarization results look similar to the observations such as Assaf *et al.* (2012) and Kemball & Diamond (1997). Moreover, we found that the flip of EVPA is dominated by a Stokes-Q flip at the MFD of 50 to 55 degrees which is similar to the result of Goldreich *et al.* (1973) in case 2a.

We made a new domain with 2 different MFDs internally to test whether the EVPA flip in one cloud that has been detected in observations can also be generated by our simulation. The MFDs are set as 15 degrees for the lower part and 75 degrees for upper part of the domain. We compare the result with the EVPA flip observed from TX Cam (Tobin *et al.* 2019) and shown in Figure 3.

We found that the EVPA from the lower part is close to 0° while in the upper part, the direction is about -23°. So, the field variation in this new domain leads to results that are consistent with the observed EVPA change. Georgiev *et al.* (2023) monitors R

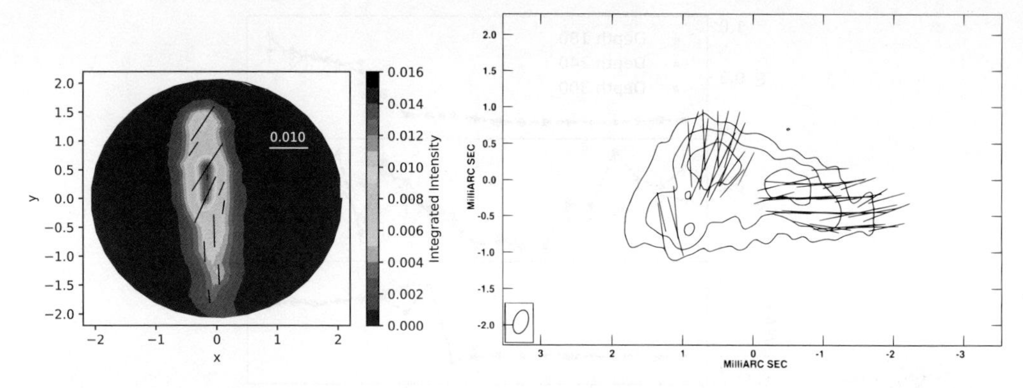

Figure 3. A new domain with different directions of a magnetic field in a single cloud compare with EVPA change in TX Cam (Tobin *et al.* 2019).

Sct in the wavelengths between 375 to 1050 nm and found that the surface magnetic field in the CSE was detected at different phases of pulsation and suggests that this may be caused by the pulsation shock wave. However, the goal of this project is to attempt to model the global magnetic field during a pulsation, so, we need to test more complicated cases, such as multi-clouds, anisotropic pumping, and the motion cloud.

References

Assaf, K. A., Diamond, P. J., Richards, A. M. S., & Gray, M. D. 2012, *IAU Symposium*, 287, 235

Assaf, K. A., Diamond, P. J., Richards, A. M. S., & Gray, M. D. 2013, *MNRAS*, 431, 1077

Diamond, P. J., Kemball, A. J. & Boboltz, D. A., 1997, *Vistas in Astron.*, 41, 175

Georgiev, S., Lèbre, A., Josselin, E., Mathias, P., Konstantinova-Antova, R., & Sabin, L. 2023, *MNRAS*, 522, 3861

Goldreich, P., Keeley, D. A., & Kwan, J. Y. 1973, *ApJ*, 179, 111

Gray, M. D., Mason, L., & Etoka, S. 2018, *MNRAS*, 477, 2628

Herpin, F., Baudry, A., Thum, C., Morris, D., & Wiesemeyer, H. 2006, *A&A*, 450, 667

Kemball, A. J., & Diamond, P. J. 1997, *ApJ*, 481, L111

Nedoluha, G. E., & Watson, W. D. 1994, *ApJ*, 423, 394

Phetra, M., Gray, M. D., Asanok, K., Kramer, B. H., Sugiyama, K., Chanapote, T., & Nuntiyakul W. 2023, *J Phys Conf Ser*, 2431, 012088

Tobin, T. L., Kemball, A. J. & Gray, M. D. 2019, *ApJ*, 871, 189

Tobin, T. L., Gray, M. D. & Kemball, A. J. 2023, *ApJ*, 943, 123

Vlemming, W. 2019, *IAU Symposium*, 343, 19

Watson, W. D. & Wyld, H. W. 2001, *ApJ*, 558, L55

Western, L. R., & Watson, W. D. 1984, *ApJ*, 285, 158

Review talk in the session Theory of Masers and Maser Sources by Martin Houde. Taken by Tomoya Hirota.

Chapter 7
New Projects and Future Telescopes

Cosmic Masers: Proper Motion toward the Next-Generation Large Projects
Proceedings IAU Symposium No. 380, 2024
T. Hirota, H. Imai, K. Menten, & Y. Pihlström, eds.
doi:10.1017/S1743921323003381

Overview of the Maser Monitoring Organisation

Ross A. Burns[1,*], Agnieszka Kobak, Alessio Caratti o Garatti, Alexander Tolmachev, Alexandr Volvach, Alexei Alakoz, Alwyn Wootten, Anastasia Bisyarina, Andrews Dzodzomenyo, Andrey Sobolev, Anna Bartkiewicz, Artis Aberfelds, Bringfried Stecklum, Busaba Kramer, Callum Macdonald, Claudia Cyganowski, Fransisco Colomer, Cristina Garcia Miro, Crystal Brogan, Dalei Li, Derck Smits, Dieter Engels, Dmitry Ladeyschikov, Doug Johnstone, Elena Popova, Emmanuel Proven-Adzri, Fanie van den Heever, Gabor Orosz, Gabriele Surcis, Gang Wu, Gordon MacLeod, Hendrik Linz, Hiroshi Imai, Huib van Langevelde, Irina Valtts, Ivar Shmeld, James O. Chibueze, Jan Brand, Jayender Kumar, Jimi Green, Job Vorster, Jochen Eislöffel, Jungha Kim, Koichiro Sugiyama, Karl Menten, Katharina Immer, Kazi Rygl, Kazuyoshi Sunada, Kee-Tae Kim, Larisa Volvach, Luca Moscadelli, Lucas Jordan, Lucero Uscanga, Malcolm Gray, Marian Szymczak, Mateusz Olech, Melvin Hoare, Michał Durjasz, Mizuho Uchiyama, Nadya Shakhvorostova, Olga Bayandina, Pawel Wolak, Sergei Gulyaev, Sergey Khaibrakhmanov, Shari Breen, Sharmila Goedhart, Silvia Casu, Simon Ellingsen, Sonu Tabitha Paulson, Stan Kurtz, Stuart Weston, Tanabe Yoshihiro, Tim Natusc, Todd Hunter, Tomoya Hirota, Willem Baan, Wouter Vlemmings, Xi Chen, Yan Gong, Yoshinori Yonekura, Zsófia Marianna Szabó and Zulema Abraham

[1]RIKEN Cluster for Pioneering Research, Wako-shi, Saitama, 351-0198, Japan.
email: rossburns88@googlemail.com

[*] The Maser Monitoring Organisation

Abstract. The Maser Monitoring Organisation is a collection of researchers exploring the use of time-variable maser emission in the investigation of astrophysical phenomena. The forward directed aspects of research primarily involve using maser emission as a tool to investigate star formation. Simultaneously, these activities have deepened knowledge of maser emission itself in addition to uncovering previously unknown maser transitions. Thus a feedback loop is created where both the knowledge of astrophysical phenomena and the utilised tools of investigation themselves are iteratively sharpened. The project goals are open-ended and constantly evolving, however, the reliance on radio observatory maser monitoring campaigns persists as the fundamental enabler of research activities within the group.

Keywords. maser emission, M2O, radio astronomy, long-term monitoring

---

1. Purpose and early beginnings

Astrophysical masers are excellent tracers of physical environments due to the sensitive relationship between maser emission brightness and physical parameters such as temperature, density and collision rates (Cragg *et al.* 2005; Hollenbach *et al.* 2013). Addressing this, several radio astronomy observatories have been conducting long-term monitoring of maser emission in order to uncover changes in the physical conditions of various astronomical phenomena, as a function of time (Volvach *et al.* 2020; Gaylard *et al.* 2002; Szymczak *et al.* 2018; Yonekura *et al.* 2016; Brand *et al.* 2007; Tanabe *et al.* 2023).

In the not-so-frequent astronomical conferences centered on maser emission, the fruits of labor of long-term monitoring campaigns have been discussed, and, occasionally, transient maser flare events which occurred in their data sets are reported. Unfortunately however, many of these flares were presented long after the subsiding of the event itself when it was too late to conduct follow-up observations to gather more information about the flare and its driver. The situation, brought to light in the IAUS 336 in Cagliari, Italy 2017, was determined to be worth addressing and in doing so the Maser Monitoring Organisation (M2O) was formed, with the goal of providing a communication platform - bringing maser monitoring observatories together with teams wishing to pursue follow-up investigations of time-domain maser activity. The membership and scientific scope of the M2O has expanded well beyond its original scope. Despite this, the back-bone of all M2O activities is derived from the reporting of maser flare events from long-term monitoring campaigns of radio astronomy observatories.

2. Operations

At its core, the M2O is a communications platform connecting maser monitoring observatories with follow-up facility users, and star-formation and maser experts. This overview proceedings will walk through the typical activities and resource management considerations that the M2O follows in relation to maser flare alerts.

2.1. *Monitoring activities*

Monitoring activities are reported to the M2O from radio astronomy observatories such as Ibaraki, Torun, Irbene, Medicina, VERA, Hartebeesthoek, Simeiz and Pushchino, which have their own long-term maser monitoring campaigns and independent research targets such as monitoring periodic sources. These observatories agree to alert the M2O to transient flare events as an additional activity. On the other hand, dedicated maser monitoring activity has also been initiated as part of the M2O effort, as in the case of the Parkes (P1073) project. Additionally, some stations contribute follow-up-like flare confirmation observations and short term, intensive monitoring in response to flare alerts. Effelsberg, Warkworth and Kuntunse are in this category. 14 single dish observatories (left panel of Figure 1) are contributing to M2O activities. Their cumulative monitored source lists (right panel of Figure 1) comprise: 562 targets at the 6.7 GHz methanol transition, 260 at the 22 GHz water transition, and 65 at various L-band hydroxyl transitions.

2.2. *Flare statistics*

At the time of the IAUS 380 symposium, and since inception at the IAUS 336 symposium, the M2O has received reports of 11 flares from 22 GHz water maser sources and 8 flares from 6.7 GHz methanol maser sources. This corresponds to a discovery rate of about 1 or 2 maser flare events per year (left panel of Figure 2).

2.3. *Establishing context*

Once a flare is reported, before deciding if/how to pursue follow-up observations, it is important to establish some context about the event in order to evaluate the appropriate amount of observing resources to commit. The first source of context comes from the

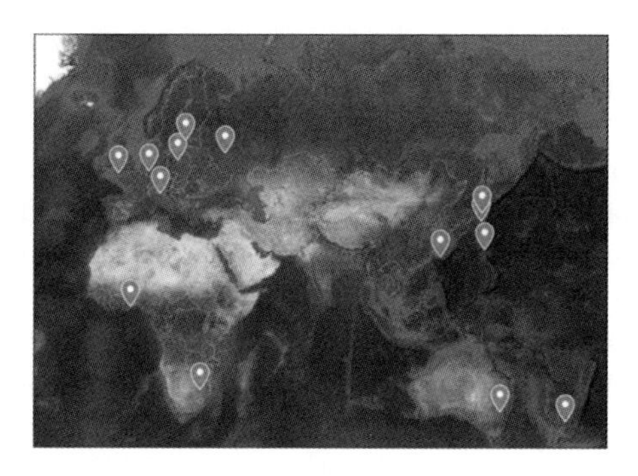

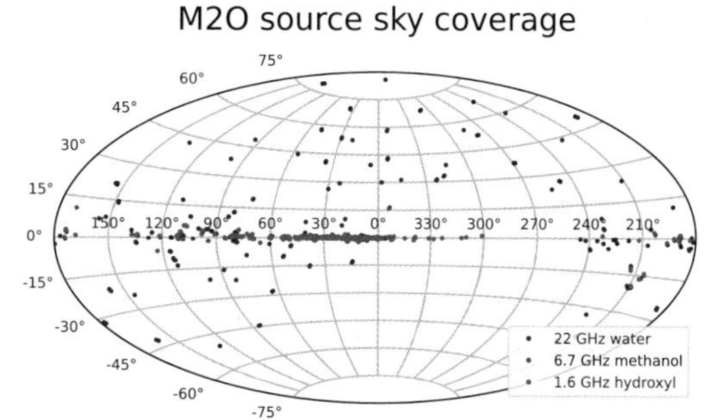

Figure 1. *Left:* Global distribution of maser monitoring observatories that have contributed monitoring data to the M2O. *Right:* Distribution of maser targets monitored by the cumulative of M2O participating stations. Up-to-date versions of these maps are accessible in the M2O website.

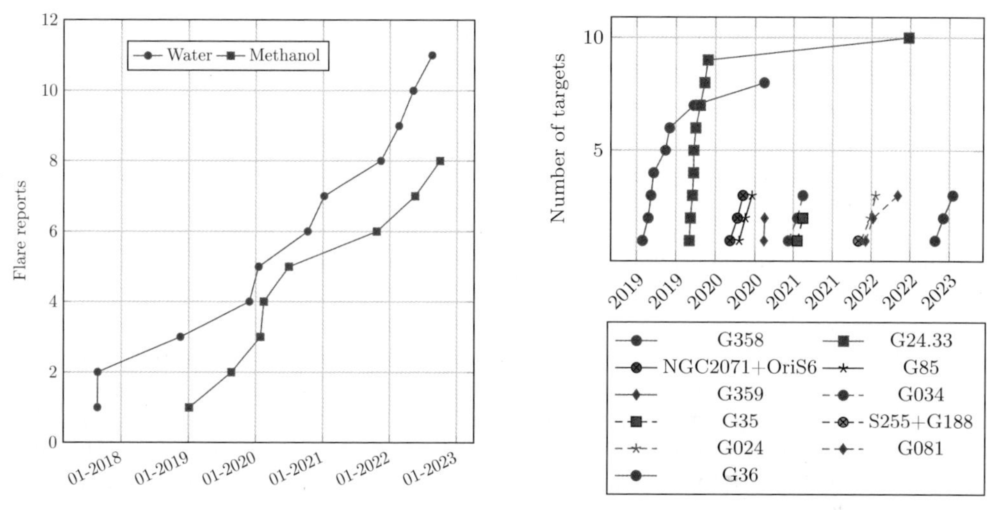

Figure 2. *Left:* Cumulative plot of the detections of 22 GHz water masers and 6.7 GHz methanol masers reported to the M2O as a function of time. *Right:* Cumulative plots of the VLBI follow-up observations conducted for maser flares in various sources.

maser itself. A rich spectrum may indicate that a large portion of the region exhibits the appropriate physical conditions to produce maser emission. Occasionally, a rich spectrum may indicate multiple progenitors within the same beam - especially telling if there are many masers at velocity separations of more than an order of magnitude, since methanol masers typically are within $\sim$ 5 km s^{-1} of their driving source (Green and McClure-Griffiths 2011).

Temporal variations then provide an additional dimension of information. Flares of only a single velocity feature may be caused by a maser superposition (Burns *et al.* 2020a) rather than any changes in physical conditions. Enhancements in the brightness of a small fraction of the maser velocity features may indicate physical changes affecting only a small region. On the other hand, accretion bursts are identified by an increase in all maser spectral features and accompanying ignition of emission in previously empty parts of the spectrum (MacLeod *et al.* 2018a). This first evaluation stage can set the tone for the later steps in looking at a maser flare.

Next, the Maser Database ("MaserDB", https://maserdb.net; Ladeyschikov *et al.* 2019) is consulted. This fantastic web search tool catalogs historical records of observational results, both detections and non-detections, of maser observations across a variety of maser transitions. The database can be queried using coordinates, names, in addition to offering filtering tools. In the case of M2O operations, the MaserDB can quickly provide context on the typical flux of a flaring maser in historical data, in addition to serving information on which other maser transitions have or have not been detected in the same target. For example, a 6.7 GHz methanol maser flare target may be revealed by the MaserDB to also have hydroxyl and water masers. Including these transitions in follow-up imaging by requesting multi-band observations can provide additional context such as the locations of shocks and jets in the region, which in turn help to interpret the flaring 6.7 GHz data. Good examples of this can be seen in the cases of G25.65+1.05 (Bayandina *et al.* 2019, 2023), G358.93−0.03 (Bayandina *et al.* 2022a,b) and G24.33+0.14 (Kobak *et al.* 2023).

Providing that the maser flare and progenitor seem interesting at the aforementioned stages then interest will likely lead to direct observations with more radio observatories of some common maser transitions, in addition to initiating high-cadence monitoring of flaring transitions. The participation of multiple observatories at this stage provides maser spectrum acquisition for a large number of maser transitions owing to the varieties of receivers equipped at each radio observatory. In addition, flaring behavior can be independently confirmed at this stage.

2.4. *Follow-up observations*

When a monitoring observatory identifies and reports a new maser flare, the decision whether or not to pursue follow-up observations, and if so, which, and what amount of resources to allocate, is made on a case-by-case basis. A crude visualisation of the allotment of VLBI follow-up resources can be seen in Figure 2, *right.* Generally most events deemed worth following up with VLBI are given three observation epochs in order to provide proper motion measurements of masers, in addition to spatially and spectrally isolating flaring features and identifying any maser emitting regions that are created/destroyed during the observing campaign. This information greatly helps in interpreting the structures and motions which are generating the maser emission.

Generally, the 6.7 GHz methanol maser flares are the most highly prised by the community since they are rarer and potentially indicate that a high-mass protostellar accretion burst is occurring (Fujisawa *et al.* 2015; Brogan *et al.* 2018; Moscadelli *et al.* 2017; Szymczak *et al.* 2018; MacLeod *et al.* 2018b, 2019). As previously mentioned, if a target exhibiting a flare in the 6.7 GHz methanol maser also exhibits other maser transitions, these other transitions are also included in follow-up observations to provide context at similar angular resolutions.

Comparatively, fewer of the 22 GHz water maser flares are followed up. This is mainly to retain observational resources which may need to be allocated to following up 6.7 GHz methanol maser flares, i.e. interest in following up water maser flares suffers from the basic principle of supply-and-demand.

In the case that a large scale follow-up campaign is to be initiated, the timing will depend on the precision of the knowledge of the maser's coordinates. Initiating high-resolution VLBI observations at the wrong coordinates can lead to potentially wrecked observations and time wasted. If previous VLBI-derived coordinates are known (these will be checked in the MaserDB) then this problem is avoided. In cases where precise coordinates are not available the VLA is an excellent instrument for constraining the maser location thanks to its large field of view, sub-arcsecond resolution, and ability to observe at several frequency bands in the same observing session (Bayandina *et al.* 2022a,b). The VLA is also less susceptible than VLBI to fatally resolving out too much of the maser emission at super milliarcsecond scales.

2.5. *Follow-up resources*

In order to obtain these observational follow-up data sets and respond quickly to new reports of maser flares, the M2O maintains triggerable target of opportunity (ToO) proposals at a number of facilities. By doing so, such pre-graded proposals can be activated once a trigger condition is met. Our trigger conditions relate to brightness increases of maser transitions.

We endeavor to maintain triggerable ToO proposals on all major VLBI arrays, in addition to shorter baseline interferometers such as the VLA, SMA and ALMA. Our observational resources in the infrared domain has included triggerable proposals on the JWST, Subaru and an ongoing partnership with the JCMT Transients team. Their project pursues long-term, regular infrared monitoring of 14 star forming regions, ten of which contain M2O targets.

2.6. *Interpretation and publication*

The diverse membership of observers, star formation researchers, data reducers and maser theorists, all share access to data, results, and a communications platform where interpretations can be presented and discussed. These discussions guide the path to conclusions about the observed phenomena in their appropriate context, either progressing ultimately toward publication of findings, or instead leading to further observations to test inconclusive hypotheses. As a rule, key members representing any monitoring station that reports a maser flare event to the M2O are subsequently included as co-authors in all publications following up the flare event. This is requested as a form of acknowledgement to maser monitoring observatories for promptly providing an essential information basis upon which all subsequent follow-up investigations are built.

At the time of the IAUS 380 symposium, and since inception at the IAUS 336 sympoisum, the M2O has 23 peer-reviewed journal publications which can be found listed on the M2O website at https://www.masermonitoring.com/#publications.

3. Case study: G358.93−0.03−MM1

Much of the aforementioned aspects of the M2O collaboration were exemplified in the identification and follow-up campaign of a 6.7 GHz methanol maser flare in high-mass star forming region G358.93−0.03.

Monitoring The maser source was one of the 488 targets monitored by the 32 m Hitachi radio telescope which is operated by Ibaraki University under the 'iMet' long-term monitoring campaign (Yonekura *et al.* 2016). A flare in the source was identified and promptly reported to the M2O via the group's mailing list and to the *astronomer's telegram* (Sugiyama *et al.* 2019). Flaring activity was seen in all of the existing maser features, in addition to the appearance and flaring of maser emission in previously empty regions of the spectrum (Breen *et al.* 2019; Brogan *et al.* 2019; MacLeod *et al.* 2019; Chen *et al.* 2020a,b), overall indicating sudden large-scale changes in physical conditions.

Establishing context Checking the entry for G358.93-0.03 in the MaserDB revealed that, prior to the excitement stoked by the accretion burst, not much was known about this star forming region aside from a few < 10 Jy maser detections and survey data (see https://maserdb.net/object.pl?object=G358.93-0.03.

To establish further context the target was observed with a large number of radio observatories and frequency bands, confirming flare activity at the 6.7 GHz transition and other methanol maser transitions, and other molecular species such as hydroxyl, prompting the mobilisation of a large-scale follow-up campaign.

Follow-up The extraordinary breadth of, and brightness of maser flaring was sufficient to gain the approval of observing time on a variety of astronomical facilities via requests of Target of Opportunity and Director's Discretionary Time (DDT). The findings of the initial follow-up and monitoring efforts influenced the observational approaches of subsequent follow-ups, such as by identifying new maser transitions which would later be mapped at ever increasing angular resolutions.

Observation time was acquired on the following instruments as part of the coordinated follow-up campaign: ATCA, VLA, SMA, ALMA, VERA, VLBA, EVN, KVN, LBA, AusSCOPE, GROND, SOFIA. Multiple epochs of data were acquired for many of these instruments. Beyond the countless single-dish radio telescope monitoring observations, the above facilities conducted more than 20 epochs of follow-up observations of G358.93−0.03.

Interpretation and publication Newly discovered maser transitions were being classified by single dish observatories (MacLeod *et al.* 2019) and interferometers (Chen *et al.* 2020a,b; Brogan *et al.* 2019; Breen *et al.* 2019), and having their brightness temperatures measured. With information from multiple molecular species and multiple maser transitions within species, attempts were made to interpret the physical conditions necessary to drive the observed maser line ratios. However, the methanol line ratios were exceptionally divergent from expectations from other star forming regions and model calculations (Cragg *et al.* 2005).

ALMA was able to identify eight millimeter cores in the region, (G358.93−0.03 MM1 through MM8) with G358−0.03 MM1 (Hereafter "G358−MM1") as the progenitor of the maser flare. A velocity gradient was revealed across the core. These observations also contributed 14 new maser discoveries in the millimeter regime (Brogan *et al.* 2019). Infrared observations confirmed that continuum dust emission had risen, pointing to the occurrence of an accretion burst (Stecklum *et al.* 2021). VLA observations produced the first maps of some of the new maser species discovered by single dish observatories and ATCA, while also uncovering sub-structures in the postulated protostellar disk which had given rise to the observed velocity gradient (Chen *et al.* 2020a,b; Bayandina *et al.* 2022a). VLBI observations traced rings of 6.7 GHz methanol masers at ever-increasing radii from the protostar, indicating that heat produced in the accretion process was traversing outward through the disk and igniting maser emission along its way (Burns *et al.* 2020b). These rings were later concatenated into a single image to provide sparse sampling of the disk at milliarcsecond resolution, revealing Keplerian rotation and a 4-arm spiral structure (Figure 3, Burns *et al.* 2020b).

Despite such a rich trove of discoveries, the entire data acquisition stage that produced these results was completed within a few months of the flare report arriving at M2O communications. It was the sharing of fresh results that provided essential context and valuable guidance which enabled subsequent discoveries to be made. Unrestricted collaboration lead to the most detailed observational account of a high-mass protostellar accretion burst to date.

To date, 11 peer-reviewed journal works have been published on the G358−MM1 flare by the M2O team, all of which crediting key members from the Ibaraki University maser monitoring programme who first provided the maser flare alert (Sugiyama *et al.* 2019). The flow of operations outlined in these proceedings, which generally describe the investigative process, for G358−MM1 and other targets, is visually summarised in Figure 4.

4. Latest developments

4.1. *The VIRAC single baseline interferometer*

In addition to cooperating with already established maser monitoring observatories and follow-up observation facilities, recent efforts have also been invested into observatory

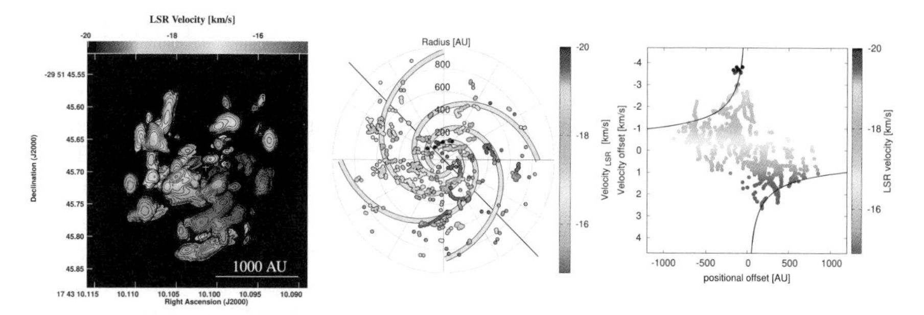

Figure 3. Figure reproduced from Burns *et al.* (2023); *Left*: Zeroth (contours) and first (colours) moment maps of the 6.7 GHz methanol maser emission in G358−MM1 which was created by combining VLBI maps. The white cross indicates the position of the G358−MM1 millimeter core (Brogan *et al.* 2019). *Middle*: Spotmap of the VLBI data sets centered on the G358−MM1 position. The black line indicates the direction of largest velocity gradient to which a position-velocity cut was taken. The spiral arms identified in Burns *et al.* (2023) are plotted as thick grey lines. *Right*: Position-velocity diagram of the maser spots. A Keplerian function for a 11.5 $M_\odot$ enclosed mass is shown as a black line.

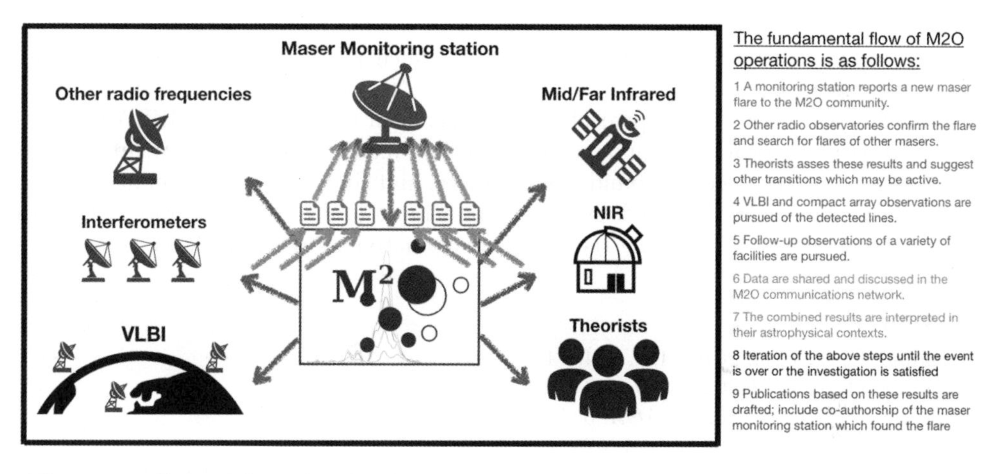

Figure 4. General flow of an M2O response to a new maser flare, coloured arrows match coloured text.

development. These efforts are necessary for addressing the very specific needs of some areas of research into star formation, specifically relating to cadence. Occasionally the observing configurations and cadences offered by established, open call observing facilities is not sufficient, and alternative route must be found or created.

One example is the single-baseline interferometer at the Irbene Radio Observatory. The specific need which this facility addresses can be explained as follows. After an accretion burst it is hypothesised that a portion of in-falling material is launched in the form of a jet of plasma from the central protostellar disk region. This hypothesis has been confirmed observationally in one high-mass protostar; S255IR-NIR3, in which a detectable increase of radio emission extending perpendicular to the disk was observed (Cesaroni *et al.* 2018). However, one limitation of the study was a low precision on the initiation of the continuum flux increase start time. A higher observing cadence could provide more details about the exact delay between the accretion event and jet launching. Additionally, the rate of increase in radio emission from ejected plasma may provide clues to the geometric distribution and flow rate of jet gas.

In order to monitor radio continuum emission with high sensitivity and high cadence, the Virac single-baseline interferometer in Irbene has stood out as an excellent instrument for conducting such an investigation. The interferometer comprises one of each of 32 m and 16 m radio telescopes separated by a 800 m baseline which provides sensitivity to structures of 15 arcseconds, matching the typical sizes of ultra-compact HII regions. Utilizing the telescopes in interferometric mode also achieves deepened sensitivity by resolving out much of the background sky noise. Since the interferometer can operate at C-band, simultaneous monitoring of the radio continuum and 6.7 GHz methanol maser can be conducted. Finally, since the telescopes are operated by Ventspils University (several members of which are members of the M2O too) operations of the interferometer can be conducted with a relative freedom in terms of cadence, within the other constraints from other commitments. A pilot study from this facility, conducted as part of a successful grant application to monitor ~ 30 high-mass protostars, is described in Steinbergs *et al.* (2022).

4.2. *A 3.7m radio telescope for astronomy in Nigeria*

The M2O cooperates with the Global Emerging Radio Astronomy Foundation (GERAF; https://www.gerafoundation.com) whose aims include sourcing funding for radio astronomy projects on various scales. One such project is the construction of a 3.7 m radio telescope which will be assembled in Nsukka, Nigeria, on the grounds of the University of Nigeria. The project funding has been raised through GERAF and the M2O will assist in providing guidance on practical maser monitoring. As time-domain maser astronomy has recently found new pastures, and since many maser emitters in the Galaxy are very bright (hundreds of Jansky and more), even a modest radio observing facility can participate in leading radio astronomy activities where cadence is king.

References

Bayandina, O. S., *et al.* 2019, *ApJ*, 884, 140
Bayandina, O. S., *et al.* 2022a, *AJ*, 163, 83
Bayandina, O. S., *et al.* 2022b, *A&A*, 664, A44
Bayandina, O. S., *et al.* 2023, *A&A*, 673, A60
Brand, J., *et al.* 2007, In *Astrophysical Masers and their Environments*, IAUS242, 223
Breen, S. L., *et al.* 2019, *ApJ*, 876, L25
Brogan, C. L., *et al.* 2018, In *Astrophysical Masers: Unlocking the Mysteries of the Universe*, IAUS336, 255
Brogan, C. L., *et al.* 2019, *ApJ* (Letters), 881, L39
Burns, R. A., *et al.* 2020a, *MNRAS*, 491, 4069
Burns, R. A., *et al.* 2020b, *Nature Astronomy*, 4, 506
Burns, R. A., *et al.* 2023, *Nature Astronomy*, 7, 557
Cesaroni, R., *et al.* 2018, *A&A*, 612, A103
Chen, X., *et al.* 2020a, *ApJ* (Letters), 890, L22
Chen, X., *et al.* 2020b, *Nature Astronomy*, 4, 1170
Cragg, D. M., Sobolev, A. M. and Godfrey, P. D. 2005, *MNRAS*, 360, 533
Fujisawa, K., *et al.* 2015, *ATEL*, 8286, 1
Gaylard, M. J., Goedhart, S. and Dhlamini, D. 2002, In *Cosmic Masers: From Proto-Stars to Black Holes*, IAUS206, 127
Green, J. A. and McClure-Griffiths, N. M. 2011, *MNRAS*, 417, 2500
Hollenbach, D., Elitzur, M. and McKee, C. F. 2013, *ApJ*, 773, 70
Kobak, A., *et al.* 2023, *A&A*, 671, A135
Ladeyschikov, D. A., Bayandina, O. S. and Sobolev, A. M. 2019, *AJ*, 158, 233
MacLeod, G. C., *et al.* 2018a, *MNRAS*, 478, 1077

MacLeod, G. C., *et al.* 2018b, *MNRAS*, 478, 1077
MacLeod, G. C., *et al.* 2019, *MNRAS*, 489, 3981
Moscadelli, L., *et al.* 2017, *A&A*, 600, L8
Stecklum, B., *et al.* 2021, *A&A*, 646, A161
Steinbergs, J., *et al.* 2022, In *European VLBI Network Mini-Symposium and Users' Meeting 2021*, 33
Sugiyama, K., Saito, Y., Yonekura, Y., and Momose, M. 2019, *ATEL*, 12446
Szymczak, M., *et al.* 2018, *MNRAS*, 474, 219
Tanabe, Y., Yonekura, Y. and MacLeod, G. C. 2023, *PASJ*, 75, 351
Volvach, A. E., Volvach, L. N. and Larionov, M. G. 2020, *MNRAS*, 496, L147
Yonekura, Y., *et al.* 2016, *PASJ*, 68, 74

Cosmic Masers: Proper Motion toward the Next-Generation Large Projects
Proceedings IAU Symposium No. 380, 2024
T. Hirota, H. Imai, K. Menten, & Y. Pihlström, eds.
doi:10.1017/S1743921323003149

Maser Science with the African VLBI Network and MeerKAT

James O. Chibueze[1,2,3]

[1]Department of Mathematical Sciences, University of South Africa, Cnr Christian de Wet Rd and Pioneer Avenue, Florida Park, 1709, Roodepoort, South Africa
email: james.chibueze@gmail.com

[2]Centre for Space Research, North-West University, Potchefstroom, South Africa

[3]Department of Physics and Astronomy, University of Nigeria, Nsukka, Nigeria

Abstract. The African VLBI Network (AVN) is slowly becoming a reality. A couple of successful fringe test observations have been conducted even as single-dish maser monitoring observations constitute the main activity on the telescopes (HartRAO 26 m and Ghana 32 m). Some of the recent observational results from the AVN telescopes includes detection of velocity drifts in masers. Although MeerKAT is largely designed for high sensitivity continuum and HI science, its bands cover some masers and is already making impressive discoveries. The need to grow the critical mass of radio astronomers in the African continent persists. The NWU 4-dish interferometer, the Nigeria 3.7 m radio telescope and the African Millimeter Telescope (AMT) are some of the initiatives that will significantly improve the statistics of radio astronomers in Africa.

Keywords. Masers, stars: formation, ISM: jets and outflows

1. Introduction

The landscape of radio astronomy research in Africa is rapidly improving with the advent of the Square Kilometer Array (SKA) and its precursor, MeerKAT. Hartebeesthoek Radio Astronomy Observatory's 26 m radio telescope has played an active rule in its involvement in the very long baseline interferometric (VLBI) observations of the European VLBI Network (EVN) and the Australian Long Baseline Array (LBA). The Ghana 32 m radio telescope came online in 2018 after a few successful VLBI fringe test experiments with the EVN and the single baseline 6.7 GHz fringe detection with HartRAO 26 m telescope. These two telescopes constitute the current operational components of the AVN.

Prior to the coming into operations of MeerKAT telescope,

2. Maser science results of the AVN telescopes and MeerKAT

HartRAO 26 m telescope has played a key in the hunt for accretion burst events in massive protostellar objects, e.g. the case of NGC 6334I (MacLeod *et al.* 2018; Hunter *et al.* 2017; Burns *et al.* 2022; Chibueze *et al.* 2021). In recently times, it has contributed to the discovery of a second period in the periodic behavior of G9.62, including the identification of Zeeman pair in the same source using the velocity drift property of the masers (MacLeod *et al.* 2022, 2021). High cadence observations of the "bunny hop" periodic maser source is being explored as a means of confirmation the existence of or not of quiescent phase in the lowest flux density phase of the periodic masers.

The Ghana 32 m telescope, on the other hand, have continued to monitor bright maser sources due to its sensitivity limitation cause by the ambient receiver system installed on the telescope. It has also been a handy facility for observational radio astronomy training of African students and cohorts of the Development in Africa through Radio Astronomy (DARA) project [PI: Melvin Hoare]. The most recent development of Ghana 32 m telescope is the acquisition of a H-maser clock which will replace the existing Rubidium clock, making it a more functional VLBI station with improved phase.

The MeerKAT debuted its maser science with the discovery of "Nkalakatha", a luminous OH megamaser at $z > 0.5$ (Glowacki *et al.* 2022). The MeerKAT Galactic Plane Legacy Survey (MGPLS) conducted in L-band focuses on continuum imaging, however, the coarse spectral resolution of the MGPLS data could still be used for OH maser detection and could be useful for mapping the most complete distribution of the OH maser in the Milky Way yet. However, this will require spectral imaging of the MGPLS data. A proof of concept for the OH maser imaging has already been conducted.

3. Growing the critical mass of African users of the SKA

Th MeerKAT and the SKA have strongly underscored the gross deficiency in the availability of human resources in the field of radio astronomy and interferometry. In fact, there are less than a dozen African professional radio astronomers. The implication of the current human capacity situation is that Africa could host revolutionary facilities but contribute minimally to the scientific output of the facility. This lack of critical mass of prospective users on future facilities like the SKA informed my desire to drive the growth of the critical mass of radio astronomers in the continent.

One way to ensure proper hands-on training that will stimulate interest among African in radio astronomy is to develop cheap radio telescope that can be used effectively for such training. Understanding interferometric data handling without proper understanding of how the individual single-dish telescopes work and the principle of radio interferometry/aperture synthesis, make the field a bit of a "blackbox" to new entrants into the field.

There are a number of initiatives toward growing the number of radio astronomers in Africa, and I will highlight three of them, namely, North-West University 4-dish interferometer, the Nigeria 3.7 m radio telescope and the African Millimeter Telescope (AMT).

3.1. *NWU 4-dish interferometer*

North-West University Potchefstroom Campus is home to the Centre for Space Research, and recently acquired and installed four 3.7 m prime focus, altitude-azimuth radio telescopes supplied by POAM Electronics United Kingdom. The installation of the telescopes took place in August, 2022. The telescope site is located ~35 km from Potchefstroom at the sparsely populated place called Nooitgedacht. The location is fairly radio quiet with only a few mobile phone/GSM and wifi potential radio frequency interference (RFI) sources.

Each of the four telescopes has C-band and L-band front-end, with a wideband band end to enable the switching of band to enhance the scientific output of the instrument. The C-band covers the 6.7 GHz methanol maser frequency, with the L-band front-end covers the 21 cm neutral hydrogen (HI) line frequency.

The science goals of the NWU 4-dish interferometer includes but not limited to the following;

(1) Intensive monitoring observations of bright 6.7 GHz methanol maser sources
(2) Focus observations of bright radio transient events
(3) High cadence monitoring of bright radio galaxies and spectral index measurement
(4) Future of observation of pulsars (pending the availability of the pulsar back-end)

Figure 1. Antenna 1 of the NWU 4-dish interferometer.

Figure 2. Aerial view of the four telescopes of the NWU 4-dish interferometer.

Beside the possibly observational work that could be done with the interferometer, it can be used for hands-on training. In fact, the assembly and installation of the four telescopes were done by DARA cohorts from the Southern African region. It is important to mention that the NWU 4-dish interferometer is already being used by postgraduate students of NWU as the solid part of their research project. The interferometer has already started attracting new collaborations for NWU. For example, the radio receiver development teach at the University of Pretoria led by Prof Tinus Stander. And we look forward to more collaboration and the improvements in the front and back-ends that this collaboration will deliver to the array. Figures 1 and 2 show a close view of one of the four 3.7 m radio telescopes and an aerial view of the array, respectively.

Figure 3. Photo of the Nigeria 3.7m radio telescope at the manufacturer's site undergoing drive tests.

3.2. *Nigeria 3.7 m radio telescope*

The NWU 4-dish interferometer has attracted a lot of interests from the astronomy community of the African continent and the demand to install similar system in many African countries have grown. South Africa has the highest number of radio astronomers in Africa, followed by Nigeria, thus it made sense to consider Nigeria as the next destination of a 3.7 m that could be used for training and some level of research.

Global Emerging Radio Astronomy Foundation (https://www.gerafoundation.com), a non-profit organization registered in Canada, in collaboration with the Centre for Basic Space Science (CBSS) of the National Space Research and Development Agency of Nigeria, raised funds and acquired 3.7 m telescope from POAM Electronic to be installed in Nigeria in the 2nd half of 2023 (see Figure 3). The telescope will be fitted with an L-band receiver for a start and more bands can be added to the front-end subsequently.

The Nigeria telescope will be used extensively for teaching and for HI mapping observations.

3.3. *African Millimeter Telescope (AMT)*

Event Horizon Telescope (EHT) imaging of the blackholes' event horizon have received a lot of publicity in recent times. The importance of the addition of a millimeter telescope on the African continent to the EHT array cannot be over-emphasized. Such addition will evidently improve the imaging capabilities of the eastern components of the array. Radboud University in collaboration with the University of Namibia are planning to install the first ever millimeter telescope on Mount Gamsberg, close to the High Energy Stereoscopic System (HESS) telescope site in Namibia.

This project is now significantly funds and we will be seeing a millimeter telescope on the continent soon. The priority project of the AMT will be participation in the annual blackhole imaging campaigns. However, there will be 1000s of hours available for single-dish observing projects, and these will include but not limited to, intensive monitoring of millimeter masers, monitoring of AGN, mapping observations of giant molecular line.

Yes, the radio astronomy landscape looks promising, but there is dire need to train more radio astronomer who will play some role in the science output of our society.

References

MacLeod, G. C., Yonekura, Y., Tanabe, Y., *et al.* 2022, *MNRAS*, 516, L96
MacLeod, G. C., Chibueze, J. O., Sanna, A., *et al.* 2021, *MNRAS*, 500, 3425
Burns, R. A., Kobak, A., Garatti, A. C., *et al.* 2022, European VLBI Network Mini-Symposium and Users' Meeting 2021, 2021, 19
Hunter, T. R., Brogan, C. L., MacLeod, G., *et al.* 2017, *ApJ*, 837, L29
Chibueze, J. O., MacLeod, G. C., Vorster, J. M., *et al.* 2021, *ApJ*, 908, 175
Glowacki, M., Collier, J. D., Kazemi-Moridani, A., *et al.* 2022, *ApJ*, 931, L7
MacLeod, G. C., Smits, D. P., Goedhart, S., *et al.* 2018, *MNRAS*, 478, 1077

Cosmic Masers: Proper Motion toward the Next-Generation Large Projects
Proceedings IAU Symposium No. 380, 2024
T. Hirota, H. Imai, K. Menten, & Y. Pihlström, eds.
doi:10.1017/S1743921323002661

Southern Hemisphere Maser Astrometry

Simon Ellingsen[1], Mark Reid[2], Karl Menten[3], Lucas Hyland[1], Jayender Kumar[1], Gabor Oroz[1,4], Stuart Weston[5], Richard Dodson[6] and Maria Rioja[6]

[1]University of Tasmania, Australia. email: Simon.Ellingsen@utas.edu.au

[2]Harvard Smithsonian Center for Astrophysics, USA

[3]Max Planck Institut für Radionastronomie, Germany

[4]Joint Institute for VLBI in Europe, The Netherlands

[5]Auckland University of Technology, New Zealand

[6]International Centre for Radio Astronomy Research, Australia

Abstract. Many astrophysical phenomena can only be studied in detail for objects in our galaxy, the Milly Way, but we know much more about the structure of thousands of nearby galaxies than we do about our own Galaxy. Accurate distance measurements in the Milky Way underpin our ability to understand a wide range of astrophysical phenomena and this requires observations from both the northern and southern hemisphere. Our ability to measure accurate parallaxes to southern masers has been hampered a range of factors, in particular the absence of a dedicated, homogeneous VLBI array in the south. We have recently made significant advances in astrometric calibration techniques which allow us to achieve trigonometric parallax accuracies of around 10 micro-arcseconds (μas) for 6.7 GHz methanol masers with a hetrogeneous array of 4 antennas. We outline the details of this new "multiview" technique and present the first trigonometric parallax measurements that utilise this approach.

Keywords. interstellar masers, astrometry, calibration, distance

1. Introduction

The BeSSeL and VERA surveys have provided accurate distances to more than 200 star formation regions in the part of the Milky Way accessible to northern hemisphere instruments (Reid *et al.* 2019). However, this excludes the majority of the fourth quadrant of the Milky Way and obtaining trigonometric parallax measurements to this region of the Galaxy has been a long-standing issue.

We know more about the structure of thousands of nearby galaxies, as our location within the plane of the Milky Way hinders our ability to determine the number and location of the spiral arms due to obscuration at most wavelengths. Trigonometric parallax is the only direct (model independent) method which can be used to measure distances to objects outside the solar systems, hence it represents the "gold standards" for determining distances. While the *GAIA* mission will measure accurate distances and proper motions for millions of stars within the Milky Way (see the talk by Rygl et al. in these proceedings), extinction at optical wavelengths prevents it from observing objects close to the Galactic Plane which are more than a few kiloparsecs from the Sun. The result is that *GAIA* will not significantly improve our knowledge of the number or location of the Milky Way's spiral arms and instead we have to utilise parallax measurements at radio wavelengths which are not hindered by obscuration. Interstellar masers are closely associated with young high-mass star formation regions (which by definition trace the spiral arms) and so are ideal targets for such measurements.

The BeSSeL and VERA surveys have primarily targeted 22 GHz H_2O and 12.2 GHz methanol masers and in the best cases are able to achieve a formal uncertainty in the parallax measurements of around 10 micro-arcseconds (hereafter, μas). This enables distances to be determined with an accuracy of better than 10% to the Galactic Center.

2. Challenges for Southern Hemisphere Trigonometric Parallax Measurements

The fundamental accuracy which can be obtained in relative astrometry measurements from very long baseline interferometry observations depends on the accuracy to which a wide range of factors have been measured or calibrated. Critical factors include the location of the antennas, the position and any structure of the reference sources, the stability of the frequency standards (clocks), electronic delays and the ionospheric and tropospheric delay contributions at each antenna (and how they change over the course of the observation). It is this final factor which prevents accurate astrometric measurements being made for sources at low-elevation (Honma *et al.* 2008), hindering measurements of fourth quadrant masers with the VLBA, VERA and other northern arrays. Good atmospheric calibration has been more difficult to achieve for southern hemisphere arrays for a number of reasons.

(1) Ionospheric calibration is traditionally obtained through the use of total electron content (TEC) maps extracted from measurements made by a network of Global Navigation Satellite System (GNSS) stations. There is a much lower density of such stations in the southern hemisphere and the result is that the TEC data for the south is often of lower accuracy (e.g. Deller *et al.* 2009).
(2) Tropospheric calibration is best achieved through observations of sources at low-elevations, sampling a range of azimuths and spanning a wide range of frequencies, often called the “geo-block” technique as it was developed for geodetic very long baseline interferometry (e.g. Reid *et al.* 2009). Hetrogeneous very long baseline interferometry arrays are generally less able to make good “geo-block” measurements due to differences in the receiver frequency coverage and antenna parameters such as the slew speeds and elevation limits between the different elements.

Furthermore, the wavelength dependence for ionospheric and tropospheric delays have the opposite signs, which further complicates the impact on astrometry when the residual uncertainty in the calibration of both terms is of comparable magnitude. The best astrometry at radio wavelengths has been achieved using homogeneous arrays (such as the VLBA and VERA) operating at frequencies above 10 GHz where the ionosphere impacts less. Furthermore, the timing of the peaks in the amplitude of the parallax signature depends on the ecliptic longitude of the target source and so can be most accurately measured if the observations are timed within a few weeks of the optimal times during the year.

The only southern hemisphere very long baseline interferometry array which has previously made trigonometric parallax observations has been the Long Baseline Array (LBA), with antennas operated by CSIRO Space and Astronomy, the University of Tasmania (e.g. Deller *et al.* 2009; Krishnan *et al.* 2015). These observations have been at frequencies less than 10 GHz, which means that they are more heavily impacted by inaccurate ionospheric calibration and the timing of the observations has not been optimal because the LBA is an ad-hoc array. The result is that the accuracy of the limited number of trigonometric parallax observations which have been made of both pulsars and masers with the LBA have had an astrometric accuracy typically of 100 μas or worse, a factor of 10 poorer than the best obtained from northern instruments. With this level of precision it is only possible to obtain sufficiently accurate distances to sources within a few kiloparsecs of the Sun (i.e. comparable to *GAIA*).

3. Multiview

The challenge for obtaining accurate relative astrometry for hetrogeneous very long baseline interferometry arrays (particularly in the southern hemisphere), has been to find an alternative atmospheric calibration approach. A range of approaches have been developed and trialed and a comprehensive review of developments can be found in Rioja and Dodson (2020). One of these techniques "multiview" involves making observations of multiple calibrators distributed around the target sources and from these data determining the time-variable two-dimensional phase-slope produced by the residual delay error. The efficacy of this approach for measuring and correcting for residual delays has been demonstrated by Hyland *et al.* (2022) who showed that it was possible to obtain a single-epoch relative astrometric accuracy of 20 μas with a four-antenna array at 8.4 GHz. Multiview is able to produce good astrometric accuracy in the presence of any residual delay errors which vary on timescales which are slower than the time it takes to determine the 2D phase-slope. This will depend on a range of factors and we have insufficient experience in implementing Multiview to be able to give clear guidance as to which of those are most critical in setting the timescale (Hyland *et al.* 2022). At frequencies between 5 and 10 GHz going around the multiple calibrators within 20-30 minutes generally allows unambiguous tracking of the changes in the phase slope. It is important to realise that for Multiview, having some of your calibrators with larger separations from the target source is advantageous, this is in contrast to standard phase referencing, or inverse phase referencing, where the closer your calibrator to the target the better. A larger separation between calibrator and target allows better measurement of the phase slope, but perhaps the most important factor for good Multiview calibration is to have an even azimuthal distribution of calibrators around the source.

The first trigonometric parallax observations towards a 6.7 GHz methanol maser obtained using inverse multiview have been obtained by Hyland *et al.* (2023). This paper reports measurements towards two sources, one (G232.62+0.99), is in the southern hemisphere but sufficiently close to the equator that it has a distance measured by the BeSSeL survey. The trigonometric parallax measured for the 6.7 GHz methanol maser in this source of 610 ± 11 μas is consistent with the measurement based on the 12.2 GHz methanol maser observations from the VLBA (596 ± 35 μas Reid *et al.* 2009). The efficacy of the Multiview calibration approach is clearly demonstrated by this result, as is has been possible to obtain a formal uncertainty in the parallax measurement with a 4-antenna hetrogeneous array that is a factor of three better than that obtained at a higher frequency with a 10-antenna homogeneous array. The other parallax measurement is for G323.74−0.26 at a declination of -56°, which is completely inaccessible to northern instruments. Hyland *et al.* (2023) measure a parallax of 364 ± 9 μas for this source, corresponding to a distance of 2.75 ± 0.07 kpc for this southern star formation region.

4. Conclusions

Application of the Multiview calibration technique is able to achieve astrometric accuracies in trigonometric parallax observations of around 10 μas with a four-antenna, hetrogeneous antenna array at a frequency of 6.7 GHz. This is comparable with the best relative astrometry which has been achieved at radio wavelengths and opens the potential for better astrometry with a range of hetrogeneous very long baseline interferometry arrays. Application of Multiview with the VLBA may enable significant improvements in its best astrometric accuracy. Additional observations covering a wider range of frequencies, times and calibrator distributions are required to better understand the critical factors which determine the accuracy of Multiview.

References

Deller, A. T., Tingay, S. J., Bailes, M., & Reynolds, J. E. 2009, *ApJ*, 701, 1243
Honma, M., Tamura, Y., & Reid, M. J. 2008, *PASJ*, 60, 951
Hyland, L. J., Reid, M. J., Ellingsen, S. P. *et al.* 2022, *ApJ*, 932, 52
Hyland, L. J., Reid, M. J., Orosz, G. *et al.* 2023, *arXiv e-prints*, arXiv:2212.03555
Krishnan, V., Ellingsen, S. P., Reid, M. J. *et al.* 2015, *ApJ*, 805, 129
Reid, M. J., Menten, K. M., Brunthaler, A. *et al.* 2019, *ApJ*, 885, 131
Reid, M. J., Menten, K. M., Zheng, X. W. *et al.* 2009, *ApJ*, 700, 137
Rioja, M. J. & Dodson, R. 2020, *A&AR*, 28, 6

Cosmic Masers: Proper Motion toward the Next-Generation Large Projects
Proceedings IAU Symposium No. 380, 2024
T. Hirota, H. Imai, K. Menten, & Y. Pihlström, eds.
doi:10.1017/S1743921323002909

The 40-m Thai National Radio Telescope with its key sciences and a future South-East Asian VLBI Network

Koichiro Sugiyama[1], Phrudth Jaroenjittichai[1], Apichat Leckngam[1], Busaba H. Kramer[2,1], Wiphu Rujopakarn[1], Boonrucksar Soonthornthum[1], Nobuyuki Sakai[1], Songklod Punyawarin[1], Nattapong Duangrit[1], Kitiyanee Asanok[1], Taufiq Hidayat[3], Zamri Zainal Abidin[4], Juan Carlos Algaba[4], Pham Ngoc Diep[5] and Saran Poshyachinda[1], on behalf of the TNRO project team and science working group members

[1]National Astronomical Research Institute of Thailand (Public Organization), 260 Moo 4, T. Donkaew, A. Maerim, Chiangmai 50180, Thailand. email: koichiro@narit.or.th

[2]Max Planck Institut für Radioastronomie, Auf dem Hügel 69, 53121 Bonn, Germany

[3]Bosscha Observatory and Astronomy Research Division, FMIPA, Institut Teknologi Bandung, Jl. Ganesha 10, Bandung 40132, Indonesia

[4]Radio Cosmology Lab, Department of Physics, Faculty of Science, Universiti Malaya, 50603 Kuala Lumpur, Malaysia

[5]Department of Astrophysics, Vietnam National Space Center, Vietnam Academy of Science and Technology, 18 Hoang Quoc Viet, Nghia Do, Cau Giay, Hanoi, Vietnam

Abstract. National Astronomical Research Institute of Thailand (Public Organization) initiated a national flagship project in 2017 for development of radio astronomy and geodesy in Thailand. In this project, a 40-m Thai National Radio Telescope (TNRT) and a 13-m VLBI Global Observing System (VGOS) radio telescope as its co-location are constructed in Chiang Mai. The 40-m TNRT is the largest telescope for radio astronomy in South-East Asia. Its flexible operation with a wide-coverage of observable frequencies 0.3-115 GHz will allow us to uniquely contribute to the time-domain astronomy as well as carry out unbiased surveys for a wide variety of science research fields, which were published in a white paper. Within the framework of collaboration with VLBI arrays in the world, TNRT will drastically improve the imaging quality and performances based on its unique geographical location, for both radio astronomy and geodetic VLBI studies in South-East Asia for the first time. On-going commissioning of TNRT particularly in the L-band system (1.0–1.8 GHz) is introduced as well as vision for establishment of forthcoming regional VLBI networks based on TNRT: Thai National VLBI Array and South-East Asian VLBI Network in collaboration with Indonesia, Malaysia, and Vietnam.

Keywords. Telescopes, Radio continuum, Radio lines, Interferometers

---

1. Introduction

Along with an essential probe of electromagnetic waves in the multi-messenger astronomy era, following the successful pathway of the 2.4-m optical telescope (Thai National Telescope) in the last decade, National Astronomical Research Institute of Thailand

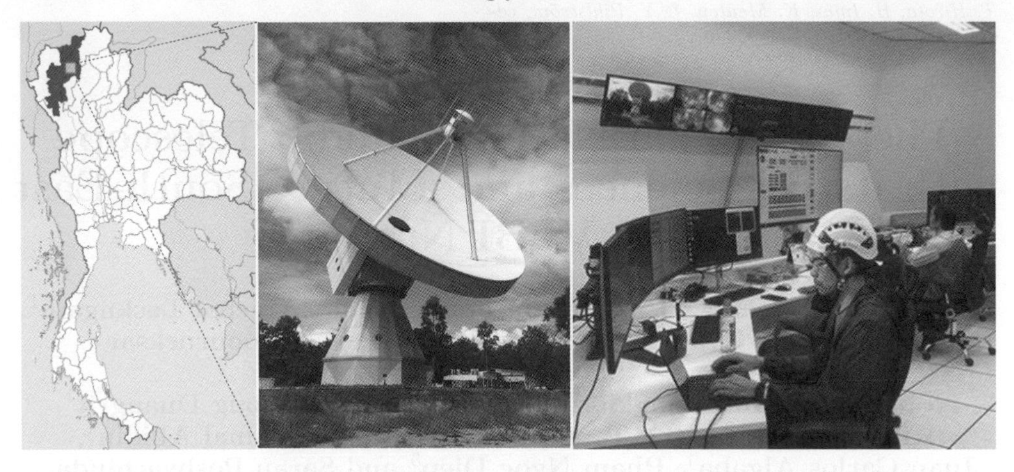

Figure 1. The 40-m Thai National Radio Telescope in Chiang Mai, Thailand, with its operation room (Image credit: Left - NordNordWest in Wikipedia; Middle and Right - TNRO/NARIT).

(NARIT: Public Organization) initiated a national flagship project in 2017 for development of radio astronomy and geodesy, known as Radio Astronomy Network and Geodesy for Development. This project was strongly motivated by the importance of the development by "Ourselves" to achieve an empyreal goal of "Capacity building through radio astronomy and geodesy", via constructing national radio telescopes in Thailand. This construction has provided precious opportunities to develop engineering / technical / instrumental skills, its technology, unique sciences achieved with these telescopes, and essential experiences on the basis of collaboration with world-wide colleagues at the world-class facilities, as well as contribute to education via cultivating potential young astronomers, engineers, and geodesists.

For this, NARIT has established the Thai National Radio Observatory (TNRO) in Huai Hong Khrai Royal Development Study Centre, Doi Saket District, Chiang Mai, in the northern part of Thailand since 2018, which is 40 km away from NARIT headquarters in the North-East direction. Given a radio quiet zone and low amount of water vapor on the basis of site investigations, this location is the best suitable site to build up radio telescopes covering low to high frequencies, consisting of a 40-m Thai National Radio Telescope (TNRT, Figure 1) and a 13-m VLBI Global Observing System (VGOS) radio telescope (TNRT co-location, in collaboration with Shanghai Astronomical Observatory (SHAO)), with a visitor center to be opened to general public (Jaroenjittichai 2018; Jaroenjittichai *et al.* 2017).

2. Overview of the 40-m Thai National Radio Telescope (TNRT)

The 40-m TNRT is located at latitude 18°51′52″ N and longitude 99°13′01″ E at 450 m above mean sea level. This is the largest telescope for radio astronomy in South-East Asia. The TNRT was started to be built in 2018, based on the National Geographical Institute (IGN) 40-m Yebes telescope (e.g. López Fernández *et al.* 2006; de Vicente *et al.* 2006) with a classical Nasmyth-Cassegrain focus optics, but is upgraded with the installation of a Tetrapod Head Unit at the prime focus. This upgrade enables the selection of either a receiver at the prime focus or the sub-reflector to be used for receivers installed at the Nasmyth focus in the receiver cabin. Given the specifications at the best performance: a pointing accuracy of 2 arcsec (no wind case) and a surface accuracy of 150 μm rms, we plan to install multiple receivers for the telescope from 300 MHz up to 115 GHz, which contains P/L/C/X/Ku/K/Q/W-bands.

The first receivers to be installed for commissioning and early sciences are the L-band (1.0–1.8 GHz, linear pol.) and K-band (18–26.5 GHz, circular pol.) receivers equipped with the Universal Software Backend (USB) system enabling multiple observation modes. These were developed in collaboration with Max Planck Institute for Radio Astronomy (MPIfR), while the Ku-band receiver (10.70–12.75 GHz) used for microwave holography was developed in collaboration with Yebes Observatory, IGN (e.g. Lopez-Perez *et al.* 2014). The Telescope Control Software (TCS) is based on the ALMA Common Software and was also developed in collaboration with Yebes Observatory, IGN. The L-/Ku-bands and K-band receivers have been installed at the prime and the Nasmyth focuses, respectively. The multiple observation modes consist of 1) pulsar with coherent dedispersion, baseband and search mode recording; 2) spectrometer / continuum; 3) polarimeter; and 4) Very Long Baseline Interferometry (VLBI) with vdif format. This will be followed-up by developing and installing forthcoming receivers step by step: C/X/Ku-bands (4.5–13.6 GHz), Q/W-bands (35–50 and 75–115 GHz) to be integrated with existing K-band into a simultaneous quasioptics triband system in collaboration with Korea Astronomy and Space Science Institute (KASI), and 0.7–2.1 GHz Phased Array Feed in the design study phase.

2.1. *The First Lights of TNRT*

The first lights of TNRT were achieved in 2022 at last, reaching a milestone, as shown in Figure 2. The intrinsic first light received was an HI emission at 1.42 GHz from our Milky Way Galaxy with the L-band receiver transiting the zenith standby position on 24 March 2022 (Figure 2 left). The first light from tracking a source was completed through receiving a pulsar signal from B0329+54 on 15 June 2022 (Figure 2 upper-right), which is the brightest pulsar in the Northern hemisphere. The latest light in L-band was detected on 25 October 2022 in the spectrometer mode for ground-state of all four OH maser lines in the high-mass star-forming region W49 North. In particular, the brightest emission at 1.665 GHz (Figure 2 lower-right) presented multiple spectral components in the LSR velocity range from 0 to +25 km s^{-1}, which is consistent with the spectrum in previous literature (e.g. Bayandina *et al.* 2021). These milestones were immediately publicized via NARIT Facebook (accessible through scanning QR codes in Figure 2) and news media with the PR divisions at NARIT. In addition, the first light at K-band was also achieved via detecting one of the brightest H_2O maser emission in the same high-mass star-forming region W49 North on 16 December 2022.

3. Key Science Cases with TNRT

To prepare for launching TNRT, a science working group was organized with worldwide collaborators to discuss key science cases achievable with TNRT for a wide-variety of research fields and the wide-coverage of observable frequencies: pulsar, fast radio burst, gravitational wave, star formation, galaxy, active galactic nucleus (AGN), evolved star, chemically peculiar star, maser, and geodesy. Based on an advantage as our own radio telescope with flexible operations and as being accessible to both the Northern and the Southern hemisphere, key sciences with TNRT as a single dish focuses on time-domain astronomy, which addresses exploration of transients / variability and achievement of high-cadence monitoring campaigns planned for known sources or as unbiased surveys for all the sky. These ideas were published as a white paper in arXiv website on 12 October 2022 (Jaroenjittichai *et al.* 2022).

Figure 2. (Cited from NARIT Facebook, accessible through scanning QR codes in each panel) The 1st lights detected with TNRT L-band receiver in 2022. (Left) HI emission from the Milky Way Galaxy. (Upper-right) Pulsar signal from B0329+54. (Lower-right) OH maser at 1.665 GHz in the high-mass star-forming region W49 North (Image credit of background: DePree, et al.; Sophia Dagnello, NRAO/AUI/NSF; Spitzer/NASA).

3.1. *Impacts on VLBI*

Moreover, TNRT will be a powerful telescope for promoting any VLBI activities, with VLBI arrays that have sufficient common-sky with TNRT: East Asia VLBI Network (EAVN), European VLBI Network (EVN), Long Baseline Array (LBA), and Giant Metrewave Radio Telescope (GMRT). For any of those cases, due to its unique geographical location, TNRT contributes to improving the UV-coverage effectively, by adding one of the longest baselines in unique directions as well as filling gaps in the coverages, such as shown in Figure 3 for K-band. The TNRT will thus enhance spatial angular resolution and imaging quality for a wide-range of frequencies for each VLBI array. Figure 4 is an example of a simulation for VLBI towards the collimated radio jet ejected from AGN M87, using EAVN without TNRT (left panel: K. Hada & Y. Cui, in private communication), and using EAVN with TNRT (right panel) in K-band, respectively. This simulation presents the drastic improvement of the imaging quality via reduced side-lobes and also demonstrates the impressive snap-shot imaging capabilities. The TNRT participation will thus enhance VLBI large programs, accelerate the synergy with essential projects in multi-messenger astronomy era, and achieve the first trial of Geodetic VLBI in South-East Asia.

4. Commissioning

To prepare TNRT for operations, performance evaluations for the telescope are currently undertaken, particularly for the general engineering commissioning of the L-band receiver.

4.1. *RFI Bird's-eye View*

Even in the radio quiet zone at the site, an essential commissioning component is to clarify the directions and the strengths of dominant radio frequency interference (RFI). This is achievable with making an RFI distribution map. It enables us to mitigate the

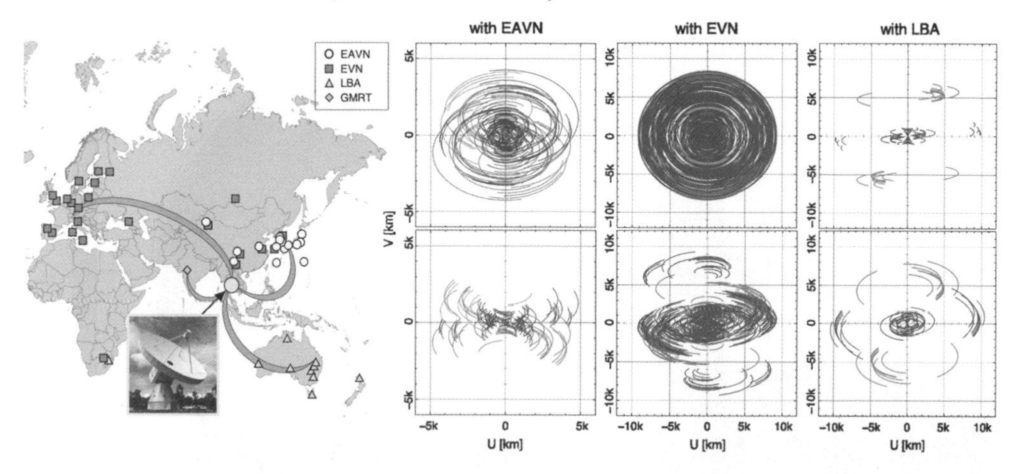

Figure 3. (Cited from the white paper, section 1 and appendix) (Left) Geographical location of TNRT and VLBI arrays that have sufficient common-sky with TNRT: EAVN (circle), EVN (square), LBA (triangle), and GMRT (diamond) (Image credit of background world-map: Illust AC). (Right) simulated results of uv-coverages in K-band for EAVN+TNRT, EVN+TNRT, and LBA+TNRT, respectively. Contributions of TNRT are shown by bold-red lines. The upper and lower panels are for source declination of +40, +60, +10, and −29, +20, −29 deg (toward the Galactic Center) from left to right, respectively.

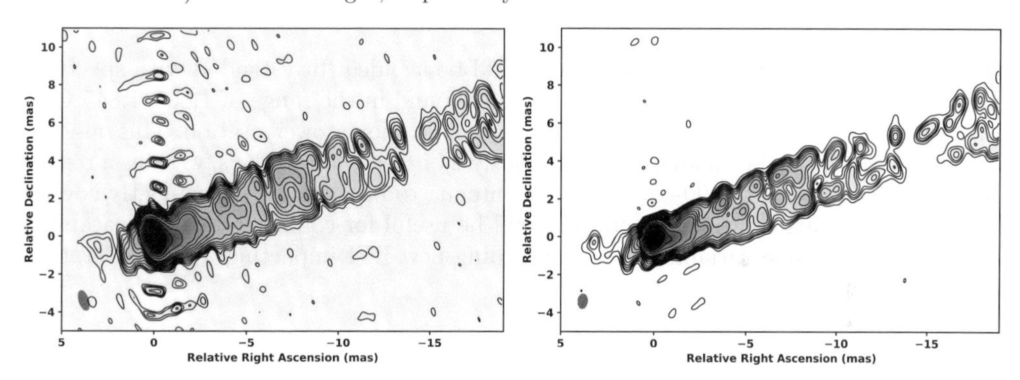

Figure 4. (Cited from the white paper, section 4) Simulated VLBI CLEAN (uniformly weighted) images of M87 at 22 GHz in K-band. (Left) image reconstructed with EAVN (KaVA+Tianma+Nanshan+Takahagi: K. Hada & Y. Cui, in private communication). (Right) image reconstructed with EAVN+TNRT. Beam size is shown as an ellipse in the bottom left corner of each panel. For both images, the contours start from −1, 1, 2, ... times 240 μJy beam^{-1} and increase by a factor of $\sqrt{2}$.

impact on TNRT in L-band and to avoid saturation of total-power in the receiver. This mapping was conducted with a 90-cm parabolic reflector antenna (Rohde & Schwarz AC008 with HL024A1: Figure 5 left) and the data were recorded with a spectrum analyzer (Keysight Fieldfox N9918B). The 90-cm antenna was installed on the top of GNSS (Global Navigation Satellite System) tower, which is 80 m away from the TNRT in roughly North-East direction and enables us to watch the same sky and the directions as for with the TNRT. The observations were designed with 3 deg steps both in AZ and EL directions with an integration time of 30 sec at each point and completed in February 2023. Integrated over 1–2 GHz for 30 sec, the first RFI bird's-eye view at the TNRO site was unveiled, as shown in Figure 5 (right). This viewing map clarified the direction and zone of the strongest RFI with an approximately 15 dB increased power level, corresponding to the South-East direction in AZ from 90 to 180 deg in EL lower

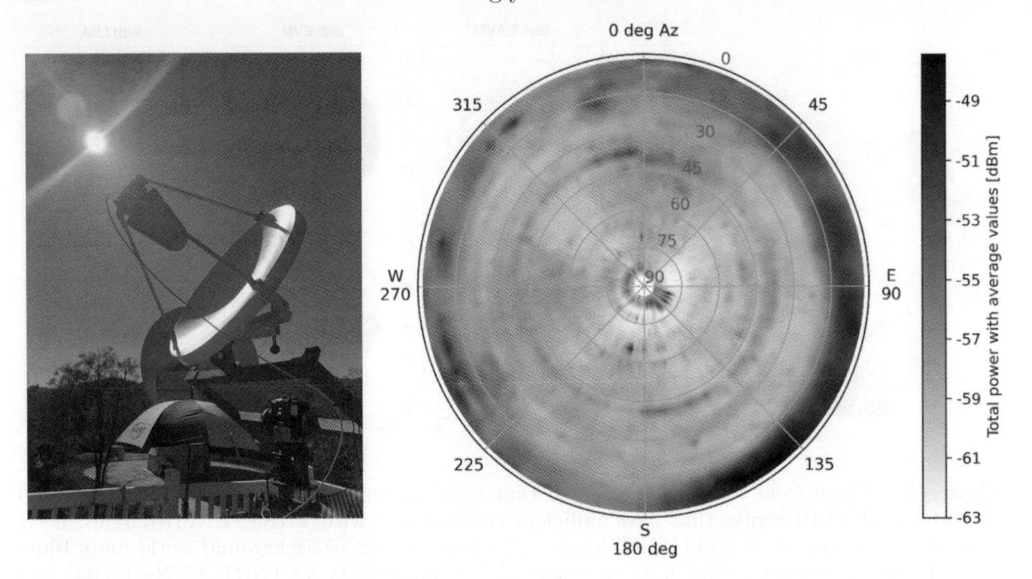

Figure 5. RFI bird's-eye view map at the TNRO site observed with a 90-cm parabolic reflector antenna. Color scale (white to black) is indicated with the right-hand side bar.

than 15 deg (shown in black). This zone should be avoided. Investigated with spectra, a major contributor of this elevated power level is found in the range of 1.805–1.845 GHz, which could be caused by transmission from a mobile phone tower/station. This on-going essential activity will be conducted continuously 24 hrs/day and 7 days/week as a regular RFI monitoring using the 90-cm dedicated antenna to be installed permanently covered by a membrane. Its continued operations will be useful for confirming the repeatability, for verifying the time variability, and for finding new RFI impacting TNRT operations immediately.

4.2. *Determining the Temperature of a Noise-source (NS)*

Another essential commissioning component is to calibrate an amplitude with its scaling into temperature. For this, the temperature of a NS ($\mathrm{T_{NS}}$) in the L-band system was determined with the R-Sky method and injecting the NS. The basic formula to determine $\mathrm{T_{NS}}$ is

$$\mathrm{T_{NS}} = (\mathrm{T_{rx}} + T_{\mathrm{amb}}) \cdot \{(P_{\mathrm{R+NS}}/P_{\mathrm{R}}) - 1\} \tag{1}$$

where $\mathrm{T_{rx}}$ and T_{amb} is the temperature of the receiver and the ambient environment, respectively, and $P_{\mathrm{R+NS}}$ and P_{R} represents the power in the case of covering the horn by an absorber with and without injecting the NS, respectively. The term $\mathrm{T_{rx}} + T_{\mathrm{amb}}$ can be estimated by the sec$Z$ method with an absorber (temporary installed on top of the L-band horn, shown in Figure 6 right). As a result, $\mathrm{T_{NS}}$ was determined to be 30.9 ±1.1 K. Verification that this $\mathrm{T_{NS}}$ works well was done by comparing the system noise temperatures T^{*}_{sys} measured by injecting the NS to the ones by R-Sky. Along the sec$Z$ axis, T^{*}_{sys} consistently ranged between 15 and 30 K using either of the methods as shown in Figure 6 (left). The best T^{*}_{sys} of around 15 K signifies that this L-band receiver has one of the best noise performances in the world.

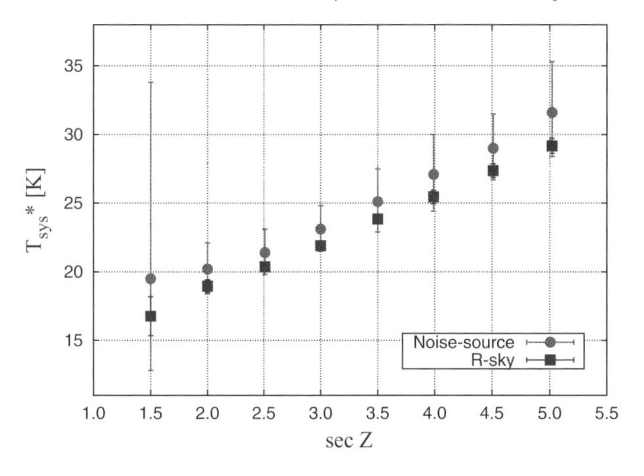

Figure 6. (Left) System noise temperature $T^*_{\rm sys}$ (included atmosphere) versus sec$Z$. $T^*_{\rm sys}$ measured with injecting the noise-source and R-sky methods are shown by circle and square symbols, respectively. (Right) scene in temporal installation of an absorber on top of the L-band horn (detached already).

4.3. *Others*

Other commissioning items are on-going for the L-band system: pointing tuning†, beam-pattern confirmation, aperture efficiency with gain-curve measurement, skyline bird's-eye view mapping, linearity investigation, and Allan variance evaluation. This will be followed by commissioning for the K-band receiver installed at the Nasmyth focus.

5. Vision for Establishment of Regional VLBI Networks

5.1. *Thai National VLBI Array (TVA)*

As introduced in section 1, NARIT is working on facilitating a 13-m VGOS station in TNRO in collaboration with SHAO, to be launched in 2024. This will be followed by building another VGOS station in Songkhla, in the southern part of Thailand, at a distance of 1,330 km from Chiang Mai, allowing extensive applications in geodesy and tectonic studies covering the two different Eurasian and Sunda plates in South-East Asia. A receiver installed at the Songkhla VGOS has been developed in collaboration with Yebes Observatory, IGN. Futhermore, there are other possible candidate antennas available for radio astronomy in Thailand: telecommunication antennas with the diameter of 32-m at the campus of CAT Telecom Public Company Limited in Chonburi and Ubon Ratchathani, in the central and eastern part of Thailand. These antennas will potentially be converted into radio telescopes with a similar technique used for the Yamaguchi 32-m, Hitachi and Takahagi 32-m, Warkworth 30-m radio telescopes, and so on (e.g. Fujisawa *et al.* 2002; Yonekura *et al.* 2016; Woodburn *et al.* 2015). Completing these plans will result in the establishment of Thai National VLBI Array (TVA: Figure 7 left), aiming to be launched since 2026 in C/X/Ku-bands and possibly in K-band.

5.2. *South-East Asian VLBI Network (SEAVN)*

The TVA will be the solid foundation for establishing a forthcoming regional VLBI network: South-East Asian VLBI Network (SEAVN). The SEAVN will be achieved in collaboration with Indonesia, Malaysia, and Vietnam. The Institut Teknologi Bandung together with the Timau National Observatory in Indonesia plan to build a new radio telescope at the Timau site, in the eastern part of Indonesia, and conduct the conversion

† Deconstructed pointing model: https://icts-yebes.oan.es/reports/doc/IT-OAN-2007-26.pdf

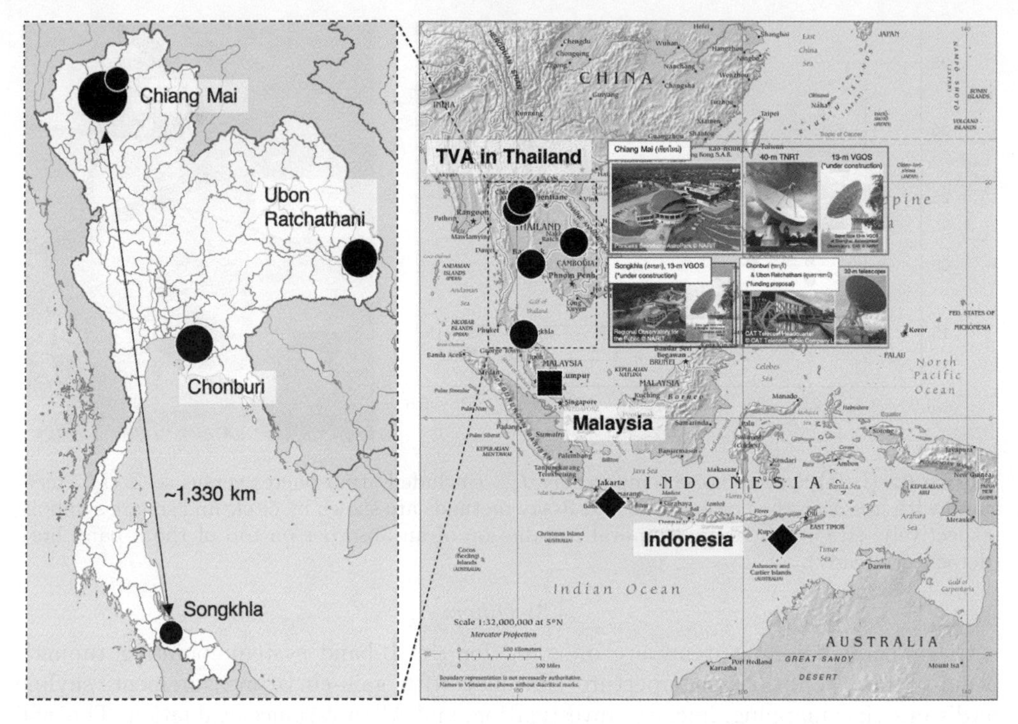

Figure 7. Forthcoming regional VLBI networks: Left - Thai National VLBI Array; Right - South-East Asian VLBI Network. In the left-panel, each radio telescope is shown by circles with the size proportional to its diameters. In the right-panel, radio telescopes (including forthcomings) are shown by circle, square, and diamond symbols with its fixed sizes for in Thailand, Malaysia, and Indonesia, respectively (Image credit of backgrounds: Left - NordNordWest in Wikipedia; Right - Monedula, CIA World Factbook, Wikimedia).

of a 32-m telecommunication antenna of Indosat Ooredoo located at Earth Station in Jatiluhur, West Java, at a distance of 1,840 km from the Timau site. Universiti Malaya (UM) in Malaysia is hosting a 13-m VGOS telescope in Jelebu in collaboration with SHAO. These stations combined with the TVA, eight stations in total, will constitute the start of SEAVN, as shown in Figure 7 (right). Key science goals of the SEAVN will be updated to reflect the fact that SEAVN will be one of the most flexible operations for VLBI accessible both in the Northern and Southern hemispheres, in collaboration with ITB, UM, and Vietnam National Space Center, Vietnam.

The development of the SEAVN will enhance VLBI networks, bringing them to the next level, through the drastic upgrade of baseline lengths, baseline / imaging sensitivities, and the UV-coverages resulting in much better synthesized-beams and improved imaging qualities. In particular, this enhancement is notable for reconstructing the Asia-Pacific Telescope, by filling in severe holes in the UV-coverage with the SEAVN telescopes located around the equator. These upgrades will result in the acceleration of an essential activity for the VLBI future led by the Global VLBI Alliance†.

We are grateful to Nikom Prasert, Spiro Sarris, Dan Singwong, Teep Chairin, and all of the engineers, technicians, and operators in the TNRO project team for excellent efforts to construct, develop, and operate TNRT. TNRO team would like to thank (1) all members of the International Technical Advisory Committee for System Integration & VLBI Development for the TNRO (TNRO-ITAC) for their advice and support; (2) Pablo de Vicente & José Antonio López-Pérez and Yebes Observatory team, IGN, for

† http://gvlbi.evlbi.org/welcome

the development of the Ku-band receivers for holographic measurement and the TCS, and advice on telescope construction with its operations; (3) Gundolf Wieching and the electronics division team in MPIfR for the development of the L- and K-band receivers, and the USB system.

References

Bayandina, O. S., Val'tts, I. E., Kurtz, S. E., *et al.* 2021, *ApJS*, 256, 7

de Vicente, P., Bolaño, R., & Barbas, L. 2006, *Astronomical Data Analysis Software and Systems XV*, 351, 758

Fujisawa, K., Mashiyama, H., Shimoikura, T., *et al.* 2002, Proc. IAU 8th Asian-Pacific Regional Meeting, Volume II, 3

Jaroenjittichai, P. 2018, *Pulsar Astrophysics the Next Fifty Years*, Proc. IAU Symposium, Vol. 337, p. 346

Jaroenjittichai, P., Punyawarin, S., Singwong, D., *et al.* 2017, *Journal of Physics Conf. Series*, 901, 012062

Jaroenjittichai, P., Sugiyama, K., Kramer, B. H., *et al.* 2022, arXiv:2210.04926

López Fernández, J. A., Gómez González, J., & Barcía Cáncio, A. 2006, *Lecture Notes and Essays in Astrophysics*, 257

Lopez-Perez, J. A., de Vicente Abad, P., Lopez-Fernandez, J. A., *et al.* 2014, *IEEE Transactions on Antennas and Propagation*, 62, 2624

Woodburn, L., Natusch, T., Weston, S., *et al.* 2015, *PASA*, 32, e017

Yonekura, Y., Saito, Y., Sugiyama, K., *et al.* 2016, *PASJ*, 68, 74

Cosmic Masers: Proper Motion toward the Next-Generation Large Projects
Proceedings IAU Symposium No. 380, 2024
T. Hirota, H. Imai, K. Menten, & Y. Pihlström, eds.
doi:10.1017/S1743921323002302

Expanded Maser Science Opportunities with the ALMA Wideband Sensitivity Upgrade

Crystal L. Brogan

National Radio Astronomy Observatory, Charlottesville, VA, USA. email: cbrogan@nrao.edu

Abstract. The ALMA Project is embarking on a partner-wide initiative to at least double, and ultimately quadruple the correlated bandwidth of ALMA by @2030. This initiative is called the ALMA Wideband Sensitivity Upgrade (WSU). In this contribution, I briefly describe the main aspects of the upgrade and status. Then I provide several examples of how the WSU will enhance (sub)millimeter maser science by affording the ability to observe more diagnostic maser transitions (and thermal lines) with a single observation.

Keywords. Radio observatories, Radio interferometers, Astrophysical masers, Interstellar molecules

---

1. Introduction

Consistent with the ALMA Development Roadmap (Carpenter *et al.* 2020), the ALMA Partnership has selected the ALMA 2030 Wideband Sensitivity Upgrade (WSU) as the highest priority development initiative for the coming decade. The primary goal of the WSU is to increase the system bandwidth by at least a factor of 2x, with a goal of 4x, with enhanced sensitivity. The WSU construction will involve replacing the majority of the ALMA Signal Chain, from the receivers to the correlator with new or improved technologies (Asayama *et al.* 2020). In order to ease the maintenance burden and cooling costs for a much more powerful correlator, the WSU ALMA Correlator called the "Advanced Technology ALMA Correlator" (ATAC) will be located at the Operations Support Facility (OSF) at 2,900 meters (rather than the Array Operations Site (AOS) at 5,000m). In addition to the ATAC, an upgraded Atacama Compact Array Total Power Spectrometer (ACAS), and the "receiving" portion (DRX) of the digital Data Transmission System (DTS) will also be housed in the new OSF Correlator Room.

The majority of the required hardware projects for the WSU are already underway. In November 2022, the ALMA Board approved the WSU Correlator Project, known as the Advanced Technology ALMA Correlator (ATAC). The ATAC is a collaboration between the National Research Council of Canada (NRC) and the National Radio Astronomy Observatory (NRAO). A project to prototype and demonstrate the WSU DTS was also approved at the November 2022 ALMA Board meeting; this project is a collaboration between National Astronomy Observatory of Japan (NAOJ) and NRAO. Projects are also underway to prototype and demonstrate WSU digitizers (and associated analog Back-end components) with the ability to process 4x bandwidth. The first upgraded wideband receiver bands are also under development: Band 2 (Yagoubov 2019), Band 6v2 (Navarrini *et al.* 2021), and Band 8v2 (Lee *et al.* 2021), led by the European Southern Observatory (ESO), NRAO, and NAOJ, respectively. In order to facilitate the transport of the significantly increased WSU data rate, ESO will lead a WSU infrastructure project to lay new fiber between the AOS and OSF; this project is due to begin in the next

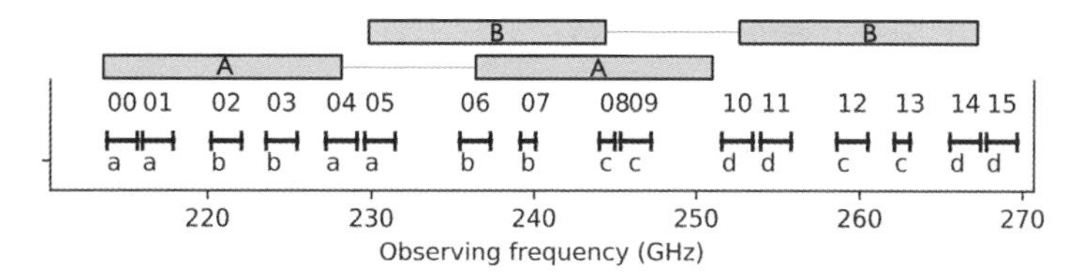

Figure 1. Demonstration of the tuning setup for the ALMA Large Program "ATOMIUM" (2018.1.00659.L, PI: Leen Decin, Decin *et al.* (2022)) that observed 17 Oxygen rich evolved stars, including a wide range of maser transitions. This program employed four spectral tunings (with spectral windows denoted by a, b, c, and d labels) in Band 6 (adapted from Gottlieb *et al.* (2022)). After the WSU, a similar spectral range can be covered by only two tunings (denoted by tunings A and B) and include additional spectral lines throughout the band. During the 2x BW era of the WSU, up to 16 GHz of the receiver's IF bandwidth (assumed to be 14 GHz per sideband for Band 6v2) can be correlated per tuning.

year. Additional efforts are underway in parallel to upgrade the downstream computing subsystems including data processing.

We hope to start science observing with the WSU, with at least 2x the current system bandwidth (16 GHz per polarization) and three wideband receivers (Bands 2, 6v2, and 8v2), before the end of this decade. Additional receiver bands will be upgraded to wideband designs over time (next in line are likely to be Bands 7, 9, and 10), and the correlated bandwidth will eventually be expanded to 4x the current bandwidth (32 GHz per polarization).

The science case and expanded capabilities for the WSU is described in detail in ALMA Memo 621 (Carpenter *et al.* 2023). In this proceeding we focus on a few examples of how the WSU will expand the ability to observe multiple diagnostic maser species simultaneously, or with many fewer tunings than previously. For demonstration purposes, we assume that upgraded WSU receivers will all have an IF bandwidth $= 4 - 18$ GHz in a 2 Sideband configuration (dual polarization); and that the maximum correlated bandwidth is 16 GHz per polarization (consistent with the start of WSU Science Operations). These assumptions are consistent with the WSU goals but the final receiver designs may be somewhat different.

2. Wider Bandwidth = More Maser Science with Less Observing Time

A significant limitation of the current Baseline Correlator used for 12 m-array observations (Escoffier *et al.* 2007) and the ACA spectro-correlator system, used for 7 m-array and total power array observations (Kamazaki *et al.* 2012) is that when high spectral resolution is required, one must give up significant bandwidth, especially in the lower ALMA bands. A key science goal of the WSU is to enable the ability to observe with a spectral resolution as fine as 0.1 km s^{-1} at full correlated bandwidth at any ALMA frequency from 35 to 950 GHz (a requirement that cannot presently be achieved at any ALMA band). This new capability will dramatically affect maser science, since high spectral resolution, especially for star formation masers, is essential.

2.1. *Band 6v2 Example for evolved stars*

The ALMA Cycle 6 Large Program: ATOMIUM (2018.1.00659.L, PI: Leen Decin) observed 17 Oxygen rich evolved stars using 4 distinct Band 6 tunings that spanned from 213.8 to 269.7 GHz with a total bandwidth of 27.19 GHz (see Figure 1). The key science goals were to study both thermal and maser emission lines from a wide range of diagnostic chemical tracers with a spectral resolution of 1.3 km s^{-1} (Gottlieb *et al.*

2022; Baudry *et al.* 2023). With the WSU, during the 2x BW era, a similar frequency range, including almost all the originally targeted maser transitions, can be covered with only two tunings, together covering a total of 32 GHz of correlated bandwidth. Taking into account the sensitivity improvements and fewer tunings, the same survey sensitivity could be accomplished with only 25% of the original observing time, or including 4x more targets. Alternatively, such a survey with same number of targets could be 2x deeper for the same observing time. This example demonstrates that after the WSU spectral scans will be much more efficient. Although it wasn't necessary for the evolved star science of the ATOMIUM survey, the same total bandwidth with two tunings (32 GHz in total) could be achieved for a spectral resolution as fine as 0.1 km s^{-1}.

2.2. *Band 6v2 Example for massive star formation*

A significant limitation of the current Baseline Correlator used for 12 m-array observations (Escoffier *et al.* 2007) and the ACA spectro-correlator system, used for 7 m-array and total power array observations (Kamazaki *et al.* 2012) is that when high spectral resolution is required, one must give up significant bandwidth, especially in the lower ALMA bands. A key science goal of the WSU is to enable the ability to observe with a spectral resolution as fine as 0.1 km s^{-1} at full correlated bandwidth at any ALMA frequency from 35 to 950 GHz (a requirement that cannot presently be achieved at any ALMA band). This new capability will dramatically affect maser science, since high spectral resolution, especially for star formation masers, is essential.

A prime example of the impact that this new WSU capability could have can be demonstrated by considering the follow-up of the recent accretion outburst event discovered toward the massive protostar G359.93-0.03-MM1 (Sugiyama *et al.* 2019; Burns *et al.* 2020; Stecklum *et al.* 2021). In (sub)millimeter follow-up aimed at constraining the expected rise in dust continuum emission, 14 never-before-seen methanol masers, primarily from $v_t = 1$ transitions, were discovered with initial peak emission in the thousands of Jy, declining in one month by factors of 4–7, and by 9 months later declining to near-thermal levels (See Figure 2, Brogan *et al.* 2019a). While a single low angular resolution but wideband observation with the SubMillimeter Array (SMA) detected 14 bright maser transitions from 198 to 360 GHz, higher angular resolution observations with ALMA required substantial observing time, including a total of four tunings (and three bands) to examine only nine of them. Moreover, very high angular resolution (30 mas) observations could only be arranged for one Band 6 tuning that could include only two of the masers (see Brogan 2019b), which revealed that the emission arose from a remarkable ring-structure surrounding the continuum emission and encompassing the 6.7 GHz masers which trace out a spiral arm structure (Burns *et al.* 2023). This lack of spatial coincidence could only be probed for two of the (sub)millimeter $v_t = 1$ maser transitions that are *A*-type pairs with $E_{upper} = 374$ K. Unfortunately, we will never know if other transitions, including $v_t = 0$ and $v_t = 2$ transitions, and E_{upper} as high as 877 K showed a similar morphology. After the WSU, many more maser transitions (and diagnostic thermal lines) can be included in a single observation – especially critical for time-variable phenomena (see Figure 3). Moreover, even for other kinds of science, the ability to include maser lines in more observations will enable more serendipitous detections, and even improved self-calibration (Brogan *et al.* 2018) in cases where the line signal-to-noise ratio is higher than the continuum.

2.3. *Band 7v2 Example*

(Sub)millimeter water masers and SiO masers (including other isotopologues of these species) are important tracers of a wide range of phenomena from star forming regions

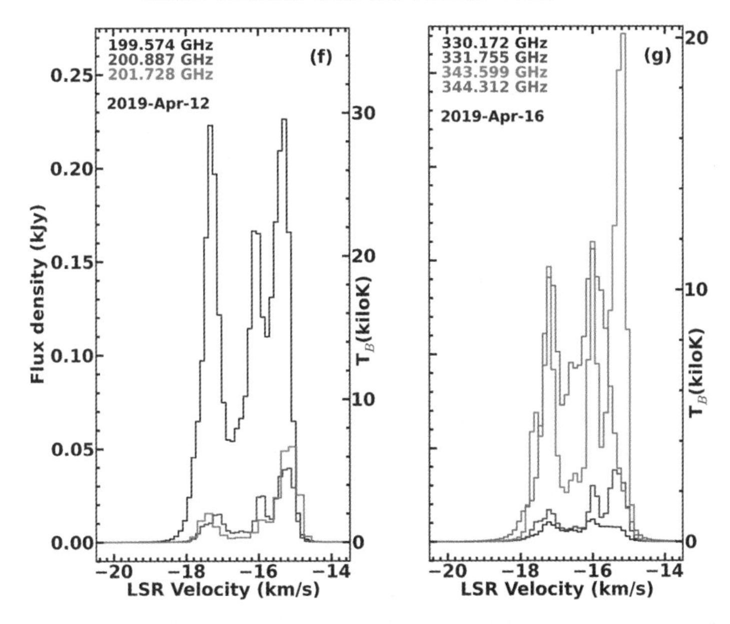

Figure 2. Spectra extracted from cubes obtained by ALMA observations in April 2019 of several newly-discovered torsionally-excited methanol maser transitions toward G358.93-0.03-MM1 (adapted from Brogan *et al.* 2019a). The 330.172 GHz line is the first detection of a maser from the $v_t = 2$ state.

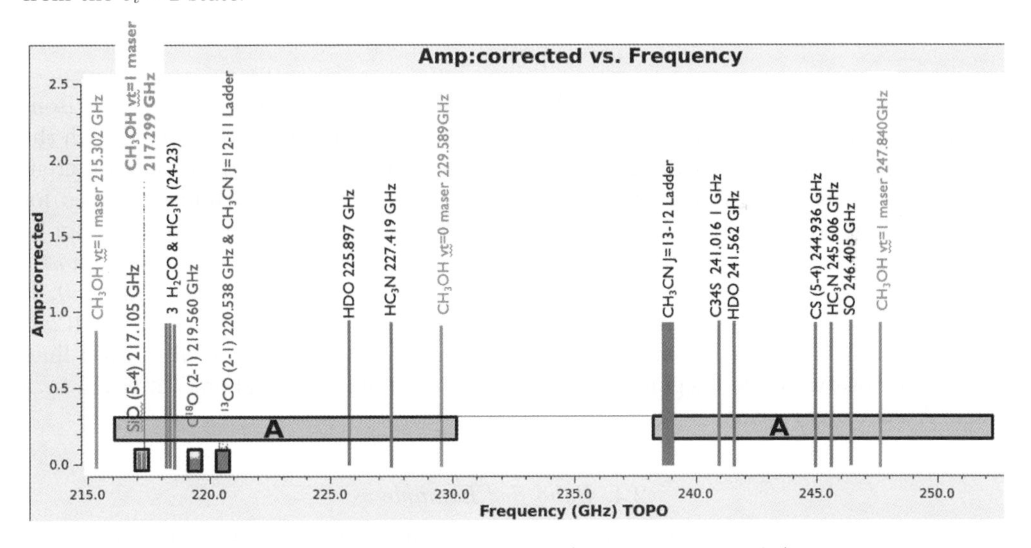

Figure 3. Example of an ALMA Band 6v2 setup (denoted tuning "A") that will be enabled by the WSU that could be used to follow-up future massive protostellar accretion outbursts for both methanol masers and diagnostic thermal transitions. Actual data from the narrow spectral regions shown along the bottom left side of the plot indicate the very limited amount of bandwidth (distributed within 217–221 GHz) that could be employed using the current ALMA system for the high angular and spectral resolution ($\sim 0.1\,\mathrm{km\ s^{-1}}$) follow-up of G358.93-0.03.

and evolved stars in our Milky Way to the nuclear disks of luminous galaxies (see for example Hirota *et al.* 2014; Wittkowski & Paladini 2014; Kameno *et al.* 2023). Band 7 covers two particularly diagnostic water maser transitions at 321 and 325 GHz with E_{lower} of 1846 K and 453 K, respectively (Neufeld & Melnick 1990; Menten *et al.* 1990). Unfortunately, as demonstrated in Figure 4, both of these transitions cannot presently

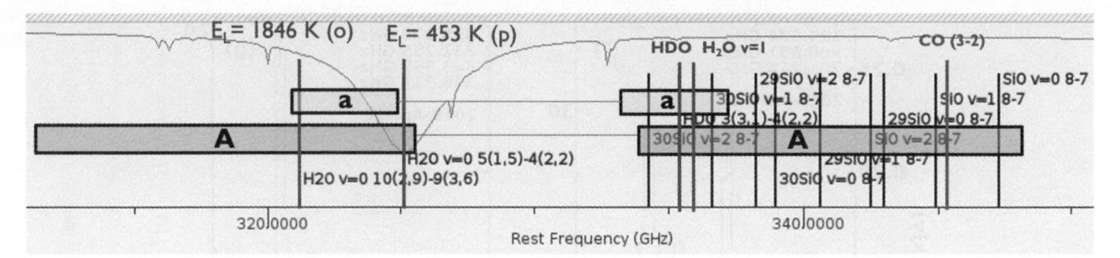

Figure 4. Example of a maser-focused current Band 7 tuning (denoted with "a") and a potential WSU tuning using an upgraded Band 7v2 tuning (denoted with "A"). During the 2x BW era of the WSU, up to 16 GHz of the receiver's IF bandwidth (assumed to be 14 GHz per sideband for Band 7v2) can be correlated per tuning.

Rest Freq (GHz)	Elower (K)	$S_{ij}\,\mu^2$	Target
437.347 (p)	1045	0.32	SF, Star
439.152 (o)	742	1.10	SF, Star
443.018 (o)	1045	0.98	thermal
470.889 (p)	742	0.36	SF, Star
474.689 (p)	488	0.41	Star

Figure 5. Previously detected water maser transitions in Band 8, where (o) or (p) denote whether it is an ortho or para transition. For the Target environment column, SF=Star Formation.

be observed simultaneously with ALMA, as a result there are few spatial comparisons between the two. Similarly, only a few SiO transitions can currently be observed with the same tuning at Band 7, though comparison between them can be used to study different envelope scale heights around evolved stars, as well as, isotopic abundances (see for example, Peng *et al.* 2013). With the 2x BW WSU, both water maser transitions can be observed in a single tuning, as well as a number of additional transitions with diagnostic value. For example, the ^{28}SiO, ^{29}SiO, and ^{30}SiO J=8–7 transitions, from the ν =0, 1, 2, vibrational states along with thermal emission from CO (3-2) and HDO 3(3,1)–4(2,2) can also be included in the same tuning. The ability to observe more diagnostic lines in a single observation will significantly improve the science "throughput" afforded by a single ALMA observation.

2.4. *Band 8v2 Example*

The WSU will also afford new simultaneous water maser tuning opportunities in Band 8. First discovered by Melnick *et al.* (1993), Figure 5 provides the rest frequencies, the energies above ground of the lower state (E_{lower}), the line strengths multiplied by the square dipole moment ($S_{ij}\mu^2$), and the typical environment(s) in which these water masers have been found to date (either "Star Forming" or AGB stars). As shown in Figure 6, presently, only the two lower frequency transitions can be observed simultaneously with ALMA. After the 2x BW WSU, four water transitions, including two ortho- and para- "pairs", can be observed simultaneously. These maser lines are of particular interest because there is evidence that they have a radiative component to their pumping in contrast to, for example 22 GHz water masers, which are thought to be pumped by collisions alone. The diagnostic value of surveying multiple of these water maser transitions was demonstrated using the 12 m telescope of the Atacama Pathfinder Experiment

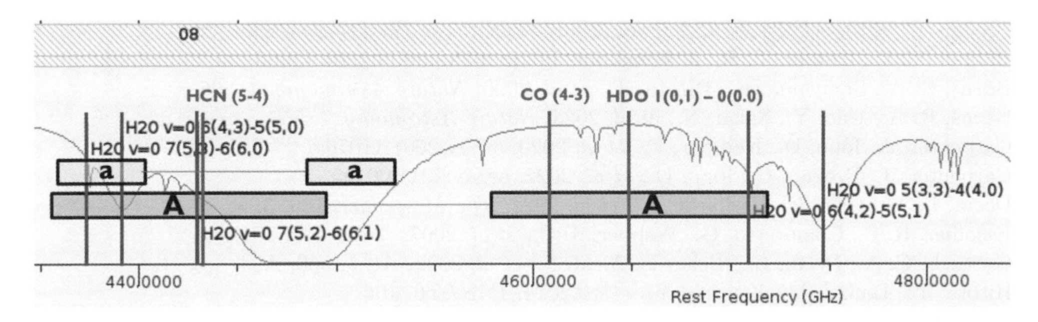

Figure 6. Example of a maser-focused current Band 8 tuning (with sidebands denoted with "a") and a potential WSU tuning using an upgraded Band 8v2 tuning (width sidebands denoted with "A"). During the 2x BW era of the WSU, up to 16 GHz of the receiver's IF bandwidth (assumed to be 14 GHz per sideband for Band 8v2) can be correlated per tuning.

(APEX) by Bergman & Humphreys (2020) toward a sample of evolved stars. At the same time, the thermal lines of CO (4-3), HCN (5-4), and HDO 1(0,1)–0(0,0) in Band 8 can also be observed to provide a more complete picture of the physical conditions that masers can provide alone. As demonstrated in NGC 6334 I by McGuire *et al.* (2018), HDO can be a powerful tracer of the thermal water reservoir associated with water masers.

3. Summary

The vision of the ALMA2030 Development Roadmap will begin to be realized with the increase in bandwidth, sensitivity, and spectral grasp that will be provided by the WSU. The new digital signal chain will raise the observing efficiency for all projects, while the increase in instantaneous receiver bandwidth and correlator capacity at high spectral resolution will increase the scientific yield in a variety of ways for many use cases. For maser science in particular, the ability to image multiple maser lines and thermal lines *simultaneously* coupled with more sensitive continuum should revolutionize our view of the formation mechanism of massive protostars. By the end of the decade, the WSU will become operational, including multiple receiver bands with expanded capability. Upgrades of the rest of the receiver bands will follow.

Acknowledgements

The National Radio Astronomy Observatory is a facility of the National Science Foundation operated under agreement by Associated Universities, Inc. ALMA is a partnership of ESO (representing its member states), NSF (USA) and NINS (Japan), together with NRC (Canada), MOST and ASIAA (Taiwan), and KASI (Republic of Korea), in cooperation with the Republic of Chile. The Joint ALMA Observatory is operated by ESO, AUI/NRAO and NAOJ. This paper makes use of the following ALMA data: ADS/JAO.ALMA#2019.A.00031.T. The SMA is a joint project between the Smithsonian Astrophysical Observatory (SAO) and the Academia Sinica Institute of Astronomy and Astrophysics (ASIAA) and is funded by the Smithsonian Institution and the Academia Sinica. This research used the https://www.splatalogue.net spectroscopy database (Remijan *et al.* 2016).

References

Asayama, S., Tan, G. H., Saini, K., *et al.* 2020, *SPIE Proceedings*, 11445, 1144575
Baudry, A., Wong, K. T., Etoka, S., *et al.* 2023, arXiv:2305.03171
Bergman, P. & Humphreys, E. M. L. 2020, *A&A*, 638, A19
Brogan, C. L., Hunter, T. R., Towner, A. P. M., *et al.* 2019, *ApJL*, 881, L39

Brogan, C. 2019, *ALMA2019: Science Results and Cross-Facility Synergies*, 28
Brogan, C. L., Hunter, T. R., & Fomalont, E. B. 2018, arXiv:1805.05266
Burns, R. A., Sugiyama, K., Hirota, T., *et al.* 2020, *Nature Astronomy*, 4, 506
Burns, R. A., Uno, Y., Sakai, N., *et al.* 2023, *Nature Astronomy*, 7, 557
Carpenter, J., Iono, D., Kemper, F., *et al.* 2020, arXiv:2001.11076
Carpenter, J., Brogan, C., Iono, D., *et al.* 2022, arXiv:2211.00195
Decin, L., Gottlieb, C., Richards, A., *et al.* 2022, *The Messenger*, 189, 3
Escoffier, R. P., Comoretto, G., Webber, J. C., *et al.* 2007, *A&A*, 462, 801
Gottlieb, C. A., Decin, L., Richards, A. M. S., *et al.* 2022, *A&A*, 660, A94
Hirota, T., Tsuboi, M., Kurono, Y., *et al.* 2014, *PASJ*, 66, 106
Kamazaki, T., Okumura, S. K., Chikada, Y., *et al.* 2012, *PASJ*, 64, 29
Kameno, S., Harikane, Y., Sawada-Satoh, S., *et al.* 2023, *PASJ*, 75, L1
Lee, J.-W., Kojima, T., Gonzalez, A., *et al.* 2021, *ALMA Front End Development Workshop*, 12
McGuire, B. A., Brogan, C. L., Hunter, T. R., *et al.* 2018, *ApJL*, 863, L35
Melnick, G. J., Menten, K. M., Phillips, T. G., *et al.* 1993, *ApJL*, 416, L37
Menten, K. M., Melnick, G. J., & Phillips, T. G. 1990, *Liege International Astrophysical Colloquia*, 29, 243
Navarrini, A., Kerr, A. R., Dindo, P., *et al.* 2021, *ALMA Front End Development Workshop*, 3
Neufeld, D. A. & Melnick, G. J. 1990, *ApJL*, 352, L9
Peng, T.-C., Humphreys, E. M. L., Testi, L., *et al.* 2013, *A&A*, 559, L8
Remijan, A., Seifert, N. A., & McGuire, B. A. 2016, *71st International Symposium on Molecular Spectroscopy*, FB11
Stecklum, B., Wolf, V., Linz, H., *et al.* 2021, *A&A*, 646, A161
Sugiyama, K., Saito, Y., Yonekura, Y., *et al.* 2019, *The Astronomer's Telegram*, 12446
Wittkowski, M. & Paladini, C. 2014, *EAS Publications Series*, 69-70, 179
Yagoubov, P. 2019, *ALMA Development Workshop*, 49

Cosmic Masers: Proper Motion toward the Next-Generation Large Projects
Proceedings IAU Symposium No. 380, 2024
T. Hirota, H. Imai, K. Menten, & Y. Pihlström, eds.
doi:10.1017/S1743921323001928

Maser science with the next generation Very Large Array (ngVLA)

Todd R. Hunter[1,2]

[1]National Radio Astronomy Observatory, Charlottesville, VA 22903, USA.
email: thunter@nrao.edu

[2]Center for Astrophysics | Harvard & Smithsonian, Cambridge, MA 02138, USA

Abstract. Imaging the bright maser emission produced by the various molecular species from 1.6 to 116 GHz provides a way to probe the kinematics of dense molecular gas at high angular resolution. Unimpeded by the high dust optical depths that affect shorter wavelength (sub)mm observations, the high brightness temperature of these emission lines have become an essential tool for understanding the process of massive star formation. Operating from 1.2–116 GHz, the next generation Very Large Array (ngVLA) of 263 antennas will provide the capabilities needed to fully exploit these powerful tracers, including the ability to resolve accretion and outflow motions down to scales as fine as ∼1-10 au in deeply embedded Galactic star-forming regions, and at sub-pc scales in nearby galaxies. I will summarize the proposed specifications of the ngVLA, describe the current status of the project, and offer examples of future experiments designed to image the vicinity of massive protostars in continuum, thermal lines, and maser lines simultaneously.

Keywords. Radio observatories, Radio interferometers, Astrophysical masers, Interstellar molecules

1. Introduction: The next generation Very Large Array

The 2020 US Decadal Survey of the National Academies (National Academies of Sciences 2021) identified the ngVLA as a high-priority large, PI-driven ground-based facility whose construction should start this decade. With scientific use cases that span imaging, beam forming, and pulsar timing (Murphy *et al.* 2018), the technical baseline of the facility (Selina *et al.* 2018) is summarized in Table 1. A key design choice was to place all antennas in fixed locations, in order to minimize operations costs. The main reflector geometry is an off-axis 18 m-diameter Gregorian with feed low to promote easier maintenance, an rms surface accuracy of 160 μm, and shaped optics to maximize the total system efficiency (>80% in bands 2-5, including digital efficiency). The panels are machined aluminum on a steel backup structure while the feed arm is mostly carbon fiber and the subreflector is composite. The servo performance specification includes a 4° slew and settle in 10 sec. The prototype is under construction in Germany. The current timeline of the project is shown in Table 2, with initial radiometric tests anticipated to begin in late 2024 at 3 cm and 7 mm.

2. Examples of science topics addressed by masers

We list some of the major scientific use cases of maser observations with the ngVLA. These topics are described in more detail in other contributions of this volume.

Table 1. ngVLA technical baseline.

Antennas			Receivers		Correlator / beamformer requirements	
Configuration group	Number	Baseline longest (km)	Band	Frequency (GHz)	Quantity	Requirement (design goal)
Core	114	4.3	1	1.2–3.5	digital efficiency	> 95%
Spiral	54	39	2	3.5–12.3	narrowest channel	< 1 kHz (0.4)
Mid	46	1070	3	12.3–20.5	total channels	>240000
Long	30	8860	4	20.5–34	total bandwidth	> 14 GHz/pol (20)
Total 18m	244	Range (m)	5	30.5–50.5	sub-band width	<250 MHz (218.75)
SBA 6m	19	11–60	6	70–116	# formed beams	10 (50)

Table 2. ngVLA major milestone dates (as of May 2023).

Milestone	Date
Conceptual Design Review (technical)	July 2022
Cost Review	December 2023
Prototype antenna delivery	Late 2024
Preliminary Design Review	Early 2026
Final Design Review (technical)	Mid 2028
Construction	2028-2037
Early Science Operations	Mid 2031
Full Science Operations	Mid 2037

2.1. *Megamaser cosmology*

The use of water megamasers to provide constraints on cosmological parameters is described in Braatz *et al.* (2018). Current precision on H_0 derived from megamaser observations is 3%. Reducing the value to 1% will require at least 50 galaxies measured with $\approx 7\%$ precision. New species of megamasers have been identified recently (Gorski *et al.* 2021).

2.2. *Evolved Stars*

Simultaneous imaging of masers and stellar photosphere with excellent uv coverage is essential to more accurately study the layering of gas properties in evolved stars (Matthews & Claussen 2018). The ability to measure the proper motion of masers over a complete stellar pulsation period will better constrain the pumping mechanism. Multi-band spectra of SiO 2-1 and SiO 1-0, observed contemporaneously in Bands 5 and 6 with ngVLA, will improve the efficiency of surveys of AGB star masers like the BAaDE survey (Sjouwerman *et al.* 2020).

2.3. *Structure of the Milky Way*

Very Long Baseline Interferometry (VLBI) enables parallax and proper motion measurements of masers in massive star-forming regions, which trace the spiral structure of the Milky Way and provide the most accurate estimates of the distance to the Galactic center (Reid *et al.* 2014). The greater sensitivity of ngVLA will provide a more complete view of the Galaxy by enabling routine measurements of these regions on the far side of the Galaxy and greatly expanding the pool of magnetically active young stars for proper motion measurements on the near side (Loinard & Reid 2018).

3. Importance of maser emission in star formation

The process of star formation leads to the concentration of molecular gas to high densities in molecular cloud cores. The potential energy released by gravitational collapse and accretion onto the central protostars heats and excites the surrounding material

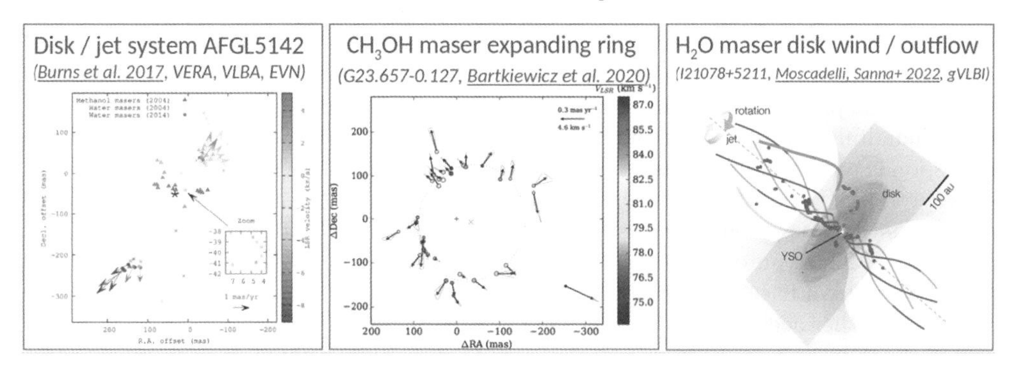

Figure 1. Three examples of kinematic structures traced by VLBI observations of masers around young massive protostars from Burns *et al.* (2017), Bartkiewicz *et al.* (2020), and Moscadelli *et al.* (2022).

through infrared radiation and high velocity bipolar outflows. Both of these feedback mechanisms (radiative and mechanical) naturally produce population inversions between specific pairs of energy levels in several abundant molecules, including H_2O, CH_3OH, OH, NH_3, SiO, and H_2CO. The resulting non-thermal maser emission in the corresponding spectral transitions provides a beacon whose brightness temperature far exceeds the more commonly-excited thermal emission lines. Consequently, maser lines at centimeter wavelengths have traditionally provided a powerful probe of star formation through single-dish surveys and interferometric imaging. In general, they trace hot, dense molecular gas, revealing the kinematics of star-forming material within a few 1000 au from very young stars, including accretion disks and their associated jets, as well as shocks in the outflow lobes where the jets impact ambient gas (Fig. 1).

Masers are generally more prevalent in regions surrounding massive protostars, due to their higher luminosities and more energetic outflows. Furthermore, masers are sensitive indicators of sudden changes in the pumping conditions near the protostars, and, recently, it has been recognized that maser flares, in those lines which are pumped by infrared photons, can be directly associated with bursts of accretion onto the stars. In this context, maser emission provides a unique tool for probing how massive stars form, allowing us to reconstruct the gas dynamics in the vicinity of young stars with tens of Solar masses, as well as to study the accretion process in the time domain.

4. Obscuration of line emission by dust

With the advent of the Atacama Large Millimeter/submillimeter Array (ALMA), imaging thermal lines at high resolution has become easier, and recent results have begun to place previous and ongoing maser studies into better physical context (see, e.g., Orion Source I, Plambeck & Wright 2016; Hirota *et al.* 2017). At the distances of more typical massive star-forming regions, however, the brightness temperature sensitivity of ALMA is still not sufficient to trace the accretion flow and accompanying jet structures that surround massive protostars, because of the high angular and spectral resolution required. Moreover, at the shorter wavelengths of ALMA, the combination of molecular line confusion and high dust opacity toward the hot cores in protoclusters will often block the most interesting details from ALMA's view (van Gelder *et al.* 2022; Nazari *et al.* 2022). In contrast, the centimeter maser transitions propagate unobscured from the innermost regions and provide a strong signal for self-calibration, which enables high dynamic range imaging on long baselines.

5. Importance of angular resolution coupled with sensitivity

Unfortunately, the angular resolution of the VLA is insufficient to resolve the details of accreting gas, particularly in the 6 GHz band where the beamsize is limited to ∼0.3 arcsec. In the handful of nearest examples of massive star formation (d∼1 kpc), this resolution corresponds to 300 au (e.g., Brogan *et al.* 2016). However, in the majority of star-forming sites across the Milky Way at several kpc from the Sun, it exceeds 1000 au, which is often more than the separation of protostars in the centers of massive protoclusters (e.g., Palau *et al.* 2013). Thus, an order of magnitude improvement in angular resolution (requiring at least ∼300 km baselines) is needed to resolve the spatial morphology and kinematics of disks, or other accretion structures, at scales of 10-100 au in a large sample of massive protostars. Such a resolution would also enable three dimensional measurements of gas velocity via multi-epoch proper motions. For instance, with an angular resolution of 10 mas in the bright H_2O maser line, it is possible to determine the maser positions with an accuracy better than 0.1 mas (assuming $S/N > 100$), and then to measure proper motions of order 10 km s^{-1} in a few months only, for sources located up to several kpc distance. Furthermore, proper motion measurements of different maser species toward the same region have the potential to trace simultaneously the complementary kinematic structures around a young star, providing a unique picture of the gas dynamics locally (Sanna *et al.* 2010; Goddi *et al.* 2011). These measurements are currently conducted by means of VLBI observations, but with the small number of available antennas their sensitivity is limited to non-thermal processes exceeding brightness temperatures of ∼ 10^7 K (e.g. Matsumoto *et al.* 2014; Bartkiewicz *et al.* 2009).

While current VLBI facilities (VLBA, EVN, eMERLIN, KVN, VERA, and LBA) have the requisite angular resolution to trace maser proper motions accurately, studies at these scales currently suffer from poor surface brightness sensitivity, which affects the science in two key ways. First, only the brightest maser spots can be detected, reducing the fidelity with which kinematic structures can be delineated in a single epoch, and reducing the number of potential spots that will persist over multiple epochs (used for proper motion studies). Second, the thermal radio continuum emission which arises in the immediate vicinity of massive protostars, with flux densities of < 1 mJy typically (Cyganowski *et al.* 2011), cannot be observed with the masers simultaneously, leading to (relative) positional uncertainties between the protostellar and maser components. The resulting ambiguity of the dynamical center severely hinders the interpretation of multi-epoch proper motion measurements, which are essential to understand the mass, momentum, and kinetic energy of the inner jet where it transitions into a bipolar molecular outflow. Studying additional objects at scales of 10-100 au in a comprehensive list of maser lines, and with sufficient brightness sensitivity to image simultaneously the associated thermal free-free continuum emission, will be an important task for the ngVLA. These studies will test and expand our current picture of massive star formation into the broader context of the Milky Way.

6. Masers as probes of accretion outbursts

With its proposed frequency span, the ngVLA will uniquely provide access to all of the most important maser transitions from OH at 1.6 GHz to CH_3OH at 109 GHz (Table 3). While each maser transition offers a unique view into any particular region massive star formation, masers can be broadly classified into major categories. The Class II CH_3OH maser lines, primarily at 6.7 GHz, 12.2 GHz, and 19.9 GHz, trace hot molecular gas that is (at least) moderately close to the youngest massive protostars ($\lesssim$ 1000 au), such that they can provide sufficiently intense mid-infrared emission to pump the maser transitions (Sobolev *et al.* 1997; Cragg *et al.* 2005). The light curves of these maser

Table 3. Detected maser lines (>94) in the ngVLA bands; see Table 1 of Menten (2007) for further details. Frequencies in boldface have been detected only during the accretion outburst in G358.93-0.03.

Band	f (GHz)	Species and line frequencies (GHz)	# lines
1	1.2–3.5	ground-state OH (1.612, 1.665, 1.667, 1.720)	4
2	3.5-12.3	excited OH (4.660, 4.750, 4.765, 6.031, 6.035, 6.049)	6
		CH_3OH Class I (9.936)	1
		CH_3OH Class II (6.668, 12.178), (**7**.682, 7.830, Breen *et al.* (2019)), CH_3OH likely Class II: (**12**.229, MacLeod *et al.* (2019))	5
		CH_3OH torsionally-excited: **6**.181, Breen *et al.* (2019)	1
		ortho-H_2CO (4.829)	1
		CH_2NH (5.291) (Gorski *et al.* 2021), CH_3NH_2 (4.364) (Xue *et al.* 2023)	2
3	12.3–20.5	CH_3OH Class II (19.967), **20.346** (MacLeod *et al.* 2019)	2
		$^{13}CH_3OH$: **14.300** (Chen *et al.* 2020)	1
		excited OH (13.441, Baudry *et al.* 1981; Caswell 2004)	1
		NH_3 (**19.757**) (McCarthy *et al.* 2023)	1
4	20.5–34	H_2O (22.235)	1
		CH_3OH J_2-J_1-series Class I (24.9-30.3) Towner *et al.* (e.g., 2017)	15
		CH_3OH Class I (23.445, Voronkov *et al.* 2011)	1
		CH_3OH Class II: 23.121, 28.970 (**27.283, 28.969**, Miao *et al.* (2022))	4
		CH_3OH torsionally-excited: **20.970** (Breen *et al.* 2019), **26.120** (Miao *et al.* 2022)	2
		ortho-NH_3 (3,3) (6,6) (9,9) (12,12) Brogan *et al.* (e.g., 2011)	4
		other NH_3 inversion lines (thermal and/or maser)	>10
		excited OH (23.817, Baudry *et al.* 1981)	1
		HDO: **20.460** (Chen *et al.* 2020)	1
		HNCO: **21.980** (Chen *et al.* 2020)	1
5	30.5–50.5	Class I CH_3OH (36.169, 44.069)	2
		Class II CH_3OH: 37.703, 38.293, 38.452, (**31.977, 34.236, 41.110, 46.558** Miao *et al.* (2022))	3
		SiO 1-0 v=1, rv=2, v=3 (43.122, 42.820, 42.519)	3
		CH_3OH torsionally-excited: **44.955** (Breen *et al.* 2019), **48.708** (Miao *et al.* 2022)	2
		CS 1-0: 48.990 (Ginsburg & Goddi 2019)	1
6	70–116	CH_3OH Class I (84.521, 95.169, 104.300)	3
		CH_3OH Class II (76.247, 76.509, 85.568, 86.615, 104.060, 107.013, 108.894)	7
		CH_3OH torsionally-excited (**99.772, 102.957**, Menten (2019))	2
		HCN J=1-0 vibrationally-excited (89.087)	1
		HCO^+ 1-0: 89.188 (Hakobian & Crutcher 2012)	1
		CS 2-1: 97.980 (Ginsburg & Goddi 2019)	1
		SiO 2-1 v=1, v=2, v=3 (86.243, 85.640, 85.038)	3

species also show intriguing properties. Quasi-periodic flares in one or more Class II CH_3OH maser lines (120–500 days) have been observed in about a dozen objects (e.g., Goedhart *et al.* 2014); in one case, the 4.83 GHz H_2CO maser also shows correlated flaring (Araya *et al.* 2010). Recently, three spectacular accretion outbursts in massive protostars have been accompanied by strong flaring of these lines simultaneously with mid-IR continuum, S255 NIRS3 (Caratti o Garatti *et al.* 2017; Moscadelli *et al.* 2017), NGC 6334 I-MM1 (Hunter *et al.* 2017, 2018; MacLeod *et al.* 2018; Hunter *et al.* 2021), and G358.93-0.03 (Brogan *et al.* 2019; Breen *et al.* 2019; MacLeod *et al.* 2019; Stecklum *et al.* 2021) supporting the idea that maser flares are caused by a variable accretion rate onto the central protostar. The first two of these extraordinary events led to the formation of the international Maser Monitoring Organization (M2O), with the goal of detecting and reporting future maser flares so that interferometers can be alerted to study the accretion event while it is still underway. Such an accretion event is also expected to yield variation in the continuum emission from the thermal jet (Cesaroni *et al.* 2018) and/or the hypercompact HII region (Brogan *et al.* 2018). Since both of these phenomena are powered by the protostar, the ability of ngVLA to perform simultaneous observations of the continuum and the masers will enable direct measurements of the correlations between them, yielding important constraints on the physics of the accretion mechanism that is currently being explored via hydrodynamic simulations (Elbakyan *et al.* 2023). Furthermore, the fixed antenna layout means that *rapid* follow-up at each necessary angular scale will *always* be possible, no longer limited by the traditional cadence of VLA configurations. This point is especially important since some outbursts rise and fall in just a few weeks (e.g., G36.11+0.55 Yonekura 2022).

The Class II CH_3OH maser lines, along with the 1.6 GHz ground state OH lines and several excited state OH lines which are radiatively pumped (at 4.66 GHz, 4.75 GHz, 4.765 GHz, 6.030 GHz, and 6.035 GHz), are also seen to trace the ionization front of ultracompact HII regions (e.g., Fish & Reid 2007), which are powered by the more evolved massive protostars and Zero-Age Main Sequence (ZAMS) OB stars. Although excited OH lines are generally considered rare, a recent unbiased survey found that the 6.035 GHz line is detected toward nearly 30% of Class II CH_3OH masers and with a similar distribution in Galactic latitude (Avison *et al.* 2016). A similar detection rate is seen in survey of the 4.765 GHz line (Dodson & Ellingsen 2002). A simple explanation is these excited OH masers always occur in the same objects that power Class II masers, but simply have a correspondingly shorter mean lifetime or duty cycle, perhaps reflecting how long they remain above current sensitivity levels following each successive accretion outburst. In rare cases, the main line OH masers can also trace outflow motion (e.g., W75N and W3OH-TW Fish *et al.* 2011; Argon *et al.* 2003).

The strong water maser line at 22 GHz also traces gas close to massive and intermediate mass protostars. Often these lines span a broad (LSR) velocity range, of several tens of km s^{-1}, about the systemic velocity of the young stars, particularly compared to both classes of methanol masers ($\lesssim 10\,\mathrm{km\,s^{-1}}$). In some cases, water masers clearly arise from gas in the first few hundred au of the jet, such as in Cepheus A (e.g., Torrelles *et al.* 2011; Chibueze *et al.* 2012), or in bow shocks somewhat further out (e.g., Sanna *et al.* 2012; Burns *et al.* 2016). With continent-scale baselines, proper motion studies of these masers reveal the 3D velocities and orientations of collimated jets and/or wide-angle winds in the inner few 1000 au from the central protostars (e.g., Torrelles *et al.* 2014; Sanna *et al.* 2010). When these studies are combined with high-resolution radio continuum observations of radio thermal jets, they can allow us to quantify the outflow energetics directly produced by the star formation process (e.g., Moscadelli *et al.* 2016; Sanna *et al.* 2016), as opposed to estimates of the molecular outflow energetics that are attainable on scales greater than 0.1 pc (typically through CO isotopologues). Long-term monitoring studies demonstrate that water masers are also highly variable (e.g., Felli *et al.* 2007), and their primary pumping mechanism is not believed to be radiative but collisional. Thus, since water masers are fundamentally produced in specific ranges of gas density and temperature within shocked gas layers, they are likely to trace different types of coherent motions at different stages of protostellar evolution. This is the case, for instance, of the star-forming region W75 N, where the 22 GHz masers (and the radio continuum) show different spatial distributions around two distinct young stars at different evolutionary stages (Carrasco-González *et al.* 2015). The 22 GHz line is also unique in exhibiting the 'superburst' phenomenon, in which brief flares reach 10^5 Jy or more. It has happened in only a few objects including Orion KL (Hirota *et al.* 2014, and references therein) and G25.65+1.05 (Lekht *et al.* 2018), but it has repeated in both, and appears to be due to interaction of the jet with high density clumps in the ambient gas, but within a few thousand au of the central protostar. In addition to Galactic studies, the detection and imaging of water masers in nearby star-forming galaxies provides a powerful probe of optically-obscured areas of star formation like the overlap region of the Antennae (Brogan *et al.* 2010).

Maser emission from the vibrationally-excited levels of SiO offers a powerful (though rare) probe of the innermost hot gas surrounding massive protostars. For example, in one spectacular nearby case (Orion KL), movies of the vibrationally-excited SiO J=1-0, v=0 and v=1 transitions at 43 GHz have revealed a complicated structure of disk rotation and outflow (Matthews *et al.* 2010). Additional massive protostars (at greater distances) have recently been detected in these lines (Cordiner *et al.* 2016; Ginsburg *et al.* 2015; Zapata *et al.* 2009). The increased sensitivity of the ngVLA will no doubt yield further detections and enable new detailed images of the inner accretion structures in these objects.

7. Connection to thermal lines

Even with the sensitivity of ngVLA, masers will provide only a partial view of circumprotostellar material. We will still rely on thermal lines to complete the picture and provide quantitative measurements of gas temperature and column density. The lines traditionally used at the VLA like NH_3, and 3 mm lines like CH_3CN 6-5 at 110 GHz will be exploited to higher angular resolution than currently possible. In addition, HDO $1_{1,0}$–$1_{1,1}$ at 80.578 GHz will provide a powerful tracer of the thermal water reservoir. Not yet observed by ALMA (it lies in the range of the Band 2 receiver, which is still in the construction phase), this transition traces the hot corino IRAS 16293-2422 B (Parise *et al.* 2005). As shown by the energy level diagram (Kulczak-Jastrzebska 2016), the E_{lower} of this transition is 47 K and its lower state is the upper state of the 893.636 GHz transition to the ground state $0_{0,0}$ in ALMA Band 10, which was recently observed by McGuire *et al.* (2018) in NGC 6334 I. This line beautifully traces a thermal gas outflow from the outbursting protostar MM1, strikingly aligned and interspersed with a linear distribution of 22 GHz maser spots, and demonstrates the potential of the 80.578 GHz line to probe similar structures. Finally, variations in the intensities of thermal lines in response to accretion outbursts is a promising line of research with initial results in the periodic CH_3OH maser G24.33+0.14 (Hirota *et al.* 2022).

8. Conclusion

In summary, with the ability to image the non-thermal and thermal processes simultaneously, it will finally be possible to link the studies of the 3D gas dynamics (using the masers) with studies of the physical conditions of the ionized and molecular gas (using the continuum and strong, compact thermal lines like ammonia, respectively) at the same spatial resolution. Also, the ability to acquire high-fidelity images of all of these maser lines in just a few tunings will promote more uniform surveys of massive protostars as well as enable rapid monitoring of protostars currently undergoing an accretion outburst. Furthermore, the broader bandwidth receivers will provide more robust measurements of the spectral index of the continuum emission by promoting the ability to obtain all the necessary observations at a common epoch. With the improved continuum sensitivity, young lower-mass T Tauri stars, which are chromospherically active stars associated with highly-variable faint (synchrotron) radio continuum emission, will be also detected in the lower frequency bands (e.g., Band 2, Forbrich *et al.* 2017), providing information about the low-mass population of protoclusters. Finally, the high spectral resolution and full polarization capability of the ngVLA will allow measurements of the magnetic field in the masing molecular gas via the Zeeman effect in methanol and water, which is a fundamental quantity for understanding the physics of star formation.

Acknowledgements

The National Radio Astronomy Observatory is a facility of the National Science Foundation operated under agreement by Associated Universities, Inc. This research used the https://www.splatalogue.net spectroscopy database (Remijan *et al.* 2016).

References

Araya, E. *et al.* 2010, *ApJ*, 717, 133
Argon, A. L., Reid, M. J., & Menten, K. M. 2003, *ApJ*, 593, 925
Avison, A., Quinn, L. J., Fuller, G. A., *et al.* 2016, *MNRAS*, 461, 136
Bartkiewicz, A., Sanna, A., Szymczak, M., *et al.* 2020, *A&A*, 637, A15
Bartkiewicz, A., Szymczak, M., van Langevelde, H. J., Richards, A. M. S., & Pihlström, Y. M. 2009, *A&A*, 502, 155
Baudry, A., Walmsley, C. M., Winnberg, A., & Wilson, T. L. 1981, *A&A*, 102, 287
Braatz, J., Pesce, D., Condon, J., *et al.* 2018, *Science with a Next Generation Very Large Array*, 517, 821
Breen, S. L., Sobolev, A. M., Kaczmarek, J. F., *et al.* 2019, *ApJL*, 876, L25
Brogan, C. L., Hunter, T. R., Towner, A. P. M., *et al.* 2019, *ApJL*, 881, L39
Brogan, C.L., *et al.* 2018, in *Proc. of IAU Symp.* 336, ed. A. Tarchi, M.J. Reid & P. Castangia, Cambridge University Press
Brogan, C.L., *et al.* 2016, *ApJ*, 832, 187
Brogan, C. L., Hunter, T. R., Cyganowski, C. J., *et al.* 2011, *ApJL*, 739, L16
Brogan, C., Johnson, K., & Darling, J. 2010, *ApJL*, 716, L51
Burns, R. A., Handa, T., Imai, H., *et al.* 2017, *MNRAS*, 467, 2367
Burns, R. A., Handa, T., Nagayama, T., Sunada, K., & Omodaka, T. 2016, *MNRAS*, 460, 283
Caratti o Garatti, A., Stecklum, B., Garcia Lopez, R., *et al.* 2017, *Nature Physics*, 13, 276
Carrasco-González, C., Torrelles, J. M., Cantó, J., *et al.* 2015, *Science*, 348, 114
Caswell, J. L. 2004, *MNRAS*, 352, 101
Cesaroni, R., Moscadelli, L., Neri, R., *et al.* 2018, *A&A*, 612, 103
Chen, X., Sobolev, A. M., Ren, Z.-Y., *et al.* 2020, *Nature Astronomy*, 4, 1170.
Chibueze, J. O., Imai, H., Tafoya, D., *et al.* 2012, *ApJ*, 748, 146
Cordiner, M. A., Boogert, A. C. A., Charnley, S. B., *et al.* 2016, *ApJ*, 828, 51
Cragg, D. M., Sobolev, A. M., & Godfrey, P. D. 2005, *MNRAS*, 360, 533
Cyganowski, C. J., Brogan, C. L., Hunter, T. R., & Churchwell, E. 2011, *ApJ*, 743, 56
Dodson, R. G., & Ellingsen, S. P. 2002, *MNRAS*, 333, 307
Elbakyan, V. G., Nayakshin, S., Meyer, D. M.-A., *et al.* 2023, *MNRAS*, 518, 791
Felli, M., Brand, J., Cesaroni, R., *et al.* 2007, *A&A*, 476, 373
Fish, V. L., Gray, M., Goss, W. M., & Richards, A. M. S. 2011, *MNRAS*, 417, 555
Fish, V. L., & Reid, M. J. 2007, *ApJ*, 670, 1159
Forbrich, J., Reid, M. J., Menten, K. M., *et al.* 2017, *ApJ*, 844, 109
Ginsburg, A. & Goddi, C. 2019, *AJ*, 158, 208.
Ginsburg, A., Walsh, A., Henkel, C., *et al.* 2015, *A&A*, 584, L7
Goddi, C., Moscadelli, L., & Sanna, A. 2011, *A&A*, 535, L8
Goedhart, S., *et al.* 2014, *MNRAS*, 437, 1808
Gorski, M. D., Aalto, S., Mangum, J., *et al.* 2021, *A&A*, 654, A110.
Hakobian, N. S. & Crutcher, R. M. 2012, *ApJL*, 758, L18.
Hirota, T., Wolak, P., Hunter, T. R., *et al.* 2022, *PASJ*, 74, 1234
Hirota, T., Machida, M. N., Matsushita, Y., *et al.* 2017, *Nature Astronomy*, 1, 0146
Hirota, T., Tsuboi, M., Kurono, Y., *et al.* 2014, *PASJ*, 66, 106
Hunter, T. R., Brogan, C. L., De Buizer, J. M., *et al.* 2021, *ApJL*, 912, L17
Hunter, T. R., Brogan, C. L., *et al.* 2018, *ApJ*, 854, 170
Hunter, T. R., Brogan, C. L., MacLeod, G. C., *et al.* 2017, *ApJL*, 837, L29
Kulczak-Jastrzebska, M. 2016, *Acta Astronomica*, 66, 239
Lekht, E. E., *et al.* 2018, *Astronomy Reports*, 62, 213

Loinard, L. & Reid, M. J. 2018, Science with a Next Generation Very Large Array, 517, 411
MacLeod, G. C., Sugiyama, K., Hunter, T. R., *et al.* 2019, *MNRAS*, 489, 3981
MacLeod, G. C., *et al.* 2018, *MNRAS*, 478, 1077
Matsumoto, N., Hirota, T., Sugiyama, K., *et al.* 2014, *ApJL*, 789, L1
Matthews, L. D. & Claussen, M. J. 2018, *Science with a Next Generation Very Large Array*, 517, 281
Matthews, L., *et al.* 2010, *ApJ*, 708, 80
McCarthy, T. P., Breen, S. L., Kaczmarek, J. F., *et al.* 2023, *MNRAS*, 522, 4728.
McGuire, B. A., Brogan, C. L., Hunter, T. R., *et al.* 2018, *ApJL*, 863, L35
Menten, K.M. 2019, private communication
Menten, K. M. 2007, *Astrophysical Masers and their Environments*, IAU Symposium 242, 496
Miao, D., Chen, X., Song, S.-M., *et al.* 2022, *ApJS*, 263, 9
Moscadelli, L., Sanna, A., Beuther, H., *et al.* 2022, *Nature Astronomy*, 6, 1068
Moscadelli, L., Sanna, A., Goddi, C., *et al.* 2017, *A&A*, 600, L8
Moscadelli, L., Sánchez-Monge, Á., Goddi, C., *et al.* 2016, *A&A*, 585, A71
Murphy, E. J., Bolatto, A., Chatterjee, S., *et al.* 2018, *Science with a Next Generation VLA*, 517, 3
National Academies of Sciences, E. 2021, *Pathways to Discovery in Astronomy and Astrophysics for the 2020s*, Washington, DC: The National Academies Press, 2021
Nazari, P., Tabone, B., Rosotti, G. P., *et al.* 2022, *A&A*, 663, A58
Palau, A., Fuente, A., Girart, J. M., *et al.* 2013, *ApJ*, 762, 120
Parise, B., Caux, E., Castets, A., *et al.* 2005, *A&A*, 431, 547
Plambeck, R. L., & Wright, M. C. H. 2016, *ApJ*, 833, 219
Reid, M. J., Menten, K. M., Brunthaler, A., *et al.* 2014, *ApJ*, 783, 130
Remijan, A., Seifert, N. A., & McGuire, B. A. 2016, *71st Intl. Symp. on Molecular Spectroscopy*, FB11
Sanna, A., Moscadelli, L., Cesaroni, R., *et al.* 2016, *A&A*, 596, L2
Sanna, A., Reid, M. J., Carrasco-González, C., *et al.* 2012, *ApJ*, 745, 191
Sanna, A., Moscadelli, L., Cesaroni, R., *et al.* 2010, *A&A*, 517, A78
Selina, R. J., Murphy, E. J., McKinnon, M., *et al.* 2018, *Science with a Next Generation VLA*, 517, 15
Sjouwerman, L. O., Pihlström, Y. M., *et al.* 2020, *Galactic Dynamics in the Era of Large Surveys*, 353, 45
Sobolev, A. M., Cragg, D. M., & Godfrey, P. D. 1997, *A&A*, 324, 211
Stecklum, B., Wolf, V., Linz, H., *et al.* 2021, *A&A*, 646, A161
Torrelles, J. M., Trinidad, M. A., Curiel, S., *et al.* 2014, *MNRAS*, 437, 3803
Torrelles, J. M., Patel, N. A., Curiel, S., *et al.* 2011, *MNRAS*, 410, 627
Towner, A. P. M., Brogan, C. L., Hunter, T. R., *et al.* 2017, *ApJS*, 230, 22
Voronkov, M. A., Walsh, A. J., Caswell, J. L., *et al.* 2011, *MNRAS*, 413, 2339
Xue, C., *et al.* 2023, in preparation
van Gelder, M. L., Nazari, P., Tabone, B., *et al.* 2022, *A&A*, 662, A67
Yonekura, Y., 2022, private communication
Zapata, L. A., Menten, K., Reid, M., & Beuther, H. 2009, *ApJ*, 691, 332

Cosmic Masers: Proper Motion toward the Next-Generation Large Projects
Proceedings IAU Symposium No. 380, 2024
T. Hirota, H. Imai, K. Menten, & Y. Pihlström, eds.
doi:10.1017/S1743921323002405

GASKAP-OH: A New Deep Survey of Ground-State OH Masers and Absorption in the Southern Sky

J. R. Dawson[1,2], S. L. Breen[3] and the GASKAP-OH Team

[1]School of Mathematical and Physical Sciences and Macquarie University Astrophysics and Space Technologies Research Centre, Macquarie University, NSW 2109, Australia.
email: joanne.dawson@mq.edu.au

[2]Australia Telescope National Facility, CSIRO Space & Astronomy, NSW 1710, Australia

[3]SKA Observatory, Jodrell Bank, Lower Withington, Macclesfield SK11 9FT, UK

Abstract. The Galactic ASKAP survey of OH (GASKAP-OH) is surveying the Milky Way Fourth Quadrant, the Galactic Centre, the Galactic Bulge and the Large Magellanic Cloud (LMC) in the 18-cm ground-state lines of the hydroxyl radical (OH), using Australia's Square Kilometre Array Pathfinder (ASKAP) telescope. With an expected per-channel rms sensitivity of 36 mJy/beam in its shallowest regions, and a velocity channel width of $0.1\,\mathrm{km\,s^{-1}}$, GASKAP-OH is expected to discover hundreds of new star-formation and evolved star OH masers, as well as extensive absorption from quasi-thermal OH throughout the Galactic Plane. We here summarise the science goals and technical specifications of the survey, and report initial detection results from test observations. GASKAP-OH is expected to run for several years and is an open collaboration. Data products will be made available to the wider community as soon as they are verified.

Keywords. masers, surveys, radio lines: ISM, radio lines: stars

---

1. Introduction

The 18-cm hyperfine transitions of ground-state ($^2\Pi_{3/2}, J=3/2$) OH are a versatile probe of a range of astrophysical phenomena – from star formation (e.g. Caswell 1998), to late-stage stellar evolution (e.g. Habing 1996), to molecular cloud evolution (e.g. Barriault et al. 2010) and stellar feedback (Wardle and Yusef-Zadeh 2002). The four transitions (two main lines at 1665.402 and 1667.359 MHz and two satellite lines at 1612.231 and 1720.530 MHz) all exhibit maser emission whose diverse pumping mechanisms and physical requirements set strong constraints on the astrophysical environments that produce them (Elitzur 1992). Even when not masing, the lines are often anomalously excited, and are observed extensively in absorption or weak emission from extended molecular gas (e.g. Dawson et al. 2022).

While a number of OH surveys have been conducted in the Southern Hemisphere, they either cover only relatively small portions of the Plane, are shallow with poor sensitivity, are low resolution, and/or cover only some of the 4 ground-state transitions (e.g. Sevenster et al. 1997; Caswell 1998; Qiao et al. 2018; Dawson et al. 2022). GASKAP-OH aims to address this deficit, by achieving uniform coverage of the Galactic Plane in the Fourth Quadrant, and carrying out multi-epoch observations of the Galactic Centre region, the Galactic Bulge, and the Large Magellanic Cloud (LMC) – all at high resolution and sensitivity.

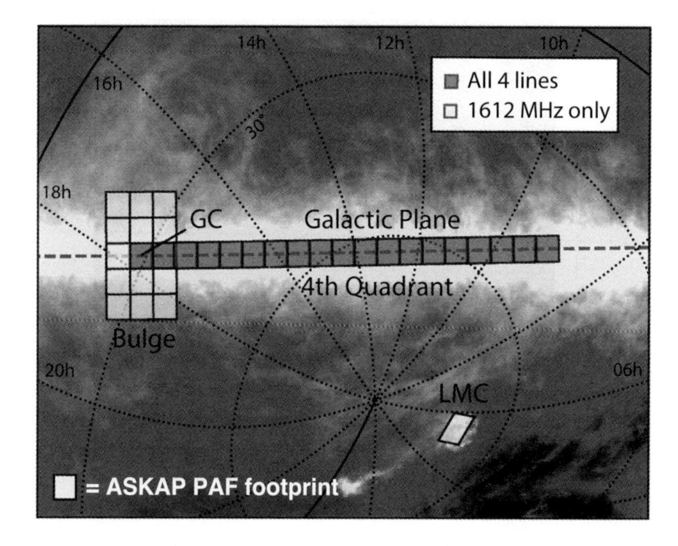

Figure 1. Planned GASKAP-OH survey coverage shown in equatorial coordinates, overlaid on an integrated HI column density map from the HI4PI survey (HI4PI Collaboration 2016). Each square corresponds to the size of a single ASKAP PAF footprint, approximately 5 × 5 degrees. The Galactic Plane coverage runs from $l = 290°$ to the Galactic Centre – the entirety of the Fourth Quadrant.

Table 1. GASKAP-OH planned survey specifications.

Region	Time/field (hrs)	No. fields	Epochs	Transitions (MHz)	Total time (hrs)	Target rms (mJy/beam)	Vel. chan. (km s^{-1})	HPBW arcsec
Galactic Plane	6	18	1	1612, 1665, 1667, 1720	342	36 (15)	0.1 (0.7)	10
Galactic Centre	18	1	6	1612, 1665, 1667, 1720	56	18 (7)	0.1 (0.7)	10
Galactic Bulge	6	13	3	1612	78	36 (15)	0.1 (0.7)	10
LMC	50	1	4	1612	50	8	0.2	10

2. Survey Specifications

GASKAP-OH is one of the nine approved Survey Science Projects (SSPs) to be carried out on the Australia Square Kilometre Array Pathfinder (ASKAP) telescope. ASKAP consists of 36 12m dishes situated in one of the most radio-quiet locations in the world – *Inyarrimanha Ilgari Bundara*, the CSRIO Murchison Radio-astronomy Observatory. Its Phased Array Feed (PAF) receivers allow it to observe a field of view of approximately 25 square degrees in a single pointing, and its backend permits a velocity resolution of up to $\sim 0.1\,\mathrm{km\,s^{-1}}$ (at the frequency of the OH lines) in a single 9 MHz band (see Hotan et al. 2021, for detailed system information). Figure 1 shows the GASKAP-OH survey region, and Table 1 shows the expected sensitivity and resolution in each of the survey subregions. The Galactic Centre, Galactic Bulge and LMC will be observed in multiple epochs to quantify maser variability and derive a more complete population of sources. In common with all ASKAP SSPs, GASKAP-OH data will be processed automatically by the observatory using the ASKAPsoft package, and released to the survey-team as a collection of datacubes, images and catalogues.

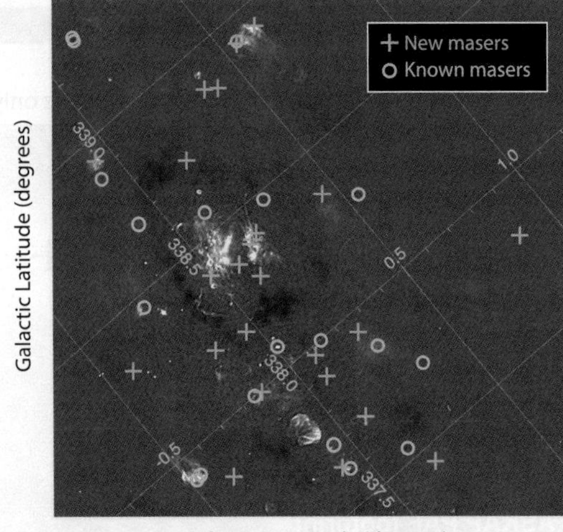

Figure 2. Newly-detected 1665 MHz masers in a portion of the G340 Test Field, overlaid on the 1665 MHz continuum image. Preliminary source-finding using SELAVY (Whiting and Humphreys 2012) recovered all 19 previously published sources in this region and detected 26 new ones, down to a 5σ detection limit of ~ 100mJy.

3. Main Science Goals

Star Formation: GASKAP-OH will provide the deepest ever unbiased survey of star-formation OH masers in the Southern Hemisphere, revealing a currently invisible population. Main line OH masers are widely regarded as a signature of the later stages of high-mass star formation (Forster and Caswell 1989), however all four ground-state OH transitions may be seen (see e.g. Qiao et al. 2018), with uncommon combinations indicating the presence of unusual physical conditions. Combining GASKAP-OH with the wealth of complementary multi-wavelength maser and thermal line surveys available in the South (e.g. class II methanol and water lines from the Methanol Multibeam Survey and the H_2O Southern galactic Plane Survey, Green et al. 2009; Walsh et al. 2014) will allow us to piece together a cohesive picture of where OH masers fit in the maser evolutionary timeline (Breen et al. 2010), both in phase and duration. Zeeman splitting of the OH maser lines will also provide a large sample of in-situ measurements of the strength and direction of the magnetic field in star forming regions.

Evolved Stars: GASKAP-OH will provide the deepest ever unbiased sky survey of circumstellar envelopes (CSE) OH masers across the Galactic Plane, Galactic Centre, Bulge and the LMC. Evolved stars, especially in the AGB and post-AGB phase, play a major role in returning metal-rich stellar mass to the ISM. However, the total rate of mass return is poorly constrained, and the precise relationship between the mass-loss rate and luminosity (which depends sensitively on the dust fraction in the outflow) is still in unknown. OH masers (primarily at 1612 MHz) associated with circumstellar envelopes (CSEs) provide an excellent tool for addressing these questions, allowing us to derive the luminosity function of CSE OH maser sources, and directly measure the CSE wind speed. This is key to understanding wind driving, to determining the dust-to-gas ratio (Marshall et al. 2004; Goldman et al. 2017), and to determining the mass-loss rates of individual stars.

Molecular Cloud Evolution: The formation of dense molecular clouds from the diffuse atomic ISM sets fundamental boundaries on the star formation rate. GASKAP-OH, in combination with GASKAP-HI (e.g. Dempsey et al. 2022) will provide a statistical

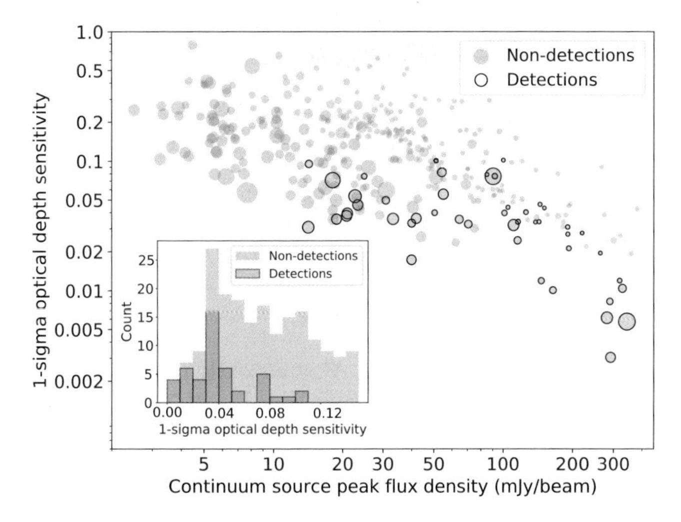

Figure 3. 1667 MHz absorption detection statistics for the G340 Test Field. The scatter plot shows the 1σ optical depth sensitivity (at a velocity resolution of $1\,\mathrm{km\,s^{-1}}$ and integrated over the entire continuum source) as a function of source peak brightness for SELAVY continuum "islands" output as part of the ASKAPSOFT processing workflow. The symbol size is proportional to the continuum island major axis. The inset panel shows optical depth sensitivity histograms for detected and non-detected continuum sightlines. 70 distinct velocity components are detected towards 47 continuum islands.

census of the atomic and molecular gas along hundreds of absorption sightlines throughout the Milky Way Disk, allowing us to probe the cold ISM pre- and post-transition to the molecular phase, and constrain the conditions required for the atomic-molecular phase transition. Modelling the non-LTE excitation of the quasi-thermal OH lines (e.g. Guibert et al. 1978) can constrain temperature, density and mass of both the molecular and atomic gas, and track how these vary with Galactic environment. Importantly, the OH lines are able to probe so-called 'dark' molecular gas – diffuse H_2 that is very poorly traced by common tracers such as CO (e.g. Li et al. 2018; Busch et al. 2021).

4. Preliminary Results & Detection Statistics

Figures 2 and 3 show preliminary detection results based on a single 8-hour integration centred on l=340°, b=0°, at a central frequency of 1666 MHz – covering both the main lines. These first test observations (the G340 Test Field) were carried out in September 2021. While the data is not of full publication-quality – in particular, the Doppler correction is not properly applied, resulting in velocity smearing across several channels – it was sufficient to verify the basic ASKAPsoft imaging parameters and explore initial detection rates. Even with compromised velocity resolution, we have more than doubled the known number of 1665 MHz OH masers in this part of the Galactic Plane (at a 5σ detection limit of ~ 100 mJy) and detected around three 1667 MHz absorption sources per square degree – triple the detection rate of the only other comparable survey THOR (Rugel et al. 2018), in the North. We expect the main survey to achieve comparable or better (despite a marginally shorter integration time on the Galactic Plane fields), once velocity corrections are properly applied.

References

Barriault, L., Joncas, G., Lockman, F. J., & Martin, P. G. 2010, *MNRAS*, 407(4), 2645–2659.

Breen, S. L., Ellingsen, S. P., Caswell, J. L., & Lewis, B. E. 2010, *MNRAS*, 401(4), 2219–2244.

Busch, M. P., Engelke, P. D., Allen, R. J., & Hogg, D. E. 2021, *ApJ*, 914(1), 72.
Caswell, J. L. 1998, *MNRAS*, 297(1), 215–235.
Dawson, J. R., Jones, P. A., & Purcell, C. *et al.* 2022, *MNRAS*, 512(3), 3345–3364.
Dempsey, J., McClure-Griffiths, N. M., Murray, C. *et al.* 2022, *PASA*, 39, e034.
Elitzur, M. 1992, *Astronomical Masers.* Astrophysics and Space Science Library. Springer Netherlands.
Forster, J. R. & Caswell, J. L. 1989, *A&A*, 213, 339–350.
Goldman, S. R., van Loon, J. T., Zijlstra, A. A. *et al.* 2017, *MNRAS*, 465(1), 403–433.
Green, J. A., Caswell, J. L., Fuller, G. A. *et al.* 2009, *MNRAS*, 392(2), 783–794.
Guibert, J., Rieu, N. Q., & Elitzur, M. 1978, *A&A*, 66(3), 395–405.
Habing, H. J. 1996, *A&AR*, 7(2), 97–207.
HI4PI Collaboration 2016, *A&A*, 594, A116.
Hotan, A. W., Bunton, J. D., & Chippendale, A. *et al.* 2021, *PASA*, 38, e009.
Li, D., Tang, N., Nguyen, H. *et al.* 2018, *ApJS*, 235(1), 1.
Marshall, J. R., van Loon, J. T. *et al.* 2004, *MNRAS*, 355(4), 1348–1360.
Qiao, H.-H., Walsh, A. J., Breen, S. L. *et al.* *ApJS*, 239(1), 15.
Rugel, M. R., Beuther, H., Bihr, S. *et al.* 2018, *A&A*, 618, A159.
Sevenster, M. N., Chapman, J. M., & Habing, H. J. *et al.* 1997, *A&AS*, 122, 79–93.
Walsh, A. J., Purcell, C. R., Longmore, S. N. *et al.* 2014, *MNRAS*, 442(3), 2240–2252.
Wardle, M. & Yusef-Zadeh, F. 2002, *Science*, 296(5577), 2350–2354.
Whiting, M. & Humphreys, B. 2012, *PASA*, 29(3), 371–381.

Cosmic Masers: Proper Motion toward the Next-Generation Large Projects
Proceedings IAU Symposium No. 380, 2024
T. Hirota, H. Imai, K. Menten, & Y. Pihlström, eds.
doi:10.1017/S1743921323002181

Introducing the MeerKAT Telescope: Studies of masers and their environment

Sharmila Goedhart

South African Radio Astronomy Observatory, 2 Fir Street, Black River Park, Observatory, Cape Town, 7925, South Africa. email: sharmila@sarao.ac.za

Abstract. The 64-dish MeerKAT telescope was inaugurated in 2018 and has been conducting regular science operations since then. In the meantime, new observation modes have been under development. Spectral line modes are available, as well as L-, UHF- and S-band receivers. MeerKAT's excellent sensitivity over a wide range of angular scales makes it an excellent choice for studies of HII regions, supernova remnants and planetary nebulae. In addition, an OH megamaser has been detected at $z > 0.5$ for the first time.

Keywords. masers; radio lines: general; radio lines: ISM; radio continuum: general; techniques: interferometric

1. Introduction

The MeerKAT (Karoo Array Telescope) is a precursor to the Square Kilometer Array (SKA) consisting of 64 offset Gregorian dishes with a diameter of 13.5 m. It is located in the Northern Cape, South Africa, in a radio-quiet environment.

The initial science programme consists of 8 Large Survey Projects (LSPs), taking up approximately 67% of allocated observing time in the first $\sim$ five years, as well as yearly open calls for proposals and Director's Discretionary Time (DDT). A wide range of science interests can be addressed with MeerKAT, including star formation, galaxy evolution, transients and pulsar timing (Jonas et al. 2016; Camilo et al. 2018).

Discoveries of note to the maser community include a luminous mainline megamaser at $z > 0.5$ in J033046.20-275518 (Glowacki et al. 2022, and this volume) and satellite lines at $z = 0.89$ towards PKS 1830-211 (Combes et al. 2021). These are just the first results of the LSPs LADUMA (Blythe et al. 2016) and MALS (Gupta et al. 2016) and many more detections from these surveys and other projects are anticipated.

2. Key specifications

More details and updates on the telescope status can be found through `https://www.sarao.ac.za/science/` on the MeerKAT link.

2.1. *Receivers*

The antennas are equipped with dual linearly polarised receivers covering the UHF, L and S bands. Table 1 gives the key parameters of the telescope. The mean system-equivalent flux density (SEFD) is 550 Jy in UHF band, 425 Jy in L-band and 365 Jy in S-band. Sensitivity calculators can be found at `https://apps.sarao.ac.za/calculators`.

2.2. *Correlator modes*

Table 2 lists the available correlator channelisation modes. At the moment, narrow-band modes are only available for L-band. Please note that only the NE107M zoom mode,

Table 1. Key MeerKAT specifications. Note that for S-band the correlator can only process 875 MHz at a time, in one of 5 pre-defined sub-bands. The effective frequency range is narrower than the full digitized range.

Number of Antennas	64
Dish diameter	13.5 m
Minimum baseline	29 m
Maximum baseline	7700 m
UHF frequency range	580 - 1015 MHz [544 to 1088 MHz digitised]
L-band frequency range	900 - 1670 MHz [856 to 1712 MHz digitised]
S-band frequency range	1750 - 3500 MHz

Table 2. Correlator channelisation modes. Note that the filter roll-off for both narrowband modes is 13.5 MHz, thus 27 MHz of the band is not usable. The narrowband modes always run concurrently with the wideband coarse mode.

Mode	Channels	L-band channel width (kHz)	UHF channel width (kHz)	S-band channel width (kHz)
Wideband coarse (4K)	4096	208.984	132.812	213.623
Wideband fine (32K)	32768	26.123	16.602	26.703
Narrowband (NE107M)	32768	3.3	n/a	n/a
Narrowband (NE54M)	32768	1.633	n/a	n/a

with a usable bandwidth of $\sim$ 80 MHz has sufficient bandwidth to observe the 1612 MHz OH line simultaneously with the mainlines at 1665 and 1667 MHz. This mode gives a velocity resolution of $\sim$ 0.6 km/s at 1665 MHz. The recently commissioned NE54M mode, which has a usable bandwidth of $\sim$ 27 MHz, has a velocity resolution of $\sim$ 0.3 km/s. The full band (4096 channel) continuum data is recorded simultaneously with narrowband.

2.3. *Array layout*

The MeerKAT array layout is optimised for observations of low-surface-brightness objects with a 'core heavy' configuration consisting of a dense inner component containing 70% of the dishes. The outer component contains 30% of the dishes distributed in a 2D Gaussian uv-distribution, with a dispersion of 2500 m and the longest baseline of 7.7 km. This layout provides roughly uniform sensitivity for angular scales ranging from 8$''$ to 80$''$ (at L-band). The best spatial resolution that can be achieved at these frequencies is $\sim$ 4$''$ (depending on declination), with $\sim$ 50% loss in sensitivity.

2.4. *Polarimetry*

The receivers have two orthogonal linear feeds. All polarisation products (XX, XY, YX, YY) are always recorded. While good results can be achieved over a limited field of view and frequency range (Cotton et al. 2020), calibration methods for widefield, wideband polarimetry are still under development (Sekhar et al. 2022) and potential users are advised to check the MeerKAT documentation† for updates. A fundamental issue due to the design of the receivers affects the polarisation leakage response across the field of view as a function of position in the primary beam as well as frequency (De Villiers 2023). This effect is much worse in the upper half of each receiver band, and it is advised to limit polarisation measurements strictly to the beam centre at frequencies above 1500 MHz in L-band. This means that detailed studies of maser polarisation and measurements of Zeeman splitting in OH masers are likely not feasible at this time unless the target is on boresight.

† `https://skaafrica.atlassian.net/wiki/spaces/ESDKB/pages/1481572357/The+MeerKAT+primary+beam`

3. Potential applications

MeerKAT's sensitivity to a wide range of angular scales make it ideal to study diffuse extended structures such as supernova remnants, star forming regions and planetary nebulae. The MeerKAT 1.28 GHz Galactic Centre mosaic (Heywood et al. 2022) is a spectacular example of MeerKAT's imaging capabilities. This is a public legacy data release, and both images and the original visibilities are available to download‡. This dataset provides an opportunity to probe the radio environment of masers seen in this region. A legacy survey of the Galactic Plane covering $250^{\circ} < l < 60^{\circ}$, $|b| < 1.5$ deg (Goedhart et al. in prep) has also been conducted, and visibilities are available from the MeerKAT archive§.

While the MeerKAT correlator does not offer very high spectral resolution, the telescope sensitivity makes searches for new masers quite viable, with e.g. a naturally weighted noise of 5.8 mJy/beam in a single channel achieved in a 5 minute integration in the NE107M correlator mode or 8.3 mJy/beam in the narrower NE54M mode. MeerKAT can also observe Targets of Opportunity, either through regular proposals with well-defined triggering criteria, or through DDT proposals, providing an opportunity for multiwavelength follow-up of maser flares or accretion bursts as seen in Bayandina et al. (2022), for example.

4. Conclusion

The newly commissioned narrow band modes of MeerKAT offer a new opportunity to observe OH masers down to a velocity resolution of 0.3 km/s while the ability to simultaneously record wideband continuum data enables high dynamic range imaging of the maser environment.

MeerKAT's high sensitivity and UHF band receivers enable the detection of OH megamasers to unprecedentedly high redshifts.

5. Acknowledgements

The MeerKAT telescope is operated by the South African Radio Astronomy Observatory, which is a facility of the National Research Foundation, an agency of the Department of Science and Innovation.

References

Bayandina, O. S., Brogan, C. L., Burns, R. A. *et al.* 2022, *AJ*, 163, 83
Blythe, S. Baker, A.J., Holwerda, B.W. *el al.* 2016, in Proc. MeerKAT Science: On the Pathway to the SKA Conf., ed. R. Taylor (Trieste: PoS),004
Camilo, F., Scholz, P., Serylak, M., it et al. 2018, *ApJ*, 856, 180
Combes, F., Gupta, N., Muller, S., *et al.* 2021, *A&A*, 648, 116
Cotton, W. D, Thorat, K., Condon, J.J., *et al.* 2020, *MNRAS*, 495, 1271
De Villiers, M. S., 2023, *AJ*, 165, 78
Glowacki, M., Collier, J. D., Kazemi-Moridani, *A. et al.* 2022, *ApJ*, 931, L7
Gupta, N., Srianand, R., Baan, W. *et al.* 2016, in Proc. MeerKAT Science: On the Pathway to the SKA Conf., ed. R. Taylor (Trieste: PoS), 014
Heywood, I., Rammala, I., Camilo, F. *et al.* 2022, *ApJ*, 925, 165
Jonas, J. & MeerKAT Team 2016, in Proc. MeerKAT Science: On the Pathway to the SKA Conf., ed. R. Taylor (Trieste: PoS), 001
Sekhar, S, Jagannathan, P., Kirk, B., Bhatnagar, S., Taylor, R., 2022, *AJ*, 163, 87

‡ `https://doi.org/10.48479/fyst-hj47`
§ `https://apps.sarao.ac.za/katpaws/archive-search`

Cosmic Masers: Proper Motion toward the Next-Generation Large Projects
Proceedings IAU Symposium No. 380, 2024
T. Hirota, H. Imai, K. Menten, & Y. Pihlström, eds.
doi:10.1017/S1743921323002752

Exploring galactic and extragalactic masers with LLAMA

Tânia P. Dominici[1] and LLAMA Collaboration

[1]Divisão de Astrofisica Instituto Nacional de Pesquisas Espaciais,
Av. dos Astronautas, 1758, São Jose dos Campos - SP, Brazil
email: tania.dominici@inpe.br

Abstract. LLAMA (Large Latin America Millimeter/submillimeter Array) is a new radio observatory that is being constructed in a collaboration between Argentina and Brazil. It will consist of a 12 meters diameter antenna that is being installed in Alto Chorrillos at 4850 m of altitude, in the Salta province of Argentina. Alto Chorrillos is a high-quality astronomical site similar to Chajnantor (Chile), where ALMA observatory operates. When completed, LLAMA will allow line, continuum and linear polarization observations between 35 and 700 GHz, approximately. For the first light, LLAMA will be equipped with ALMA-like receivers at bands 5 (163 - 211 GHz), 6 (211 - 275 GHz) and 9 (602 - 720 GHz). LLAMA is being planned to be a versatile astronomical facility that will serve the scientific community for the exploration of scientific topics as diverse as the molecular evolution of the Universe, black holes and their accretion disks, astrophysical jets, stellar formation and evolution, the structure of our galaxy and the Sun, planetary atmospheres and extragalactic astronomy. In this work, I will present the LLAMA project and the perspectives for this new astronomical facility in the context of the investigation of galactic and extragalactic masers.

Keywords. radio observatory, LLAMA, millimeter/submillimeter astronomy

---

1. Introduction

Between 2004 and 2015, the Argentine Institute of Radio Astronomy (IAR) carried out a series of atmospheric transparency monitoring campaigns in the Northwest of Argentina (Salta and Jujuy provinces), in order to select suitable high altitude sites for the installation of radio astronomical instruments that could operate at millimeter and submillimeter wavelengths (Bareilles *et al.* 2011). In 2009, based on the fact that there were sites with comparable quality to Chajnantor (Chile), Brazil and Argentina began to articulate the project of a binational radio observatory.

Thus, the Large Latin American Millimeter/submillimeter Array - LLAMA - is the joint scientific and technological undertaking of Argentina and Brazil whose goal is to install and to operate an observing facility capable of performing observations of the Universe at the higher frequencies of the radio atmospheric window (Lepine *et al.* 2021). At the end of the site search, the new observatory is being constructed at 4850 m above sea level in the Puna Saltea, northwest region of Argentina, where is the selected site, called Alto Chorrillos.

LLAMA was planning to be an appropriate (sub)millimeter wave single-dish telescope (similar in performance and instrumentation to an individual ALMA antenna and/or to APEX), with the ability to be integrated as an element in Very Large Baseline Interferometric networks (VLBI), and with the room for harboring new cutting-edge instruments. Despite its binational base, LLAMA is being developed with support and collaboration of many international institutions, what will be briefly described here. As a general purpose and very sensitive radio telescope, conceived and designed to be a long-lived user observatory, LLAMA can be used to study the emission from a wide variety of astronomical sources, including galactic and extragalactic masers.

2. Overview of the LLAMA Observatory

LLAMA will consist of a 12-m ALMA-like antenna with the addition of two Nasmyth cabins, similar to APEX telescope. The requirement for the shape of the parabolic disk is to reach less than 15 μm rms adjusting its panels by using holography (e.g. Baars *et al.* 2007). The roughness of the disk surface was planned to allow to point the antenna towards the Sun, making LLAMA a unique facility to explore high frequency solar emissions at several timescales. The antenna pointing precision was specified to be better than 2 arcseconds.

As said before, the observatory is being constructed in a new astronomical site at Alto Chorrillos, 4850 m above sea level. When completed, LLAMA will be equipped with six ALMA-like receivers covering bands 1, 2+3, 5, 6, 7, and 9 (35 - 700 GHz), which will populate the two Nasmyth cabins. Besides, invited instruments can be installed at the Cassegrain cabin. The concept of LLAMA cover the capability to obtain continuum, spectral, polarization and Solar observations, including the possibility for carrying out simultaneous observations at two or more bands. Each Nasmyth cabin can be equipped with one cryostat housing three ALMA-like receivers.

At the beginning of the scientific operations, it is expected to have three receivers at bands 5 (163 - 211 GHz), 6 (211 - 275 GHz) and 9 (602 - 720 GHz). While bands 6 and 9 receivers were made possible mainly in collaboration with NOVA-Groningen (Netherlands), band 5 receiver is a former prototype for ALMA Observatory refurbished by GARD (Gothenburg, Sweden), NOVA and offered by European Southern Observatory (ESO) for LLAMA as a permanent loan. Besides, a band 2+3 receiver (67 - 116 GHz) is being developed by the Universidad de Chile (Reeves *et al.* 2023), but it is not expected to be available for the first light.

One of the most complex developments for LLAMA is the Nasmyth Cabin Optical System (NACOS). It is the optical-electrical-mechanical instrument that will guide the mm/sub-mm waves collected by the antenna to the receivers. For the first light, only one of the cabins will be populated with a cryostat and three receivers (bands 5, 6 and 9). NACOS includes a robotic arm with three calibration loads (two hot and one at room temperature), subsystem being developed in collaboration with Universidad de Concepcin (Chile, Reeves *et al.* 2023). In Fig. 1 we can see a schematic view of NACOS for the first light, with the components of the Cassegrain and of one of the Nasmyth cabins. While NACOS is not complete, enabling the use of the second Nasmyth cabin, band 2+3 receiver is planned to be installed in the Cassegrain cabin for commissioning and initial science operations. Besides the construction of the expansion for the second Nasmyth Cabin, the long-term completion of NACOS implies in a redistribution of the receivers between the cabins and the replacement of flat mirrors by dichroics, what will allow simultaneous observations among some of the receivers in both and within a single Nasmyth cabin.

The site infrastructure has been evolving (Fig. 2). A road to access the site was constructed, as well as a flat area for the antenna assembly. The antenna foundation and anchor ring installation were successfully concluded at the beginning of 2023. All components of the antenna are in the site and its assembly must be completed by the end of this year. After that, the Assembly, Integration and Verification (AIV) phase will start. Some of the most important milestones of the project are the so-called "Eyes Opening",

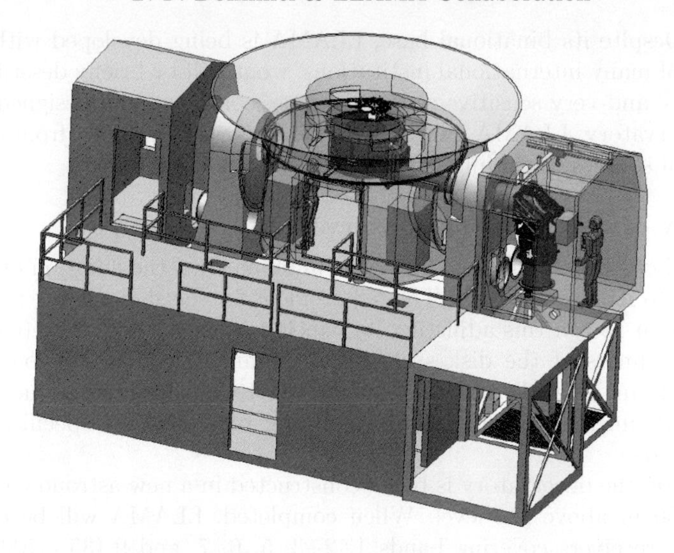

Figure 1. Schematic view of NACOS for LLAMA's first light, for which only one Nasmyth cabin will be populated.

Figure 2. View of LLAMA site at Alto Chorrillos (Argentina) from the approximated position where the holographic tower is now installed.

defined as the preliminary uncalibrated measurement of the continuum and spectrum of a standard object in one of the receivers, and the "First Light", the preliminary calibrated measurement of the continuum flux and spectrum of a standard object in one of the receivers. "Eyes Opening" is expected from the end of 2024 and the "First Light" should be reached six to twelve months after "Eyes Opening".

3. Cosmic Masers with LLAMA

Eight key scientific topics and niches were identified for LLAMA: planetary atmospheres, LLAMA as part of VLBI networks, time domain (sub)mm astronomy, solar physics, interstellar medium, astrochemistry, magnetic fields and extragalactic astronomy. Most of these subjects are related with or can be investigated through maser emission. Particularly interesting is the the variability detected in several maser species in different astronomical scenarios. Time domain submillimeter astronomy is a window of opportunities for future facilities like LLAMA, along with the importance of the improvement of submillimeter calibration techniques. LLAMA will be able to contribute by monitoring a large number of masers species with high temporal resolution and and large temporal coverage. It will probe different transitions and vibrational states simultaneously, which is crucial to find out the physical conditions and investigate the pumping mechanism producing the maser effect. Even as a single dish, LLAMA can be used to study the inner parts of AGNs orbited by clouds showing water megamaser emission.

During Cosmic Masers IAU Conference, several speaker reinforced the relevance of a new facility operating at LLAMA's frequencies. A few examples include SiO maser monitoring at 86 GHz (J=2-1; band 2+3) in AGB/RGB stars (Lorant Sjouwermans talk); HCN (around 176 GHz, band 5) to explore inner regions of CSE in Carbon-rich AGB stars (Lynn Matthews); H_2O at 183 GHz (band 5) in ULIRGs, including linear polarization measurements (Masatoshi Imanishis talk) and methanol masers at 349 GHz (band 7, Alberto Sannas talk). Surveys as Nearby Evolved Star Survey (NESS, Scicluna *et al.* 2022), using single dish observations obtained with JCMT and APEX, and Bulge Asymmetries and Dynamical Evolution (BAaDE, Stroh *et al.* 2019), based on ALMA observations at 86 GHz (band 2+3) are good examples of large survey programs that can be implemented and or extended using single dishes like LLAMA or through interferometric experiments at those and higher radio frequencies.

In fact, LLAMA is a promising observatory in the context of VLBI networks. In particular, the collaboration intends to be part of the next generation Event Horizon Telescope (ngEHT), as LLAMA could be a key station for that experiment (e.g. Raymond *et al.* 2021). The ngEHT has among its science cases the observation of maser spectral lines in Galactic star-forming regions and evolved stars at 86, 230 and 345 GHz (Richard Dodson's talk).

The LLAMA project is made possible mainly by Ministerio de Ciencia, Tecnologa e Innovacin (MinCyT, Argentina) and So Paulo Research Foundation (FAPESP, Brazil, through the grants 2011/51676-9, 2015/50360-9 and 2015/50359-0). It also has investment from CNPq/Finep/MCTI/BRICS-STI (grant 402966/2019-8), Government of Salta, Dutch Research Council (NWO, Netherlands) and FONDEF/Conicyt (Chile). MinCyT, CONICET, Government of Salta, INVAP, IAR, IAFE, OAC and UNSa participate in the development of the project in Argentina. In Brazil, USP, CRAAM/Mackenzie and INPE/MCTI are part of the collaboration.

References

Baars, J. W. M., Lucas, R., Mangum, J. G., Lopez-Perez, J. A. 2007, *IEEE Antennas Propag Mag*, 49, 24

Bareilles, F. A., Morras, R., Hauscarriaga, F. P., Guarrera, L., Arnal, E. M., Lepine, J. R. D. 2011, *BAAA*, 54, 427

Lepine, J. R. D., Abraham, Z., Castro, C. G. G. de, *et al.* 2021, *An. Acad. Bras. Cienc.*, 93, e20200846

Raymond, A. W., Palumbo, D., Paine, S. N., Blackburn, L., Córdova Rosado, R., Doeleman, S. S., Farah, J. R., *et al.* 2021, *ApJS*, 253, 5

Reeves, R., Bovino, S., Bronfman, L., *et al.* 2023, in: V. Ossenkopf-Okada *et al.* (eds.), *Physics and Chemistry of Star Formation: The Dynamical ISM Across Time and Spatial Scales*, Proceedings of the 7th Chile-Cologne-Bonn Symposium (Puerto-Varas, Chile), p. 318

Scicluna, P., Kemper, F., McDonald, I., Srinivasan, S., Trejo, A., Wallström, S. H. J., Wouterloot, J. G. A., *et al.* 2022, *MNRAS*, 512, 1091

Stroh, M. C., Pihlström, Y. M., Sjouwerman, L. O., Lewis, M. O., Claussen, M. J., Morris, M. R., Rich, R. M. 2019, *ApJS*, 244, 25

Cosmic Masers: Proper Motion toward the Next-Generation Large Projects
Proceedings IAU Symposium No. 380, 2024
T. Hirota, H. Imai, K. Menten, & Y. Pihlström, eds.
doi:10.1017/S1743921323003058

Sub-mm spectral astrometric VLBI with the ngEHT

Richard Dodson[1] and Maria J. Rioja[1,2,3]

[1]ICRAR, M468, The University of Western Australia, 35 Stirling Hwy, Crawley, Western Australia, 6009. email: richard.dodson@icrar.org

[2]CSIRO Astronomy and Space Science, PO Box 1130, Bentley WA 6102, Australia

[3]Observatorio Astronómico Nacional (IGN), Alfonso XII, 3 y 5, 28014 Madrid, Spain

Abstract.
We present the prospects from astrometric spectral line VLBI in the era of ngEHT. We review the potential targets, that span many interesting science cases. We summarise the approaches that have been demonstrated to work at lower frequencies and touch on the simulations that give us great confidence that these same approaches will continue to work at sub-mm wavelengths. We conclude that this is a worthwhile pursuit with a high probability of success.

Keywords. Astrophysics - astrometry; Astrophysics - masers; Astrophysics - Instrumentation and Methods for Astrophysics

1. Introduction

We present our conclusions on the opportunities provided by the Next Generation Event Horizon Telescope (ngEHT), which include the ability to perform astrometric VLBI observation at millimeter and sub-millimeter wavelengths.

The challenges for Astrometry at high frequencies is all about the atmosphere; as the frequency get higher the coherence time also gets shorter as the same extra atmospheric pathlength represents an increasing fraction of the wavelength. But additionally the system temperatures of the radio telescopes gets worse, as the engineering challenges become more difficult. The trigger for this paper was the work towards a new sub-mm VLBI array, the next-generation Event Horizon Telescope, led by the Black Hole Institute out of CfA in Harvard (Johnson *et al.* (2023)). The ngEHT will eventually be 10 antenna sites, all capable of performing simultaneous observations at 86, 230 and 340GHz bands, with new small high performance parabolic dishes. It is envisaged to operate both in conjunction with the Event Horizon Telescope (EHT) and in isolation. To ensure sensitivity when observing without the massive collecting area of ALMA it is possible that there will be a small number of high sensitivity sites with multiple antennas that will be phased up together.

2. Maser targets for ngEHT

There has been a recent study on the range of spectral line VLBI that would be possible with the ngEHT (Kim & Fish (2023)), as summarised by their figure reproduced in Figure 1. They point to the myriad of detections of unexpected mazing species made by the ALMA connected array, at lower resolutions. This provides confidence that there are maser targets for ngEHT.

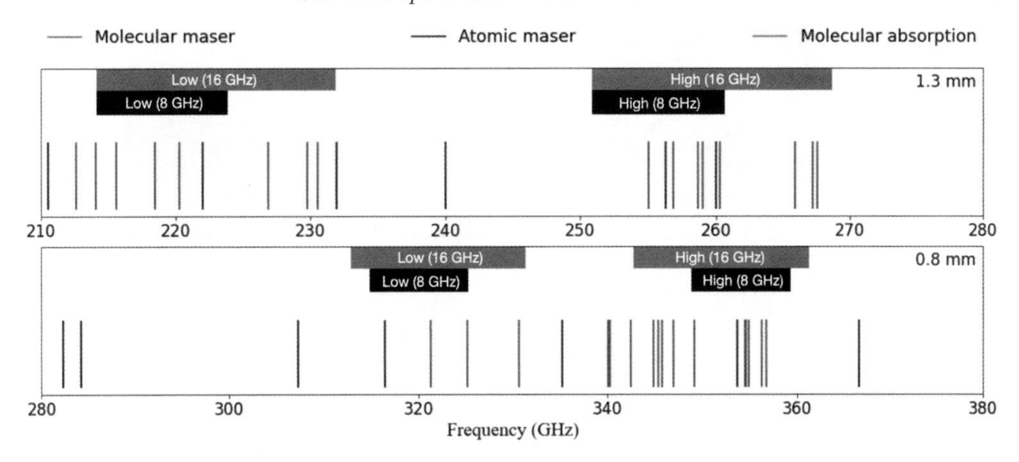

Figure 1. Distribution of spectral lines over the frequency range of 1.3 mm and 0.8 mm. Each color indicates the type of spectral lines. Frequency coverage of 8 GHz and 16 GHz observations is marked with two different frequency tunings (low and high).

These can be summarised as:

- Molecular lines

These trace distinct excitation conditions in the circumstellar envelopes of AGB stars and star forming regions. Asymptotic Giant Branch (AGB) are the final stage of the majority of the stars on the main branch and their pulsations seed the ISM with molecular species generated in the stellar body. Thus these are of vital importance for our understanding of galactic evolution. They act as delicate probes of the astro-chemistry (i.e. temperature, the elements present, and pressure). Some of the most common maser species are SiO, which forms in the outer atmospheres of the star, and 22GHz H_2O, which form via shock interactions in the accelerated outflows. The sub-mm H_2O (e.g. 345GHz) will also form closer to the stellar surface in inner regions and connect the two most common species, allowing the seamless connection of the matter transfer. The sub-mm H_2O is expected to form and this would be about 2-5$R_\odot$ (which is half of the radius at which 22GHz emission is to be found).
On the other hand carbon rich, and thus oxygen poor, AGBs lack water and SiO. Thus the masing species of HCN and SiS, which both have sub-mm transitions, will fill similar roles as H_2O and SiO in those AGB stars.
Massive Star Forming Regions (SFR) host OH, H_2O and CH_3OH masers, which have been widely used to measure the proper motion, parallax and outflows of these regions (e.g. Reid *et al.* (2019)). The distances to these relatively rare objects has been vital to, for example, measuring their individual physical conditions (e.g. absolute brightness) as well as global parameters such as the Galactic Rotation curve. Lower mass SFRs host sub-mm masers of similar character, as has been observed by ALMA. High sub-mas resolution and precision is required to resolve the proper motion, parallax and outflows of these objects.

- Atomic lines:

Hydrogen Radio Recombination Lines (RRL) can maser in sub-mm range; first observation were reported in Cox *et al.* (1995). For example the H_α-30 line has been used to probe post-AGB and Planetary Nebulae for bipolar outflows using VLBI. Of particular interest are ultra-compact H-II regions, where the low resolution observations suggest bipolar outflows, but VLBI is required to confirm this and to image the details.

- Molecular Absorption Lines:

It is important in the study of Gravitational Lens to identify the distance to the lensing

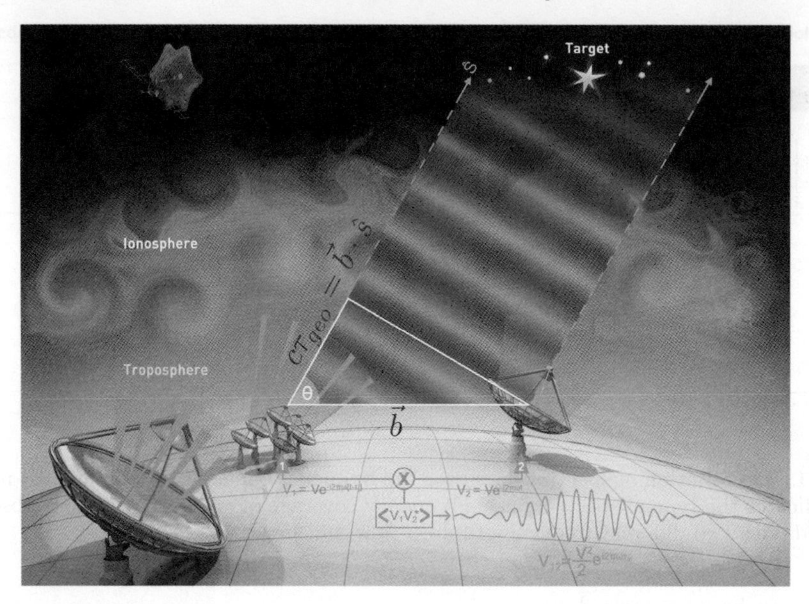

Figure 2. Schematic from Rioja & Dodson (2020) that shows all the possible approaches for sub-mm astrometric VLBI. Next generation methods include: Source/Frequency Phase referencing using different frequencies to solve for non-dispersive terms, and MultiView which use multiple sources to solve for spatially varying terms. Next generation technology options for SFPR include rapid frequency switching (order of seconds) and simultaneous multi-band receivers. Next generation technology options for MultiView include moderate sized single dishes combined with: large single dishes with multiple beam technologies, arrays of smaller dishes with multiple Tied Array Beams and paired antennas of smaller dishes with multiple pointing centres.

galaxy, which can be very hard. For example for the gravitational lens 1830-211 it took a long time to identify the distance to the lensing galaxy (Winn *et al.* (2008)). Absorption is one of the tools that will give a distance to the lensing galaxy directly. Furthermore, the detection of rare molecular species is easier in absorption, as high resolution is required to resolve the clouds, improve the SNR of the detection and to get accurate mass estimates of the absorbers. Similarly absorption from the circumnuclear gas in AGN allows studies of the structure and chemical composition of these disks surrounding and feeding the central black hole.

3. Approaches for astrometry with mm-VLBI

We have completed a through and deep analysis of the paths to ultra-precise astrometry, as reported in Rioja & Dodson (2020). This was focused on the lower frequencies and SKA, but the conclusions are still applicable to the higher frequencies, and potentially ngEHT. For example we have completed a study of the expected performance of the Frequency Phase Transfer (FPT) between the 86- and 340-GHz frequency bands (Rioja *et al.* (2023)), and a survey of the major science applications (Jiang *et al.* (2023)). Figure 2, reproduced from Rioja & Dodson (2020), shows schematically the possible station combinations for sub-mm astrometric VLBI.

For sub-mm VLBI astrometry the most well established method is that of Source/Frequency Phase Referencing (Rioja & Dodson (2011)), where the calibration from lower frequency is scaled and applied to the higher frequency. This implicitly assumes that the calibration errors are non-dispersive, which include the antenna position or the source position, but are dominated by the troposphere residual path-length. Additionally we are coming to realise that MultiView (Rioja *et al.* (2017)) will also be applicable to sub-mm VLBI, even though it was developed for the lower frequencies. This is because MultiView uses multiple calibrators to solve for all contributions for the errors, interpolated to the line of sight of the target. Thus all (linear) systematic contributions are removed. In the sub-mm domain the one additional requirement is that observations of the sources are performed simultaneously, at least to some extent. This is discussed further in the next section.

4. Prospects

Firstly it is important to note that no one has performed sub-mm astrometric VLBI. The requirements have long been considered impossible. Nevertheless we now believe we can see a path to achieving this goal.

The Frequency Phase Transfer method is expected to work up to the sub-mm regime; similar demonstrations have been performed in band-to-band calibration with ALMA (Asaki *et al.* (2020)). Source/Frequency Phase Referencing tests between 86-GHz and 215-GHz have been performed, but are yet to be released. However the data seems good with potential simultaneous observations at Pico Valeta and the combined SMA/JCMT site, which will demonstrated a 'paired antenna' configuration. Unfortunately simultaneous observations at KVN-Yebes failed; the experiment will be repeated.

Demonstrations of the MultiView method at 86GHz are under consideration. Ideally MultiView would be performed with multiple beams from one antenna, but a dish small enough to see multiple 86GHz compact sources would be too small to be sensitive enough. Thus the only option is to have pairs of antennas, which are sufficiently physically close such that the atmospheres can be considered common. Using the typical wind speed of 10m/s would imply that the same formulae for the errors due to the 'temporal sampling' of the atmosphere could be used with the distance between the paired antennas standing in as the time step. See Walker *et al.* (2020) for discussions. Groups of paired antennas were a suggested concept for some of the stations of the ngEHT, to provide an increased collecting area for crucial hub stations that would then be the reference antennas for the other less sensitive sites.

To investigate the possibilities and requirements we have started to do some testing in this direction, using the fastest possible switching times on the VLBA. Currently we have made 43GHz observations switched at sub-minute timescales over angular scales as much as 5 degrees. These seem to be promising.

5. Conclusions

In conclusion, there are multiple interesting targets for spectral line sub-mm astrometric VLBI. These provide unique probes of different aspects of astrophysics, such as the various regions around AGB stars. There are routes to perform sub-mm astrometric VLBI; for a single source the reference of emission between transitions and molecules is a delicate and precise probe of the thermal and chemical conditions in the pumping regions of astrophysical masers. The frequency phase transfer method is well tested and can be expected to work between the 86 and 340GHz bands. The potential for relative spatial astrometry is under investigation, and has rigorous requirements, such as multiple paired antennas per site. On the other hand these seem to be achievable, if a sufficient strong science case can be made to address the additional costs.

In summary, it is worthwhile to attempt spectral line astronomical VLBI at sub-millimeter wavelengths, we are confident that it will be possible to perform trans-frequency astrometry and we are hopeful it will be possible to perform trans-source astrometry.

References

Asaki, Y., Maud, L. T., Fomalont, E. B., *et al.* 2020, *ApJS*, 247, 23
Cox, P., Martin-Pintado, J., Bachiller, R., *et al.* 1995, *A&A*, 295, L39
Jiang, W., Zhao, G.-Y., Shen, Z.-Q., *et al.* 2023, *Galaxies*, 11, 3
Johnson, M. D., Akiyama, K., Blackburn, L., *et al.* 2023, *Galaxies*, 11, 61
Kim, D.-J. & Fish, V. 2023, *Galaxies*, 11, 10
Reid, M. J., Menten, K. M., Brunthaler, A., *et al.* 2019, *ApJ*, 885, 131
Rioja, M. & Dodson, R. 2011, *AJ*, 141, 114
Rioja, M. J., Dodson, R., Orosz, G., *et al.* 2017, *AJ*, 153, 105
Rioja, M. J. & Dodson, R. 2020, *AARv*, 28, 6
Rioja, M. J., Dodson, R., & Asaki, Y. 2023, *Galaxies*, 11, 16
Walker C., Dodson, R., Rioja, M., Reid, M., in prep, *ngVLA memo*
Winn, N. *et al.* 2008, *AJ*, 575, 103

Chapter 8
Concluding Remarks

Cosmic Masers: Proper Motion toward the Next-Generation Large Projects
Proceedings IAU Symposium No. 380, 2024
T. Hirota, H. Imai, K. Menten, & Y. Pihlström, eds.
doi:10.1017/S1743921323003253

Closing Remarks of the International Astronomical Union Symposium 380

Anna Bartkiewicz[1] and **Ylva Pihlström**[2]

[1]Insitute of Astronomy, Faculty of Physics, Astronomy and Informatics, Nicolaus Copernicus University, Grudziadzka 5, 87-100 Torun, Poland. email: annan@astro.umk.pl

[2]Department of Physics and Astronomy, University of New Mexico, Albuquerque, NM 87131, USA

Abstract. We are pleased to summarise the IAU Symposium 380, *Cosmic Masers: Proper Motion toward the Next-Generation Large Projects* held on March 2023 March 20-24 in Kagoshima City, Japan. It is the sixth symposium focusing on astrophysical masers broadly used in research of star formation and evolved stars, astrometric measurements of the Milky Way galaxy, as well as studies of extragalactic environments.

Keywords. Masers, Milky Way galaxy, Extragalactic

---

1. Introduction

For the first time, the cosmic maser meeting series was held in hybrid mode gathering in total more than 172 participants; 70 online and 102 in-person admiring Kagoshima city and Mt. Sakura-jima – an active stratovolcano. In total, we represented 28 countries counted by the location of the main institute affiliation of each participant. This sixth astrophysical maser symposium finished the "continent cycle" (North America, South America, Australia, Africa, Europe, and Asia) which began with the first symposium taking place in USA (1992), followed by meetings in Brazil (2001), Australia (2007), South Africa (2012), and Italy (2017). The IAUS maser meetings are recurrent approximately every 5 years, and we can therefore be considered to be super-periodic considering the conference locations! This pattern will be analysed in more detail at the proper time.

In his closing remarks of the 2017 IAUS 336 in Cagliari, Italy, Prof. Phil Diamond stated (Diamond 2018): *First, we are seeing the culmination of major, long-term monitoring programmes; secondly, we are seeing the massive impact of ALMA and the JVLA; thirdly, it is clear that Gaia is a game-changer for galactic science; and, finally, the panchromatic information that is now available is enabling a much deeper view of the physical conditions and overall environments in which masers exist than was previously available. And, it is only going to get better.* It is clear that after six years significant progress has been made in all aspects of maser astrophysics, and in particular ALMA and Gaia have made deep impacts. As evidenced by the results reported on in these proceedings, ALMA served the wide astrophysics community as a supplier of maser as well as counterpart data with high-sensitivity and high-angular resolution, covering topics from planetary nebulae, star-forming regions, to Ultraluminous Infrared Galaxies (ULRIGs). The access to Gaia DR3 has enabled cross-checking opportunities for maser astrometry projects (e.g., Quiroga-Nunez†). While in many cases the Gaia parallaxes

† When none citation is given, we link the proceedings from this volume by the name of a speaker/first author

markedly enhance the scientific output through distance estimates, during the meeting it was emphasised that for evolved stars Gaia parallaxes need to be treated with caution due to the obscuration at optical wavelengths.

Seven major topics on maser sciences were presented and discussed: theory, cosmology, galaxies, Milky Way, star-formation, evolved stars, and future projects. Just as in previous meetings, the details of high-mass star-formation (HMSF) continues to stimulate extensive research through primarily methanol and water maser studies. The importance of monitoring programs over more than 10 years using mid-sized antennas was stressed, as it has enabled new discoveries in the regime of maser variability. The flare spectra and light curves may indicate the underlying mechanisms of the variability. Houde proposed that superradiance is responsible for the fast variability, while Gray pointed out effects like catastrophic release of saturation capable of causing fast intensity variations at a given velocity. The new and strongly time-dependent features reported thus encourage the community to continue, or to start, monitoring programs with frequent sampling for a diverse set of maser transitions. Progress has also been made in understanding the polarization of masers. Lankhaar further expressed the need for 183 GHz galactic and extragalactic water maser observations.

2. New insights from maser research

In recent years, accurate Galactic astrometry has been done and the Milky Way rotation curve has been verified (e.g., Rygl, Honma, Reid, Ellingsen). It is clear that we now can study the "unreachable" – e.g., the Bulge (Sjouwerman, Lewis), the Long Bar (Kumar), the Galactic Centre (Paine, Sakai), and we can learn about kinematics in extremely obscured LIRGs (e.g. Aalto). New hypotheses were posed to be verified in the futur: *1) Has the gap in the Perseus-arm originated in a cloud collision?* (Sakai), and *2) Do kilomasers indicate supermassive stars?* (Nowak)

Concerning HMSFRs, in this volume we find multiple new close-ups providing milliarcsecond details of individual star-forming regions like the discovery of the magnetohydrodynamic disc wind in IRAS 1078+5211 (Moscadelli), the 3D structure reconstruction of masing regions in periodic methanol masers (Olech), the magnetic field estimation at a few 100s AU scale using 6 GHz OH masers (Kobak), and tracing rotation of magnetic field in W75N (Surcis).

Since the 2017 maser meeting high-resolution observational results have also been added for evolved stars, for example the VLBI results presented on OH/IR (Nakagawa). We observe wind motions in circumstellar envelopes (CSE) of red-supergiants (RSG) (Brand), and they are a signpost of transitional phases along the Asymptitic Giant Branch (AGB) shell (Etoka). Water fountains also have been found to contain SiO masers (Amada, Uscanga).

In the cosmology domain, we are closer to resolving the Hubble tension problem (e.g. Pesce, Kuo, Nakai) where the water megamaser project plays a role. Detailed kinematics of Acitve Galactic Nuclei (AGN) is possible (e.g. Impellizzeri, Nakai).

Finally, new maser species and transitions were discovered: a methanimine maser at 5.29 GHz (Xue) and mm methanol maser transitions (Brogan).

3. The power of teams

Considering the relatively small number of researchers in the cosmic maser community, it is important to appreciate the worldwide efforts under collaborations such as M2O (www.masermonitoring.com), GASKAP–OH (gaskap.anu.edu.au), BeSSeL (bessel.vlbi-astrometry.org), and BAaDE (leo.phys.unm.edu/˜baade/). These teams aim to examine the outbursts at the cm maser transitions in the time and frequency domain by combining results from a diverse set of instruments, to provide astronomers with an unprecedented view of the neutral gas content of the Milky Way and nearby Magellanic system via H I and OH line observations, to estimate parallaxes to methanol and water masers in HMSFRs to discover the structure of the Milky Way, to map the positions and velocities of up to $\sim 30,000$ evolved stars via SiO masers along the full Galactic plane, respectively. These are ambitious project goals that can only be achieved through team efforts.

4. New capabilities and Challenges for future

New instruments always open up for new research programs. We gleaned results from SKA precursors as ASKAP and MeerKAT. New OH megamasers were discovered and we may expect more progress in our understanding of these systems (Glowacki, Roberts). Hopefully, in the next symposium, we will see the first results from the Italian INAF radio telescopes and from the Nobeyama 45-m radio telescope that are about to be equipped with new receivers. We also look forward to access data from the extended Korean VLBI Network and 40-m Thai National Radio Telescope.

Despite the new monitoring programs, time-sampling of masers on a larger scale still is a challenge, requiring substantial amounts of telescope time. We still need multi-epoch and multi-frequency observations to see details and to infer physical conditions of obscured regions from where we can detect masing clouds: interiors of star-forming regions and environments around evolved stars. The maser community also plan to do further astrometry to obtain more accurate measurements at the μas level including parallax estimation of long-period OH/IR stars since Gaia will not be able to probe the optical emission from their obscured central stars. A noted challenge is to obtain more parallaxes in the $3^{\rm rd}$ and $4^{\rm th}$ quadrants, and therefore Australia – New Zealand VLBI is requested.

Science questions raised during the Symposium include: *Are water fountains common and are they periodic?* (Imai), *What are the longest and shortest periods of periodic methanol sources?* (Olech, Tanabe, van den Heever, Kenta), and *What is the mass-loss history in evolved stars?* (Yun). To find answers to these questions, high-sensitivity surveys are requested for water megamasers (Castangia) and gigamasers (Pesce, Kuo). Similarly, increased sensitivity is needed for research of sub-mJy water masers in HMSFRs and radio-continuum counterparts (Sanna). There are hopes that ngVLA will help with at least some of the above challenges.

5. Summary

The Symposium was an amazing (a-masing!) opportunity to catch up on the newest science that has been done in last years, including during the difficult time of 2020-2022, when the pandemic time due to the COVID-19 forced our community to be separated. However, the science progress since the Cagliari meeting in 2017 is remarkable and exciting. The first results from MeerKAT and ASKAP have been presented, and we wait for more: from SKA and the upgraded ALMA. However, we need to keep operating the single-dish telescopes, especially mid-sized, as they have proven crucial for monitoring programs.

On behalf of all the attendees, in person and online, we thank the Local Organisers and 14 young volunteers for all their hard work before the Symposium and during it. We acknowledge the social program with the excursion to a black vinegar factory at Kakuida Black Vinegar and Arimura Lava Observatory to have a view for Sakura-jima volcano and Kinko-wan Bay. Also, the banquet that took place in The Peak Premium Terrace in Tenmonkan area, was enjoyable time with delicious food and dances. This IAUS gathered the largest number of speakers, the idea of flash-talks should be continued. It was a great opportunity for younger researchers to meet in person "more advanced/evolved" colleagues and start collaborations since the power of maser studies is crossing borders - national and institutional.

Acknowledgment

AB acknowledges support from the National Science Centre, Poland through grant 2021/43/B/ST9/02008.

Reference

Diamond, P.J. 2018, *Astrophysical Masers: Unlocking the Mysteries of the Universe*, Proc. IAU Symposium No. 336, p. 451-454

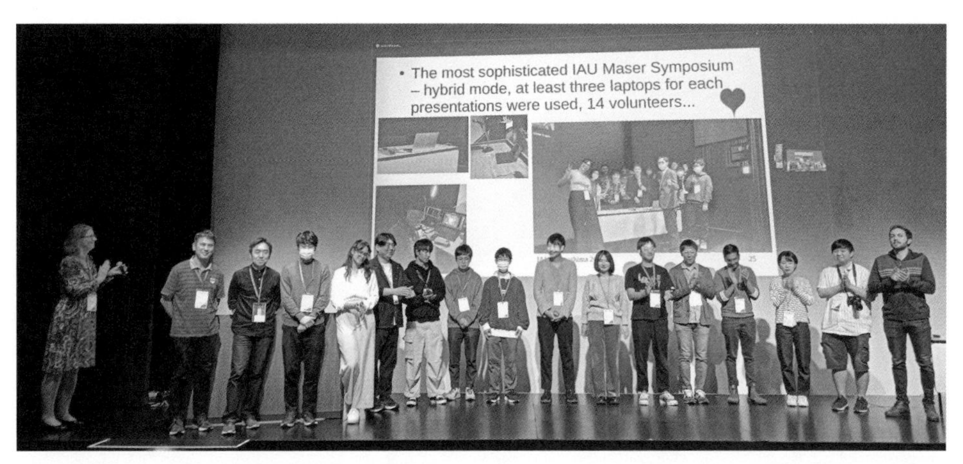

Group photo of the volunteer staff. From left to right, Anna Bartkiewicz (speaker of the concluding remarks), Hiroshi Imai and Akiharu Nakagawa (LOC co-chairs), Yuichi Sakamoto, Tatiana M. Rodríguez, Kaito Kawakami, Ryosuke Watanabe, Yosuke Shibata, Koki Tanaka, Kei Amada, Rina Kasai, Keisuke Nakashima, Daisuke Takaishi, Jayender Kumar, Nao Ikeda, Ka-Yiu Shum, and Roldán A. Cala. Taken by Tomoya Hirota.

Author index